A Manual of

Mammalogy

A Manual of Mammalogy with Keys to Families of the World

Second Edition

Anthony F. DeBlase
Field Museum of Natural History

Robert E. Martin
University of Mary Hardin-Baylor
and Field Museum of Natural History

wcb

Wm. C. Brown Company Publishers
Dubuque, Iowa

Library of Congress Catalog Card Number: 79–53828

ISBN 0–697–04591–9

Printed in the United States of America

dedicated to
Alyce Myers DeBlase 1939–1976

Contents

Preface ix
Acknowledgments xi

1. **Introduction** 1
2. **The Skull** 7
3. **Teeth** 15
4. **The Integument** 27
5. **Horns and Antlers** 35
6. **Claws, Nails and Hoofs** 41
7. **Appendicular Skeleton** 43
8. **Locomotor Adaptations** 47
9. **Reproduction** 59
10. **Populations** 73
11. **Systematic Methods** 81
12. **Keys and Keying** 95
13. **The Orders of Mammals** 101
    Key to the Orders of Living Mammals 102
14. **The Monotremes** 109
    Order Monotremata 109
    Key to Living Families 111
15. **The Marsupials** 113
    Order Marsupicarnivora 117
    Key to Living Families 118
    Order Peramelina 122
    Key to Living Families 123
    Order Paucituberculata 123
    Order Diprotodonta 125
    Key to Living Families 126
16. **The Insectivores** 133
    Order Insectivora 133
    Key to Living Families 135
17. **The Colugos** 143
    Order Dermoptera 143
18. **The Bats** 145
    Order Chiroptera 145
    Key to Living Families 148
19. **The Primates** 159
    Order Primates 159
    Key to Living Families 161

20. The Anteaters, Sloths and Armadillos 171
Order Edentata 171
Key to Living Families 175
21. The Pangolins 177
Order Pholidota 177
22. The Whales 181
Order Mysticeta 184
Key to Living Families 186
Order Odontoceta 188
Key to Living Families 190
23. The Carnivores 199
Order Carnivora 199
Key to Living Families 201
24. The Rabbits, Hares and Pikas 215
Order Lagomorpha 215
Key to Living Families 217
25. The Rodents 219
Order Rodentia 219
Key to Living Families of the World 224
Key to Living Families of North America 246
26. The Aardvark and the Subungulates 251
Order Tubulidentata 251
Order Proboscidea 253
Order Hyracoidea 255
Order Sirenia 256
Key to Living Families 258
27. The Perissodactyls 259
Order Perissodactyla 259
Key to Living Families 262
28. The Artiodactyls 263
Order Artiodactyla 263
Key to Living Families 266
29. Statistical Analysis and Representation of Data 275
30. Age Determination 289
31. Diet Analysis 297
32. Identifying Mammal Sign 305
33. Recording Data 311
34. Collecting 319
35. Specimen Preparation and Preservation 333
36. Collecting Ectoparasites of Mammals 353
37. Analysis of Spatial Distribution 365
38. Estimation of Relative Abundance and Density 375
39. Performing a Literature Search 383

Glossary 389
Literature Cited 413
Index 429

Preface

Mammals occur on every major land mass and in every major body of water in the world. We believe that with today's rapid communication and global travel it is essential for students of mammalogy to develop a familiarity with mammals of the world rather than only with those of a particular county, state or nation. Therefore, we have designed this manual to be worldwide in scope. Recognizing that most institutions will not have examples of many of the world's mammals, we attempted to select North American representatives as examples for particular characteristics whenever possible. However, when North American examples will not suffice we have written instructions using specimens from other parts of the world. We feel it is better for the student to examine illustrations of mammals in the revised manual or in one of the several well-illustrated books now available than to ignore a particular group or characteristic simply because specimens are not readily available. We realize that few North American institutions have many non-North American mammals to use with the keys. However, we believe that the students' use of keys covering all families will increase their awareness of the world's mammalian fauna even though they work only with North American forms.

Mammalogy is a broad field including a wide variety of subject areas. Since it is impossible to touch upon all of these areas in a single volume designed for use in a one semester course, we have concentrated our focus on the methods for identifying mammals and determining their evolutionary and ecological relationships. We hope the use of the manual will be augmented by library and field research projects and encourage its use in conjunction with a basic mammalogy text or a carefully planned selection of readings from various texts and the original literature.

Prior to beginning the revision of this manual, we solicited comments from many colleagues. Based on many of the suggestions received and our current research interests, we have added six new chapters: Populations, Statistical Analysis, Age Determination, Diet Analysis, Analysis of Spatial Distribution, and Estimation of Relative Abundance and Density. In addition, the chapters on Reproduction, Recording Data, Specimen Preparation (including a new section on Cleaning Skeletal Material), and Collecting Ectoparasites (including a new Key to Arthropod Ectoparasites of Mammals) were extensively revised and rewritten. All of the order chapters have been greatly augmented with illustrations and substantially rewritten to include more information on the biology of mammals and their characteristics, and to reflect current understanding of the classification of various groups.

Although the manual is designed for an undergraduate or graduate course in mammalogy it will also be of use in other courses such as wild-life management and vertebrate zoology. It is assumed that the student will have a background in general zoology and comparative anatomy but review material is included.

The keys in this manual are unique in their coverage of the cranial (and many external) characters of *all* the world's living families of mammals. These will be of interest and value to the professional mammalogist as well as the interested student.

AFD
REM

Acknowledgments

First Edition

This manual has grown from a set of laboratory exercises written by participants in graduate seminar at Oklahoma State University. The seminar was directed by Bryan P. Glass and the participants included Alberto Cadena, Andrew Chien, Anthony DeBlase, Barbara Garner, Robert Ingersol, John Jahoda, Pegge Luken, Robert Martin, George Rogers, and Stanley Rouk. We particularly wish to thank Bryan Glass and the participants for providing the seed from which this manual has grown.

Karl Liem, Joseph Curtis Moore and Luis de la Torre, all of Field Museum of Natural History, allowed us to examine specimens in their care for preparation and testing of the keys. Walter W. Dalquest, Midwestern University, Bryan P. Glass, Oklahoma State University Museum, and Robert L. Packard, The Museum, Texas Tech University, also granted us access to their study collections for the purpose of examining and photographing specimens. J. E. Hill kindly supplied us with notes on mammals of the world which were useful in constructing several keys.

Many of the keys to orders and families were tested by André Dixon, Edward Gray, Williams Mohs, and Steven Rissman, students of DeBlase at Roosevelt University, who also helped rewrite certain sections. Various versions of the manual have been used and tested from 1967 to 1973 in the mammalogy classes of Bryan P. Glass at Oklahoma State University. Stephen R. Humphrey used an early version of the manual in his mammalogy class at the University of Florida in 1971. Students in Robert J. Baker's mammalogy class (1970) at Texas Tech University made critical comments on the keys in Chapters 15 and 17, and students in DeBlase's mammalogy class (1971) at Roosevelt University offered critical comments on many aspects of the manual.

Many people kindly read all or portions of the manuscript, corrected errors, and made suggestions on the content of the manual. Their suggestions, in most instances, were incorporated, but the authors are responsible for the final content and accuracy of the text. General comments on the outline or entire manual were made by Walter W. Dalquest, Bryan P. Glass, Stephen R. Humphrey, Thomas H. Kunz, George A. Moore, Guy G. Musser, Hans N. Neuhauser and Terry A. Vaughn. Their comments were very helpful in determining the final scope of the manual. Selected chapters or portions of chapters were read and criticized by Sydney Anderson, Dale L. Berry, Hugh H. Genoways, Bryan P. Glass, David L. Harrison, Theodore A. Heist, Philip Hershkovitz, Stephen R. Humphrey,

Karl F. Koopman, Thomas H. Kuntz, John A. Morrison, Guy G. Musser, J. Mark Rowland, C. David Simpson, James P. Webb, Robert W. Wiley, and Daniel R. Womochel. Preparation of Chapter 3 was aided by discussions with Craig C. Black. George A. Moore provided helpful suggestions on many occasions and kindly proofread the final draft of the manuscript. Donna Womochel and Alyce M. DeBlase assisted in reading proofs.

The original illustrations were prepared by Janet Blefeld, L. Patricia Martin, Kenneth G. Matocha, George A. Moore, Richard Roesner, John Whitesell, and the authors. Photographs processed by Lee A. Jones, Milton R. Curd and the Photography Department at Field Museum of Natural History facilitated the preparation of many original drawings.

Helen Walker and the Department of Zoology, Oklahoma State University, were especially helpful in the formative stages of the manuscript. Patricia Gaddis and Marge Bell typed portions of the manuscript. Alyce M. DeBlase assumed the major burden of typing the various drafts of all chapters and offered a critical editorial eye to the entire manuscript. Special thanks go to our wives, Alyce and Patty, for patience and help during the preparation of this manual.

We greatly appreciate the efforts of all of these people, with the realization that our task would have been extremely difficult, if not impossible, without their generous assistance.

Second Edition

Many persons have contributed to this second edition. We are grateful to Ronald H. Pine who provided detailed comments on the entire first edition and to all of you who have written to us, talked to us at meetings, or returned our survey forms offering comments, student reactions, and constructive criticism.

Eric H. Smith revised Chapter 36 on ectoparasites of mammals and wrote the key and group descriptions in this chapter. Laurie Wilkins wrote the section on cleaning skeletal material in Chapter 35, and Keith Carson made contributions to the section on skinning large mammals in the same chapter. Patricia Freeman, present Curator of Mammals at the Field Museum, facilitated our work on this project in many ways. We are grateful to all of these persons and to those listed below who have read and criticized or have made other contributions to one or more of the chapters in the current new edition: Sharon Adams, Elmer C. Birney, Sara Derr, Jerran T. Flinders, George Fulk, Robert J. Izor, Laurel E. Keller, Karl F. Koopman, Cliff A. Lemen, Larry G. Marshall, L. Patricia Martin, Chris Maser, Peter L. Meserve, Dale Osborn, Pamela Parker, Ronald H. Pine, Linda Porter, James D. Smith, Mike Smolen, Sandra L. Walchuk, and Laurie Wilkins.

We especially appreciate the assistance and cooperative attitude of Mary Ann Cramer in preparing a large number of original illustrations and the cooperation of family and colleagues during the period of this revision. But most of all we want to thank all of you who have continued to use the Manual of Mammalogy in your courses. Without you the second edition would never have come to be.

AFD
REM

1 Introduction

Mammals are animals that possess a hollow dorsal nerve tube throughout life, and that, in early stages of embryonic development, possess gill pouches and a notochord. These three characters place mammals in the phylum **Chordata.** The brain and spinal cord of mammals are enclosed in, and protected by, a bony skeleton. Thus mammals are members of the subphylum **Vertebrata.**

Embryos of the class **Mammalia,** like those of the classes Reptilia and Aves, are surrounded by protective amnionic and allantoic membranes. As in crocodilians and birds, the mammalian heart has four distinct chambers, and as in birds, there is no mixing of oxygen-rich and oxygen-poor blood. Homoiothermy is thus generally well developed.

Characters Defining Mammals

In defining the class Mammalia, it is necessary to use characters that may be applicable to fossil remains as well as to living animals. Thus, the major defining characters are skeletal. The lower jaw of mammals is unique among vertebrates in being composed of only a single pair of bones, the **dentaries,** which articulate directly with the cranium. In other vertebrates the dentary is only one of several bones in the lower jaw and it usually does not articulate directly with the cranium. The presence of an articulation between the dentary and squamosal is the characteristic used to define Mammalia. The articular and quadrate bones, elements of the lower jaw and cranium respectively in other vertebrates, are modified in mammals to form two of the three middle ear bones (=ossicles). The articular and quadrate of other vertebrate classes become the **malleus** and **incus** respectively in mammals. The columella, the only middle ear bone of other vertebrates, becomes the **stapes** in mammals. Thus, mammals are comonly defined by the presence of a single bone in the lower jaw and by the presence of three ossicles in the middle ear (Fig. 1-1).

In addition to these skeletal characters, living mammals have several characteristics of the soft anatomy

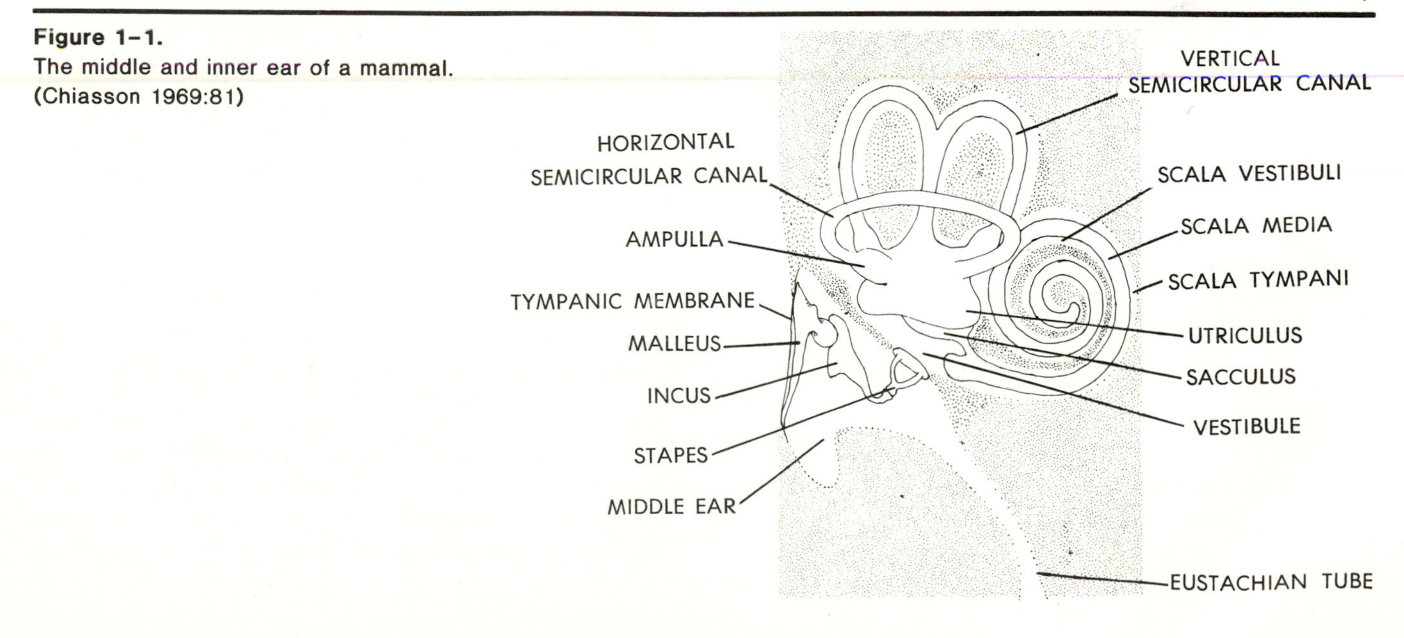

Figure 1–1.
The middle and inner ear of a mammal.
(Chiasson 1969:81)

Figure 1–2.
Anterior or ventral view of the mammalian heart with the chambers exposed. Note that the aorta exits from the heart and arches to the left.
(Chiasson 1969:45)

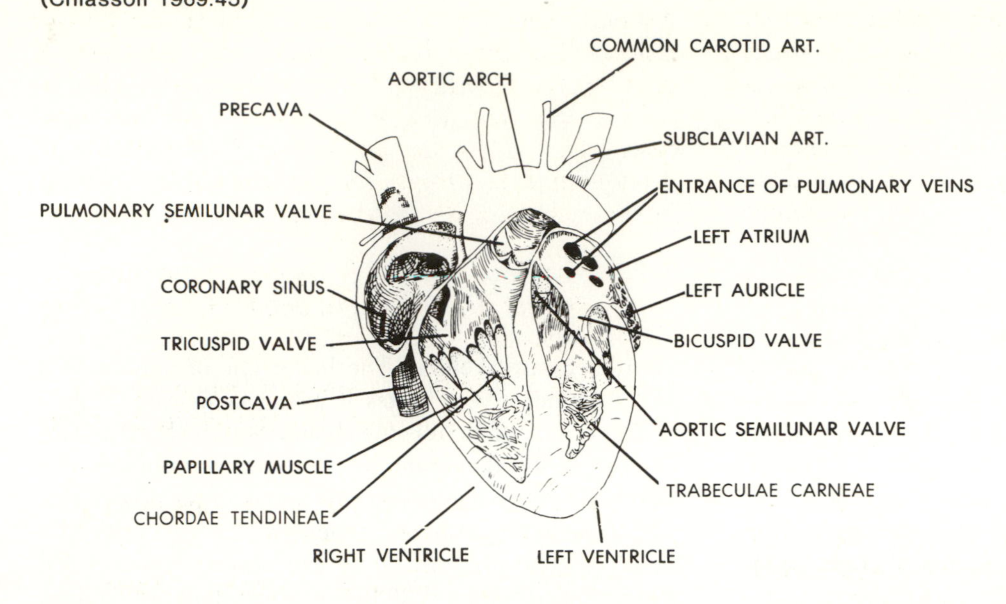

that are unique to mammals and common to all living mammalian species. Some of the more conspicuous ones are given below.

All mammals possess **hair** at some stage of their life cycle. Most have hair throughout life, but several aquatic forms have only a few stiff bristles. In some whales even these bristles are lost in the adult.

All female mammals possess **mammary glands** that produce milk to nourish the young. In males these glands are present but rudimentary. In monotremes the milk flows from pores in the skin but all other mammals have nipples and the young suckle.

The thoracic and abdominal cavities of mammals are separated by a **muscular diaphragm.** Some other vertebrates have a membranous septum between these cavities but only in mammals is this structure fully muscular.

Only the **left aortic arch** is present in adult mammals (Fig. 1-2). The right aortic arch is lost during early embryonic development. Both arches persist in reptiles while only the right aortic arch persists in birds.

Mammals have **enucleate erythrocytes.** No nuclei are observable in mature red blood cells. Other vertebrates have nuclei in mature erythrocytes.

The **neopallium,** or roof of the forebrain, is proportionately larger in mammals than in other vertebrates and the **corpora quadrigemina,** an elaboration of the midbrain, is found only in mammals.

1-A. Compare the lower jaw of a dog or coyote (genus *Canis*) with those of a bird, reptile, amphibian, and bony fish. How many bones are present in each jaw? With which bone(s) of the cranium does each mandible articulate?

1-B. In demonstration material locate the middle ear bones in a mammalian skull and compare with the middle ear structure of a bird, reptile, and amphibian.

1-C. On a demonstration dissection of a cat or other mammal examine the structure of the heart and locate the left aortic arch and muscular diaphragm. Compare with dissections of a frog and bird.

1-D. Compare a prepared slide of mammalian blood with blood of a bird, reptile and/or amphibian.

1-E. Compare the structure of the mammalian brain with that of other vertebrates.

Distribution

Mammals were originally four-footed terrestrial animals and most living mammals retain this basic plan. But over the millennia of their development, mammals have diversified to fill a great variety of niches. They are now found underground, in marine and fresh water, and in the air as well as on the ground. They exist on all continents including Antarctica (some seals) and in all oceans. Ignoring man and the commensals that follow him, mammals are absent only from a few remote oceanic islands.

Because of this widespread distribution, it is convenient to refer to ranges of particular mammal groups in terms of **faunal regions** (Fig. 1-3). The faunal regions are based upon broad similarities in animal life. The boundaries of these regions are generally formed by barriers such as the Himalaya Mountains and the Sahara Desert which have restricted the distribution of mammals.

The **Neotropical** Region includes all of the South American continent and extends north to central Mexico. The **Nearctic** Region includes the remaining portion of the North American continent. The **Palearctic** Region includes all of Europe, Africa to the south edge of the Sahara, and Asia north of the south slope of the Himalayas. Because of the great similarities between the Nearctic and Palearctic faunas, these regions are frequently grouped as a single region, the **Holarctic.** The **Ethiopian** Region includes Africa south of the Sahara and most of the Arabian Peninsula. Madagascar, here considered a portion of the Ethiopian, is sometimes considered a distinct region. The **Oriental** Region extends south and east of the Himalayas to Wallace's Line which passes through Indonesia. The **Australian** Region includes the Australian continent, New Guinea, and the Indonesian islands south and east of Wallace's Line.

New Zealand and other islands of Oceania, and the Antarctic continent are not placed in named faunal regions. The desert belt extending from the western edge of the Sahara in Morocco through southwest Asia to the Sind area of northwest India has been referred to as the Saharo-Sindian Region. While it is unquestionable that this Region has a unified fauna, its true relation to the above faunal regions is not clear.

1-F. What are the major barriers separating each of the faunal regions from neighboring regions?

1-G. List at least three kinds of mammals that may be considered characteristic of each region.

1-H. In which faunal region or regions is each of the following situated?

Bolivia, Iceland, Japan, Tahiti
China, India, Malaya, Tunisia
Greenland, Indonesia, Mexico, Uganda
Guatemala, Iran, Spain, Yemen
Haiti

Figure 1–3.
The major Faunal Regions of the world.
(A.F. DeBlase)

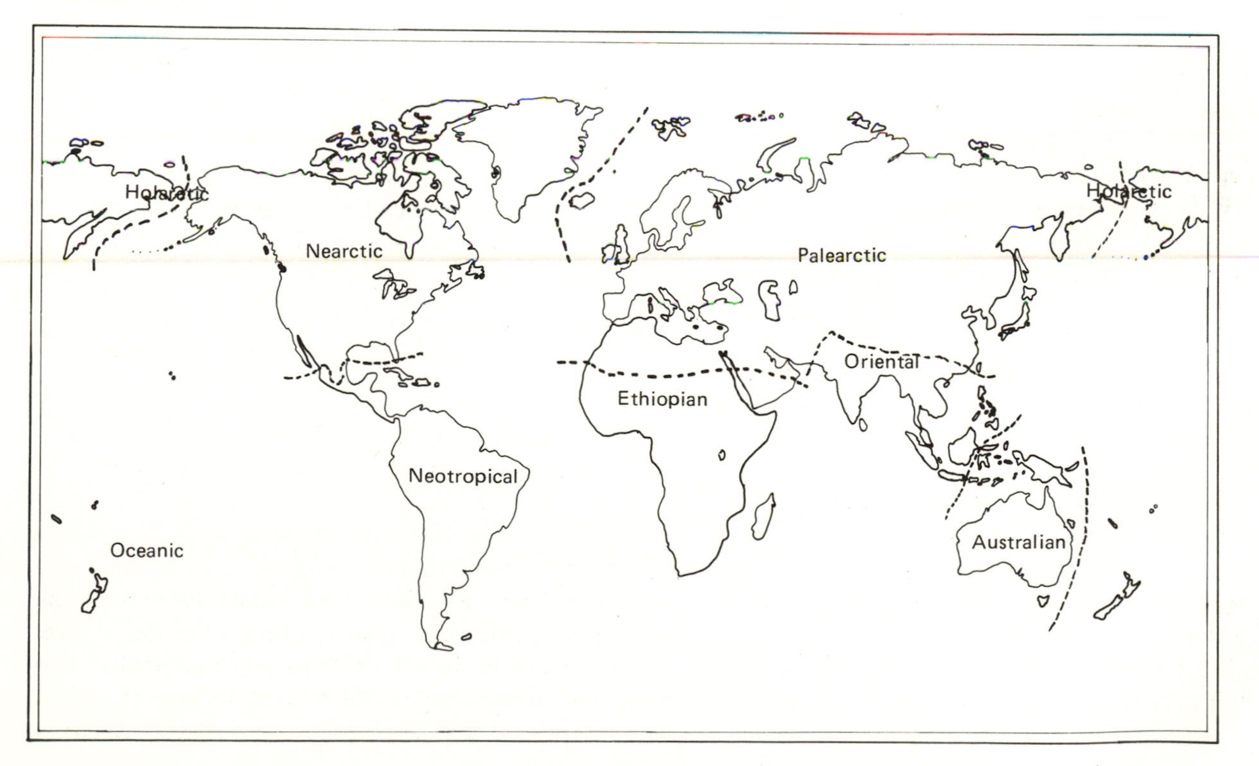

Table 1-1
Geologic Time Divisions

ERA	PERIOD	EPOCH	Approximate beginning time (millions of years ago)*	
CENOZOIC	Quaternary	Recent	.006	Appearance of modern species.
		Pleistocene	2	The "Ice Age." Appearance of many modern genera.
	Tertiary	Pliocene	10	
		Miocene	23	Appearance of grasslands and associated mammals.
		Oligocene	34	
		Eocene	50	Origins of most modern orders of mammals.
		Paleocene	70	Beginning of great diversification of eutherian mammals.
MESOZOIC	Cretaceous		135	First eutherian and metatherian mammals.
	Jurassic		180	
	Triassic		200	The oldest known fossil mammals from late Triassic of Europe.
PALEOZOIC	Permian		270	The first therapsid reptiles, ancestors of mammals.
	Pennsylvanian		320	The first known mammallike reptiles, the †Synapsida.
	Mississippian		350	
	Devonian		400	"Age of Fishes."
	Silurian		430	
	Ordovician		490	
	Cambrian		600	First vertebrates.
	Precambrian		4,500	

*Council of Biology Editors, Committee on Form and Style (1972:120).
†Extinct

Fossil History

Mammals arose from therapsid reptile ancestors in the early Mesozoic. The oldest fossils definitely referable to the class Mammalia are from the late Triassic of Europe. Several extinct mammalian groups are known from the Mesozoic; and metatherian and eutherian mammals appeared in the late Cretaceous. The Cenozoic is known as the "Age of Mammals" and during this era the great diversification of therian mammals occurred. The fossil history of mammals will be discussed in greater detail in chapters 3 through 28, but the geologic time scale (Table 1-1) is presented here for your reference and review.

Continental Drift

Evidence accumuated largely since the mid-1960's suggests that the present configuration of continental land masses is the result of large-scale movements of segments or plates of the earth's crust (Dietz and Holden 1970, Tarling and Tarling 1971, Marvin 1973, Windley 1977). These plate movements and additions to the crustal surface (by liquid magma from the fluid-interior of the earth) by the spreading of ocean basins has been termed **plate tectonics. Continental drift,** the shifting of the relative position of the continental units on various plates, is a component of plate tectonics.

Near the end of the Paleozoic Era (about 200 myBP*), the continental land masses were joined together (or in close apposition) into a supercontinent termed **Panagea** (Fig. 1-4A). During the early Jurassic (180 myBP), the northern landmass, **Laurasia,** began to separate from the southern landmass, **Gondwanaland** (Fig. 1-4B). At the beginning of the Cretaceous Period (135 myBP), a rift developed between the African and South American landmasses (Fig. 1-4C). The spreading of the Atlantic ocean basin and the contraction of the Pacific basin continued into the Tertiary (Fig. 1-4D). Further discussion of the history, processes, and results of continental drift can be found in Dietz and Holden (1970), Tarling and Tarling (1971), Marvin (1973), Windley (1977), and Vaughan (1978).

Although most extant families of mammals appeared during the Tertiary after a considerable amount of continental drift had occurred, the continental configurations existing during that time had a profound

*million years Before Present. The dates for various continental arrangements are the subject of current debate; see Marvin (1973) for further details.

Figure 1-4.
The composition and fragmentation of Pangaea as postulated by Dietz and Holden (1970). A, Pangaea about 200 myBP; B, Laurasia and East and West Gondwanaland, about 180 myBP; C, initial rift between South America and Africa, about 135 myBP; D, isolation of Antarctica from Australia and separation of Africa from Madagascar, about 65 myBP. The distributions of the mammalian faunas are those hypothesized by Fooden (1972).
(Breakup of Pangaea and isolation of relict mammals in Australia, South America and Madagascar., "Fooden, J., *Science,* Vol. 175, pp. 894-898, Fig. 1, 25 February 1972. Copyright 1972 by the American Association for the Advancement of Science. Reprinted with permission.)

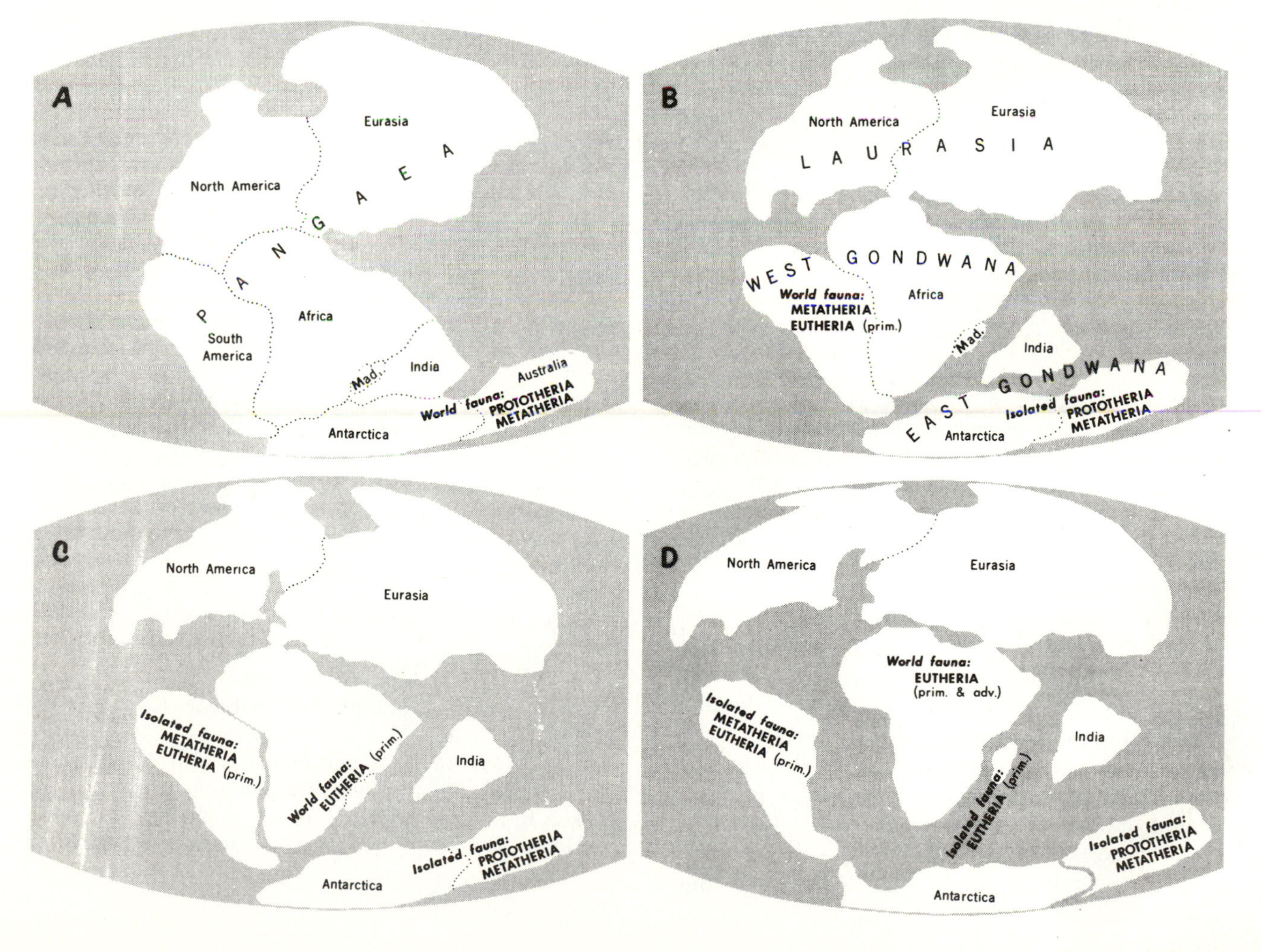

effect on the development of the mammalian faunas in many areas of the world. Fooden (1972) and Cracraft (1974b) suggested that the composition of continental mammalian faunas was significantly influenced by the isolation of primitive placental, marsupial, and prototherian mammals during the separation of the landmasses. Fooden (1972) suggested that the land mammal faunas of Australian and New Guinea, South America, and Madagascar represent successively detached samples of the evolving mammal fauna as Pangaea fragmented into various units (Fig. 1-4). Fooden's hypothesis was challenged by Cracraft (1974) since fossil evidence is lacking for certain groups and areas —e.g., no Cretaceous marsupials known for Africa; no prototherians and metatherians known for Antarctica.

Further discussion of the biological implications of plate tectonics and continental drift can be found in Jardine and McKenzie (1972), McKenna (1972), and Marvin (1973).

Care of Specimens

Throughout this manual you will be called upon to closely examine and handle numerous mammal specimens. Whenever possible, we have recommended the use of species most readily available and most easily replaced. However, whether the specimen is expendable or rare, it should always be handled and treated with care.

Skulls and skeletal elements should be handled gently and never picked up by slender processes or other portions that are likely to break. When possible, place skulls on a pad. Never drop them on a hard surface such as a tabletop. Be particularly careful when handling small or delicate skulls. If a tooth falls out of its socket or a bone breaks or becomes disarticulated, report it to the instructor at once so that pieces are not lost.

Study skins require even more careful handling than most skulls. Never pick up a study skin by the feet or tail. A dried skin can be very brittle, and projections such as the feet, ears, or tail can be easily broken off. Most study skins are made to be placed belly down. Do not leave them resting on their backs or sides and do not stack them on top of one another.

Specimens preserved in liquid must be kept moist and not allowed to dry out. If these are being studied for an extended period, moisten them occasionally with the preservative or water. Pay special attention to thin structures such as ears and bat wings.

Take particular care not to detach labels from any specimen. As will be emphasized in a later chapter, the catalog numbers and data on these labels are exceedingly important to the scientific value of the specimen. If a label should become loose notify the instructor at once.

Whenever live mammals are studied in the laboratory, they should be treated humanely. Wild mammals need particular attention with respect to housing and diet. Use caution whenever handling live animals.

Supplementary Readings

Anderson, S., and J. K. Jones (Eds.) 1976. *Recent mammals of the world: a synopsis of families.* Ronald Press, New York. 435 pp.

Bourliere, F. 1964. *The natural history of mammals,* 3rd Ed. Alfred A. Knopf, New York. 387 pp.

Cockrum, E. L. 1962. *Introduction to mammalogy.* Ronald Press, New York. 455 pp.

Cracraft, J. 1974. Continental drift and vertebrate distribution. *Ann. Rev. Ecol. Syst.* 5:215-261.

Davis, D. E. and F. B. Golley. 1963. *Principles in mammalogy.* Reinhold, New York. 335 pp.

Fooden, J. 1972. Breakup of Pangaea and isolation of relict mammals in Australia, South America, and Madagascar. *Science* 175:894-898.

Grzimek, B. (Ed.). 1972. *Animal life encyclopedia, Vol. 10, Mammals I,* 627 pp.; 1975. *Vol. 11, Mammals II,* 635 pp.; 1975. *Vol. 12, Mammals III,* 657 pp.; 1972. *Vol. 13, Mammals IV,* 566 pp. Van Nostrand Reinhold, New York.

Gunderson, H. L. 1976. *Mammalogy.* McGraw-Hill Book Co., New York. 483 pp.

Jones, J. K. Jr., S. Anderson, and R. S. Hoffman (Eds.). 1976. Selected readings in mammalogy. *Mus. Nat. Hist. Univ. Kansas Monogr.* No. 5. 640 pp.

Keast, A.; F. C. Erk, and B. Glass, (Eds.) 1972. *Evolution, Mammals and Southern Continents.* State Univ., New York, Albany. 543 pp.

Kowalski, K. 1976. *Mammals, an outline of theriology.* [English Translation] Nat. Cent. Sci. Tech. Econ. Info., Springfield, Va. 617 pp.

Marvin, U. 1973. *Continental drift: the evolution of a concept.* Smithsonian Institution Press, Washington. 239 pp.

Matthews, L. H. 1969. *The life of mammals, Vol. 1.* 340 pp.; 1971, *Vol. 2,* 440 pp. Universe Books, New York.

Morris, D. 1965. *The mammals, a guide to the living species.* Harper and Row, New York. 448 pp.

Romer, A. S. 1966. *Vertebrate paleontology.* 3rd Ed. Univ. of Chicago Press, Chicago. 468 pp.

Sclater, W. L. and P. L. Slater. 1899. *The geography of mammals.* Kegan Paul, Trench, Trubner and Co., London. 335 pp.

Stumpke, H. 1967. *The snouters, form and life of the Rhinogrades.* Natural History Press, Garden City, N. Y. 92 pp.

Vaughan, T. A. 1978. *Mammalogy,* 2nd Ed., W. B. Saunders Co., Philadelphia. 522 pp.

Walker, E. P., *et al.* 1975. *Mammals of the world,* 3rd ed. Johns Hopkins Univ. Press, Baltimore. 2 vols., 1500 pp.

2 The Skull

The mammalian skull is a complex structure. It houses and protects the brain and the receptors for four major senses: smell, vision, hearing, and taste. Through the skull enter oxygen, water, and food—all essential for life. The braincase has adapted to the changes in the size and proportions of the brain. Specializations in the senses of hearing, smell, and sight have frequently resulted in corresponding changes in the skull, as have various adaptations for gathering food and preparing it for digestion. It is not surprising then, that the skull is one of the most important anatomical units used in mammalian classification. A knowledge of its anatomy is essential for the identification of mammals. The keys included in this volume have been constructed primarily on the basis of skull and tooth characters. Teeth will be discussed in the next chapter.

The skull is composed of two easily disarticulated elements: the **cranium,** or skull proper, and the **mandible,** or lower jaw. Only those bones that are visible externally on a cleaned skull are discussed below.

The Cranium

For convenience two major regions of the mammalian cranium may be recognized: the **braincase** and the **rostrum.** The braincase is a box of bone protecting the brain. Attached to it or associated with it are the **auditory bullae,** which house the middle and inner ears; the **occipital condyles,** which articulate with the first vertebra; and numerous processes and ridges which serve as points of attachment for muscles. Several **foramina** and **canals** penetrate the bone and allow for the passage of nerves and blood vessels. The rostrum is composed of the group of bones projecting anteriorly from a vertical plane drawn through the cranium at the anterior edges of the orbits. It includes the upper jaws and the bones that surround the nasal passages and divide these passages from the oral cavity.

2-A. On a wolf, coyote or dog skull (genus *Canis*) locate each of the bones or structures listed below in bold face type. Label each of these on the various views of the coyote skull in Figure 2-1. All terms are listed in the glossary. A key to the numbers on Figure 2-1 is located at the end of this chapter. Use it only to verify your identifications.

The dorsal part of the cranium is composed of a series of paired bones that meet along the midline. The long slender **nasal** bones roof the **nasal passages.** Posterior to these are the paired **frontals.** Each of these extends down the side of the cranium to form the inner wall of the **orbit** or eye socket. The **postorbital process** is a projection of the frontal that marks the posterior margin of the orbit. Posterior to the frontals are the paired **parietals.** A small, unpaired **interparietal** is located between the posterior edges of the parietals; in *Canis* this is fused posteriorly with the supraoccipital. Low **temporal ridges** arise on the frontals near the postorbital processes and continue posteriorly until they converge to form the **sagittal crest.** These ridges (including the "crest") increase the area available for attachment of jaw muscles. The posterior portion of the skull is formed by a single bone, the **occipital.** The **foramen magnum,** through which the spinal cord passes, is located near the center of the occipital and is flanked by two knobs, the **occipital condyles,** which articulate with the atlas, the first of the neck vertebrae. In embryonic mammals four bones fuse to form the single occipital bone of the adult. The names for these are used to designate regions of the occipital. Around the foramen magnum these are the ventral **basioccipital,** the dorsal **supraoccipital,** and the lateral **exoccipitals.** The **occipital crests** extend laterally from the sagittal crest. Branches of the exoccipitals, the **paroccipital processes,** extend ventrally in

Figure 2–1.
A *Canis* cranium. A, dorsal view; B, ventral view; and C, lateral view.
(A.F. DeBlase)

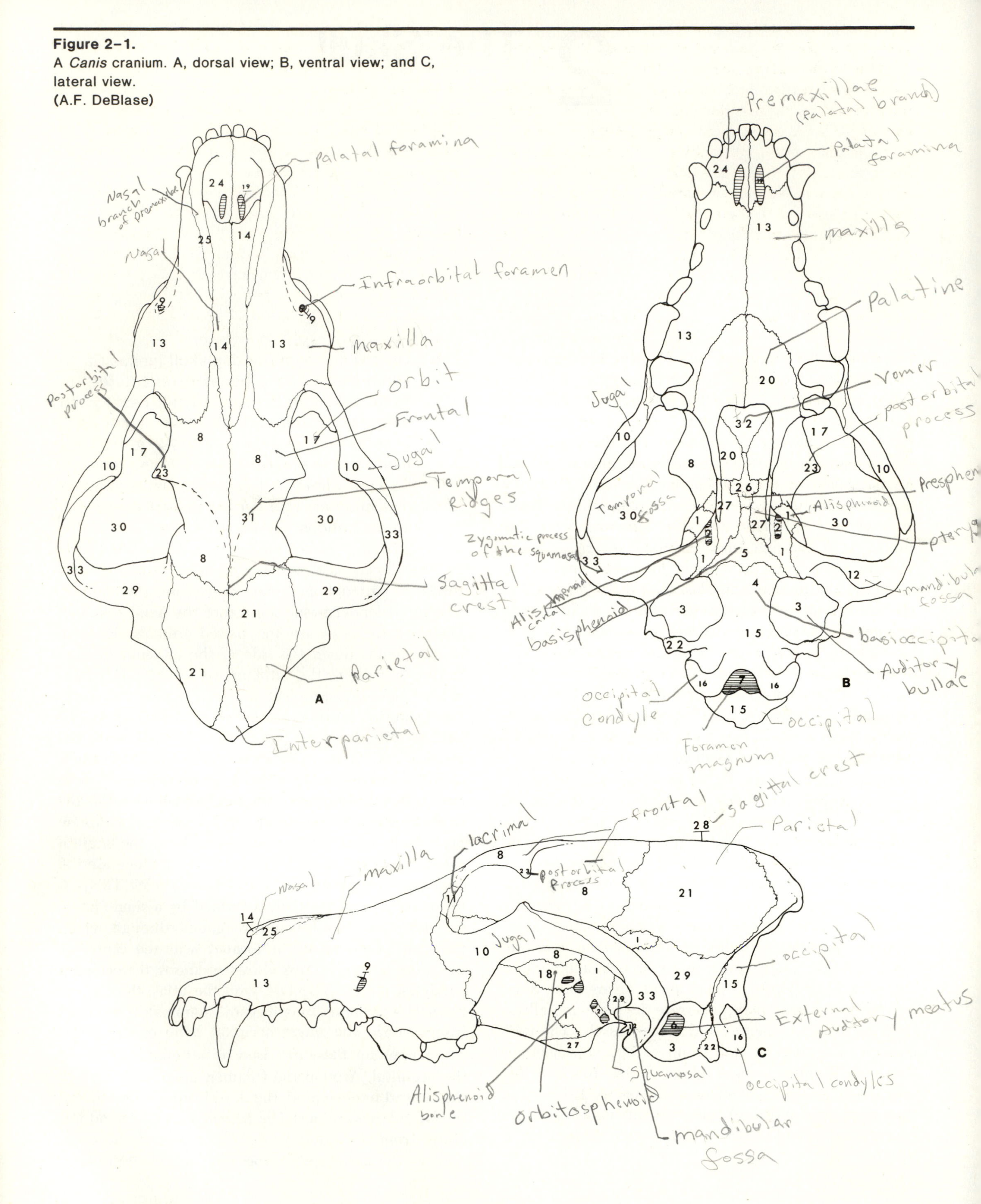

close association with the auditory bullae. The entire posterior region of the cranium is termed the **occiput.**

The tooth-bearing bones of the cranium are the paired premaxillae and maxillae. The **premaxillae,** which meet at the anterior end of the cranium, have two major branches. The **palatal branches** of the premaxillae meet along the midline of the skull and form the anterior end of the **hard palate;** the **nasal branches** of the premaxillae project dorsally and posteriorly to form the sides of the **anterior nares.** Posterior to the premaxillae, the **maxillae** form the major portions of the sides of the rostrum. A large foramen in each maxilla is the anterior opening of the **infraorbital canal.** Each canal terminates in the orbit and serves for passage of blood vessels and nerves. In some mammals this opening is not elongated into a canal and is termed the **infraorbital aperature** or **infraorbital foramen.**

The palatal branches of the premaxillae and maxillae together with the paired **palatine bones** form the hard **palate** that separates the mouth from the nasal passages. A pair of openings at the suture between the premaxillae and maxillae are the anterior **palatal foramina** (also termed the **incisive foramina).** Posterior and dorsal to the palatine bones are the proximal openings of the nasal passages, the **internal nares.** The **vomer** is an unpaired bone forming a septum between the two nasal passages. The highly convoluted bones within these passages are the **turbinals.** Posterior to the internal nares and the palatine bones are the paired **pterygoids.** Between the paired pterygoids and posterior to the vomer is the unpaired **presphenoid.** This complex bone passes beneath the pterygoid, palatine, and maxillary bones to reappear dorsally in the wall of each orbit where it is termed the **orbitosphenoid** and is perforated by the **optic foramen.** The medial **basisphenoid** lies between the basioccipital and the ventral visible portion of the presphenoid.

The conspicuous bony arches forming the lateral borders of the orbits and **temporal fossae** are the **zygomatic arches.** Three bones contribute to each zygomatic arch. Anteriorly the **jugal** bone articulates with the **zygomatic process** of the **maxilla.** Posteriorly the jugal articulates with the **zygomatic process** of the **squamosal** bone. A short process on the dorsal edge of the zygomatic arch marks the posterior edge of the orbit. In some mammals (but not in *Canis*), this process is continuous with the postorbital process of the frontal, forming a **postorbital bar.** The postorbital bar separates the orbit or eye socket from the temporal fossa, through which some of the muscles of the lower jaw pass. On the ventral side of the base of each zygomatic process of the squamosal, the **mandibular fossa** provides an articulation point for the lower jaw.

Between the jugal and frontal bones, at the anterior root of each zygomatic arch, is the small **lacrimal** bone. The foramen in this bone is for passage of the tear or lacrimal duct. Anterior to the squamosal and posterior to the frontal and orbitosphenoid is the **alisphenoid** bone. Ventrally on this bone, near its suture with the basisphenoid, is a small arch of bone forming the **alisphenoid canal.**

The bulbous structures between the mandibular fossae and the occipital condyles are the **auditory bullae.** The foramen in the side of each bulla is the **external auditory meatus** across which the tympanic membrane or eardrum is stretched. In *Canis* the **tympanic** bone is the only bone visible on the external surface of the bulla, but in some mammals the **entotympanic bone** is also visible externally. Within each bulla is the **middle ear** chamber containing the three ossicles, the **incus, malleus,** and **stapes.** The **otic capsule,** which houses the structure of the inner ear, is covered by the tympanic in *Canis,* but is visible in primitive mammals that have incomplete auditory bullae. A portion of the **periotic,** one of the bones forming each otic capsule, is frequently exposed between the squamosal and occipital bones. The distal exposed portion of the periotic forms a distinct **mastoid process** in many mammals, but this is not a conspicuous structure in *Canis.* In some mammals (including cats and higher primates), the tympanic and squamosal bones fuse to form a single structure termed the **temporal bone.**

The Mandible

Compared to the cranium, the mandible is a very simple structure. It is composed of left and right **dentary** bones. The anterior point of contact between the paired dentaries is the **mandibular symphysis.** This suture is fused firmly in *Canis* and most other Carnivora and in many other mammals. But in rodents, most artiodactyls, and many other forms the two dentaries are easily disarticulated. The horizontal portion of each dentary, the portion that normally bears teeth, is termed the **ramus** (see the glossary for other definitions of **ramus**). The **mandibular condyle** is the portion of the mandible that articulates with the mandibular fossa of the cranium. Dorsad to the condyle, the **coronoid process** extends up to fit into the temporal fossa and provides a surface for muscle attachment. Ventrad to the condyle, the **angular process** protrudes posteriorly. The shallow depression near the bases of these processes is the **masseteric fossa.** In some mammals (but not in *Canis*), this depression is very deep and occasionally completely penetrates the mandible forming a **masseteric canal.**

Figure 2–2.
Lateral view of a *Canis* mandible.
(A.F. DeBlase)

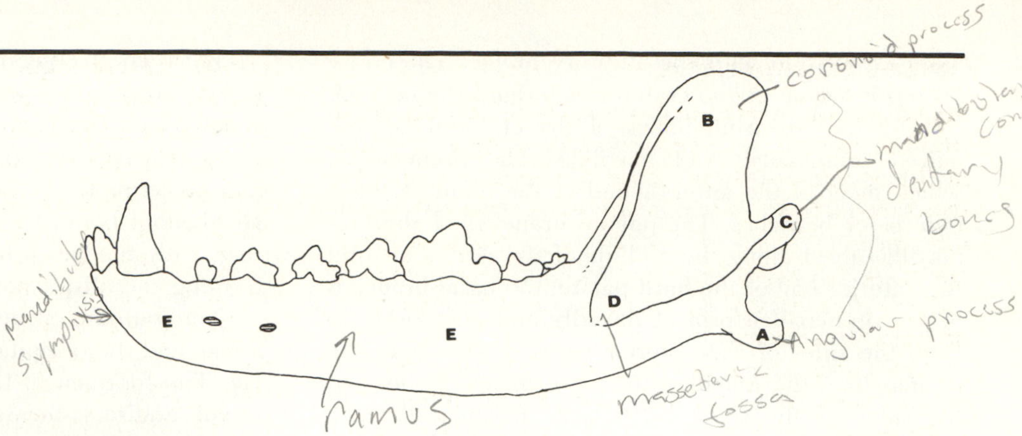

2-B. Locate on a *Canis* mandible each of the structures listed above in boldface type. Label these on Figure 2-2. Check your identifications with the key at the end of this chapter.

Cranial Variations

The skull of *Canis* may be considered to represent a "typical" mammal skull. From this typical structure many variations occur. The postorbital bar, mastoid process, and other structures conspicuous in some mammals, but absent in *Canis*, have already been mentioned.

The relative lengths of braincase and rostrum vary considerably. Mammals such as the whales and anteaters have relatively short braincases and long rostra, while other species, such as man, have large braincases and virtually no rostra.

The orbits may be directed anywhere from laterally, as in the pronghorn, to anteriorly, as in man. They may be positioned low on the head, as in raccoons, or high on the skull, as in woodchucks.

Nasal bones may be short and broad, or long and narrow. Palatal, nasal, or both branches of the premaxillae may be enlarged, reduced, or lost.

Zygomatic arches may be incomplete, weak, or amazingly robust. Auditory bullae may be complete, incomplete, inflated, or compressed.

Many other such variations can and do exist but are far too numerous to list.

2-C. To get an idea of the range of variation that exists in mammalian skulls, make as many of the following comparisons as possible.
   a. Compare the degree of separation of the orbit and temporal fossa in shrew, man or monkey, raccoon, cat, and horse.
   b. Compare the bone structure of the temporal region in *Canis*, cat, and man.
   c. Compare the relative lengths and sizes of the rostrum and braincase in an opossum, shrew, man, coyote, cat, horse, and elephant.
   d. Compare the position of the orbits in a man, raccoon, otter, woodchuck, and deer.
   e. Compare the size and proportion of the nasal bones in an opossum, man, porpoise, elephant, horse, tapir, and moose.
   f. Compare the zygomatic arches of an opossum, shrew, man, *Canis*, rat, porcupine, porpoise, and horse.
   g. Compare the structure of the auditory bullae in a hedgehog, man, *Canis*, kangaroo rat, bear, porpoise, and deer.
   h. Compare the placement of the foramen magnum of an opossum, monkey, and deer.

Determination of Maturity

There are several methods of determining the age of an individual. These are discussed in detail in Chapter 30. Since most identification keys, including the ones in this manual, are for adult mammals only, it is necessary for you to be able to distinguish between immature and adult animals. Two cranial characteristics help to identify immature specimens but neither of these is perfect. An individual in which it is evident that teeth are not yet fully erupted is usually an immature specimen. The degree of fusion of cranial sutures is generally an indication of age. An immature specimen will have poorly fused sutures and an old adult will have sutures that are almost indiscernible. If a skull has fully erupted dentition and fully fused cranial sutures, it should be possible to identify it using the keys in this manual. The keys may or may not correctly identify a specimen that does not meet these criteria.

Cranial Measurements

Several measurements that are more or less standardized are used in gaining information about variation in the mammalian skull. Since the skull is a complex structure that can vary in many ways, different sets of measurements are used for different groups of mammals. The ten most frequently taken measurements for a *Canis* skull would not be the same as the ten for a porpoise or a rodent.

Skull measurements are taken in a straight line between two points and are recorded in millimeters. **Calipers** are customarily used for taking these measurements. Dial calipers are the easiest and most efficient type to use. While various brands and models differ in design, in most models the centimeters are read directly from the bar and millimeters and tenths of millimeters are read directly from the dial mounted on the movable slide. Vernier calipers are equally accurate, but are slightly more difficult to read. Again models vary in precise design, but in most models centimeters and millimeters are read directly from the bar, and tenths of millimeters are determined by the best match between gradations on the bar and one of the lines on the sliding scale (Fig. 2-3).

If calipers are not available, measurements can be taken by spanning the desired skull dimension with a divider and then measuring the distance between the points of the divider with a rule. This method is not as precise as the use of a caliper.

Whenever handling specimens, take care not to damage them. Calipers should be closed to fit snugly but be careful not to crush the bone.

The measurements listed below are a selection of those most frequently taken. An asterisk (*) indicates those that are taken on most species.

Figure 2–3.
Reading a vernier caliper. The beam is graduated in millimeters and the number of full millimeters is read at the "0" line of the slide or vernier scale. (In this instance 37. mm.) Tenths of millimeters are determined by the closest match between lines on the beam and the slide. In this example the fifth line on the slide coincides most closely with one of the lines on the beam, thus the reading is 37.5 mm. If the seventh line on the slide had coincided with one of the lines on the beam the reading would be 37.7 mm. (J. & S. Precision Scientific Instrument Mfrg. Co.)

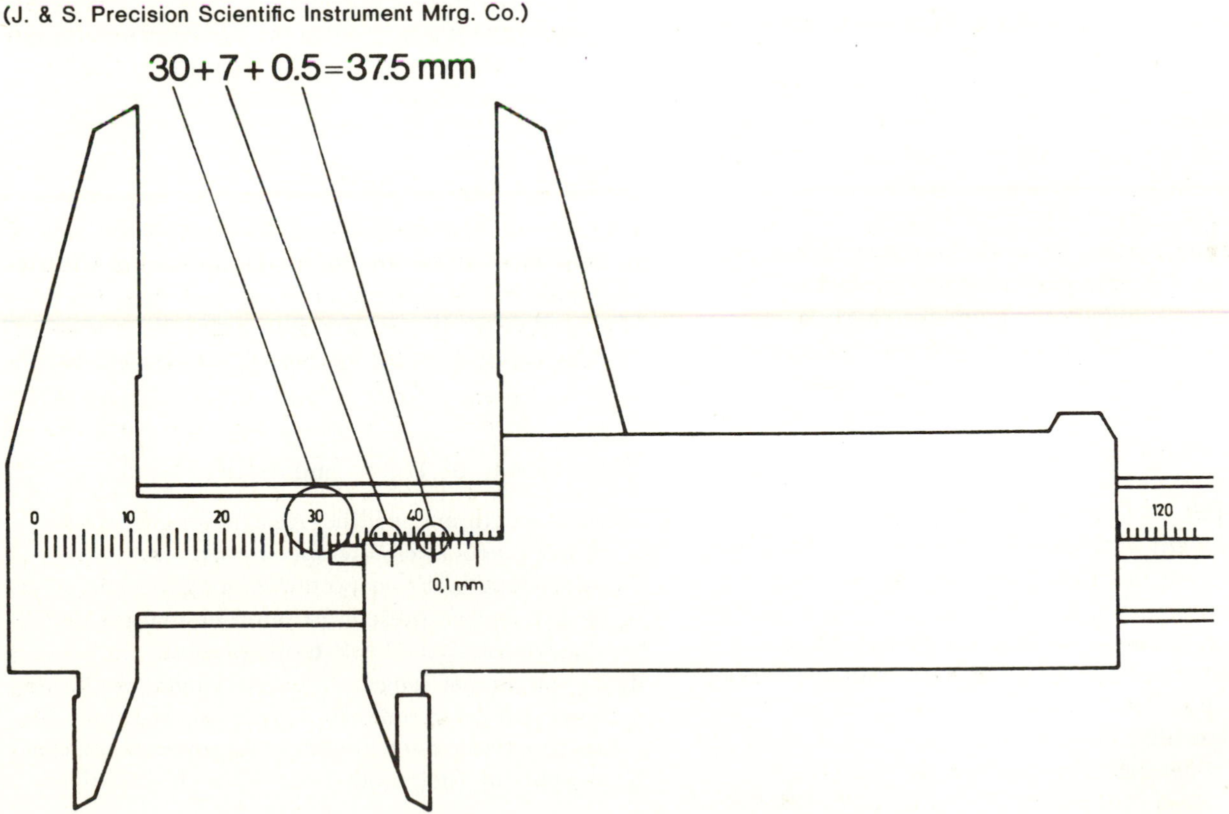

Fig. 2–4.
Felis skull showing points for taking measurements of the ventral side.
(A.F. DeBlase)

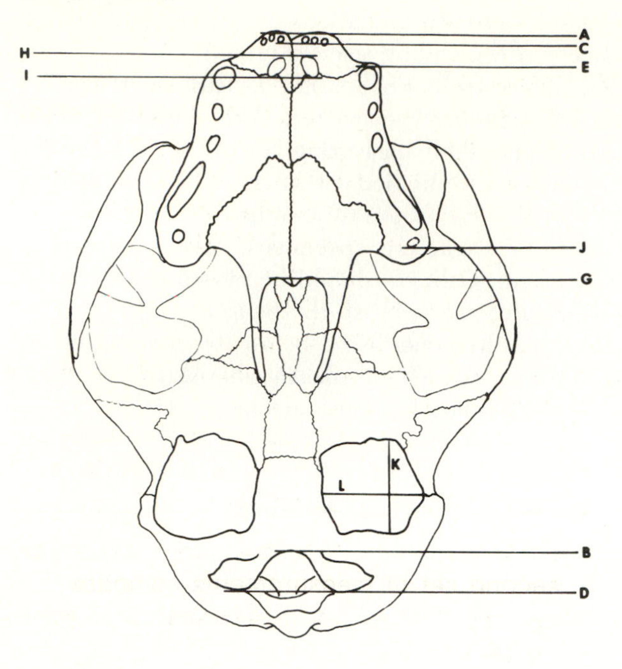

Figure 2–5.
Felis skull showing points for taking measurement of the dorsal side.
(A.F. DeBlase)

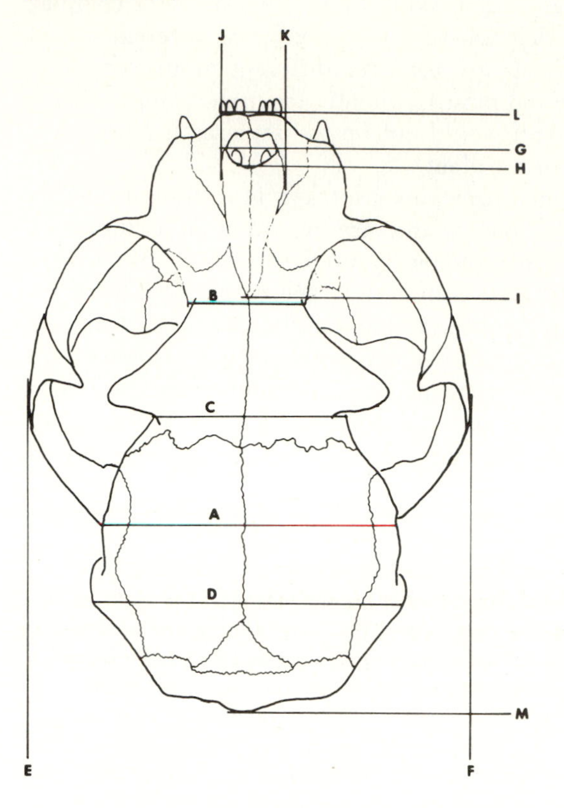

Measurements of the Entire Skull

All measurements of length are taken along the midline of the skull.

Basal length. From the anterior edge of the premaxillae to the anteriormost point on the lower border of the foramen magnum, Figure 2-4, A-B.

Basilar length. From the posterior margins of the alveoli of the upper incisors to the anteriormost point on the lower border of the foramen magnum, Figure 2-4, C-B.

***Condylobasal length.** From the anterior edge of the premaxillae to the posteriormost projection of the occipital condyles, Figure 2-4, A-D.

Condylocanine length. From the anterior edges of the alveoli of the upper canines to the posterior edges of the occipital condyles. (Usually taken instead of condylobasal length in forms in which the premaxillae are frequently lost.) Figure 2-4, E-D.

***Greatest length of skull.** From the most anterior part of the rostrum (excluding teeth) to the most posterior point of the skull, Figure 2-5, L-M.

***Breadth of braincase.** Greatest width across the braincase posterior to the zygomatic arches, Figure 2-5, A.

***Least interorbital breadth.** Least distance dorsally between the orbits, Figure 2-5, B.

Mastoid breadth. Greatest width of skull including the mastoid, Figure 2-5, D.

***Postorbital constriction.** Least distance across the top of the skull posterior to the postorbital process, Figure 2-5, C.

Rostral breadth. Breadth of rostrum usually taken at suture between premaxillae and maxillae, but may be taken at any designated point, e.g., rostral breadth at M^1.

***Zygomatic breath.** Greatest distance between the outer margins of the zygomatic arches, Figure 2-5, E-F.

Measurements of Palate and Upper Teeth

Alveolar length and width. Greatest length or width of the alveolus of any specified tooth.

Diastema length. From posterior margin of alveolus of last incisor present to anterior margin of alveolus of first cheek tooth present.

Incisive foramina length. Greatest length of the anterior palatal foramina, Figure 2-4, H-I.

***Maxillary tooth row.** Length from anterior edge of alveolus of first tooth present in the maxillae to posterior edge of alveolus of last tooth, Figure 2-4, E-J.

***Palatal length.** From anterior edge of premaxillae to anteriormost point on posterior edge of palate, Figure 2-4, A-G.

Palatilar length. From posterior edges of alveolae of first incisors to anteriormost point on posterior edge of palate, Figure 2-4, C-G.

Palatal width. Usually width of palate between alveoli of any specified pair of teeth. Occasionally includes alveoli.

Measurements of Other Portions of the Skull

Nasal length. From anteriormost point of nasal bones to posteriormost point taken along midline of skull, Figure 2-5, G-I.

Nasal width. Greatest width across nasals, Figure 2-5, J-K.

Nasal suture length. Greatest length of suture along midline of paired nasal bones, Figure 2-5, H-I.

Postpalatal length. From anteriormost point on posterior edge of palate to anteriormost point on lower edge of the foramen magnum, Figure 2-4, G-B.

Tympanic bullae length and width. Greatest length and width of bulla, Figure 2-4, K and L.

Measurements of Mandible and Lower Teeth

Mandibular diastema. Same as for maxillary diastema, Figure 2-6, A-B.

***Mandible length.** Greatest length of the mandible, usually excluding teeth, Figure 2-6, D-E.

Fig. 2–6.
Rodent jaw showing points for taking the most commonly used measurements.
(A.F. DeBlase)

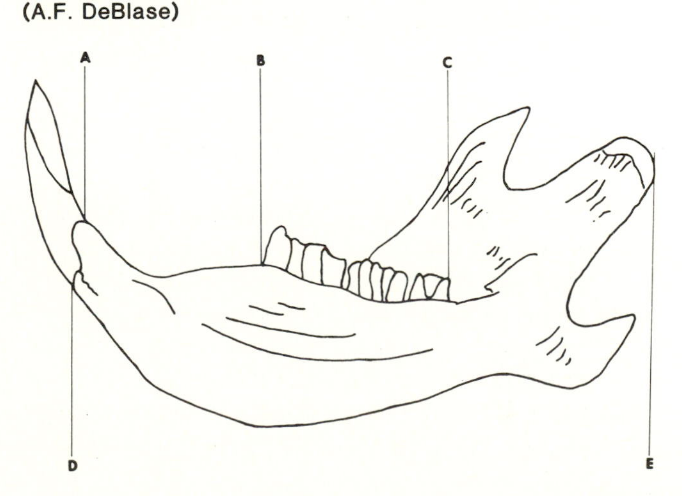

***Mandibular tooth row.** Length from anterior edge of alveolus of canine or first cheek tooth to posterior edge of alveolus of last tooth. The incisor is not included in this measurement, Figure 2-6, B-C.

2-D. Take each of the measurements listed above on the following mammals (not all can be made on all skulls; e.g., length of diastema cannot be taken on an animal without a diastema). Record measurements to nearest tenth of a millimeter.

- *Canis* or other carnivore
- Rat or other rodent
- Man or other primate

Compare your figures with those of others in the class.
How closely do you agree with others who measured the same specimen? (A measurement of **accuracy.**)

2-E. Remeasure one specimen. How does your second set of measurements compare with your first? (A measurement of **precision.**)

Key to Labeling of Figures

Figure 2-1

1. alisphenoid bone
2. alisphenoid canal
3. auditory bulla (tympanic bone)
4. basioccipital
5. basisphenoid
6. external auditory meatus
7. foramen magnum
8. frontal
9. infraorbital foramen
10. jugal
11. lacrimal
12. mandibular fossa
13. maxilla
14. nasal
15. occipital bone
16. occipital condyle
17. orbit
18. orbitosphenoid
19. palatal (=incisive) foramen
20. palatine
21. parietal
22. paroccipital process
23. postorbital process (of the frontal)
24. premaxilla, palatal branch
25. premaxilla, nasal branch
26. presphenoid

27. pterygoid
28. sagittal crest
29. squamosal
30. temporal fossa
31. temporal ridge
32. vomer
33. zygomatic process of squamosal
34. zygomatic process of maxilla

Figure 2-2

A. angular process
B. coronoid process
C. mandibular condyle
D. masseteric fossa
E. ramus

Supplementary Readings

Crompton, A. W. 1963. The evolution of the mammalian jaw. *Evolution* 17:431-439.

Hildebrand, M. 1974. *Analysis of vertebrate structure.* John Wiley and Sons, New York. 710 pp.

Mystkowska, E. T. 1966. Morphological variability of the skull and body weight of the red deer. *Acta Theriol.* 11:129-194.

*Thomas, O. 1905. Suggestions for the nomenclature of cranial length measurements and of the cheek teeth of mammals. *Proc. Biol. Soc. Washington* 18:191-196.

*Included in Jones, Anderson and Hoffman (1976).

3 Teeth

Although mammalian teeth are similar in basic components, they exhibit great diversity in number, size and shape. The radiation of mammals into virtually every macrohabitat has resulted in evolutionary adaptations in tooth morphology to cope with varied diets. Teeth are readily fossilized, and many extinct mammals are known only from teeth. Thus, teeth are valuable tools in classifying, identifying, and studying mammals.

Tooth Anatomy and Replacement

The **crown** is the portion of a tooth exposed above the gumline; the **root** is the portion fitting into the **alveolus** or socket in the jaw. Teeth with a particularly high crown are termed **hypsodont,** and those with a particularly low crown are **brachydont.** Points and bumps on the crown of the tooth are generally termed **cusps** (see Premolar-Molar Section for nomenclature of crown elements). Teeth may be unicuspid, bicuspid, tricuspid, etc. The side of a tooth closest to the tongue is termed the **lingual** side, and the side closest to the cheek is the **labial** or **buccal** side. The surface of a tooth that meets with a tooth in the opposing jaw is termed the **occlusal** surface.

The major portion of each tooth is made up of a bonelike material called **dentine** (Fig. 3-1). The crown has a thin layer of hard, usually white, **enamel** covering the dentine, and the root is covered by a layer of bonelike **cementum.** The central, living portion of a growing tooth, the **pulp,** is supplied with blood vessels and nerves through one or more openings in the base. In most species when the tooth has reached a certain size, this opening constricts, the blood supply is much reduced, and growth ceases. Such teeth are termed **rooted.** In some groups the opening does not constrict and growth of the tooth continues throughout the life of the mammal. Such evergrowing teeth are termed **rootless.**

Figure 3–1.
Diagrammatic cross section of a mammalian tooth. A, enamel; B, dentine; C, pulp; D, root canal; E, cementum; F, crown; G, root.
(L.P. Martin)

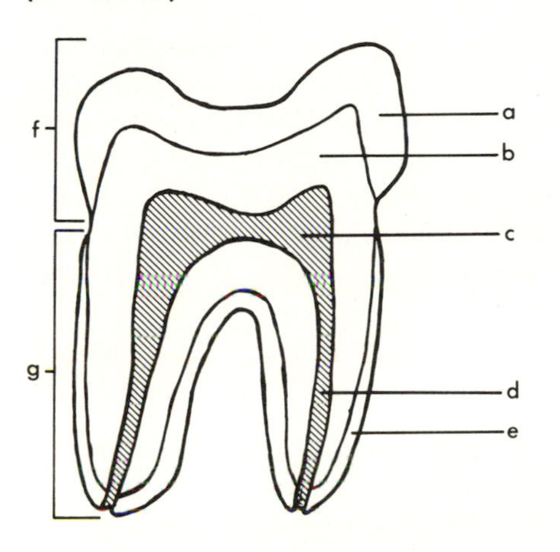

Most mammals are **diphyodont,** having only two sets of teeth. The **deciduous** or **milk teeth** present in immature mammals are usually replaced by a set of **permanent teeth** that are retained for life. Toothed whales, Odontoceta, and a few other mammals are **monophyodont,** having only one set of teeth. Marsupials and some other mammals have only some of the milk teeth replaced and others remain as a part of the adult dentition.

Some mammals, such as elephants, manatees, and kangaroos, have a slightly different system of tooth replacement. These mammals feed primarily upon harsh vegetation and this diet makes for considerable wear on their teeth. In elephants, the alveoli of the cheek teeth converge into a groove and tooth replacement occurs only at the posterior end of the toothrow. As the anterior tooth is worn away, a new tooth develops from the rear and the entire row moves forward (Fig. 3-2). A total of six cheek teeth are avail-

Figure 3–2.
Diagram of tooth replacement in an elephant jaw. A, portion of tooth worn away; B, portion exposed; C, portion still embedded in the jaw. Arrow indicates direction of tooth replacement.
(A.F. DeBlase)

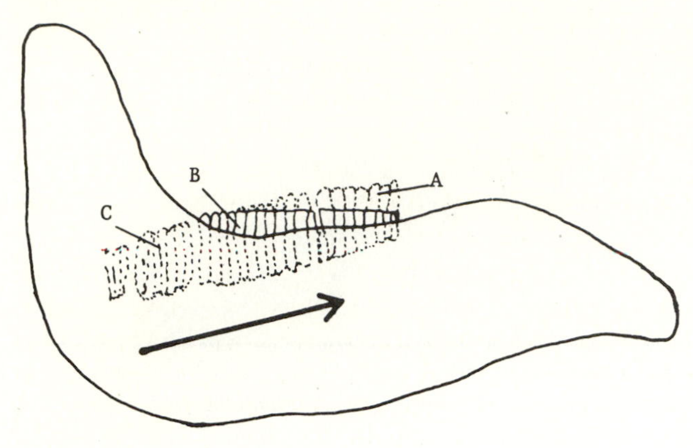

able to each quadrant, but only one or parts of two teeth are functional at any one time. Manatees have a similar system of tooth replacement with a potential number of twenty teeth per jaw, but only six to eight function at one time. In kangaroos, tooth replacement is primarily from the rear of the jaw, but the anterior two deciduous cheek teeth are replaced from below by a single tooth.

3-A. Examine the internal structure of a sectioned tooth and compare with Figure 3-1.

3-B. Examine a coyote or dog skull (genus *Canis*) and note the placement of teeth in the alveoli. Examine a similar specimen from which the teeth have been removed. How many roots does each tooth have? How many cusps?

3-C. Examine skulls of mammals that are in the process of shedding their deciduous teeth. How does replacement occur?

3-D. Examine an elephant jaw. How many cheek teeth are present? Can you notice any difference in wear between the first and last tooth in each jaw (excluding tusks)?

The Kinds of Teeth

An individual mammal usually has two or more morphologically different kinds of teeth, a condition termed **heterodont.** This contrasts with the **homodont** dentition of lower vertebrates, in which all teeth in an individual resemble each other in shape. In mammals four basic kinds of teeth are recognized: **incisors, canines, premolars,** and **molars.**

Incisors

Incisors are the teeth rooted in the premaxillary bone and the corresponding teeth in the lower jaw. Placental mammals never have more than three incisors in each jaw quadrant, but marsupials may have up to five in each half of the upper jaw and up to four in each half of the lower jaw. These are usually unicuspid teeth with a single root, but in some groups of mammals accessory cusps, additional roots, or both may be present.

Incisors are generally chisel-shaped teeth that function primarily for nipping; e.g., a human biting an apple, or a horse cropping grass. In cattle, deer, and their relatives, this nipping action has been modified by the loss of the upper incisors. Instead of nipping the vegetation between upper and lower incisors, these animals use their highly mobile lips and prehensile tongue to draw vegetation across the lower teeth, which cut it off in much the same way that a tape dispenser cuts tape. In rodents and certain other specialized forms, the number of incisors has been reduced, but those remaining are stout chisel-edged teeth used in gnawing. These incisors are rootless and grow continually as they are worn away at the tips. In vampire bats the first pair of incisors has a long, sharp edge. These teeth are used to shave away a layer of skin to expose blood vessels. The blood that flows to the surface is then lapped up. Elephants have incisors that are enlarged to form tusks. These are rootless and evergrowing and may be used for digging and removing bark from trees. Shrews have incisors that project anteriorly and act as forceps in catching and holding insects and other prey.

3-E. Examine the incisors of a shrew, vampire bat, monkey, rodent, horse, and cow, sheep, or deer. What can you deduce about the diet or feeding habits of each of these mammals?

3-F. Examine the pectinate (comblike) lower incisors of a colugo (Dermoptera: Cynocephalidae). Compare these incisors with those of the ringtail lemur (Primates: Lemuridae). In what way are the incisors similar? How do they differ? What is their function?

Canines

Canines are the most anterior teeth rooted in the maxillae and the corresponding teeth of the lower jaw. They never number more than one per quadrant. Canines are usually long, conspicuous, unicuspid teeth with a single root. However, some mammals, particularly some primitive forms, may have canines with accessory cusps, additional roots, or both.

Canines are usually used to capture, hold, and kill prey. In herbivorous species they are frequently reduced or absent. In some groups, such as the hogs and some primitive deer, they are very long and sharp and used for fighting. Pig "tusks" are rootless and in some species arranged so they do not fully occlude. This minimizes wear and allows at least the upper tusks to grow very long. Walruses have been said to use their elongated canines to scrape the mollusks that they feed upon from the ocean floor, but recent evidence indicates that these conspicuous teeth are not used in this way (Ray 1973, Miller 1975).

Frequently canines and/or other teeth are absent leaving a wide space between the anterior teeth and the cheek teeth. Any such wide gap between teeth is termed a **diastema.**

Note! In some species the most conspicuous unicuspid tooth in the anterior part of the jaw is not the canine. Occasionally the last incisor is large and **caniniform** and the canine is absent or small and resembles a premolar. Conversely, the first premolar is occasionally caniniform and the canine is small and **incisiform.** In the upper jaw these teeth are easily identified by locating the suture between the premaxilla and maxilla.

3-G. Examine the canines of the following pairs of mammals. Can you suggest functions (if any) for the specializations?
- a. Peccary (*Tayassu)* and wart hog (*Phacochoerus*)
- b. *Canis* and *Felis*
- c. Man and baboon (*Papio*)
- d. Extinct sabre-toothed marsupial (†*Thylacosmilus)* and sabre-toothed cats (e.g., †*Smilodon).* See Romer (1966, pp. 203 and 234 respectively).

Premolars and Molars

Premolars are situated posterior to the canines and differ from molars in having deciduous predecessors in the milk dentition. Molars are situated posterior to the premolars and do not have deciduous predecessors. Authorities disagree as to whether molars are permanent teeth for which there are no corresponding milk teeth, or whether they are milk teeth which erupt late and are not replaced. Premolars are usually smaller than molars and have fewer cusps. However, without embryological investigation or a knowledge of the milk dentition of the species being studied, it is frequently impossible to distinguish between premolars and molars in an adult mammal. Therefore, these two tooth types frequently are referred to together as **cheek teeth, postcanine teeth,** or **molariform teeth.**

Placental mammals generally have a maximum of four postcanine teeth in the milk dentition and seven postcanine teeth in the adult dentition. Thus they are regarded as having a maximum of four premolars and three molars. Marsupials have only a single postcanine tooth in the milk dentition. This milk tooth corresponds to the third postcanine tooth in the adult dentition. Thus marsupials are regarded as having a maximum of three premolars and four molars. Teeth are absent in adult monotremes.

Since cheek teeth do the major job of masticating food, they are the teeth that exhibit the greatest diversity correlated with diet. Cheek teeth occur that are adapted for such a variety of foods as mollusks, meat, soft vegetation, tough grasses, hard-bodied insects, worms, and krill. The structure of the cheek teeth is one of the most important criteria in mammalian classification.

A standardized terminology for dental crown elements that is acceptable to all paleontologists and mammalogists is not presently available. The greatest obstacles to the development of a generally accepted terminology are questions of homology of cusps between early and later groups of mammals. Some workers (Vandebroek 1967) advocate an examination of tooth crests to determine homologies. Others (Patterson 1956) maintain that only complete, interconnecting fossil material will provide the final solution. Our approach in this section is to discuss tooth topography of Mesozoic mammals without reference to formal cusp names. Modified Cope-Osborn cusp names are used for certain Cretaceous and Cenozoic mammals. Our terminology is derived, in part, from information presented by Patterson (1956), Van Valen (1966), Szalay (1969), and Hershkovitz (1971).

Cheek Teeth of Triassic-Jurassic Mammals

The earliest known mammals, members of the late Triassic families †Eozostrodontidae and †Kuehneotheriidae, had teeth with four main cusps arranged in a longitudinal or slightly triangular row (Figs. 3-3 and 3-4). A shelflike ridge, termed the **cingulum,** bor-

†Extinct groups are indicated by a dagger (†) preceeding the name.

Figure 3–3.
Upper molars of †*Eozostrodon*, a late Triassic mammal that is frequently classified in the order †Triconodonta. Lingual (A) and occlusal (B) views showing the essentially linear arrangement of the major cusps.
(Modified from Crompton and Jenkins 1968)

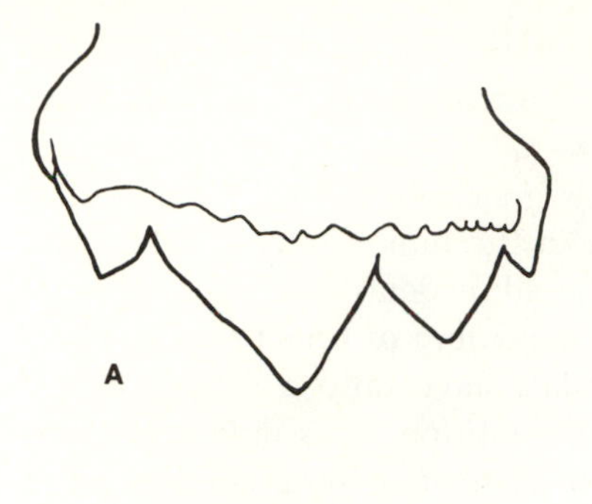

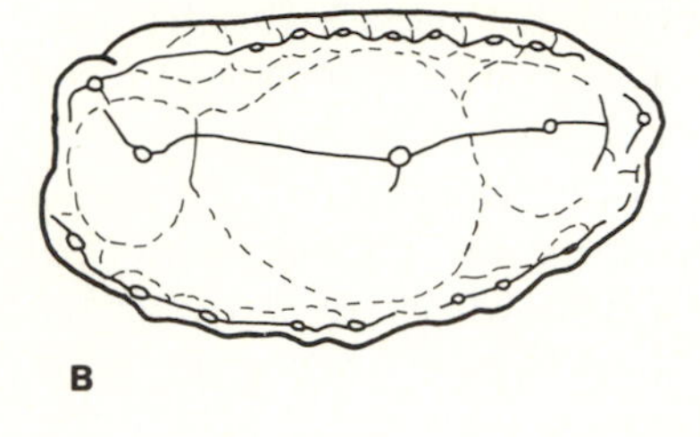

Figure 3–4.
Upper molars of †*Kuehneotherium*, a late Triassic mammal, sometimes classified as an early member of the †Symmetrodonta and sometimes as an early member of the †Pantotheria. Lingual (A) and occlusal (B) views, showing partially triangular outline formed by the major cusps.
(Modified from Crompton and Jenkins 1968)

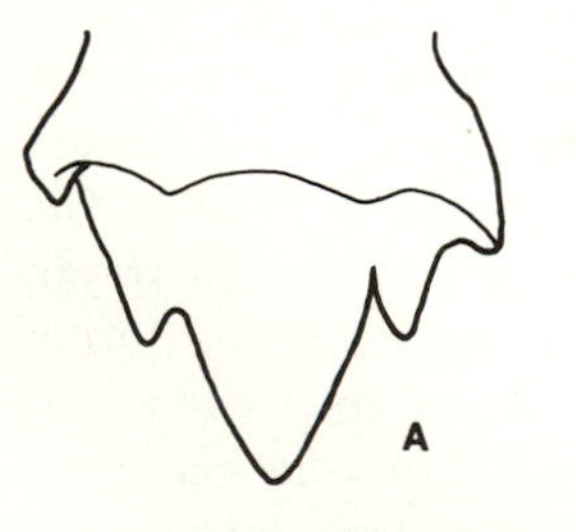

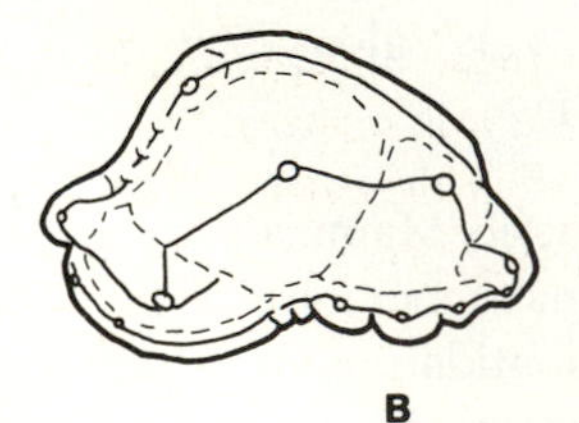

Figure 3–5.
Upper and lower cheek teeth of a triconodont, †*Priacodon*, an upper Jurassic member of the order †Triconodonta. The cusp arrangement on the occlusal surface is similar, although not identical, to that of †*Eozostrodon.*
(Modified from Gregory 1934)

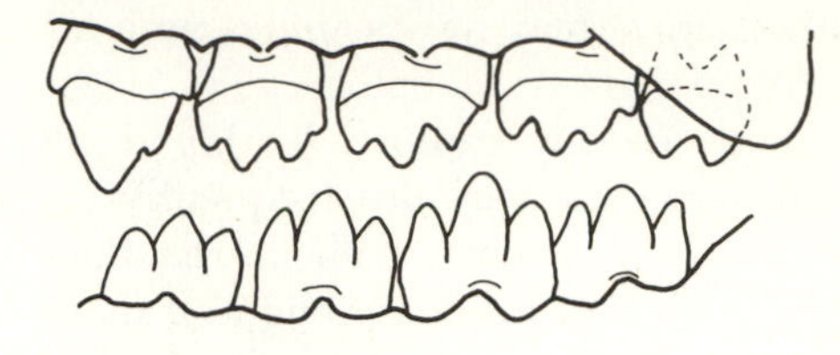

Figure 3–6.
Occlusal view of right upper (A) and left lower (B) molars of †*Docodon*, a member of the upper Jurassic order †Docodonta.
(Modified from Butler 1941)

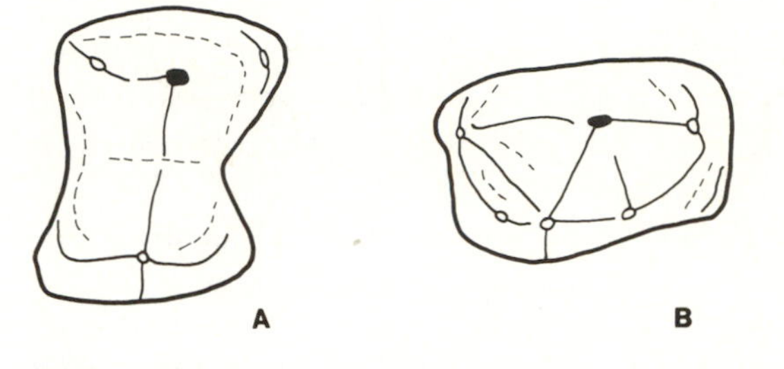

dered the labial and lingual base of the tooth crown. A single **cingulid**, the lower tooth equivalent of the cingulum, was present on the lingual side of the crown. The main, highest cusp of the upper molar occluded in the **embrasure**, or space, between cusps of two immediately adjacent lower molars. Presumably, the molars sheared together in a cutting or scissorslike action known as **embrasure shearing.** Members of the †Triconodonta (†Triconodontidae and †Amphilestidae) had teeth (Fig. 3-5) similar to those of the †Eozostrodontidae (early members of the order †Docodonta) which are generally termed **triconodont.** Some living mammals (e.g., *Lobodon*) show a tooth form that parallels that of the triconodonts, but this undoubtedly represents a secondary acquisition of this tooth form.

The †Docodonta (not including †Eozostrodontidae) have rectangular lower, and hourglass-shaped upper molars (Fig. 3-6). The main cusps are situated near the labial border of the tooth with additional cusps present on the expanded internal cingulum and cingulid. During occlusion, these teeth were capable of shearing and crushing food.

The molars of the order †Multituberculata had two (upper and lower) or three (some upper) rows of cusps. The lower premolars were frequently enlarged for a

Figure 3-7.
Cheek teeth of two members of the order †Multituberculata. Lower dentition of †*Ptilodus douglassi*, occlusal (A) and lingual (B) views. Upper dentition (C) of †*Ptilodus* sp., occlusal view. Note the distinctive fourth lower premolar and the two or three rows of cusps on each molar.
(A and B, modified from Simpson 1937; C, modified from Osborn 1907)

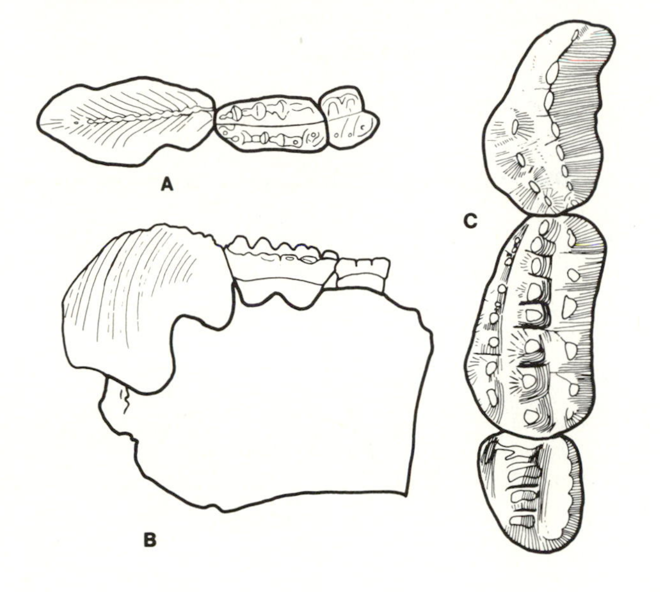

shearing occlusion with the corresponding upper teeth, while the upper molars occluded directly with their lower counterparts. The multituberculate dentition (Fig. 3-7) was presumably adapted for an herbivorous diet.

The earliest therian mammals, members of the orders †Symmetrodonta and †Pantotheria (=†Eupantotheria of recent authors), had essentially triangular-shaped molars. In †Symmetrodonta, there were three main cusps arranged in the form of an isosceles triangle (Fig. 3-8). In the upper molar this arrangement of cusps is termed a **trigon;** in the lower molar a **trigonid.** The occlusion was essentially embrasure shearing in that the molars did not directly occlude. That is, when the jaws closed, the teeth of the upper jaw fitted into the spaces between the teeth of the lower jaw. No grinding surface was present.

The upper molars of the †Pantotheria were essentially like those of the symmetrodonts (Fig. 3-9). However, a posterior extension of the cingulum, termed the **talonid,** was present in the lower molars. Pantothere molars were capable of embrasure-shearing occlusion and direct occlusion by action of trigon on trigonid. The small talonid served primarily as a stopping device for the upper molar but permitted some crushing action with the upper tooth.

Figure 3-8.
Occlusal (A) and labial (B) views of symmetrodont molars. The upper molars are based on †*Peralestes*, the lower molars, on †*Spalacotherium*. In A, the lower molars are indicated by the dotted outlines. The action of these molars was primarily shearing and cutting, with little "cusp on cusp" occlusion.
(Modified from Gregory 1934)

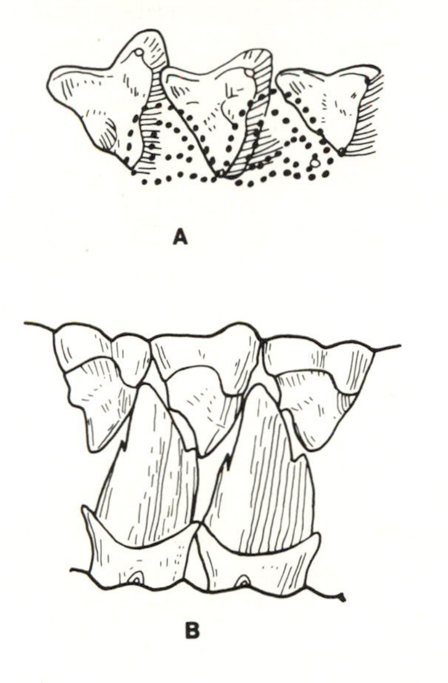

Figure 3-9.
Occlusal view of dryolestid pantothere molars. The upper molars, based on †*Melanodon*, are indicated by light lines, the lower molars, based on †*Laolestes*, by bold lines. On the lower molars, note the shelf-like extension, the talonid, that occludes with the apex of the trigon.
(Modified from Gregory 1934)

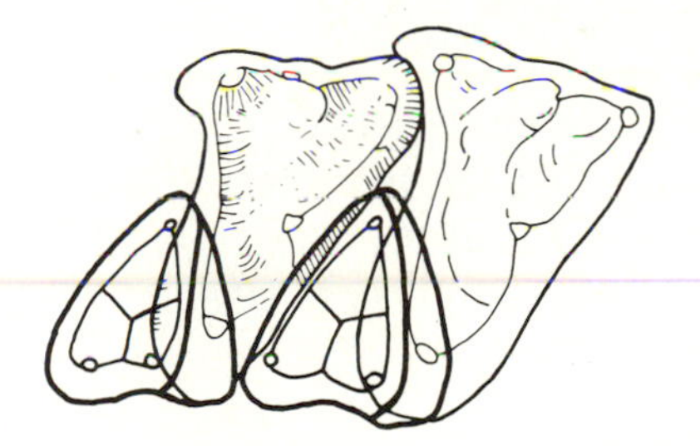

3-H. Examine the cheek teeth (models, photographs, or drawings, if necessary) of Triassic and Jurassic mammals. Can you identify the major groups of early mammals?

Simple Tribosphenic Cheek Teeth

The earliest known tribosphenic cheek teeth were present in primitive marsupials and placentals of the Cretaceous. A simple **tribosphenic** upper molar (Fig.

3-10A) has a trigon with three major cusps: a lingual cusp, the **protocone,** at the apex of the trigon; an anterior **paracone,** and a posterior **metacone.** The **stylar shelf,** a broad ledge situated labial to the paracone and metacone, has a stylocone, parastyle, and other stylar cusps. The **parastyle,** the most anterior stylar cusp, provides a convenient directional reference point to orient a tooth for study. The lower tribosphenic molar (Fig. 3-10B) consists of a high-cusped trigonid and lower-cusped talonid. The talonid is much larger than that found in pantotheres and helps to square the outline of the crown. The posteriormost cusp of the talonid is the **hypoconulid.** The main cusps of the talonid are the labial **hypoconid** and the lingual **entoconid.** The latter three cusps enclose a depression (sometimes absent) known as the **talonid basin,** that receives the protocone of the trigon during occlusion. The anteriormost cusp of the trigonid is the **paraconid.** Two additional cusps, a labial **protoconid** and a lingual **metaconid** are situated at the posterior portion of the trigonid.

The tribosphenic molar has been modified in many ways to produce the varied dentition of all modern mammals. This molar type is important because it permits two main types of occlusion: a shearing or cutting action between trigon and trigonid; and a mortar and pestle, or crushing action between protocone and talonid. Some of the premolars, particularly the most posterior ones, show the same crown specializations as the molars.

Two terms sometimes encountered in association with tribosphenic dentitions are **tritubercular** molar and **tuberculosectorial** molar. These terms are essentially synonyms of the upper and lower tribosphenic molars respectively.

The simple tribosphenic molar was present in Cretaceous therian mammals. Simple tribosphenic molars are also found in certain living marsupials and placentals, although some authors utilize different terminology to describe these (see next section).

3-I. Examine the upper and lower tribosphenic molars of an opossum (Didelphidae). Identify the trigon, trigonid, talonid, talonid basin, and the major cusps. Observe the shearing and crushing actions that occur as the upper and lower jaws are brought into occlusion.

Modified Tribosphenic Cheek Teeth

The simple tribosphenic cheek tooth has been modified in various lineages of mammals. For convenience, some authors (Butler 1941; Hershkovitz 1971; Turnbull 1971) divide the simple and derived tribosphenic molars into three main groups: zalambdodont, dilambdodont, and euthemorphic. Although these modifications apply to molars and some premolars, particularly the most posterior premolars in a series, the discussion that follows is based on molars. A **zalambdodont** upper molar (Fig. 3-11) is characterized by a V-shaped ectoloph. An **ectoloph** is a series of **cristae,** or crests, connecting the paracone (and sometimes the metacone) with cusps on the stylar shelf. Typically the zalambdodont molar lacks a protocone, and the paracone (sometimes combined with the metacone) is located at the lingual apex of the crown. This type of molar is found in many Insectivora and in the marsupial mole, *Notoryctes.* A dilambdodont upper molar (Fig. 3-12) has a W-shaped occlusal surface with the protocone near the lingual apex of the trigon. The

Figure 3–10.
Occlusal views, somewhat diagrammatic, of simple tribosphenic left upper (A) and left lower (B) molars. *Upper crown elements*: c, cingulum; cr, crista; mt, metacone; mtc, metaconule; pa, paracone; pr, protocone; prc, paraconule; pst, parastyle. *Lower crown elements:* cd, cristid; end, entoconid; hyd, hypoconid; hycd, hypoconulid; mtd, metaconid; pad, paraconid; prd, protoconid; tlb, talonid basin. Major cusps in solid black. Based on information in Van Valen (1966) and Szalay (1969).
(Modified from Van Valen 1966)

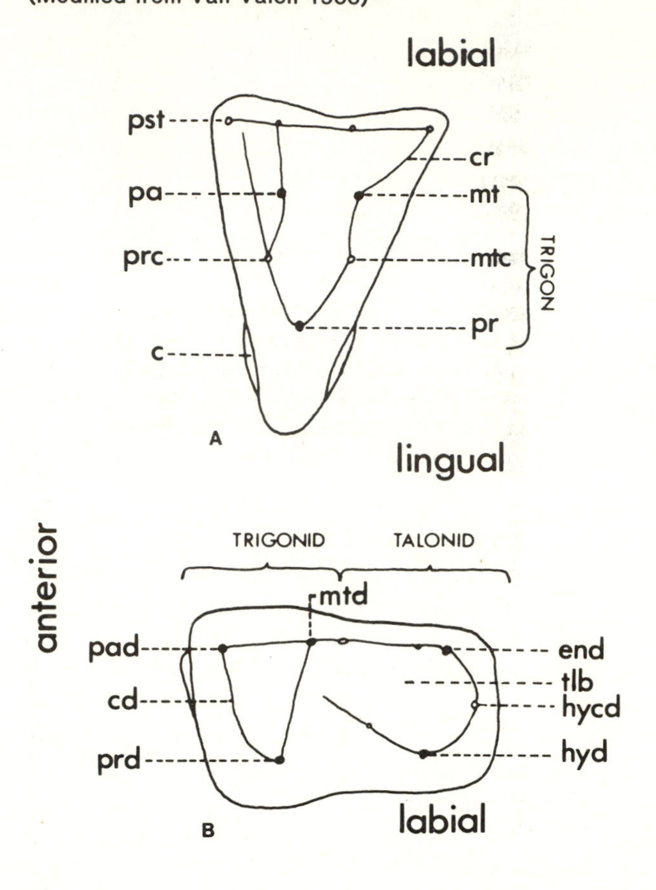

Figure 3–11.
Occlusal view of zalambdodont right upper (A) and left lower (B) molars of the otter shrew, *Potamogale.*
(Modified from Butler 1941)

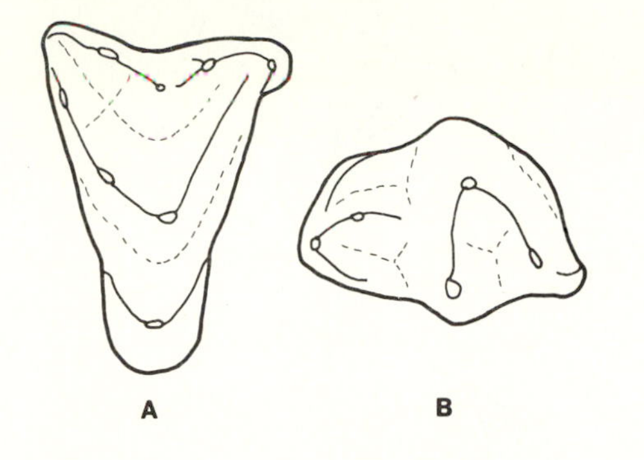

Figure 3–12.
Occlusal view of dilambdodont right upper (A) and left lower (B) molars of a tree shrew, *Tupaia.*
(Modified from Butler 1941)

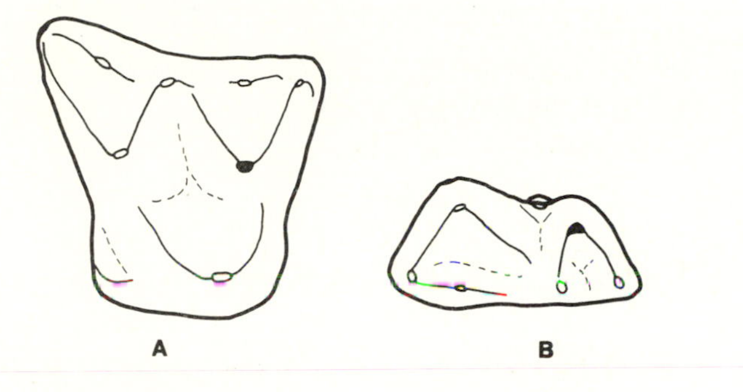

W-shaped pattern is formed by an ectoloph connecting the metacone and paracone with cusps on the stylar shelf. The molars of the opossums, (Didelphidae) and tree shews (Tupaiidae) are examples of the dilambdodont type. A **euthemorphic** upper molar (Fig. 3-13) usually has a square or quadrate crown. The square outline results from the addition of a main cusp, the **hypocone,** to the posterior lingual side of the crown. In certain molars, the hypocone area is identified as the **talon.**

A euthemorphic upper molar with four main cusps is termed **quadritubercular.** Upper and lower molars may become fully **quadrate,** or square, by loss or reduction of some cusps (e.g., in the lower dentition the paraconid is generally lost). Most living mammals have basically euthemorphic molars, although the teeth may be modified in several ways (see next section).

3-J. Examine a zalambdodont upper molar of Chrysochloridae, Solenodontidae or Tenrecidae. Locate the ectoloph and position of the paracone (or fused paracone-metacone).

Figure 3–13.
Bunodont left upper (A) and right lower (B & C) molars of man, *Homo sapiens.* A and B are occlusal views, C is a labial view. *Upper:* hy, hypocone; mt, metacone; pa, paracone; pr, protocone. *Lower:* end, entoconid; hyd, hypoconid; hycd, hypoconulid; mtd, metaconid; prd, protoconid.
(Modified from Osborn 1907)

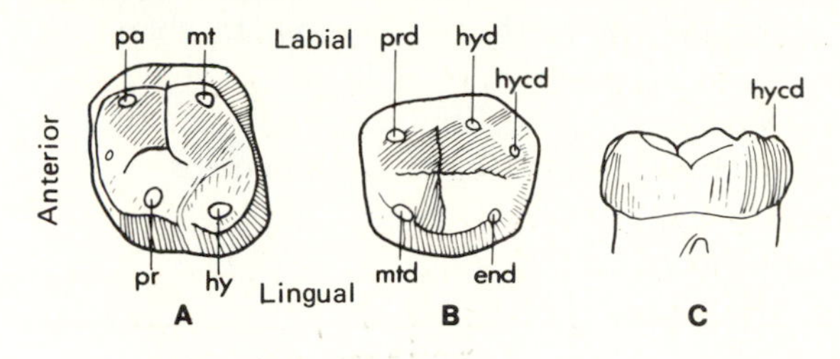

3-K. Examine a dilambdodont upper molar of Didelphidae, Talpidae or Tupaiidae. Locate the ectoloph, paracone, metacone, and protocone.

3-L. Examine upper and lower euthemorphic molars of a pig, man, various rodents, and a horse. In which species do molars show well-defined cusps? Which are quadritubercular? Which quadrate?

Specializations of Cheek Teeth

The bunodont tooth is found in many mammals that are basically omnivorous. The bunodont tooth is euthemorphic, quadrate, frequently brachydont, and has four major rounded cusps (Fig. 3-13). It is considered to have developed from a tribosphenic tooth by the bulging out of the side between the protocone and metacone and the development of a new cusp, the hypocone, in this area (hypoconid in the lower teeth). Other smaller cusps may develop between the larger ones. For example, a small **paraconule (=protoconule)** may develop between the protocone and paracone and a small **metaconule** may develop between the metacone and hypocone. In the lower cheek teeth a **hypoconulid** is situated on the posterior margin of the talonid between the hypoconid and entoconid. The crowns of bunodont teeth oppose each other directly and the paraconid is lost. Men and hogs are examples of mammals with bunodont teeth used for an omnivorous diet.

In mammals that tend toward an herbivorous diet, the cheek teeth frequently are hypsodont. The abrasive action of plant material quickly erodes teeth; so the higher the crown, the longer the tooth will last. Some herbivorous mammals (particularly grazers) have cheek teeth that are rootless and continue to grow throughout life as they are worn away at the top. Many herbivorous mammals have **lophodont** teeth in which cusps fuse to form elongate ridges termed **lophs** (Fig. 3-14). These ridges create elongated abrasive surfaces for the grinding of plant materials. A **selenodont** tooth functions in much the same manner but in it each ridge is formed by the elongation of a single cusp. The ridges of selenodont teeth are always crescent-shaped and longitudinally oriented (Fig. 3-15), while those of lophodont teeth are variable in shape and may be transversely oriented (Fig. 3-14). Some mammals such as the horses, Equiidae, have complex selenolophodont teeth that combine aspects of both of these types.

Figure 3–14.
Lophodont molar tooth of a rhino (order Perissodactyla), occlusal view. Note the fusion of cusps into transverse and longitudinal lophs.
(Modified from Osborn 1907)

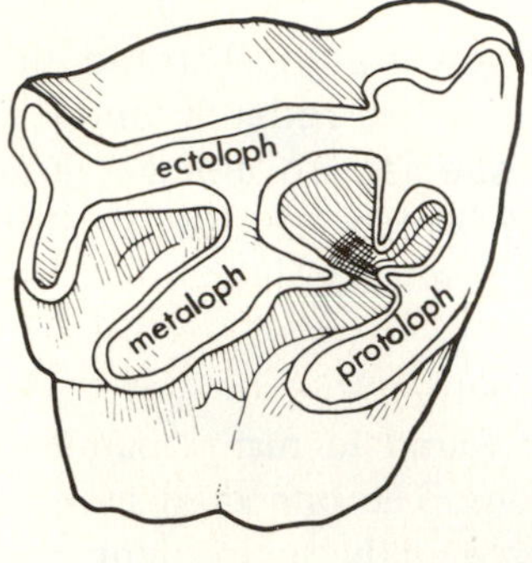

Figure 3–15.
Selenodont upper molars of *Capreolus capreolus*, the roedeer. Note the crescent-shaped patterns on the occlusal surface of each molar.
(Modified from Gromova 1962)

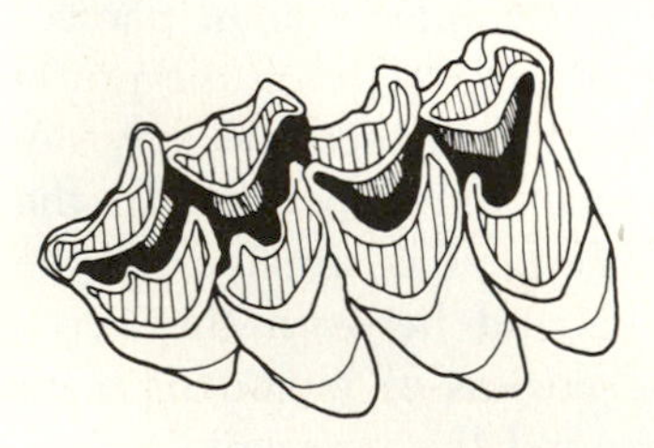

Enamel is the hardest known animal substance and resists wear well. However, when the enamel finally is worn away from the top of a bunodont tooth the softer dentine below wears away much more rapidly leaving a cavity (Fig. 3-16) encircled by a ridge of enamel. As the occlusal surface becomes folded into ridges the wear surface of the tooth has numerous enamel ridges interspersed through the dentine (Fig.

Figure 3–16.
Bunodont tooth (right M_3) of a pig (*Sus scrofa*) that is unworn (A) and worn (B).
(Mary Ann Cramer)

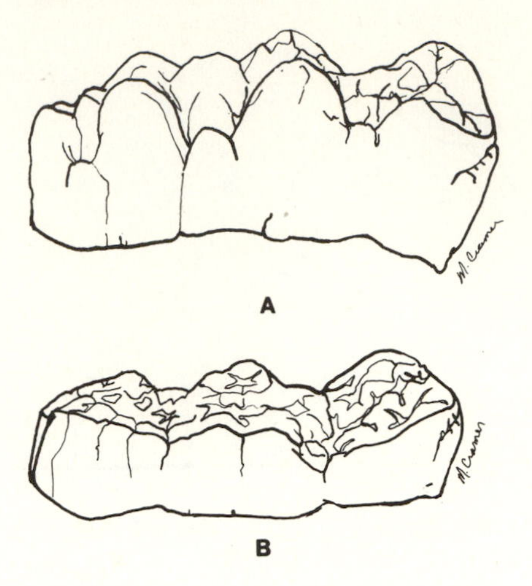

Figure 3–17.
Lophodont tooth (left M^1) of a bison (*Bison bison*) that is unworn (A) and one that is worn (B).
(Mary Ann Cramer)

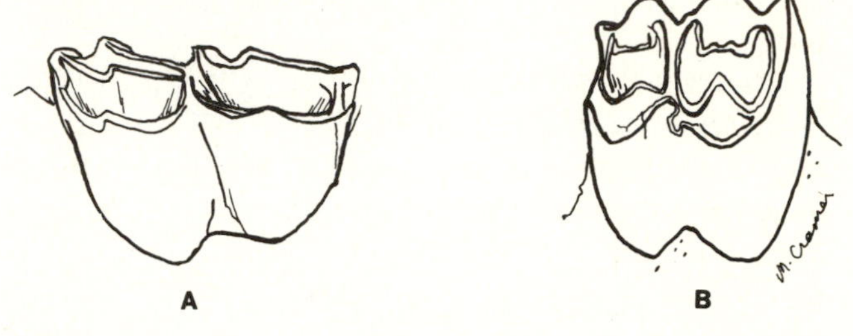

3-17). These hard ridges of enamel wear away more slowly than the surrounding tissues and provide a grinding surface similar to that of a millstone.

The cheek teeth of rodents show numerous modifications from the basic quadritubercular plan. These may include simplification of the occlusal pattern, fusion of cusps, or infoldings along the margins of the teeth. Many of these modifications are described and illustrated in chapter 25. Taxonomic studies of certain rodents, particularly members of the families Cricetidae and Muridae, require a worker to have a detailed knowledge of cusp and crown morphology. A schematic diagram of molar crown elements is presented in Figure 3-18. Many of the features shown in this diagram will not be present in a particular specimen or species. Likewise, a thorough knowledge of this diagram is not necessary for using the keys in this manual. These terms, however, are widely used in other reference books and publications on mammals.

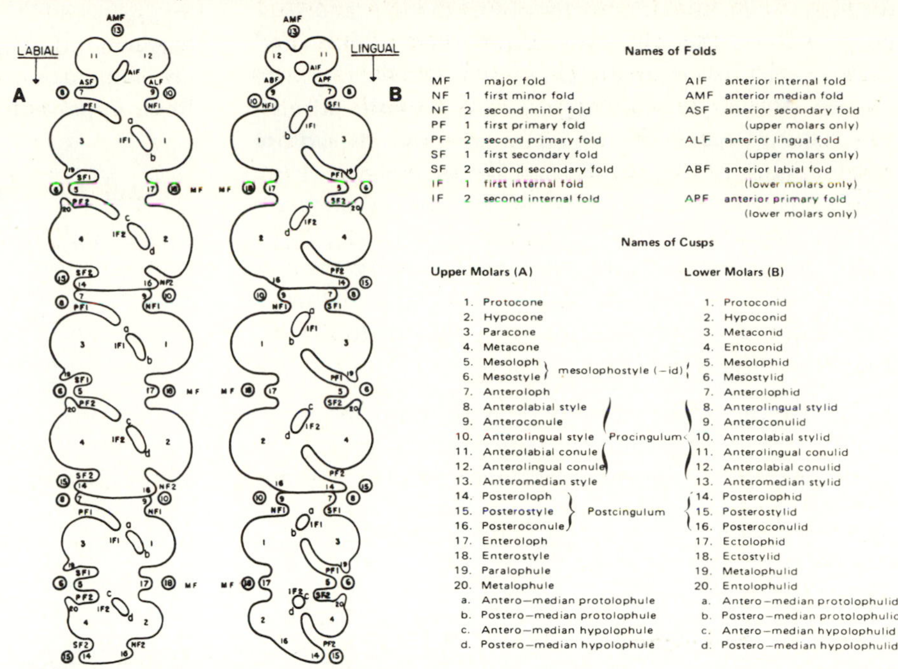

Figure 3–18.
Diagrammatic representation of crown elements that may be present in the molar teeth of cricetid and murid rodents. A given species may not show all of the possible specializations.
(Hershkovitz 1962)

A familiarity with the diagram and information presented in Hershkovitz (1962) and Reig (1977) will serve as a starting point for more detailed studies of rodent crown patterns.

Teeth modified for a carnivorous diet generally are reduced secondarily to two major cusps. The upper and lower teeth, working together, provide a scissors action for shearing flesh (Fig. 23-3). *Note!* The term carnassial has two meanings: **carnassial or secodont dentition** is the general type of dentition found in mammals whose principal diet is flesh. The **carnassial pair** or **carnassial teeth,** found only in the order Carnivora, are the two teeth on each side which do most of the shearing. In living carnivores these teeth are the fourth upper premolar and the first lower molar in the adult dentition and the third upper and fourth lower premolars in the milk dentition.

Many bats have modified tribosphenic teeth (dilambdodont or quadritubercular) in which the three cusps elongate into sharp crescent-shaped cristas (Fig. 18-20, 18-28, etc.), sometimes termed **commissures.** These cristas are useful in cutting and crushing the hard chitinous exoskeletons of insects. Similar specializations are present in the teeth of many Insectivora.

Many fish-eating mammals such as sea lions and porpoises have cheek teeth reduced to a series of sharp unicuspids for holding their slippery food (Figs. 22-28 and 23-9). The sea otter, *Enhydra lutris,* and walrus, *Odobenus rosmarus,* feed primarily on mollusks and echinoderms and have flat brachydont cheek teeth that crush their food (Figs. 23-5 and 23-6). A highly specialized cheek tooth is found in the Antarctic crab-eating seal, *Lobodon carcinophagus.* This species feeds upon krill, small planktonic shrimp, in the cold Antarctic waters. Each cheek tooth of *L. carcinophagus* has three to five long, curving cusps in a straight line reminiscent of the teeth of members of the Triconodonta. These teeth collectively form a sieve (Fig. 3-19) for straining krill from the ocean.

Many diverse groups of mammals have adapted to diets in which teeth serve little or no major function. In many of these, including bats (Fig. 18-11B) and marsupials (Fig. 15-24) that feed upon pollen and/or nectar, sloths that feed upon soft buds (Fig. 20-9) and armadillos (Fig. 20-11) and aardwolves (Fig. 23-21) that feed upon soft-bodied insects, the entire dentition is degenerate; and frequently the teeth are reduced to a series of simple flat-topped or unicuspid pegs. The

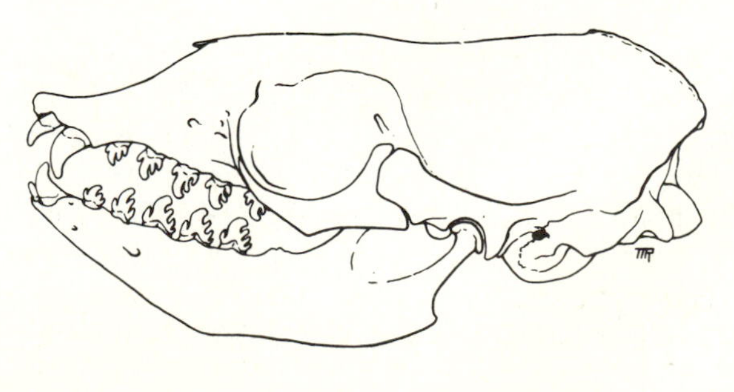

Figure 3–19.
Skull of a crab-eating seal, *Lobodon carcinophagus,* showing teeth that resemble those of the triconodonts.
(Hatt 1946)

echidnas (Fig. 14-5), anteaters (Fig. 20-10), and pangolins (Fig. 21-3), all of which feed upon soft-bodied insects, and the platypus (Fig. 14-4) which feeds on aquatic worms, are **edentulate** (i.e., lack teeth entirely). The baleen whales are also edentulate and instead use **baleen** plates to filter krill from the ocean water (Figs. 22-7 and 22-8).

3-M. Examine a *Canis* skull and identify the incisors, canines, premolars, and molars. For what function is each type of tooth modified?

3-N. Examine the molars of a primate, hog, or both. Locate and identify the four major cusps on each molar. Which, if any, smaller cusps are present? Compare the occlusion of these teeth with those of an opossum.

3-O. Examine the dentition of as many of the following mammals as possible. Identify the kinds of teeth in each species and ascertain the probable diet associated with each dentition.

shrew	sea otter
fruit bat	mink
nectar feeding bat	cat
vampire bat	crab-eating seal
vespertilionid bat	sea lion
anteater	walrus
armadillo	porpoise
sloth	aardvark
rabbit	elephant
rat	horse
deer mouse	deer
vole	cow

Theories of Dental Evolution

Numerous theories exist as to the evolution of mammalian cheek teeth. One of the most widely held theories, the Cope-Osborn or Tritubercular Theory, contends that the upper tribosphenic or tritubercular tooth evolved from a triconodont-type tooth through rotation of the cusps to form a triangular outline; and that the lower tribosphenic or tuberculosectoral tooth evolved from a tritubercular-type tooth. Further consideration of this and related theories, and the controversies surrounding them, are beyond the scope of this manual. However, the subject provides fascinating, if often complicated, reading. General reviews of some or all of the theories may be found in Osborn (1907), Gregory (1934), Butler (1941), Patterson (1956), Hershkovitz (1971), and Reig (1977).

Most theories agree that the teeth of the multituberculates are highly specialized forms. Whatever theory of tooth evolution is accepted, the pantotheres are generally considered to be ancestral to most modern mammals and these later mammals have cheek teeth apparently derived from the tribosphenic type.

Dental Formulas

The **dental formula** is a shorthand method used by mammalogists to indicate the numbers of each kind of tooth in a particular mammal. The complete dental formula for *Canis*, showing the number of each kind of tooth in each half of each jaw, is written:

$$\text{incisors } \frac{3\text{-}3}{3\text{-}3} \text{ canines } \frac{1\text{-}1}{1\text{-}1} \text{ premolars } \frac{4\text{-}4}{4\text{-}4} \text{ molars } \frac{2\text{-}2}{3\text{-}3} = \frac{20}{22} = 42.$$

The numbers above the line represent the teeth in the upper jaw and those below the line represent those in the lower jaw. Since the two halves of each jaw normally have identical numbers of teeth, the dental formula is usually written to show only one side. The total number of teeth is found by multiplying by two. Thus the above formula may be abbreviated as:

$$I\,\frac{3}{3}\quad C\,\frac{1}{1}\quad P\,\frac{4}{4}\quad M\,\frac{2}{3} = 42.$$

or, since the four kinds of teeth are always recorded in the order shown above, the formula can be further abbreviated by deleting the initials for the tooth types:

$$\frac{3}{3}\quad \frac{1}{1}\quad \frac{4}{4}\quad \frac{2}{3} = 42.$$

If a particular type of tooth is not represented in a species, a zero is used. Thus, the dental formula for a Norway rat, *Rattus norvegicus*, is:

$$I\,\frac{1}{1}\quad C\,\frac{0}{0}\quad P\,\frac{0}{0}\quad M\,\frac{3}{3} = 16$$

$$\text{or}\quad \frac{1}{1}\quad \frac{0}{0}\quad \frac{0}{0}\quad \frac{3}{3} = 16.$$

Primitive Dental Formulas

Placental mammals have a maximum of three incisors and one canine per quadrant, and usually have no more than four premolars and three molars per quadrant. Marsupials have a maximum of five upper and four lower incisors on each side and one canine per quadrant. They usually have no more than three premolars and four molars per quadrant. These tooth numbers are considered to be like the ancestral condition. The following dental formulas represent the primitive tooth numbers for marsupials and placentals:

marsupial	I 5/4	C 1/1	P 3/3	M 4/4 = 50.
placental	I 3/3	C 1/1	P 4/4	M 3/3 = 44.

While reduction in tooth number from the primitive formula is common, an increase in this number is rare. Among marsupials only the banded anteater, *Myrmecobius fasciatus* (Fig. 15-10), has more than fifty teeth. Among placental mammals only the giant armadillo, *Priodontes giganteus* (Fig. 20-11A), the African bat-eared fox, *Otocyon megalotis*, and most of the toothed whales (e.g., Figs. 22-27 and 22-32) have more than forty-four teeth. *Priodontes* has up to 100 unicuspid, peglike teeth. *Otocyon* has additional molars to make a total of forty-six teeth (occasionally forty-eight to fifty). Some toothed whales have up to 260 unicuspid, essentially homodont teeth. In the manatees, Trichechidae, up to eighty teeth develop during the life of the animal but usually only twenty-four (occasionally up to thirty-two) are visible at any one time (Fig. 26-10).

3-P. Examine the teeth in a hog (*Sus,* a placental) skull and write the dental formula. How do you know which cheek teeth are premolars and which are molars?

3-Q. Examine the teeth of an opossum (*Didelphis,* a marsupial) skull and write the dental formula. How do you know which cheek teeth are premolars and which are molars?

3-R. Examine the teeth in a cat (*Felis* or *Lynx*) skull and write the dental formula. How do you know which cheek teeth are premolars and which are molars?

3-S. Examine the dentition of a porpoise. How many incisors are present? Can you identify the canines? What is the total number of teeth?

Grouped Dental Formulas

Since premolars and molars are frequently impossible to distinguish in the skull of an adult, these two kinds of teeth are sometimes grouped in writing a dental formula. Such a grouped formula for the common harbor seal, *Phoca vitulina,* is:

$$\text{I } 3/2 \quad \text{C } 1/1 \quad \text{P+M } 5/5 = 34.$$

A similarily grouped formula for a typical nine banded armadillo, *Dasypus novemcinctus,* is:

$$\text{I } 0/0 \quad \text{C } 0/0 \quad \text{P+M } 7/7 = 28.$$

However, since the nine banded armadillo is a placental mammal with a maximum potential of four premolars and three molars, and since seven postcanine teeth are present, it is possible to write a standard dental formula:

$$\text{I } 0/0 \quad \text{C } 0/0 \quad \text{P } 4/4 \quad \text{M } 3/3 = 28.$$

In shrews and some other primitive groups the posterior incisors, the canines, and the anterior premolars may all be simple, single-cusped teeth which are difficult to distinguish from one another. These are collectively termed the "unicuspids" and a particular tooth may be referred to as, for instance, the third upper unicuspid. Some authors write a standard dental for these animals but Choate (1975) has recommended using a formula that identifies the known incisors and premolars and lumps the remaining incisors, canines and premolars as unicuspids. His formula for *Cryptotis* is "first incisor, 1/1; unicuspids, 4/1; fourth premolar, 1/1; molars, 3/3."

3-T. Examine a beaver (*Castor*) skull. There are a total of twenty teeth and only the last premolar is present. Write a dental formula combining the cheek teeth. Write a standard dental formula.

3-U. Write a dental formula grouping cheek teeth for the porpoise examined in 3-S above.

3-V. Examine a shrew skull. How many unicuspids are present? You will need a binocular microscope or a hand lens to see them clearly. Write a dental formula grouping the unicuspids.

Formulas Identifying Missing Teeth

Occasionally a dental formula is written to indicate exactly which teeth have been lost. For *Canis* this type of formula is:

$$\text{I } \frac{123}{123} \quad \text{C } \frac{1}{1} \quad \text{P } \frac{1234}{1234} \quad \text{M } \frac{120}{123} = 42.$$

Here each number represents a particular tooth. The zero indicates that the last upper molar is absent. A similar dental formula for the Norway rat is:

$$\text{I } \frac{100}{100} \quad \text{C } \frac{0}{0} \quad \text{P } \frac{0000}{0000} \quad \text{M } \frac{123}{123} = 16.$$

The last incisors and last molars are usually lost before the first of either of these kinds of teeth, while the first premolars are usually lost before the last premolars. Thus, if a mammal has only one incisor or one molar it is usually I 100/100 or M 100/100 rather than I 003/003 or I 020/020. If a mammal has only two premolars these are usually P 0034/0034 rather than P 1200/1200 or P 0230/0230. However, this is a usual trend and not a rule! Teeth are sometimes lost from the opposite end or from the middle of a series.

3-W. Write a dental formula for a cat (*Felis* or *Lynx*) that shows which teeth are absent. (Remember that the carnassial pair in adults is always the last upper premolar and the first lower molar.)

Notation for Single Teeth

P^2 is a shorthand method of saying "the second upper premolar" and M_3 is a shorthand method of saying "the third lower molar." This use of the tooth type initial combined with a superscript or subscript is in common usage in scientific literature and is used throughout this manual. Some authors use a capital letter to represent a tooth in the upper jaw and a lower case letter to represent a tooth in the lower jaw. By this system P2 is the second upper premolar and m3 is the third lower molar. A lower case letter may also be used to refer to a tooth in the deciduous dentition, but the letter "d" usually accompanies such a designation. For example, the fourth upper deciduous premolar could be termed p^4, dp^4, or pd^4.

Note! If a mammal has a dental formula of:

I 123/123, C 1/1, P 0234/0234, M 120/123 = 38

and an author refers to P^3, he may be referring to the third upper premolar present in the specimen, or he may be referring to the third upper premolar that is potentially present based upon the primitive dental formula. In this manual a shorthand note such as P^3 will always refer only to the teeth actually present in the particular species.

Variation in Formulas

The dental formula is generally considered to be a characteristic of a genus or a higher taxon, but there are certain genera in which there is variation in dental formulas. For instance, the gray squirrel, *Sciurus carolinensis*, has a tiny upper premolar giving a dental formula of:

I 1/1, C 0/0, P 2/1, M 3/3 = 22,

while the closely related fox squirrel, *Sciurus niger*, lacks this tiny premolar giving a dental formula of:

I 1/1, C 0/0, P 1/1, M 3/3 = 20.

There is also variation of dental formulas within certain species. This variation is frequently associated with secondary sexual differences, as when canines are developed in the male but are absent in the female or with age as with the late eruption of the last molars (wisdom teeth) in man. In certain other species, usually, though not always, those with a degenerate homodont dentition, the number of teeth may vary among individuals without regard to sex or age.

3-X. Compare the dentition of a gray squirrel with that of a fox squirrel.

3-Y. Write and compare the dental formulas of a male (ungelded) and female horse.

3-Z. Compare the tooth counts in a series of armadillo and/or black bear skulls.

Supplementary Readings

Butler, P. M. 1956. The ontogeny of molar pattern. *Biol. Rev. Cambridge Philosoph. Soc.* 31:30-70.

Crompton, A. W. 1971. The origin of the tribosphenic molar. pp. 65-87 *In* D. M. Kermack and K. A. Kermack. (Eds.) Early Mammals. *J. Linn. Soc. (London) Zool.* 50:165-180.

Hiiemac, K. 1967. Masticatory function in mammals. *J. Dent. Res.* 46:883-893.

Patterson, B. 1956. Early Cretaceous mammals and the evolution of mammalian molar teeth. *Fieldiana: Geology* 13:1-105.

Peyer, B. 1968. *Comparative odontology.* Univ. of Chicago Press. Chicago. 347 pp. (Translated by R. Zangerl)

4 The Integument

The skin of mammals and other vertebrates is composed of two layers, the **dermis** and **epidermis** (Fig. 4-1).

The epidermis is composed entirely of epithelial cells. Only the basal layer of these cells, the **stratum germinativum,** is living and reproducing. Progressing toward the surface, successive layers of epithelial cells become more flattened and cornified, or keratinized. This surface layer of dead cells, the **stratum corneum,** receives the brunt of environmental wear and tear and continually flakes off the skin and is replaced by growth from below. Thickened portions of this keratinized epithelium form the **pads** on the feet of most mammals and the **friction ridges** on the digits and palms of primates. A fingerprint is an impression of these ridges. **Calluses,** which form where the skin is subjected to constant friction, are further thickenings of this cornified layer. Epidermal scales, hair, horn, and claws are all modifications of cornified epithelial cells.

The dermis lies below the epidermis and is a thick layer of fibrous connective tissue with associated muscular, adipose, vascular, and nervous tissue. All blood vessels and sensory receptors associated with the skin are in the dermal layer. In fishes, **dermal scales** are present but in extant higher vertebrates these are almost always absent.

4-A. Examine a slide of mammalian skin under the compound microscope. Differentiate between epidermis and dermis. Note the change in shape and degree of cornification of epithelial cells from base to surface. In the dermis differentiate between connective, vascular, muscular, nervous, and adipose tissues.

4-B. Examine the friction ridges of your fingers under a binocular microscope. Compare these with the pads on the foot of a rodent, a carnivore, or both, and with the ischial callosities on the buttocks of a primate.

Scales

Dermal bone is true bone formed within the dermal layer of the integument. This bone formed the protective shells of ancient fishes and in modern animals occurs in the scales of some fish, the shells of turtles, and the skins of many lizards and crocodilians. Dermally derived bones contribute portions of the skulls of most vertebrates (e.g., the frontals and parietals are dermal bones) and teeth are believed to be derivatives of dermal scales. In many lower vertebrates dermal bones compose much of the pectoral girdle, but in mammals only the interclavicle (found only in Monotremata) and the clavicle are of dermal origin. Except for these elements, dermal bone is absent in all living mammals except the armadillos.

Epidermal scales are modifications of the cornified epithelium and are never bony. Reptiles are usually completely covered by epidermal scales. Birds have epidermal scales on their legs, and their bodies are covered with modified epidermal scales, termed feathers. Several species of mammals retain epidermal scales on the tail, feet, or both (e.g., Anomaluridae, Castoridae). However, only two groups of mammals, the armadillos and pangolins, have a major portion of the body covered with scales of epidermal derivation.

The armadillos (Edentata, Dasypodidae) have series of bony dermal scales embedded in the skin over the back and sides, and most species have rings of dermal scales encircling the tail. Overlying these dermal scales are thin horny epidermal scales that usually abut one another but do not overlap.

The pangolins (Pholidota) lack dermal scales but have most of the body covered with large, leaf-shaped, imbricate (overlapping) keratinized scales (Figs. 21-1 and 21-2) of epidermal origin.

4-C. Examine a fragment of armadillo shell and a whole armadillo. Note the arrangement of dermal bone and epidermal scales. How do the size and shape of the pieces of the two layers compare? How are the sutures of the bony shell positioned relative to the lines of contact between the epidermal scales? How is the shell constructed to allow for flexibility?

4-D. Examine a pangolin and note the arrangement of scales over the body. What is the function of the scales in this group?

4-E. Examine the scaly tail of a beaver or rat and note the placement of hairs in relation to the placement of scales. Compare this to the arrangement of hair follicles on the inner surface of a pigskin. How do these arrangements support the belief that hair is not a derivative of epidermal scales but is of separate origin?

Hair Anatomy

Though hairlike structures may be found on birds, insects, and even plants, true epidermal hair is unique to mammals. In most mammals it is conspicuous, in others it is sparse, and in some, notably some whales, it is represented only by a few bristles on the embryo.

A hair first develops as a thickening of the epidermis that pushes into the dermis to form a **follicle** (Fig. 4-1). Directly under the follicle a tiny cup, the **papilla,** is formed. The papilla is richly supplied with blood vessels which nourish the growing hair. The epidermal cells at the base of the follicle proliferate to form a column of dead, keratinized cells which pushes out through the neck of the follicle. An outgrowth of epidermal cells from the side of the follicle forms a **sebaceous** (oil) **gland;** the secretions of which keep the hair from becoming brittle and render a degree of water-proofing (see Integumentary Glands section below).

Hairs do not emerge vertically from the skin but project at an angle. Associated with each follicle is a small, involuntary **arrector pili muscle.** Contraction of this muscle erects the hair. In man this brings about "gooseflesh." In other mammals it raises the hair, thereby increasing insulating properties or serving as a threat posture.

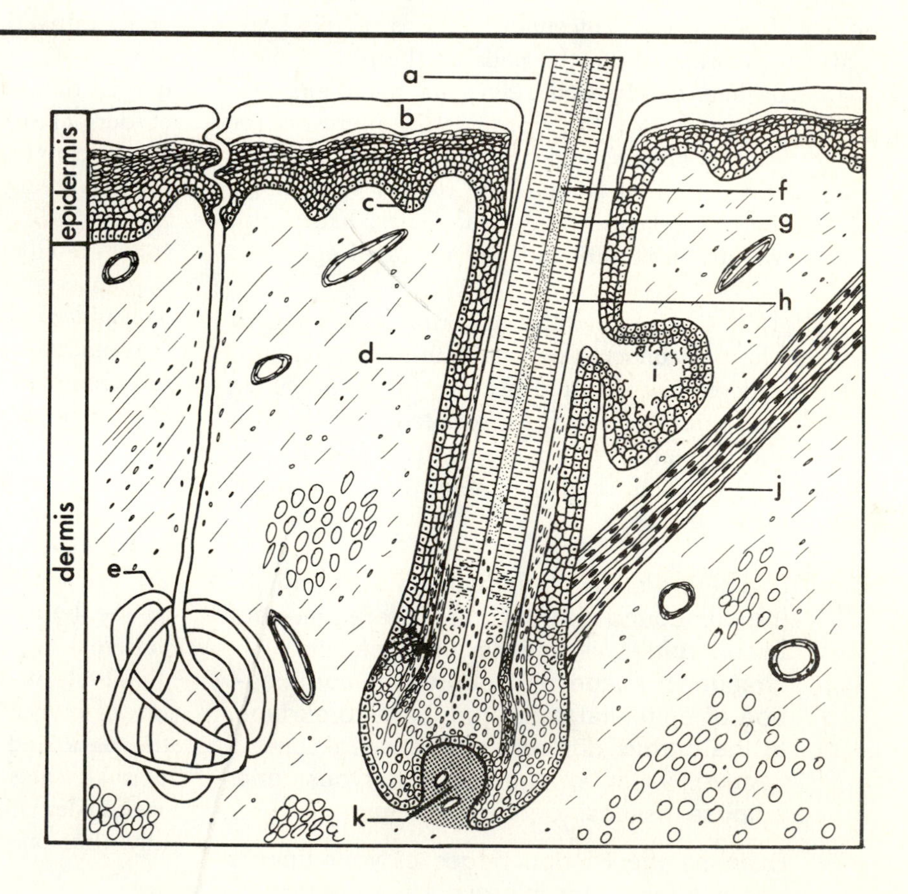

Figure 4–1.
Sectional view of skin showing hair and various structures in dermis and epidermis. a, hair; b, stratum corneum; c, stratum germinativum; d, hair follicle; e, eccrine sweat gland; f, medulla; g, cortex; h, cuticle; i, sebaceous gland; j, arrector pili muscle; k, papilla.
(R. E. Martin)

4-F. On a slide of mammalian skin locate a follicle. Identify the papilla, hair, sebaceous gland, and arrector pili muscle.

4-G. The arrector pili muscle raises the hair to increase insulation or to make the animal seem larger in threat postures. In man "gooseflesh" is usually produced by cold or fear. Is this consistent with what occurs in other mammals?

Each hair normally consists of three well-defined layers (Fig. 4-1). The central **medulla,** present in all but the smallest hairs, may be continuous or may have regularly or irregularly spaced air cavities. In some mammals the medulla is absent leaving a central air column. The **cortex** immediately surrounds the medulla and makes up the bulk of a hair. The **cuticle** is a thin outer layer of scales covering the cortex. The scales may form a relatively smooth surface or may overlap in various distinctive patterns (Fig. 4-2). Pigment granules may be located in the medulla, the cortex, or both but are never found in the cuticle. In cross section a hair may be circular or somewhat flattened. Flattened hairs are curly while cylindrical hairs are normally straight.

Combinations of variations in hair structure are distinctive for certain mammalian groups. Keys can be constructed to aid in identifying hairs found in dens, fecal material, owl pellets, etc. (e.g., Benedict 1957; Day 1966; Mayer 1952; Miles 1965). Hair size, the character of the medulla, the amount and distribution of pigments, and the type of cuticular scale patterns are all considered in writing these keys.

4-H. Remove a hair from your head and from at least three other species of mammals. Mount each hair in a drop of water on a clean microscope slide and cover with a cover slip. With a compound microscope compare the structure of the medulla and the location and relative abundance of pigment granules in each of the hairs.

Figure 4–2.
The basic types of cuticular scales found on mammalian hair.
(Nason 1948)

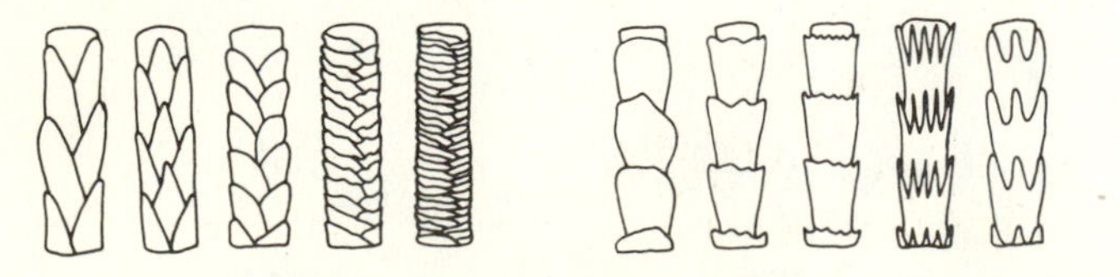

4-I. Place a sample of each of the kinds of hair used above on separate microscope slides. Cover the hairs with plastic cover slips and then pass each slide through a flame until the cover slip has melted somewhat. Allow it to cool before removing hairs. Remove the hair from the impressions, mount the cover slips on microscope slides and examine the cuticular scale patterns under the compound microscope. See Chapter 30 for additional references on making slide preparations of hair.

Classification of Hair

On the basis of growth, hair may be classified as either **definitive** or **angora.** Hair with definitive growth reaches a certain length typical for a species and body location and growth ceases. These hairs are shed and replaced periodically, such as human eyelashes. Angora hair grows almost continuously and reaches a considerable length before being shed. Some angora hairs, such as a horse's mane, are never shed but continue to grow through the life of the animal. Growth patterns and hair functions are combined to classify hair into the various types recognized on mammals.

Vibrissae

Vibrissae are long stiff hairs with well innervated bases. They primarily serve as tactile receptors. The best known vibrissae are the "whiskers" on a mammal's face (Fig. 4-3), but vibrissae may also be located on the legs and elsewhere on the body.

Figure 4–3.
Locations of vibrissae on the head of a gray fox (*Urocyon cinereoargenteus*).
(Hildebrand 1952:422)

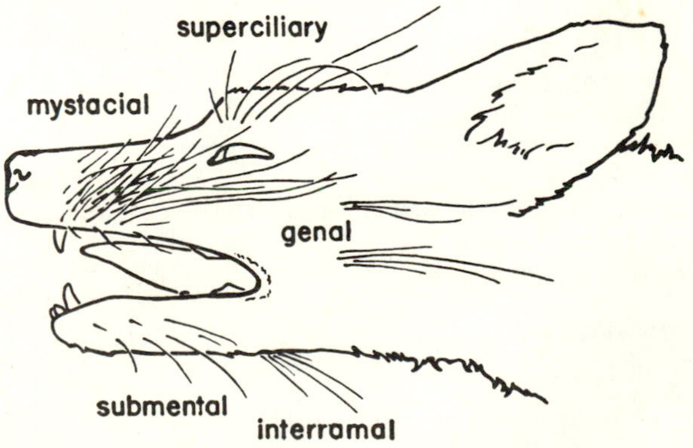

superciliary
mystacial
genal
submental
interramal

Guard Hairs

Guard hairs or overhairs are the most conspicuous hairs on most mammals. They serve primarily for protection. The three major types of guard hair recognized are given below. However, intermediate types do occur. **Spines** are greatly enlarged, stiff guard hairs with definitive growth. They serve primarily as defense from predators. The New World porcupines, Erethizontidae and the tenrec genus *Hemicentetes* (Eisenberg and Gould 1970:49), have **barbs** on the tips of the spines (Fig. 4-4). Once embedded in a predator's skin, such a spine cannot be easily removed, and the victim's muscular action will actually cause the spine to embed more deeply. If it works its way in deeply enough to penetrate a vital organ, it can be fatal. Spines on other groups of mammals (e.g., Old World porcupines, Hystricidae; hedgehogs, Erinacidae; and echidnas, Tachyglossidae) do not have barbs. **Bristles** are long, firm hairs with angora growth (e.g., horse and lion manes). **Awns** are hairs with definitive growth that have a firm, expanded distal portion and a smaller, weaker base. Awns are the most noticeable hairs on most mammals.

Underhairs

Underhairs function primarily for insulation. Three major types are generally recognized but intermediate types do occur. **Wool** is angora underhair. It is usually long, soft, and curly. **Fur** is fine, relatively short, hair with definitive growth that grows densely over the body. **Velli,** down or fuzz, are very fine, short hairs that are velvety in appearance. Embryonic hair, lanugo, is a type of velli.

4-J. Note the location of vibrissae on a variety of mammals. Explain these locations considering the habits and/or habitat of each species. Observe the action of the vibrissae in a live mammal.

4-K. Under a binocular microscope compare the spines of a New World porcupine with those of an Old World porcupine, hedgehog, or other spiny mammal. How do they differ? Under what circumstances may a porcupine be said to "throw its quills"?

4-L. Examine a variety of mammals. What types of hair are found on each? What is the function of each type of hair?

Figure 4–4.
Quill tip of a New World porcupine (*Erethizon dorsatum).*
Note presence of barbs.
(Shadle and Chedley 1949:172)

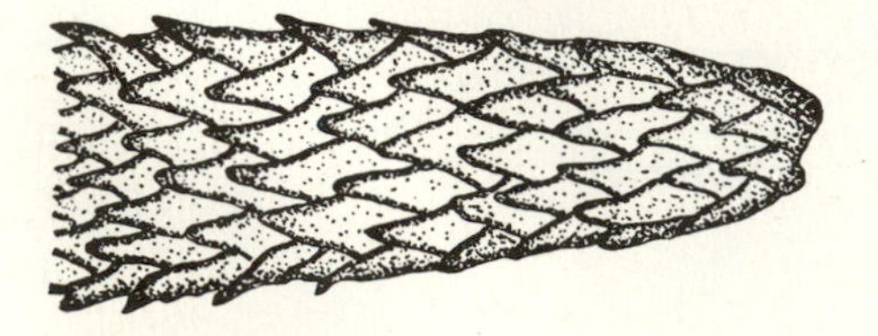

Hair Replacement

Some angora hairs grow continuously for the life of the mammal and continuously are worn away at the tips. Most hair, however, is shed and replaced periodically in a process termed **molting.**

Molts may occur continuously with some hairs being replaced at all times, e.g., human eyelashes. But most mammals, particularly those living in temperate or arctic climates, have an **annual molt** during which all hairs are replaced in a short period of time. Such molts usually begin in a specific region of the body and progress in orderly fashion until all hairs have been replaced. The **molt pattern** varies with the species and occasionally with the age of an individual (Fig. 4-5). Some mammals have **seasonal molts** with more than one molt per year. This is most conspicuous in species that change from a brown summer **pelage** to a white winter pelage. In northern populations of the long-tailed weasel, *Mustela frenata,* for example,

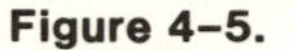

Figure 4–5.
Postjuvenal molt pattern in the wood rat, *Neotoma cinerea.*
Stippled areas represent appearance of new pelage as the molt progresses.
(Egoscue 1962:335)

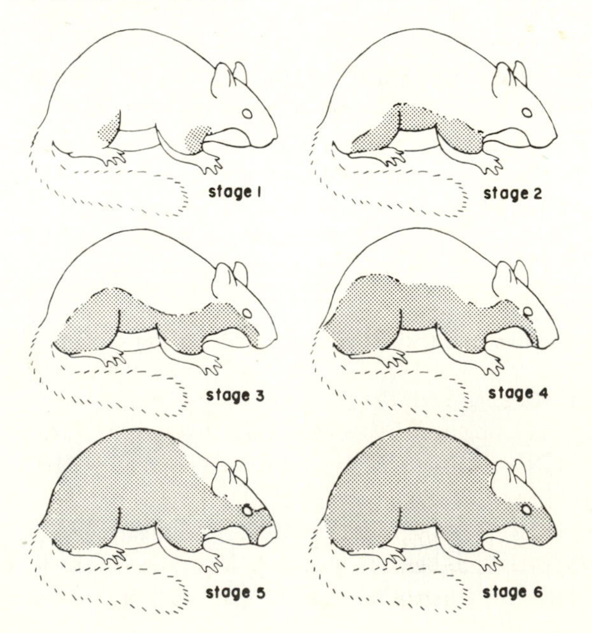

the spring molt begins along the dorsal surface and brown hairs replace the white winter hairs over the dorsal parts of the animal. In the fall the molt pattern is reversed with replacement of brown hairs by white ones progressing dorsally from the perpetually white ventrum. In the southern parts of the range of this species, the molts occur but do not result in a color change, and the weasel has a brown dorsal pelage throughout the year. The change from winter to summer pelage is influenced by hormones, photoperiod, and temperature.

A distinctly **juvenal pelage** is recognizable in many mammals. This is usually grayer and duller than typical **adult pelage.** The juvenal pelage can also be variously striped or spotted while the adult is more or less uniformly colored. In some groups a distinctive **subadult pelage** occurs between the juvenal and adult pelages.

4-M. Examine series of skins of several species collected while the animals were in the process of molting. How does the molt pattern differ among the species examined?

4-N. Compare winter and summer pelts of *Mustela frenata, Lepus americanus,* or some other species that has seasonal alteration of colors. Compare with pelages of specimens collected in spring and fall.

4-O. Examine a large series of deer mice (*Peromycus*) and identify juvenile individuals on the basis of pelage. Compare pelages of juvenile and adult specimens of other species of mammals.

Color

The color of an individual hair is affected by numerous factors. The kind, amount, and distribution of pigment granules in a hair can all vary to produce different effects. In addition, hair surface texture, the thickness of the hair, and the amount of air space in the medulla can all alter the way in which light is reflected by the hair and, therefore, change the appearance of the pigments present.

The overall coloration of a mammal is determined by the coloring of individual hairs and the relationships among these hairs in the pelage. An animal may, therefore, be red and brown speckled because each hair has red and brown color bands or because the pelage has a mixture of red hairs and brown hairs.

There are two types of pigment in mammalian hair. **Eumelanin** in various concentrations produces blacks and browns. **Pheomelanin** in various concentrations produces reds and yellows. White is the complete lack of pigment. Each hair usually has a series of color bands. The **agouti** hair has a black tip followed by successive bands of pheomelanin and eumelanin.

4-P. Examine individual hairs from a cottontail rabbit (*Sylvilagus*) or other animal with agouti hair and note the sequence of color bands.

4-Q. Examine the color banding on hairs of at least ten species of mammals whose color could be described as brown. How do these compare in number, width, sequence, and hue of the bands? Do all of the hairs in a given body region exhibit the same banding?

Mammalian hair and skin coloration serves several functions. Among these are concealment, communication, and protection from ultraviolet (UV) radiation. Many predators and many prey have **concealing** or **cryptic coloration** that allows them to blend with their habitat to avoid detection. The primitive agouti pattern provides a coloration that is usually very similar to the color of the earth and dead vegetation. Mammals such as the tiger and okapi are strikingly marked with sharply defined light and dark colors. But in their normal habitats these examples of **disruptive coloration** obscure the body contours and cause the animal to blend into the patterns of light and shadow caused by sunlight penetrating the vegetation. Facial stripes, present in many mammals, are also disruptive coloration, usually intended to conceal the eye. Most mammals have a ventral surface that is more lightly colored than the dorsum—an arrangement termed **countershading.**

Coloration is also important for intra- and interspecific communication. The color pattern typical of a species serves to identify each individual to others of his species. In several species, display of distinctive color regions, such as the red genital area of some baboons, is an important part of the courtship ritual. Conspicuously colored "flags," such as the white underside of the tail in many rabbits and deer, may be exposed to alert others of a group or herd to dangerous conditions. These flags may also offer protection since a predator's eyes will frequently focus on the white tail of his fleeing prey and will lose sight of the prey when it quickly stops and lowers the flag. Warning coloration is present in some species that have other special means of defense. The striking black and white patterns of skunks are examples.

Prolonged exposure to ultraviolet (UV) radiation from the sun can cause burns and in other ways be deleterious to an animal's health. The melanin pigments in mammalian hair and skin filter this harmful UV radiation from the sunlight. Ultraviolet radiation is highest in intensity in equatorial zones and becomes progressively less intense toward the poles. Thus, if other influencing factors such as concealment and communication are ignored, mammals of equatorial regions need to be more heavily pigmented than do those of the polar regions (Hershkovitz 1968, 1970; Lawlor 1969). Color can also be an important aid in thermoregulation (Hamilton 1973).

4-R. Examine skins of all of the mammal species known from your state. Which exhibit concealing coloration? Do any have areas of the body where disruptive coloration patterns are present? Which exhibit countershading? Which are equipped with white "flags" or other color signals? Do any exhibit warning coloration?

4-S. Do any mammals other than skunks exhibit warning coloration? Examine illustrations in Walker et al. (1975), Burt and Grossenheider (1964), van den Brink (1967), Dorst and Dendelot (1970), etc.

4-T. Which mammals other than the tiger and okapi mentioned above exhibit a marked degree of disruptive coloration? Check the same sources listed in 4-S above.

4-U. Consider the coloration of peoples inhabiting central Africa and compare this with the successively lighter colors of people from the Mediterranean regions of northern Africa and southern Europe, central Europe, and Scandinavia. What factors might have played an important role in bringing about this clinal variation from equator towards the pole?

Terrestrial vertebrates inhabiting arid regions are usually paler in coloration than closely related forms inhabiting more humid regions. This phenomenon is known as **Gloger's Rule.** While the rule links color and aridity, the influencing factor is most likely background color of the habitat. As the habitat becomes more arid, vegetation becomes more sparse and soil color becomes generally lighter. Thus light-colored pelages are adaptations for concealment. In desert areas where there are large expanses of black volcanic rock, the small mammals are usually black like the rock rather than light as Gloger's Rule would dictate.

4-V. Examine a series of rabbits (*Sylvilagus*) or wood rats (*Neotoma*) including specimens collected in the eastern deciduous forests, the Great Plains, and the southwestern deserts. Do differences in color correlate with annual precipitation? If possible, compare these with specimens collected from an arid lava field.

Albinism, the complete lack of integumentary pigments, results from genetic mutation and has been found in many species of mammals. Since albinos are not well adapted for camouflage, communication, or ultraviolet radiation protection, they generally do not survive to establish themselves as part of a wild population. But geographically localized populations of albinistic individuals have become established in several species. True albinos, which lack pigments in the irises and thus are pink-eyed (due to the red blood pigments visible through the transparent, colorless tissues of the eye), have been propagated in captivity, and albino rats and rabbits, in particular, are common. Partial albinos, individuals having patches of white on the body, are not as radically different from the typical coloration of a species and are, therefore, better able to survive in the wild. The white patches of a partial albino are usually caused by somatic mutations and thus cannot be passed on to the offspring.

Melanism, a tendency toward completely black color, is also a genetic mutation frequently encountered in wild populations. While melanistic individuals differ from the normal coloration of their species with respect to concealment and communication, they do have ample protection against UV radiation. In some species melanistic tendencies are common and melanistic individuals commonly occur. Examples of these are the silver and cross varieties of the red fox, *Vulpes vulpes;* black fox squirrels, *Sciurus niger;* and black grey squirrels, *Sciurus carolinensis.*

4-W. Examine a live albino animal. Note color of hair, skin, and eyes. Why aren't the eyes also white? Compare the color of an individual hair with one from a normally colored individual of the same species. Is the albino really white?

4-X. Examine mammal skins exhibiting partial albanism. (Such irregular white patches are common in the Eastern mole, *Scalopus aquaticus*, and the Mexican freetailed bat, *Tadarida brasiliensis*.)

4-Y. Examine a series of squirrel or fox skins showing the typical color for the species and a variety of melanistic shades. Compare several hairs from the melanistic specimens with one from a normally colored individual of the same species. How is the melanism produced? Does each hair have fewer or smaller red bands. Is each hair completely black? Are all hairs black?

A few species of mammals have coloration that is not dependent upon pigmentation. Sloths, for example, have coarse overhair with numerous external grooves. Algae grow in these grooves and give the animals a greenish color that allows them to better blend in with their forest environment. The color of many mammals is frequently altered in appearance by layers of dust.

4-Z. Examine sloth hair under a compound microscope.

Integumentary Glands

There are two basic types of glands in the skin of mammals, sebaceous glands and sweat glands. All other integumentary glands are considered to be modifications of one of these two types.

Sweat glands are found only in mammals, although several kinds of mammals—echidnas, moles, sirenians, cetaceans, etc.—have none. Sweat glands consist of two basic types, sudoriferous and eccrine. **Sudoriferous (=apocrine sweat) glands** are highly coiled and usually located in the vicinity of hair follicles. Sudoriferous glands produce the odorous component of perspiration. In humans, apocrine sweat glands are concentrated in the axillae, naval, ano-genital areas, nipples, and ears. **Eccrine sweat glands,** are also highly coiled (Fig. 4-1) but open directly onto the skin surface, independent of the hair follicles. Eccrine sweat glands are responsible for most of the fluid portion of sweat and also excrete some metabolic wastes and salts. Evaporation of sweat from the surface of the skin is a cooling mechanism for the body and perspiration also improves tactile sensitivity and grip when secreted onto the palms and soles. The wax-producing glands of the external ear, the glands of Moll, are modified apocrine sweat glands that protect the tympanic membrane from becoming dry and loosing flexibility.

Sebaceous glands (Fig. 4-1), which are usually associated with hair follicles, serve primarily to keep the hair from becoming too dry and brittle. In many mammals they are also important in waterproofing the pelage. Sebaceous secretions in the hair of otters and fur seals, for instance, keep cold water from contacting the skin and thereby retard heat loss. Some sebaceous glands, as in the upper lip, nose, and upper cheek areas of humans, open directly onto the skin surface rather than into hair follicles. Meibomian glands, located on the eyelid, and Hartner's glands, located behind the eyeball, are modified sebaceous glands that lubricate the eyelid and nictitating membrane respectively.

4-AA. Examine prepared slides of mammal skin. Locate and compare the structures of a sebaceous gland and a sweat gland.

4-BB. Keep your hand tightly closed until it begins to sweat, then examine the palm under a binocular microscope.

Scent glands are complex odor-producing glands of variable composition. They may be predominately sebaceous or sudoriferous or divided about equally between these two types of glands. While much remains to be learned about the function and significance of scent glands and their secretions (Ralls, 1971, Eisenberg and Kleiman, 1972, Doty, 1976), the functions can be divided into three general categories: defense, recognition of territory, and social interactions.

In defense, skunks (Mustelidae: Mephitinae) will discharge a mercaptan-based **musk** from anal glands. Wolverines (Mustelidae: *Gulo*) and peccaries (Tayassuidae: *Tayassu*) are examples of other mammals that emit a musk when they are in danger.

Many mammals establish a scent trail in their territory to guide their travels. This process of labeling an area with scent is termed **marking.** An area marked may even be recognizable as a visible signpost (see chapter 32). The interdigital glands of deer (Cervidae, e.g., *Odocoileus*) and the musk glands of badgers (Mustelidae, e.g., *Taxidea*) apparently provide the scent for trail marking in these mammals. In rodents and some primates, urine and preputial gland (see chapter 10) secretions are important for marking territory.

A **pheromone** is an odor or musk that has a behavioral or physiological effect on another individual or individuals of the same species. For example, pheromones in the urine of some rodents are known to influence the onset of estrus and other phenomena of the reproductive cycle. An **alarm pheromone** is released when an animal is in danger. The secretions of the metatarsal glands of deer are thought to be alarm pheromones.

An individual encountering a scent mark may have several possible responses. A mark made by the individual, or in social species any member of the group, may reduce anxiety. A foreign mark, one made by another species or member of another group, may result in increased aggression or a readiness to flee (Ewer 1968). The position of an individual in a social group may be influenced by the amount of scent production. In some rabbits (Leporidae: *Oryctolagus*), individuals with a high social status have well-developed chin glands and engage in more territorial marking than individuals with smaller chin glands. The tarsal glands of deer serve a similar social function.

The secretions of scent glands are sometimes utilized by man's commercial enterprises. Some scent glands of mammals are removed by trappers and used to prepare scent baits or attractants (see chapter 34). The musk, or civet, produced by the anal glands of certain species of civets (Viverridae, principally *Viverra, Viverricula,* and *Civettictis*) is utilized as a base in the manufacture of fine perfumes.

4-CC. Examine preserved material of a variety of mammals. How many scent glands can you find? Where on the animal's body are these located?

4-DD. Examine a demonstration dissection of the anal glands of a skunk (e.g., *Mephitis*). How are these animals able to propel their musk for such great distances?

4-EE. Observe live deer, rabbits, or rodents for evidence of marking behavior. In what behavioral or social situations do the animals mark?

Mammary glands, unique to mammals, are considered to be modified sweat glands. They are present in both sexes until puberty, at which time they degenerate in males and develop in females. Since these glands are so closely associated with the reproductive cycle, they are discussed in greater detail in chapter 10.

Claws, nails, hoofs, and horns, which are also composed of keratinized epithelial cells, are discussed in chapters 5 and 6.

Supplementary Readings

Cott, H. B. 1966. *Adaptive coloration in animals.* Methuen and Co., London. 508 pp.

Ebling, F. J., D. Bellamy, and P. Hale (Eds.). 1964. *The mammalian epidermis and its derivatives. Symp. Zool. Soc. London* 12:1-133.

Hamilton, J. B., *et al.* 1951. The growth replacement and types of hair. *Ann. New York Acad. Sci.* 53:461-752.

Hamilton, W. J. III. 1973. *Life's Color Code.* McGraw Hill Book Co., New York. 238 pp.

Hart. J. S. 1956. Seasonal changes in insulation of fur. *Canadian J. Zool.* 34:53-57.

Hershkovitz, P. 1968. Metachromism or the principle of evolutionary change in mammalian tegumentary colors. *Evolution* 22:556-575.

*Linzey, D. W., and A. V. Linzey. 1967. Maturational and seasonal molts in the golden mouse, *Ochrotomys nuttalli. J. Mammal.* 48:236-241.

Maderson, P. F. A. (Ed.) 1972. The vertebrate integument. Symposium in *American Zool.* 12:12-171.

Muller-Schwarze, D. 1971. Pheromones in the black-tailed deer (*Odocoileus hemionus columbianus*). *Animal Behav.* 19:141-152.

*Noback, C. R. 1951. Morphology and phylogeny of hair. *Ann. New York Acad. Sci.* 53:476-492.

Parnkkal, P. F. and N. J. Alexander. 1972. *Keratinization; a survey of vertebrate epithelia.* Academic Press, New York. 59 pp.

Quay, W. B. and D. Muller-Schwarze. 1970. Functional histology of integumentary glandular regions in black-tailed deer. (*Odocoileus hemionus columbianus*). *J. Mammal.* 51:675-694.

Ralls, K. 1971. Mammalian scent marking. *Science* 171: 443-449.

*Smith, M. H., R. W. Blessing, J. L. Carmon and J. B. Gentry. 1969. Coat color and survival of displaced wild and laboratory reared old-field mice. *Acta Theriol.* 14:1-9.

*Included in Jones, Anderson and Hoffman (1976).

5 Horns and Antlers

In modern mammals head ornamentation in the form of horns and antlers is confined to the ungulate orders Artiodactyla and Perissodactyla. However, the fossil record includes horned mammals in other orders, even Rodentia. Head ornamentation of living mammals may be divided on the basis of structure and method of formation into five major types.

True Horns

True **horns,** which occur only in the family Bovidae (buffalo, sheep, goats, cattle, antelopes, etc.), are unbranched and permanent. They are composed of an inner bony core, formed from a frontal bone, and an outer layer of true **horn,** formed from keratinized epidermis (Fig. 5-1). *Note!* "Horn" can refer either to the entire structure (e.g., a cow's horns) or to the keratinized material that forms the sheath. A true horn grows from its base throughout the adult life of the animal, but neither the bony core nor the keratinized portion is shed. Portions of the sheath are often worn away and in some species parts or layers of the sheath may regularly break away (O'Gara and Matson 1975) but the entire sheath is not shed. In many horned bovids each season's growth produces a ring at the base of the sheath (Fig. 5-2) and counts of these annual rings have proven useful in determining age of wild sheep and certain other species.

Horns may be present on both sexes or may occur only on males .When present on both sexes they are usually larger on the male. A few breeds of domestic

Figure 5–1.
Diagrammatic section of a bovid horn: a, horn or keratinized epidermis; b, epidermis; c, dermis; d, bone. (L.P. Martin)

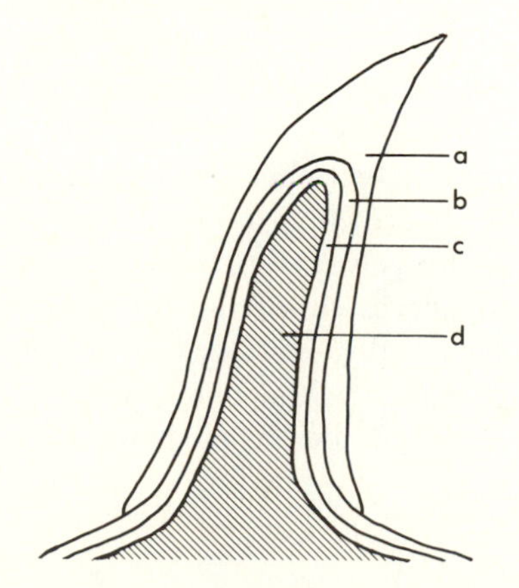

Figure 5–2.
Head of Grant's gazelle, *Gazella granti.* Note the growth rings at the base of the horns. (Flower and Lydekker 1891)

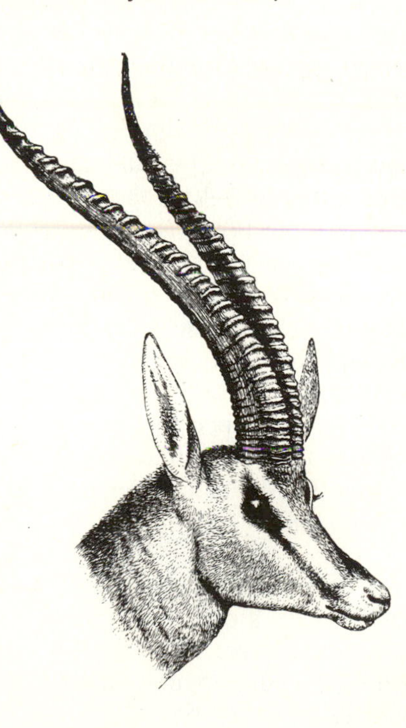

Figure 5–3.
Skull of the four horned antilope, *Tetracerus quadricornis,* the only living bovid with more than one pair of horns. (Owen 1868:625)

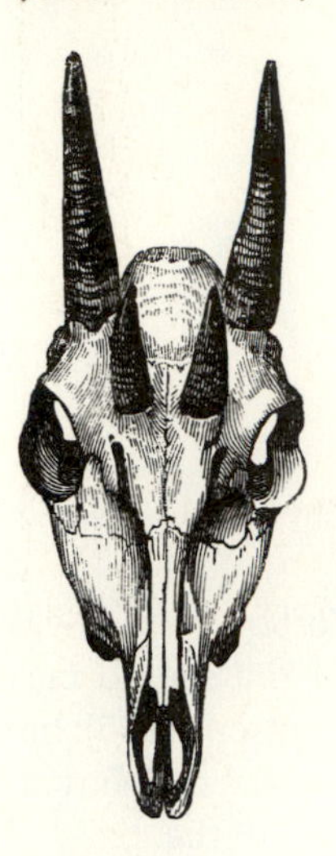

bovids (e.g., Aberdeen Angus, Polled Hereford) are hornless. Horns usually occur as a single pair, however one living bovid, the four-horned antelope, *Tetracerus quadricornis,* has four well developed horns (Fig. 5-3).

5-A. Examine horns and horn cores of a variety of bovids. Note differences in size, length, and curvature.

Pronghorns

Pronghorns are found among modern mammals only on the North American pronghorn, *Antilocapra americana,* the single living species of the family Antilocapridae. Their basic structure (Fig. 5-4) is similar to that of the bovine horn, consisting of a permanent, unbranched, bony core projecting from a frontal bone and an epidermal, horny sheath. However, in pronghorns the horny sheath is shed annually and is branched, having a small anterior projection or prong. When the sheath is about to be shed, it becomes loose and a new one begins to form on the bone core. O'Gara and Matson (1975) have described this process in detail. Female pronghorns are sometimes hornless and frequently lack prongs. The horns of the male are larger than those of the female.

5-B. Compare the horns (both cores and sheaths) of a pronghorn with those of the bovid series examined above. How do the horn cores differ? (Be sure to compare to a goat as well as to other bovids.)

Figure 5–4.
Diagrammatic section of a pronghorn: a, keratinized epidermis; b, epidermis; c, dermis; d, bone.
(L.P. Martin)

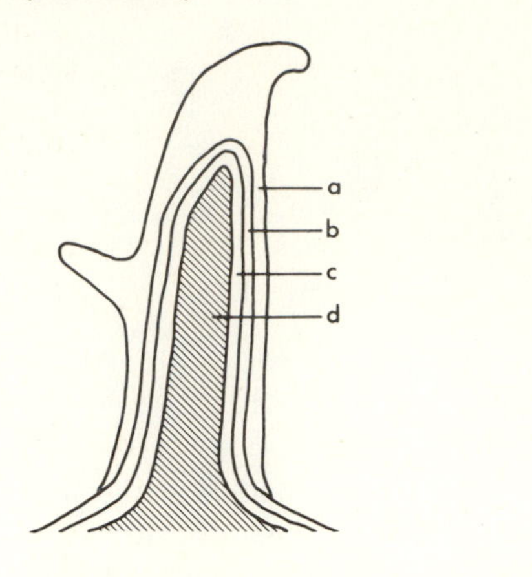

Figure 5–5.
Diagrammatic section of antler with (A) and without (B) velvet: a, hair; b, epidermis; c, dermis; d, bone (or antler); e, abscission line at region of burr.
(L.P. Martin)

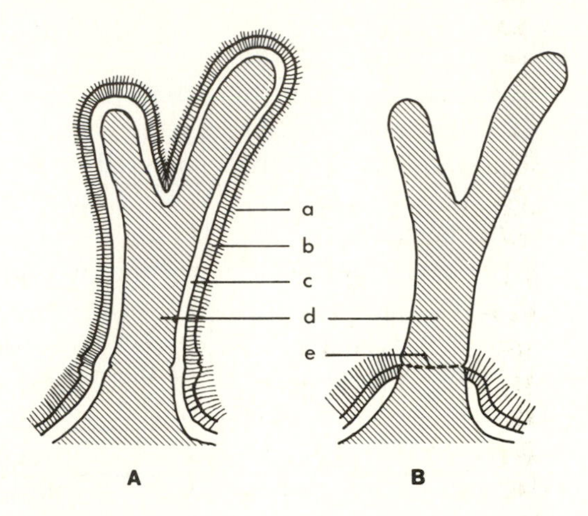

Antlers

Antlers occur on males of all but the most primitive genera of the family Cervidae, the deer, and are found in both sexes of the genus *Rangifer*, the reindeer and caribou. Fully developed antlers are entirely bony structures (Fig. 5-5B) that are branched and are shed annually. While the antler is growing, the bone is covered with a layer of skin, the **velvet** (Fig. 5-5A), which carries blood vessels and nerves supplying the growing bone. When the bone is fully ossified, this velvet is shed. After each mating season the bony antler is shed and in the spring a new set begins to grow (Fig. 5-6).

Figure 5-6.
Stages in the growth of antlers of the red deer, *Cervus elaphus.*
(Sokolov 1959)

Figure 5-7.
Skull of a male Indian muntjak, *Muntiacus muntjak,* a primitive deer with unusually long pedicels: a, pedicel; b, burr; c, antler.
(Owen 1866)

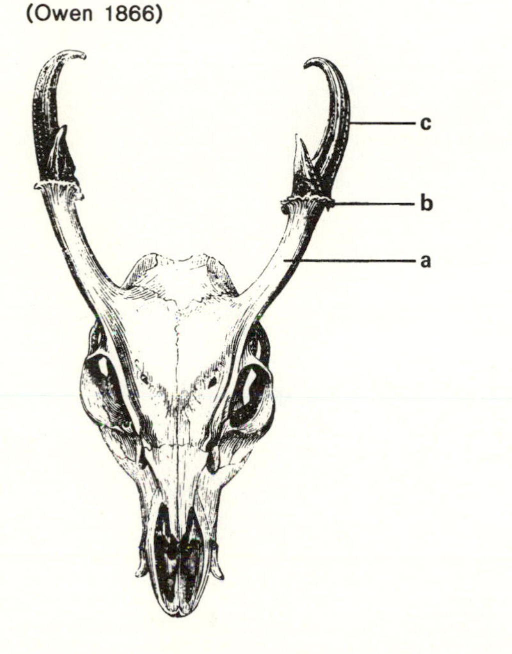

Figure 5-8.
Red deer, *Cervus elaphus,* antler: a, burr; b, beam; c, brow tine; d, bez tine; e, crown.
(Sokolov 1959)

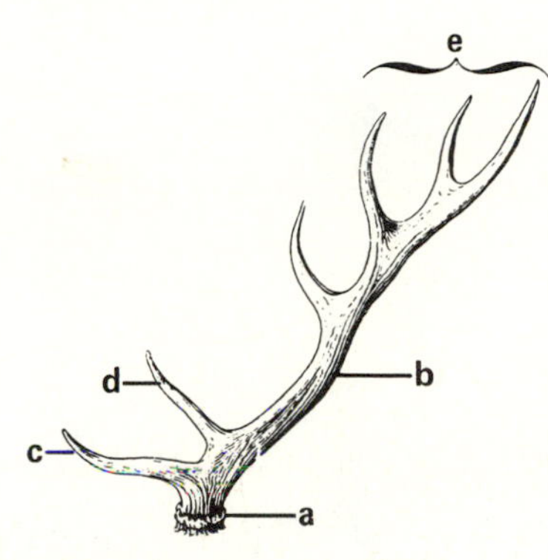

The antler forms from the **pedicel,** an extension of the frontal bone. A **burr** marks the point of separation between the permanent pedicel and the deciduous antler. The pedicel is usually very short but in some primitive deer it is as long or longer than the antler (Fig. 5-7). The antler usually consists of a main stem, the **beam,** with a variable number of branches or **tines.** The first tine to arise from the beam, immediately over the forehead, is termed the **brow tine,** and the second is termed the **bez tine.** The points at the summit of the antler are collectively termed the **crown** (Fig. 5-8). The pair of antlers together are termed the **rack.** Commonly all of the branches of the antler are essentially cylindrical but in some species, as in the moose, *Alces,* they are more or less expanded and flattened. Such flattened antlers are termed **palmate** (Fig. 5-9B).

In the white-tailed deer, *Odocoileus virginianus,* the cycle of antler growth begins at about nine months when small, paired bulges first appear on the frontal bones and rapidly develop into the pedicels. The first antlers grow from the pedicels at about eighteen months. These begin to ossify at the base and ossification progresses toward the tip. The growth is very rapid and may be completed in fourteen weeks. By late summer or early fall growth is completed, circulation in the velvet becomes sluggish, and the skin dies. The antlers are rubbed against trees and brush until the velvet hangs in shreds and falls off. After the mating season, absorption just under the burr results in a plane of weakness across the burr and the antlers fall off. In the spring the cycle begins again with new growth from the burr.

The number of points or tines displayed by an individual increases with age until the optimum for the individual is reached when the animal is in its prime. Antler development is influenced by several factors other than age, thus it is never possible to determine the absolute age of any animal merely by counting the number of tines (see chapter 30). Nutrition plays

Figure 5–9.
Examples of the variation possible in the configuration of antlers: A, Schomburgh's deer, *Cervus schomburghi;* B, moose, *Alces alces;* C, reindeer, *Rangifer tarandus;* D, wapiti, *Cervus canadensis.* (Not all to same scale.) (Flower and Lydekker 1891:309, 327, 325, 322)

an important role in antler development, and undernourished individuals will never have as many tines, or as well-developed racks as properly nourished individuals of the same age. In most deer, antlers are secondary sex characters, thus their formation is also controlled by male hormones. Injury to the testes or other factors influencing hormonal production may result in stunted or deformed antlers. In very old males poor nutrition or reduced hormonal production will frequently result in antlers consisting of a single spike or a rough burr. Deformed antlers are frequently seen. While these may be the result of hormonal or nutritional deficiencies, they are frequently due to mechanical injury to the antler while it is in velvet and still growing.

5-C. Examine a series of skulls or racks from *Odocoileus* males, collected in various seasons. Compare the structure of the bone in each. What has caused the small grooves visible on a mature antler?

Figure 5–10.
A giraffe head, *Giraffa camelopardalis,* and a giraffe "horn." a, hair covered epidermis; b, dermis; c, bone. (Head, Giebel 1859:369; horn, L.P. Martin.)

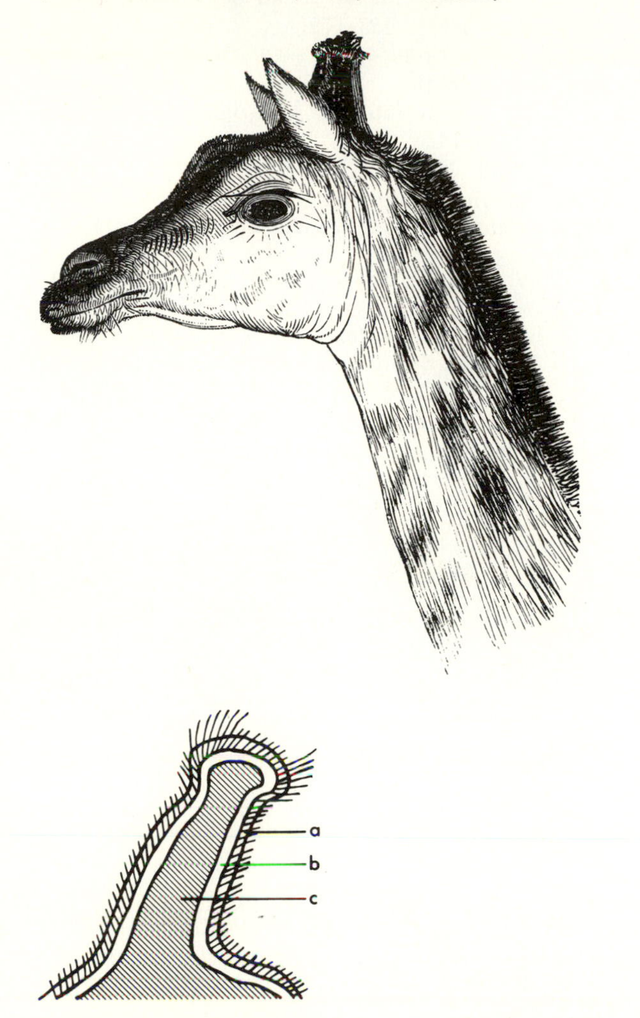

5-D. Examine a series of racks from *Odocoileus* of different ages. What factors influence the differences in antler complexity? Which, if any, can you age absolutely?

5-E. Locate the burr, beam, brow tine, bez tine, and crown on a white-tailed deer, mule deer, wapiti, moose, and caribou. How do the general size and arrangement of these racks vary?

Giraffe "Horns"

The head ornamentation of giraffes and okapis, Giraffidae, consists of a pair of short, unbranched, permanent, bony processes, that are situated over the sutures between the frontal and parietal bones and are covered with skin and hair (Fig. 5-10). They ossify from distinct centers and then fuse to the skull. Thus they

Figure 5–11.
Diagrammatic section of a rhinoceros "horn": a, epidermis; b, dermis; c, bone; d, dermal papilla; e, matrix of epidermal cells; f, fiber.
(L.P. Martin)

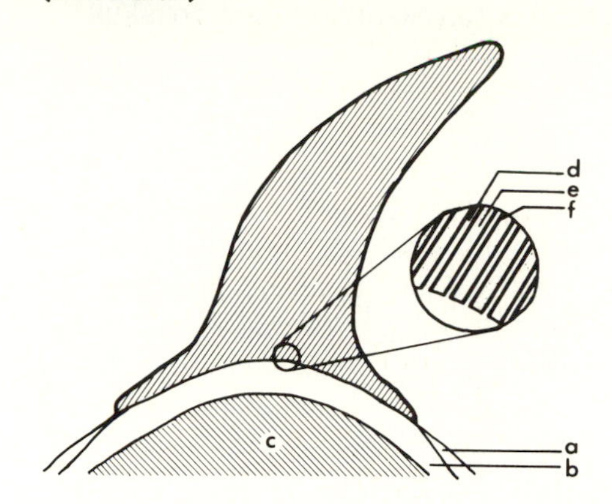

are not projections of the frontal bone as are other types of artiodactyl headgear. "Horns" are present in both sexes of giraffids and even in newborn animals. Anterior to the paired "horns" a median protuberance of the frontal bone is frequently present (Fig. 28-10). This "third horn" increases in size with the age of the individual.

5-F. Examine the headgear of a giraffe or okapi and compare the structure to that of a true horn and an antler in velvet.

Rhinoceros "Horns"

The only living nonartiodactyls to possess keratinized head ornamentation are the rhinoceroses of the order Perissodactyla (Fig. 27-4). The rhino "horn" does not have a distinct core and sheath, but is a solid mass of hardened epidermal cells that are formed from a cluster of long **dermal papillae** (Ryder 1962). The cells formed around each papilla constitute a distinct horny fiber resembling a thick hair. These fibers are cemented together by a mass of epidermal cells which grow up from the spaces between the fibers. They differ from true hairs in growing from a dermal papilla which extends up into them, rather than from a follicle extending down into the dermis (Fig. 5-11). The skin bearing the horn is situated over the fused nasal bones. These bones are generally enlarged and have a very rough surface which insures firm seating of the skin and horn (Fig. 27-3). In species with two horns the second is positioned over the frontal bones. The horns are conical and frequently curve posteriorly. In some species they may reach a length of four feet.

5-G. Examine a rhino horn and note its fibrous texture. What is the reason for the deep pores in the base?

5-H. Examine a rhinocerous skull. Where is/are the horn(s) situated in relation to skull bones?

Function

Horns and antlers of all the above mammals frequently serve as offensive and defensive weapons. They are sometimes used to fend off attacking predators and are frequently used during the rutting season in battles between males (Schaffer and Reed 1972). Rhinoceros horns are also used in the courtship ritual preceding mating. In some species the horns seem to be expressions of sexual dimorphism. In species with hierarchial social systems, the relative size of horns and antlers usually influences the relative ranking of the individual in the group or herd. See Ewer (1968) for further discussion of this subject.

Supplementary Readings

Bronson, W. S. 1942. *Horns and antlers.* Harcourt, Brace & Co., New York. 143 pp.

Chapman, D. I. 1975. Antlers—bones of contention. *Mammal Review* 5:121-172.

Ewer, R. F. 1968. *Ethology of mammals.* Plenum Press, New York. 418 pp.

Geist, V. 1966. The evolution of horn like organs. *Behaviour* 27:175-214.

Modell, W. 1964. Horns and antlers, pp. 71-79. *In* Scientific American Readings, *Vertebrate structures and functions.* W. H. Freeman and Co., San Francisco.

O'Gara, B. W., and G. Matson. 1975. Growth and casting of horns by pronghorns and exfoliation of horns by bovids. *J. Mammal.* 56:829-846.

Ryder, M. L. 1962. Rhinoceros horn. *Turtox News* 40: 274-277.

Schaffer, W. M., and C. A. Reed. 1972. The co-evolution of social behavior and cranial morphology in sheep and goats (Bovidae, Caprini). *Fieldiana: Zoology* 61:1-88.

6 Claws, Nails and Hoofs

The extremities of the digits of mammals are protected by plates or sheaths formed of keratinized epidermal cells. Only the whales and most sirenians lack digital keratinizations.

Claws

In their most primitive form these structures are **claws** (Fig. 6-1), similar to those of birds and most reptiles. A claw is composed of a dorsal, scalelike plate, the **unguis,** and a ventral plate, the **subunguis.** The unguis, the better developed and harder of the two, is curved longitudinally and transversely, and encloses the subunguis between its lower edges. The subunguis is continuous with the **pad** at the end of the digit. The claw encases the last phalanx of the digit.

In addition to protection for the ends of the digits, claws serve several other functions. In some running mammals they aid in increasing traction and stability. Mammals that dig have long stout claws to aid in excavation. Arboreal species frequently have sharp claws that enable them to scamper up the side of a tree. The sloths have long curving claws from which they hang suspended in trees. Many carnivorous mammals use their claws to help hold or kill prey.

Claws are usually fixed in position, but most cats have retractile claws that may be drawn back into protective bony sheaths. This is an arrangement that minimizes wear of the claws.

Figure 6–1.
Digit with a claw. A, lateral section; B, ventral view; unguis shaded, subunguis stippled, and pad dotted.
(R. Roesener)

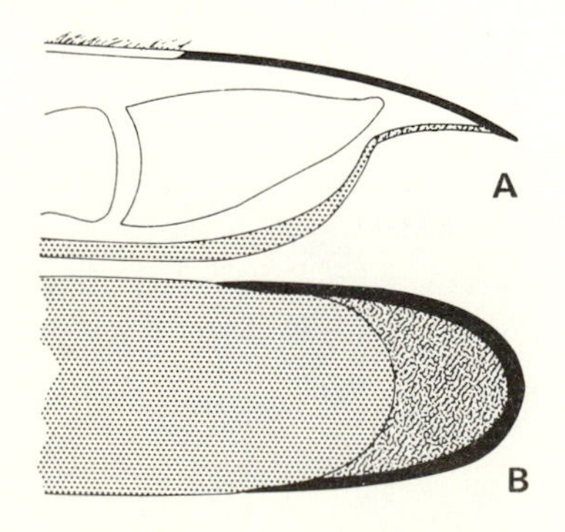

6-A. Examine the claws of a tree squirrel, cat, dog, and badger or mole. Locate the unguis and subunguis on each. What is the main function of the claws in each of these mammals?

6-B. Compare the structure and arrangement of claws in a cat or cat skeleton with those of some other carnivore. How does the retractile mechanism work?

Nails

A **nail** (Fig. 6-2) is a modified claw that covers only the dorsal surface of the end of the digit. In nails the unguis is broad and flattened and the subunguis is reduced to a small remnant that lies under the tip of the nail. Nails provide less protection for the ends

Figure 6–2.
Digit with a nail. A, lateral section; B, ventral view; unguis shaded, subunguis stippled, and pad dotted.
(R. Roesener)

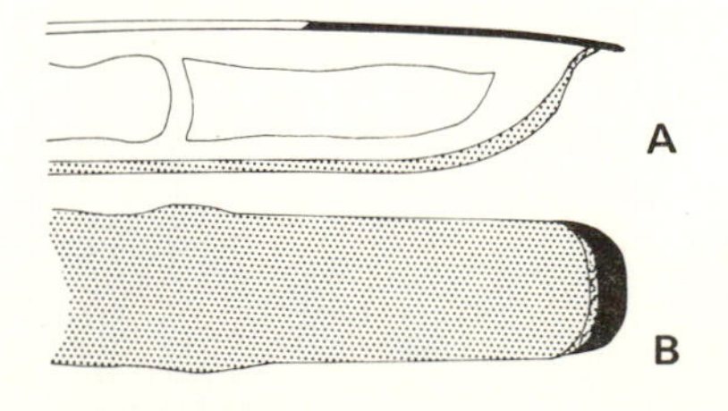

of the digits than claws, but they also allow for greater precision in manipulation of objects and for increased tactile perception.

6-C. Examine your fingernail and locate the unguis and subunguis. Compare with the nails of other primates.

Hoofs

In the **hoof** (Fig. 6-3) of an ungulate mammal the unguis curves almost completely around the end of the digit and encloses the subunguis within it. The pad lies just behind the hoof and is called the **frog.** Since the unguis is harder than the subunguis, it wears away more slowly and a rather sharp edge is maintained. Ungulates have only their digital keratinizations, hoofs, in contact with the ground. These durable structures provide good traction and protect the digits from wear.

6-D. Examine the hoofs of a horse and bovid (cow or sheep). Locate the unguis, subunguis, and frog.

Some mammals have keratinized structures that are very similar to hoofs but are not the only part of the animal in contact with the substrate. These **subungulates** also have pads on the bottoms of the feet.

Figure 6–3.
Digit with a hoof. A, lateral section; B, ventral view; unguis shaded, subunguis stippled, and pad dotted.
(R. Roesener)

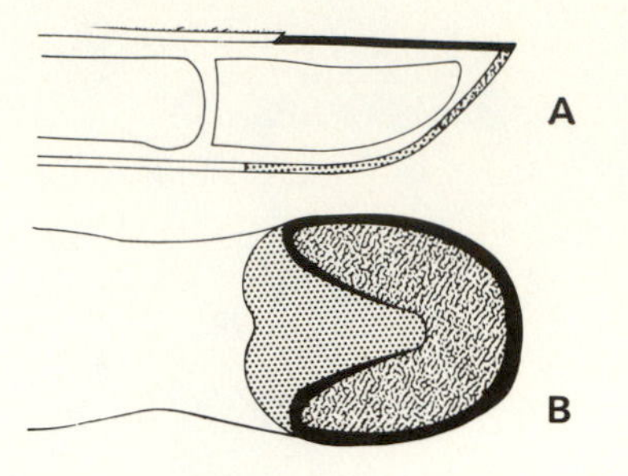

6-E. Examine the feet of an elephant, aardvark, or hyrax. How do the digital keratinizations compare with true nails and hoofs?

Spurs

The male monotremes are unique among mammals in possessing a hollow horny spur on each hind leg (Fig. 14-3). This is not a digital keratinization since it projects from the ankle rather than from the end of a digit, but it is made of keratinized epithelium. There are poison glands associated with the spurs.

Supplementary Readings

See supplementary readings sections of chapters 7 and 8.

7 Appendicular Skeleton

Mammals are **tetrapods** and with few exceptions (cetaceans and sirenians) have four conspicuous locomotor appendages, an anterior or pectoral pair and a posterior or pelvic pair. Like the skull and teeth, the limbs of mammals have a consistent general anatomy, but show considerable variation in size and proportion because of numerous adaptations for life in diverse habitats.

The Pectoral Limb

The shoulder or **pectoral girdle** (Fig. 7-1) of most mammals is composed of two elements: **scapula** and **clavicle.** The scapula is embedded in muscles dorsal to the ribs and has no direct articulation with any of the bones of the axial skeleton. The clavicle extends from the **glenoid fossa,** the shoulder socket, to the sternum and provides a firm brace for the anterior limb. Most mammals that run on hard ground have the clavicle reduced or absent. In these mammals there is no direct contact between the pectoral and axial portions of the skeleton, so that the shock of the body striking the ground is absorbed by the muscles and other soft tissues. The **coracoid,** the third main bone of the pectoral girdles of lower vertebrates, is rudimentary and fused with the scapula in marsupial and placental mammals. Monotremes have a pectoral girdle that is much more reptilian in structure with well-developed coracoids, **precoracoids,** and **interclavicle.**

The skeleton of the **pectoral limb** (Fig. 7-1) or forelimb is composed of a series of articulating elements. The single proximal element, the **humerus,** has a large head that articulates in the glenoid fossa of the scapula forming a ball-and-socket joint. This type of joint generally allows great mobility of the limb in several planes. Distally the humerus articulates with the two bones of the lower forelimb, the **radius** and **ulna.** This joint is termed the elbow. The ulna articulates with the humerus as a hinge joint, allowing movement in only one plane. The **olecranon process** of the ulna extends proximally beyond the humerus and serves as the short arm of the lever for attachment of the muscles that extend the forearm. In most mammals this process prevents the forelimb from being completely straightened. The radius, the more medial of the two forearm bones, articulates at both ends in a manner that allows the two bones to rotate around one another. It is this action that allows man to turn his

Figure 7–1.
Left pectoral girdle and limb skeleton of a rat, *Rattus norvegicus.*The clavicle is omitted.
(Chiasson 1969:9)

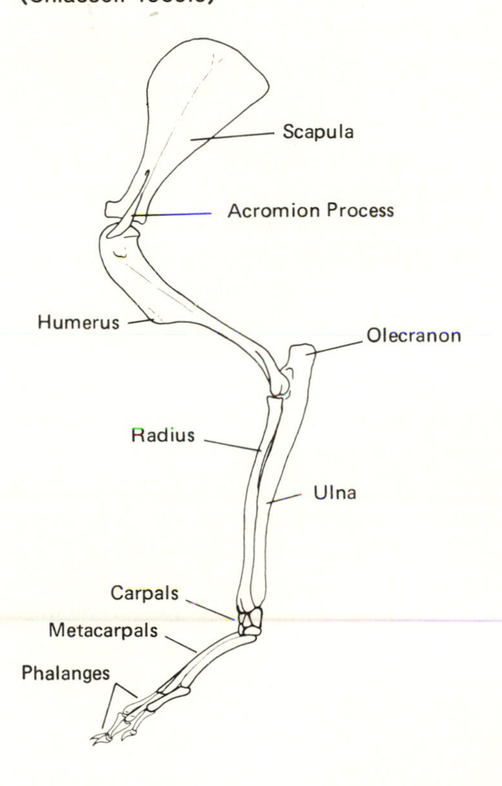

Figure 7–2.
Left manus of a rat, *Rattus*. Names of the individual carpal bones are given.
(Chiasson 1969:10)

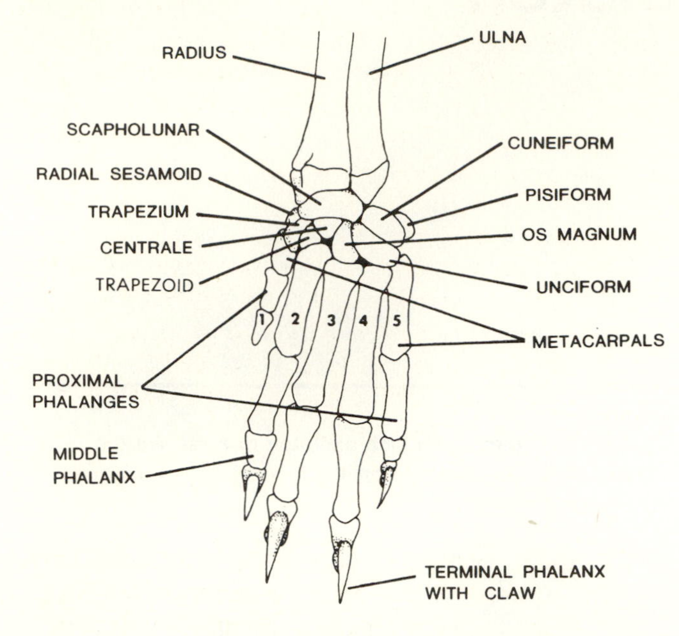

Figure 7–3.
Left pelvic girdle and limb skeleton of a rat, *Rattus norvegicus*. The patella is omitted from this figure. The tibia and fibula are fused distally in this species.
(Chiasson 1969:11)

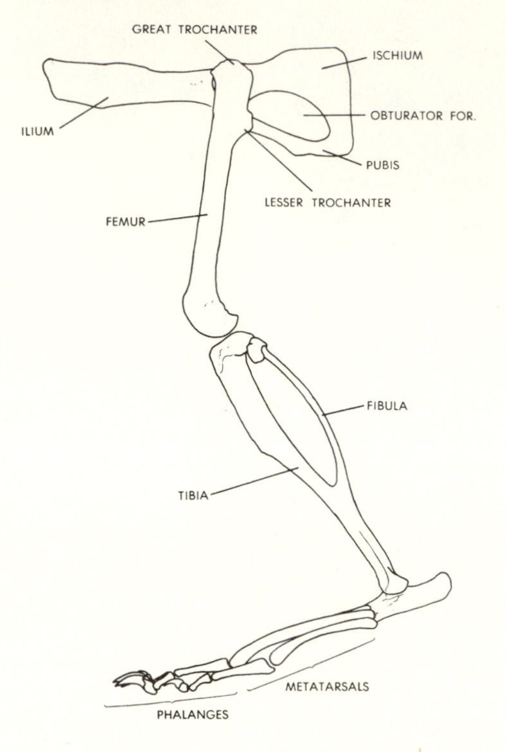

hand either palm up or palm down. Distal to the radius and ulna are a group of small bones, the **carpals** or wrist bones, which provide the great, yet sturdy, flexibility of the wrist (Fig. 7-2).

Primitively, mammals are **pentadactyl** or five-digited. While reduction in the number of digits frequently occurs, an increase in the number is very rare and does not normally occur in any mammalian group. Following the carpals are five elongate **metacarpals,** one for each digit. The metacarpals are enclosed within the body of the forefoot or palm of the hand. A series of **phalanges** (singular, phalanx) extends from the distal end of each metacarpal to form each **manual digit.** The first or most medial digit of the forefoot, termed the **pollex,** has two phalanges. In many mammals, including man, the pollex is opposable. The second through fifth digits typically have three phalanges each. With modifications of the limb for various environments, the number of phalanges may be reduced, or, in a few groups, increased. The forefoot or hand, including the carpals, metacarpals, and phalanges, is termed the **manus.**

The Pelvic Limb

The hip or **pelvic girdle** (Fig. 7-3) is a single structure formed by the fusion of three pairs of bones. The dorsal **ilia** (singular **ilium**) articulate with the sacral vertebrae. The **ischia** (singular **ischium**) are directed posteriorly and form the bony part of the rump. The paired **pubic bones** project anteriorly and ventrally and are joined at their distal ends. These three pairs of bones, together with the sacral vertebrae, form a ring through which the digestive, urinary, and reproductive tracts exit from the body. In females of some species the junction between the pubic bones, the **pubic symphysis,** is somewhat elastic to allow for passage of a large fetus.

The skeletal structure of the hind limb is similar to that of the forelimb. At the point where the three pelvic bones meet, a large socket, the **acetabulum,** receives the head of the **femur,** the proximal element of the **pelvic limb.** The femur is followed by a pair of bones, the **fibula** and **tibia.** The more medial of the two, the tibia, is larger in most mammals and the fibula is often greatly reduced. While the elbow bends to **project the forelimb anteriorly,** the knee bends to project the hind limb posteriorly. A sesamoid bone, the **patella** or knee cap, develops in a tendon on the anterior side of this joint.

Distal to the tibia and fibula is a complex of **tarsal bones** or ankle bones that correspond to the carpals of the forelimb (Fig. 7-4). The largest of the tarsals, the **calcaneum** or heel bone, extends posteriorly from the joint with the tibia and serves for attachment of the Achilles tendon. The lever arrangement is similar

Figure 7–4.
Left pes of a rat, *Rattus norvegicus.* Names of the individual tarsal bones are given.
(Chiasson 1969:12)

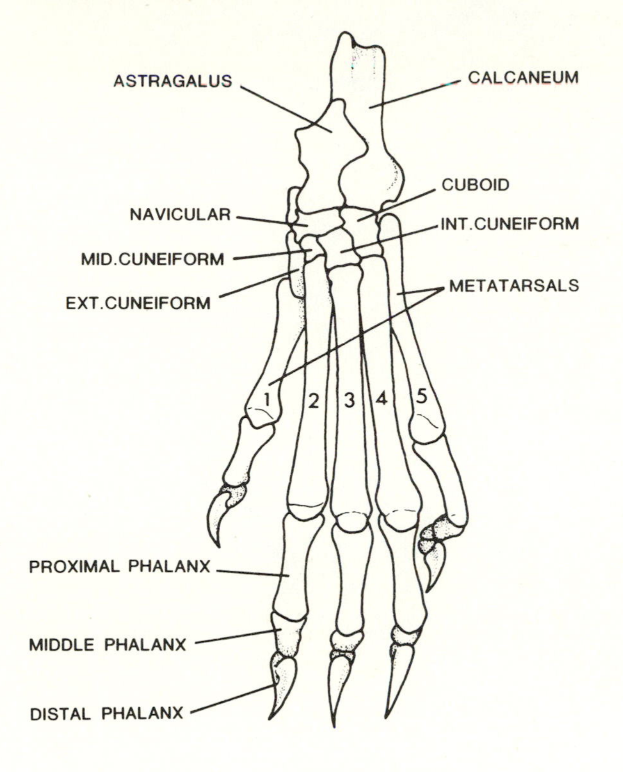

to that of the olecranon process of the ulna. The **astragulus** is the large tarsal bone adjacent and medial to the calcaneum. Five elongate **metatarsal bones** extend from the tarsals and correspond to the metacarpals of the forelimb. A series of **phalanges** extends from each metatarsal and forms each **pedal digit.** The first or most medial digit, the **hallux** or big toe, contains two phalanges. The remaining digits have three each. The hind foot, including the tarsals, metatarsals, and phalanges, is termed the **pes.**

7-A. Examine the skeleton of a raccoon, an opossum, or another plantigrade mammal. Locate each of the bones or structures given in boldface type above. (Cat or human skeletons may be used but each of these has modifications from the primitive ambulatory limb.)

7-B. Draw a patella in the correct position in Fig. 7-3.

7-C. Examine the pectoral girdle of a monotreme. (Use an illustration if necessary.) Identify each of the bones and compare the structure to that examined in 7-A above.

Supplementary Readings

Eaton, T. H. 1951. Origin of tetrapod limbs. *American Midland Naturalist* 46:245-251.

Hildebrand, M. 1974. *Analysis of vertebrate structure.* John Wiley and Sons, New York. 710 pp.

Also see the supplementary readings section of chapter 8.

8 Locomotor Adaptations

The generalized limb structure discussed in chapter 7 is well suited for walking or **ambulatory** locomotion. The metacarpals and metatarsals are unmodified, the pectoral and pelvic limbs are about equal in length, and the joints allow movement of the limb in several planes. The feet are **plantigrade** (Fig. 8-1A) and five digits are usually present. This primitive ambulatory limb structure has been modified in several ways to produce the diversity of limb types found in mammals.

Terrestrial Locomotion

Most mammals with the ambulatory limb are capable of running or **cursorial locomotion.** However, species that rely heavily upon running to catch prey or escape predators have specially adapted limbs. As the length of limb increases, so does the length of stride. When walking, man is plantigrade, resting his weight on the entire sole of the foot; but when running, he lengthens his limbs by raising the metatarsals off the ground and placing his weight only on the digits. Many cursorial mammals are permanently **digitigrade** (Fig. 8-1B) with metatarsals and metacarpals which never contact the surface during locomotion. Digitigrade mammals frequently exhibit reduction in the number of toes and elongation of the metacarpals and metatarsals. Toe reduction can occur in two basic ways, leaving either a large single central digit to carry the main axis of weight, the **mesaxonic foot,** or with the main axis of weight passing between a pair of similarly sized digits, the **paraxonic foot.** Perissodactyls, raccoons and humans are examples of mammals

Figure 8–1.
Skeletons of the plantigrade right hindfoot of a bear (A) and the digitigrade right hindfoot of a lion (B).
(Owen 1866)

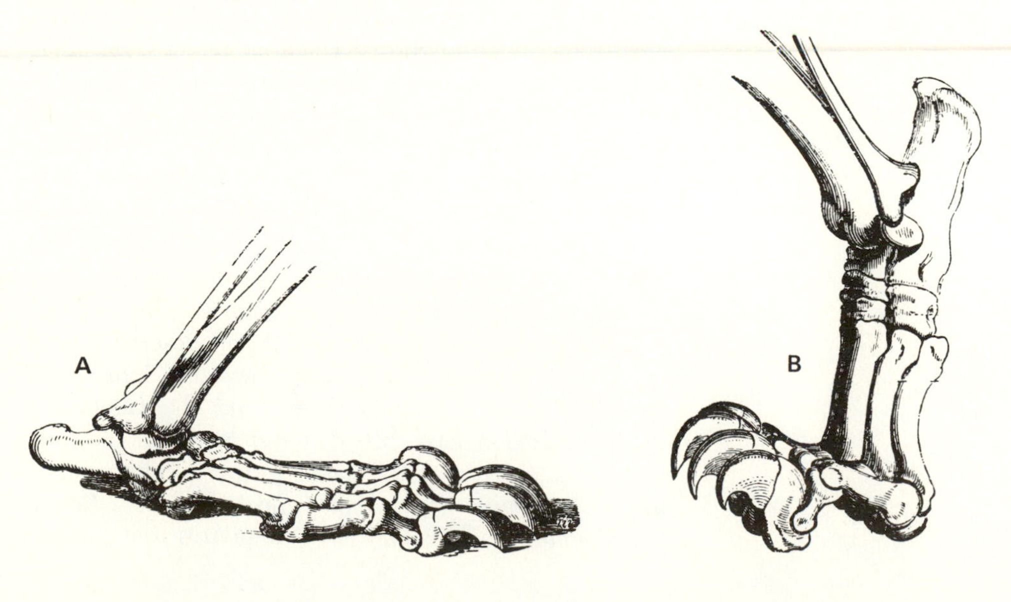

with mesaxonic feet. Artiodactyls, Tasmanian "wolves" and hyaenas are examples of mammals with paraxonic feet. A few mammals have paraxonic front feet and mesaxonic hind feet (e.g., the capabara, *Hydrochoerus hydrochoeris*) and at least one (the pig-footed bandicoot, *Choeropus eucaudatus*) has paraxonic hind feet and mesaxonic forefeet (Brown and Yalden 1973:115).

Most digitigrade mammals have limbs capable of moving in several planes, e.g., cats. This freedom of movement allows great agility and allows the limbs to aid in the capture of prey. Several digitigrade carnivores also have great suppleness of the back that allows the hind feet to be placed well in front of the forefeet when the animal is in full gallop. This **lordosis** is found, for example, in the cheetah, the fastest living mammal.

Hoofed animals illustrate the ultimate of cursorial limb structure, **unguligrade** limbs (Fig. 8-2). These have the phalanges elevated so that only the hoofs, the modified digital keratinizations, are in contact with the substrate. The proximal portions of the limbs are shortened and heavily muscular. The radius is usually fused with the ulna and the fibula with the tibia. These bones, as well as the metacarpals, metatarsals, and phalanges are usually greatly elongated and the number of digits is usually greatly reduced. The ultimate of reduction is reached in horses that run on a single digit. Most of the "cloven-hoofed" mammals, order Artiodactyla, have two functional digits. The pairs of metacarpals or metatarsals associated with these two digits may be fused to form the **cannon bone.** Unguligrade mammals generally have limbs capable of little lateral motion. The resulting restricted agility is usually compensated for by a long neck with a large head which may be used as a counterweight in maintaining balance.

Figure 8–2.
The skeleton of the unguligrade pectoral limb of a horse. (Owen 1866)

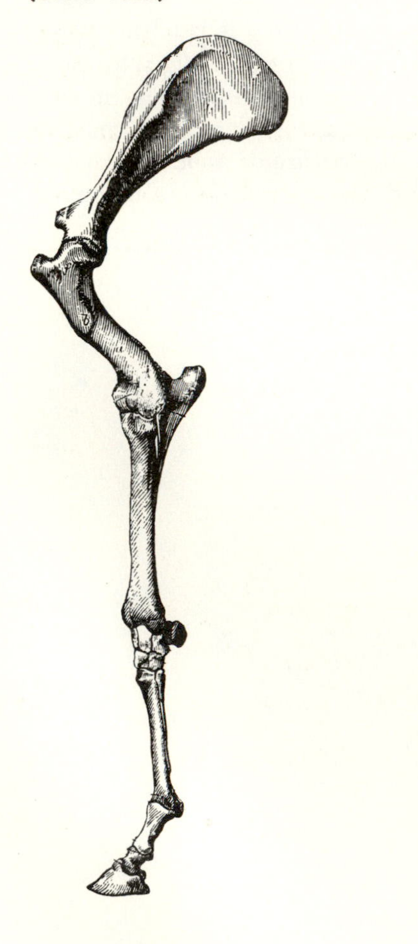

8-A. Examine the limbs of a mounted specimen and a skeleton of a cat or other digitigrade mammal. Compare the lengths and relative positions of the various limb elements with those of a plantigrade mammal. Compare the placement of pads on the feet of a cat with those of a raccoon or opossum.

8-B. Examine the limbs of mounted unguligrade animals and skeletons. Compare the lengths and arrangement of elements with those of digitigrade and plantigrade mammals. What causes the restriction of lateral movement in the unguligrade limb?

8-C. Observe a live cat and horse running. Why can a cat turn more sharply than a horse?

Mammals that normally walk on particularly soft surfaces such as snow, sand, or mud have special limb modifications. For example, the lemmings, *Dicrostonyx*, and the hare, *Lepus americanus*, inhabit areas that have snow much of the year. These and several other mammals that occupy similar habitats have feet that are larger than those of their relatives in more temperate climates. Arctic mammals also frequently have the soles of their feet well covered with hair which offers insulation and further increases the size of the foot. The enlarged foot distributes the weight of the animal over a greater surface area and functions in the same manner as man's snowshoes. The jerboas, e.g., *Jaculus*, of the African and Asian deserts are examples of mammals that have similarly hairy feet for walking on soft sand (Fig. 8-3). The camels of the same region have broad, tough pads on the bottoms of the feet for the same purpose. The swamp deer, *Blastocerus dichotomus*, and waterbuck, *Kobus ellipsipymnus*, are examples of ungulates that inhabit wet areas. Their elongated and widely spread hoofs facilitate movement on mud.

8-D. Compare the feet of small mammals that inhabit regions of snow with those from areas of loose sand. Why are they so similar?

8-E. Compare the hoofs of the caribou, *Rangifer,* with those of the similarly-sized deer, *Odocoileus.* How do you explain the difference in terms of habitat?

8-F. Compare the feet of a jerboa with those of a camel. How do these different feet serve the same function?

Several mammals are **saltatorial;** they progress by a series of leaps with the hind limbs providing the main propulsive force. Some of these, such as the rabbits and hares, are quadrupedal. Their leap is termed a **spring** and all four feet are involved.

Many only distantly related saltatorial mammals (kangaroos, jerboas, kangaroo rats, jumping mice, etc.) are bipedal. Their leap is termed a **ricochet** and the forefeet are not utilized (Fig. 8-4). Ricochetal species have greatly elongated and very muscular hind limbs and reduced forelimbs. The pes is particularly long (Fig. 8-5). The tail, which is usually long and frequently tufted at the tip, functions as a counterbalance when the animal is moving and provides support when the animal is resting in a bipdal position. Ricochetal locomotion is particularly valuable for rapid movement over soft substrates and is common in sand-dwelling rodents.

8-G. Examine mounted specimens and skeletons of a kangaroo, kangaroo rat, or jumping mouse. Compare the structures of hind and forelimbs and compare these with typical plantigrade and digitigrade mammals.

8-H. Observe rapid locomotion of a live kangaroo rat (*Dipodomys*) or a jumping mouse (*Zapus*). How does the tail aid in maneuvering?

8-I. Examine a mounted or live rabbit and a rabbit skeleton. How does the limb structure

Figure 8–3.
The hindfoot of *Salpingotus,* a small rodent inhabiting arid, sandy areas of Asia.
(Vinogradov 1937)

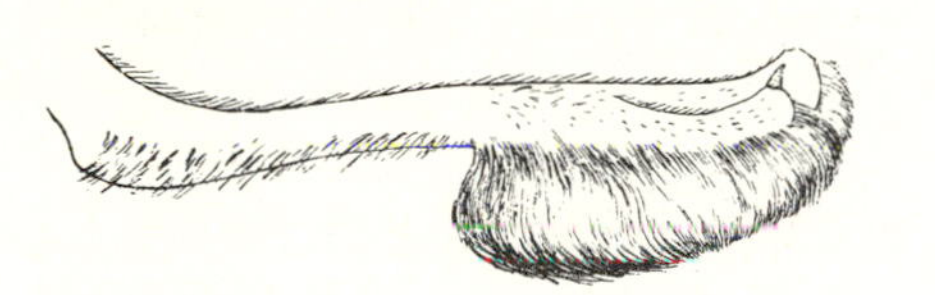

Figure 8–4.
A kangaroo exhibiting richochetal locomotion.
(Brazenor 1950:14)

Figure 8–5.
Skeleton of a richochetal rodent, the three—toed jerboa, *Dipus sagitta.*
(Owen 1866)

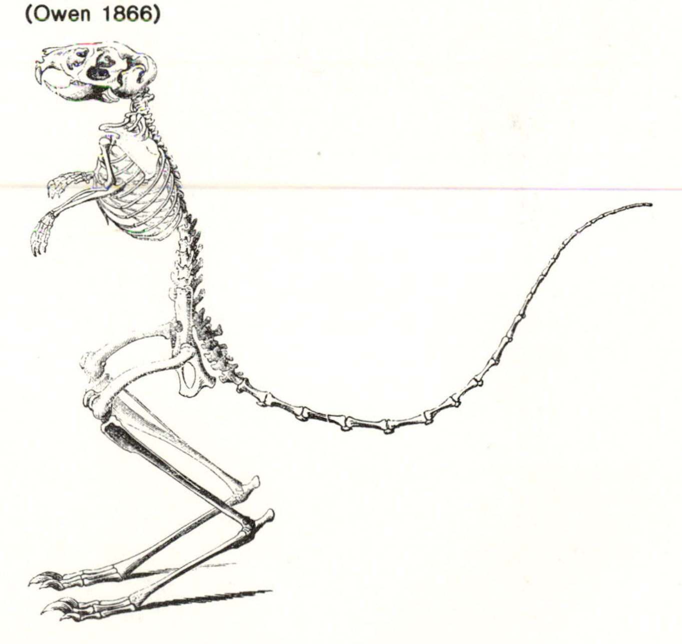

Figure 8–6.
The skeleton of an Indian elephant, *Elephas maximus*. Note the columnar nature of the graviportal limbs. (Owen 1866)

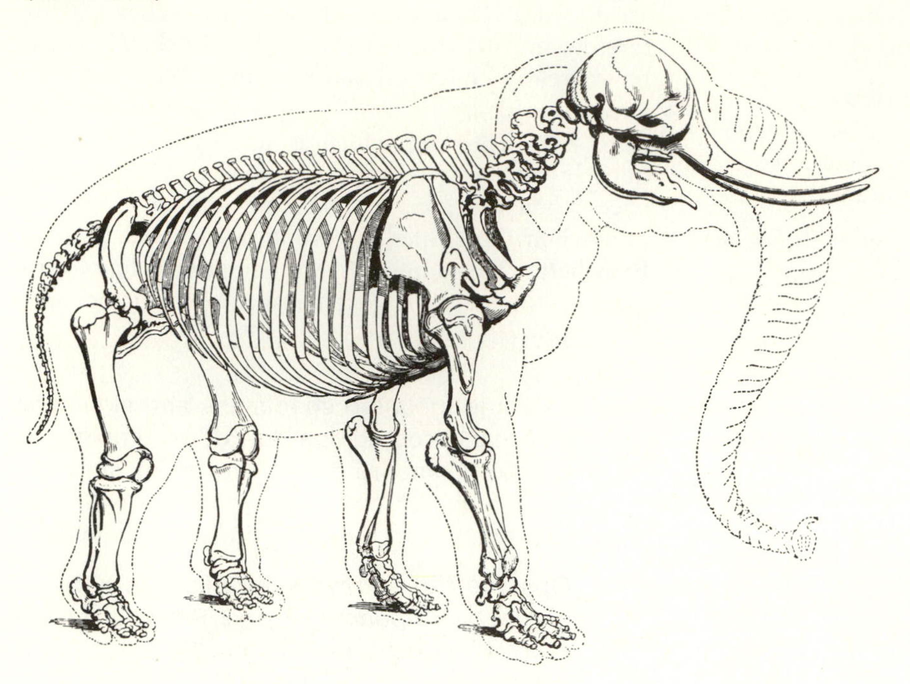

compare with that of an ambulatory or cursorial species? With that of a ricochetal species?

Some terrestrial mammals (e.g., the elephants) are extremely heavy-bodied and have a limb structure that is dictated more by the great weight of the animal than by a particular method of locomotion. This **graviportal** limb (Fig. 8-6) is essentially columnar, with each element situated directly above the one below it. The digits radiate out to form a series of arches that are incorporated in the flesh to form the base of a massive column. The bottom of the foot has a thick cushioning pad.

8-J. Examine specimens or illustrations of an elephant or hippopotamus leg. Compare the leg skeleton with the complete leg. Is it digitigrade or plantigrade?

Fossorial Adaptations

Many species of mammals have habits that involve digging. These range from species that dig only to obtain food and seldom go entirely beneath the surface, to species that feed, live, and breed beneath the ground and seldom come to the surface. Tendencies associated with life underground include reduction of external body projections, development of valvular body openings, reduction of vision, increase in number and sensitivity of tactile receptors, enlargement of the forefeet and claws, and reduction in length of the tail and neck.

Semifossorial mammals are those that burrow into the ground but also spend a great deal of time moving around on the surface. Ground squirrels, marmots, prairie dogs, kangaroo rats, grasshopper mice, pigmy rabbits, and many other rodents and lagomorphs live in burrows that they themselves dig, but gather food primarily on the surface. As prey species they are constantly alert and their eyes are frequently positioned high on the head so that they may observe their surroundings before exposing themselves. Badgers, some armadillos, and many other mammals both den and gather food below the surface but still spend a great deal of time above ground.

Fully **fossorial** mammals live underground and only rarely move about on the surface. In all of these animals the body is very compact; the tail is rudimentary; the neck is very short; pinnae are tiny or absent; eyes are frequently vestigal (Fig. 8-7); and the forefeet, pectoral girdle, and associated musculature are large and powerful (Fig. 8-8).

Figure 8–7.
A golden mole, *Chrysochloris,* a fully fossorial mammal.
(Flower and Lydekker 1891:639)

8-K. Compare the limb structure and general body shape of a badger with that of a raccoon. What modifications for semifossorial life are present in the badger?

8-L. Compare the position of the orbits in skulls of a raccoon and a marmot. Explain the reasons for the difference observed.

8-M. Examine a skinned mole carcass and a mole skeleton. Compare the structures of the hind and forelimbs.

8-N. Compare forefeet of a mole (Talpidae), golden mole (Chrysochloridae), pocket gopher (Geomyidae), molerat (Bathyergidae), and marsupial mole (Notoryctidae). Use photographs when necessary. How are the limbs of each of these distantly related fossorial mammals modified to serve the same function? Compare their eyes, external ears, and tails.

8-O. Examine a skin of a mole (Talpidae). Brush the hairs in various directions. What is the advantage of this type of pelage to a fossorial mammal?

8-P. Examine a specimen or photograph of the burrowing rodent, *Heterocephalus glaber* (Fig. 25-46B). How do you explain this unusual pelage?

8-Q. The lemmings (*Lemmus* and *Dicrostonyx*) live in Arctic areas where they must spend a considerable portion of the year burrowing through snow. Examine the claws on the fore-feet of a winter-caught lemming. How do these compare with the claws of a vole (*Microtus*) from a temperate region?

Figure 8–8.
Skeleton of a mole, *Scalopus aquaticus,* a fully fossorial mammal.
(Hatt 1946)

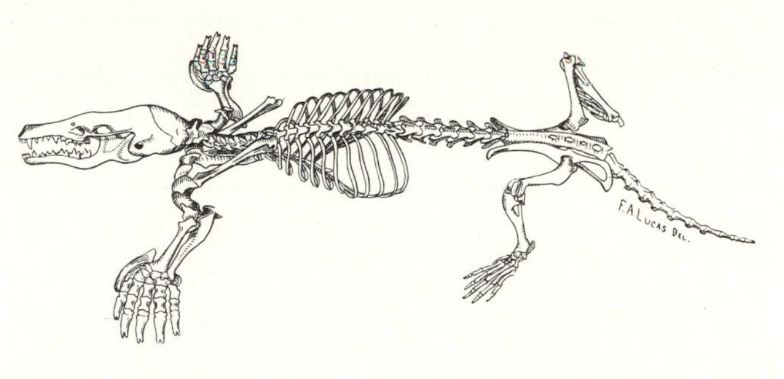

Figure 8–9.
Hindfoot of a water shrew, *Neomys fodiens.* Note the fringe of stiff hairs that increase the surface area of the foot.
(Stroganov 1957:231)

Aquatic Adaptations

Mammals that live in water, like those that dwell underground, show a variety of adaptations ranging from infrequent swimming to perpetual existence in water.

Semiaquatic mammals, such as the beaver, otters, water shrews, platypus, polar bear, etc., occur in several orders. They all exhibit some degree of aquatic adaptation. With the exception of the sea otter, all of these spend a good deal of their lives out of water either on land or in their islandlike dens. Many have **webbed feet** but none have true flippers. Some have a fringe of stout hair along the edge of the foot to increase the surface area (Fig. 8-9). Almost all have a generally **fusiform** body streamlined for easy movement through the water. The neck and tail are often thick at the base so that there is no sudden change of size where they join the body. The tail is frequently

Figure 8–10.
Tail of a water shrew, *Neomys fodiens.* Note the ventral keel of stiff hairs.
(Stroganov 1957:87)

Figure 8–11.
Forefoot of a platypus, *Ornithorynchus anatinus.* A, with webbing extended in the swimming position; B, with membrane folded back so that the stout claws may be used for burrowing.
(Ognev 1951)

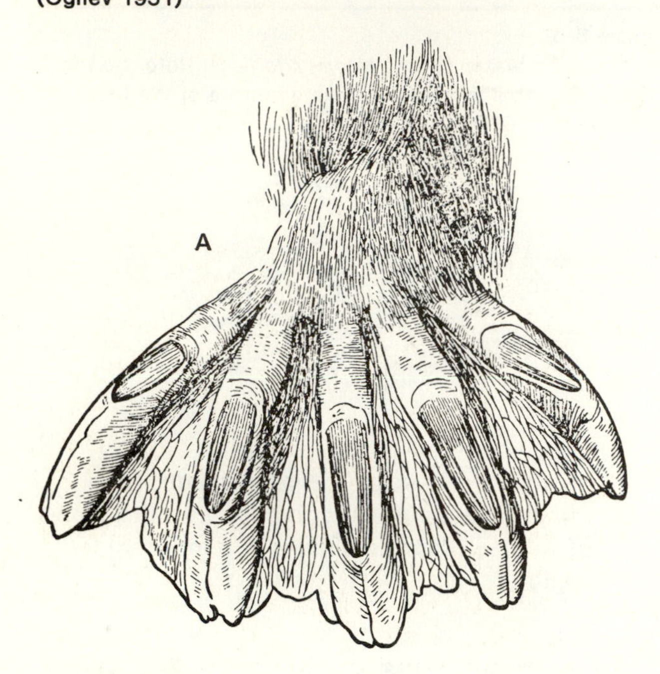

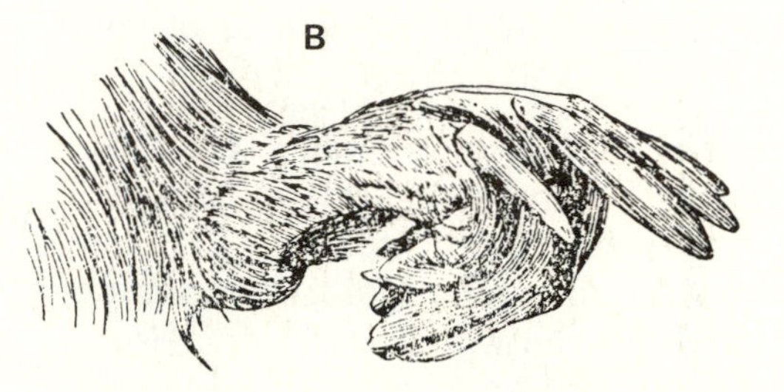

flattened or equipped with a keel of stiff hairs (Fig. 8-10) to assist in propulsion, maneuvering or both. The neck may be elongated to provide a counterbalance or shortened to provide a compact torpedo shape. The ears and nostrils are usually **valvular** and each eye is protected by a **nictitating membrane.** Pinnae and other external projections are reduced.

8-R. Examine a platypus, water opossum, water shrew, otter shrew, beaver, muskrat, polar bear, river otter, and sea otter for the above modifications. Use pictures when necessary. In what ways are these distantly related forms very similar in structure? How are the feet of each modified to increase the surface area for swimming?

8-S. How are the forefeet of a platypus well adapted for both swimming and digging its nest burrows (Fig. 8-11)?

8-T. The hippopotamus is also a semiaquatic mammal. How does its structure compare with other semiaquatic species? In what ways is it similar? In what ways is it different? Why are the limbs so different?

Pinniped mammals, the seals, sea lions, and walruses, spend a large part of their lives in the water but bear their young ashore. The forelimbs are modified into **flippers** that can be rotated under the body and used for locomotion on land. Seals, Phocidae, have hind flippers that are permanently rotated backwards; while sea lions and walruses, Otariidae, have hind flippers that can be rotated under the body and used

Figure 8–12.
Skeleton of a seal, Phocidae.
(Hatt 1946)

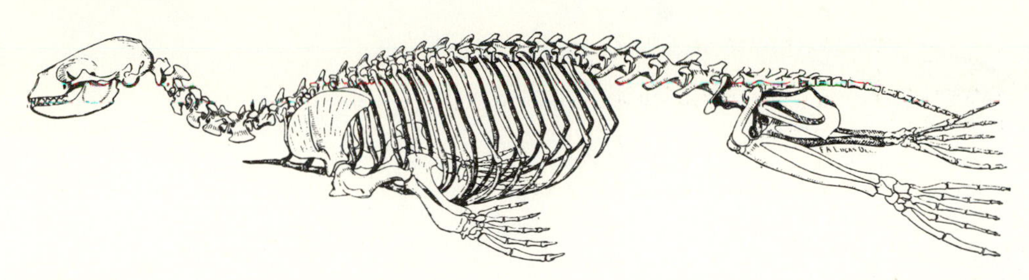

Figure 8–13.
Skeleton of a porpoise.
(Hatt 1946)

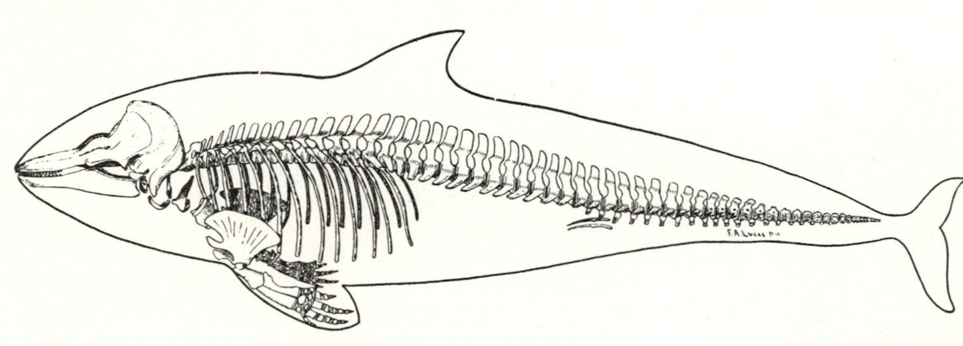

for locomotion on land. The main propulsive force of seals in water is produced by a lateral undulation of the posterior portion of the body, including the hind limbs. The neck is shortened to provide a torpedo shape and the forelimbs are relatively small and used primarily for maneuvering (Fig. 8-12). Sea lions and walruses use their relatively large forelimbs for propulsion. The neck is long to provide a counterbalance.

8-U. Examine the ears, nostrils, and flippers of a seal and a sea lion. Compare the positions and sizes of the flippers and the lengths of the necks.

8-V. Observe photographs or films of seals and sea lions resting and moving on land and in water. Compare the methods of propulsion and maneuvering.

8-W. Compare photographs or specimens of polar bears with those of other bears. How do the neck lengths compare? Do these indicate the method of propulsion in water?

The fully **aquatic** mammals, the whales and sirenians, do not come onto land. Pinnae are absent and the ears are valvular. The body is extremely fusiform with a very short, thick neck and an even taper from trunk to tail. The forelimbs are modified as flippers and the hind limbs are absent externally (Fig. 8-13). The tail tip is laterally expanded and dorsoventrally flattened to form a paddle (manatees) or **flukes** (dugong and whales). These mammals swim by undulating the posterior part of the body in a vertical plane. The expanded flukes provide increased surface area for propulsion, while the pectoral appendages are used primarily for maneuvering. A **dorsal fin,** present in many whales, aids in stabilization in the water. The flukes and fin are composed of fibrous connective tissue without bony support (Fig. 8-13).

8-X. Examine models or pictures of whales and sirenians (live animals if possible). Note the structure of the tail, fin, and pectoral appendages. Locate the nostril(s).

8-Y. Examine the skeletons of the flippers of a whale and of a pinniped. How many digits are present? How many phalanges are present in each digit? How do these flippers compare with the feet of terrestrial mammals?

8-Z. Locate the pelvic rudiments on whale and sirenian skeletons. Which bone ele-

Figure 8–14.
White handed gibbons, *Hylobates lar,* brachiating.
(Flower and Lydekker 1891:729)

ments remain? What skeletal support is present in the flukes?

8-AA. Compare the tail structure of a whale with that of a fish. The fish undulates laterally. Of what advantage is dorsoventral undulation to an air-breathing mammal?

8-BB. Compare the placement of the orbits of several aquatic and semiaquatic mammals with those of several terrestrial and fossorial ones. Are they more similar to the terrestrial ones or the fossorial ones? Why?

Arboreal Adaptations

Locomotion in and between trees has taken many diverse forms ranging from the rapidly scampering of squirrels to the slow-motion of sloths. Modifications to the limbs and to the tail are many and diverse. **Prehensile** tails have evolved in many groups of arboreal mammals and are particularly common in the

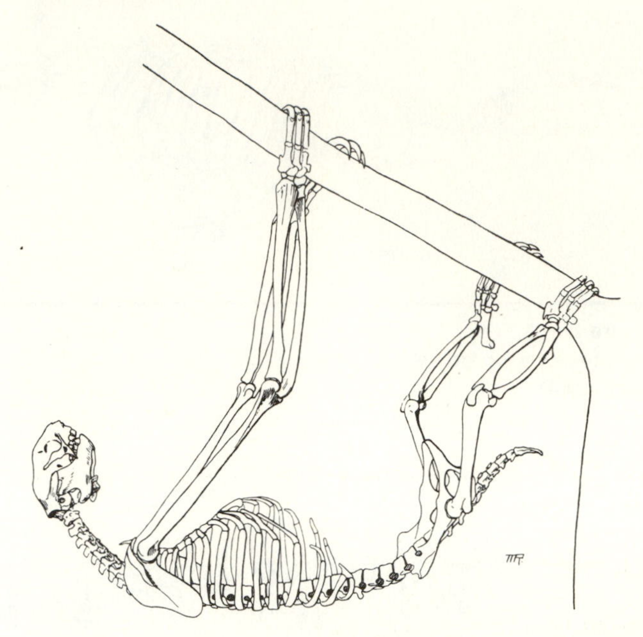

Figure 8–15.
Sloth skeleton, Bradypodidae.
(Hatt 1946)

Neotropical Region. The mammals that move rapidly through trees, particularly those such as the gibbons and flying squirrels that swing or jump from point to point, have well-developed binocular vision and depth perception. Some animals that commonly move about in trees, such as the tree hyraxes, *Dendrohyrax,* have no striking adaptations unique to their habitat, but they have good balance and move carefully.

Scansorial mammals, such as the tree squirrels, e.g., *Sciurus,* are essentially terrestrial in structure. But they have sharp, strong claws that allow them to scamper up vertical surfaces and long fluffy tails that aid in balance.

Arboreal mammals, such as arboreal mice, most monkeys and most opossums, cling to branches by prehensile, opposable digits, prehensile tails, or both.

Brachiating mammals, such as the gibbons, swing through the trees with their hands (Fig. 8-14). This movement may be described as a bipedal walk with the arms. The olecranon process of the ulna is very small and thus allows the arm to be extended perfectly straight to support the weight of the body. The pollex of brachiating mammals is frequently reduced or absent and the fingers are very long.

Sloth movement, found in sloths and colugos, is a quadrupedal walk by an animal hanging suspended from all four limbs. Sloths hang from strong recurved claws on all four limbs (Fig. 8-15).

The various flying squirrels, colugos or flying "lemurs," and certain Australian phalangers have a thin

Figure 8–16.
Ventral view of a colugo, *Cynocephalus*, showing the extent of the gliding patagia.
(Pocock 1926)

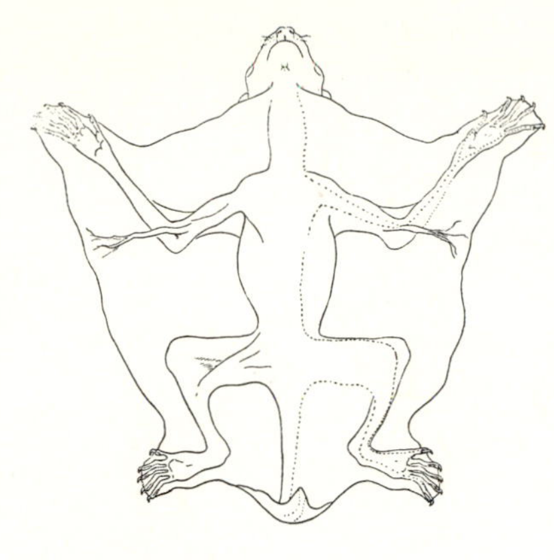

Figure 8–17.
A "flying" squirrel, *Hylopetes*.
(Hsia et al. 1964:31)

flap of skin, a **patagium,** extending between the hind limbs and forelimbs. The colugo has additional patagia extending from neck to forelimbs and from hindlimbs to tail (Fig. 8-16). These flaps offer maximum surface area for gliding through the air from tree to tree. The tails of the other gliding species are not enclosed in the patagia, but are long and have hairs arranged to provide an overall structure that is dorsoventrally flattened and that further increases the surface area (Fig. 8-17). Lifting and lowering of the limbs and tail facilitates maneuvering in midair. When not gliding, flying squirrels and flying phalangers move in trees by scansorial locomotion, but the colugo uses sloth-like movement.

While the common names of these animals are *flying* lemur, *flying* squirrel, and *flying* phalanger, it must be emphasized that they *do not fly*. They merely glide. Additional confusion on this point has been brought about by the use of the term "volant" for gliding locomotion. Volant is derived from the Latin, *volare,* meaning *"to fly."* Thus, volant is not a proper term to be used for mammals that glide but do not fly. We recommend that the term volant no longer be used in this context and that this type of locomotion be termed simply **gliding** locomotion.

8-CC. Examine the limbs and tails of a tree squirrel, arboreal opossum, cebid monkey, arboreal cercopithecid monkey, marmoset, gibbon, tree hyrax, raccoon, tree mouse, sloth, flying squirrel, and colugo. Into which of the above categories (scansorial, etc.) do each of these fit? Use photos where necessary.

8-DD. How do the gliding patagia and tails of a flying squirrel and a colugo compare? How do the foot structures of a sloth and a colugo compare?

8-EE. Examine a series of skulls of arboreal mammals. Which have the orbits directed anteriorly to provide overlapping visual ranges and, therefore, binocular vision? What is the selective advantage of depth perception in these forms?

8-FF. Prehensile tails have evolved in diverse mammalian groups. List mammals having a prehensile tail and the continent(s) on which each is found.

Aerial Adaptations

True flight has developed in only one mammalian order, the Chiroptera. The **wing** is a naked or nearly naked patagium that extends between the greatly elongated digits, the body, and the hind foot (see Fig. 18-1). The **uropatagium** is the continuation of this membrane between the hind legs and tail. The major skeletal modifications for flight include: elongated distal portions of the arm; elongated metacarpals and phalanges of the 2nd, 3rd, 4th, and 5th digits; a **keeled sternum** to which powerful flight muscles are attached; a knee that is directed outward and back; and a cartilaginous spur, the **calcar,** on the ankle to help support the uropatagium. All long bones are very slender and light in weight (Fig. 8-18 and 8-19).

Associated with the nocturnal habits of one of the two suborders of bats, the Microchiroptera, is the ability to echolocate. For this activity the ears of these bats are generally enlarged. A small structure in the inner edge of the pinna, the **tragus,** is very much enlarged in many bats and is believed to aid in echolocation. Elaborate structures are found on the noses of many bats and these are also believed to aid in

Figure 8–18.
Skeleton of a fruit bat, *Pteropodidae.*
(Hatt 1946)

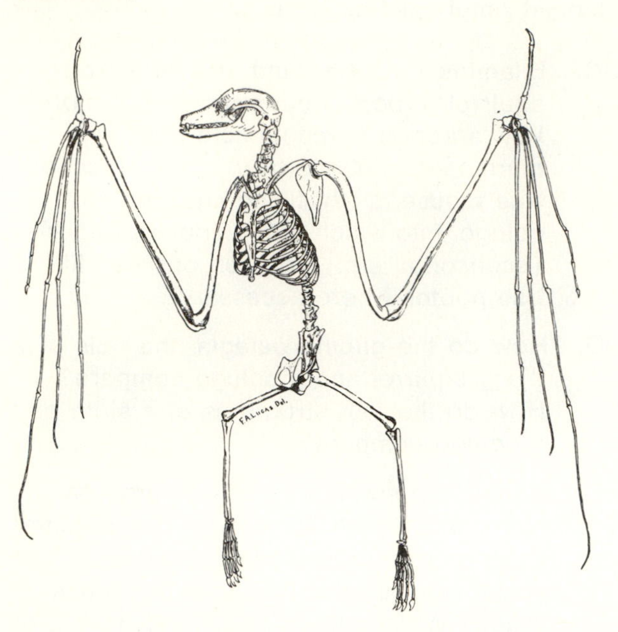

echolocation. The eyes of microchiropterans are present and functional but are generally very much reduced. In the suborder Megachiroptera the eyes are generally large, the tragus is absent, there is no nasal ornamentation, and there is echolocation in only one genus (*Rousettus*) utilizing a mechanism quite different from that of microchiropterans.

Bats differ greatly in their habits and habitats. Many are relatively sedentary spending their entire lifetime in a few square miles of cave and forest. Others migrate great distances annually. Due to these factors and others, there are different requirements for the general wing structure. Long narrow wings provide swift, sustained flight at the expense of maneuverability, while broad, relatively short wings provide maneuverability at the expense of speed and sustained flight.

8-GG. Examine a bat skeleton and note the modifications for flight mentioned above. How does the structure of the bat wing differ from those of the wings of the other flying vertebrates, the birds and pterosaurs?

8-HH. Examine a series of microchiropteran bats of the families Vespertilionidae, Phyllostomatidae, Rhinolophidae, Molossidae, and Emballonuridae. Compare the structures of the ears, noses, and tails of these forms. Do you find any relation-

Figure 8–19.
A bat, *Nyctalus,* with one wing extended.
(Ognev 1951)

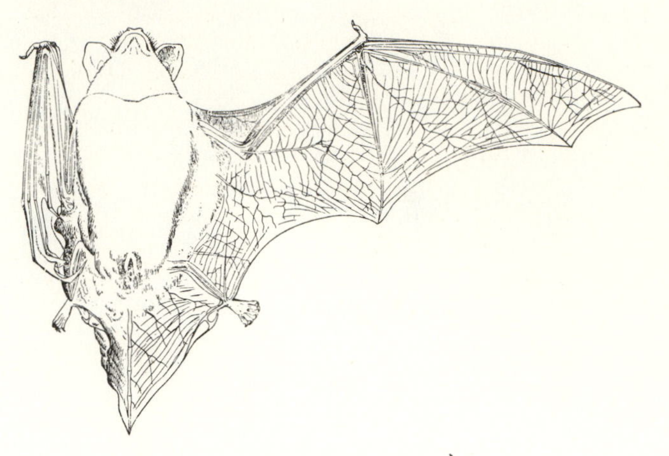

ship between size of tragus and development of nasal ornamentation?

8-II. Compare a megachiropteran with the above series. Note particularly the differences in eyes and ears.

8-JJ. Compare the extended wing of a free-tailed bat (*Tadarida*) with that of a vespertilionid bat such as *Plecotus, Myotis,* or *Pipistrellus.* Does this reveal anything about their respective habits?

Supplementary Readings

Brown, J. C. and D. W. Yalden. 1973. The description of Mammals—2 limbs and locomotion of terrestrial mammals. *Mammal Review* 3:107-134.

Dagg, A. I. 1973. Gaits in mammals. *Mammal Review* 3:135-154.

Dagg, A. I. and D. E. Windsor. 1972. Swimming in northern terrestrial mammals. *Canadian J. Zool.* 50: 117-130.

Ellerman, J. R. 1959. The subterranean mammals of the world. *Trans. Roy. Soc. of South Africa* 35:11-20.

Erikson, G. E. 1963. Brachiation in New World monkeys and in anthropoid apes. *Zool. Soc. London Symposium* 10:135-164.

Griffin, D. R. 1958. *Listening in the dark.* Yale Univ. Press, New Haven. 413 pp.

Harris, J. E. (Ed.). 1961. Vertebrate locomotion. *Symp. Zool. Soc. London* 5:1-134.

*Hildebrand, M. 1959. Motions of the running cheetah and horse. *J. Mammal.* 40:481-495.

Hildebrand, M. 1960. How animals run. *Scientific American* 202:148-157.

Hildebrand, M. 1974. *Analysis of vertebrate structure.* John Wiley and Sons, New York. 710 pp.

*Included in Jones, Anderson and Hoffman (1976).

Jenkins, F. A. 1971. Limb posture and locomotion in the virginia opossum (*Didelphis marsupialis*) and in other non-cursorial mammals. *J. Zool. London* 165: 303-315.
Lawlor, T. E. 1973. Aerodynamic characteristics of some Neotropical bats. *J. Mammal.* 54:71-78.
Marshall, L. G. 1974. Why kangaroos hop. *Nature* 248:174-176.
Muybridge, E. 1899. *Animals in motion.* Reprinted 1957, Dover Press, New York.
Slijper, E. J. 1961. Locomotion and locomotory organs in whales and dolphins (Cetacea). *Symp. Zool. Soc. London* 5:77-94.
Tarasoff, F. J.; A. Bisaillon; J. Pierard; and A. P. Whitt. 1972. Locomotory patterns and external morphology of the river otter, sea otter, and harp seal (Mammalia). *Canadian J. Zool.* 50:915-929.
Vaughan, T. A. 1959. Functional morphology of three bats. *Eumops, Myotis, Macrotus. Univ. Kansas Publ., Mus. Nat. Hist.* 12:1-153.
Vaughan, T. A. 1970. The skeletal system, the muscular system, flight patterns and aerodynamics. pp. 97-216. *In* W. A. Wimsatt (Ed.) *Biology of bats.* Academic Press, New York.
Yalden, D. W. 1966. The anatomy of mole locomotion. *J. Zool., London* 149:55-64.

9 Reproduction

Living mammals are divided into infraclasses primarily on their method of reproduction. Prototherian mammals lay eggs. Metatherians give birth to partly developed embryos that complete development in the female's marsupium. Eutherian mammals possess a placenta for providing nourishment to the embryos in the uterus. However, the reproductive cycle of all mammals is similar in that all have internal fertilization and females nourish their young on the secretions of mammary glands. Further details can be found in Asdell (1964; 1965), Nalbandov (1976), Perry and Rowlands (1969; 1973), Rowlands (1966), Sadleir (1969; 1973), and Weir and Rowlands (1973).

Anatomy of the Female Genital System

The female genital systems of all eutherian mammals are similar in basic plan (Fig. 10-1). Paired **ovaries** are the site of gamete production. Upon discharge from an ovary, the female gametes, **ova** (singular **ovum**), enter the **oviducts** or **fallopian tubes** which carry them to the **uterus** where, in the event of fertilization, development occurs. Upon parturition, the fetus exits through the **cervix**, or "neck" of the uterus, and through the **vagina.**

In the Monotremata (Prototheria) and in the Marsupialia (Metatheria), there is no true vagina since the uteri (Monotremata) or vaginal canals (Marsupiala) and the ureters empty into a common chamber termed the **urogenital sinus** (Figs. 9-2, 9-3). In the Monotremata, the urogenital sinus and the digestive tract empty into a common chamber, the **cloaca** (Fig. 9-2). A shallow cloaca is also present in some marsupials.

In monotremes, the egg passes through the uterus to the urogenital sinus and cloaca. In marsupials, the fetus passes down a temporary medial vaginal canal that connects to the urogenital sinus. This medial canal may remain open (many macropodids and *Tarsipes*) or seal shut (Fig. 9-3) until time for a subsequent parturition. See Sharman et al. (1966), Tyndal-Briscoe (1973), and Barbour (1977) for further details on marsupial reproduction.

Figure 9–1.
The female urogenital system of a rat, *Rattus norvegicus*. (Chiasson 1969:59)

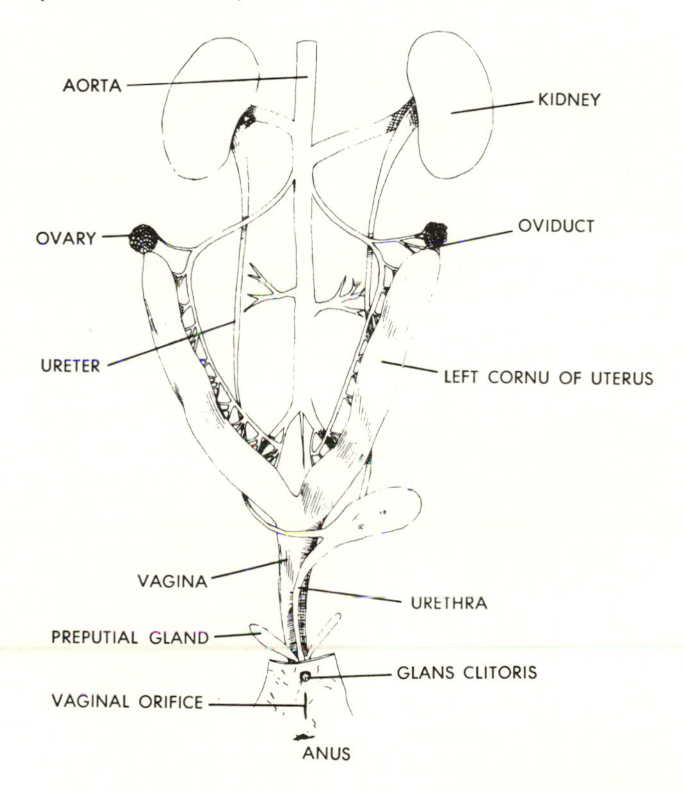

The female reproductive tracts of eutherian mammals have uteri, cervixes, and vaginas more or less fused. The **duplex** system (Fig. 9-4A) found, for example, in lagomorphs and most rodents has a single vagina, two cervixes, and two uteri. The **bipartite** system (Fig. 9-4B) has two essentially distinct uteri but

Figure 9–2.
The reproductive tract of a female monotreme (diagrammatic). Note the presence of the urogenital sinus and cloaca.
(L.P. Martin)

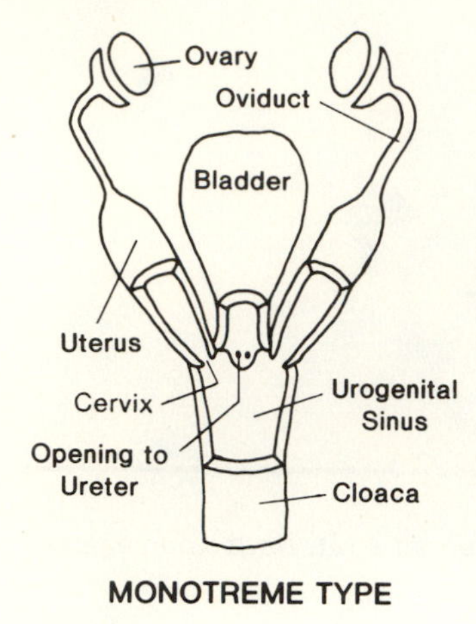

Figure 9–3.
The reproductive tract of a female marsupial (diagrammatic). The lateral vaginal canals receive the penis of the male. The vaginal sinus opens and connects with the urogenital sinus to serve as a passageway for delivery of the young.
(L.P. Martin)

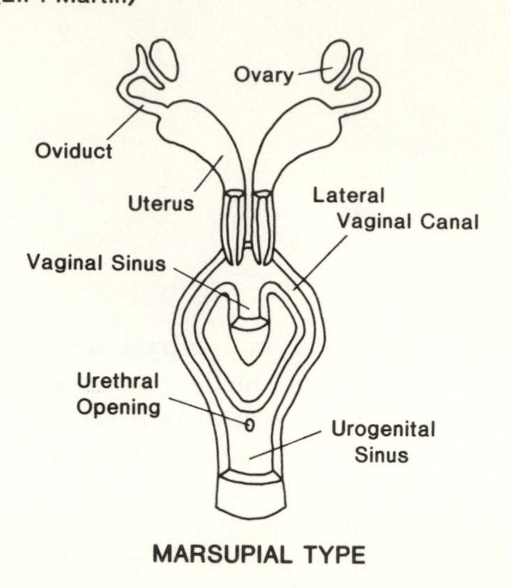

Figure 9–4.
The four types of eutherian uteri (diagrammatic): A, duplex, with two cervixes; B, bipartite, with one cervix and unfused uteri; C, bicornuate, with partially fused uteri; D, simplex, with completely fused uteri.
(L.P. Martin)

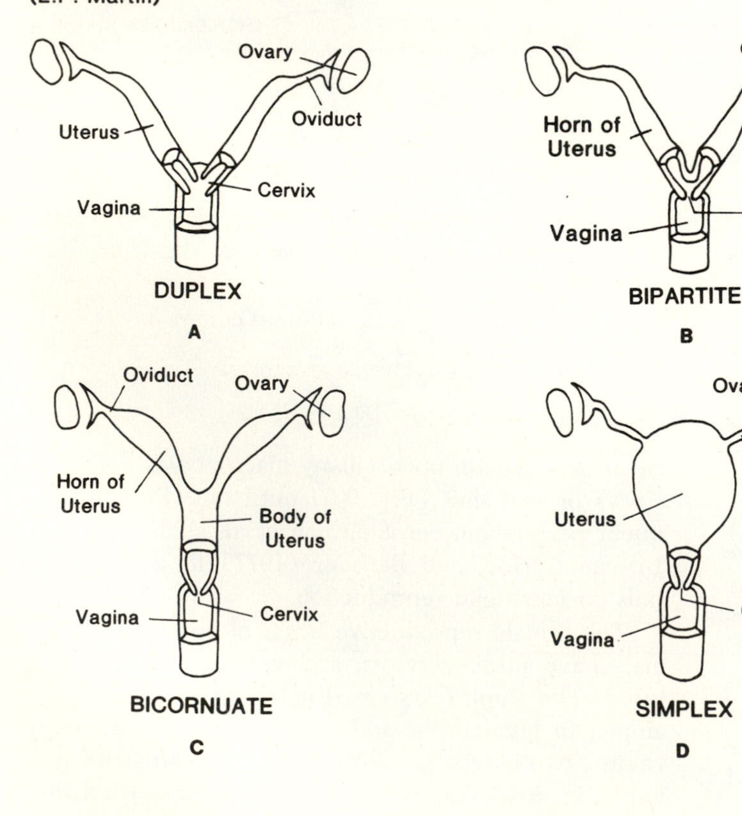

Figure 9–5.
The male urogenital system of a rat, *Rattus norvegicus*. (Chiasson 1969:57)

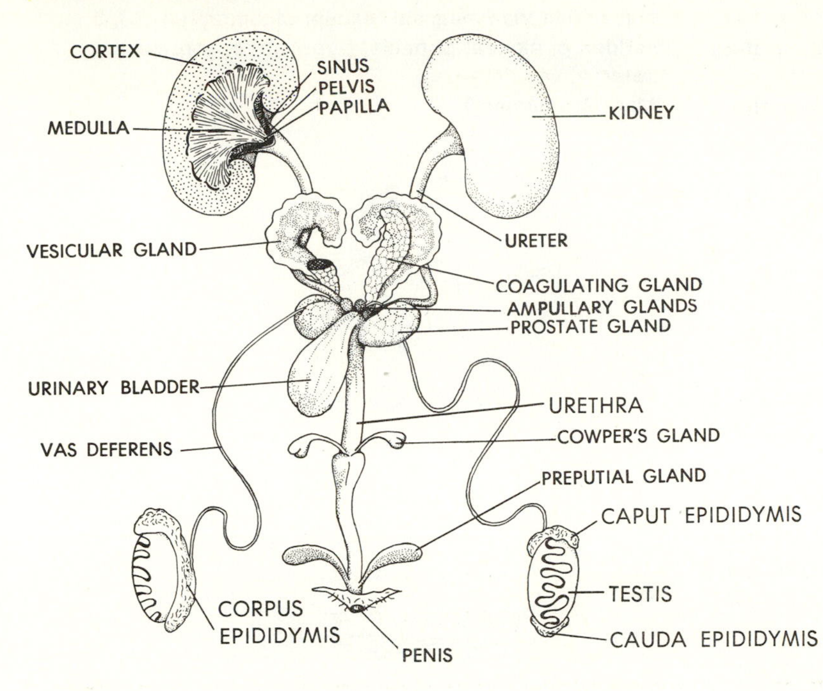

a single cervix and vagina. This system occurs, for example, in pigs and most carnivores. The **bicornuate** system (Fig. 9-4C), found, for example, in insectivores and most ungulates, bats, and primitive primates, is further fused with only about one-third to two-thirds of each uterus remaining distinct; the unfused portion of each of these partly-fused uteri is termed the **cornu** or **horn.** In the **simplex** system (Fig. 9-4D), found, for example, in edentates and higher primates, the fusion is complete and only a single uterus, cervix, and vagina are present.

The **clitoris,** the female homolog of the glans penis, is located anterior to the urethral opening. In some mammals, it may be supported by a small bone, the **os clitoris.** In many rodents, the clitoris is perforated by the urethral canal. In the spotted hyaena (*Crocuta crocuta*) the clitoris is large and resembles the penis of the male. It is difficult to determine sex in these mammals on the basis of external genitalia.

In adult females, the vaginal opening is generally open although in many rodents (particularly hystricomorphs) and insectivores, the opening is closed by a membrane during the nonbreeding season and, in some species, during pregnancy (Sadleir 1973).

In primates and most ungulates there are one or more pairs of fleshy folds of skin, the **labia,** that secrete lubricating fluids at the time of mating. The larger of these folds, the **labia majora** are the female homolog of the male scrotum. In certain primates, these folds, and additional areas known as **ischial callosities,** may become brightly colored during the period of estrus (Sadleir 1973).

9-A. Examine a demonstration dissection of a female cat or other eutherian mammal. Locate the ovaries, oviducts, uterus, cervix, and vagina. What type of uterus is present?

9-B. Compare the above reproductive system with that of a female opossum.

9-C. Examine an assortment of preserved eutherian uteri. Identify each of the four types.

Anatomy of the Male Genital System

The male gametes, **sperm,** are produced in the paired **testes.** From each testis the sperm pass through the ducts of the **epididymis,** a storage structure, into the **vas deferens** (plural **vasa deferentia).** The vasa deferentia empty into the **urethra,** a common duct for urine and gamete flow. Near the junction of the vasa deferentia and urethra (Fig. 9-5), the **prostate glands** (the anterior prostate glands are also termed the **coagulating glands), Cowper's glands** (=**bulbo-urethral glands),** and particularly, the **vesicular glands** (=**seminal vesicles),** when present, add secretions to the spermatozoa to produce **semen.** Additional accessory reproductive glands include the **preputial glands,** which are modified sebaceous glands, and the **ampullary glands,** located near the ampulla of the vas deferens (Fig. 9-5). In some mammals the coagulating glands produce a substance that coagulates the secretions of the vesicular gland. Following **copulation** or **coitus,** this coagulated fluid may form a **copulation plug** that fills the lumen of the vagina. Presence of a copulation plug indicates that coitus has occurred. Fertilization may or may not be correlated with plug formation. These plugs persist for a few minutes, hours, or days. Copulation plugs have been noted in many, but not all, species of rodents and are reported for some bats, insectivores and marsupials.

Some mammals (e.g., monotremes, edentates, cetaceans, subungulates, some insectivores and pinnipeds) have testes that are permanently abdominal. Other mammals (e.g., cricetine rodents in temperate areas) have testes that are abdominal through most of the year, but descend during the mating season through the **inguinal canals** in the muscular abdominal wall and into the **scrotum,** an external pouch of skin. The presence of the testes in a scrotum is a thermoregulatory mechanism that protects the developing sperm from high body temperature. Elephants, with abdominal testes, have spermatozoa that are able to tolerate the temperatures of the body cavity proper (Short 1972b). Testes are permanently scrotal in most marsupials, ungulates, carnivores, and primates. In marsupials and lagomorphs, the scrotum is situated anterior to the penis.

The **penis** of monotremes is posteriorly directed and situated in a sheath within the cloaca (Fig. 9-6A). The penis of marsupials is situated in a sheath and is posteriorly directed (Fig. 9-6B; see Biggers 1966). The penis of eutherian mammals is anteriorly directed (except in lagomorphs) and either pendant (higher primates and bats) or withdrawn into an external sheath (other mammals, Fig. 9-6C).

The sensitive distal end of the penis, the **glans penis,** is bifurcated in monotremes and marsupials and thus is compatible with the paired vaginas of the females. The glans of eutherian mammals is not divided, nor is the vagina, but in may species it may be variously adorned with lobes or spines.

All carnivores; most primates, rodents, and bats; and some insectivores have a bone, the **baculum** or **os penis,** within the penis. The structure of the baculum varies greatly from group to group (Fig. 9-7) and is frequently used as a major taxonomic character. (A much smaller bone, the **os clitoris,** is found in the **glans clitoris** (Fig. 9-7) of females of most species in which the males possess a baculum.)

Figure 9–6.
Lateral view (diagrammatic) of the male reproductive systems of a monotreme (A), a marsupial (B), and a typical eutherian mammal (except lagomorphs) (C), b, bladder; c, cloaca; p, penis; r, rectum; s, scrotum; u, ureter; v, vas deferens.
(Mary Ann Cramer)

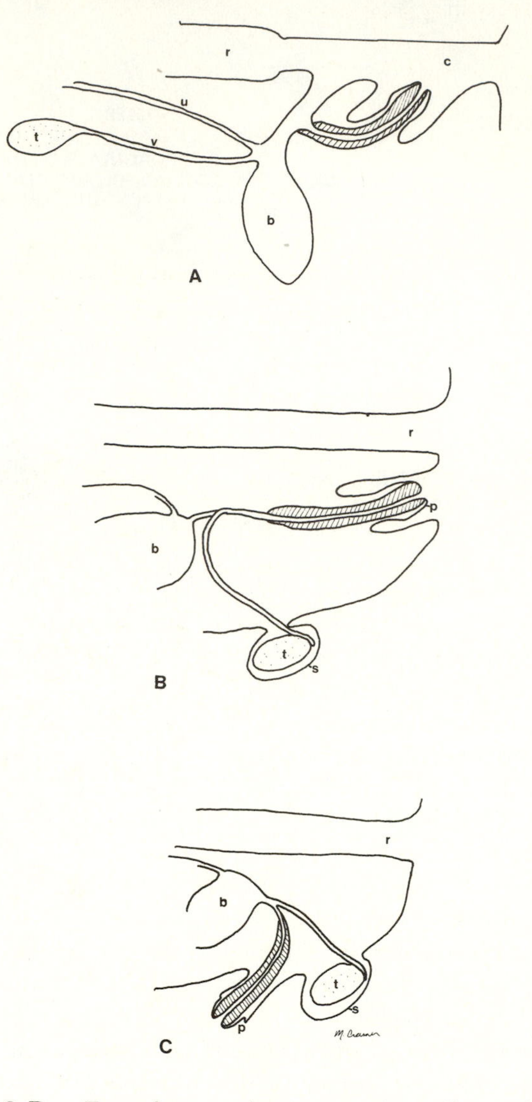

9-D. Examine a demonstration dissection of a male cat or other carnivore. Locate the testes, epididymides, vasa deferentia, prostate glands, urethra, penis, and glans penis. Is a baculum present? Are vesicular glands present? Compare with a demonstration dissection of a rat, *Rattus,* or other rodent.

9-E. Examine the penis of an opossum. What do you suppose gave rise to the old belief that opossums copulate through the nose?

9-F. Examine a series of bacula from various mammals. Use Burt (1960) to identify them.

Figure 9–7.
Bacula of seven species of *Mustela*. Note the variation that occurs among species. The shape and size of the baculum also varies greatly among the orders and families of mammals from a simple rod to a complex structure with numerous radiating parts. See Burt (1960) for illustrations of the bacula of North American mammals.
(Novikov 1956:130)

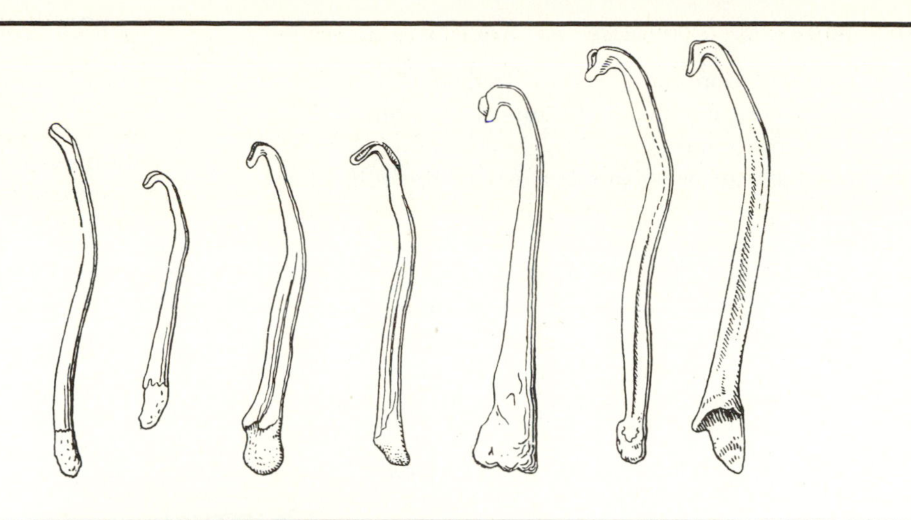

Figure 9–8.
Section through ovary and oviduct. a, antrum; b, ovum; c, Graafian follicle; d, developing follicles; e, corpus luteum; f, oviduct.
(W.J. Bleier)

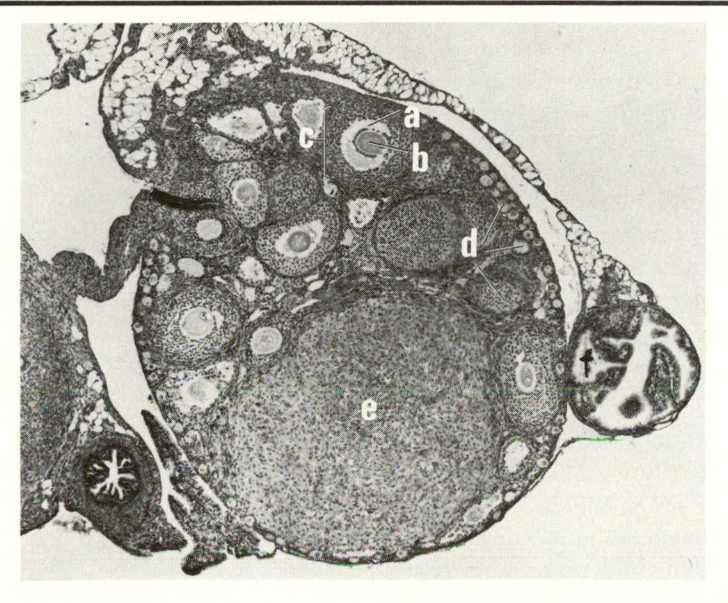

Oogenesis, Ovulation, and Corpora Lutea

Oogenesis

The ovary (Fig. 9-8) is composed of a central **medulla** surrounded by a thick **cortex.** The medulla and cortex are enclosed in a capsule of connective tissue, the **tunica albuginea,** which is in turn covered by a single layer of cube-shaped epithelial cells, the **germinative epithelium.** During the embryonic development of the female, the cells of the germinative epithelium undergo meiosis to produce **primary oocytes.** The oocytes penetrate the tunica albuginea and become situated in small **primary follicles** in the outer portions of the cortex. Well before the female has been born, the total number of eggs that she may ovulate during her lifetime lies dormant in the ovaries.

After sexual maturity and upon the proper hormonal stimulation, certain primary follicles and the enclosed eggs begin to develop. The follicle and egg grow into a large **Graafian follicle** that may occupy the entire thickness of the cortex. The Graafian follicle consists of several layers of cells surrounding a liquid-filled space, the **antrum,** in which the egg (which is small in contrast to most vertebrates) is suspended.

The vast majority of follicles that begin developing never mature and degenerate into **atretic follices** by a process that is not fully understood. In the human female, for example, of some seven million primordial germ cells that exist about five months after her conception, only about one-half million remain by birth and only some 60,000 by the time the woman reaches her middle twenties (Adams 1972). Since some 400 oocytes are ovulated during the reproductive life of the female, the number is sufficient despite the preciptious decline with age (Baker 1972).

Ovulation

Upon **ovulation** the Graafian follicle ruptures at the point where it touches the tunica albuginea. The tunica albuginea and germinative epithelium also rupture, and the egg oozes out along with the follicular liquid. The timing of ovulation within the reproductive cycle is variable and may be dependent on environmental factors (e.g., diurnal rhythms) or mating behavior. Generally, ovulation is **spontaneous** and occurs at fairly regular intervals during each estrus cycle (see below). In the Norway rat, *Rattus norvegicus,* ovulation occurs about 8 to 11 hours after the start of estrus while in the domestic sheep, *Ovis aries,* ovulation occurs some 18 to 24 hours after the onset of estrus. In Primates, and particularly in humans, the onset of ovulation is much more variable than in other mammals (Sadleir 1972). In mammals (Primates excluded) with extended periods of sexual receptivity, the stimulus of copulation may cause **induced ovulation.**

Corpora Lutea

After ovulation, while the egg is passing down the oviduct, a hormone (**lutenizing hormone, LH**) causes the ruptured Graafian follicle to fill first with fluid and then with luteal cells from the epithelian lining of the follicle. The yellowish glandular tissue that results is termed the **corpus luteum.** In the absence of mating, the corpora lutea remain virtually inactive and secrete very little **progestogens,** the hormones of pregnancy. Should fertilization occur, the corpora lutea become very active and secrete progestogens (principally progesterone) that prepare the uterus for implantation of the zygote and subsequent gestation. In a few species, cervical stimulation or mating without fertilization may induce **pseudopregnancy** in which the corpora lutea remain active for most of the "pregnancy." Inactive corpora lutea degenerate into **corpora albicans.**

Examination of the number of corpora lutea and corpora albicans in the ovaries may reveal much about the female's history of reproductive activity. The presence of corpora lutea indicates fairly recent ovulation. Accessory corpora lutea are found in many species of rodents (particularly hystricomorphs) that ovulate many (e.g., 200) oocytes but implant only one or two in the uterine wall (Perry 1972).

9-G. Examine a sectioned, stained, and mounted mammalian ovary under a compound microscope. Locate the medulla, cortex, tunica albuginea, and germinative epithelium. Locate a primary follicle. How many Graafian follicles are present? Are any corpora lutea present? How many? How many corpora albicans? What can you say about the reproductive history of this individual?

Estrous and Menstrual Cycles

In most species of mammals, the period of **estrus*** or heat is the time when females may accept males and copulate. During this state, females frequently exhibit special behavioral traits or show alterations in the external genitalia (e.g., swelling of the labia).

The **estrous* cycle** is the sequence of reproductive events, including hormonal, physiological, and behavioral, that typically occurs at fairly regular intervals in a mature female mammal. In many mammals (but to a lesser extent in Primates), the changes in the epithelial lining of the vagina provide a means to monitor the stages of the estrus cycle. In the **vaginal smear technique,** cells are collected from the vaginal lining, placed on a slide, and their condition noted and compared with standards developed for particular species. In the Norway rat (*Rattus norvegicus*), nucleated epithelial cells are found in the smears of **proestrus** when estrogen, progesterone, and lutenizing hormone (LH) levels reach their peak. During **estrus** proper, when ovulation occurs, the epithelial cells are cornified, the female is generally receptive to the male and copulation may occur. In this stage, LH levels drop dramatically along with a lesser decline of estrogen levels. Progesterone levels remain high initially then drop to base levels followed by a steady rise. In **metestrus,** leucocytes appear among the cornified cells of the smear, the corpora lutea are fully formed along with an increase of progesterone levels and a decrease of estrogen and LH levels. In **diestrus,** cornified cells occur in lesser numbers and some mucus may be present in the smear. Progesterone levels continue to rise, reach a peak, and then decline as the corpus luteum of nonpregnancy wanes (Nalbandov 1976; Short 1972a).

Many female primates, including humans, have a special sequence of events termed the **menstrual cycle.** In human females the menstrual cycle averages about 28 days but it varies widely among different women and even within the same individual. Ovulation in humans normally occurs about the midpoint of the cycle and is followed by a higher basal body temperature. Should fertilization fail to occur, the corpus

*Estrus is the noun, estrous is the adjective.

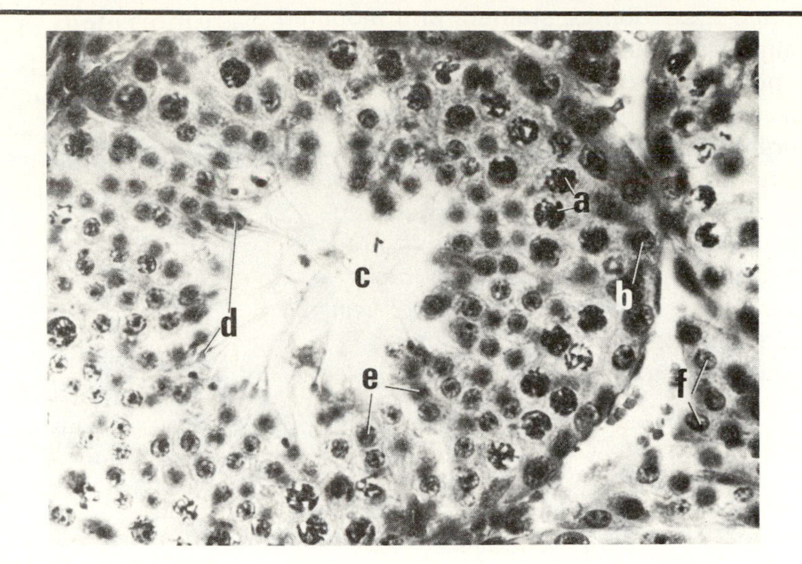

Figure 9–9.
Section through seminiferous tubule and interstitial tissue of testis. a, spermatocytes; b, spermatogonium; c, lumen of seminiferous tubule; d, spermatozoa; e, spermatids; f, interstitial cells.
(W.J. Bleier)

luteum regresses, progesterone levels drop, and the start of the **menstruation** (the sloughing of the endometrium) signals the start of a new cycle (Sadleir 1972).

In **monestrous species,** (e.g., dog, bear) only one estrus or heat occurs per year. In **diestrous species** (e.g., some species of squirrels), estrus occurs twice a year. In **polyestrous species,** estrous cycles occur at intervals throughout the year. In reality, however, many species of mammals are **seasonally polyestrous** since reproductive activity is restricted to certain times of the year. Females of species with **continuous estrus** (e.g., lagomorphs, camels) are receptive to males throughout the year and mating may induce reproductive and hormonal changes (Sadleir 1972).

9-H. Secure several live, mature, female *Rattus norvegicus*. Assemble several clean microscope slides, clean blunt-tipped pipettes with rubber bulbs, and a small quantity of physiological saline. Holding the female firmly by the nape of the neck, gently infiltrate a small quantity (several ml) of saline solution into the vagina (2-3 mm) and aspirate several times. Withdraw a small quantity of the aspirated fluid and place on a clean labeled slide. Examine the slide under the microscope. Repeat the operation for several individuals using clean slides and pipettes. What types of cells do you observe? What are the states of estrus of the different females? Would this be a valuable field technique?

Spermatogenesis and Fertilization

Spermatogenesis

Male Gamete Production

Each testis (Fig. 9-9) is composed of numerous **seminiferous tubules** in which sperm are produced and masses of hormone-producing **interstitial cells** situated in the spaces between the tubules. The testis is enclosed in a tough capsule of connective tissue, the **tunica albuginea.**

The seminiferous tubules are composed of triangular-shaped **Sertoli cells** arranged around the central lumen. Embedded near the outer edges of the Sertoli cells are the **spermatogonia** or primary germ cells. The large dark-staining nuclei of these cells divide to produce **primary spermatocytes** that undergo further division into **secondary spermatocytes.** The secondary spermatocytes divide quickly into **spermatids** and thus are frequently not seen in testis sections. The spermatids near the inner edge of the Sertoli cells elongate and grow tails. The resulting tailed cells are the **spermatozoa** or sperm.

Complete spermatogenesis takes about 30 days in the mouse (*Mus musculus*), 50 days in the Norway rat (*Rattus norvegicus*), and 74 days in humans (Sadleir 1973).

Upon completion of development, the spermatozoa move through the lumens of the seminiferous tubules, epididymal ducts, and vasa deferentia to the urethra. Upon copulation, the **semen** (spermatozoa plus secretions from the seminal vesicle and other accessory glands) is ejaculated into the vagina of the female (or directly into the uterus in horses, pigs, and in some species of rodents). In males that have had a vasectomy, the output of semen is not affected other than that it lacks spermatozoa.

9-I. Examine a sectioned, stained, and mounted mammalian testis under low power of a compound microscope. Locate the seminiferous tubules, interstitial cells, and tunica albuginea. Under high power examine a seminiferous tubule and locate the Sertoli cells, spermatogonia, primary spermatocytes, secondary spermatocytes (if present), spermatids, and spermatozoa. Where in this series is meiosis occurring? Can you locate and identify cells undergoing meiotic or mitotic division?

Fertilization

Some spermatozoa pass through the cervix and may reach the site of fertilization (generally lumen of fallopian tube) in a very few minutes. Several hours may elapse prior to actual fertilization until certain physiochemical changes, termed **capacitation,** take place in the spermatozoa to enable them to penetrate the protective covering of cells surrounding the oocyte (Austin 1972). After union of the spermatozoan and egg, a **zygote** is formed.

In some species of hibernating bats (e.g., in various species of *Eptesicus, Myotis, Plecotus,* and *Pipistrellus*), mating occurs in late summer but fertilization of the ovum does not occur until the following spring (Asdell 1964; Sadleir 1972, 1973). This mechanism, termed **delayed fertilization,** is an adaptation to time copulation during the period of greatest mixing of populations of these species and still to permit young to develop during the spring months and then be born in early summer when food resources, principally insects, are in greater abundance. During the winter months, the spermatozoa remain in the uterus or upper portion of the vagina and apparently receive nutrients from uterine glands (Asdell 1964).

9-J. Place a drop of physiological saline (0.9 g NaCl, 100 ml distilled H_2O) on a clean microscope slide. Remove an epididymis from a freshly killed adult male mammal and squeeze the contents onto the drop of physiological saline. Examine under a compound microscope. Are any live spermatozoa present? What can you say about the number and activity of the spermatozoa?

9-K. How does the timing of meiosis in sperm production significantly differ from the timing of meiosis in egg production?

Embryology and Implantation

Following fertilization, the zygote begins to divide or cleave while still in the oviduct. Division continues and the early embryo or **blastocyst** (8-16 cells) arrives in the uterus. The embryo is nourished initially by secretions of uterine glands (Nalbandov 1976).

In mammals having bipartite and bicornuate uteri, embryos resulting from eggs produced by one ovary may migrate to the uterine horn of the opposite side. This phenomenon, known as **uterine migration,** apparently aids in equalizing the number of embryos in each horn (Nalbandov 1976). In bats, a developing embryo (twins are rare) is generally found in the right uterine horn although maturing follicles and corpora lutea are approximately evenly distributed in right and left ovaries (Asdell 1964).

The attachment or **implantation** of the embryo to the uterine wall generally occurs within one to two weeks following fertilization (Nalbandov 1976). In **delayed implantation,** many months elapse between fertilization and implantation. This phenomenon is common for most temperate species of the Mustelidae, the roe deer (*Capreolus*), otarid and phocid seals, bears and armadillos (Asdell 1964). The review edited by Enders (1963) gives details on this fascinating adaptation. A related phenomenon, termed **delayed development** (Fleming 1971) occurs in the Neotropical fruit bat, *Artibeus jamaicensis.* Delayed development is apparently a mechanism to permit births to occur during the period when fruit is most plentiful (Fleming 1971). It differs from delayed implantation in that development slows down markedly after implantation of the blastocyst. In mammals with a postpartum estrus (many rodents), should copulation and fertilization occur, implantation may be delayed several weeks due to the effect of lactating young from the previous litter. In eutherian mammals, this latter mechanism is termed **aseasonal delayed implantation** while in marsupials it is generally referred to as **embryonic diapause** (Sadleir 1973).

Development and Placentation

In monotremes, the developing zygote is covered by a leatherly shell produced by shell glands. The shell in monotremes differs from that of reptiles and birds in that it continues to increase in size while in the uterus. When laid, the eggs are about thirteen to fifteen millimeters in diameter. The female echidna develops a transitory pouch on her abdomen into which a single egg is placed (Griffiths 1968). After approximately seven to ten days the egg hatches but the naked young remains in the pouch for an additional six to eight weeks. At this time, the developing spines of the young cause the female to eject it from the pouch permanently although it may continue to suckle. The platypus has no pouch but deposits eggs (usually two, occasionally one or three; Burrell 1927) in a nest deep inside an extensive underground burrow system. The naked young suckle milk from two patches of hair at the site of the mammary glands and do not leave the incubation chamber for about four months.

In marsupials, the young are born after a very short gestation period (12 1/2 days in the opossum, *Didelphis virginiana*). At birth, the hind limbs, eyes, and many other regions are only poorly differentiated, but the forelimbs are well developed and capable of grasping. Marsupials usually possess only a yolk-sac or **choriovitelline placenta** for intrauterine nourishment of the unborn fetus. In some bandicoots, Order Peramelina, true chorioallantoic placenta (see section on Eutheria) are also present and functional (Sharman 1959).

The **marsupium** is a pouch of skin on the abdomen of female marsupials. It is present in some species but absent in many other species (e.g., lacking in species of *Marmosa;* R. H. Pine, personal communication). In members of some genera, the marsupium opens ventrally (e.g., in *Didelphis*); in others it opens anteriorly and transversely (e.g., kangaroos) or posteriorly (e.g., koalas).

At birth the newborn animal climbs from the vagina through the female's belly hair to the opening of the marsupium (Sharman et al. 1966). The female licks a path for the young who apparently rely on olfaction to reach the pouch area. Once in the pouch the young locates a nipple and takes it into its mouth. The nipple then swells, securing the mouth of the young to the nipple. The female produces two kinds of milk: one for pouch young and one for independent joeys (Tyndal-Briscoe 1973). The developing young hangs there for a period of weeks or months, depending on the species, and then it detaches and continues its development more independent from the mother.

9-L. Examine a live or fluid-preserved female opossum *(Didelphis virginiana)* and locate the marsupium. How many nipples are present? If possible, examine preserved material with the developing young attached to the nipples.

In eutherian mammals, a yolk-sac or choriovitelline placenta, the typical placenta of marsupials, may persist for a short to moderate time after implantation of the embryo (Mossman 1937). This type of placenta is replaced by a **chorioallantoic placenta** (Fig. 9-10) composed of the extraembryonic membranes termed the **chorion** (outer layer) and the **allantois** (inner vascularized layer). The **placenta,** as defined by Mossman (1937), is an apposition or fusion of the fetal membranes to the uterine mucosa for physiological exchange. The actual contact between the maternal and fetal tissues is quite variable and may involve one to seven layers of tissue. The significance of this variability is not fully understood (Mossman 1937) but may relate to the efficiency of the physiological exchange of nutrients and waste products.

In gross morphology, the chorioallantoic placenta may be **diffuse** (Fig. 9-10A) and completely surround the embryo (e.g., in nonruminant artiodactyls, pangolins, and primitive primates). In ruminants and some other mammals, a **cotyledonary** placenta (Fig. 9-10B) is arrayed into discrete patches. The **zonary** placenta (Fig. 9-10C) of carnivores surrounds the embryo in a band, the size and thickness of which may vary depending on the species. Many mammals, including humans, possess a **discoid** placenta (Fig. 9-10D) in which the placenta resembles a disc or plate. Most rodents possess variations of the discoid placenta that may be cup-shaped (Fig. 9-10E) or spheroidal.

In some mammals, the vascularization of the placenta may produce **villous** or finger-like projections that facilitate physiological exchange. In other mammals, the projections are more elaborate and complex and are termed **labyrinthine.**

The nondeciduate placenta of some eutherian mammals (e.g., nonruminant artiodactyls) pulls free from the uterine wall at birth or **parturition** without leaving a tear in the uterine tissue. In species that possess a deciduate placenta, a portion of the uterine wall is torn away with each parturition. The resulting **placental scar** on the uterine wall remains and counts of placental scars can be useful in studying an individual's reproductive history (Rolan and Gier 1967; Lidicker 1973). **Nuliparous** females show no evidence of scars. A **parous** female is pregnant or has placental scars that indicate prior parturition. A **multiparous** female has placental scars of different ages thus indicating evidence of several litters.

9-M. Examine a sectioned, stained, and mounted placenta under a compound microscope. Differentiate between maternal and embryonic tissues.

9-N. Examine examples or drawings (Mossman 1937) of the various types of placentas described above. In what groups of mammals is each found?

Figure 9–10.
Sectional views through diffuse (A), cotyledonary (B), zonary (C), discoid (D) and cup-shaped discoid (E) types of placentas. Views A, B and D are saggital; views C and E are transverse. alv, allantoic vesicle; amc, amniotic cavity; cot, cotyledon; emb, embryo; exo, exocoelom; lum, uterine lumen; uep, uterine epithelium; ubc, umbilical cord; vtc, vitelline cavity.
(Redrawn by Mary Ann Cramer from Mossman 1937)

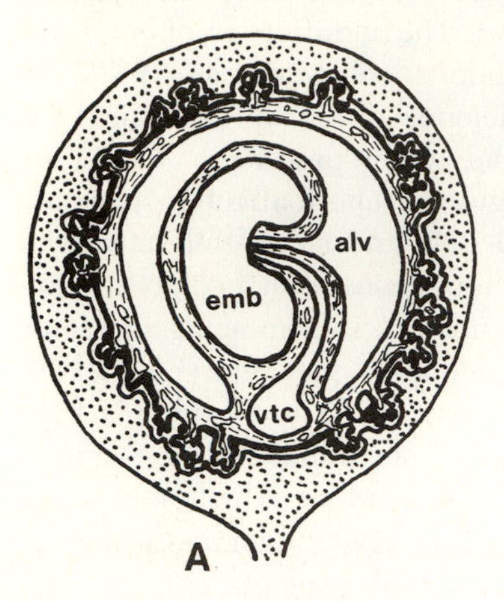

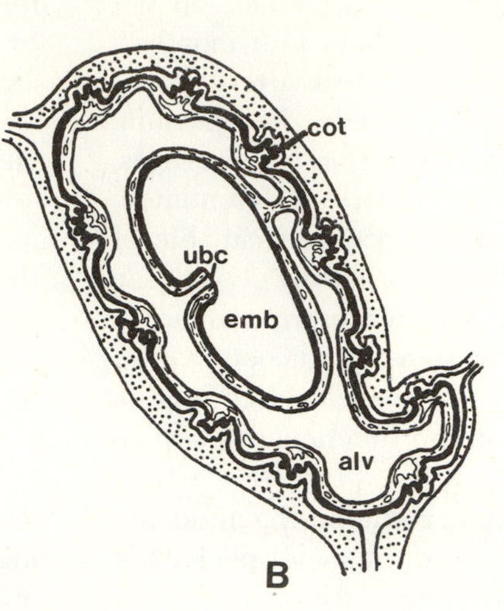

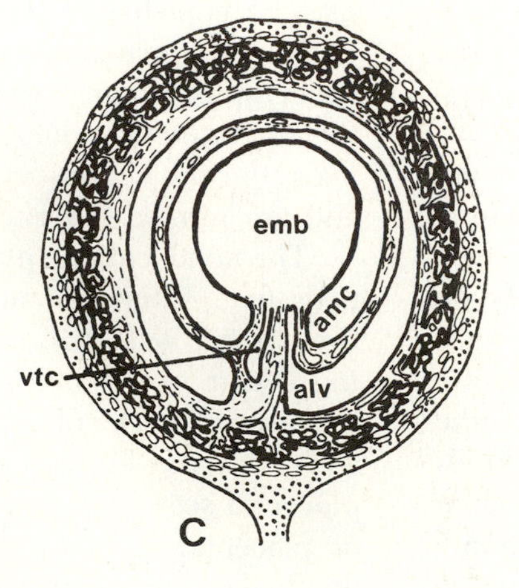

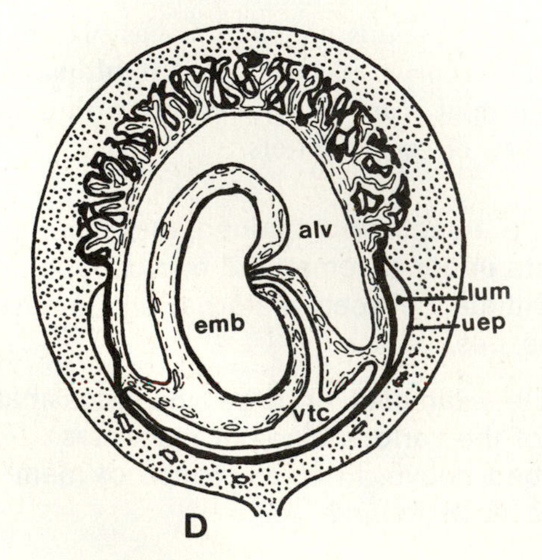

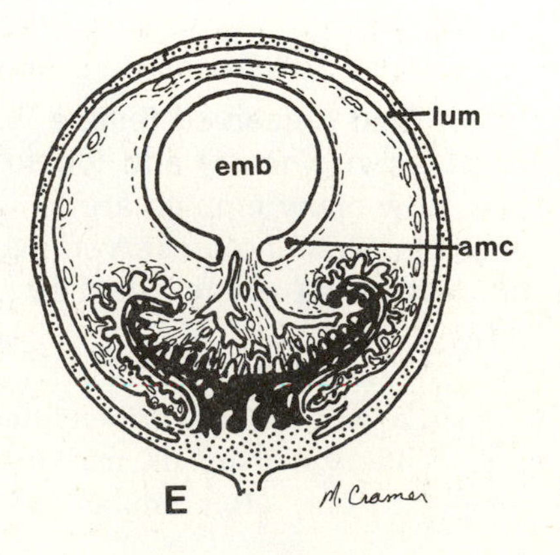

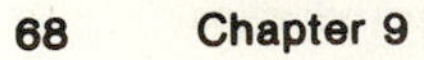

Figure 9-11.
Gestation lengths in several genera of cricetine rodents. (Adapted from Helm 1975:585)

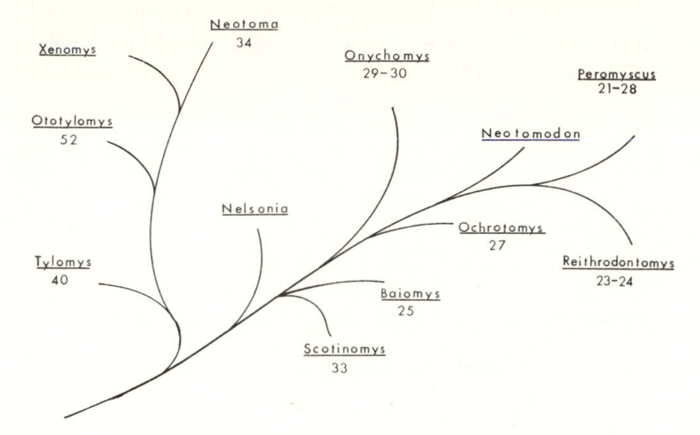

9-O. Examine the uteri of old and young females of a species that has a deciduate placenta. Locate and count the placental scars. What does this tell you about the reproductive history of these individuals?

Gestation, Litter Size, and Reproductive Strategies

Gestation

Gestation is the length of time from fertilization until the birth of the fetus. In monotremes, this period of time is apparently relatively short (in the echidna supposedly 27 days, Griffiths 1968) although the evidence is sketchy. Similarly, the gestation period in marsupials is short and, in contrast to eutherian mammals, is typically less than or approximately equal to the length of the estrus cycle (Sharman et al. 1966) The gestation lengths of eutherian mammals are quite variable (Fig. 9-11) and range from about 16 1/2 days in the golden hamster, *Mesocricetus auratus,* to some 22 months in the African elephant, *Loxodonta africana,* (Sadleir 1973).

Since monotremes are hatched and marsupials are born at a very early stage of development (no metanephric kidney, no dentary-squamosal joint, eyes closed, little hair, etc., Tyndal-Briscoe 1973) the young are termed **altricial.** The altricial newborn of eutherian mammals are more developed than young monotremes and marsupials. The **precocial** young of other eutherian mammals such as hyraxes, hares, ungulates, and many rodents (particularly, hystricomorphs) may be born with their eyes open (or they open in one or two days) and are capable of getting up and walking around shortly after birth.

The overall trend in mammals is for the length of gestation to increase with body size. But marsupials, cetaceans, and probably monotremes have much shorter gestation lengths than would be expected for their body sizes (Asdell 1965; Sadleir 1973). Higher primates have gestation periods significantly longer than expected for their size and the newborn are generally fully dependent on extended parental care. Hystricomorph rodents have gestation lengths three to four times longer than expected for their size and the young are generally born in a very precocial state (Weir 1974).

9-P. What factors do you think contribute to the lack of definitive data on gestation lengths for monotremes and for many marsupials and eutherians?

9-Q. What sort of evidence is necessary for establishing a definitive gestation length for a species? Why is gestation length in humans often recorded from the date of the last menstruation to the date of birth? Is this latter assumption reasonable in view of the wide variability in the "gestation lengths" computed by this method?

Litter Size and Reproductive Strategies

Litter size is the number of young delivered by a female at birth. Estimates of litter size can be made from counts of embryos in the uterus or by counting placental scars. The latter two methods may not be entirely accurate since **resorption** of embryos can occur; also, a placental scar may persist at the site of resorption and may be indistinguishable from a placental scar of a term embryo (see Rolan and Gier 1967).

Smaller species tend to have larger litter sizes than larger species and within a species the larger females usually have larger litters than their smaller conspecifics (Fig. 9-12). In species having more than one litter per breeding season, the second and subsequent litters tend to have fewer young than the first litter (Sadleir 1973).

Litter size also tends to increase in species living at higher altitudes and at greater distances from the equator. Spencer and Steinhoff (1968) proposed an hypothesis to explain the observed variations in litter size in mammals. Under this hypothesis, short seasons limit the maximum number of times that a female can reproduce in her lifetime, giving an advantage to phenotypes producing large litters. Correspondingly, long seasons favor females that produce small litters. In contrast, Millar (1973) found that these trends did not hold for pikas (*Ochotona princeps*).

Figure 9–12.
Relationship between female weight at parturition and litter size in *Peromyscus yucatanicus*.

(Lackey 1976:642)

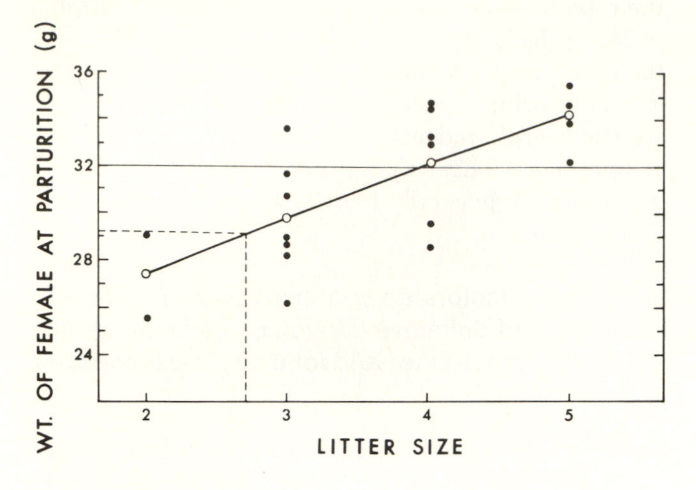

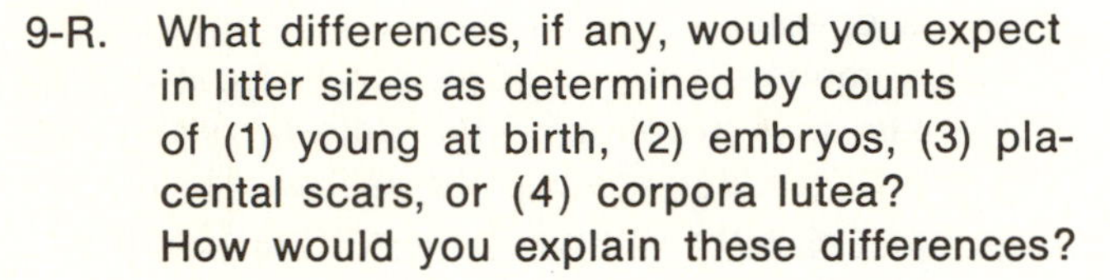
9-R. What differences, if any, would you expect in litter sizes as determined by counts of (1) young at birth, (2) embryos, (3) placental scars, or (4) corpora lutea? How would you explain these differences?

Species that maximize the production of relatively short-lived young, among other parameters, are often referred to as ***r*-selected** species while those that have fewer numbers of young that live longer are termed ***K*-selected** species (MacArthur and Wilson 1967). Not all species can be cast into these two categories of selection nor are these selection pressures static with time. Refer to Chapter 10 (this book), Nichols et al. (1976), and recent issues of *Ecology* for further information on this subject.

Studies of the reproductive tactics of mammals are some of the most exciting, yet difficult to quantify, aspects of mammalian ecology. One important measure, **reproductive value,** is the relative number of female offspring that remain to be born to each female of age *x*. This value is highest for females at the age they begin to reproduce and is zero for females that have ceased reproduction (Wilson 1975). **Reproductive effort** is the energy investment of an individual in a particular set of offspring and can apply to both males and females (Pianka and Parker 1975). Refer to Chapter 10 (this book), Wilson and Bossert (1971), Wilson (1975), Pianka and Parker (1975) and Pianka (1976) and Parker (1977) for further information and/or methods for calculating these values.

Many species of mammals have a seasonal period of reproductive activity when females are receptive to males and mating takes place. Following the season of mating, males may become sexually inactive. These seasonal trends are frequently strongly correlated with rainfall or other environmental variables such as length of the growing season or photoperiod. Refer to the reviews by Perry and Rowlands (1969; 1973), Sadleir (1972, 1973), and Fretwell (1972) for further information. Methods for gathering quantitative data on the reproductive condition of mammals can be found in Chapter 35 (Preservation—section on autopsy), Chapter 33 (Recording Data), and Twigg (1977).

Postparturitional Care and Development

Mammary Glands and Lactation

After parturition (or in the case of monotremes, hatching) all mammals feed upon **milk** secreted from the **mammary glands.** In monotremes there are two mammary glands in the abdominal wall that secrete milk into abdominal depressions. The hatchings then suck milk (Griffiths 1968) from hairs associated with these mammary glands (Fig. 9-13A). All therian mammals have mammary glands that are equipped with either nipples or teats and the young suckle. **Nipples** are found in most mammals and have numerous small glandular ducts that exit from the tip of a small, fleshy projection (Fig. 9-13B). **Teats,** such as those found in artiodactyls, have ducts that lead from the glands into a common reservoir that in turn is connected to the exterior through a single duct (Fig. 9-13C). See Cowie (1972) for further details.

Mammary glands are thought to be highly specialized sweat glands. They are found in both sexes but generally degenerate in males and develop in females.

The number and placement of mammary glands varies greatly and is correlated with the typical litter size for a species. Marsupials generally have a somewhat circular arrangement of nipples in the pouch. The opossum, *Didelphis virginiana,* for instance has thirteen nipples, twelve arranged in a large U-shape and the thirteenth centrally located. Eutherian mammals usually have the nipples arranged in two longitudinal, ventral rows. In species such as the hog, *Sus scrofa,* which have a relatively large litter size, each row extends from a point between the pectoral limbs to a point between the pelvic limbs. Higher primates, which usually have a single young, have a single pectoral pair of mammae while horses, which have the same litter size, have a single abdominal pair. Some rodents, e.g., species of *Mastomys,* that have an extremely large litter size, have the nipples extending out onto the backs of the thighs. Some hystricomorph rodents, e.g., *Myocastor,* have the nipples located rela-

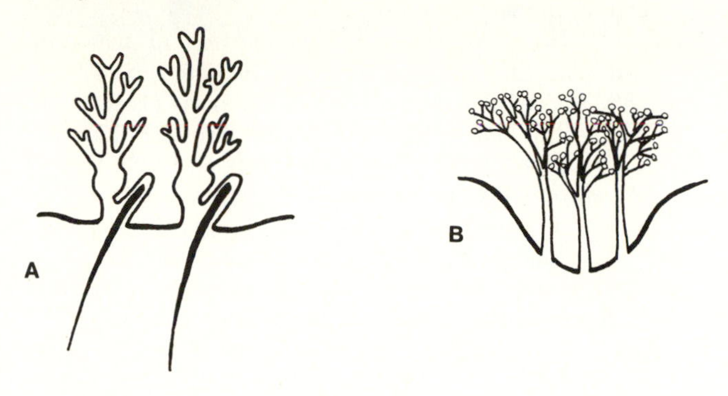

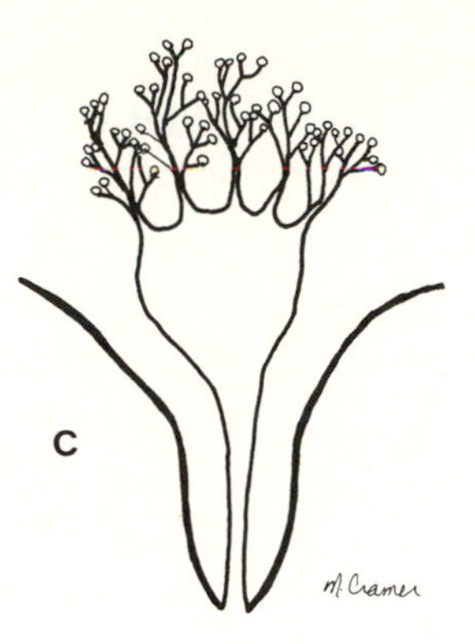

Figure 9–13.
Lateral view (diagrammatic) of mammary gland of monotreme (A) and structure of nipple (B) and teat (C). (Mary Ann Cramer)

tively high on the sides, apparently an adaptation for nursing the large, precocial young of this group of mammals.

9-S. Examine a variety of species of female mammals and compare the number and location of mammary glands.

9-T. Examine preparations of a teat and a nipple. How do the structures compare?

During pregnancy, various hormonal and physiological changes (e.g., increase in estrogen and prolactin levels) stimulate the development of mammary tissue in preparation for **lactation** (see Peaker 1977). The suckling of young stimulates nerve endings in the nipple or teat and these impulses travel to the neurohypophysis that in turn releases the hormone oxytocin (Sadleir 1973). This hormone travels via the blood stream to the mammary gland where it results in milk ejection or "letdown" of milk from the alveoli of the mammary glands. Milk is actually **expressed** rather than sucked from the nipple or teat through pressure by the tongue and hard palate of the nursing mammal (Cowie 1972).

Colostrum, a fluid rich in antibodies, is generally the first product released by the mammary gland following the birth of young (Sadleir 1973). Colostrum is important for transfering antibodies from the mother to young since many of these proteins are unable to pass through the placental barrier.

The duration of lactation varies among and within species depending on environmental and behavioral factors. In marine mammals, the duration of lactation is relatively short, varying from two to six weeks in seals to about seven months in some species of baleen whales (Sadleir 1973). In some mammals (e.g., seals), the body weight of the female drops markedly during the period of lactation while in others it is maintained by a greater intake of food during this period. The energy demands on the female are generally much greater during lactation than during pregnancy.

In humans, the nutritional and health benefits to breast-fed babies are well documented and striking (Cowie, 1972; Gunther 1977). In addition, breast feeding offers substantial psychological and emotional advantages to mother and child (Cowie 1972).

Secondary Sex Characteristics

The gonads of mammals are primary sex characters. Many species of mammals exhibit characters that developed under hormonal control during development (e.g., external genitalia; mammae). Other sexually dimorphic characteristics may be present or appear seasonally due to hormonal or environmental effects. Such **secondary sex characteristics** are frequently related to sexual recognition, courtship behavior, or both. **Sexual dimorphism** is thus the marked difference in the characteristics of males and females.

Most deer, for instance, have antlers that normally develop only in males. Their growth is directly controlled by the hormone, testosterone, produced by the interstitial cells of the testes. Antlers serve primarily for sexual recognition, social hierarchial status, as weapons in rutting season fights between males and for thermoregulation (see Chapter 5). In many herbivorous species (e.g., horses and some deer) canines are present in males but absent in females. Pelage variations also are frequently correlated with sexual dimorphism. The color patterns of the sexes or the degree of development of various areas of hair (e.g., a lion's mane or the human beard) may vary. Sexual dimorphism may be more subtle and be reflected only by size or by proportions (e.g., the relatively thicker layer of subcutaneous adipose tissue in the hips of human females).

9-U. Examine the dentition of a stallion and a mare. Compare these to that of a gelding. Which does the gelding resemble most closely? Why?

9-V. Compare the pelage and coloration of male and female red bats, lions, or mandrills.

9-W. Compare the external or cranial measurements of the two sexes of an assortment of mammalian species (use only adult individuals). Which species have larger males? Do any have larger females?

9-X. Examine species of bovids that have horns in both sexes. What differences between the sexes do you note?

9-Y. List as many kinds of secondary sex characters as you can. Give an example of a species exhibiting each.

Supplementary Readings

Asdell, S. A. 1964. *Patterns of mammalian reproduction,* 2nd ed. Cornell Univ. Press, Ithaca. 670 pp.

Austin, C. R., and R. V. Short (Eds.). 1972. *Reproduction in mammals.* Cambridge Univ. Press, London. Vols. 1-4.

Daniel, J. C. (Ed.). 1978. *Methods in mammalian reproduction.* Academic Press, New York. 584 pp.

Nalbandov, A. V. 1976. *Reproductive physiology of mammals and birds.* 3rd ed. W. H. Freeman and Co., San Francisco. 334 pp.

Parker, P. 1977. An ecological comparison of marsupial and placental patterns of reproduction, pp. 273-286. *In* B. Stonehouse, and D. Gilmore (Eds.). *The biology of marsupials.* Univ. Park Press, Baltimore.

Perry, J. S., and I. W. Rowlands (Eds.). 1969. Biology of reproduction in mammals. *J. Reprod. Fertility Suppl.* 6:1-531 pp.

Perry, J. S., and I. W. Rowlands (Eds.) 1973. The environment and reproduction in mammals and birds. *J. Reprod. Fert. Suppl.* 19:1-613 pp.

Rowlands, I. W. (Ed.). 1966. Comparative biology of reproduction in mammals. *Symp. Zool. Soc. London* 15:1-559.

Sadleir, R.M.F.S. 1969. *The ecology of reproduction in wild and domestic mammals.* Methuen & Co., London. 321 pp.

Sadleir, R.M.F.S. 1973. *The reproduction of vertebrates.* Academic Press, New York and London. 227 pp.

Sharman, G. B. 1959. Marsupial reproduction. *Monog. Biol.* 8:332-368.

Weir, B. J., and I. W. Rowlands. 1973. Reproductive strategies of mammals. *Ann. Rev. Ecol. Syst.* 4:139-163.

10 Populations

A **population** may be defined as a group of interbreeding individuals of the same species that occupy an area in a given time interval. Cole (1957) defined a population as "a biological unit at the level of ecological integration where it is meaningful to speak of a birth rate, a death rate, a sex ratio and an age structure in describing the properties of the unit."

Studies of populations are basic for understanding many ecological and life history phenomena and for managing most species of wildlife. In this chapter, basic information is presented on fundamental population parameters. Detailed information on growth of populations, rates of increase, life history strategies, and other components of population biology can be found in Deevey (1947), Cole (1954), Slobodkin (1961), Caughley (1966, 1977), Wilson and Bossert (1971), and Hutchinson (1978).

Basic Concepts

Population Growth

Natality, mortality, emmigration, and immigration are the four main parameters that affect population size. **Natality** is the increment of young into a population through births while **mortality** is a decrement to the population through death of individuals. Animals may also be added to a population by **immigration** from areas outside of the population. Similarly, animals may leave the population by **emmigration.** The rate of increase in a population is the difference between the rate that individuals are added to the population through birth and immigration and subtracted from the population through death and emigration.

In many theoretical and empirical studies, it is useful to quantify natality and mortality in very small or instantaneous units of time. The relationship of instantaneous natality rate (b), instantaneous mortality rate (d), and population size (N), is sometimes depicted in the form of an exponential equation of population growth:

$$\frac{dN}{dt} = bN - dN$$
$$= (b-d)N = rN. \text{ Then,}$$

the **intrinsic** (instantaneous) **rate of increase** (Malthusian parameter) of the population, symbolized by r, can be defined according to the formula:

$$r = b - d.$$

By knowing r and e (the base of natural log) the solution of population size at any point in time is the following:

$$N = N_o e^{rt}.$$

The values of r vary enormously among different species (Wilson 1975). For example, in many human populations, the rate of increase is about 3 percent or less per year ($r = 0.03$ per year). In the Norway rat (*Rattus norvegicus*), the rate of increase is about 1.5 percent per day ($r = 0.015$ per day).

10-A. Examine demographic data on population parameters from various countries (see Ehrlich and Ehrlich 1970:330-334). In what countries are r-values low? High?

There are several types of r that can be calculated (Caughley and Birch 1971; Caughley 1977). The rate of increase termed r_m or the *intrinsic rate of increase* (as presented above) is the maximum rate that a population with a stable age distribution can increase when no resource is limiting. This rate is not a constant for a species but is specific to the particular environment in which it was measured (Caughley and Birch 1971). The *survival-fecundity rate of*

Figure 10–1.
Examples of exponential (A) and logistic (B) population growth at different rates of increase (*r*), initial population size (N_o), and carrying capacities of the environment (K). (Brower & Zar 1974: 113,115)

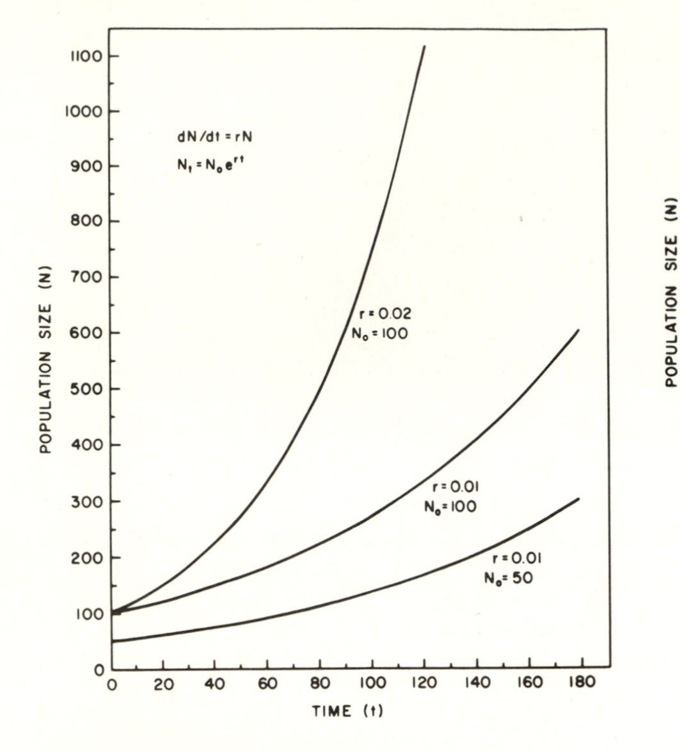

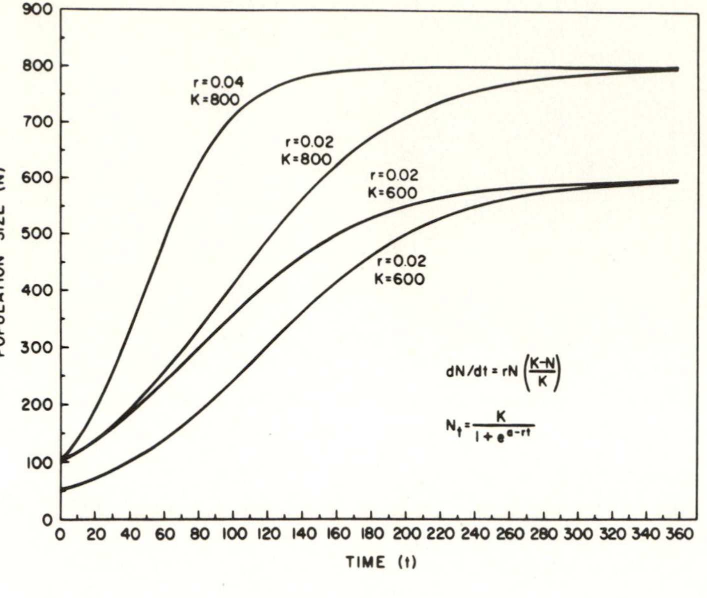

increase (r_s) is computed in exactly the same manner as r_m except that the conditions of the rate of increase need not necessarily be optimal. Thus, r_m is a special case of r_s. The *observed rate of increase* (r) is a very general measure of exponential increase in a population. To compute this rate, the conditions of a stable age distribution, a constant rate of increase, or the presence of abundant resources need not be met (Caughley 1977).

The simplest measure of a population's rate of increase is the ratio of population densities in two successive years (N_t and N_{t+1}) (Caughley 1977). This **finite rate of increase,** sometimes termed the growth multiplier (λ, lambda), can be symbolized as follows:

$$\lambda = \frac{N_{t+1}}{N_t}$$

Values less than 1 indicate a population decline while those greater than one indicate a population increase.

Population Growth Curves

In the absence of behavioral or environmental restrictions on population growth (Figure 10-1A), the number (N) in a population can be represented by the **exponential growth** equation developed by Lotka (Hutchinson 1978):

$$\frac{dN}{dt} = rN \text{ and the solution } N = N_o e^{rt}.$$

In these equations, *d* indicates the instantaneous rate of change with time (*t*) of a population of an initial (N_o) and current (*N*) size. The Malthusian parameter, *r*, is the intrinsic rate of increase and may be computed by a variety of means (for summary, see Caughley 1977; May 1976; this chapter). This population growth model is frequently unrealistic (Wilson and Bossert 1971, Hutchinson 1978) because the environment ultimately limits the growth of the population. Consequently, some form of the **logistic growth** equation (Figure 38-1B) is more reasonable since *K*, an upper population limit determined by the limits of environmental conditions (e.g., food, space, etc.), is considered. The logistic growth equation can be symbolized by

$$\frac{dN}{dt} = rN \frac{(K-N)}{K} \text{ or } N = \frac{K}{1 + e^{-rt}}.$$

The term $(K-N)/K$ is the effect of the upper limit of the population on the size of the population (see Wilson and Bossert 1971, Wilson 1975, or Hutchinson 1978 for details).

10-B. Calculate the finite rate of increase (λ) for the populations listed and record your results in the column on the right.

Species	Population Densities Year 1 No/ha	Year 2 No/ha	λ
Rodent A	119	124	1.042
Rodent B	14	21	
Rodent C	0.5	2.0	
Elephant	0.00150	0.00134	
Insectivore	1.40	1.45	

Life Tables, Survivorship, and Mortality Curves

Often it is instructive to summarize various life history parameters in the form of a **life table** (Table 10-1). The construction of a life table is straightforward mathematically although data for its preparation may be difficult to obtain (see Caughley 1966, 1977 for specifics).

To prepare one type of life table, a group or **cohort** of individuals of the same (or nearly same) age is followed through time and information gathered on deaths, ages of death, or the proportion remaining alive. This type is termed an *age-specific* (because individuals are of same age), *dynamic* (because population is followed through time), or a *horizontal* (in contrast to vertical) life table. When the life table is prepared using data on ages at death (e.g., animals aged from hunter-killed samples), the actual cohort used in the table is imaginary since these animals were probably born at different times (Caughley 1977). A life table constructed in this manner or from data or a sample where the ages of the population are known directly (e.g., human population) is termed *time-specific* (since only one point in time is measured), *static* (since the population is not followed with time), or a *vertical life table.* A *composite life table* is composed of data obtained by time-specific or age-specific procedures and may be useful in demonstrating the differences in values due to changing natality and mortality (Hutchinson 1978). In studies of mammalian populations, time-specific life tables are more commonly utilized with studies of game animals while age-specific tables, what few are available, are more commonly seen (in nonhuman species) in studies of smaller and shorter-lived mammals (Caughley 1977). Eberhardt (1969) believes that the time-specific method results in less bias. Small sample sizes can also invalidate a life-table analysis. Caughley (1977) recommends that an N of 150 should be the minimum for any life table calculations.

Survivorship (designated l_x, see column three of Table 10-1) is the probability that any individual in a population will survive to any particular age (x). **Mortality** (d_x, column four of Table 10-1) is the fraction of the population that dies during the age interval x to $x + 1$. It is calculated by finding the difference between two l_x values for successive age intervals (typically years but may be months for short-lived animals). Since the initial l_x value (survivorship at birth) has a value of 1.0, all d_x values in the age schedule total 1.0 by definition (an important check on arithmetic). The **mortality rate** (q_x, column six in Table 10-1) is the proportion of individuals alive at age x that die before age $x + 1$. It is calculated as

$$q_x = d_x/l_x.$$

The **survival rate** (p_x, last column of Table 10-1) is the complement of the mortality rate and represents the proportion of animals alive at age x that survive to age $x + 1$. It is calculated as

$$p_x = 1 - q_x.$$

The mean expectation of life (e_x, not shown in table), frequently shown in actuary tables of insurance companies, is rarely utilized in studies of population biology. It is the mean expectation of life for individuals alive at the start of age x.

Fecundity and Reproductive Strategies

Most mammals are **iteroparous** and thus produce a series of young or litters during their lifetime. Rarely, individuals of short-lived species may be functionally **semelparous** whereby only one litter is produced during the life of the female or the life of the male (Braithwaite and Lee 1979). Caughley (1977) indicated that birth frequencies in mammalian populations consist of two main types: birth-flow and birth-pulse. In the *birth-flow* model, which humans, red kangaroos (*Macropus rufa*), and a few other species follow, the rate of breeding is constant throughout the year. In the *birth-pulse* model, which includes most mammalian species, the period of births is seasonal. Several population statistics, including mean generation length, are computed by different methods depending on whether the birth pulse or birth flow models are applicable (Caughley 1977).

Fecundity (m_x) is a measure of the number of live young produced during an interval of time. Since fecundity usually changes with time, these schedules are often summarized in the form of a **fecundity table** (Table 10-1, column five) that lists m_x values for

Table 10-1
Time-Specific Life Table for Female Thar *(Hemitragus jemlahicus).**

Age (years) x	Frequency f_x	Survival l_x	Mortality d_x $(l_x - l_{x+1})$	Fecundity offspring (m_x)	Mortality Rate q_x (d_x / l_x)	Survival Rate p_x $(! - q_x)$
0	205	1.000	0.553	0.000	0.553	0.447
1	96	0.467	0.006	0.005	0.013	0.987
2	94	0.461	0.028	0.135	0.061	0.939
3	89	0.433	0.046	0.440	0.106	0.894
4	79	0.387	0.056	0.420	0.145	0.855
5	68	0.331	0.062	0.465	0.187	0.813
6	55	0.269	0.060	0.425	0.223	0.777
7	43	0.209	0.054	0.460	0.258	0.742
8	32	0.155	0.046	0.485	0.297	0.703
9	22	0.109	0.036	0.500	0.330	0.670
10	15	0.073	0.026	0.500	0.356	0.644
11	10	0.047	0.018		0.382	0.618
12	6	0.029		0.470		
12	11			0.350		

*Caughley (1966:911)

different ages. For simplicity, a fecundity table is produced only for females listing their female offspring (a fecundity table for males is logistically difficult to prepare; only one table is needed for life history studies). In most mammals, fecundity rises rapidly after puberty, reaches a plateau, and then may taper off prior to death of the female (Caughley 1977).

Once survivorship (1_x) and fecundity (m_x) schedules are available for a particular population, it is possible to calculate other useful statistics. One of these, the **net reproductive rate** (R_o), is the multiplication rate per generation (in terms of female offspring of a female). More specifically, it is computed according to the following equation (Wilson 1975; Caughley 1977; Hutchinson 1978):

$$R_o = \sum_{x=o} l_x m_x = \Sigma V_x,$$

where V_x (termed the reproductive function) gives the total (Σ) expected number of offspring (in this case female) for a given schedule of fecundity (m_x) and survivorship (l_x). Values of R_o near 1.0 indicate a stable population while values of < 1.0 and > 1.0 indicate populations that are decreasing and increasing in size, respectively, during each generation.

Another useful population parameter is **mean generation length** (T). This is the mean period between the birth of parents and the birth of offspring. Specifically, this parameter can be approximated according to the following formula:

$$T = \frac{\Sigma l_x m_x x}{\Sigma l_x m_x} = \frac{\Sigma l_x m_x x}{R_o}$$

The intrinsic rate of increase (r_m) can now be approximated once estimates are available on the net reproductive rate (R_o) and the mean generation length (G):

$$r_m = \frac{\log_e (R_o)}{T}$$

With the aid of a computer (e.g., using Fortran program in Caughley 1977) or a programmable calculator, the value of r_m can also be determined by an iterative process using the following formula (Wilson 1975; Caughley 1977; Hutchinson 1978):

$$\sum_{x=o} e^{-rmx} l_x m_x = 1$$

Trial values of r_m (e.g., starting with the initial one obtained by the approximation formula) are then substituted into the formula until the solution approximates 1.0. Since r_m is an instaneous rate, it can be converted into a finite rate by the following formula (Krebs 1972):

$$\lambda = e^{rm}, \text{ where } e = 2.71828\ldots$$

(e.g., if $r_m = 0.881$, then $\lambda = 2.413$ offspring per individual per year).

The intrinsic rate of increase (r_m) can also be computed with the use of a Leslie Matrix although Caughley (1977) suggests that it is generally not feasible to obtain field data in the form needed for the matrix solution.

Knowledge of the value of r_m (including also r_x and r) for various species and environmental conditions is useful for summarization and in answering many important theoretical and empirical questions in ecology, evolution, and management (Cole 1954; Wilson 1975; Caughley 1977; Hutchinson 1978). It is a useful parameter to use in simulations of natural populations for management purposes (Caughley 1977).

10-C. Inspect Table 38-1 and verify (thru your own calculations) that $d_x = 0.046$, $q_x = 0.106$, and $p_x = 0.894$ for a 3-year old female. Compute these statistics once more after you have added 0.067 to the l_x value for 3-year old females. How did this change affect the d_x, q_x, and p_x-values?

10-D. Given the following hypothetical (but realistic) life table, verify for each age that the values for d_x are correct.

Age					Compute	
x	f_x	l_x	d_x	m_x	$l_x m_x$	$(x)(l_x m_x)$
0	200	1.00	0.50	0.05		
1	100	0.50	0.10	0.80		
2	80	0.40	0.15	1.00		
3	50	0.25	0.25	0.80		
4	0	0.00				
		$R_o = 1.05$ $T = 1.71$		$\Sigma =$		

10-E. Compute the values of the net reproductive rate (R_o) and mean generation length (T) using the values in the life table of exercise 10-D. Do your calculations agree (rounded to 3 significant figures) with those listed?

10-F. Compute a trial estimate of r_m using the formula $r = \log_e (R_o)/T$. Use the estimates for these parameters obtained from exercises 10-D and 10-E. Did you obtain a value of 0.087? What does this value tell you?

10-G. Compute the intrinsic rate of increase (r_m) for the population described in the life table of exercise 10-D. Use the following table to record your intermediate results. *Hint.* Start with the value of r (0.087) computed in Exercise 10-F to see how close this comes to solving the equation listed below (see text for details):

$$\sum_{x=o} e^{-rmx} / {}_{xmx} = 1$$

If not, try values of r above and below this number until equality (or near equality) is reached. The calculations for computing r_s and r are the same as those for r_m. What makes these values different (if any) are the conditions under which the population is experiencing increase (or decrease). See text and Caughley (1977) for details.

x	$l_x m_x$	e^{-rx}	$e^{-rx}(l_x m_x)$
0	0.05		
1	0.40		
2	0.40		
3	0.20		
4	0.00		

In recent years, increasing attention has focused on the study of life history strategies (see reviews by Wilson 1975; Pianka and Parker 1975; Pianka 1976; Charlesworth and Leon 1976; and Hutchinson 1978) although the subject, as Cole (1954) amply demonstrated, is not new. Central to these studies are the concepts of reproductive value and reproductive effort. **Reproductive value** (v_x) is the age-specific expectation of all present and future offspring or, more practically, the relative number of female offspring that remain to be born to each female of age x. Reproductive value can be calculated from the age-specific survivorship (l_x) and fecundity (m_x) schedules using the following formulas (Wilson 1975; Pianka and Parker 1975):

$$v_x = \sum_{t=x}^{\max age} \frac{l_x m_x}{l_x} \text{ for a population not changing in size } or$$

$$v_x = \frac{e^{rx}}{l_x} \sum_{t=x} e^{-rt} l_t m_t \text{ for a population that may be changing in size.}$$

Of what practical significance is a measurement of reproductive value? In seeking to maximize the production of a species of game animal, the population biologist would recommend against harvesting methods that kill females during their most productive years. The same approach has been used, with only limited success, to stop the precipitious decline in the populations of baleen whales.

Reproductive effort is the investment (e.g., in energy) that an individual places in any current act of reproduction. In most animals, reproductive effort increases with age of the individual. Individuals expending great amounts of energy in reproduction may reduce their individual survivorship and ultimately reduce the fitness of their offspring (i.e., if the female is exposed to a greater risk of death because of pregnancy or care of young). For mammalian populations, good quantitative data on reproductive effort are scarce; data in Randolph *et al.* (1977) are a notable exception.

The upper population size (*K*) for a particular environmental condition and the rate of increase (***r***) in this population are both subject to selection. This led MacArthur and Wilson (1967) to define an ***r*-strategist** or *opportunistic species* as one that relies on a high intrinsic rate of increase (***r***) to survive in fluctuating environments with ephemeral resources. A ***K*-strategist** or *equilibrium* (*stable*) *species* is one that is more likely to live in an environment subject to less fluctuation. Individuals of such species are likely to be keen competitors, have slower developmental rates, greater longevity, and greater ability to utilize resources efficiently (Wilson 1975:101). In some instances, it appears that species exhibit typically *r*-selected or *K*-selected traits but recent work suggests that rigid classification of such species into these categories fails to account for the temporal nature of these strategies. Nichols *et al.* (1976) pointed out that populations of a species may shift from one or the other of these strategies with time to compensate for different environmental conditions.

Population Regulation

The subject of population regulation in mammals is a complex subject that has been amply reviewed by a number of sources (Sladen 1969; Southwick 1969; Krebs *et al.* 1973; Lidicker 1973; Delany 1974; Golley *et al.* 1975; Snyder 1978; Tamarin 1978). Some general comments on this subject are present below.

Population densities of mammals may be regulated by *extrinsic factors* of the environment or by mechanisms *intrinsic* to the individuals (e.g., genetic factors). Similarly, factors affecting population regulation may be classified according to whether they are affected by the densities of the individuals in the population. In populations regulated by *density-dependent* factors, a decline in population size (*N*) is due to a decline in the birth rate or an increase in the death rate as *N* increases. In populations regulated by *density-independent factors,* the size of *N* is regulated not by the size of the population but by factors independent of population size. In many instances, climatic factors (rainfall, drought) function as density-independent factors but in other situations (e.g., individuals killed by cold weather because no adequate refuge sites available) they may be density-dependent factors (Krebs 1972). Typical density-dependent factors include such events as predation, disease, parasitism, and starvation.

Other theories of population regulation suggest that species are capable of regulating their own population densities by intrinsic means without destroying the renewable resources of their environment or requiring various extrinsic factors for this purpose (Chitty 1960). What mechanism or combination of mechanisms regulate mammal populations is not known (see Krebs *et al.* 1973; Golley *et al.* 1975; Snyder 1978). But multi-factor models of population regulation will most likely approach the situations observed under natural conditions (Lidicker 1973).

Supplementary Readings

Berry, R. J. and H. N. Southern (Eds.). 1970. Variation in mammalian populations. *Symp. Zool. Soc. London* 26:1-404.

*Caughley, G. 1966. Mortality patterns in mammals. *Ecology* 47:906-918.

Caughley, G. 1977. *Analysis of vertebrate populations.* John Wiley and Sons, New York. 234 pp.

Humphrey, S. R., and J. B. Cope. 1976. Population ecology of the little brown bat, *Myotis lucifugus,* in Indiana and north-central Kentucky. *Spec. Publ., Amer. Soc. Mammalogists* 4:1-81.

*Included in Jones, Anderson and Hoffman (1976).

Hutchinson, G. E. 1978. *An introduction to population ecology*. Yale Univ. Press, New Haven. 260 pp.
Krebs, C. J., M. S. Gaines, B. L. Keller, J. H. Meyers, and R. H. Tamarin. 1973. Population cycles in small rodents. *Science* 179:35-41.
Lidicker, W. Z. 1973. Regulation of numbers in an island population of the California vole: a problem of community dynamics. *Ecol. Monog.* 43:271-302.
Martinka, C. J. 1976. Population characteristics of grizzly bears in Glacier National Park, Montana. *J. Mammal.* 55:21-29.
Pianka, E. R. 1976. Natural selection of optimal reproductive tactics. *Amer. Zool.* 16:775-784.
Sladen, B. K., and F. B. Bang (Eds.). 1969. *Biology of populations: the biological basis of public health.* Amer. Elsevier Publ. Co., New York. 449 pp.
Snyder, D. P. (Ed.). 1978. Populations of small mammals under natural conditions. *Univ. Pittsburgh Pymatuning Lab. Ecol., Spec. Publ. Ser.* 5:1-237.
Tamarin, R. H. (Ed.). 1978. *Population regulation.* Benchmark Papers in Ecology, Vol. 7. Academic Press, Inc., New York. 416 pp.
Tanner, J. T. 1978. *Guide to the study of animal populations.* Univ. Tennessee Press, Knoxville. 186 pp.
Wilson, E. O. 1975. *Sociobiology: the new synthesis.* Harvard Univ. Press, Cambridge. 416 pp.
Wilson, E. O., and W. H. Bossert. 1971. *A primer of population biology.* Sinauer Associates, Sunderland, Mass. 192 pp.

11 Systematic Methods

Our understanding of the biology of mammals is partly dependent upon knowledge of the diversity of organisms. The field of science concerned with the diversity of mammals and other organisms is termed **systematics,** and a scientist who studies the diversity of organisms is termed a **systematist.** A major subdiscipline of systematics is **taxonomy** defined by Mayr (1969) as ". . . the theory and practice of classifying organisms" and practiced by a **taxonomist. Classification** is the process by which animals are arranged into various groups, these groups of organisms, termed **taxa** (singular, **taxon**) are then assigned to a category or level of classification. Each taxon is assigned a distinctive name according to rules governing **nomenclature. Identification** is the assigning of particular specimens to already established and named taxa.

A portion of taxonomy is devoted to descriptions of taxa not previously known to science. This aspect is especially important in areas where extensive faunal surveys have not been made or with little studied higher taxa. Another important aspect of taxonomy is the assigning of taxa to groups and the definition of boundaries between groups. But taxonomy is not limited to descriptions and classification schemes; it is also concerned with patterns of geographic variation, gene flow, mechanisms that maintain reproductive isolation between species, and various other phenomena. Thus systematics involves both the mechanisms and the contemporary products of evolution.

Another major subdiscipline of systematics is the attempted determination of phylogenies. A **phylogeny** is the evolutionary history of an organism or groups of related organisms. The **phylogenist** studies extant and/or fossil materials and attempts to determine their phylogenetic history.

Figure 11–1.
Categories of classification generally used for mammals, and the taxa to which the spotted ground squirrel, *Spermophilus spilosoma,* belongs as they are compared to these categories. Standardized endings for the names of taxa in the superfamily, family, subfamily, and tribe categories are shown in boldface type. Names of taxa in categories at the generic level and below are italicized. Categories always used in the classification of animals are printed here in upper case.

Categories	Examples of Taxa
KINGDOM	Animalia
PHYLUM	Chordata
Subphylum	Vertebrata
CLASS	Mammalia
Subclass	Theria
Infraclass	Eutheria
ORDER	Rodentia
Suborder	Sciuromorpha
Superfamily	Sciur**oidea**
FAMILY	Sciur**idae**
Subfamily	Sciur**inae**
Tribe	Marmot**ini**
GENUS	*Spermophilus*
SPECIES	*Spermophilus spilosoma*
Subspecies	*Spermophilus spilosoma pallescens*

Hierarchies of Classification

Animals and plants are arranged into taxa. These taxa, in turn, are organized into a hierarchy of categories. The categories generally used for animals are shown in Figure 11-1. The categories printed in all upper case letters are the major categories used in the classification of all animals. In this list the smallest group is at the bottom, thus a species may be composed of several subspecies, a genus may be composed of many species, etc. If a taxon in any category contains only a single taxon at the next lower category the taxon is termed **monotypic,** thus the genus *Antilocapra* is monotypic because it contains only one species, *A. americana*, and the order Dermoptera is monotypic because it includes only one family, Cynocephalidae. A taxon containing more than one taxon in the next lower category is **polytypic.**

Species and Subspecies

To most systematists the **species** is the basic unit of taxonomy. The most widely accepted current defintion of species is that of Mayr (1969), "Species are groups of interbreeding natural populations that are reproductively isolated from other such groups." Thus species are naturally occuring groups of organisms that a systematist attempts to define and delimit but does not create.

Reproductively isolated individuals are any two animals of opposite sexes which cannot mate to produce offspring that will contribute to the gene pool of the species. Such isolation can be produced by premating mechanisms that prevent mating, such as isolation in different habitats in the same geographic area, isolation in nonoverlapping seasons of reproductive activity, isolation by different fixed behavioral patterns, or isolation by incompatability of the genitalia. Reproductive isolation can also be produced by postmating mechanisms that prevent production of young, such as inability of the gametes to join to produce a zygote or inability of the zygote to develop into a viable offspring. Even if an offspring is produced, reproductive isolation is present if the offspring is sterile or if it is for any other reason unable to reproduce and contribute to the gene pool of the species.

If two populations are **sympatric,** i.e., have overlapping geographic ranges (Fig. 11-2A) and are reproductively isolated, they are indeed distinct species. Similarly if they are **parapatric,** i.e., have contiguous but not overlapping ranges, (Fig. 11-2B) and are reproductively isolated along this line of contact they are considered distinct species. However, if the populations are fully **allopatric,** i.e., have nonoverlapping and noncontiguous geographic ranges (Fig. 11-2C), potential reproductive isolation is not easily demonstrated. If it is hypothesized that the allopatric populations would interbreed if they were sympatric or parapatric they are considered to be **conspecific.** But if it is hypothesized that they would not interbreed they are usually considered to be distinct species.

The concept of species discussed above often does little to help taxonomists forced to make decisions on the basis of anatomical, physiological, or behavioral characteristics that may not be related to the presence or absence of reproductive isolation. And thus, in the absence of data on reproductive isolation, taxonomic decisions are based upon similarities and differences in characters. **Sibling species** are species that are reproductively isolated but morphologically indistinguishable (or very difficult to distinguish).

A **subspecies** is a relatively uniform and genetically distinct portion of a species representing a separately or recently evolved lineage with its own evolutionary tendencies, definite geographic range, and a narrow zone of intergradation (i.e., interbreeding, usually inferred by a shift in character gradient) with adjacent subspecies (Lidicker 1962). A subspecies may be an important unit of evolution, but **speciation,** i.e., formation of a new species, occurs *only* when the subspecies becomes reproductively isolated from other related subspecies. Usually, the ranges of subspecies are parapatric, the populations are interfertile and thus should not automatically be considered incipient species. In some instances the delimitation of subspecies may not be an adequate expression of complex or diverse patterns of variation within a species (see comments in Wilson and Brown 1953; Lidicker 1962.)

A **cline** is a gradual change in the attribute of a character (e.g., pale to dark color; small to large size) along a geographical gradient. The designation of subspecies as an expression of clinal variation is usually inappropriate. General reviews of the subspecies ques-

Figure 11–2.
Diagrammatic representation of A, sympatric; B, parapatric; and C, allopatric populations.
(Mary Ann Cramer)

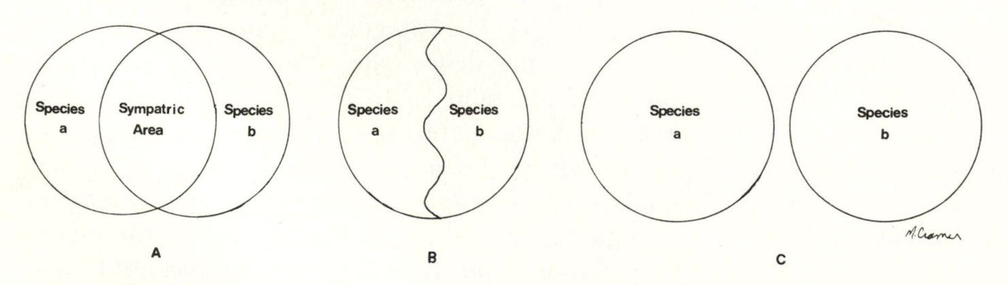

tion can be found in Mayr (1963:347-350) and Simpson (1961b:171-176).

Circular overlap occurs when a polytypic species exists as a chain of contiguous and intergrading, i.e., interfertile, populations that curve back until the terminal populations overlap one another. These terminal populations are generally reproductively isolated from one another and have frequently been considered to be different species until their true relationships were determined by tracing of the complete chain.

11-A. Examine the distribution maps in Hall and Kelson (1959). Find examples of closely related species pairs that are sympatric, examples that are parapatric and ones that are allopatric. Find a monotypic species. Can you locate any species that have sympatric subspecies and which may indicate circular overlap?

11-B. Could a classification based entirely on morphological characters distinguish sibling species? If not, what biological criterion could be invoked to solve the problem?

Higher Categories

A higher taxonomic category consists of one or more related species that differ sufficiently from other such categories and share a common lineage. Just what constitutes a significant difference, especially at the generic level, is somewhat arbitrary. Practically, a **genus** is a category containing a single species or group of species that differ from species in other genera by marked discontinuities, e.g., different morphological features, behavior, adaptations, or other features. A **family** is a group of closely related genera that share a common evolutionary origin and generally exhibit marked differences from species in other families. An **order** is an assemblage of one or more related families, and a **class** contains one or more related orders. In the class Mammalia there are approximately twenty-two living orders and 131 living families. (The reason for the word "approximately" will become apparent in the Taxonomic Remarks sections of chapters 13 through 28.)

Attempts were made in the past to record characters or attributes of characters that would define a genus, family, or a higher category in all situations. These characters were then referred to as generic characters, family characters, and so on. But evolutionary events do not always lead to the same characters or attributes being important criteria for defining various taxa. The delimitation of a taxon depends on an overall evaluation of different characters or attributes. Thus, a strictly defined generic character usually does not exist.

Hybrids

While true species do not normally hybridize in the wild, **hybrids** are often encountered among domestic and zoo animals where human influences have directly or indirectly induced a mating that would not occur in the wild. Such human influenced hybrids are often sterile. Hybridization does occasionally occur in the wild (Birney 1973:127) often as a result of major disturbances to the environment. VanGelder (1977) discussed mammalian hybridization and recommended that the ability to produce hybrids be a criterion for placing species in the same genus.

The scientific names of hybrids are generally written as the names of the two parent species with a cross (X) between them. Thus the coy-dog, a hybrid between a coyote (*Canis latrans*) and a dog (*Canis familiaris*) is written as *Canis latrans* X *C. familiaris*.

Philosophies of Systematics

Three basic philosophies are in use currently to develop classifications and/or phylogenies. Debates over the relative merits of the three have, on occasion, become quite heated.

Evolutionary Systematics

Evolutionary or **traditional systematics** seeks to discover the natural groups that are current products of evolutionary lines. In this regard evolutionary systematists are not totally at odds with the advocates of phenetics. Students of evolutionary systematics state that members of a taxon should be classified together because they share in a common evolutionary heritage. In contrast, the pheneticists state that OTUs, and/or taxa, should be grouped together on the basis of their overall similarity. In practice, the phenetic and evolutionary taxonomists utilize similar, if not identical, characters (i.e., those that are observable) and often similar numerical techniques. However, the evolutionary systematist uses weighted characters (i.e., certain characters are thought to have greater value for determining relationships than other characters) and considers information from fossil history and other parameters to make inferences on classification schemes and hypothesized phylogenies. **Phylograms** or

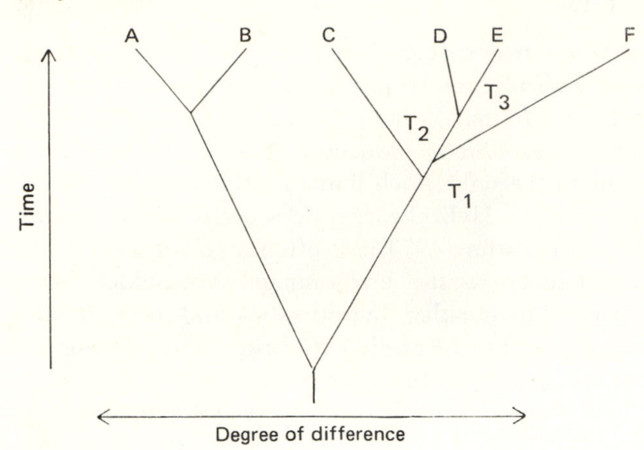

Figure 11-3.
Two-dimensional phylogram (or phylogenetic dendogram) of supposed relationships between taxa. D and E are considered to be more closely related to C than to F, based on time sequence and degree of divergence. (Mayr 1965)

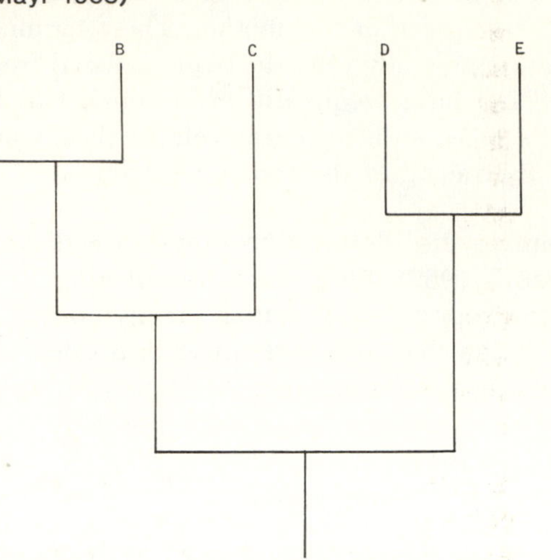

Figure 11-4.
Two-dimensional phenogram. The nearness of the terminal points indicates degree of similarity, not phylogeny. Note that D is more similar to E than to C. (Mayr 1965)

phylogenetic dendrograms (Fig. 11-3) are graphical representations of hypothesized phylogenies based upon the data evaluated by the evolutionary systematist.

The terms **phyletic** or **phylogenetic systematics** and **phyleticist** are often used by persons engaged in evolutionary systematics. However these descriptive terms are also claimed by the cladists. We have therefore, avoided using the terms for either group. Evolutionary or traditional classification is reviewed by Simpson (1961b), Mayr (1963, 1965, 1969) and Bock (1963, 1974 and 1977).

Phenetics

Phenetics or **numerical phenetics** is a philosophy of classification based upon the degree of overall phenotypic similarity between taxa or between operational taxonomic units (**OTUs**). The practice of numerical phenetics is commonly called numerical taxonomy, taxometrics, taximetrics, or Neo-Adansonian taxonomy. Typically, the procedure involves the selection of a large number of operational (i.e., definable) characters that are not assigned any *a priori* weight (i.e., each character is initially assumed to be equal to each other character in taxonomic importance). A data matrix is then accumulated and usually processed with the aid of digital computers and multivariate statistical techniques. The results of these analyses produce measures of total phenetic correlation and distance between the OTUs. Phenetic similarity and distance between the OTUs is frequently illustrated with a two-dimensional **phenogram** (Fig. 11-4) or a three-dimensional projection (Fig. 11-5). The numerical taxonomist does not negate the value of time relationships, e.g., fossil history, and branching of lineages in evolution; but he does believe that such criteria are not essential for producing classification schemes. The pheneticist is generally not interested in producing phylogenies. Sokal and Sneath (1963), Sokal and Camin (1965), Jardine and Sibson (1971), Sneath and Sokal (1973), Clifford and Stephenson (1975) and Farris (1977) present comprehensive summaries of the phenetic method. The journal *Taxometrics* is an important source of bibliographic references for this field.

Cladistics

Cladistics or **cladisim** is a method of developing phylogenies based upon the branching sequences of evolution. A **cladist** divides characters into those that are primitive (**plesiomorphic**) and those that are recently derived (**aptomorphic**). The presence of shared primitive characters (**synplesiomorphy**, Fig. 11-6 A and B) does not tell the researcher when groups branched but the presence of shared derived characters (**synaptomorphy**, Fig. 11-6C) indicates that the taxa sharing the characters are more closely related to each other than to a taxon without the derived characters. While **convergence**, which could result in the same derived characters arising independently in two, only distantly related, groups (Fig. 11-6D), is considered possible, the principle of parsimony makes the probability of

Figure 11–5.
Three-dimensional projections of sixteen taxa of Chiroptera, representing the family Phyllostomatidae. (Gerber and Leone 1971:163)

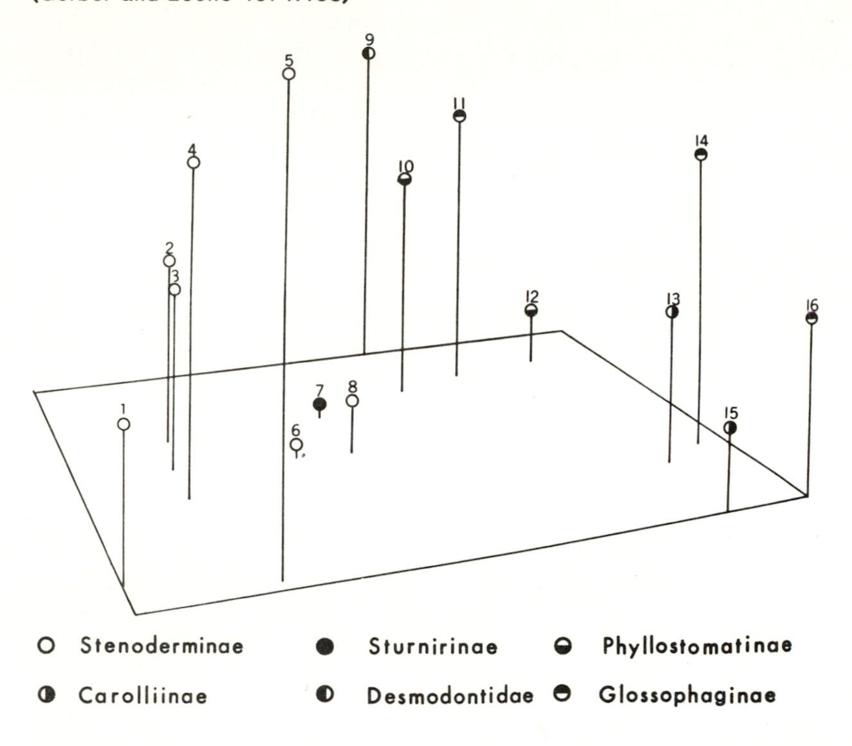

Figure 11–6.
The types of similarities of characters that can occur as viewed by a cladist. a and a' indicate two expressions of a character with a' representing the more advanced and a the more primitive state. The arrow (———>) indicates where in the evolutionary history of the group the change of state from a to a' occurred. A and B show two of the three taxa sharing the primitive character (Synplesiomorphy), C and D show two taxa sharing the derived character, C by a single change (Synaptomorphy) and D by convergence.
(A.F. DeBlase after Bonde 1977:764)

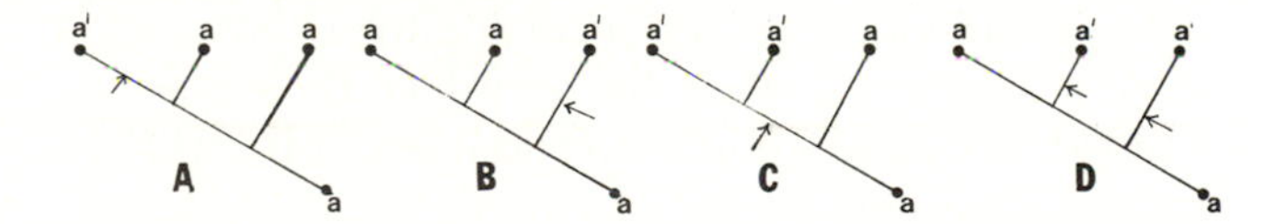

synaptomorphy outweigh the probability of convergence. In constructing cladograms (Fig. 11-7) the cladist uses **sister groups,** the two groups resulting from each bifurcation of the phylogeny. The cladist is usually not particularly concerned about the division of his cladogram into classical classification categories. For discussions of cladistic (phylogenetic) systematics see Hennig (1966), Camin and Sokal (1965), Nelson (1971, 1972, 1974a,), Ashlock (1971), Cracraft (1974a), McKenna (1975), Sneath (1975), Lovtrup (1975, 1977), and Bonde (1977).

Figure 11–7.
A cladogram illustrating the hypothesized phylogenies of taxa A through E. Characters m through q occur in one to three variants each with the plain letter the most primitive form (e.g., m, m', and m'' in order from most primitive to most recently derived) A and B are sister groups; A and B together are a sister group to C, D and E; C is a sister group to D and E; etc.
(A.F. DeBlase after Bonde 1977:764)

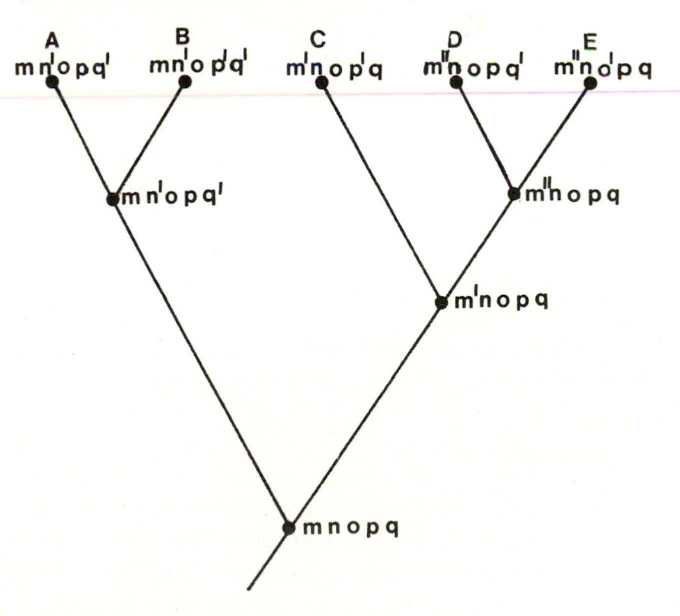

Synthesis

The philosophies espoused by evolutionary, cladistic and phenetic systematists overlap to some extent and their methods are even more similar. They are all looking at the same or very similar data and differ mainly in the conclusions they draw from them. At present the evolutionary systematist is more prone to utilize the statistical methods of the pheneticists. The pheneticist, in turn, is seeking operational ways to make reconstructions of evolutionary history. For further discussions of the relative merits of these philosophies see Johnson (1968), Throckmorton (1968), Rohlf (1974), Nelson (1974b), Mayr (1974), Hennig (1975), Sokal (1975), and Farris (1977).

Most classification schemes employed in this manual are based on an evolutionary approach. This scheme is adopted to conform with most of the available literature on mammalian classification. We recognize the important contributions of phenetic and cladistic approaches to systematics and encourage studies that seek a synthesis of these philosophies.

11-C. Examine Figure 11-4. Is taxon C more closely related to A-B or D-E? In general what is the basis for such similarity in a phenetic study?

11-D. When would a phenetic OTU not be equal to a phyletic taxon?

11-E. If A through E are taxa, m through q are plesiomorphic characters, and m' through q' are aptomorphic characters, then which of the following taxa are sister groups?

A,	m	n'	o	p	q
B,	m	n'	o	p	q'
C,	m	n'	o'	p	q
D,	m'	n	o	p'	q
E,	m'	n	o	p	q

Methods of Systematics Research

The purpose and goals of any research project should be clearly in mind before specimens are examined or before specimens or data are collected. In practice, the systematist may conduct preliminary investigations to test new techniques or aid in formulating specific goals.

Collection of Samples

Systematic studies rely on available museum specimens or other specimens (samples) that the systematist collects. The specimens may be live mammals for behavioral studies; whole individuals preserved in alcohol; or skins, skulls, or other body portions such as postcranial skeletons, blood, hair, or internal organs. The specimens should represent a random sample of individuals in the population. The samples ideally should also include a sufficient number of individuals of each sex and age class to allow for the validity of statistical computations and tests (see Sokal and Rohlf 1969). In actual practice most taxonomic data are obtained from adult individuals, although immatures may be needed for certain specialized studies.

Selection of Characters

Systematic studies involve a variety of techniques and characters. The taxonomist must decide whether characters will be weighted (in previous studies have certain characters been shown to be of greater importance in separating taxa?) or unweighted (no character is assumed *a priori* to have greater importance than any other) and randomly selected from a pool of possible characters. The assumptions of most techniques of phenetic taxonomy require the use of unweighted characters. Mayr (1969) presents an argument for the value of using weighted characters and Sokal and Sneath (1963) present an alternative argument for the use of unweighted characters. The weighted character approach is still widely used in mammalian systematic studies.

11-F. Make a list of possible taxonomic characters of each of the following specimens: live mammal, study skin, skull, tooth, and microscope slide of chromosomes. How would you select characters? How would you decide which are to be weighted more heavily?

Recording Data

The information or data available for various characters of specimens must be recorded. Data may be recorded on discontinuous, continuous, or attribute variables (see chapter 29).

Procedures for measuring cranial and external features of mammals are presented in chapters 2 and 35, respectively. These measurements are widely used in systematic literature on mammals.

Mensural and coded data must be tabulated onto forms to allow subsequent inspection or statistical analysis. For multivariate statistical procedures it is helpful, and often necessary, to prepare data for transfer to tabulating cards or other computer input systems.

Sorting Methods

Specimens must be sorted to analyze different patterns of variation. A heterogenous sample may be sorted into groups of morphologically similar individuals. These groups may or may not be related to discrete taxonomic units, i.e., taxa. In numerical taxonomy each group can represent an OTU for computational purposes. The groups can then be divided into subgroups according to age, sex, and locality and examined again. The groups and sub-groups should be compared to published descriptions and/or previously identified species of taxa known to occur, or which might be reasonably assumed to occur, in the area from which the group was collected.

Data can then be recorded and statistically compared by hand or coded for groups, transferred to tabulating cards, and sorted by the computer. A measure of **intrapopulation variation** or variation within a population or sample must be obtained before inferences can be made on variation between populations, **interpopulation variation,** or taxa. An intrapopulation analysis will permit an evaluation of differences between sex and age groups and aid later interpretations and methodology. For example, an analysis revealing no significant difference in certain measurements between males and females permits pooling of individuals from both groups to obtain larger samples for further statistical analyses.

11-G. Examine a series of skins and skulls of both sexes of a single taxon. Is sexual dimorphism present in any character? Select four cranial measurements (see chapter 2) and measure the series of skulls. What is the degree of overlap in cranial measurements between male and female specimens?

11-H. Examine a sample of specimens (skins and skulls) of a single taxon collected from several widely separated localities. Segregate the specimens by localities. Can you detect any differences in cranial measurements, pelage colors, or other characters when specimens from the separate localities are compared with one another?

Presentation of Data

A frequency **histogram** indicates the number of specimens in a sample that possess certain characters or measurements (e.g., number that are black, gray, or white). Different samples can be visually compared with one another if each character is plotted on the same scale. Frequency histograms can also be plotted using multivariate discriminant scores (Fig. 11-8).

Scatter diagrams (Fig. 11-9) are helpful in illustrating differences between populations or taxa. One character is plotted on the abscissa, X-axis, and the other character on the ordinate, Y-axis.

Descriptive statistics, such as mean, range, standard deviation, and standard error, can be presented in the form of **Dice-Leraas diagrams** or **Dice-grams** (Fig. 29-5). These diagrams are helpful in showing overall patterns of variation but should not be used for extensive testing of differences between means (see chapter 29, Statistics).

The results of multivariate cluster analyses are frequently presented as two-dimensional phenograms (Fig. 11-4) that show similarities between taxa or OTU's. The results of discriminant functions and principal components analyses are sometimes depicted as two- or three-dimensional projections (Fig. 29-8).

11-I. Obtain a heterogeneous sample of skulls representing several taxa or one taxon from several localities. Select one measurement of cranial length and one of cranial width from the list in chapter 2. Measure the specimens and record the results. Plot your results in a two-dimensional scattergram. Are these characters helpful in separating the taxa or localities? If not, would other characters provide better differentiation?

Taxonomic Interpretations and Decisions

Interpretation of patterns of inter- and intrapopulation variation is one of the first steps in delimiting taxa at the species level. A broad range of variation within a single "taxon" from one locality, when compared with a closely related "taxon" from another locality, may indicate several things. It may indicate that the first taxon may be composed of several discrete species or, that the taxon is a polymorphic species. Similarity between what are suspected to be two species must also be carefully investigated. These two "species" may represent one taxon or may actually be two sibling species. Determination of the presence or absence of isolating mechanisms may be necessary to correctly delimit the two taxa, or careful inspection (frequently with the aid of multivariate discriminant analysis) may reveal discrete characters or attributes that will be useful in separating the taxa.

Figure 11–8.
Histograms showing frequencies of linear discriminant scores (X-axis) in various Nebraska populations of *Blarina* (number of individuals on Y-axis). The bottom histogram refers to reference samples of two subspecies of *B. brevicauda.* The upper histogram refers to test samples of individuals collected near zones of contact.
(Genoways and Choate 1972)

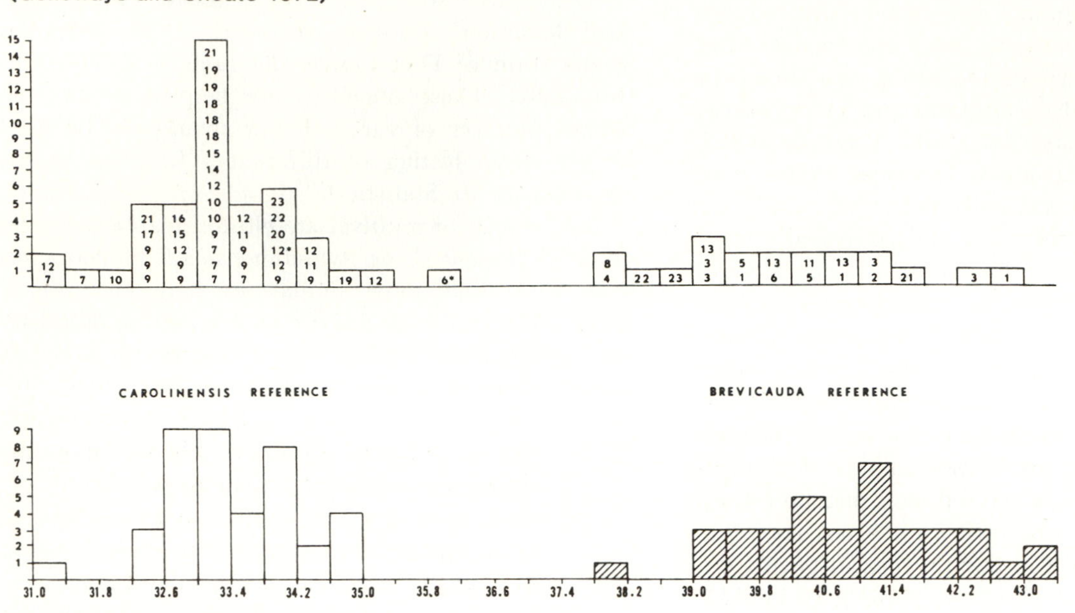

Figure 11–9.
Scatter diagram to illustrate separation of two subspecies of *Myotis thysanodes.* Solid symbols, *M. t. pahasapensis;* open symbols, *M. t. thysanodes.*
(Jones and Genoways 1967)

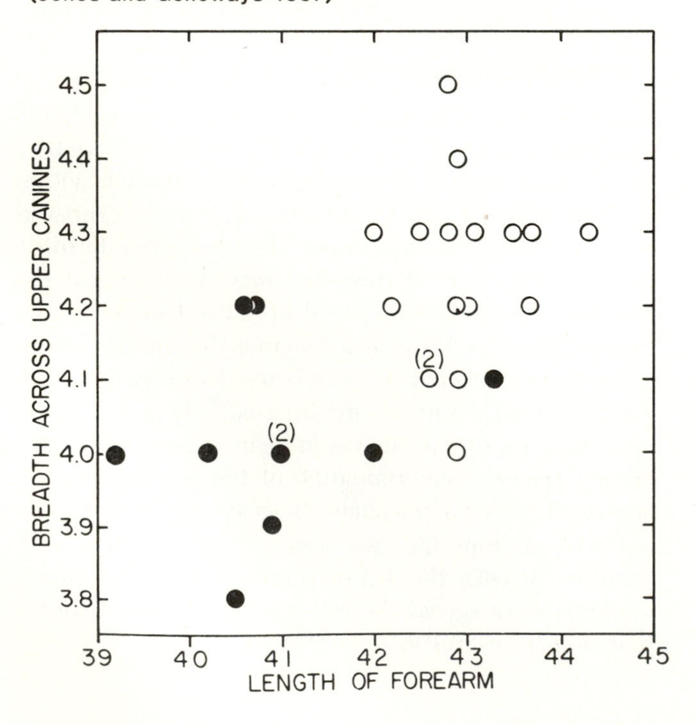

Allopatric populations, as discussed earlier, make these decisions more difficult. Allopatric taxa may be assigned, somewhat arbitrarily, to different species or subspecies, depending on the degree of difference.

At the subspecific level the systematist must determine whether the differences between populations are clinal or marked by definite character discontinuities. If clinal variation is present, the allocation of subspecific status to various populations is usually discouraged. However, definite steps in the cline (i.e., abrupt changes in the values of a character) may merit subspecific recognition on either side of the discontinuity. Suggestions for delimiting species and subspecies may be found in Mayr (1963; 1969:181-197).

Taxonomic decisions at the generic and higher levels of classification are difficult and somewhat arbitrary. Species included within a particular taxon should show similar trends in adaptive features. A knowledge of fossil history is also important in delimiting families and higher categories. Frequently, however, the fossil record is sparse and the systematist must make inferences based on similarities and discontinuities in characters. Refer to Simpson (1961b) for suggestions on delimiting higher categories.

11-J. Examine several series of mammal skins, skulls, or both representing the following: two subspecies in the same species, two species in the same genus, two genera in the same family. What characters or measurements seem most useful for separating the subspecies? The species? The genera?

11-K. Examine a series of specimens representing an unknown (to you) number of taxa (at least two but no more than six). How many different "taxa" are represented? Check your determinations with the "correct" identifications provided by the instructor.

Zoolological Nomenclature

Zoological nomenclature is the system of scientific names applied to taxa of animals, living and extinct. Ideally, any system of nomenclature should promote names that are *unique,* only one name for a given taxon; *universal,* written in a single language accepted by all zoologists; and *stable,* free of unnecessary or arbitrary name changes.

The set of rules governing zoological nomenclature is the current edition (1964) of the *International Code of Zoological Nomenclature* adopted by the 15th International Congress of Zoology (1958) with amendments by the 16th International Congress of Zoology (1963). Between successive Congresses the International Commission on Zoological Nomenclature may issue Declarations on proposed changes in the rules. The Declarations remain in force until a succeeding Congress either accepts, modifies, or rejects them. The Commission is charged with the additional duties of (1) rendering Opinions and Directions on questions of zoological nomenclature that do not involve changes in the Code, (2) compiling the Official Lists of accepted names and the Official Indexes of rejected names and works in zoology, (3) performing other actions requested by the Congresses. An official copy of the Code may be obtained from the International Trust for Zoological Nomenclature, 14 Belgrave Square, London S. W. 1, England. A partial text of the Code with comments can be found in Mayr (1969). Some of the important provisions of the Code (with comments) are presented below.

Binominal Nomenclature

The 10th edition (1758) of *Systema Naturae,* written by the Swedish botanist Carolus Linnaeus (the Latinized version of his name, Carl von Linné) is the starting point for zoological nomenclature. Linnaeus consistently used in the 10th and later editions a system of two Latin or Latinized words, the **binomen,** for the name of a species. The Linnaean system, termed **binominal nomenclature,** has been adopted as the standard for the formation of scientific names of species. The name of a species is a binomen consisting of a capitalized **generic** name, first word of the binomen, and an uncapitalized **specific** name, second word of the binomen. Single-word names are termed **uninominals.** The *species* name, as opposed to the *specific* name, is always a binomen. For example, the species name of humans is *Homo sapiens,* not just the specific *sapiens.*

The name of a subspecies is a **trinomen** consisting of the generic, specific, and subspecific names. The subspecific name, like the specific name, is never capitalized, e.g., *Lynx rufus baileyi.*

A **subgeneric** name, if used, is placed in parentheses between the generic and specific names, e.g., *Microtus (Pitymys) pinetorum.* This uninominal is capitalized but is not considered a part of the binomen or trinomen. The names of taxa in higher categories (e.g., families, orders) are also uninominals (Fig. 11-1).

Generic, subgeneric, specific, and subspecific names are always printed in italics (or underlined to indicate italics) and the formation and emendation of these names must conform to the rules of the Code and to Latin grammar. A generic name is always a noun in the nominative singular, e.g., *Homo, Ursus.* Specific names must consist of more than one letter and be (1) an adjective in the nominative singular, e.g., *Sciurus niger, Sylvilagus aquaticus;* (2) a noun in the nominative singular, e.g., *Felis leo;* (3) a noun in the genitive, e.g., *Eutamias merriami* named for C. Hart Merriam or *Rhinophylla fisherae,* named for Sigrid Fischer; or (4) an adjective used as a substantive in the genitive case, e.g., *Fonsecalges saimirii,* a psoroptic mite, described from *Saimiri sciureus,* the squirrel monkey. **Patronyms** are names based upon the names of persons and follow the rules for a noun in the genitive case. (A specific name based upon the name of a man ends in *i,* on that of a woman in *ae,* on those of men or combinations of both sexes in *orum,* on those of women in *arum.* For example, *Talpa streetorum* is named for William S. and Janice K. Street.)

The name of a species or subspecies may be abbreviated if such usage does not lead to misinterpretation. For example, in a paper on *Sciurus niger,* it would be permissible to refer to the species as *S. niger* if the combination *Sciurus niger* was used earlier in the paper and the abbreviation would not be confused with any other generic name mentioned in the paper. Likewise, a subspecific name that is identical to the specific name may be listed with the specific name abbreviated as a single letter, e.g., *Sciurus n. niger* or *S. n. niger.* In many publications the names of species and subspecies are abbreviated with one word, e.g., *niger,* but this practice can lead to confusion and should be avoided except in nomenclatural papers that are discussing the name *per se* rather than the animal.

11-L. From the list below determine whether the specific name is based upon a noun, adjective, or a noun in the genitive case:
Felis catus
Mormopterus minutus
Spermophilus variegatus
Peromyscus truei
Molossus major
Peromyscus californicus
Molossus ater
Vulpes fulva

Names of Taxa in Other Categories

The names of families and subfamilies are formed by the addition of the suffixes **-idae** and **-inae** respectively, to the stem word of the type-genus (see section on Types). In addition, the Code recommends that the terminations **-oidea** and **-ini** be added to the stems of type-genera to form the names of superfamilies and tribes, respectively. Thus, the family and subfamily names for the type-genus *Sciurus* would be Sciuridae and Sciurinae. The names for the corresponding superfamily and tribe would be Sciuroidea and Sciurini. Standardized endings for the names of other taxonomic categories such as phyla, classes, and order have not been established. The names of taxa in categories larger than genus are capitalized but not written in italics. A writer may thus refer in lower case letters to the sciurid rodents or the sciurids, but must capitalize the family name Sciuridae.

11-M. Using the information presented above write the correct names for the missing taxa in the spaces provided in the table below. Unless otherwise indicated, assume that the same stem name applies to all of the taxa on each line.

Availability of Names

Several rules govern the **availability** of a name for use in association with a taxon being described. An author must propose the new name in a publication that is (1) reproduced in ink on paper by some method that assures numerous identical copies, (2) obtainable by purchase or free distribution, and (3) issued for the purpose of a scientific, public, permanent record. The new name must be accompanied by a description (see below), definition, or indication. The new name must be written in Latin or must be Latinized (see Appendix D of Code). Descriptions and new names published after 1930 must give one or more characters distinguishing the taxon. After 1950 no name proposed by an anonymous author is available. Neither is any name (after 1960) proposed in recognition of an infrasubspecific category such as a form or variety. A name that does not meet the provisions of availability is declared a **nomen nudum** and has no status in nomenclature. In addition, the author of a new name should make every effort to insure that the name he proposes is not a synonym or homonym of a previously published name (see Validity section below).

Dates and Authorship

The date of publication must be listed in the work with a new name and description. If only a month or year are listed, the date is considered to be the last day in the stated month or year. The date listed on the publication is assumed to be correct unless convincing contrary evidence is later brought forth. The author of a new name for a taxon is frequently listed immediately following the name with no intervening punctuation, e.g., *Dipodomys merriami* Mearns or *Sylvilagus aquaticus* (Bachman); see the Synonymy section for explanation of parentheses around Bachman. The year of publication of the name may also be listed after the author and must be separated from the author's name by a comma, e.g., *Lepus americanus* Erxleben, 1777.

Validity of Names

Under the **law of priority** the valid name of a taxon is the oldest available name applied to it, provided that the name is not invalidated by any provision of the Code or suppressed by the Commission. A statute

Genus	*Subfamily*	*Family*	*Superfamily*
Sciurus		Sciuridae	
Abrocoma	Abrocominae		Octodontoidea
Vespertilio			Vespertilionoidea
Sorex		Soricidae	

of limitations (Article 23b) was adopted by the Fifteenth International Congress of Zoology in 1960 to prevent the revival of a long-forgotten name that would replace a well-known one. The wording of this article was somewhat ambiguous and caused difficulty in interpretation and application. Therefore, the Seventeenth International Congress of Zoology, meeting in 1972, adopted a revision (Article 23a-b) to insure that the Law of Priority will be applied in a manner that will insure stability of nomenclature (Corliss 1972). If a taxonomist determines in a given instance that application of the Law of Priority would threaten stability, he may refer the matter to the Commission for a decision under the Plenary Powers (Article 79).

A **homonym** is one of two or more identical names proposed for the same or different taxa (identical specific or subspecific names included in different genera are not considered homonyms). A **primary homonym** can be, (1) one of two or more identical names at the generic or a higher level, or (2) identical subspecific or specific names included in the same genus when the names were first proposed. A **secondary homonym** is a homonym resulting from the transfer of a specific or subspecific name from the genus in which it was originally described to a genus which already contains an identical name. The Law of Homonymy states that a **junior homonym,** a name published or transferred at a later date, must be rejected and the earlier (older) name, the **senior homonym,** retained. In primary homonymy, for example, *Xenurus* Wagner, 1830, an armadillo, is a junior homonym of *Xenurus* Boie, 1826, a genus of birds. *Xenurus* Boie is thus the senior homonym and the valid generic name for the bird taxon and *Xenurus* Wagner is an invalid name for the armadillo.

A **synonym** is one of two or more different names that have been applied to the same taxon (see Synonymy section below). A **senior synonym** is an available name that has the earliest date of publication. Under the Law of Priority the senior synonym will be considered the applicable name unless the Commission rules otherwise. When taxa are described simultaneously, the **first reviser** principle permits a taxonomist to select a name that will best insure stability and universality of nomenclature. For example, if two synonyms are equally available under the Law of Priority, the first reviser (taxonomist) selects one of the names. When the synonymous names first appear in the same publication, the first name to appear has "page priority" and the first reviser may select this name as the valid name for the taxon. According to the present Code, however, the taxonomist is not required to follow page priority in selecting a valid name, since stability and universality of usage may be more important than page priority.

Types

The type is the standard of reference for the application of a name. The type of a species or subspecies is a specimen so designated. The type of a genus is a species designated as a **type-species.** The type of a subfamily or family is a genus designated as a **type-genus.**

For subspecific and specific names the author designates a single specimen as the **type** or **holotype.** Several other forms of types have nomenclatural standing. A **neotype** is a specimen selected as the type, subsequent to the original description, to replace the original type that has been destroyed, lost, or suppressed by the Commission, i.e., found to be invalid. A neotype is generally designated only in exceptional circumstances in connection with revisionary work where such a designation will best insure stability of nomenclature. A **lectotype** is a specimen subsequently selected as the type from a series of syntypes. A **syntype** (= **cotype**) is one of a series of specimens, collectively designated as the type. It is strongly suggested (recommendation 73A of the Code) that an author designate a single specimen as the holotype and not use syntypes.

There are also several kinds of types that have no official status in nomenclature. A **paratype** is a specimen other than the holotype that was used by the author in preparation of the description and designated or indicated as such by the original author. A **topotype** is any specimen collected at the type locality. A topotype is of no greater importance for the designation of a species or subspecies than any other specimen of these taxa. An **allotype** is a specimen of the opposite sex from the holotype. Designation of allotypes is sometimes utilized in groups where there is considerable sexual dimorphism.

Types with nomenclatural standing are generally conspicuously labeled with a red tag, housed separately from the general collection, and given special care. These types (and other specimens as well) should be carefully labeled with data on locality, date of collection, sex, and other information that is pertinent, such as a bibliographic reference to the published description. Type specimens should be made available for scientists to study. Generally types are studied in the museums where they are housed and are not mailed or shipped for loan because of the danger of loss or damage in transit.

11-N. Why should the naming of syntypes be discouraged?

11-O. In order to insure stability a taxonomist must have a type specimen for a given

taxon, however, the original type is known to be nonexistent.
What would be his course of action?

Descriptions

A **description** is a statement of characters and supplementary information that is associated with the proposal of a new name for a taxon. Descriptions generally consist of the following information: (1) designation of type, (2) geographic range or distribution of taxon, (3) diagnosis and/or description, (4) comparison or differential diagnosis, (5) measurements and/or illustrations, (6) remarks, and (7) list of specimens examined.

The designation of the type may refer to a specimen (holotype) or a taxon. For a holotype, the collector, date and location collected, method of preservation, and museum where housed, are listed. The geographic range of the taxon is described briefly and may be supplemented by a map. The **diagnosis** is a statement of the characters that serve to distinguish the taxon from similar or closely related taxa. (A general description of the taxon may be used instead or in addition.) A **comparison** or **differential diagnosis** compares the named taxon with specifically mentioned equivalent taxa (i.e., taxa of same rank, such as species versus species). Measurements of the holotype and paratypes are frequently listed. An illustration or a bibliographic reference to such an illustration should be included for descriptions of species and subspecies. The author of the description may include additional comments in the remarks section (e.g., why taxon was recognized as new, derivation of name). The specimens examined section provides a list of specimens examined by the author as a basis for the written description. Consult the *Journal of Mammalogy*, *Mammalia*, or the publications of major natural history museums for examples of published descriptions of new taxa.

Synonymy

A **synonymy** is a list of the synonyms and other names, arranged in chronological order, that have been applied to a given taxon. Examples of partial synonymies of the genus *Lynx* and of the bobcat, *Lynx rufus rufus* are shown below (after Miller and Kellogg, 1955:777-778).

Genus LYNX Kerr

1792. *Lynx* Kerr, The animal kingdom, . . . , vol. 1, systematic catalogue inserted between pages 32 and 33, description on p. 157.

1867. *Cervaria* Gray, Proc. Zool. Soc. London, pt. 2, p. 276, October 1867. (Not of Walker, 1866.) (Type, *Felis pardina* Temminck = *Lynx pardellus* Miller.)

1903. *Eucervaria* Palmer, Science, new ser., vol. 17, p. 873, May 29, 1903. (Substitute for *Cervaria* Gray.)

Lynx rufus rufus (Schreber)

1777. *Felis rufa* Schreber, Die Säugtiere . . . , Theil 3, Heft 95, pl. 109b. (For use of the name *rufus* Schreber 1777 in place of *ruffus* Guldenstaedt 1776 (not a scientific name) see J. A. Allen, Journ. Mamm., vol. 1, No. 2, p. 91, Mar. 2, 1920.)

1817. *Lynx rufus* Rafinesque, Amer. Monthly Mag., vol. 2, No. 1, p. 46, November 1817.

1884. *Lynx rufus* True, Proc. U.S. Nat. Mus., vol. 7 (App., Circ. 29), p. 611, Nov. 29, 1884. (Part).

Note in the above synonymies that the specific name was based on *Felis rufa* described by Schreber. But *Felis*, was based upon a type species quite different from the cats now included in the genus *Lynx*. The oldest and present name for this genus is thus *Lynx*. The generic name *Cervaria* Gray, 1867, is not presently recognized since it lacks priority and also proved to be a junior homonym (*Cervaria* Walker, 1866, a lepidopteran, is the senior homonym). *Eucervaria* was an attempt to correct the junior homonym *Cervaria* Gray, but it likewise, had no priority over *Lynx*. The binomen presently used is attributed to Rafinesque. Note, however, that since the specific name, *rufus*, relates back to Schreber, the name "Schreber" is listed after *Lynx rufus rufus*. Since Schreber did not use *rufus* in a binomen that included *Lynx*, his name is enclosed in parentheses. If Schreber had described this cat as *Lynx rufus rufus*, his name would be listed after the subspecific name without parentheses.

11-P. Why is the accepted name *Lynx rufus* as Rafinesque used it rather than *Lynx rufa* as Schreber gave the specific name?

11-Q. Given the information listed below prepare a synonymy for the hypothetical species known (currently) as *John johnsoni* (Smith), 1811. List only the year, name of taxon, and author.
   a. R. J. Smith described *Fred johnsoni* in 1811.
   b. A. B. Black described the genus *John* based on *Bill whitei* in 1800.
   c. F. A. Clark described *Fred kingi* in 1830 as a new form that later proved

to be conspecific (the same species as) *John johnsoni.*

d. A. B. Black described *John blackeyi* in 1812, a synonym of *John johnsoni.*

Alternative Systems of Nomenclature

Several authors (Michener 1964; Little 1964; Hull 1966; and others) have advocated a replacement of the Linnean system of nomenclature with a numerical system. The numerical system, they contend, would increase stability and facilitate electronic data processing and information retrieval. Where space considerations are a factor, these proposals may have merit. Randal and Scott (1967), however, suggest that the Linnean system is superior to numbers since typographical errors would be recognized more easily. Hennig (1966); Nelson (1971), Lovtrup (1973, 1975) and Farris (1976) discussed classification based upon cladograms. Some of these closely resemble classical systems discussed above, others are radically different.

Supplementary Readings

ASC Newsletter, published six times per year by the Association of Systematics Collections. Lawrence, Kansas.

Blackwelder, R. E. 1972. *Guide to the taxonomic literature of vertebrates.* Iowa St. Univ. Press, Ames. 259 pp.

Clifford, H. T. and W. Stephenson. 1975. *An introduction to numerical classification.* Academic Press, New York. 229 pp.

Hecht, M. K., P. C. Gody, and B. M. Hecht. (Eds.). 1978. *Major patterns in vertebrate evolution.* Plenum Press, New York. 908 pp.

Hennig, W. 1966. *Phylogenetic systematics.* Univ. of Illinois Press, Urbana. 263 pp.

International Commission on Zoological Nomenclature. 1964. *International Code of Zoological Nomenclature adopted by the XV International Congress of Zoology.* International Trust for Zool. Nomenclature, London. 176 pp.

Jones, J. K., S. Anderson, and R. S. Hoffman (Eds.). 1976. Section I—Systematics, pp. 7-103. *In Selected Readings in Mammalogy.* Mus. Nat. Hist. Univ. Kansas Monogr. 5.

Mayr, E. 1969. *Principles of systematic zoology.* McGraw-Hill, New York. 428 pp.

McKenna, M. C. 1975. Toward a phylogenetic classification of the mammalia, pp. 21-46. *In* W. P. Luckett and F. S. Szalay (Eds.) *Phylogeny of the Primates.* Plenum Press, New York.

Ross, H. H. 1974. *Biological systematics.* Addison Wesley Pub. Co., Reading, Mass. 345 pp.

Simpson, G. G. 1945. The principles of classification and a classification of mammals. *Bull. Amer. Mus. Nat. Hist.* 85:1-350.

Simpson, G. G. 1961. *Principles of animal taxonomy.* Columbia Univ. Press, New York. 245 pp.

Sneath, P. H. H. and R. R. Sokal. 1973. *Numerical taxonomy,* 2nd Ed. W. H. Freeman, San Francisco. 573 pp.

Systematic Zoology, Published quarterly by the Society of Systematic Zoology, Washington, D. C.

Van Gelder, R. G. 1977. Mammalian hybrids and generic limits. *Amer. Mus. Novitates.* 2635:1-25.

Yablokov, A. V. 1974. *Variability of mammals.* Amerind Publ. Co., New Delhi. 350 pp. [English Edition]

12 Keys and Keying

A biological **key** is a tool for the identification of a specimen to a particular taxonomic unit such as order, family, genus, or species. It consists of a series of pairs of mutually exclusive statements (Fig. 12-1). Each pair of statements is termed a **couplet** and the couplets are numbered or lettered consecutively on the left side of the page.

The person using the key reads both parts of the first couplet and judges which of the two statements best describes the specimen he is trying to identify. Once a decision is made, the number at the right-hand margin indicates the next couplet to be considered. In Figure 12-1, for instance, if the user decides that the second set of statements in the first couplet (1′) best matches his specimen, he then proceeds to couplet 3. If his specimen best matches the first part of this couplet (3), he is told that he has a mole of the genus *Parascalops*. However, if this specimen best matches the second part (3′), he is directed to couplet 4. He thus proceeds, by the process of elimination, to identify his specimen.

Figure 12–1.
Example of a dichotomous bracket key using some of the characters in Table 12-1.

1		Auditory bullae complete; nostrils open on superior surface	2
1′		Auditory bullae incomplete; nostrils open on lateral or superior surface	3
2	(1)	Total number teeth 36; forefeet webbed	*Scalopus*
2′		Total number teeth 44; forefeet not webbed	*Scapanus*
3	(1′)	Accessory basal cusp on first upper incisor; tail length less than 1/4 total length	*Parascalops*
3′		Accessory basal cusp lacking on first upper incisor; tail length greater than 1/4 total length	4
4	(3′)	Total number teeth 36; nostrils open laterally	*Neurotrichus*
4′		Total number teeth 44; nostrils open anteriorly	*Condylura*

While the above paragraph may seem basic for inclusion in an advanced manual, it is essential that practicing biologists be able to use a key. No one can be an expert on the identification of all of the kinds of living organisms in any given area, let alone in the world as a whole. All biologists, including taxonomists, ecologists, physiologists, ethologists, etc., must identify the organisms with which they are working. A key makes the task of identification much simpler.

The essentials of key construction were known to Aristotle (Mayr 1969) but keys were not widely used for identification purposes until the 17th and 18th centuries (Voss 1952). Modern keys usually serve solely for straightforward and accurate identification of a specimen. They are not a **synopsis** (summarized descriptions of particular taxonomic groups, Metcalf 1954) and do not necessarily illustrate phylogenetic relationships among the groups included. This last statement cannot be overemphasized. The characters used in a key are those most easily observed by the user. While occasionally they are the same characters that indicate relationships between various groups, it should never be assumed that this is the case.

Selection of Key Characters

Once the taxonomic and geographic limitations of the key have been defined, the author's first step is the selection of the **key characters** that will be used to distinguish the organisms to be covered. The characters selected should be valid for both sexes and all age classes; if this is not possible, the limitations of the key with respect to sex and age should be clearly stated.

The author of the key must consider the portion(s) of the organism that the reader of the key will be able to examine. For instance, if the key is intended for use by a field ecologist or ethologist who will be identifying live animals, an internal structure such as length of intestine or presence or absence of an interparietal bone is useless. Similarly, a behavioral or physiological character is useless to a taxonomist trying to identify a museum skin or skull.

Key characters should be easily observable with a minimum of manipulation and should be described or stated clearly and concisely. In addition, the characters selected should exhibit little individual variation. Absolute conditions such as the presence or absence of a structure are best. Meristic or discontinuous characters that compare counts of discrete objects, e.g., two incisors versus three incisors, are useful as long as individual variation does not result in overlapping of the counts. Continuous characters, e.g., width of zygomatic arch, should be used only when there is no possibility of overlap between the groups being separated or when these characters offer support to other mutually exclusive characters.

Characters that call for a value judgment on the part of the user are least desirable. Judgment against a generally recognized standard may be used when all else fails. For instance, the alternatives may be "blond pelage" versus "brown pelage." Most people have an idea of what "blond" and "brown" mean, but who can decide precisely when "dark blond" becomes "light brown"? Comparative value judgments (e.g., "size large" versus "size small") should never be used unless they are tied to extremes of measurements (e.g., "size large, over 100 mm" versus "size small, never exceeding 75 mm").

Ideally each statement in the first part of a couplet is completely contradicted by a corresponding statement in the second part of the couplet. In practice, however, it is sometimes not possible for *each statement* to be contradicted. But it is essential that the two parts of a couplet be mutually exclusive. Thus, if each part of a couplet contains three statements, at least one must exhibit no overlap. The other two might exhibit minimal overlap that will render them useless with some specimens, but will be highly useful with others. If, for instance, the following couplet were given and you had only a skull, you could identify

1 Hair present on soles of feet; incisors 2-3/3; canines present ....**Taxon A**
1′ Hair absent on soles of feet; incisors 3/3; canines present or absent ....**Taxon B**

the specimen as Taxon A if it had two upper incisors or as Taxon B if it had no canines. But if it had three upper incisors and had canines, you could not identify it without also examining the skin.

Occasionally it is impossible to use any single character to distinguish between two groups and it is necessary to write a couplet such as the following:

1 Rostrum long and pointed; cheek teeth bunodont .... **Taxon C**
1′ Rostrum short and broadly rounded, or if long and pointed, cheek teeth lophodont .... **Taxon D**

Specimens identified as Taxon C must have both a long pointed rostrum and bunodont teeth. But specimens identified as Taxon D may have either a short and rounded rostrum and any kind of teeth or a long pointed rostrum and lophodont teeth.

Once the group of characters to be used have been decided upon it is helpful to summarize them in a table (Table 12-1). The table can then be used to organize the series of couplets in the keys. Methods have been developed for programming data on key characters into a computer and using the computer to organize these data into a key (Hall 1970, Morse 1971, 1974, Pankhurst 1971, Wilcox et al. 1973).

Key Construction

Thus far we have described keys that are composed of **dichotomous** couplets. Each couplet has only two parts (1 and 1′). Keys have been written with three or more alternatives in each "couplet," but such keys are now discouraged.

Each couplet should be written with two series of parallel and mutually exclusive statements. The following example is *not* an acceptable couplet:

1 Dental formula 1/3 1/1 3/3 2/3 = 34; tail long and bushy .... **Taxon E**
1′ Cheek teeth unicuspid; tail sharply bicolored .... **Taxon F**

While a dental formula of 1/3 1/1 3/3 2/3 = 34 may be exclusive to Taxon E and unicuspid teeth may be exclusive to Taxon F, this fact is not made clear by the key. The two parts of the couplet are not "parallel." To be placed in parallel, the couplet must be rewritten so that the condition of each key character is clearly stated in each part of the couplet. Thus the above couplet could be rewritten as follows:

1 Dental formula 1/3 1/1 3/3 2/3 = 34; cheek teeth bunodont; tail long, bushy, and uniformly colored .... **Taxon E**
1′ Dental formula 1/1 1/1 3/3 2/2 = 28; cheek teeth unicuspid; tail short, sparsely haired, and sharply bicolored .... **Taxon F**

Table 12-1
Some Key Characters* of Five Genera of Moles (Talpidae) Based on Data in Jackson (1915).

Genus	Total Number Teeth	Auditory Bullae	Accessory Basal Cusp on First Upper Incisor	Tail Length	Nostril Opens	Forefeet
Scalopus	*36*	*complete*	absent	< 1/4 total length	*superiorly*	*webbed*
Scapanus	*44*	*complete*	absent	< 1/4 total length	*superiorly*	*not webbed*
Parascalops	44	*incomplete*	*present*	*< 1/4 total length*	*laterally*	not webbed
Condylura	*44*	*incomplete*	*absent*	*> 1/4 total length*	*anteriorly*	not webbed
Neurotrichus	*36*	*incomplete*	*absent*	*> 1/4 total length*	*laterally*	not webbed

*Characters used in key examples (Figs. 12-1, 12-2) are italicized.

A key is usually written in telegraphic style, i.e., unnecessary articles and verbs are eliminated, to economize on text space and simplify reading. The most important or most easily used character is presented first with the remaining characters arranged in the same order in each portion of the couplet. The positive condition for a character (e.g., canines present) is usually placed in the first member of a couplet with the negative condition (e.g., canines absent) given in the second member. Such an arrangement allows for a more efficient and rapid use of the key.

There are several ways of arranging couplets into a key. (Metcalf 1954, and Mayr 1969, present reviews of the types of keys used in biology.) Figure 12-1 illustrates a **bracket key** format in which the two portions of each couplet are presented in immediate succession to one another. All of the keys in the following chapters of this manual are bracket keys.

The second frequently used format is the **indented key** (Fig. 12-2). This style of organization can, and frequently does, result in wide separation between the two parts of each couplet.

Occasionally a key is written as a flow chart (Fig. 12-3) with lines indicating the appropriate succession of couplets. Flow-chart keys are useful only for very short keys and are most commonly found in popular or semipopular works. Sometimes flow-chart keys are prepared using illustrations rather than printed descriptions. Construction of such picture keys is not feasible if a character is variable because a word description can fit a variety of conditions more easily than can a figure.

Figure 12–2.
Example of an indented key using some of the characters in Table 12-1. Note that the characters utilized are identical to those in Figure 12-1.

A. Auditory bullae complete; nostrils open on superior surface
    B. Total number teeth 36; forefeet webbed .... *Scalopus*
    BB. Total number teeth 44; forefeet not webbed .... *Scapanus*
AA. Auditory bullae incomplete; nostrils open on lateral or superior surface
    B. Accessory basal cusp on first upper incisor; tail length less than 1/4 total length .... *Parascalops*
    BB. Accessory basal cusp lacking on first upper incisor; tail length greater than 1/4 total length
        C. Total number teeth 36; nostrils open laterally .... *Neurotrichus*
        CC. Total number teeth 44; nostrils open anteriorly .... *Condylura*

Occasionally it will be necessary for the user to move backward in the key in order to locate the point at which an incorrect decision was made. By the nature of their construction, the indented and flow-chart keys are easily reversible. To facilitate reverse use of a bracket key the number of the preceding couplet

Figure 12–3.
Example of a flow-chart key using characters in Table 12–1.
(R.E. Martin)

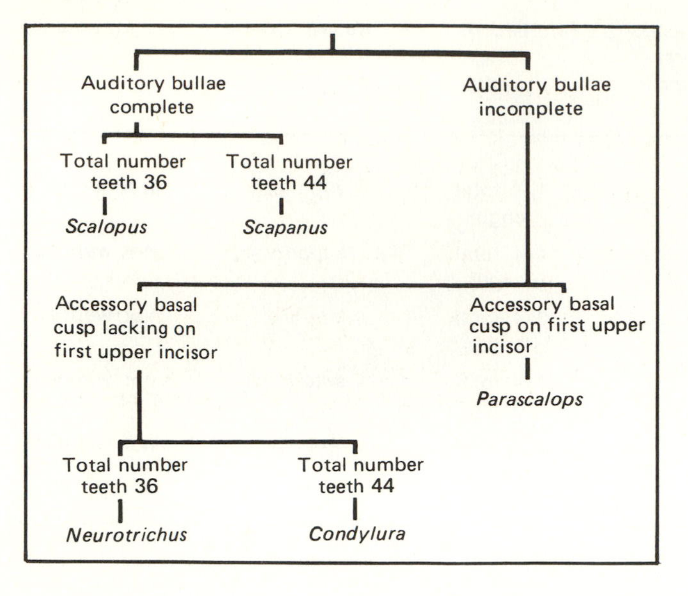

is placed in parentheses after the number of the couplet. In Figure 12-1 the number "3" is followed by "(1′)" indicating that it was "1′" which referred the user to couplet 3.

12-A. Study Table 12-1 and Figures 12-1, 12-2 and 12-3.
Why are some characters omitted in preparing the keys? Which key is easiest to use? Why? Which would be easiest to use if the key were longer and continued for several pages? Which key is easiest to use in reverse? Which key is most economical of space? Which least? Do you see other ways that the keys could be constructed using any of the characters in the table? How would you proceed?

12-B. Prepare a dichotomous bracket key using the information presented in Table 12-2. Are your couplets mutually exclusive and parallel? Rearrange your couplets into an indented key.

12-C. Assemble a selection of at least ten different "taxa" of common office supplies including various kinds of rubber bands, thumb tacks, staples, paper clips, etc. Write a key which can be used to identify these "taxa."

12-D. Using an assortment of at least ten different kinds of identified mammal skulls, skins, or both, write a dichotomous key.

Table 12-2
Key Characters of Eight Imaginary Taxa of Rodents.

Taxon	*Dominant Color of Dorsum*	*Tail Pattern*	*Number Toes on Hindfoot*	*Cheek Tooth Formula*	*Grooves on Incisors*	*Postorbital Process*	*Paroccipital Process*
A	black	unicolor	4 or 5	4/3	none	present	present
B	black or brown	unicolor	4	3/3	1	absent	present
C	gray	bicolor	5	3/3	none	absent	present
D	brown	unicolor	5	4/4	none	present	absent
E	brown	bicolor	4 or 5	3/3	2	absent	present
F	brown	unicolor	3	4/4	none	absent	absent
G	brown and white	bicolor	4	3/3	1	absent	absent
H	gray	bicolor	4	3/3	none	absent	present

The Keys in This Manual

It is relatively easy to say how a key is ideally constructed, but if you have tried the above exercises, you will know that application of these principles is not always easy. In writing the keys in the following chapters, we have attempted to practice what we preach.

The keys in the next sixteen chapters are designed to identify to order (Chapter 13) or family (Chapters 14 through 28) the skulls of extant mammals of the world. All adult mammals should be identifiable but immature specimens without fully erupted dentition or without fully ossified cranial sutures may or may not key out correctly.

External features are included in the keys wherever possible, but none of the keys are designed to work with skins alone. Naturally it is impossible for any author or team of authors to examine the complete range of individual variation that occurs in any species or the complete range of species variation in many families. If you find species or specimens that are not identifiable, we will appreciate hearing from you.

Supplementary Readings

Hall, A. V. 1970. A computer-based system for forming identification keys. *Taxon* 19:12-18.

Metcalf, Z. P. 1954. The construction of keys. *Syst. Zool.* 3:38-45.

Morse, L. G. 1974. Computer programs for specimen identification, key construction and description printing using taxonomic data matrices. *Publ. Mus. Michigan State Univ.* 5:1-128.

Voss, E. G. 1952. The history of keys and phylogenetic trees in systematic biology. *J. Sci. Labs. Denison Univ.* 43:1-25.

13 The Orders of Mammals

Mammals are the dominant animals on the earth today. They range in size from the smallest shrews weighing a few grams to the blue whale, at 112,500 kg, the largest animal that has ever lived. Mammals run over the earth's surface, burrow underground, swim in the waters, climb in the trees, and fly and glide through the air. They feed upon plankton, green vegetation, fruit, flowers, nectar, pollen, mollusks, annelids, arthropods, fish, and other vertebrates. Some provide the power to plant man's crops, and others steal the food from his granaries. Some provide flesh, milk, and hides for man's consumption, while others feed upon man himself. And one mammal, man, has developed the power to destroy all life on earth

Fossil History

Mammals evolved from an extinct group of synapsid reptiles, the †Therapsida. But the precise origin or origins of mammals within this group is not known despite intensive investigation. Some paleontologists maintain that mammals arose from several lines of therapsids, indicating a **polyphyletic** origin for the class Mammalia as currently defined (Olson 1959; Simpson 1961a). This polyphyletic origin for mammals is suggested by the presence of certain mammalian characters in several lines of therapsids. These characters include double occipital condyles, a secondary hard palate, heterodont dentition, and a phalangeal formula of 2-3-3-3-3. At least one paleontologist (Van Valen 1960) has suggested that mammals be redefined to include therapsids.

The view of a polyphyletic origin for the class Mammalia is not shared by all investigators of early mammal history. Hopson and Crompton (1969) interpreted the evidence to suggest that the origin of mammals probably lies within one family of cynodont therapsids. Their interpretation supports a **monophyletic** origin for the class Mammalia. The arrangement of orders and the classification scheme adopted for the discussion of fossil history below follow Hopson (1970) for the subclass Prototheria and Romer (1966) and Dawson (1967) for the subclass Theria.

The subclass Prototheria includes one living and three extinct orders. The living order Monotremata is known only from the Tertiary, Pleistocene and Recent of the Australian Region (Woodburne and Tedford 1975). Fossil documentation is lacking but it is probable that the monotremes have existed since the Mesozoic. The order †Multituberculata is known from the late Jurassic and did not become extinct until the late Eocene. This order existed on earth for a longer period of time than any other mammalian order. Multituberculates were small, rodentlike mammals with enlarged incisors, and molars with two or three longitudinal rows of cusps (Fig. 3-7). The orders †Triconodonta and †Docodonta are apparently closely related to each other and are classified in the infraclass †Eotheria. The order †Triconodonta includes mammals that had the principal cusps arranged in an anteroposterior row with reduced cingular cusps (Fig. 3-5). The earliest triconodonts, the †Eozostrodontidae [†Morganucodontidae], are known from the late Triassic and are thought to be ancestral to the Jurassic triconodonts (Hopson and Crompton 1969). One family of triconodonts ranged into the middle Cretaceous. †Docodonta is the remaining order of the subclass Prototheria. The docodonts had teeth similar to those of triconodonts although the cusps on the internal cingula were greatly enlarged (Fig. 3-6). Docodonts are probably derived from the Triassic morganucodontids (Hopson and Crompton 1969). Docodonts are known only from the upper Jurassic.

The subclass Theria is conventionally divided into three infraclasses. The infraclass †Trituberculata includes two orders, †Symmetrodonta and †Pantotheria,

†Extinct taxa.

both of which had essentially triangular-shaped upper molars (Figs. 3-8, 3-9). The symmetrodonts ranged from the late Triassic to early Cretaceous. The pantotheres appeared in the Jurassic and ranged into the early Cretaceous. Traditionally, the pantotheres were said to be ancestral to the other therian infraclasses. This conclusion is under close scrutiny by some paleontologists (e.g., Crompton and Jenkins 1967) and could be modified or changed with new fossil evidence or interpretations.

The infraclass Metatheria is now regarded to include four orders. (See the Taxonomic Remarks section of chapter 15 for a discussion of this point.) The oldest known marsupial is from the late Cretaceous of North America. Marsupials account for approximately eight per cent of the living genera and six per cent of the living species of mammals.

The infraclass Eutheria is first known from the late Cretaceous. Eutheria is here considered to include seventeen living and twelve extinct orders. Eutherian or placental mammals were the dominant terrestrial vertebrates throughout the Cenozoic and with the exception of the Australian Region are the dominant wild mammals on the earth today. Eutherians account for 91.5 per cent and 94 per cent, respectively, of the living genera and species of mammals.

List of Mammalian Orders

The list below is modified from Simpson (1945), Dawson (1967), Ride (1964, 1970), Rice (1967, 1977), and Hopson (1970). Refer to the Taxonomic Remarks section of this chapter for a discussion of this and other classification schemes. The names of extinct groups in the list and elsewhere in this manual are preceded by a dagger (†). Only extant forms are included in the key in this chapter and those in the remainder of the manual.

Class **Mammalia**

- Subclass **Prototheria**
  - Infraclass †Eotheria
    - Order †Triconodonta
    - Order †Docodonta
  - Infraclass **Ornithodelphia**
    - Order **Monotremata**
  - Infraclass †Allotheria
    - Order †Multituberculata
- Subclass **Theria**
  - Infraclass †Trituberculata
    - Order †Symmetrodonta
    - Order †Pantotheria
  - Infraclass **Metatheria**
    - Superorder **Marsupialia**
      - Order **Marsupicarnivora**
      - Order **Peramelina**
      - Order **Paucituberculata**
      - Order **Diprotodonta**
  - Infraclass **Eutheria**
    - Order **Insectivora**
    - Order **Dermoptera**
    - Order **Chiroptera**
    - Order **Primates**
    - Order †Tillodontia
    - Order †Taeniodonta
    - Order **Edentata**
    - Order **Pholidota**
    - Order **Lagomorpha**
    - Order **Rodentia**
    - Order **Mysticeta**
    - Order **Odontoceta**
    - Order **Carnivora**
    - Order †Condylarthra
    - Order †Litoptera
    - Order †Notoungulata
    - Order †Astrapotheria
    - Order †Xenungulata
    - Order **Tubulidentata**
    - Order †Pantodonta
    - Order †Dinocerata
    - Order †Pyrotheria
    - Order **Proboscidea**
    - Order †Embrithopoda
    - Order **Hyracoidea**
    - Order **Sirenia**
    - Order †Desmostylia
    - Order **Perissodactyla**
    - Order **Artiodactyla**

Key to the Orders of Living Mammals

This key is based primarily on skull characters. In some instances external features are included as supplementary information.

1 Teeth present 6

1′ Teeth absent 2

2 (1′) Greatest length of skull more than 500 mm 3

2′ Greatest length of skull less than 500 mm 4

3 (2) Skull essentially symmetrical, nasal bones form part of roof of nasal passages (Fig. 22-9); two external nostrils; baleen present .... **Mysticeta**
baleen whales

3′ Skull asymmetrical, particularly in the region of the external nares, nasal bones form no part of roof of nasal passages (Fig. 22-16); one external nostril; baleen absent .... **Odontoceta** (in part)
some female toothed whales

4 (2′) Zygomatic arch complete (Figs. 14-4, 14-5); body with spines or well furred; if well furred, feet webbed ....**Monotremata**
echidnas and platypus

4′ Zygomatic arch incomplete (Figs. 20-10, 21-3); body with large, imbricate scales or well haired; feet never webbed 5

5 (4′) Palate with conspicuous medial, longitudinal depression (Fig. 21-3); body covered with large imbricate scales (Figs. 21-1, 21-2) .... **Pholidota**
pangolins

5′ Palate flat, without conspicuous medial, longitudinal depression; body furred or haired, never with large scales ....**Edentata** (In part)
anteaters

6 (1) Incisors never number more than 3/3; angular process of dentary only rarely inflected, usually directed outward or backward; jugal only rarely participating in mandibular articulation 7

6′ Incisors frequently more than 3/3; angular process of dentary inflected; jugal usually participates in mandibular articulation 31

7 (6) Incisors 1/1, long and usually strongly curved (Figs. 19-4, 25-2C) 8

7′ Incisors variable, if 1/1 not strongly curved 9

8 (7) Postorbital bar present (Fig. 19-4); foramen magnum opens ventrally ....**Primates** (In part)
aye-aye

8′ Postorbital bar usually absent, if present foramen magnum opens posteriorly .... **Rodentia**
rodents

9 (7′) Incisors 1/2, upper incisors triangular in cross section (Fig. 26-8) .... **Hyracoidea**
hyraxes

9′ Incisors variable, if 1/2, upper incisors not triangular in cross section 10

10 (9′) Incisors 2/1; first pair of incisors large and strongly curved, second pair small and peglike and situated immediately behind first pair (Fig. 24-3); anterior portion of maxilla perforated .... **Lagomorpha**
rabbits, hares, and pikas

10′ Incisors variable, if 2/1, size and arrangement not as above and anterior portion of maxilla not perforated 11

11 (10′) Upper incisors forming tusks (much longer than other teeth); cheek teeth lophodont (Fig. 26-6); flexible proboscis (trunk) longer than greatest length of skull (Fig. 26-5) .... **Proboscidea**
elephants

11′ Upper incisors variable, if tusklike, then cheek teeth (if present) not lophodont; flexible proboscis absent, or much shorter than greatest length of skull 12

12 (11′) External nares displaced posteriorly, opening at or behind the anterior margin of the orbits; nasal bones reduced or absent; pelvic limbs absent externally 13

12′ External nares not, or only slightly displaced posteriorly, open well anterior to the anterior margin of the orbits; nasal bones usually well developed; pelvic limbs present externally 14

13 (12) Skull somewhat asymmetrical, particularly in the region of the external nares (Fig. 22-16); rostrum long and pointed or broadly rounded (Fig. 22-23); teeth conical .... **Odontoceta**
toothed whales

13′ Skull not asymmetrical; rostrum short and blunt (Figs. 26-9, 26-11); teeth not conical .... **Sirenia**
dugong and manatees

14 (12′) Orbit and temporal fossa separated by postorbital bar or postorbital plate 15

14′ Neither postorbital bar nor postorbital plate present 22

15 (14) Upper incisors (or incisiform teeth) absent 16

15′ Upper incisors (or incisiform teeth) present 17

16 (15) Greatest length of skull less than 100 mm; narrow diastema between premaxilla-maxilla suture and first molariform tooth; horns or antlers never present; manus with five functional digits; nails present .... **Primates** (In part)
some lemurs

16′ Greatest length of skull usually more than 100 mm; wide space between premaxilla-maxilla suture and first molariform tooth (Figs. 28-15, 28-17); horns or antlers frequently present; two or four functional digits on manus; hoofs present .... **Artiodactyla** (In part)
suborder Ruminantia

17 (15′) Incisors 2/1, 2/2 or 2/3 18

17′ Incisors 1/3 or 3/3 20

18 (17) Incisors 2/3 (Fig. 16-8); W-shaped cusp pattern present on occlusal surface of molars .... **Insectivora** (In part)
tree shrews

18′ Incisors 2/1 or 2/2; molars with cusps present or absent, if present not arranged in W-shaped pattern 19

19 (18′) Cheek teeth lack cusps, have sharp lateral edges and a median longitudinal furrow (if cusps present the canine is bicuspid) (Fig. 18-5); forelimbs modified as wings ....**Chiroptera** (In part)
some Megachiroptera

19′ Cheek teeth cuspidate, never as above; canine not bicuspid; forelimbs not modified as wings ....**Primates** (In part)
all primates except aye-aye and some lemurs

20 (17′) Incisors 1/3 (Fig. 28-10); molars selenodont; two digits present ...... **Artiodactyla** (In part)
camels

20′ Incisors 3/3; molars secodont or with complex infolded pattern of cusps and ridges, never strictly selenodont; digits one, or four or more, never two 21

21 (20′) Molars secodont, carnassial pair well developed (Fig. 23-3); digits number four or five; hoofs absent .... **Carnivora** (In part)
some cats and viverrids

21′ Molars with complex folded pattern of cusps and ridges (Fig. 3-14); one hoofed digit on each foot .... **Perissodactyla** (In part)
horses

22 (14′) First two lower incisors pectinate (resembling a comb) (Fig. 17-3); patagium extending from side of neck to manus to pes to side of tail .... **Dermoptera**
colugos

22′ Incisors never as above; patagium usually absent, if present, forelimb modified as a wing 23

23 (22′) Postcanine teeth homodont, never with complex folds and ridges 24

23′ Postcanine teeth heterodont, if homodont possessing complex folds and ridges 26

24 (23) Canines or caniniform teeth conspicuously longer than other teeth; one to three teeth in each premaxilla (Figs. 23-9, 23-21); .... **Carnivora** (In part)
aardwolf, walrus, seals, and sea lions

24′ Canines or caniniform teeth absent or not conspicuously longer than other teeth (Figs. 20-9B, 20-11, 26-3); if long, caniniform teeth are present, no teeth are present in premaxillae (Fig. 20-9A) 25

25 (24′) Incisors and canines absent (Fig. 26-3); posterior cheek teeth constricted medially to

form distinct anterior and posterior lobes (Fig. 26-4); each tooth composed of numerous hexagonal prisms of dentine surrounding tubules in the pulp cavity (Fig. 26-4) [hand lens or binocular microscope required] .... **Tubulidentata**
aardvark

25′ Incisors and/or canines or caniniform teeth present or absent (Figs. 20-9, 20-11); cheek teeth cylindrical, never bilobed; never having dentine prisms as above .... **Edentata** (In part)
armadillos and sloths

26 (23′) Size large, greatest length of skull more than 200 mm 27

26′ Size moderate to small, greatest length of skull less than 200 mm 29

27 (26) Cheek teeth secodont, carnassial pair usually well developed (Fig. 23-3); digits with claws .... **Carnivora** (In part)
large carnivores

27′ Cheek teeth not secodont, carnassial pair never developed; digits with hoofs 28

28 (27′) Canines triangular in cross section, usually with sharp edges; upper canines large, often curving upward and outward (Figs. 28-7, 28-9); pes with 2, 3, or 4 digits; if 3, two of about equal size, the third considerably smaller .... **Artiodactyla** (In part)
hogs, peccaries, and hippos

28′ Canines (if present) not triangular in cross section, not sharp edged; upper canines (if present) small, never curving upward or outward (Fig. 27-5); pes with three digits, central digit larger than the two lateral digits which are about equal in size .... **Perissodactyla** (In part)
rhinos and tapirs

29 (26′) Forearm adapted for sustained flight; phalanges of manus greatly elongated, second through fifth digits wholly enclosed in patagium; upper incisors 0-2, the total number of teeth never exceeding 38 .... **Chiroptera** (In part)
Microchiroptera and some Megachiroptera

29′ Forearm not adapted for flight; phalanges of manus not greatly elongated, and digits never wholly enclosed in a patagium; upper incisors 0-3, total number of teeth variable, up to 48 to 50 30

30 (29′) Canine or caniniform tooth (most anterior tooth in maxilla) small and not clearly differentiated from premolars; incisors small or greatly enlarged (may be larger than canine or cheek teeth); carnassial pair never developed; zygomatic arch often incomplete, auditory bullae complete or incomplete .... **Insectivora** (In part)
most insectivores

30′ Canines large and clearly differentiated from premolars; incisors always much smaller than canines or cheek teeth; carnassial pair usually well developed; zygomatic arch and auditory bullae both complete .... **Carnivora** (In part)
smaller carnivores

31 (6′) Incisors diprotodont (Fig. 15-2 right), I_1 very large, usually projecting nearly horizontally forward, I_2 much smaller or absent; canines present or absent 32

31′ Incisors polyprotodont (Fig. 15-2 left); I_1 not particularly larger than I_2, does not protrude horizontally forward; canines present 33

32 (31) Dental formula 4-5/3 1/1 3/3 4/4 = 46-48; upper incisors roughly uniform in size; C^1 widely separated from both last incisor and first premolar (Figs. 15-16, 15-17); second and third pedal digits syndactylous; tail long, never prehensile .... **Peramelina**
bandicoots

33′ Dental formula diverse; I^1 frequently larger than other upper incisors; C^1 not widely separated from both last incisor and first premolar; no syndactylous digits; tail variable, frequently prehensile .... **Marsupicarnivora**
carnivorous marsupials

33 (31′) Dental formula 4/3-4 1/1 3/3 4/4 = 46-48; no syndactylous digits ...... **Paucituberculata**
caenolestids

32′ Dental formula variable but upper incisors always fewer than four and C^1 never pres-

ent; second and third pedal digits syndactylous .... **Diprotodonta**
Australian diprotodont marsupials

Comments and Suggestions on Identification

The characters used in the above key are not necessarly those used to define an order. This key is based primarily upon cranial characters that can be easily seen by the student in the lab. Mammalian orders may be grouped on the basis of cranial characters but also may be based upon limb structure, methods of reproduction, or numerous other characters or combination of characters.

Below are some comments that will help with rapid identification of a specimen to order.

Monotremes are mammals that lay eggs. Of course this cannot be observed in a study skin or skull. There are two basic types of monotremes, the platypus and echidna; learn what the skin and skull of each looks like.

Marsupicarnivores are a diverse order of marsupials. Most are primarily carnivorous or insectivorous. The teeth are polyprotodont and the feet have no syndactylous digits. The number of incisors frequently exceeds the eutherian number of 3/3. To avoid confusion with eutherian carnivores also look for the inflected angle of the ramus and the presence of the jugal in the mandibular fossa.

Peramelinans are a rather small group of rather similar marsupials. The incisor teeth are polyprotodont and the second and third pedal digits are syndactylous. Both externally and cranially bandicoots share a basic resemblance. To distinguish them from the larger insectivores, which they resemble, look for the inflected angle of the ramus.

Paucituberculatans are small, shrew-like marsupials. They have diprotodont incisor teeth and no syndactylous digits. Externally they could be confused with shrews and their skulls could be confused with those of small marsupicarnivores or those of insectivores. Check the diprotodont lower incisor to confirm identification.

Diprotodonts are a diverse assemblage of Australian herbivorous marsupials. All share two basic characteristics. The incisors are diprotodont and the second and third digits of the hind foot are syndactylous.

Insectivores are very difficult to characterize. They are generally small mammals with long, pointed snouts and numerous, cuspidate teeth. They can best be identified by eliminating other orders.

Dermopteran skulls are recognized by the wide, pectinate (comblike), lower incisors and the skins by the extensive patagium which runs from the side of the neck to the manus, to the pes, to the side of the tail.

Chiropterans are indentified by the presence of a forelimb modified as a wing. The skulls are, however, highly diverse. Some could be confused with those of insectivores or with those of small carnivores. Only practice will help.

Primates are usually monkeylike in general external appearance, but some of the more primitive forms (e.g., some lemurs and galagos) may seem squirrel-like. The skulls always have a postorbital bar or plate separating the orbit and temporal fossa. No other small mammals have a postorbital plate, a few have a postorbital bar, but these generally can be identified by other characteristics.

Edentates include three very different appearing groups. The anteaters have long, conical, toothless skulls. While the three anteater genera differ greatly in size, they resemble each other in general shape and proportions. The sloths have squarish skulls with cylindrical, homodont cheek teeth. Externally they are easily identified by their general appearance. Armadillos have long, conical skulls with numerous cylindrical, homodont cheek teeth. Externally they are easily identified by their characteristic armor.

Pholidotes are identified externally by their armor of overlapping epidermal scales. The skull is conical and toothless, but with practice, can be distinguished readily from those of anteaters or echidnas.

Mysticetes include the largest living mammals. They and their skulls are easily identified as whales by their general form. They can be distinguished from toothed whales by the absence of teeth and the symmetry of the skull in the region of the nares. Most species are larger than all species of odontocetes but there is overlap.

Odontocetes are easily identified as whales and are readily distinguished from the baleen whales by the presence of teeth and the asymmetry of the skull in the region of the nares.

Carnivores almost always have large, well developed canines. Fissiped carnivores often have well developed carnassials and pinniped carnivores have homodont postcanine dentition. Externally carnivores are variable but most types (dogs, bears, cats, otters, weasels, skunks, hyenas, raccoons, seals, sea lions, etc.) are familiar to most students.

Lagomorps have incisors 2/1 with the first pair long like those of rodents. The second pair of upper incisors are small pegs situated directly behind the first incisors. Externally there are two main body forms: the pikas and the rabbits or hares. The latter are familiar to all students.

Rodents all have incisors 1/1 and lack canines. The incisors are long, usually strongly curved and frequently pigmented on the anterior surface. Unfortunately a few other mammals have similar incisors. Learn these. Externally rodents are diverse, but there is a basic resemblance.

Tubulidentates have long, conical skulls with flat-crowned, cylindrical or bilobed teeth composed of numerous hexagonal prisms of dentine. Since there is only one species, all aardvarks resemble each other closely.

Proboscideans are identified readily both externally and by skulls. Once you have seen an elephant or an elephant skull, they cannot be confused with any other mammal.

Hyracoideans are small mammals that superficially resemble rodents. Externally, the species bear close resemblance to each other. The unique foot structure is diagnostic. The upper incisors are long and somewhat resemble those of rodents but they are distinctly triangular in cross section.

Sirenians have distinctive skulls with large complex teeth and posteriorly displaced external nares. Manatees and dugongs lack external hindlimbs and could only be confused with dolphins. However, their flexible necks and blunt rostra distinguish them from these small whales.

Perissodactyls are hoofed mammals that have the axis of weight passing through the central digit. They usually have an odd number of toes (1 or 3), but even when four toes are present (on the forelimbs of tapirs) one toe is larger than the others. Skull and skin characters readily separate this order into three groups, the horses, tapirs, and rhinos, each of which may be rapidly identified once the basic patterns are learned.

Artiodactyls are hoofed mammals that have the axis of weight passing between the third and fourth digits. They almost always have an even number of toes (2 or 4). (On peccaries a rudimentary third toe is present on each hind leg.) Externally the various body forms (pig, camel, deer, cow, sheep, antelope, etc.) are familiar to most students. Horns or antlers are frequently present.

Taxonomic Remarks

In recent years several new classification schemes have been proposed for all or part of the class Mammalia (Turnbull 1971, McKenna 1975, Szalay 1978, Butler 1978). Some of these classifications (Romer 1966; Dawson 1967) are little modified from Simpson's (1945) system. The classifications of Romer (1966) and Dawson (1967) recognize three subclasses: Prototheria, †Allotheria, and Theria, but leave the orders †Triconodonta and †Docodonta assigned to uncertain subclasses. Turnbull's (1971) classification does not remedy the uncertain status of †Triconodonta and †Docodonta but proposes a new treatment for members of the subclass Theria. Hopson's (1970) classification of the expanded subclass Pantotheria assigns †Triconodonta and †Docodonta to this subclass and reduces the subclass †Allotheria to an infraclass within the Prototheria. The major divisions of these various classification schemes are presented in Table 13-1.

Table 13-1
Three Alternative Classification Schemes for the Class Mammalia.

Romer (1966), Dawson (1967)	*Hopson (1970)*	*Turnbull (1971)*
Subclass *Prototheria*	Subclass *Prototheria*	Subclass *Prototheria*
	Infraclass Ornithodelphia	
Monotremata	Monotremata	Multituberculata [misprint for Monotremata]
Subclass Uncertain	Infraclass †Eotheria	Subclass Uncertain
†Triconodonta	†Triconodonta	†Triconodonta
†Docodonta	†Docodonta	†Docodonta
Subclass *†Allotheria*	Infraclass †Allotheria	Subclass *†Allotheria*
†Multituberculata	†Multituberculata	†Multituberculata
Subclass *Theria*	Subclass *Theria*	Subclass *Theria*
Infraclass †Trituberculata		Infraclass †Pantotheria
†Symmetrodonta		†Symmetrodonta
†Pantotheria		†Dryolestoidea
		Infraclass Eutheria
		Cohort †Tribosphenata
		†Zalambdadonta
		†Tribospena
Infraclass Metatheria		Cohort Marsupiata (=Old Metatheria or Marsupialia) [4 orders]
Marsupialia		
Infraclass Eutheria		Cohort Placentata (=Old Eutheria or Placentalia) [Including remaining placental orders]
[Including 17 living and 8 extinct orders]		

Twenty-two orders of living mammals are given in the List of Mammalian Orders above, but there is no universal agreement among mammalogists on this arrangement. The four marsupial orders were frequently considered to be a single order, Marsupialia, and Marsupicarnivora is sometimes split into two orders of living marsupials. The Insectivora should probably be split into two or three distinct orders and Dermoptera is, on occasion, included within the Insectivora. The Mysticeta and Odontoceta are often combined into the order Cetacea. The seals, sea lions and walrus are often considered to constitute the order Pinnipedia, distinct from Carnivora.

Supplementary Readings

Butler, P. M. 1978. A new interpretation of the mammalian teeth of tribosphenic pattern from the Albian of Texas. *Brevioria.* 446:1-27.

Dawson, R. M. 1967. Fossil history of the families of Recent mammals, pp. 12-53. *In* S. Anderson and J. K. Jones (Eds.) *Recent mammals of the world: a synopsis of families.* Ronald Press, New York.

Hopson, J. A. 1970. The classification of nontherian mammals. *J. Mammal.* 51:1-9.

Lawlor, T. E. 1976. *Handbook to the orders and families of living mammals.* Mad River Press, Eureka, CA. 244 pp.

McKenna, M. C. 1975. Toward a phylogenetic classification of the Mammalia, pp. 21-46. *In* U. P. Luckett, and F. S. Szalay (Eds.). *Phylogeny of the primates: a multidisciplinary approach.* Plenum Press, New York.

Romer, A. S. 1966. *Vertebrate palaeontology.* 3rd ed. Univ. Chicago Press, Chicago. 468 pp.

Simpson, G. G. 1945. The principles of classification and a classification of mammals. *Bull. Amer. Mus. Nat. Hist.* 85:1-350.

Szalay, F. S. 1978. Phylogenetic relationships and a classification of the eutherian mammalia, pp. 315-374. *In* M. K. Hecht, *et al.* (Eds.). *Major patterns in vertebrate evolution.* Plenum Press, New York.

Turnbull, W. D. 1971. The trinity therians: their bearing on evolution in marsupials and other therians, pp. 151-179. *In* A. A. Dahlberg (Ed.) *Dental morphology and evolution.* Univ. Chicago Press, Chicago.

Van Valen, L. 1960. Therapsids as mammals. *Evolution* 14:304-313.

14 The Monotremes

Order Monotremata

ORDER MONOTREMATA

The name Monotremata, meaning "one hole," refers to the cloaca, a common chamber into which the digestive, excretory, and reproductive tracts open and from which the products of these tracts leave the body. This is only one of the distinctive characters found in these unique mammals.

Monotremes are very primitive mammals, the only living representatives of the subclass **Prototheria.** They possess all of the characteristics considered to be diagnostic of mammals (see chapter 1), but they also possess many features that are very reptilian in nature. The most striking of these is the structure of their reproductive systems (see chapter 10). Monotremes are the only mammals that lay eggs.

The two families differ in many ways. The duck-billed platypus, Ornithorynchidae, is both semiaquatic and semifossorial. It feeds on aquatic invertebrates and spends most of its time in the water (Fig. 14-1), but it also digs burrows and dens in the banks of streams and ponds. The eggs are laid in these dens and incubated until they hatch. Upon hatching, the young feed on the thick milk secreted from the mammary glands. The platypus is covered with soft fur.

Figure 14–1.
The platypus, *Ornithorhynchus anatinus.* (Brazenor 1950:11).

The echidnas, Tachyglossidae, are terrestrial and also semifossorial. They feed upon termites and other insects and insect larvae and frequently dig to obtain their food. When frightened, they can rapidly dig into the ground to escape predators and are known to excavate burrows. A transitory pouch develops on the abdomen of the female during the breeding season. The eggs are laid and placed in this pouch where they are carried until they hatch. Initially the young remain in the pouch and feed upon the mothers milk. When the young become too large the mother's pouch disappears, but the mammary glands continue to secrete and the young continue to feed. The echidnas have a pelage of coarse hair and spines (Fig. 14-2) and are sometimes called "spiny anteaters."

None of the monotremes are of major economic importance. Both are zoo attractions, but the platypus does not adapt well to captivity and has rarely been exhibited successfully outside of Australia. Echidnas do well in captivity and have been exhibited by numerous zoos around the world.

Distinguishing Characters

Teeth are absent in adults. Tooth buds form in embryonic Ornithorhynchidae and some cheek teeth may erupt in young animals, but all traces of teeth disappear soon after birth. These teeth are quite differ-

Figure 14–2.
A, the Australian echidna, *Tachyglossus aculeatus*; and B, a New Guinea echidna, *Zaglossus bruijnii*.
(A, Brazenor 1950:13; B, Cabrera 1919: pl. I)

A

B

Figure 14–3.
Right hind feet of A, *Tachyglossus aculeatus;* B, *Zaglossus bruijnii;* and C, *Ornithorynchus anatinus.*
(Cabrera 1919: Pls. I and II)

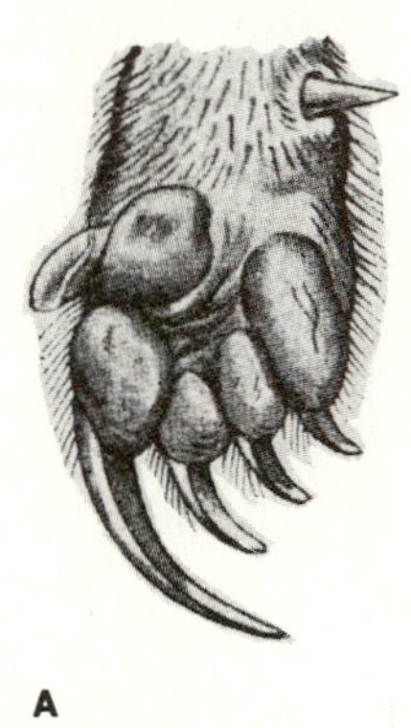

A

B

C

ent from those of other mammals and cannot be directly compared to those of other mammals (Woodburne and Tedford 1975). The mandibular fossa is located entirely within the squamosal. The lower jaws are reduced. Only a vestige of the coronoid process remains and no true angular process is present. Lacrimal bones and auditory bullae are absent. The jugals are reduced or absent but the zygomatic processes of the maxilla and squamosal meet to form complete zygomatic arches.

Large precoracoids, distinct coracoids, and an interclavicle are present in the pectoral girdle (see Vaughan 1978:44, Fig. 5-3). Large claws are present on the digits. Cervical ribs are present as are large epipubic bones. Males possess large horny spurs, each equipped with a poison gland on their ankles (Fig. 14-3). The jaws are covered with rubbery, hairless skin. Vibrissae are lacking. A cloaca is present.

The uteri are completely unfused and poorly developed. Shell glands are present and the mammary glands lack nipples. The penis is bifurcate at the tip, attached to the ventral wall of the cloaca, and used only for the passage of sperm. The testes are permanently abdominal and a baculum is absent.

Distribution

Monotremes are known only from the Australian Region. *Ornithorhynchus anatinus*, the single living ornithorhynchid, inhabits streams and lakes in Tasmania and eastern Australia. In Tachyglossidae the genus *Tachyglossus* is represented by one species in Australia, Tasmania and New Guinea, and the genus *Zaglossus* includes two living species both occuring only on the island of New Guinea (Griffiths 1968).

Fossil History

A few fossil forms of each of the two extant families are known from the Pleistocene of Australia and an ornithorhynchid has been described from the Australian Tertiary (Woodburne and Tedford 1975).

The history of this order is otherwise unknown. The monotremes are believed to represent a line distantly related to the other living mammals and may have evolved from a group of therapsids quite distinct from those ancestral to marsupials and placentals.

Key to Living Families

1 Rostrum with widely flaring premaxillae (Fig. 14-4); snout broad, "duck-billed" (Fig. 14-1); tail well developed; pelage of soft hair; no pinnae .... **Ornithorhynchidae** platypus

1′ Rostrum and snout slender, terete (Fig. 14-5); tail vestigial; pelage of coarse hair with spines (Fig. 14-2); pinnae well developed .... **Tachyglossidae** echidnas

Table 14-1
Living Families of Monotremata.

Family	Common Name	Number of Genera	Number of Species	Distribution
Ornithorhynchidae	platypus	1	1	Australian
Tachyglossidae	echidnas	2	3	Australian

Figure 14–4.
Skull of the platypus, *Ornithorhynchus anatinus.* A and B, dorsal and ventral views, respectively, of the skull; C, lateral view of skull and lower jaw in occlusion; D, dorsal view of lower jaw.
(Giebel 1859:324)

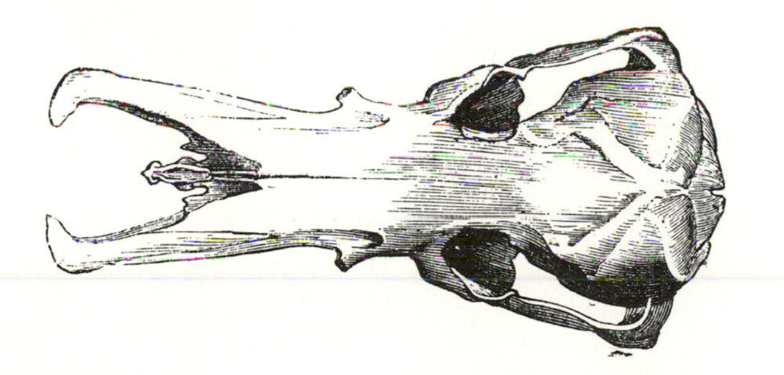

A

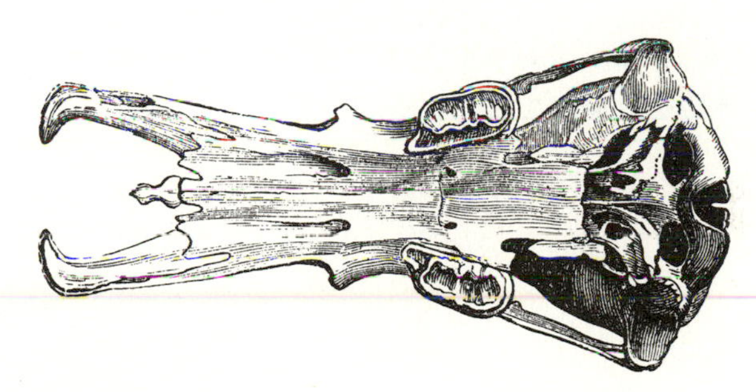

B

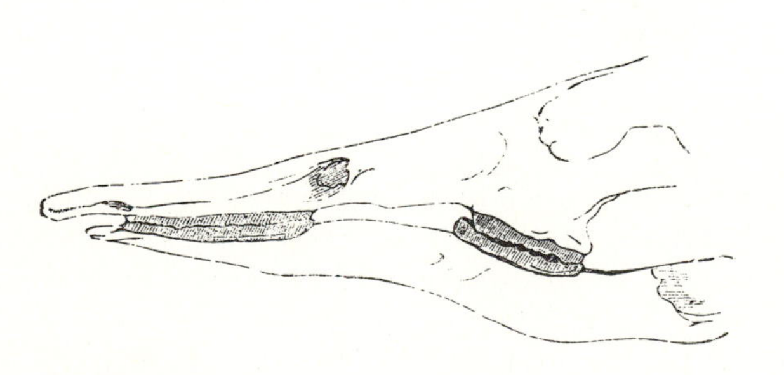

C

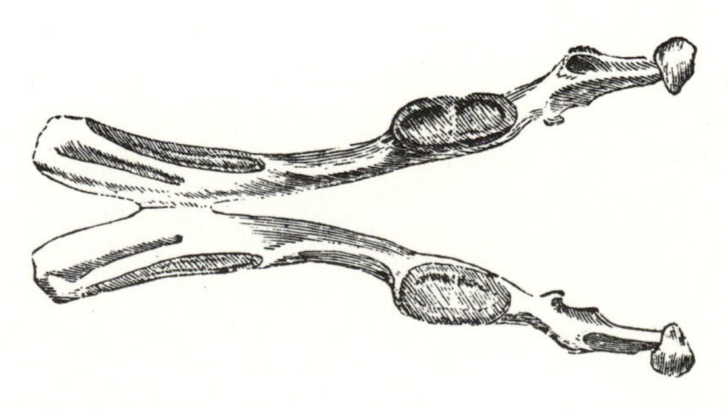

D

Figure 14–5.
Skulls of the echidnas (A) *Zaglossus bruijnii* and (B) *Tachyglossus aculeatus.*
(Cabrera 1919: pl. I)

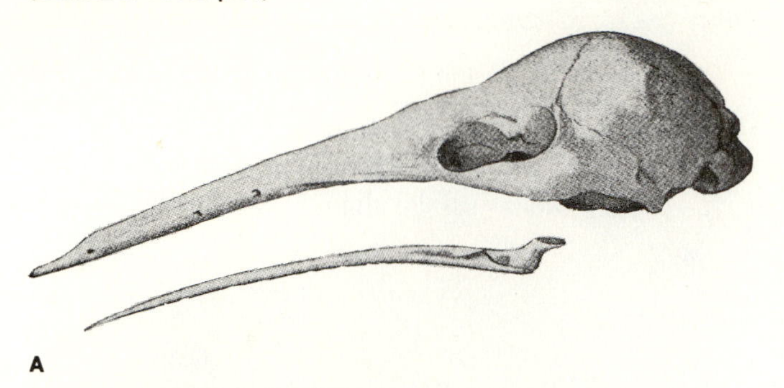

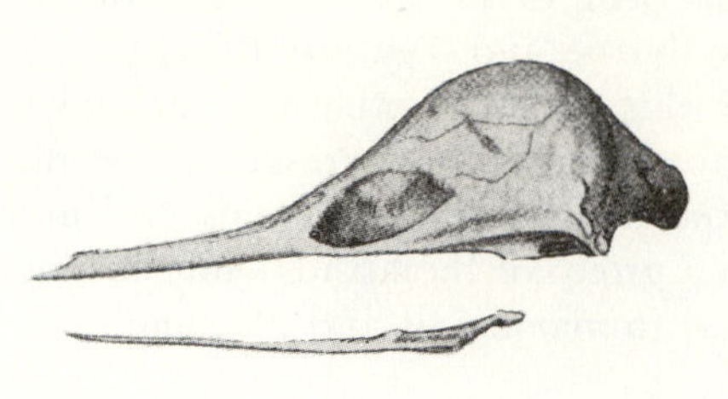

Comments and Suggestions on Identification

The platypus and echidnas are distinctive animals and, except for the echidna skull, should present no identification problems. The skull of the echidnas resembles those of the small anteaters and the pangolins in being toothless and cone-shaped. However, it has a more elevated braincase and, in the more common genus, *Tachyglossus,* the tips of the premaxillae are bent slightly upwards. The anteaters possess well-developed lacrimal bones that are absent in the echidnas and pangolins, but the presence or absence of these bones is not always easily detected. The pangolin skull is more robust than the others but its lower jaw is weaker and lacks an angular process.

Several groups of small mammals, e.g., hedgehogs, tenrecs, and certain rodents, possess spines and in this way superficially resemble echidnas. Check the snout, teeth, and skull.

Supplementary Readings

Burrell, H. 1927. *The platypus.* Angus and Robertson Ltd. Sydney. 227 pp.

Collins, L. R. 1973. *Monotremes and marsupials.* Publ. No. 4888, Smithsonian Institution Press, Washington, 323 pp.

Griffiths, M. 1968. *Echidnas.* Pergamon Press, Oxford. 282 pp.

Griffiths, M. 1978. *The biology of the monotremes.* Academic Press, New York. 368 pp.

Grizimek, B. 1975. Egg-laying mammals. pp. 38-49. *In* B. Grizimek (Ed.). *Animal Life Encyclopedia, Volume 10, Mammals I.* Van Nostrand Reinhold, New York.

Parrington, F. R. 1974. The problem of the origins of the Monotremes. *J. Nat. Hist.* 8:421-426.

Ride, W. D. L. 1970. *A guide to the native mammals of Australia.* Oxford Univ. Press, Melbourne. 249 pp.

Woodburne, M. O. and R. H. Tedford. 1975. The first Tertiary monotreme from Australia. *American Museum Novitates* 2588:1-11.

15 The Marsupials

Order Marsupicarnivora
Order Peramelina
Order Paucituberculata
Order Diprotodonta

Figure 15–1.
A wallaby with a "joey" in its marsupium.
(Brazenor 1950:49)

Figure 15–2.
Polyprotodont (left) and diprotodont (right) marsupial skulls.
(Brazenor 1950: 18 and 19)

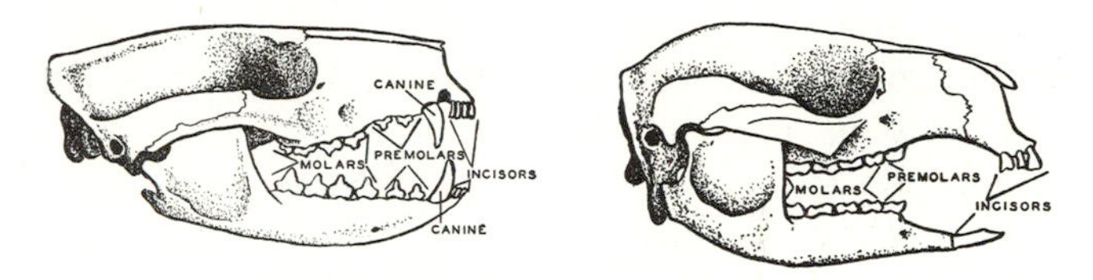

The supraordinal name Marsupialia refers to the abdominal pouch or marsupium (Fig. 15-1) in which most newborn marsupials complete embryonic development (chapter 10). Although similar structures are present in the echidnas and a marsupium is absent in many marsupial species, this structure is one of the diagnostic characters of the Marsupialia.

Marsupialia was long regarded as a single, extremely diverse order including insectivorous, carnivorous, omnivorous, nectivorous, browsing and grazing animals. They occupied diverse habitats in the Australian and Neotropical regions, ranging from the dry deserts of central Australia and central Chile to the wet tropics of New Guinea and Brazil. Marsupials have evolved fossorial, semifossorial, ambulatory, cursorial, saltatorial, semiaquatic, arboreal, and gliding forms.

The number of incisors in the upper jaw is higher than the number in the lower jaw in all living forms except Vombatidae. The primitive marsupial dental formula, with premolars 3/3 and molars 4/4, is the reverse of the premolar and molar numbers in placentals. The total number of teeth often exceeds the basic eutherian number of forty-four. In many families the incisors have been modified to form a **diprotodont** dentition (Fig. 15-2). In these the lower jaw is shortened and the first pair of lower incisors is greatly enlarged and elongated to meet the first pair of upper incisors. The upper incisors may be similarly enlarged but are usually unspecialized. Canines and the first premolars are frequently incisiform. In some macropodids and in some fossil paucituberculids the anterior premolar is plagiaulacoid (Fig. 15-20). It is an elongated blade-like tooth similar to that of the multituberculate *Plagiaulax*.

The brain is relatively small and a corpus callosum is absent. The braincase is small in relation to skull size and the auditory bullae, when present, are formed principally or entirely by the tympanic process of the alisphenoid. Large palatal vacuities are often present. The angular process of the dentary is inflected and the jugal contributes to the formation of the mandibular fossa in all except *Tarsipes* and is only weakly developed in *Phascolarctos* and *Myrmecobius*.

Limb structures vary considerably among the marsupials. Most are plantigrade but some are digitigrade.

Figure 15–3.
Skeletons of the right hind feet of three marsupials. A, a kangaroo and B, a koala, both with syndactylous digits. C, an opossum without syndactylous digits.
(Flower and Lydekker 1891:159, 156, 133)

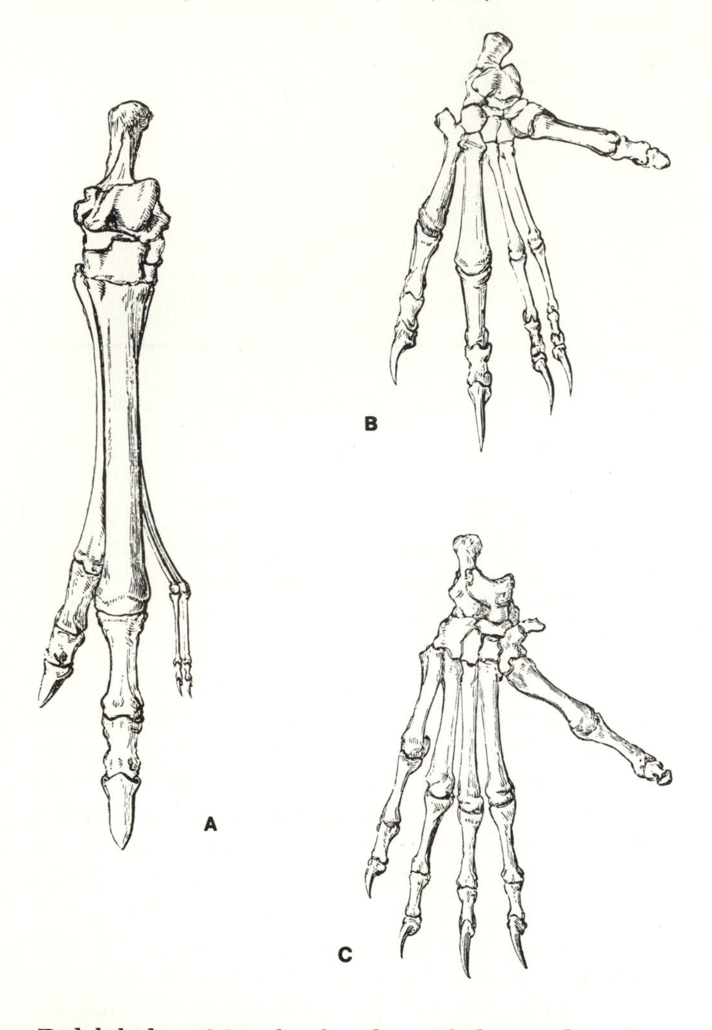

Figure 15–4.
Ventral views of the hind feet of two marsupials with syndactylous digits.
(Weber 1928:54)

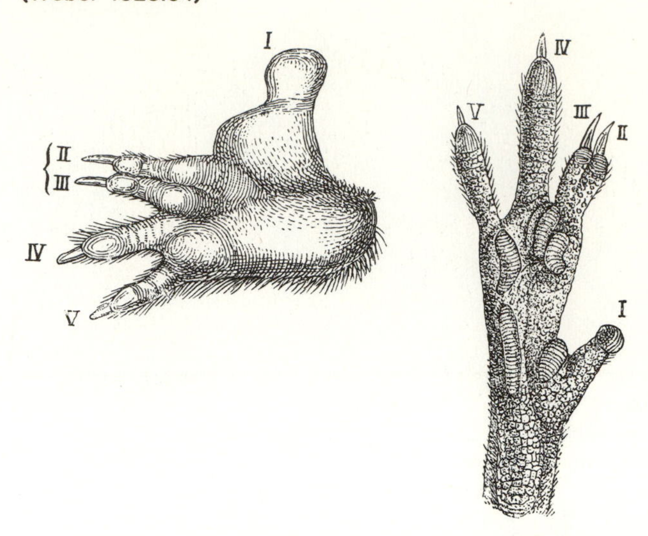

Didelphidae, Microbiotheridae, Phalangeridae, Burramyidae, Petauridae, Phascolarctidae, and Tarsipedidae are well adapted for arboreal life, and one species of Didelphidae, *Chironectes minimus*, has webbed digits and is semiaquatic. Vombatidae and some Dasyuridae are semifossorial and Notoryctidae is fully fossorial. All Peramelina are hoppers and the Macropodidae are highly modified for ricochetal locomotion (However some macropodids are actually arboreal). In arboreal forms the hallux is typically opposable, but in others it is frequently reduced or absent. In the Peramelina and Diprotodonta, the second and third digits of the hind foot are syndactylous (Figs. 15-3 and 15-4). The toes are fused so that the skeletal elements of the two toes are encased within a single skin sheath. Two claws, one for each toe, project from the end of this syndactylous digit (Fig. 15-4).

A simple yolk sac placenta is present in most forms. A chorio-allantoic placenta, similar to that of eutherian mammals but lacking villi, is found in only one order, the Peramelina. The gestation period is short and the young are relatively undeveloped at birth in comparison to placental mammals. A marsupium or abdominal folds are present on the abdomen of most species. These pouches contain the young as they complete development after birth. Epipubic bones are present in both sexes of all species (Fig. 15-5) but are vestigial in Notoryctidae and Thylacinidae. The uteri are completely separate with two distinct lateral vaginal canals. The penis is bifurcated at the tip and lacks a baculum. The scrotum is situated anterior to the penis. A shallow cloaca is present in young, but the adults typically have separate urogenital and anal openings.

Several species of marsupials are of minor economic importance. The North American opossum, *Didelphis marsupialus,* for instance, is of minor importance as a fur-bearing animal and a food source. Kangaroos are hunted for hides and for meat which is used to some extent for human consumption and to a greater extent for pet food. The brush-tailed possum, *Trichosurus vulpecula,* of Australia is an important furbearer and also a nusiance where it lives as a commensual with man (Gewalt 1975:112).

Many species of marsupials are used for food by Australian aborigines and by native peoples of New Guinea. Many Australian species, particularly the carnivorous forms and the smaller, less conspicuous herbi-

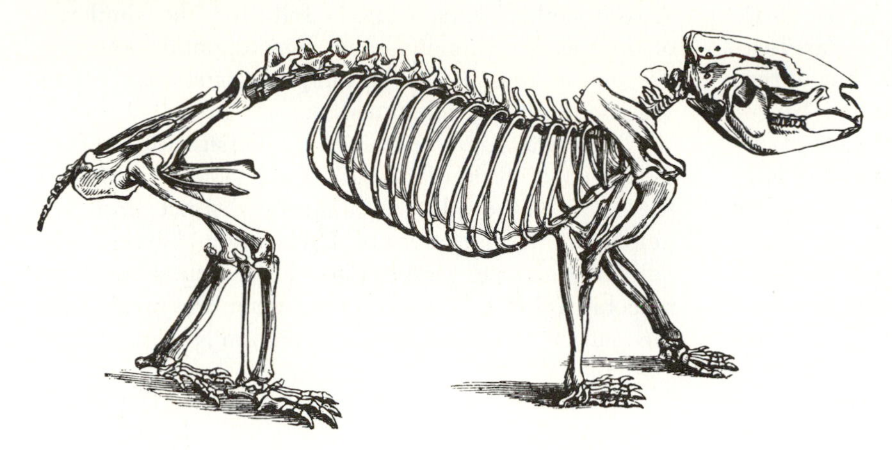

Figure 15–5.
Skeleton of a wombat. Note the epipubic bones projecting anteriorly from the pelvis.
(Owen 1866:330)

vorous forms, are endangered or extinct, probably due to competition with introduced placental species such as rabbits, dingos, cats, foxes, etc., or due to destruction of their habitat. Other species were apparently never very common. Leadbeaters possum, *Gymnobelideus leadbeateri,* was known from only five specimens collected between 1867 and 1909 in Victoria, Australia. It was considered to be extinct until a small population was discovered living at Marysville, Australia in 1961. The genus *Burramys* was known only as a Pleistocene fossil until a representative was discovered alive at a ski lodge near Mt. Hotham, Victoria, in 1966.

Forms very similar to living opossums are known from the late Cretaceous of North America, and in the early Tertiary similar forms are known from the Americas and Europe. No specimens are known from Europe or from North America after the Miocene though they existed in South America until the Recent. The single modern species in the Nearctic, *Didelphis virginiana,* probably reinvaded this area from the Neotropical.

South America was separated from North America during most of the Tertiary. During the period of isolation several distinct marsupial groups developed in South America and many large marsupial carnivores, including a marsupial version of the saber-toothed cat, are known. Toward the end of the Tertiary the Panamanian land bridge united the American continents and placentals from North America invaded the south.

The Australian Region was separated from the rest of the world even earlier than the Neotropical had been. At the time of its isolation only monotreme and marsupial mammals were present. Until the arrival of man and his dog in relatively recent times, only a few rodents and bats had managed to cross the water barriers separating the Australian Region from the rest of the world. During this period of isolation Australian marsupials radiated into almost all of the habitats available to mammals. The Dasyuridae are believed to have included the form ancestral to other Australian marsupials but Tertiary fossils from Australia are virtually unknown. During the Pleistocene several large marsuipials are known including a wombat-like animal the size of a small rhinoceros.

There is considerable variation in the way mammalogists have grouped marsupial families into higher taxa. Many older works divided the families into two suborders either on the basis of the structure of the incisors or on the basis of the structure of the feet. Using teeth as a criterion, those families with small subequal incisors and large canines were grouped in the suborder Polyprotodonta, while those with large first lower incisors and the other incisors and the canines small or absent, were grouped in the suborder Diprotodonta (Fig. 15-2).

Using foot structure as a criterion, those families without syndactylous toes were grouped in the suborder Didactyla while those with syndactylous toes were grouped in the suborder Syndactyla (Figs. 15-3 and 15-4).

These two systems agreed except for the reverse placement of the families Peramelidae and Caenolestidae.

Simpson (1930A, 1945) abandoned both of the systems above as being too simplistic since they were based on only one character, and most later authors followed him. He (Simpson 1945) organized the living families into five superfamilies (Table 15-1). Several recent authors have argued that the differences that exist among several marsupial families are as great or greater than those existing among the several orders of placentals. Ride (1964, 1970) recommended splitting the Marsupialia into four orders (Table 15-1). More recently Kirsch (1977:45) presented a classification of marsupials that is similar to Ride's, but differs in one important aspect—Kirsch included the bandicoots in the same order as the didelphids and dasyurids, and thus recognized only three marsupial orders.

If the peramelids are indeed more closely related to the dasyurids than to the diprotodontids, the origin of syndactyly of the second and third digits of the hind foot must have occured twice in the evolution of Australian marsupials. Simpson (1961C) conceded that the presence of syndactyly in the African otter shrew, *Potomogale*, a member of the order Insectivora, proves that syndactyly has evolved more than once, however he did not advocate grouping the peramelids and dasyurids together. Kirsch (1977) felt that the similarity of the pes in peramelids and diprotodontids was convergent but Marshall (1972 and pers. comm.) and others argue against this likelihood.

Other ways in which Kirsch's classification differs from that of Ride (Table 15-1) are (1) recognition of the Microbiotheriidae as a family distinct from Didelphidae, (2) placement of Thylacinidae closer to Didelphidae than to Dasyuridae, (3) recognition of Myrmecobiidae as a family distinct from Dasyuridae, (4) recognition of Thylacomyidae as a family distinct from

Table 15-1
Classifications of marsupials.

Kirsch (1977)	*Ride (1964, 1970)*	*Simpson (1945)*
Superorder Marsupialia	Superorder Marsupialia	
Order Polyprotodontia	Order Marsupicarnivora	Order Marsupialia
Suborder Didelphimorphia		
Superfamily Didelphoidea	Superfamily Didelphoidea	Superfamily Didelphoidea
Family Didelphidae	Family Didelphidae (implied)	Family Didelphidae
Family Microbiotheriidae		
Family Thylacinidae		
Suborder Dasyuromorphia		
Superfamily Dasyuroidea	Superfamily Dasyuroidea	Superfamily Dasyuroidea
Family Dasyuridae	Family Dasyuridae	Family Dasyuridae
Family Myrmecobiidae		
	Family Thylacinidae	
Suborder Notoryctemorphia		Family Notoryctidae
Superfamily Notoryctoidea		
Family Notoryctidae		
	Order Peramelina	
Suborder Peramelemorphia		
Superfamily Perameloidea		Superfamily Perameloidea
Family Peramelidae	Family Peramelidae	Family Peramelidae
Family Thylacomyidae		
Order Paucituberculata	Order Paucituberculata	
Superfamily Caenolestoidea		Superfamily Caenolestoidea
Family Caenolestidae	Family Caenolestidae	Family Caenolestidae
Order Diprotodonta	Order Diprotodonta	
Superfamily Phalangeroidea		Superfamily Phalangeroidea
Family Phalangeridae	Family Phalangeridae	Family Phalangeridae
Family Petauridae	Family Petauridae	
Family Burramyidae	Family Burramyidae	
Family Macropodidae	Family Macropodidae	Family Macropodidae
Superfamily Vombatoidea		
Family Vombatidae	Family Vombatidae	Family Phascolomidae (=Vombatidae)
Family Phascolarctidae	Family Phascolarctidae	
Superfamily Tarsipedoidea		
Family Tarsipedidae	Family Tarsipedidae	
	Marsupialia *incertae sedis*	
	Family Notoryctidae	

Peramelidae, (5) placement of Phascolarctidae closer to Vombatidae than to Phalangeridae and (6) inclusion of Notoryctidae in the same order as the dasyurids and perameloids.

In the following treatment we have followed the family classification given by Kirsch (1977) but have recognized the ordinal arrangement used by Ride (1964). Our arrangement differs from Kirsch (1977) only in the following two points. We use the name Marsupicarnivora instead of Polyprodontia for the order including Didelphidae, Dasyuridae and Notoryctidae, and we recognize Kirsch's suborder Peramelemorphia as a full order, the Peramelina.

ORDER MARSUPICARNIVORA

The Marsupicarnivora include those marsupials that are generally carnivorous. Some eat other vertebrates, many are highly insectivorous and most are quite omnivorous. But all include a significant amount of animal matter in their daily diets. They occupy terrestrial and arboreal habitats, one is semiaquatic, some are semifossorial, and one is fully fossorial.

Distinguishing Characters

Incisors number 5/4 or 4/3, small pointed or bladelike. Canines well developed (except in *Notoryctes*). Toes didactylous. Hallux clawless, well developed and opposable, or vestigial to absent. Marsupium present or absent. Stomach simple, caecum small or absent.

Distribution

The Marsupicarnivora include groups in both the Neotropical and Australian Regions. Two families are Neotropical: Didelphidae and Microbiotheriidae, and one species of didelphid, *Didelphis virginiana,* extends well into the Nearctic. The other three living families are found only in the Australian region. Dasyuridae occurs widely in Australia, New Guinea and adjacent islands. Myrmecobiidae is found only in southwestern Australia, and Thylacinidae, if it is still extant, occurs only in Tasmania. The marsupial mole is known only from central and western Australia.

Fossil History**

The oldest known marsupials, from the late Cretaceous of North America, have been referred to the several families of Didelphoidea. Except for Eocene-Miocene remains of Didelphidae from Europe the Didelphoidea are known only from the Nearctic and Neotropical. The didelphoid family †Borhyaenidae ranged from the late Paleocene to the Pliocene of South America. It included several large carnivorous forms including some similar to the thylacine (Fig. 15-6A) and others with teeth and jaws remarkably like the sabre-toothed cats (Fig. 15-6B). The Dasyuridae and Thylacinidae range from the Miocene to Recent in Australia. Myrmecobiidae and Notoryctidae are known only from the Australian Recent.

**Kirsch 1977

Table 15-2
Living Families of Marsupicarnivora.

Family	Common Name	Number of* Genera	Number of* Species	Distribution
Didelphidae	Opossums	11	70	Neotropical and Nearctic
Microbiotheriidae	Monito del Monte	1	1	SC Chile
Thylacinidae	Tasmanian "wolf" or "tiger"	1	1	SE Australia
Dasyuridae	Native "cats," marsupial "mice," Tasmanian devil	14	49	Australian
Myrmecobiidae	Numbat or banded anteater	1	1	SW Australia
Notoryctidae	Marsupial "mole"	1	1	W Australia

*Kirsch and Calaby 1977

Figure 15–6.
Skulls of two fossil Marsupicarnivora: A, †*Borhyaena* from the Miocene of Patagonia and B, †*Thylacosmilus* from the Pliocene of Argentina.
(A, Sinclair 1906; B, Riggs 1933)

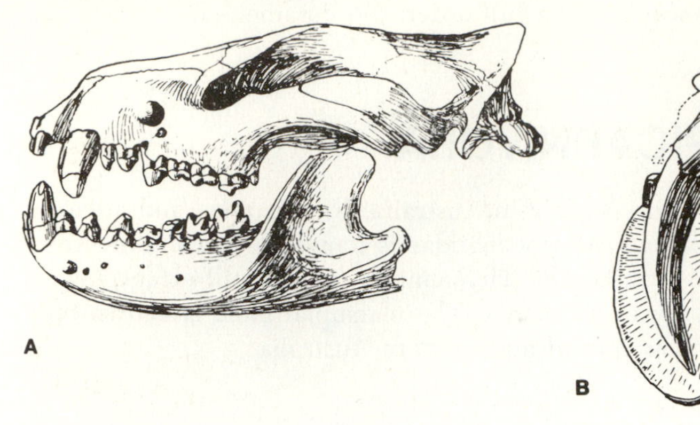

Figure 15–7.
Skull of Monito del monte, *Dromiciops australis*.
(Osgood 1943:49)

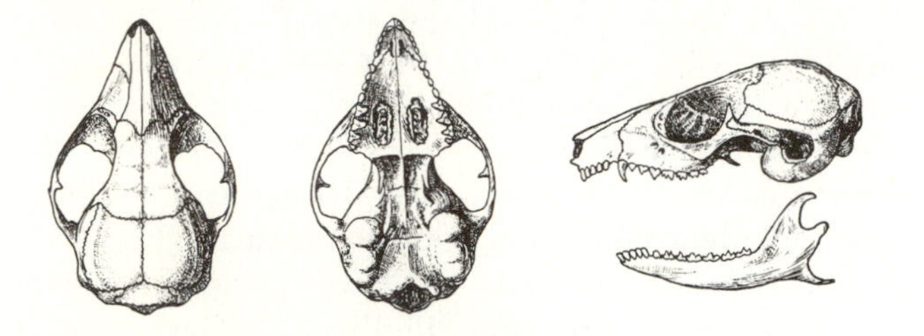

Figure 15–8.
Skull of a didelphid, the coligrueso, *Lutreolina crassicaudata*.
(Marshall, L.G. 1978. *Lutreolina crassicaudata*. Mammalian Species, 91:1–4, fig. 1. Copyright 1978 American Society of Mammalogists. Reprinted with permission.)

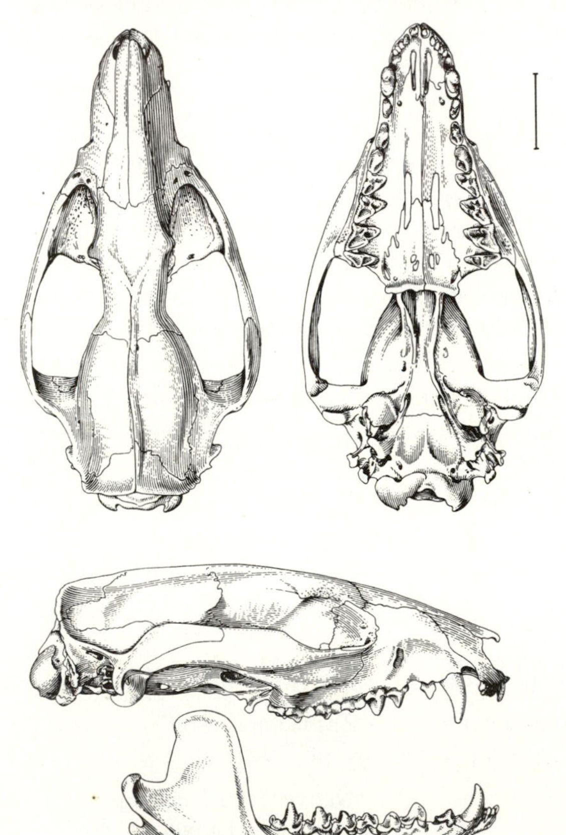

Key to Living Families

1 Incisors 5/4 (Figs. 15-7 and 15-8); hallux well developed and opposable; tail prehensile 2

1′ Incisors 4/3 (Figs. 15-10 to 15-12); hallux reduced or absent, never opposable; tail not prehensile 3

2 (1) Size small, condylobasal length less than 30 mm; first upper premolar and last upper incisor approximately equal in width (Fig. 15-7); total length lesss than 300 mm; dorsal pelage dark brown with several greyish-white patches along sides .... **Microbiotheriidae**
Monito del monte

2′ Size variable, condylobasal length more than 30 mm or if less than 30 mm then first upper premolar at least twice as wide as last upper incisor (Fig. 15-8). Size variable; pelage without greyish white patches along sides or if with such patches then total length more than 500 mm (Fig. 15-9) .... **Didelphidae**
Opossums

Figure 15–9.
A didelphid opossum, the yapok or water opossum *Chironectes minimus.*
(Duncan 1877-83:223)

3 (1′) Postcanine teeth 7-8/8-9 for a total of 48 to 50 teeth, reduced and widely separated (Fig. 15-10); dorsal pelage dark with 6 to 9 transverse light bands (Fig. 15-10); total length less than 500 mm .... **Myrmecobiidae**
Numbat or banded anteater

3′ Postcanine teeth 6-7/6-7 for a total of 42 to 46 teeth, well developed for cutting insects or shearing flesh, not widely separated (Fig. 15-11); dorsal pelage without bands, or if with bands then with more than 7 dark bands on a lighter general pelage color (Fig. 15-12); total length variable 4

4 (3′) Canines small, no larger than last incisor and smaller than first premolar; skull blunt and squarish in shape (Fig. 15-11); animal mole-like in general appearance (Fig. 15-11); no external eyes; central claws on manus huge (Fig. 15-11) .... **Notoryctidae**
marsupial "mole"

4′ Canines larger than last incisor and first premolar; skull tapered and conical (Figs. 15-12 and 15-13); not mole-like in general appearance; external eyes present; central claws on manus not greatly enlarged **5**

Figure 15–10.
The banded anteater, Myrmecobiidae, *Myrmecobius fasciatus,* feeding and lateral view of skull.
(anteater, Flower and Lydekker 1891:140; skull, Gregory 1910)

Figure 15–11.
The marsupial "mole" and three views of its skull, Notoryctidae, *Notoryctes typhlops.*
("Mole," Beddard 1902:159; skull, Cabrera 1919: pl. 9)

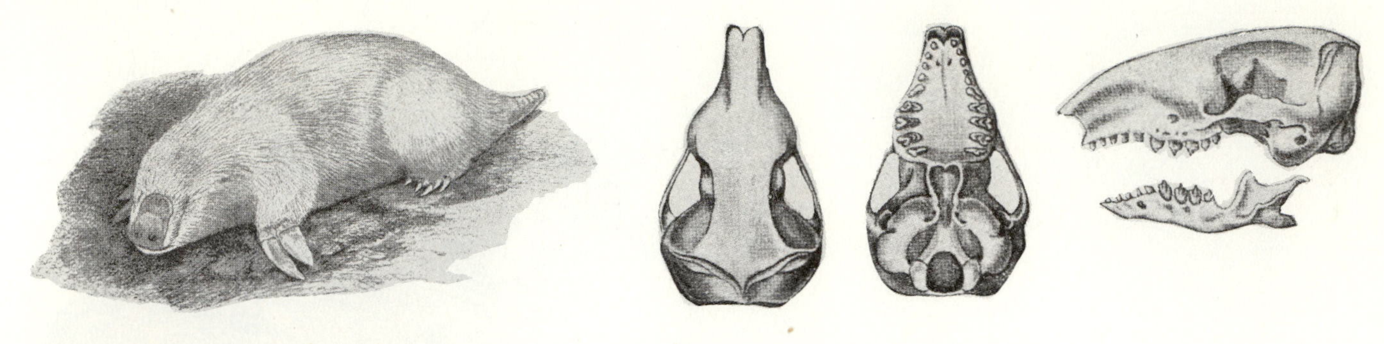

Figure 15–12.
The Tasmanian "wolf" and a lateral view of its skull. Thylacinidae, *Thylacinus cynocephalus.*
(Flower and Lydekker 1891; 136 and 137)

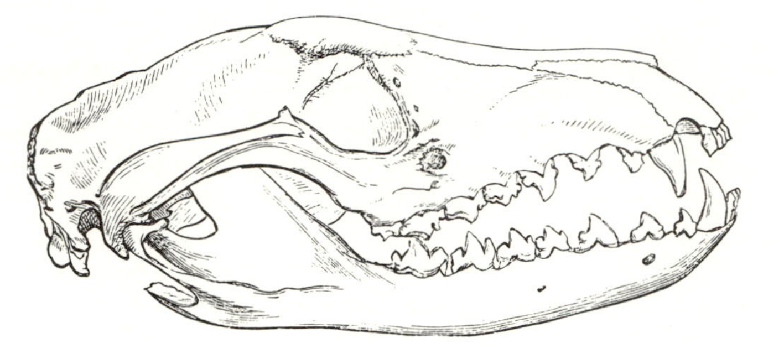

5 (4′) Size larger, condylobasal length greater than 120 mm; dental formula 4/3, 1/1, 3/3, 4/4 = 46 (Fig. 15-12); dorsal pelage with several dark transverse stripes over the rump (Fig. 15-12) .... **Thylacinidae**
Tasmanian "wolf"

5′ Size smaller, condylobasal length less than 120 mm, usually considerably less, if 120 mm or more; dental formula 4/3, 1/1, 2/2, 4/4 = 42 (Fig. 15-13); dorsal pelage plain or spotted (Fig. 15-14) but not striped as above .... **Dasyuridae**
native "cats," marsupial "mice," Tasmanian devil

Comments and Suggestions on Identification

The skulls of New World marsupicarnivores are distinctive and not easily confused with other mammals. The thylacinid skull and those of the larger dasyurids look like those of the Carnivora at first glance. The numbat and smaller dasyurid skulls resemble certain members of the Insectivora. Check the number of upper incisors and the angular process of the ramus.

The skins of marsupicarnivores frequently resemble members of the Insectivora, Carnivora or Rodentia. Most of the new world forms are easily identified by the prehensile tail and opposable, clawless hallux, but the Australian forms are more difficult to identify and require practice. The marsupial "mole" resembles the moles and golden moles of the Insectivora, but it can be readily distinguished from these by the shape of the teeth, by the size of the manual claws and by the ringed nature of the tail.

Figure 15–13.
Skulls of representative Dasyuridae. A, the fat-tailed marsupial "mouse," *Sminthopsis crassicaudata;* B, the brush-tailed phascogale, *Phascogale tapoatafa;* C, the Tiger "cat," *Dasyurus maculatus;* D, the Tasmanian devil, *Sarcophilus harrisii.* Not all to same scale.
(A–C, Brazenor 1950:97 to 99; D, Gregory 1910)

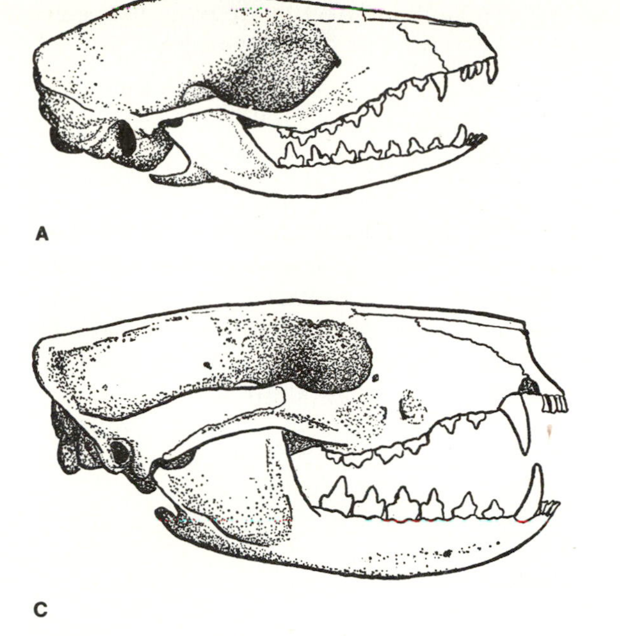

A

B

C

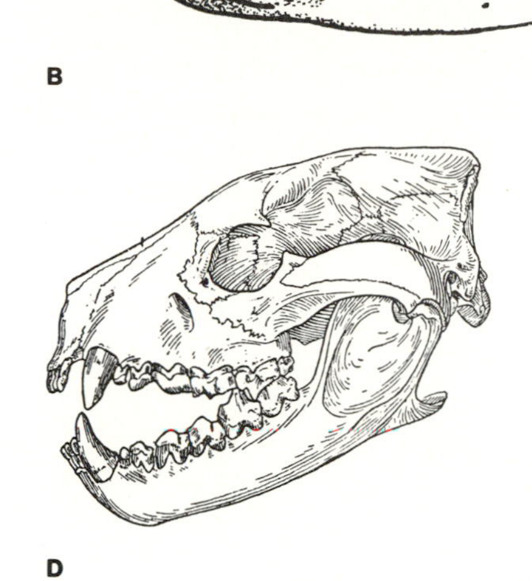

D

Figure 15–14.
Representative Dasyuridae: A, the common pouched "mouse," *Sminthopsis murina*; B, the brush-tailed phascogale, *Phascogale talpoatafa; C, the native "cat," Dasyurus viverrinus.*
(Brazenor 1950:22, 26 and 27)

A

B

C

Taxonomic Remarks

Simpson (1945) included Microbiotheriidae within the family Didelphidae but Kirsch (1977) regarded it as a distinct family on the basis of serology and Reig (Pers. Comm. *in* Kirsch and Calaby 1977) reached similar conclusions on the basis of dentition. Simpson (1945) included Thylacinidae and Myrmecobiidae in Dasyuridae. Ride (1964, 1970) placed Thylacinidae in a distinct family and Kirsch (1977) concluded that it is more closely related to Didelphidae and particularly to the extinct †Borhyaenidae than it is to Dasyuridae. Notoryctidae is frequently listed as a family *incertae sedis* (e.g., Ride 1964) but most authors place it, at least tentatively, near the Dasyuridae.

ORDER PERAMELINA

The bandicoots are terrestrial, primarily insectivorus, Australian marsupials that are frequently compared to rabbits in size and appearance (Fig. 15-15). The bandicoots are usually considered to be phylogenetially intermediate between the Australian Marsupicarnivora and the Diprotodonta. They share the nondiprotodont dentition with the Marsupicarnivora and share syndactylous toes with the Diprotodonta.

Distinguishing Characters

Dental formula 4-5/3, 1/1, 3/3, 4/4 = 46-48. Incisors with flattened, not pointed, crowns and the canines widely spaced from both the last incisor and first premolar. Rostrum elongate and skull conical in general shape. Hind limbs longer than forelimbs. Only medial two or three manual digits well developed and clawed. Lateral digits on manus rudimentary or absent. First pedal digits rudimentary or absent, second and third pedal digits slender and syndactylous, fourth pedal digit largest, and fifth reduced in size but usually functional. Marsupium present, opens to the rear. Stomach simple, moderately sized caecum present. Chorio-allantoic placenta present but unlike eutherian mammals no villi are present. Clavicle rudimentary or absent.

Distribution

Members of the Peramelidae are found throughout Australia, Tasmania, New Guinea and several adjacent islands. The Thylacomyidae are found only in southern and western Australia.

Figure 15–15.
A peramelid, the striped bandicoot, *Perameles bougainville.* Some forms of Peramelidae have shorter ears than this species but the species of Thylacomyidae have much longer ears.
(Brazenor 1950:32)

Table 15-3
Living Families of Peramelina.

Family	Common name	Number of* Genera	Number of* Species	Distribution
Peramelidae	bandicoots	7	16	Australian
Thylacomyidae	rabbit-eared bandicoots	1	2	SW Australian

*Kirsch and Calaby 1977

Fossil History

Peramelidae is known from the late Miocene and Thylacomyidae from the late Pliocene of Australia (Kirsch 1977).

Key to Living Families

1 P^3 larger than P^2; bullae usually quite small (Fig. 15-16); incisors 4/3 or 5/3; ears short to medium in length (Fig. 15-15), usually under 65 mm, never extending beyond muzzle when laid forward; pelage variable but never long and soft; hallux reduced but present and manus with five digits or hallux absent and manus with two large digits and one vestigial digit .... **Peramelidae**
bandicoots

1′ P^3 smaller than or equal to P^2; bullae large and well developed (Fig. 15-17); incisors 5/3; ears long, usually over 70 mm, extending beyond muzzle when laid forward; pelage long and soft over entire body, including tail; hallux absent, manus with five digits .... **Thylacomyidae**
rabbit-eared bandicoots

Comments and Suggestions on Identification

Bandicoot skulls resemble those of some Insectivora but they can be distinguished from these readily by the number of upper incisors and the inflected angular process of the ramus. The skins can resemble large rodents and insectivores or small rabbits but differ from all of these in the presence of the syndactylous digits on the hind feet.

Figure 15–16.
Skull of a peramelid, the eastern striped bandicoot, *Perameles bougainville.*
(Brazenor 1950:105)

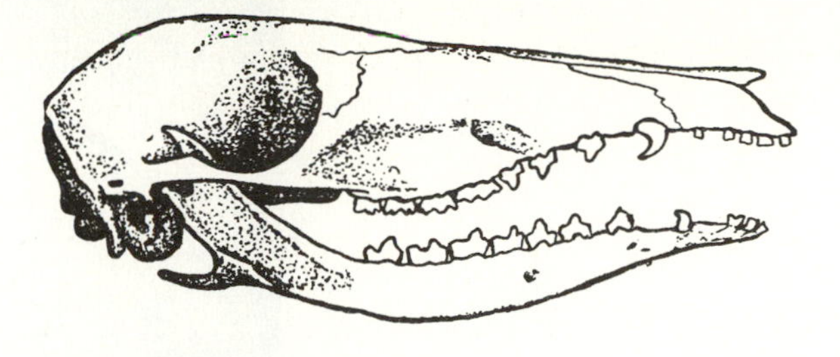

Figure 15–17.
Skull of a thylacomyid, the lesser rabbit-eared bandicoot, *Macrotis leucura.*
(Brazenor 1950:106)

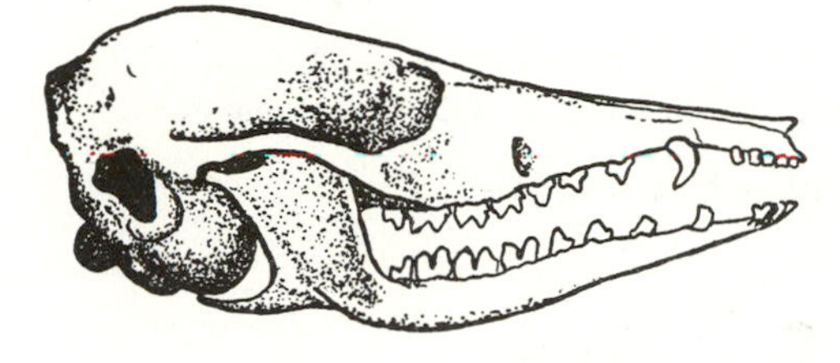

Taxonomic Remarks

Thylacomyidae was recognized as a full family by Kirsch and Calaby (1977). Prior to this the rabbit-eared bandicoots had been included in the family Peramelidae.

ORDER PAUCITUBERCULATA

The caenolestids are small and rather shrew-like in appearance (Fig. 15-18). They are forest dwellers that feed on insects and small vertebrates.

Distinguishing Characters

Diprotodont incisors, usually 4/3 in number (Fig. 15-19). Preorbital vacuity between nasal, maxillary and frontal bones. Mastoid large and broadly exposed lateraly. Marsupium absent. Tail long and haired to the tip. Stomach divided into three distinct parts. Limbs subequal, hind foot didactylous.

Distribution

The only living family is confined to western South America. Two genera, *Caenolestes* and *Lestoros,* live in high, wet, cool forests on the westren slope of the northern Andes. *Rhyncholestes* is known from only a few specimens from Chiloé Island and adjacent coastal areas of central Chile.

Fossil History*

The Caenolestidae are known from the early Oligocene to Recent. †Polydolopidae, known from the late Paleocene to early Oligocene, had a more highly specialized dentition than the living Caenolestidae. Some

*Marshall (pers. comm.)

Figure 15–18.
A caenolestid, *Caenolestes obscurus.*
(Osgood 1921 pl. 1)

Figure 15–19.
Skulls of the three genera of caenolestids, A, *Lestoros inca;* B, *Caenolestes obscurus;* and C, *Rhyncholestes raphanurus.*
(Osgood 1924:173)

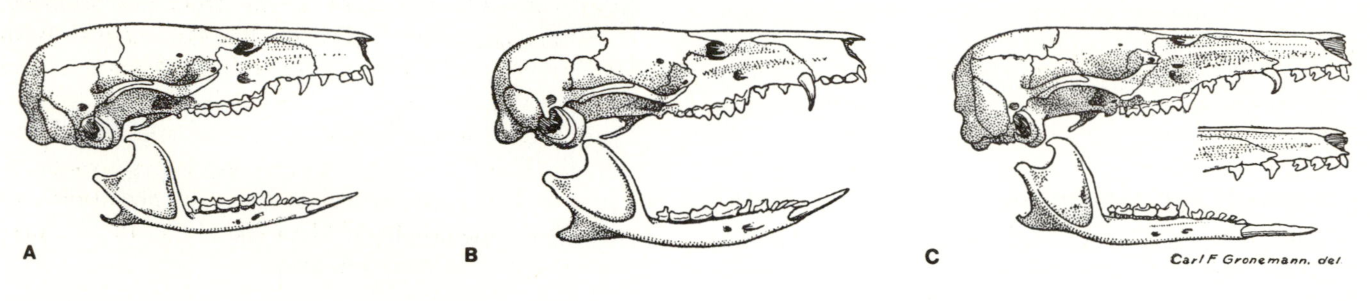

Table 15-4
Living Families of Paucituberculata.

Family	Common name	Number of* Genera	Number of* Species	Distribution
Caenolestidae	shrew opossums or rat opossums	3	7	W Neotropical

*Kirsch and Calaby 1977

of these fossil forms had a large plagiaulacoid cheek tooth (Fig. 15-20).

Figure 15–20.
The lower jaw of a Miocene caenolestid, †*Abderites* from Patagonia, with a plagiaulacoid P_4.
(Weber 1928:80)

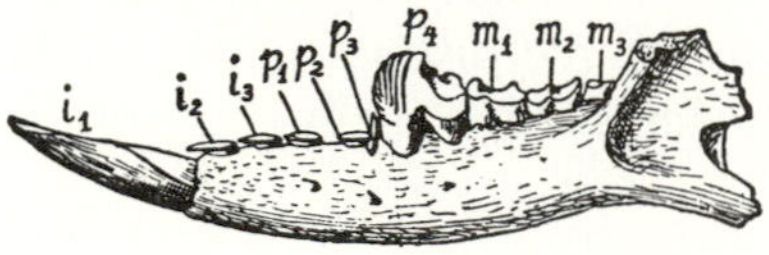

Comments and Suggestions on Identification

The diprotodont dentition separates the caenolestid skull from all except the Diprotodonta. Caenolestids are much smaller than most diprotodonts and differ from them in having four upper incisors. The skins resemble mice or shrews and can be distinguished from these only with practice.

ORDER DIPROTODONTA

The ordinal name refers to the diprotodont dentition this group share with the preceeding order, Paucituberculata. The Diprotodonta includes the primarily herbivorous marsupials of the Australian region. They vary considerably in body size and form, ranging from small mouse-like creatures to large kangaroos. They occupy diverse habitats and include terrestrial, semifossorial and arboreal animals.

Distinguishing Characters

Diprotodont incisors may be 3/2-3, 3/1, 2/1, or 1/1 in number, second and third lower incisors minute when present. Marsupium present, opening anteriorly or posteriorly. Stomach simple in most forms, complex in Macropodidae in which it functions much like that of ruminants. Caecum absent (Tarsipedidae) to very long and complex (Phascolarctidae). Second and third digits of hind foot syndactylous.

Distribution

The Diprotodonta are confined to the Australian faunal Region. The Phalangeridae range throughout the Australian region from Celebes east to the Solomon Islands and south to Tasmania. The Macropodidae and Petauridae are both known from Australia, Tasmania, New Guinea and some adjacent islands. Macropodids also have been introduced into New Zealand and elsewhere in the world. The Burramyidae occur in New Guinea, eastern, southern and western Australia and in Tasmania. Vombatidae is restricted to Tasmania and southeastern and southcentral Australia and Tarsipedidae occurs only in southwestern Australia. Phascolarctidae is found only in eastern Australia but fossils are known from the southwestern portion of the country.

Fossil History

All diprotodontid families except Tarispedidae are known from the Miocene. Tarispedidae is unknown before the Recent. Among the most interesting of the fossil forms known for the extant families are the huge browsing kangaroos of the Pleistocene (see *Procoptodon* in Vaughn 1978:50, Fig. 5-8 or Kirsch 1977:14, Fig. 5-9). Four extinct families of diprotodontids also are recognized (Kirsch 1977). †Wynyardiidae and †Ektopodontidae are known only from the Miocene. †Thylacoleonidae and †Diprotodontidae ranged from Miocene to Pleistocene. The †Diprotodontidae were quadrupedal, herbivorous marsupials (Fig. 15-21) including Pleistocene forms that were as large as a rhinocerous. †*Thylacoleo* was the size and general shape of a lion. It had reduced molars but each posterior premolar was elongated and blade-like (Fig. 15-22), presumably for a carnivorous diet.

Table 15-5
Living Families of Diprotodonta.

Family	Common Name	Number of* Genera	Number of* Species	Distribution
Phalangeridae	phalangers and cuscuses	3	11	Australian
Burramyidae	pigmy phalangers	4	7	Australian
Petauridae	ringtails and gliding phalangers	5	22	Australian
Macropodidae	kangaroos and wallabies	17	56	Australian
Phascolarctidae	koala	1	1	E Australia
Vombatidae	wombats	2	3	SE Australia
Tarsipedidae	honey possum	1	1	SW Australia

*Kirsch and Calaby (1977)

Figure 15–21.
Skull of †*Diprotodon* from the Pleistocene of Australia.
(Flower and Lydekker 1891:171)

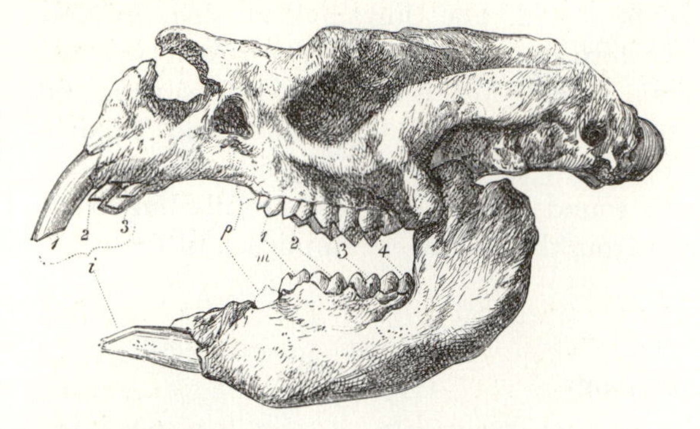

Figure 15–22.
Lateral (A) and anterior (B) views of the skull of †*Thylacoleo*.
(Weber 1928:76)

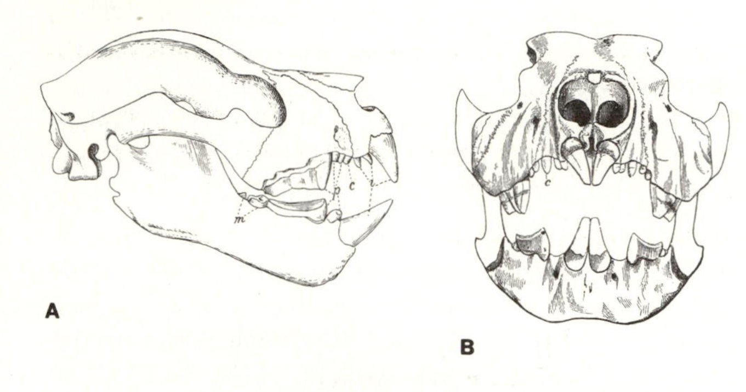

Key to Living Families

1 Dental formula 1/1, 0/0, 1/1, 4/4 = 24; incisors resemble those of rodents (Fig. 15-23); limbs subequal; hallux vestigial; tail vestigial (Fig. 15-23) .... **Vombatidae**
wombats

1′ Dental formula variable but never as above; upper incisors number 2 or 3, not like those of rodents; limbs subequal or with hind limbs longer than forelimbs; hallux absent or well developed, or if vestigial, hind limbs larger than forelimbs; tail long, or if vestigial, then hallux well developed and opposable 2

Figure 15–23.
The common wombat, *Vombatus ursinus*, and a skull of the same species.
(Wombat, Brazenor 1950:45; skull, Gregory 1910)

2 (1′) Incisors 2/1; canines 1/0; cheek teeth never more than 3/3, peglike and vestigial (Fig. 15-24); lower jaw and zygomatic arch slender; tail long and sparsely haired; feet with nails on all digits except syndactylous pedal digits that are clawed (Fig. 15-24); dorsal pelage with three dark longitudinal stripes (Fig. 15-25) .... **Tarsipedidae**
honey possum

2′ Incisors 3/1-3; canines 0-1/0; cheek teeth usually more than 3/3 and never peglike or vestigial; lower jaw and zygomatic arch well developed; tail long and fully haired for at least part of its length, long and naked but scaly, or short and vestigial; only claws present on digits, nails absent; dorsal pelage coloration variable but never with three longitudinal dark stripes 3

Figure 15-24.
Dorsal and ventral views of the right hind foot and lateral and dorsal views of the skull of the honey possum, *Tarsipes spencerae.*
(Cabrera 1919: pl. 13)

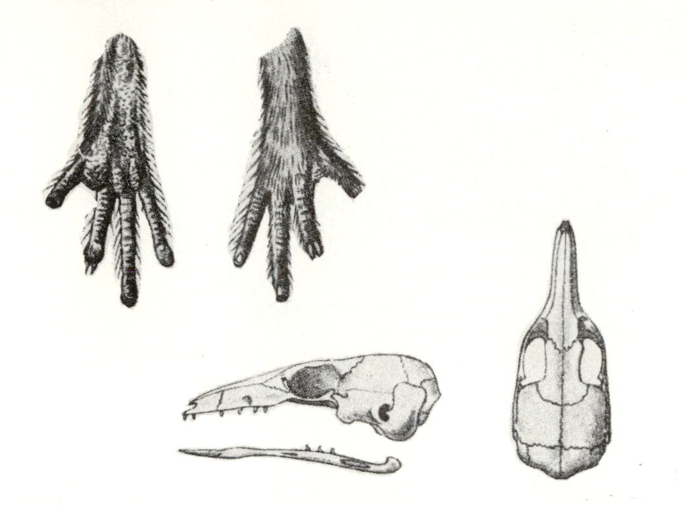

Figure 15-25.
The honey possum, *Tarsipes spencerae.*
(Flower and Lydekker 1891:148)

3 (2′) Dental formula 3/1, 0-1/0, 0-2/0-2, 3-4/3-4 = 20-34; upper incisors approximately equal in size, lower incisor very procumbent; masseteric fossa deep, masseteric canal present but not always obvious (Fig. 15-26); P^1 and P_1, if present, plagiaulacoid (Fig. 15-26A); hind legs much larger than forelegs; tail long and usually tapering from a stout base (Fig. 15-27) .... **Macropodidae**
kangaroos and wallabies

3′ Dental formula variable but if incisors 3/1 then I^1 larger than I^2 and I^3, and I_1 only slightly procumbent (Fig. 15-28); P^1 and P_1 if present, never plagiaulacoid; masseteric fossa shallow, masseteric canal absent (Figs. 15-28, -30, -32 and -34); limbs subequal; tail short and vestigial or long and slender, never tapering from a stout base (Figs. 15-30, 15-31 and 15-32) 4

Figure 15-26.
Skulls of two macropodids. A, a rat kangaroo, *Bettongia,* note the bladelike, plagiaulacoid premolars; B, a wallaby, *Macropus,* note the deep masseteric fossa and the masseteric canal.
(Flower and Lydekker 1891:163-164)

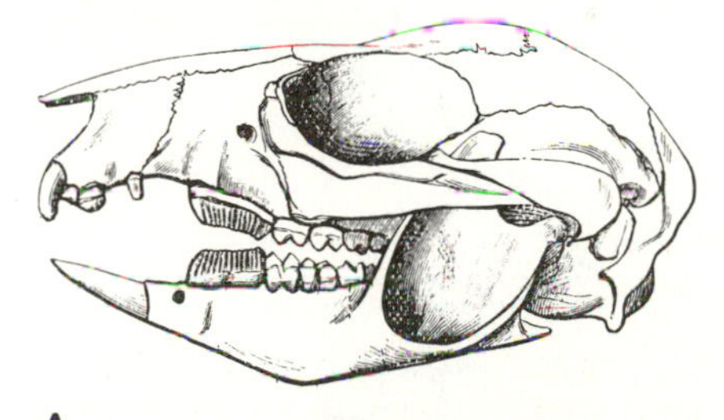

A

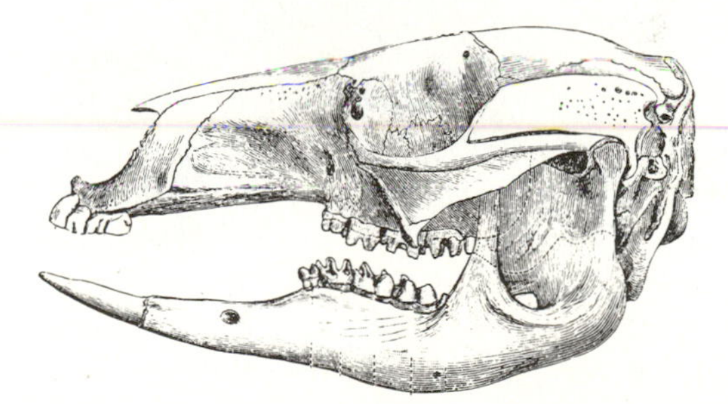

B

Figure 15–27.
Two macropodids of quite different size. A, the great grey kangaroo, *Macropus giganteus,* has a head and body length of up to 140 cm (Grzimek and Heineman 1975) and B, the rat-kangaroo, *Potorous tridactylus,* has a head and body length of about 30 cm.
(A, Flower and Lydekker 1891:160; B, Brazenor 1950:47)

A

B

Figure 15–28.
(A) the koala, *Phascolarctus cinereus,* and (B) a lateral view of its skull.
(Skull, Gregory 1910; koala, Kingsley 1884:30)

A

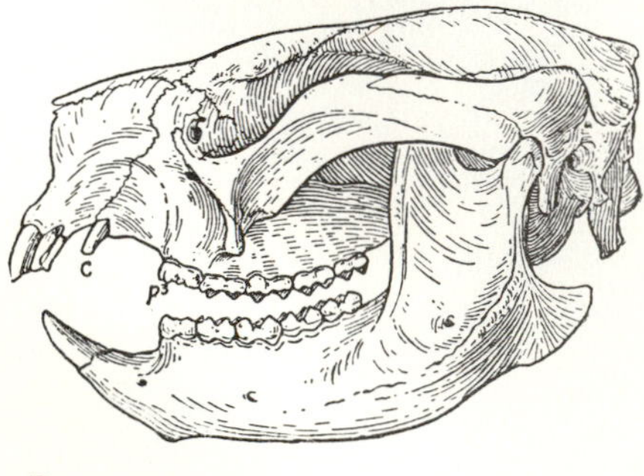

B

4 (3′) Incisors 3/1 (Fig. 15-28); molars with crescent shaped ridges (Vaughn 1978:64, Fig. 5-24B); tail rudimentary; marsupium opens posteriorly .... **Phascolarctidae** koala

4′ Incisors usually 3/2-3, second and third lower incisors tiny or sometimes absent (Figs. 15-29, 33, and 34); molars usually with rounded cusps; tail long (Figs. 15-30, 31, and 32); marsupium opens anteriorly 5

Figure 15–29.
Skull of a burramyid, the pygmy glider, *Acrobates pygmaeus*
(Brazenor 1950:108)

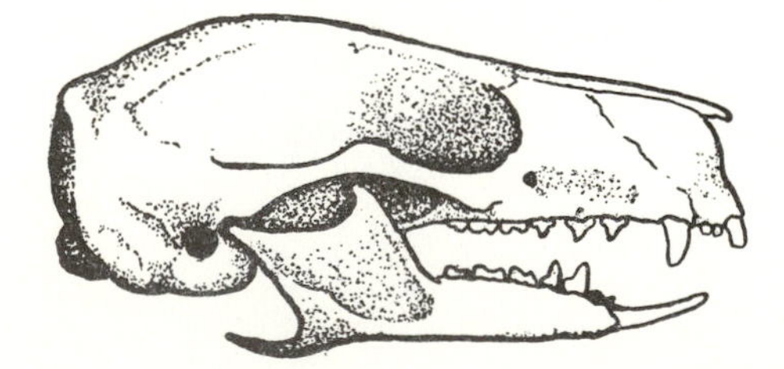

Figure 15–30.
Two representatives of the Burramyidae. A, the pygmy glider, *Acrobates pygmaeus,* and B, the dormouse possum, *Cercartetus manus.*
(Brazenor 1950:34 and 39)

A

B

5 (4′) *Molars 3/3 (Fig. 15-29), 4/4 in one group; mouse sized, total length 120-265 mm; tail either furred at base and naked for remainder of length (Fig. 15-30B) or with lateral fringes of stiff hairs (Fig. 15-30A); gliding membranes may be present (Fig. 15-30A), if present tail with lateral fringes .... **Burramyidae**
pygmy phalangers

5′ Molars 4/4; rat sized or larger, total length at least 270 mm; tail variable but never as above; gliding membranes present or absent 6

*Burramyidae and Petauridae have recently been split from Phalangeridae on the basis of physiological characters. Teeth are variable and descriptions and specimens of several species are rare. These couplets are based upon rather tenuous characters.

Figure 15–31.
Two representatives of the Petauridae. A, the greater glider, *Schoinobates volans,* and B, the ringtail, *Pseudocheirus peregrinus.*
(Brazenor 1950:38 and 41)

A

B

Figure 15–32.
Two representatives of the Phalangeridae. A, the brush-tailed possum, *Trichosurus vulpecula* and B, the spotted cuscus, *Phalanger maculatus.*
(Brazenor 1950:42 and 116)

A

B

6 (5′) *Tail fully furred to tip, or if underside of tip naked then gliding membranes present (Fig. 15-31A); fourth manual digit may be greatly elongated and slender ...... **Petauridae** (in part) gliders and some ringtails

6′ Tail with at least underside of tip naked (Fig. 15-32A), no gliding membranes present; fourth manual digit never as above 7

7 (6′) *Tail naked and scaly or tail mostly furred with tip naked on all sides .... **Phalangeridae** (in part) cuscuses and some phalangers

7′ Tail fully furred except for underside of tip only (Fig. 15-14A) 8

8 (7′) *Upper premolars usually 2 (Fig. 15-33A) .... **Phalangeridae** (in part) some phalangers

8′ Upper premolars usually 3 (Fig. 15-34A) .... **Petauridae** (in part) most ringtails

*Burramyidae and Petauridae have recently been split from Phalangeridae on the basis of physiological characters. Teeth are variable and descriptions and specimens of several species are rare. These couplets are based upon rather tenuous characters.

Figure 15–33.
Skulls of two phalangerids. A, the brush-tailed phalanger *Trichosurus caninus,* and B, the gray cuscus, *Phalanger orientalis.*
(A, Brazenor 1950:111; B, Flower and Lydekker 1891:150)

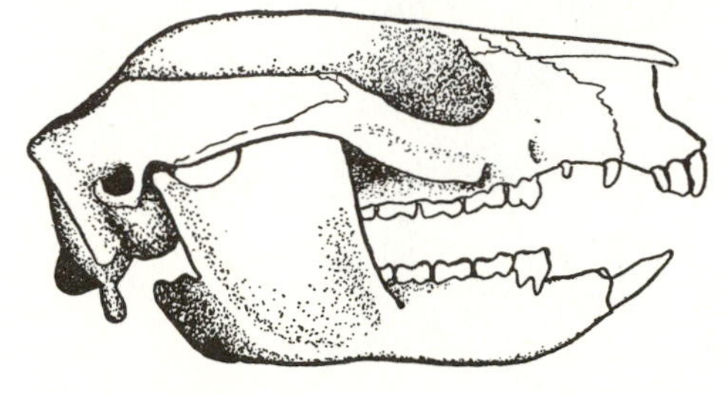

A

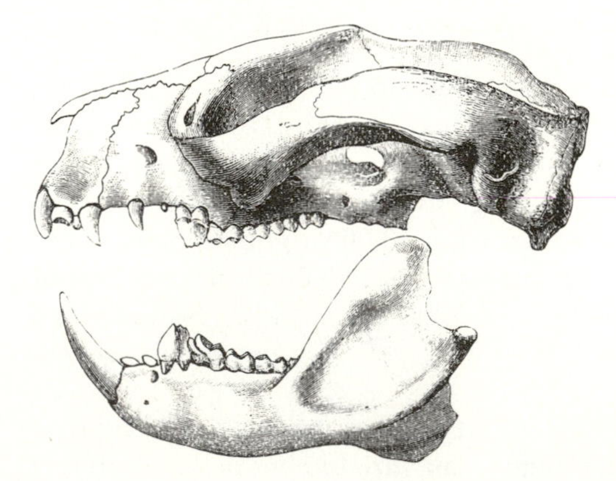

B

Figure 15–34.
Skulls of two petaurids. A, the ringtail *Pseudocheirus peregrinus,* and B, the striped possum, *Dactylopsida trivirgata.*
(Brazenor 1950:110 and 109)

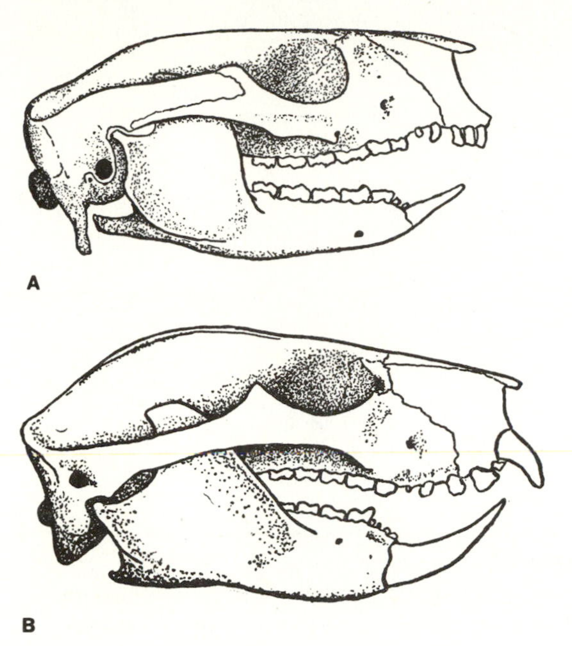

Comments and Suggestions on Identification

The Vombatidae, Phascolarctidae and Tarsipedidae are each distinctive animals, unique from all others in many ways. The wombat and koala skulls could be mistakenly identified as rodents but the inflected angular process of the ramus identifies them as marsupials. The *Tarsipes* skull somewhat resembles that of nectivorous bats but the combination of reduced dentition mal. The Macropodidae and the other three phalan-
The Macropodidae and the other three phalangeroid families can be distinguished from all other orders except Caenolestidae by the procumbent lower incisors. They are usually larger than caenolestids and differ further in the number of upper incisors. The macropodids can be differentiated from the other three phalangeroids by the masseteric fossa and canal.

Taxonomic Remarks

The Vombatidae have alternatively been known as the Phascolomidae (Simpson 1945). These and Macropodidae and Phalangeridae were recognized by Simpson (1945). All other extant families included in this order have been split out of Simpson's Phalangeridae. Ride (1970) divided Petauridae, Burramyidae, Tarsipedidae and Phascolarctidae from the three genera left in Phalangeridae. Kirsch (1977) agreed with this division and gave further evidence. He (Kirsch 1977) also concluded that Phascolarctidae is more closely related to Vombatidae than it is to the phalangeroids.

Supplementary Readings

Brazenor, C. W. 1950. *The mammals of Victoria.* Brown, Prior, Anderson Pty. Ltd., Melbourne. 125 pp.

Collins, L. R. 1973. *Monotremes and marsupials.* Publ. No. 4888. Smithsonian Institution Press, Washington. 323 pp.

Frith, H. J. and J. H. Calaby. 1969. *Kangaroos.* C. Hurst & Co., London. 209 pp.

Gardner, A. L. 1973. The systematics of the genus *Didelphis* (Marsupalia: Didelphidae) in North and Middle America. *Spec. Publ. Museum, Texas Tech* Univ. 4:1-81.

Heineman, D. (and others). 1975. The marsupials, pp. 50-173. *In* B. Grzimek (Ed.). *Animal Life Encyclopaedia, Vol. 10, Mammals I.* Van Nostrand Reinhold Co., New York. 627 pp.

Hunsaker, D. (Ed.). 1977. *The biology of marsupials.* Academic Press, New York. 537 pp.

Kirsch, J. A. W. 1977. The comparative serology of Marsupialia and a classification of marsupials. *Aust. J. Zool.* Supp. Ser. 52:1-152.

Morcombe, M. 1972. *Australian marsupials and other native mammals.* Lonsdowne Press Pty. Ltd., Melbourne. 100 pp.

Ride, W. D. L. 1970. *A guide to the native mammals of Australia.* Oxford University Press, Melbourne. 249 pp.

Serventy, V. and C. Serventy, 1975. *The koala.* E. P. Dutton & Co., New York. 80 pp.

Stonehouse, B. and D. Gilmore (Eds.) 1977. *The biology of marsupials.* University Park Press, Baltimore. 486 pp.

Tyndale-Biscoe, H. 1973. *Life of marsupials.* Edward Arnold Ltd., London. 254 pp.

Wood-Jones, F. 1923-1925. *The mammals of South Australia,* Parts I, II and III. Photo Reprint 1969. Government Printer. Adelaide, 458 pp.

16 The Insectivores

Order Insectivora

ORDER INSECTIVORA

The name Insectivora refers to the diet of most members of this order. Although a diet of insects and other small invertebrates is common among the diverse mammals included in Insectivora, the ordinal name does not indicate a diagnostic character. Some Insectivora are quite omnivorous or carnivorous, while numerous species in other orders have diets that are almost exclusively insectivorous.

Insectivores are generally small, rather primitive mammals. The order has been used as a "wastebasket taxon" into which many kinds of living and fossil mammals of dubious relationship have been placed (Findley 1967; Butler 1972). It is very likely that the order as presently recognized contains two or more groups that deserve ordinal ranking (see Taxonomic Remarks below).

Some insectivores (e.g., the moles) are considered vermin in some parts of their range but none are of major economic importance. One species, *Suncus murinus*, is a commensal with man through much of the Orient. Several forms, such as the solenodons of the West Indies, are very rare or occupy greatly reduced ranges. Most species are terrestrial but fossorial, semiaquatic, and arboreal forms are not uncommon.

Distinguishing Characteristics

No single character or simple combination of characters can be given that will distinguish all insectivores from all other mammals. They are small animals ranging in size from the smallest known mammals, the shrews, *Suncus etruscus* and *Microsorex hoyi*, to species the size of a rabbit. The pelage usually consists of only one kind of hair other than vibrissae. The feet usually are plantigrade. The dentition is generally simple and the cheek teeth are frequently zalambdodont.

Members of the suborder Lipotyphla have olfactory capsules that are longer than the brain and largely interorbital. The maxillae extend into the orbital walls and separate the lacrimal bones from the palatines. The jugals are reduced or absent and the zygomatic arches are sometimes incomplete. Postorbital bars are never present.

Figure 16–1.
Representatives of the three most widespread genera of Soricidae. A, *Sorex araneus;* B, *Suncus murinus;* and C, *Crocidura suaveolens.*
(Hsia et al. 1964:8–11)

Members of the suborder Menotyphla have olfactory capsules that are shorter than the brain and usually do not extend between the orbits. The maxillae do not extend into the medial walls of the orbits and there is no contact between the maxillae and frontals in the orbits. The jugals are well developed and the zygomatic arches are always complete. The postorbital processes of the frontal are well developed and postorbital bars are frequently present.

A baculum has been reported in some insectivores and may exist in others. The testes are never scrotal and the uterus is bicornuate.

Distribution

Insectivores range over most of the world's land surface except for the Australian and southern Neotropical Regions, Antarctica, and most oceanic islands. The European hedgehog, *Erinaceus europeas,* has been introduced into New Zealand, the tenrec, *Tenrec eucaudatus,* has been introduced onto Reunion, Mauritius, and islands in the Comoro and Seychelles groups in the Indian Ocean (Herter 1975:187), and one shrew, *Suncus caeruleus,* has been introduced into New Guinea.

Fossil History**

Insectivora is an ancient group and is generally considered to be ancestral to all other placental orders. Many living insectivores are only distantly related ends of diverse radiations. The late Cretaceous †*Kennalestes* (see Vaughan 1978:75, Fig. 6-1B) may be close to the ancestoral insectivore stock (Kielan-Jaworowska 1975). The most primitive living insectivore family, Erinaceidae, is common from the Oligocene in North America and Eurasia. The Soricidae and Talpidae both date from the upper Eocene and are widely distributed, but the Solenodontidae is unknown from before the Pleistocene and is confined to the West Indies even as fossils.

Modern tenrecs are confined to Madagascar except for the otter shrews of west central Africa, but fossil members of recent Madagascaran genera of Tenrecidae are known from the Miocene of East Africa. Fossil Chrysochloridae also date from the Miocene of East Africa.

The Macroscelididae date from the Oligocene of Africa but no certain fossil Tupaiidae are known (Butler 1972:262).

**Based upon Romer (1966) and Butler (1972).

Table 16-1
Living Families of Insectivora.

Family	Common Name	Number of* Genera	Number of* Species	Distribution
Suborder Lipotyphla				
Erinaceidae	hedgehogs and moon rats	10	14	Ethiopian, Palearctic, Oriental
Talpidae	moles and desmans	15	*ca.* 22	Holarctic, Oriental
Tenrecidae	tenrecs and otter shrews	15	32	Madagascar (tenrecs) and west central Africa (otter shrews)
Chrysochloridae	golden moles	7	18	Central and southern Africa
Solenodontidae	solenodons	2	2	Cuba, Hispaniola
Soricidae	shrews	23	*ca.* 291	Holarctic, Ethiopian, Oriental, northern Neotropical
Suborder Menotyphla				
Macroscelididae	elephant shrews	4	15	Africa
Tupaiidae	tree shrews	5	15	Oriental

*Meester and Stezer (1971) for Tenrecidae, Chrysochloridae and Macroscelididae; Diersing (1980) and Findley (1967) for Soricidae; Findley (1967), for other families.

Key to Living Families

1 I^1 large, protruding forward and hooked, small cusp present behind main cusp (Fig. 16-2); teeth may or may not be pigmented; small animals with short dense fur (16-1) .... **Soricidae**
shrews

1′ I^1 may be large, but if hooked, no accessory cusp present; teeth never pigmented; size and pelage variable, frequently spiny 2

Figure 16–2.
Skulls of a pigmented—toothed shrew, A, *Sorex araneus* and an unpigmented—toothed shrew; B, *Crocidura sauveolens.*
(Stroganov 1957:117 and 244)

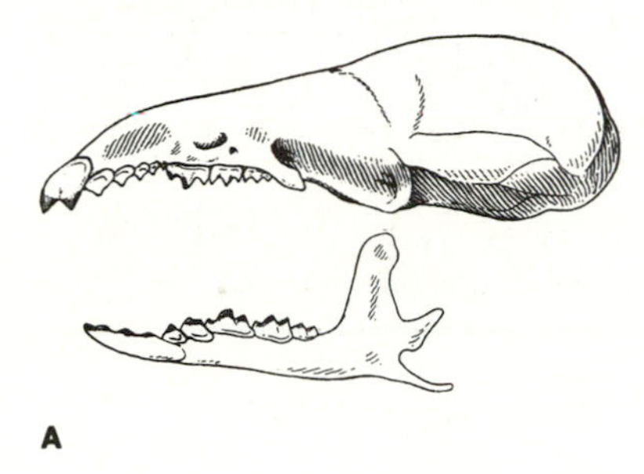

A

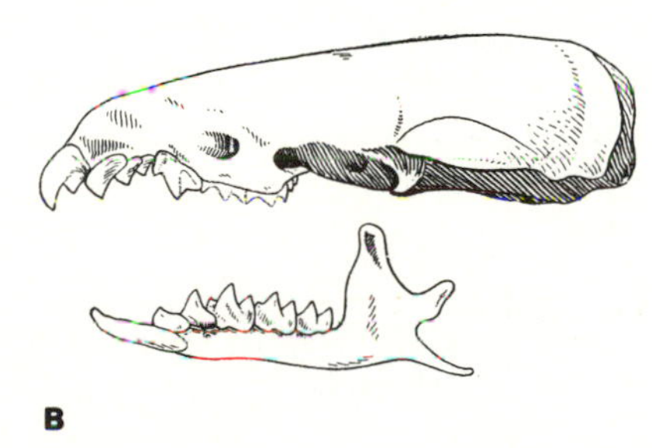

B

2 (1′) Zygomatic arch incomplete 3

2′ Zygomatic arch complete 4

3 (2) Dental formula 3/3 1/1 3/3 3/3 = 40; I^1 large (Fig. 16-3); I_2 large with longitudinal, lingual groove (Fig. 16-4); pelage long and lax; tail long and scaly (Fig. 16-5) .... **Solenodontidae**
solenodons

3′ Dental formula variable; I^1 small or large; I_2 may be large but never grooved (Fig. 16-6); pelage variable, usually very short or with spines (Fig. 16-7); tail variable ...... **Tenrecidae**
tenrecs and otter shrews

Figure 16–4.
A lingual view of the anterior portion of the mandible of a *Solenodon.*
(John D. Whitesell)

Figure 16–3.
Dorsal (A) and lateral (B) views of the skull of a solenodon, *Solenodon paradoxus.*
(Dorsal, Cabrera 1925: pl. 14; lateral, John D. Whitesell)

A

B

Figure 16–5.
The Cuban solenodon, *Solenodon paradoxus.*
(Flower and Lydekker 1891:636)

4 (2′) Dental formula 2/3 1/1 3/3 3/3 = 38; zygomatic arch usually perforated; postorbital bar present (Fig. 16-8); resembles a long-nosed, arboreal squirrel (Fig. 16-9) .... **Tupaiidae**
tree shrews

4′ Dental formula variable; zygomatic arch never perforated; no postorbital bar; body form variable but never squirrel-like 5

5 (4′) Crowns of upper molars quadrate (Figs. 6-10 and 6-12); eyes and pinnae large (Figs. 6-11 and 6-13); pelage may include spines (Fig. 6-13) 6

5′ Crowns of upper molars triangular (zalambdodont) (Fig. 16-14) or with cusps in a W-shaped pattern (dilambdodont) (Figs. 16-16, and 16-17); eyes and pinnae small or absent (Figs. 16-15, and 16-18); pelage never spiny 7

Figure 16–6.
Skulls of four tenrecids illustrating some of the diversity that occurs in this family. A, *Tenrec eucaudatus;* B, *Setifer setosus;* C, *Hemicentetes semispinosus;* D, *Potomogale velox.* Not all to same scale.
(Cabrera 1925: pl. 16 and 17)

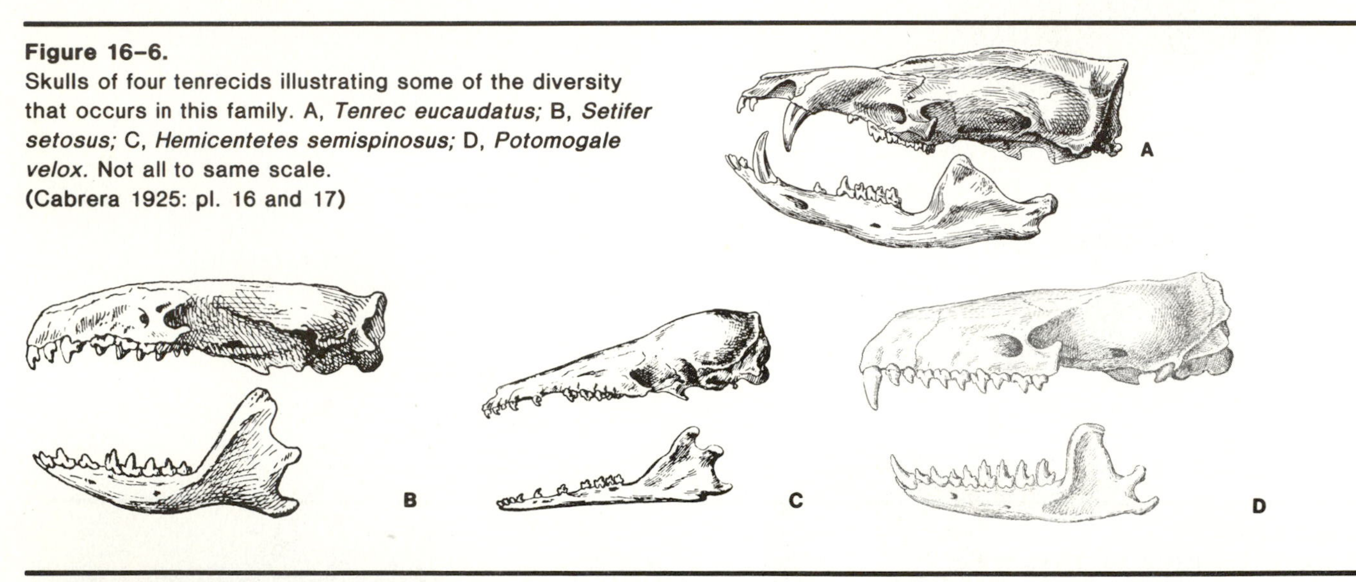

Figure 16–7.
Representative tenrecids. A, an otter shrew, *Potomogale velox* and B, a tenrec, *Tenrec eucaudatus.*
(Cabrera 1925: pl. 16 and 17)

Figure 16–8.
Skulls of two tupaiids. A, common tree shrew, *Tupaia glis* and B, the Indian tree shrew, *Anathana ellioti.*
(Cabrera 1925: pl. 1)

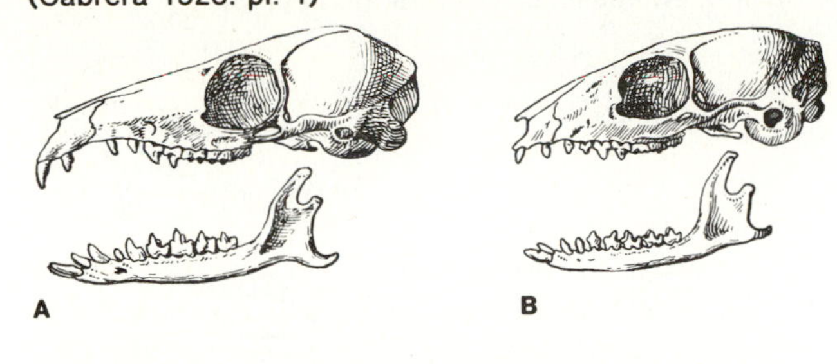

Figure 16–9.
Representative tree shrews. A, the pentail tree shrew, *Ptilocercus lowii* and B, the large tree shrew, *Tana tana.*
(Cabrera 1925: pl. 1 and 2)

Figure 16–10.
Skull of an elephant shrew, *Elephantulus rozeti.*
(Cabrera 1932:38)

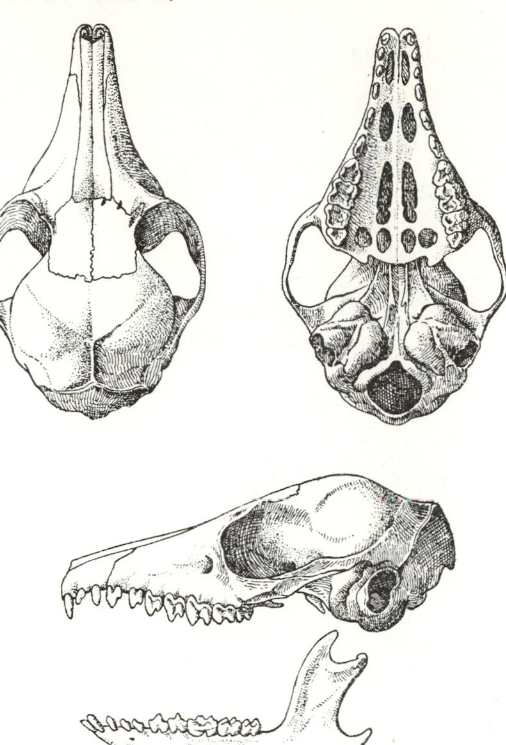

Figure 16–11.
An elephant shrew, *Elephantulus fuscipes.*
(Dekeyser 1955:98)

6 (5) Auditory bullae complete (Fig. 16-10); hind limbs much longer than forelimbs (Fig. 16-11); spines never present .... **Macroscelididae**
elephant shrews

6′ Auditory bullae incomplete (Fig. 16-12); limbs subequal; spines usually present (Fig. 16-13) .... **Erinaceidae**
hedgehogs and moon "rats"

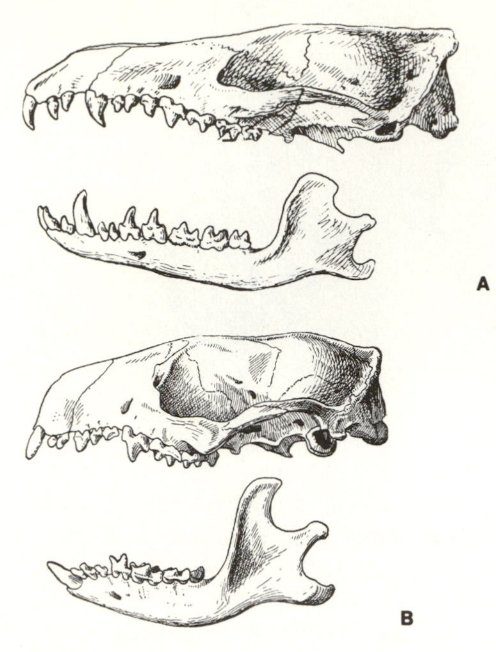

Figure 16–12.
Skulls of representative erinaceids. A, a moon rat, *Echinosorex gymnurus;* and B, a hedgehog, *Erinaceus algirus.*
(Cabrera 1925: pl. 6 and 7)

A

B

Figure 16–13.
Representative erinaceids. A, a hedgehog, *Erinaceus algirus* and B, a moon rat *Hylomys suillus.*
(Cabrera 1925: pl. 6 and 7)

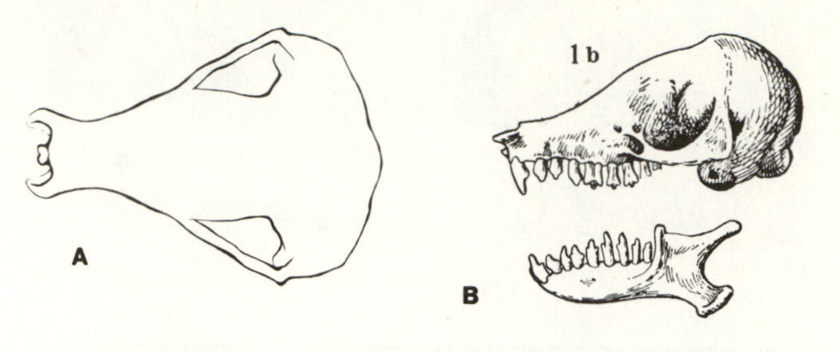

Figure 16–14.
Skull of a golden mole, *Chrysochloris asiatica.* A, diagrammatic dorsal view, B, lateral view.
(A, John D. Whitesell; B, Cabrera 1925: pl. 5)

Figure 16–15.
A golden mole, *Chrysochloris leucorhina.*
(Dekeyser 1955:95)

7 (5′) Upper molars zalambdodont (Fig. 16-14B); dental formula 3/3 1/1 3/3 2/2 or 3/3 = 36 or 40; I^1 enlarged; skull a short cone (Fig. 16-10A); zygomatic arches broad; forefoot with four digits, two central claws much larger than others; tail rudimentary (Fig. 16-15) .... **Chrysochloridae**
golden moles

7′ Upper molars dilambdodont (Fig. 16-16); dental formula variable; I^1 may or may not be enlarged (Fig. 16-17); skull generally long and conical; zygomatic arches generally weak; forefoot with five digits; claws on digits two and three not particularly larger than others; tail length variable (Fig. 16-18) .... **Talpidae**
moles and desmans

Comments and Suggestions on Identification

Unfortunately, there is no single good characteristic distinguishing all insectivores from all other mammals. Familiarize yourself with the general appearance of each family and the other groups with which it might be confused. Only the Tenrecidae has great diversity within the family, and except for the subfamily Pota-

Figure 16–16.
Skull and upper (left) and lower (right) tooth rows of the European mole, *Talpa europea*.
(Cabrera 1925: pl. 9)

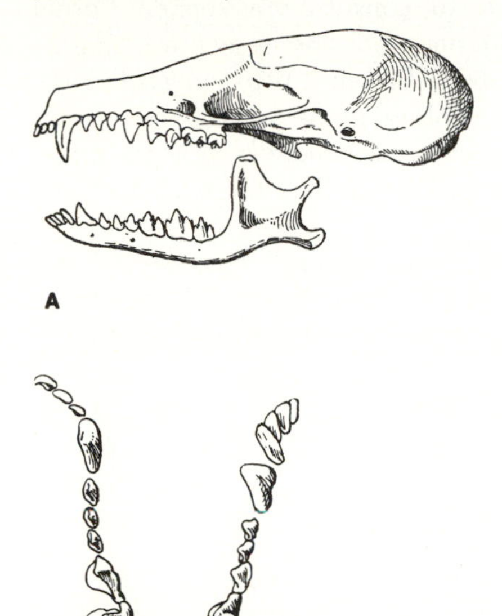

Figure 16–17.
Skull of a desman, *Desmana moschata*, Talpidae.
(Gromov et al. 1963:67)

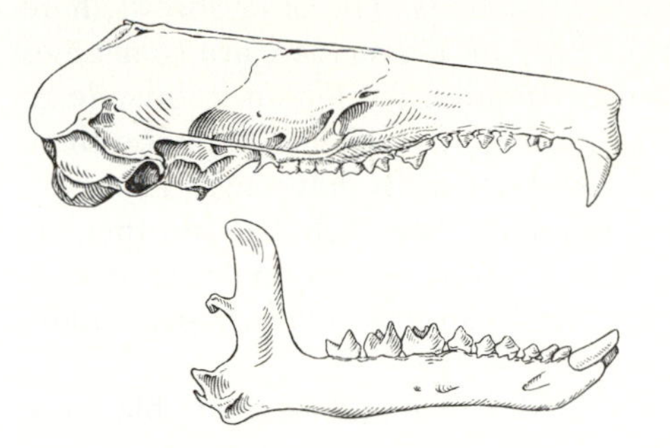

mogalinae, the otter shrews, tenrecids are confined to Madagascar. Also keep in mind that:

1. Moles, golden moles and marsupial "moles" resemble each other. Check and compare forefeet, skulls, and localities collected.
2. Hedgehogs, tenrecs, and echidnas resemble each other. Check skulls and localities. There are no hedgehogs in Madagascar and spiny Tenrecidae are confined to that island.
3. Certain shrews and certain moles resemble each other. Check the skulls for the distinctive first upper incisor of shrews.

Figure 16–18.
Representative species of Talpidae. A, a mole, *Scalopus aquaticus* and B, a desman, *Desmana moschata*.
(A, Kingsley 1884:152; B, Flower and Lydekker 1891:629)

Taxonomic Remarks

Butler (1972:253) has reviewed the frequent changes in the classification of the insectivores. At the family level there are few arguments. The otter shrews, here included as a subfamily of Tenrecidae, are sometimes regarded as a distinct family, the Potamogalidae (e.g., Walker et al., 1975 and Herter 1975). Otherwise there is general agreement on family groupings. However, various families frequently are included in other orders, are split off as separate orders, or are variously organized as taxa between the ordinal and family levels.

The suborders Lipotyphla and Menotyphla have frequently been regarded as distinct orders since Gregory (1910) first divided them. The tree shrews, Tupaiidae, resemble primitive Primates and frequently have been included in that order (Simpson 1945, Napier and Napier 1967, Chiarelli 1972, and Walker et al. 1975). Campbell (1974) reviewed the evidence that the Tupaiidae are not closely related to primates. Butler (1972:263) regarded them to be a separate order, Scandentia. The elephant shrews, Macroscelididae, are sometimes regarded to constitute a separate order sometimes called Menotyphla and sometimes Macroscelidea (Butler 1956). Broom (1916) regarded the golden moles, Chrysochloridae, to be a distinct order, Chrysochlorida, but otherwise there is general agreement that the lipotyphlan families listed above constitute a single order.

The colugos or "flying lemurs," Cynocephalidae, are closely related to the Insectivora and sometimes are considered to be members of this order (e.g., Romer 1966).

The arrangement of families into suborders given in Table 16-1 follows Findley (1967). Table 16-2 gives three of the many alternative arrangements of the living families:

Table 16-2
Classifications of Insectivora.

Simpson (1945)	Romer (1966)	Butler (1972)
Order Insectivora	Order Insectivora	
	Suborder Lipotyphla	Order Lipotyphla
Superfamily Erinaceoidea	Superfamily Erinaceoidea	Suborder Erinaceomorpha
Erinaceidae	Erinaceidae	Erinaceidae
	Talpidae	
Superfamily Soricoidea	Superfamily Soricoidea	Suborder Soriciomorpha
Soricidae	Soricidae	Soricidae
Talpidae		Talpidae
	?Solenodontidae	Solenodontidae
	?Suborder Zalambdodonta	
Superfamily Tenrecoidea	Superfamily Tenrecoidea	Suborder Tenrecomorpha
Solenodontidae		
Tenrecidae	Tenrecidae	Tenrecidae
Potamogalidae	[including Potamogalidae]	[including Potamogalidae]
Superfamily Chrysochloroidea	Superfamily Chrysochloroidea	Suborder Chrysochloroidea
Chrysochloridae	Chrysochloridae	Chrysochloridae
Superfamily Macroscelidoidea	Suborder Macroscelidea	Order Macroscelidea
Macroscelididae	Macroscelididae	Macroscelididae
Order Primates		
Suborder Prosimii	Suborder Protoeutheria	
Superfamily Tupaioidea	Superfamily Leptictoidea	Order Scandentia
Tupaiidae	Tupaiidae	Tupaiidae
Order Dermoptera	Suborder Dermoptera	Order Dermoptera
	Superfamily Plagiomenoidea	
Cynocephalidae	Galeopithecidae [=Cynocephalidae]	

Supplementary Readings

Allen, G. M. 1910. *Solenodon paradoxus. Mem. Mus. Comp. Zool. Harvard* 40:1-54 + 9 plates.

Butler, P. M. 1972. The problem of insectivore classification, pp. 253-265. *In* K. A. Joysey and T. S. Kemp (Eds.). *Studies in vertebrate evolution.* Oliver and Boyd, Edinburgh.

Campbell, C. B. G. 1974. On the phyletic relationships of the tree shrews. *Mammal Review* 4:125-143.

Cozens, A. B. 1950. The otter shrew. *Nigerian Field* 15: 76-83.

Crowcroft, P. 1957. *The life of the shrew.* Max Reinhardt, London: 166 pp.

Eisenberg, J. F. and E. Gould. 1970. The tenrecs: a study in mammalian behavior and evolution. *Smithsonian Contrib. Zool.* 27:1-137.

Evans, F. G. 1942. The osteology and relationships of the elephant shrews (Macroscelididae). *Bull. Amer. Mus. Nat. Hist.* 80:85-125.

Godfrey, G. and P. Crowcroft. 1960. *The life of the mole.* Museum Press, Ltd., London. 152 pp.

Herter, K. 1965. *Hedgehogs.* Phoenix House, London. 69 pp.

Herter, K. 1975. The insectivores, pp. 176-257. *In* B. Grzimek (Ed.). *Animal life encyclopedia, Vol. 10, Mammals I.* Van Nostrand Reinhold, New York.

Mellanby, K. 1973. *The mole.* Taplinger Publishing Co. Inc., New York. 159 pp.

Pearson, O. P. 1954. Shrews. *Scientific American* 191 (2):66-70.

Van Valen, L. 1965. Tree shrews, primates and fossils. *Evolution* 19:137-151.

17 The Colugos

Order Dermoptera

ORDER DERMOPTERA

The ordinal name, Dermoptera, means literally "skin-winged" and refers to the extensive patagia of these animals. Dermopterans are known commonly as **colugos** or "flying lemurs," but "flying" is inaccurate since these animals glide but do not fly, and "lemur" is incorrect since true lemurs are in the order Primates. Colugo is the preferred common name.

Colugos are herbivorous, feeding upon the young leaves, buds, and fruits of the trees in which they live. Adept gliders, they can cover great distances with little loss of altitude. When moving about in trees they hang from their large curved claws and move in a slow, suspended quadrupedal fashion similar to that of sloths.

In some parts of their range colugos are used for meat and hides, but they are of little economic importance. No zoo has yet kept one alive for more than a few months (Schultze-Westrum 1975:66).

Distinguishing Characteristics

The dental formula is 2/3 1/1 2/2 3/3 = 34. The first upper incisors are small and separated medially by a wide diastema (Fig. 17-2). The second upper incisors are caniniform. The first two lower incisors are pectinate (Fig. 17-3), each with from five to twenty long slender cusps resembling the teeth of a comb. The third lower incisor has five or six cusps that decrease in length posteriorly. The canines are incisiform and the cheek teeth are brachydont. The skull is broad and dorsoventrally flattened.

Well-developed and completely furred patagia extend from the sides of the neck, to the manual phalanges, to the pedal phalanges, and to the tail (Fig. 8-17). The caecum is large. The uterus is bicornuate and testes are either scrotal or in inguinal pouches. A baculum has not been reported.

Figure 17–1.
A colugo, *Cynocephalus.*
(Ognev 1951)

Figure 17–2.
Ventral and dorsal views of the skull of a colugo, *Cynocephalus volans.*
(Giebel 1859:94)

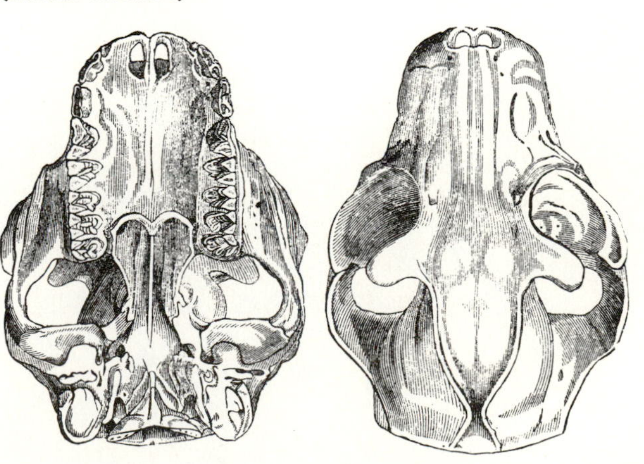

Table 17-1
Living Families of Dermoptera.

Family	Common Name	Number of Genera	Number of Species	Distribution
Cynocephalidae	colugos	1	2	Oriental

Figure 17–3.
The lower incisors of a colugo, *Cynocephalus variegatus.* (Cabrera 1925)

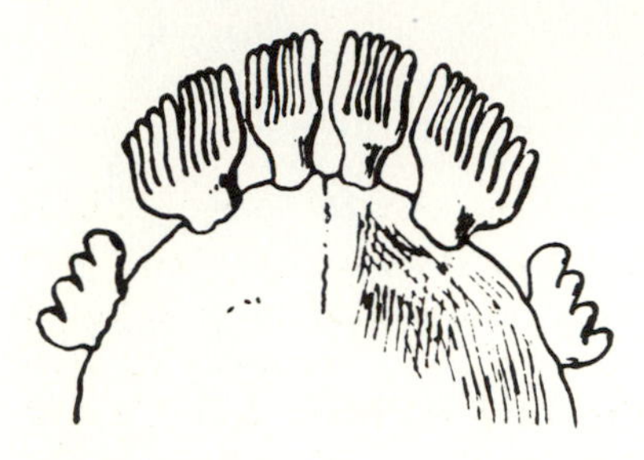

Distribution

Living dermopterans are confined to the Oriental Region, ranging through tropical forests of southern Thailand and Vietnam, Malaysia, Indonesia, and the southernmost Philippine Islands.

Fossil History

Dermopterans are poorly known as fossils. Some fragmentary specimens from the late Paleocene and Lower Eocene of North America have been tentatively assigned to this order (Romer 1966).

Comments and Suggestions on Identification

The colugo skin and skull are each distinctive and not easily confused with those of any other mammal. Certain lemurs have pectinate lower incisors, but in these primates each tooth of the "comb" is a single incisor while in colugos each tooth of the comb is only a single cusp of a large incisor. Several gliding marsupials and rodents have patagia between fore and hind limbs, but only colugos have these folds of skin extending to the sides of the neck and tail.

Taxonomic Remarks

Colugos have, in the past, been included with the bats in the order Chiroptera, with the lemurs in the order Primates, and with the insectivores in the order Insectivora (see Taxonomic Remarks section, chapter 15), but most authorities now recognize these unique mammals as comprising a distinct order. The family name Galeopithecidae was formerly used for the colugos (Simpson 1945:179).

Supplementary Readings

Lim Boo Liat. 1967. Observations on the food habits and ecological habitat of the Malaysia flying lemur *Cynocephalus variegatus. Int. Zoo. Yearbook* 7:196-197.

Schultze-Westrum, T. 1975. Colugos or flying lemurs, pp. 64-66. *In* B. Grzimek (Ed.) *Animal Life Encyclopedia Vol. 11, Mammals II.* Van Nostrand Reinhold Co., New York.

Wharton, C. H. 1950. Notes on the life history of the flying lemur. *J. Mammal.* 31:269-273.

18 The Bats
Order Chiroptera

ORDER CHIROPTERA

The ordinal name Chiroptera, meaning literally "hand winged," refers to the bat's most distinctive characteristic. Bats are the only mammals that have their forelimbs modified as wings and are the only mammals capable of sustained flight.

Most bats are insectivorous. Many feed upon flying insects that they capture in the air. Others glean insects from foliage, from the ground or from the surface of water. In tropical and semitropical areas insectivorous bats exist together with species specialized for diets of fruit, nectar, pollen, fish, small vertebrates, or blood (Glass 1970, Gardner 1977). The three families of bats that range widely into temperate regions, Rhinolophidae, Vespertilionidae, and Molossidae, are insectivorous. Since their food supply is generally unavailable in winter, these temperate species either hibernate (most Rhinolophidae and Vespertilionidae) (Davis 1970) or migrate seasonally to warmer climates (Molossidae, some Rhinolophidae and Vespertilionidae) (Griffin 1970).

Most bats are nocturnal and most navigate with the aid of echolocation. Ultrasonic sounds are emitted through the nose or mouth and the reflected sounds are received by the ears. The detailed mechanisms of this are highly varied (Henson 1970, Novick 1977). Many groups have simple to elaborate fleshy flaps, ridges or other projections associated with the nose, the ears, or both. These structures are believed to aid in echolocation (Novick 1977).

Constantine (1970) has provided a detailed review of the way in which bats affect man. Many nectar and pollen feeding bats provide service in pollination of flowers and many fruit bats aid in seed dispersal. Deposits of bat guano (feces) in caves are, or have been, mined in many parts of the world for use as fertilizer and insectivorous bats in temperate regions play an important role in insect control. Several species of bats are known to be carriers of diseases infectious and sometimes fatal to man and his domestic animals. These diseases include rabies, which bats frequently survive (Turner 1975). Vampire bats in South and Central America are responsible for the deaths of many domestic animals from disease and from loss of blood (Wimsatt and Guerriere 1962). Many species of bats are popular laboratory animals for medical research and for research on echolocation (Rasweiler 1977).

Distinguishing Characteristics

The forelimb is modified to form a wing. The metacarpals and phalanges of digits two through five are elongate and enclosed in thin webs of skin (Fig. 18-1). The **propatagium** extends from shoulder to the wrist anterior to the upper arm (brachium) and forearm (antebrachium). The **dactylopatagia** fill the spaces between digits with the *dactylopatagium minus* between digits 2 and 3, the *dactylopatagium longus* between digits 3 and 4, and the *dactylopatagium latus* between digits 4 and 5. The **plagiopatagium** extends from digit five to the side of the body and the hind leg. The **uropatagium** extends between the hind legs and usually incorporates the tail. A cartilaginous spur, termed the **calcar,** is sometimes present. It extends medially from the ankle region of each hind foot and helps support the uropatagium. The degree of development of the tail and uropatagium and the relationships between these two structures are variable (Fig. 18-3). The first manual digit is clawed, the second bears a claw only in some Pteropodidae, and the remainder are always clawless. The pedal digits all bear claws. The clavicle is well developed and the sternum is frequently keeled.

Figure 18–1.
The wing and tail structure of a bat. Numbers indicate digits one through five. pro, propatagium; dm, dactylopatagium minus; dlo, dactylopatagium longus; dla, dactylopatagium latus; pla, plagiopatagium; uro, uropatagium; and c, calcar.
(Modified from Ognev 1951 and Quay 1970)

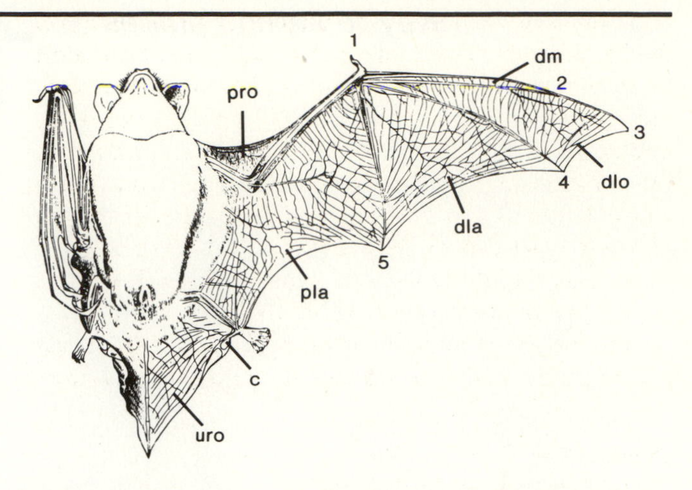

Figure 18–2.
Diagrams of the various tail and uropatagium arrangements found in bats. A, and B, tail enclosed in uropatagium; C, tail extending beyond posterior margin of uropatagium; D, tail protruding from dorsal surface of uropatagium; E, tail absent, uropatagium short; F, tail and uropatagium essentially absent.
(R. Roesener)

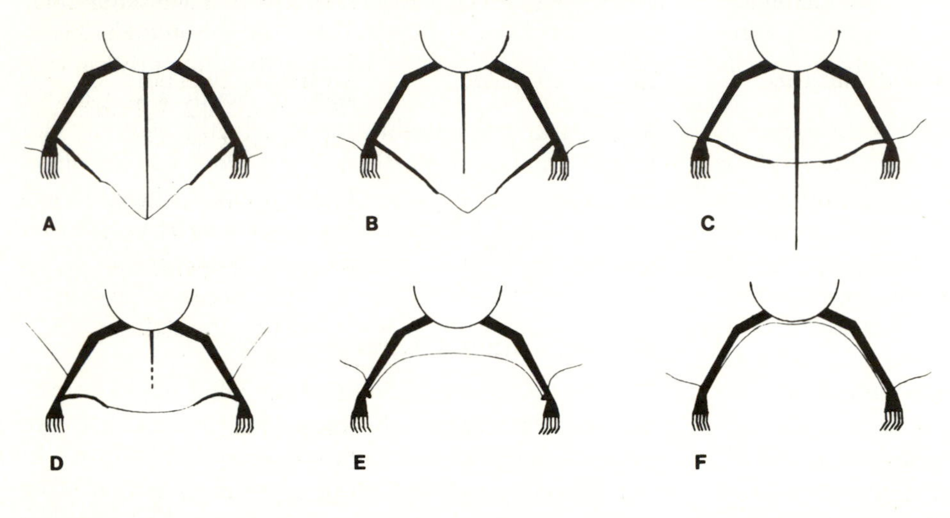

Figure 18–3.
The ears of some representative Microchiropera (A through D) and Megachiroptera (E). A, B and C, all vespertilionids, show some of the many variations in tragus size and shape. D, a rhinolophid, lacks a tragus but has the antitragus enlarged. E, a pteropodid, has a continuous inner ear margin and completely lacks both tragus and antitragus. Not all to same scale.
(A–D, Barrett-Hamilton 1910; E, R. Roesener)

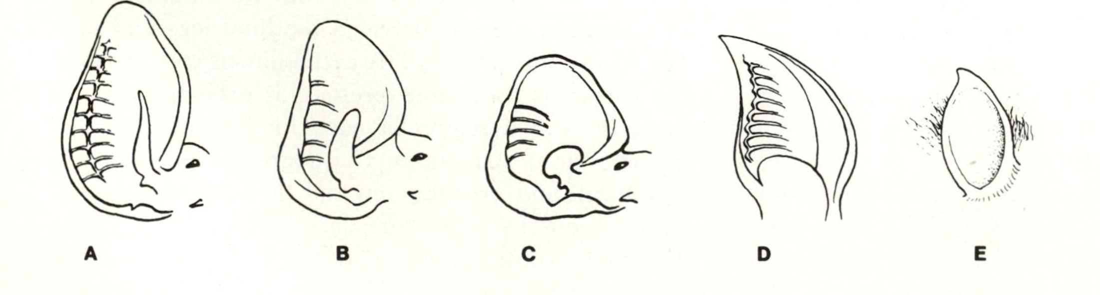

The **tragus,** a fleshy lobe projecting from the lower medial corner of the ear pinna, is present in most microchiropterans (Fig. 18-3A to 3C) but is absent in the megachiropterans (Fig. 18-3E). A flap, termed the **antitragus,** on the lower edge of the outer margin of the ear pinna is particularly well developed in microchiropterans that lack a tragus (Fig. 18-3D). The noses of many microchiropteran bats are ornamented with nose leaves that vary from simple flaps of skin to highly complex structures (Fig. 18-4).

The uterus is bicornuate (simplex in Phyllostomatidae) and in many species only the right ovary is functional (Carter 1970:235). The penis is pendant and a baculum usually is present. The testes are abdominal and in some species descend seasonally through the inguinal canal in the abdominal wall (Carter 1970:234).

Distribution

Bats occur on most of the earth's land surface except the permanently cold polar regions and a few small, oceanic islands. Even such remote islands as New Zealand and the Hawaiian Islands have endemic bats.

Table 18-1
Living Families of Chiroptera.

Family	Common Name	Number of* Genera	Number of* Species	Distribution
Suborder Megachiroptera				
Pteropodidae	Old World fruit bats	38	*ca.* 150	Australian, Oriental, Ethiopian, south Palearctic, some oceanic islands
Suborder Microchiroptera				
Rhinopomatidae	mouse-tailed bats	1	3	North Ethiopian, south Palearctic, west Oriental
Emballonuridae	sac-winged bats, sheath-tailed bats	12	44	North Neotropical, Ethiopian, south Palearctic, Oriental, Australian
Craseonycteridae	bumble bee bat	1	1	Thailand
Noctilionidae	bulldog bats	1	2	Neotropical
Mormoopidae	leaf-chinned bats	2	8	Neotropical
Nycteridae	hispid bats, hollow-faced bats	1	13	Ethiopian, Oriental
Megadermatidae	false vampire bats	4	5	Ethiopian, Oriental, Australian
Rhinolophidae	horseshoe bats	11	*ca.* 128	Most of Eastern Hemisphere
Phyllostomatidae	New World leaf-nosed bats, vampire bats	49	*ca.* 137	Neotropical, south Nearctic
Natalidae	funnel-eared bats	1	4	Neotropical
Furipteridae	smoky bats	2	2	Neotropical
Thyropteridae	disk-winged bats	1	2	Neotropical
Myzopodidae	sucker-footed bat	1	1	Madagascar
Vespertilionidae	common bats	34	*ca.* 283	Worldwide
Mystacinidae	short-tailed bat	1	1	New Zealand
Molossidae	mastiff bats	14	*ca.* 82	All Regions except northern Holarctic

*Rhinopomatidae (Hill 1977), Craseonycteridae (Hill 1974), Mormoopidae (Smith 1972), Phyllostomatidae (Jones and Carter 1976), and Molossidae (Freeman 1977). All others Koopman and Jones 1970.

Figure 18–4.
Bat heads showing varying degrees of development of flaps and projections on the nose. A, plain nose of a *Pipistrellus* B, simple leaf of a *Rhinopoma* and C, the complex noseleaf of *Rhinolophus*. Not all to same scale. (A, Kuzyakin 1950; B, Harrison 1964, used with permission; C, Gromov et al. 1963)

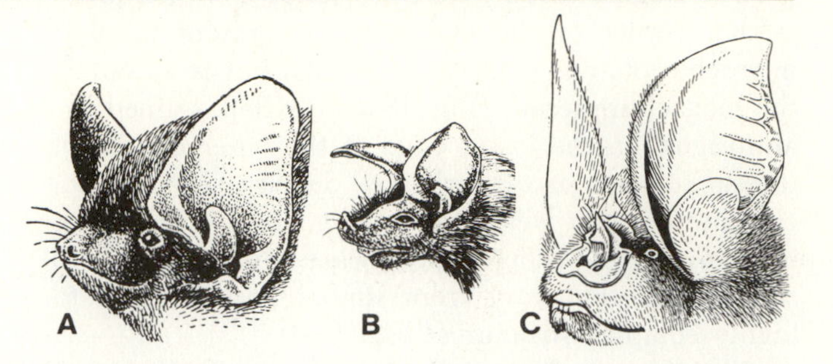

Figure 18–5.
Skulls of representative Pteropodidae: A, *Pteropus lepidus;* B, *Nyctimene major;* C, *Macroglossus minimus.* (A, Miller 1907:58; B and C, Andersen 1912:684, 748)

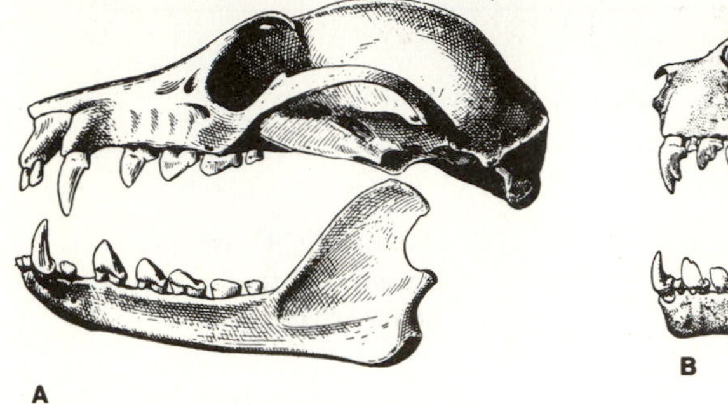

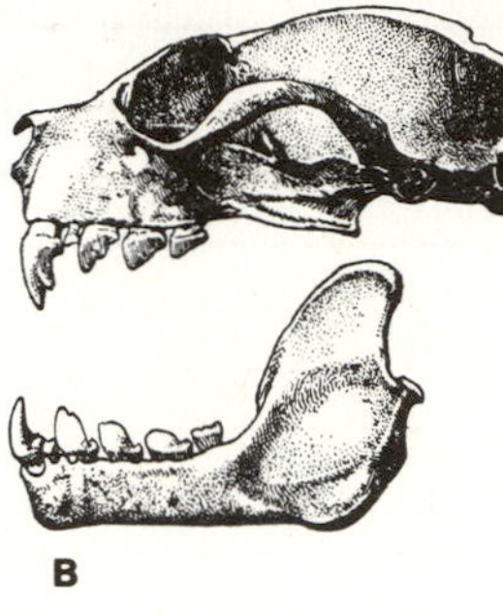

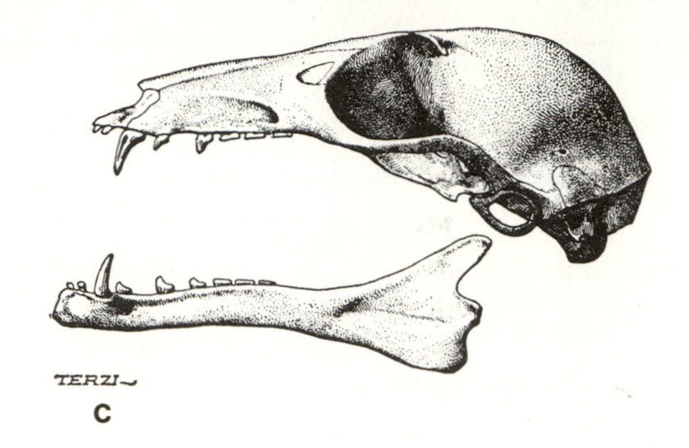

Fossil History

Fossil Chiroptera are very poorly known. Existing evidence has been reviewed by Jepson (1970) and Smith (1976). Megachiropterans are known only from the Oligocene and Miocene of Europe. The oldest known bat is *Icaronycteris index* from the early Eocene in Wyoming. This and other more recent fossils from various parts of the world are all true bats with fully developed wings and associated structures of flight. *Icaronycteris* (for photo see Jepson 1970: Figs. 4 through 17, or Vaughn 1978; Fig. 6-31) includes characteristics of both extant suborders but Jepson (1966) has included it as a very primitive microchiropteran.

Key to Living Families

1 Second finger retaining evident degree of independence from third, usually with a claw; rostrum long; postorbital process well developed, occasionally forming postorbital bar (Fig. 18-5); ears simple, the inner margin forming a continuous ring, tragus and antitragus absent (Fig. 18-3E) ...... Suborder Megachiroptera .... **Pteropodidae**
Old World fruit bats

1′ Second finger scarcely, if at all, independent from third, without a claw; rostrum varies from very long to very short; postorbital process usually absent (if present, rostrum short and second finger without claw); ears may be complex, inner margin never a complete ring, tragus usually present, if absent, antitragus well developed (Fig. 18-3) .... Suborder Microchiroptera 2

2 (1′) Postcanine teeth fewer than 4/4; upper incisors elongated to form blade edge (Fig. 18-6); lower jaw with a groove to receive large upper incisors; no tail, uropatagium short, small noseleaf present (Fig. 18-6) .... **Phyllostomatidae** (in part)
Desmodontinae, vampire bats

2′ Postcanine teeth 4/4 or more; upper incisors not large and bladelike; lower jaw without socket; tail, uropatagium, and nose variable 3

Figure 18–6.
The vampire bat, *Desmodus rotundus,* Phyllostomatidae. A, lateral view of the skull; B, anterior view of upper jaw showing broad incisors; C, face.
(A, Ruschi 1953; B, Dobson 1878: pl. 30; C, Goodwin and Greenhall 1961:267)

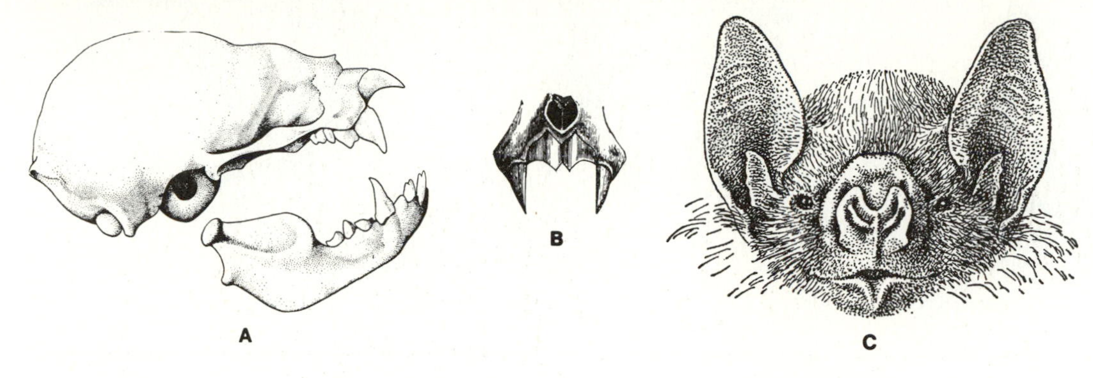

Figure 18–7.
Skull and face of *Noctilio leporinus,* Noctilionidae.
(Face, Goodwin and Greenhall 1961:220; skull, Giebel and Leche 1874: pl. 8)

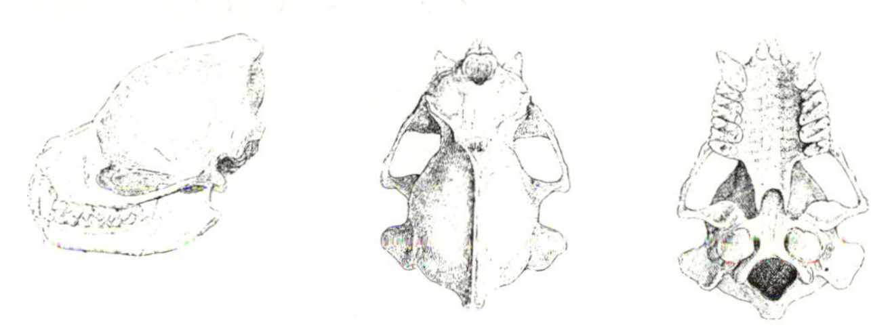

3 (2′) Incisors 2/2, 2/1, or 2/0 4

3′ Incisors not as above 6

4 (3) Incisors 2/1 (Fig. 18-7); nose leaf absent (Fig. 18-7); feet usually elongated and claws enlarged and sharp .... **Noctilionidae**
bulldog and hair-lipped bats

4′ Incisors 2/2, 2/1, or 2/0; nose leaf present or absent (if absent, incisive formula is 2/2); feet not as above 5

5 (4′) Nose leaf absent, chin-leaf present extending ventrally from lower lip (Fig. 18-8); tail protruding from dorsal surface of uropatagium; rostrum tilted up anteriorly giving a dorsal profile of skull that is concave in lateral view (Fig. 18-9); dental formula 2/2 1/1 2/3 3/3 = 34 .... **Mormoopidae**
leaf chinned bats

5′ Nose leaf present (Fig. 18-10); but sometimes rudimentary (Fig. 18-10D); chin-leaf absent; tail variable but not protruding from dorsal surface of uropatagium; rostrum not as above, dorsal profile of skull usually convex or flat (Fig. 18-11); dental formula variable including above ....
.... **Phyllostomatidae** (in part)
all New World leaf nosed bats
except Desmodontinae

Figure 18–8.
Faces of representatives of the three genera of Mormoopidae. A, *Chilonycteris rubiginosa;* B, *Pteronotus davyi;* and C, *Mormoops megalophylla.*
(Goodwin and Greenhall 1961:222, 224, 225)

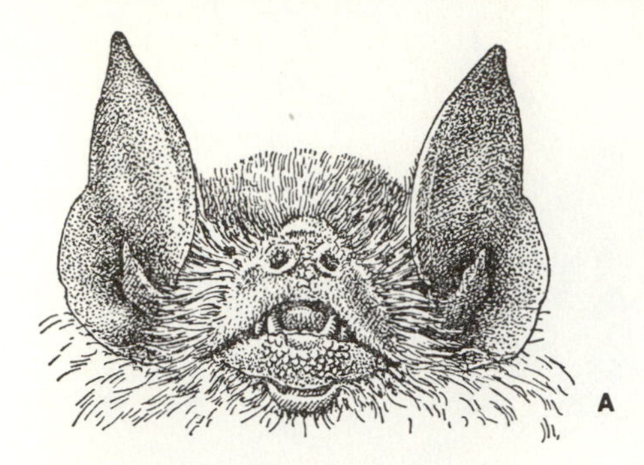

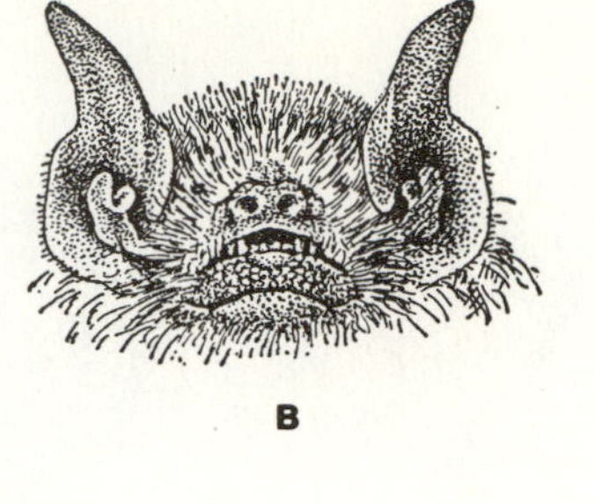

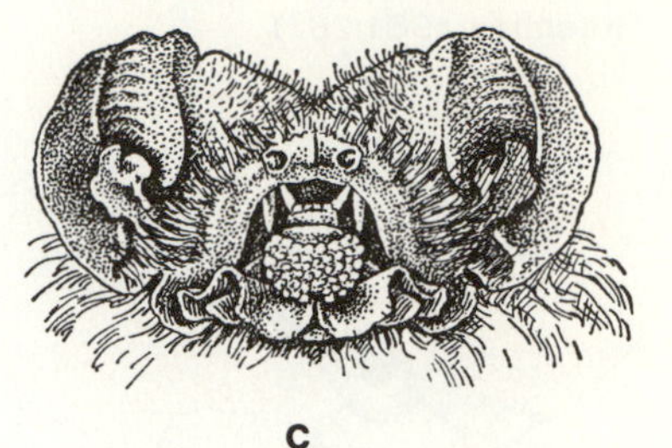

Figure 18–9.
Lateral views of skulls of representatives of two genera of Mormoopidae: A, *Pteronotus parnelli* and B, *Mormoops megalophylla.*
(Smith 1972: 6 and 109)

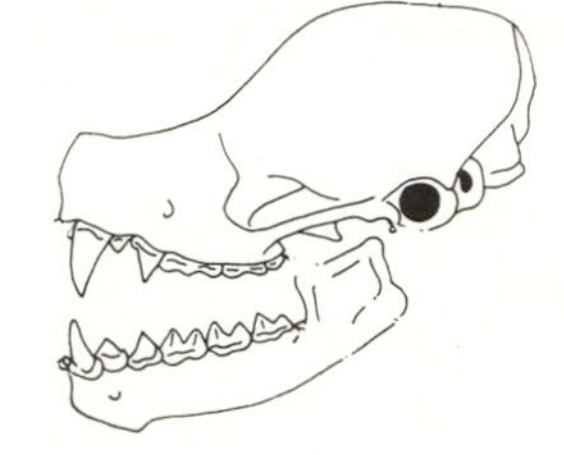

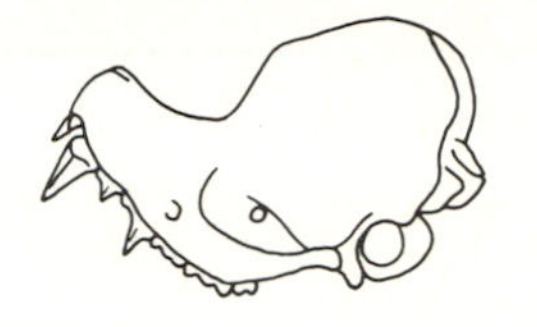

Figure 18–10.
Faces of representative phyllostomatids: A, *Tonatia bidens;* B, *Artibeus jamaicensis;* C, *Vampyrops helleri;* D, *Ectophylla alba;* E, *Lonchorhina aurita;* F, *Sturnira lilium.*
(D, Dobson 1878: pl. 30; all others, Goodwin and Greenhall 1961:234-259)

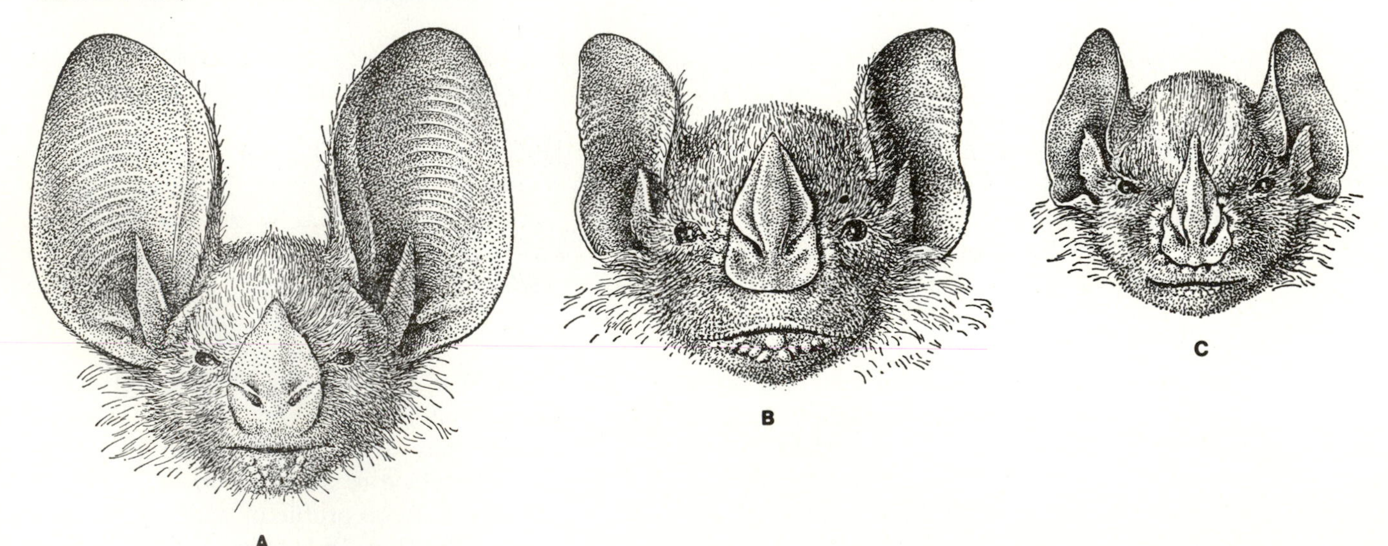

Figure 18–10.—*Continued.*

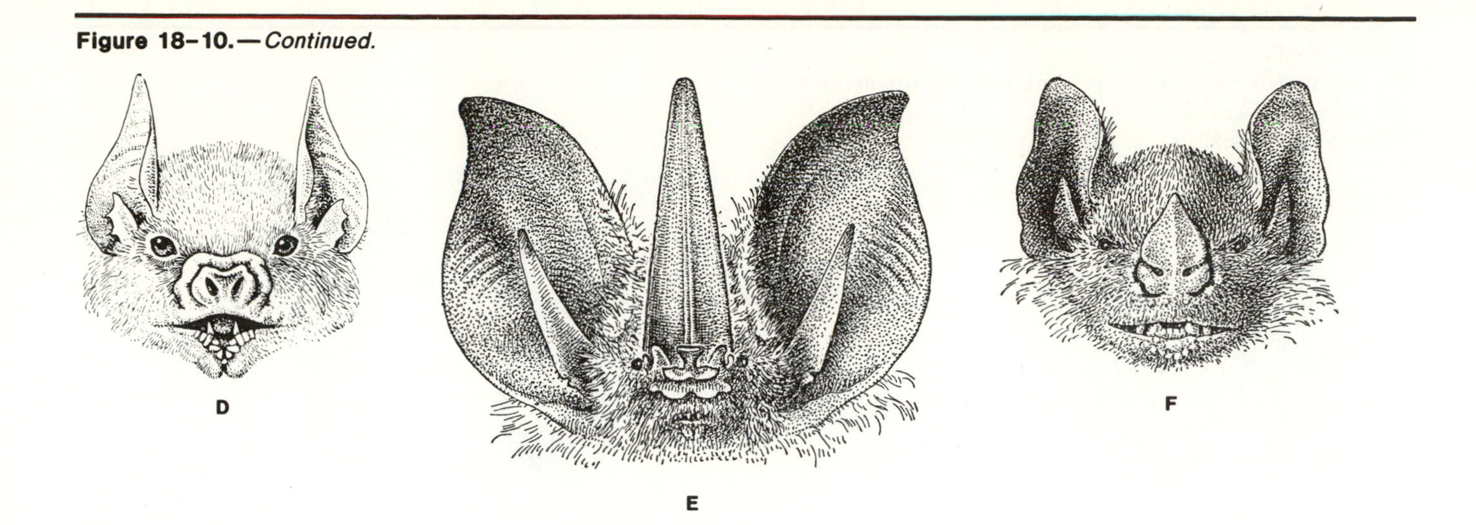

Figure 18–11.
Skulls of representative phyllostomatids, A, *Phyllostoma perspicillata;* B, *Leptonycteris nivalis.*
(Giebel and Leche 1874: pl. 8)

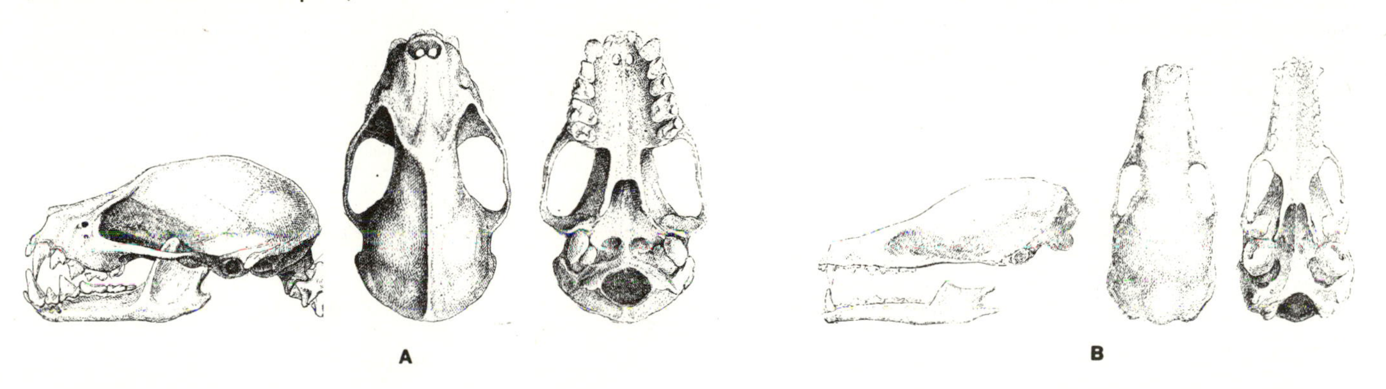

Figure 18–12.
Thumb and foot of a mystacinid bat. Note the basal talons on the claws.
(Dobson 1876:671)

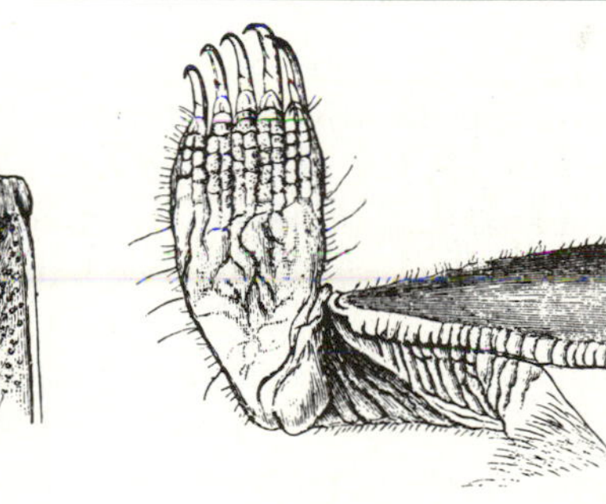

6 (3′) **Dental formula 1/1 1/1 2/2 3/3 = 28;** claws on thumb and toes with a basal talon (Fig. 18-12); short tail projects from midpoint of dorsal surface of uropatagium .... **Mystacinidae**
short-tailed bat

6′ Dental formula not as above; incisors not 1/1, or if 1/1, total number of teeth 26; claws without basal talon; tail variable 7

7 (6′) Premaxillae and upper incisors absent (Fig. 18-13); nose leaf present; tail very short or absent; tragus bifurcated (Fig. 18-15) .... **Megadermatidae**
false vampire bats

7′ Premaxillae and upper incisor(s) present; nose leaf present or absent; tail variable; tragus, if present, not bifurcated 8

Figure 18–13.
Representative megadermatids. Face of *Megaderma lyra* and skull of *M. spasma.*
(Face, Finn 1929:32; skull, Miller 1907:104)

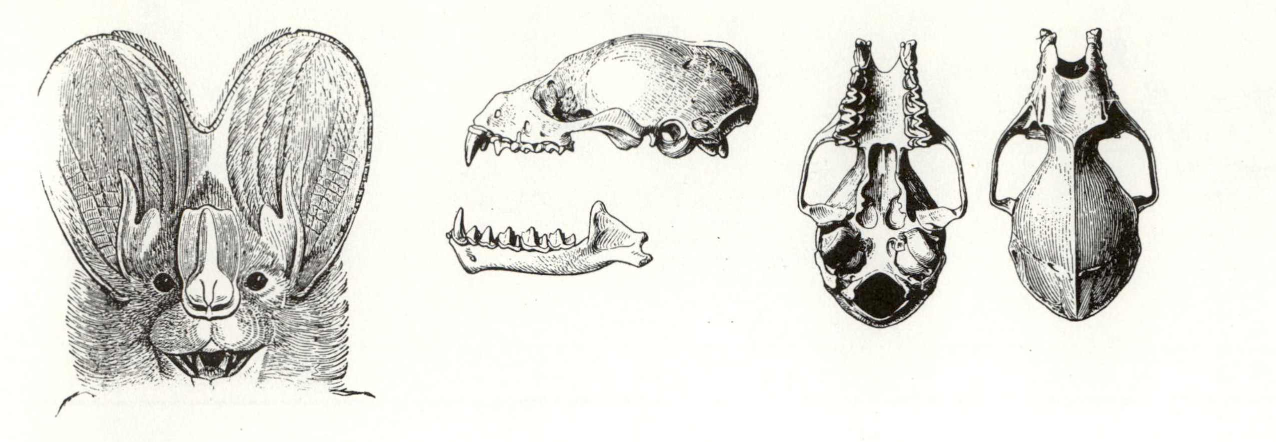

Figure 18–14.
Face and skull of *Craseonycteris thonglongyai.*
(Hill 1974) (A new family, genus and species of bat (Mammalia: Chiroptera). Bull. British Museum [Nat. Hist.], 32[2]:29–43, figs. 1-4. Copyright 1974 Trustees of the British Museum [Natural History]. Reprinted with permission.)

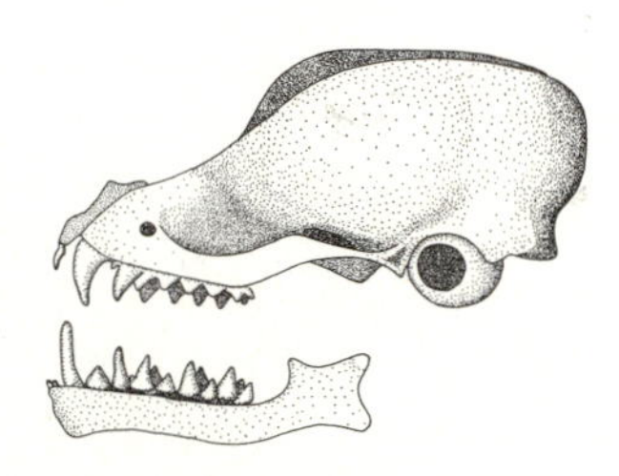

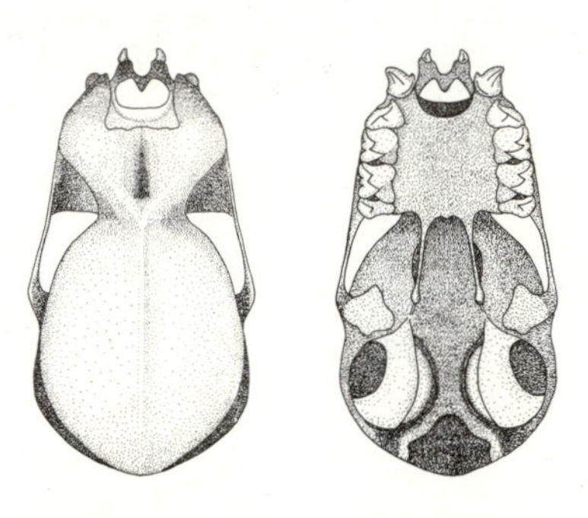

8 (7) Premaxillae form a complete ring around the external nare (Fig. 18-14); dental formula 1/2 1/1 1/2 3/3 = 28; tail absent, uropatagium well developed .... **Craseonycteridae** bumble bee bat

8′ Premaxillae not forming a complete ring around external nare; dental formula variable; tail present, uropatagium variable 9

9 (8′) Postorbital processes present, sometimes incorporated into wide supraorbital ridge (Figs. 18-15 and 18-19); tail completely enclosed in uropatagium or protruding from dorsal surface of uropatagium 10

9′ Postorbital processes and supraorbital ridges absent; tail variable but never protruding from dorsal surface of uropatagium 11

Figure 18–15.
Skull of a hollow-faced bat, *Nycteris javanica.*
(Miller 1907:100)

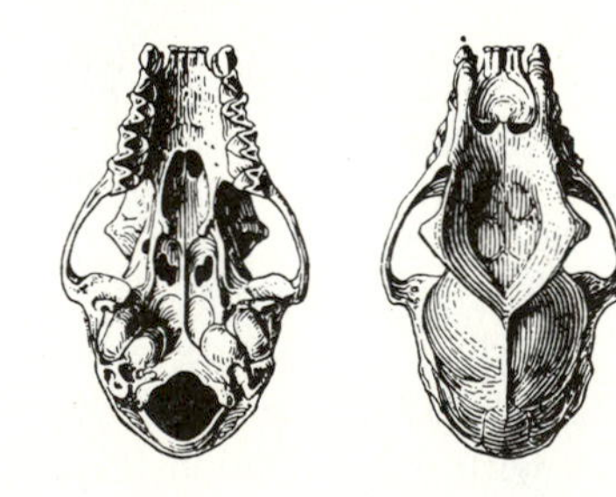

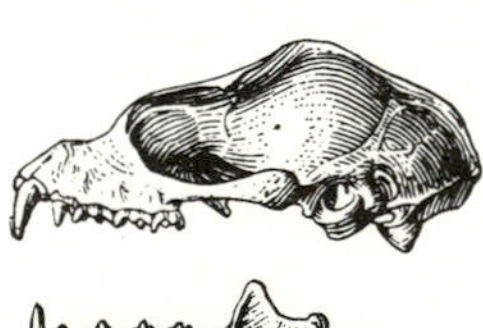

10 (9) Premaxillae with palatal branches only, nasal branches absent; skull with a deep concavity on the anterior surface between the orbits (Fig. 18-15); tip of terminal tail vertebra bifurcate, completely enclosed in uropatagium (Fig. 18-16) .... **Nycteridae**
hispid or hollow-faced bats

10′ Premaxillae with nasal branches only, palatal branches absent; usually no interorbital concavity (Fig. 18-17); tail protrudes from dorsal surface of uropatagium .... **Emballonuridae**
sheath-tailed or sac-winged bats

11 (9′) Nasal branches of premaxillae absent, palatal branches fused to each other medially, but widely separated from the maxillae laterally (Fig. 18-18); complex nose leaf present (Fig. 18-19); tragus absent, antitragus well developed (Fig. 18-19) ....**Rhinolophidae**
horseshoe bats

11′ Nasal branches of premaxillae present, palatal branches, if present, not as above; nose leaf absent, or a simple flap or projection; tragus present 12

Figure 18–16.
Head of *Nycteris avakubia* and tail and uropatagium of *N. arge*, Nycteridae.
(Lang and Chapin 1917)

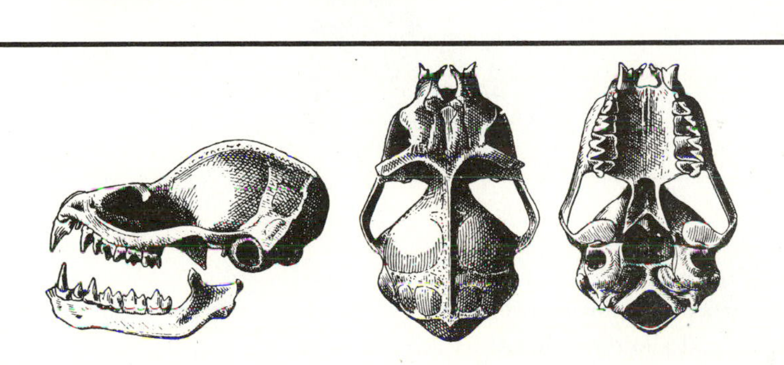

Figure 18–17.
Skull of a sac-winged bat, *Saccopteryx bilineata*.
(Miller 1907:89)

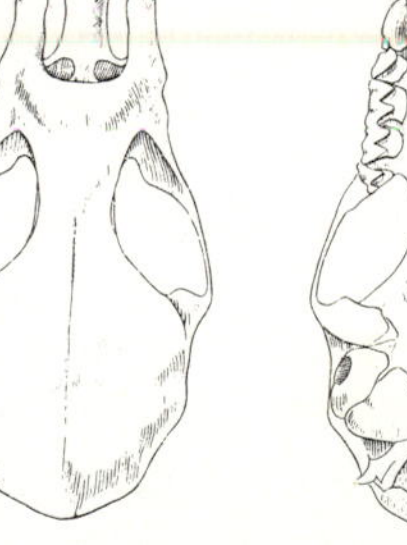

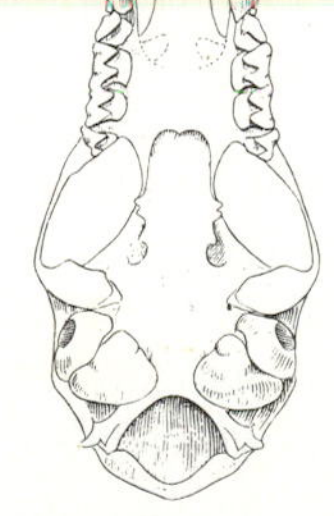

Figure 18–18.
Skull of the greater horseshoe bat, *Rhinolophus ferrumequinum*.
(Gromov et al. 1963)

Figure 18–19.
Heads of representatives of the family Rhinolophidae. A, *Rhinolophus mehelyi;* B, *Triaenops persicus;* C, *Hipposideros caffer.*
(A, Gromov et al. 1963; B, Dobson 1878, C, Cabrera 1932:86)

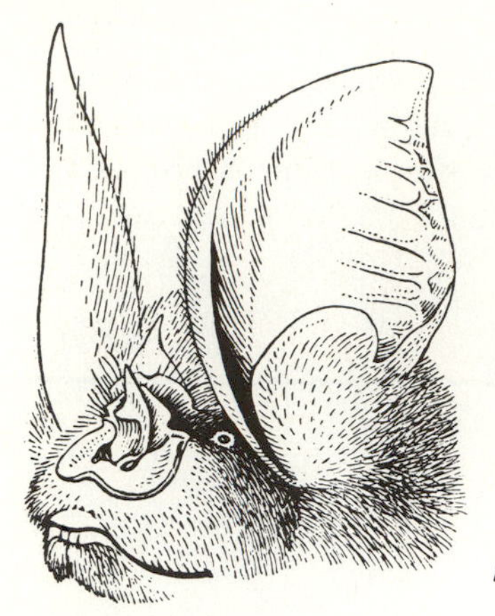

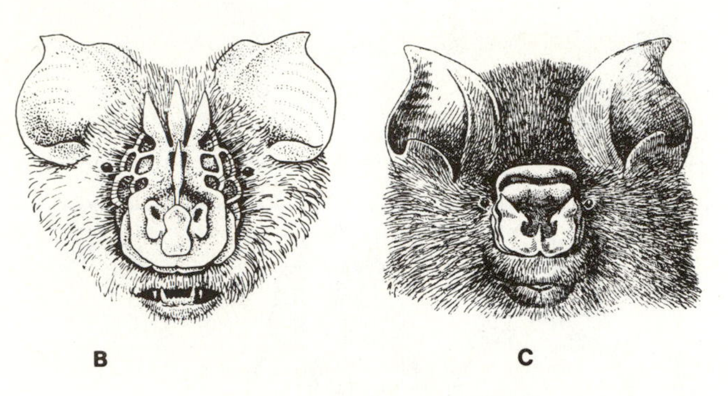

12 (11′) Dental formula 1/2 1/1 1/2 3/3 = 28; nasal region of skull greatly inflated (Fig. 18-20); tail long and slender, almost as long as head and body; uropatagium short, enclosing less than one-third of the length of the tail (Fig. 18-21) .... **Rhinopomatidae**
mouse-tailed bats

12′ Dental formula variable, teeth usually number more than 28; nasal region of skull not inflated; tail enclosed in membrane for more than one-third its length; tail usually less than two-thirds the length of head and body 13

13 (12′) Posterior bases of pterygoids connected by a distinct ridge on the basisphenoid; dental formula 2/3 1/1 3/3 3/3 = 38 14

13′ Posterior bases of pterygoids not connected by a ridge; dental formula variable 15

Figure 18–20.
Skull of a mouse-tailed bat, *Rhinopoma hardwickei.*
(R. Roesener)

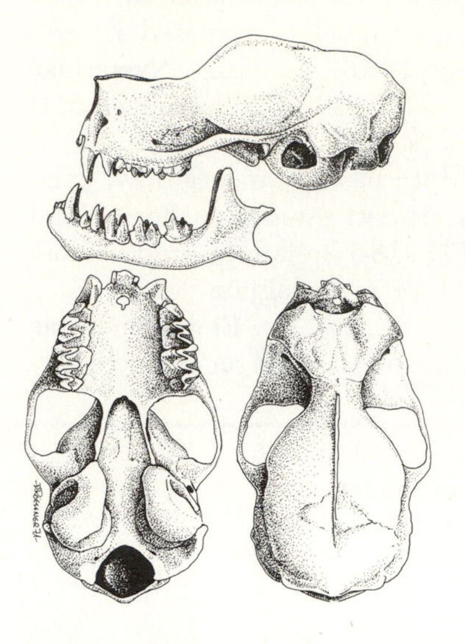

Figure 18–21.
Tail of a mouse-tailed bat, Rhinopomatidae.
(Lang and Chapin 1917)

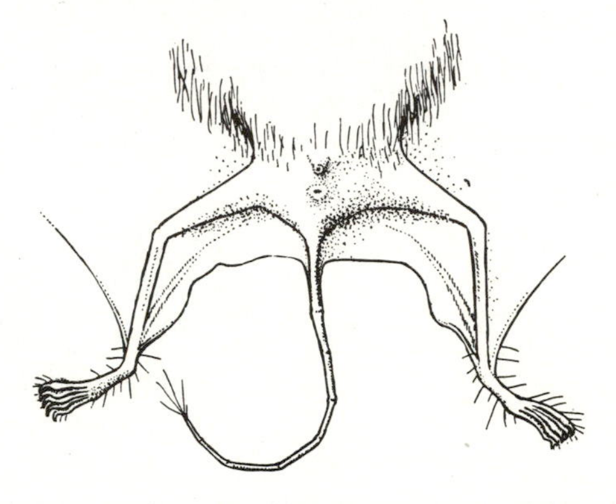

14 (13) Full length of third and fourth toes and claws fused; sucking disks present on feet and thumbs (Fig. 18-22); anterior palatal emargination, if present, V-shaped, incisive foramina absent (Fig. 18-23); anterior of rostrum somewhat pointed in dorsal view (Fig. 18-23 .... **Thyropteridae**
disk-winged bats

14′ Third and fourth toes and claws not fused; no sucking disks present; anterior palate never emarginate, two small incisive foramina usually present (Fig. 18-24); anterior of rostrum squared in dorsal view (Fig. 18-24) .... **Natalidae**
funnel-eared bats

Figure 18–22.
Sucking discs of a disc-winged bat, Thyropteridae. A, wrist, lateral view; B, wrist, ventral view; C, foot.
(Dobson 1876)

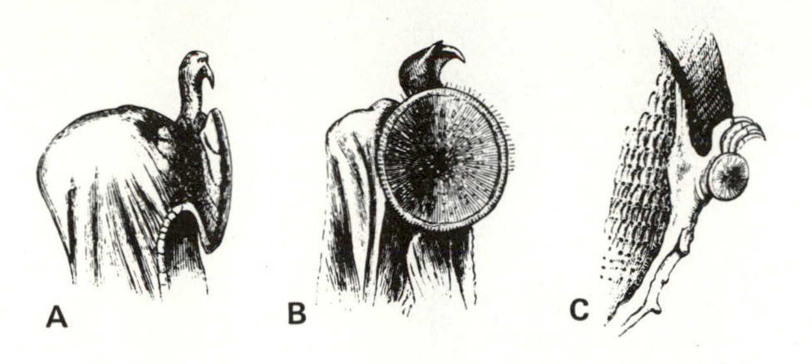

Figure 18–23.
Skull of *Thyroptera discifera* and face of *T. tricolor*, Thyropteridae.
(Skull, Miller 1907:192; face, Goodwin and Greenhall 1961:275)

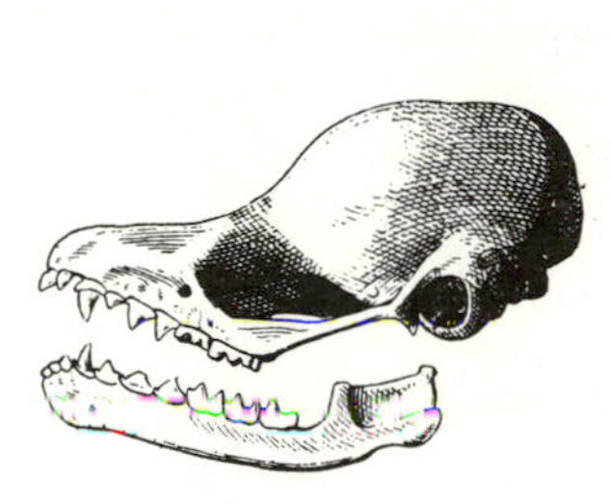

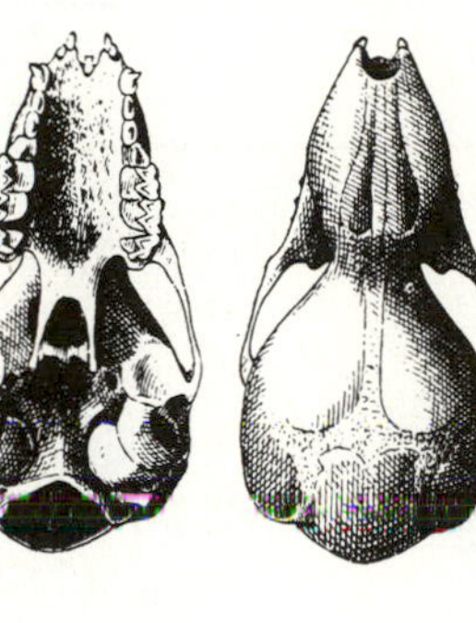

15 (13′) Palate entire or with shallow anterior emargination and two small incisive foramina (Fig. 18-24) 17

15′ Palate with deep anterior emargination and no incisive foramina (Figs. 18-26, 18-28) 16

Figure 18–24.
Skull of *Natalus mexicanus* and face of *N. tumidirastris*, Natalidae. The skull has a shallow palatal emargination and two incisive foramina.
(Skull, Miller 1907:184; face, Goodwin and Greenhall 1961:272)

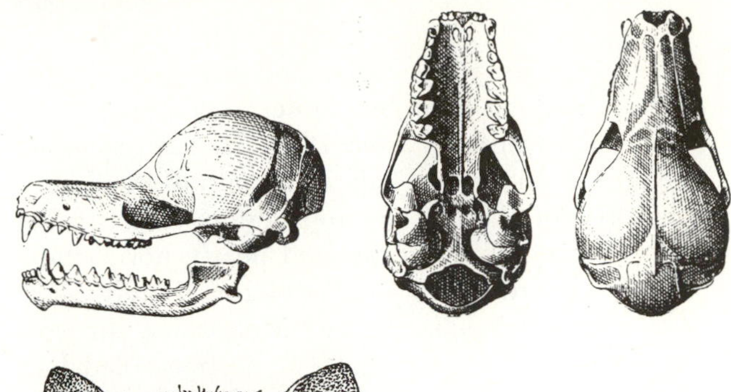

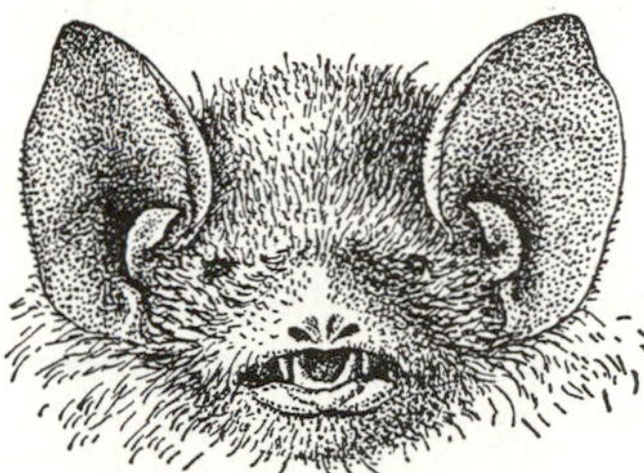

Figure 18–25.
Skull of a molossid that has no palatal emargination, *Molossus rufus*.
(Miller 1907:260)

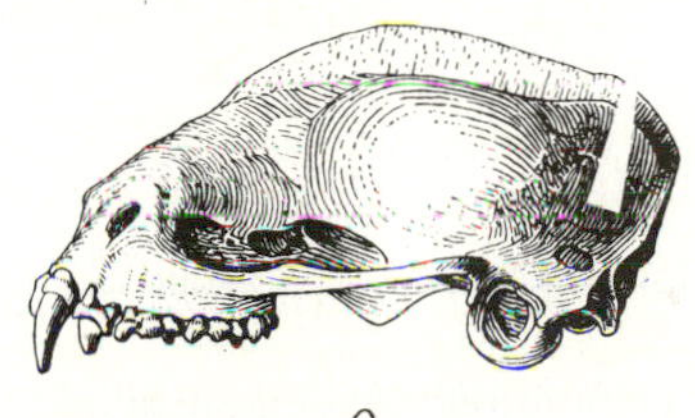

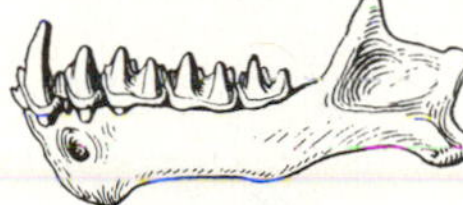

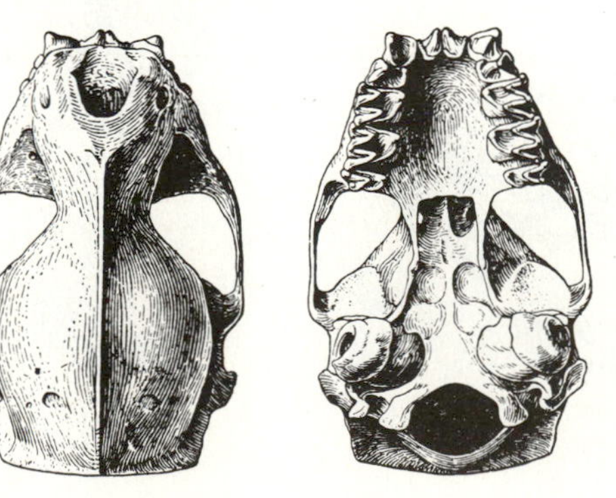

16 (15′) Incisors 1-2/2-3, upper incisors widely separated at base and tip (Fig. 18-26); teeth number from 28 to 38; tail completely or almost completely enclosed in uropatagium; ears usually more or less upright; tragus usually elongate (Fig. 18-27) .... **Vespertilionidae**
common bats

16′ Incisors 1/2-3, upper incisors widely separated at base but closer together at tips; teeth number from 28 to 32; tail extending for about one-half its length beyond posterior margin of uropatagium (Fig. 18-29); ears frequently directed nearly horizontally forward (Fig. 18-29); tragus usually short and rounded .... **Molossidae** (In part)
free-tailed bats

Figure 18–26.
Skull of a vespertilionid bat, *Myotis myotis.*
(Popov 1956)

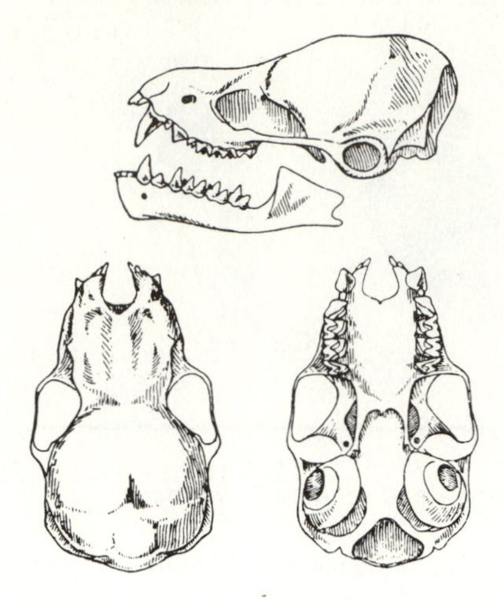

Figure 18–27.
Heads of representative Vespertilionidae. A, *Myotis nattereri;* B, *Eptesicus serotinus;* C, *Pipistrellus pipistrellus;* D, *Nyctalus noctula;* and E, *Miniopterus schreibersi.*
(Cabrera 1914:104–142)

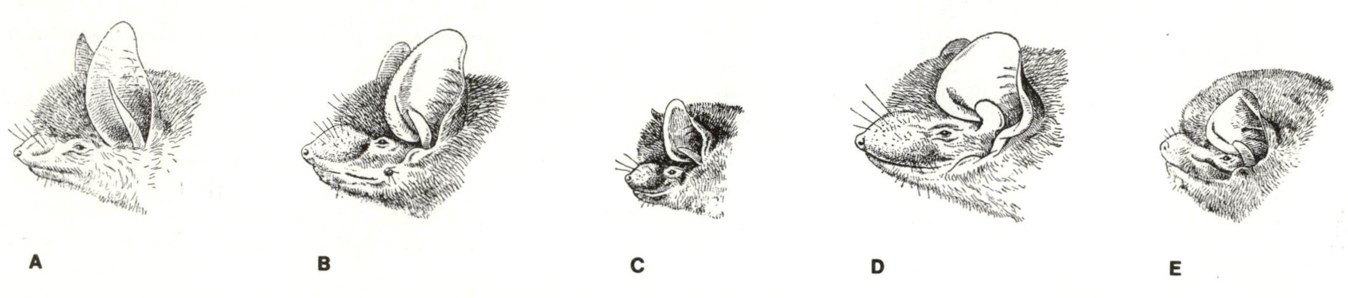

Figure 18–28.
Skull of a molossid with an anterior palatal emargination, *Tadarida macrotis.*
(Miller 1907:253)

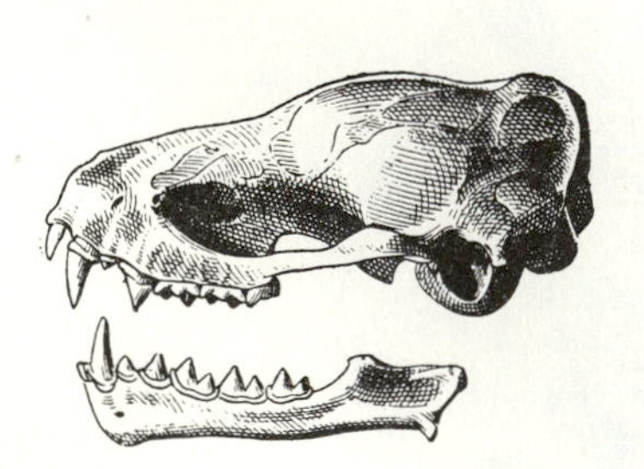

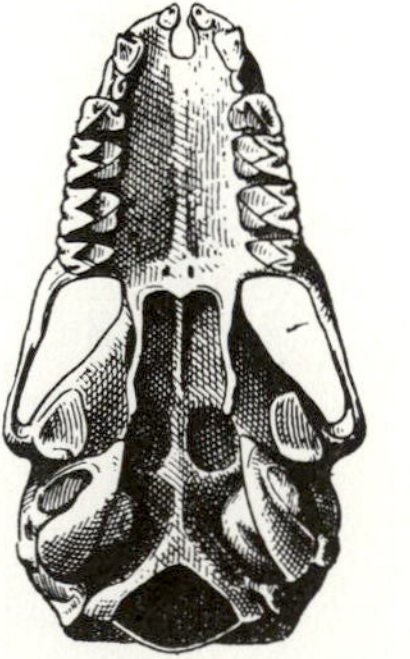

Figure 18–29.
Head and tail of the free-tailed bat *Tadarida teniotis,* Molossidae.
(Head, Cabrera 1914:146; tail, Lang and Chapin 1917)

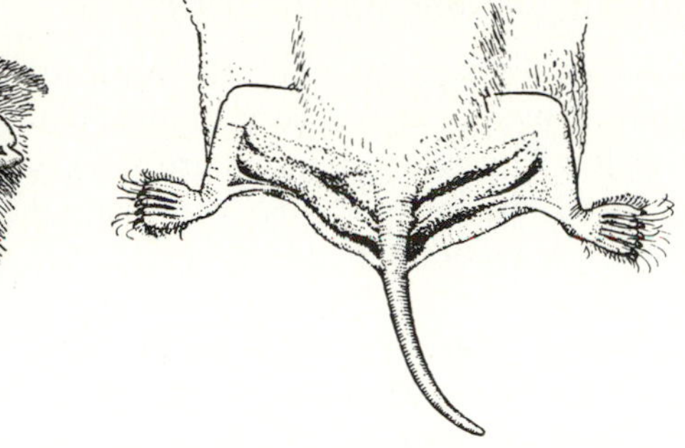

17 (15) Dental formula 1/1-3 1/1 1-2/2 3/3 = 26 to 32; tail with approximately distal one-half extending beyond posterior edge of uropatagium (Fig. 18-29) .... **Molossidae** (In part) free-tailed bats

17′ Dental formula 2/3 1/1 2-3/3 3/3 = 36 or 38; tail enclosed in uropatagium for all or most of its length 18

18 (17′) Dental formula 2/3 1/1 2/3 3/3 = 36; thumb reduced, enclosed in patagium to base of minute claw (Fig. 18-30 .... **Furipteridae** smoky bats

18′ Dental formula 2/3 1/1 3/3 3/3 = 38; thumb free of patagium or, if enclosed, a large claw present 19

Figure 18–30.
Face and wrist region of the smoky bat, *Furipterus horrens,* Furipteridae. Note the patagium enclosing the thumb to the base of its minute claw.
(Goodwin and Greenhall 1961:274)

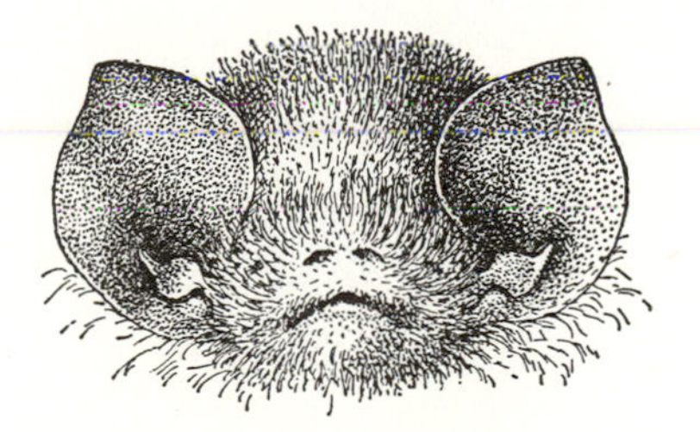

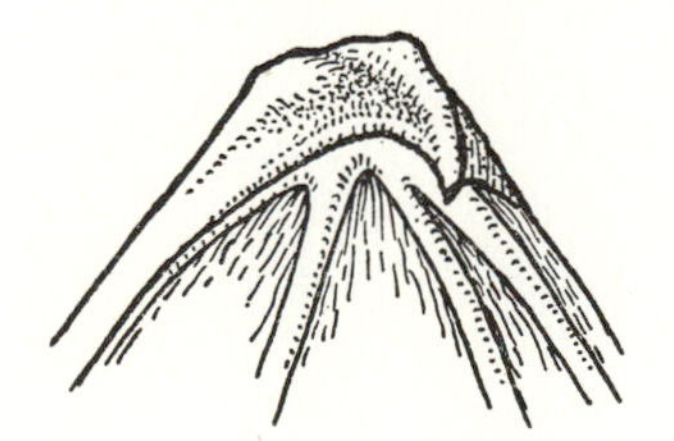

19 (18′) Posterior palatal emargination reaches anteriorly to last molar; canine in contact with adjacent incisor; rostrum broader than long; unique mushroom-shaped structure in ear partially blocking auditory canal (Fig. 18-31); thumb free .... **Myzopodidae** sucker-footed bat

19′ Palate extending posteriorly well beyond last molar (Fig. 18-24); a distinct space present between canine and adjacent incisor; rostrum as long or longer than broad; no mushroom-shaped structure in ear (Figs. 18-24 and 18-30; thumb enclosed in membrane .... **Natalidae** funnel-eared bats

Figure 18–31.
Head of *Myzopoda aurita,* Myzopodidae. Note the mushroom shaped structure partially blocking the external auditory meatus.
(Thomas 1904)

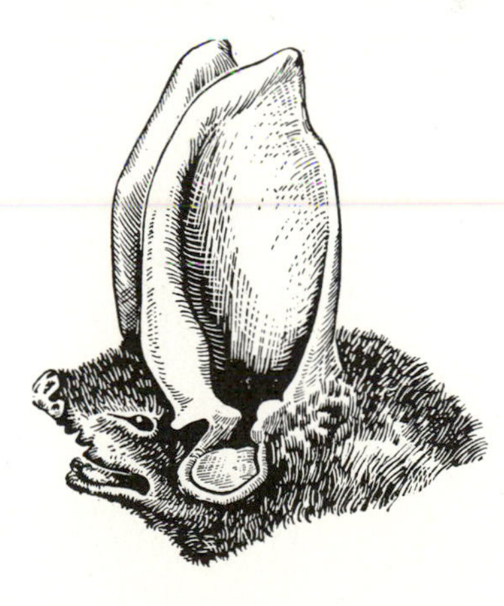

Comments and Suggestions on Identification

Externally, bats are easily recognized by the presence of wings. There are, however, many very similar families within this order. The two suborders are usually easily separated, but keep in mind that the smallest megachiropterans are smaller than the largest microchiropterans.

The families Craseonycteridae, Mystacinidae, Rhinolophidae, Rhinopomatidae, Molossidae, Noctilionidae, Nycteridae, Megadermatidae, Emballonuridae, Furipteridae, Myzopodidae, Thyropteridae, and Natalidae are each recognized by one or two distinctive characteristics. The family Vespertilionidae is very large and homogeneous but is lacking any one good distinctive character. The family Phyllostomatidae is large and very heterogeneus; become familiar with the many different types.

Taxonomic Remarks

The classification scheme used here corresponds with that presented by Koopman and Jones (1970) except that Mormoopidae is recognized as distinct from Phyllostomatidae (Smith 1972) and Craseonycteridae (Hill 1974), a family new to science and based upon material collected in 1973 in Thailand, is included. The two suborders and most of the families are universally recognized. The subfamily Hipposiderinae of the Rhinolophidae is frequently recognized as a distinct family, the Hipposideridae (Romer 1966, Eisentraut 1975, Walker et al. 1975). The vampire bats, Desmodontinae, are frequently considered to be a family distinct from the Phyllostomatidae and until recently the Mormoopidae were considered to be a subfamily of the Phyllostomatidae (Romer 1966, Eisentraut 1975, Walker et al. 1975).

Supplementary Readings

Baker, R. J., J. K. Jones Jr., and D. C. Carter, (Eds.). 1976. Biology of Bats of the New World Family Phyllostomatidae. Part I. *The Museum, Texas Tech. Univ., Spec. Pub.* 10:1-218; Part II, 1977, 13:1-36; Part III, 1978, 1:441.

Barbour, R. W. and W. H. Davis. 1969. *Bats of America.* Univ. Press of Kentucky, Lexington. 286 pp.

Eisentraut, M. 1975. Bats, pp. 67-92. *In* B. Grzimek (Ed.) *Animal life encyclopedia. Vol. 11, Mammals II.* Van Nostrand Reinhold, New York.

Greenhall, A. M. and J. L. Paràdiso. 1968. *Bats and bat banding.* Resource Publ. No. 72, U.S. Bureau of Sports Fisheries and Wildlife, Washington. 47 pp.

Leen, N. and A. Novick. 1969. *The world of bats.* Holt, Rinehart and Winston, New York. 171 pp.

Slaughter, B. H. and D. W. Walton (Eds.) 1970. *About bats.* Southern Methodist Univ. Press, Dallas. 339 pp.

Turner, D. C. 1975. *The vampire bat.* Johns Hopkins Univ. Press. Baltimore. 145 pp.

Wimsatt, W. A. (Ed.). 1970. *Biology of bats,* Vol. I. Academic Press, New York. 406 pp.; Vol. II, 1970, 477 pp.; Vol. III, 1977, 651 pp.

Yalden, D. W. and P. A. Morris. 1975. *The lives of bats.* David and Charles Ltd. Newton Abbot, Devon England. 247 pp.

19 The Primates
Order Primates

ORDER PRIMATES

The ordinal name, Primates, means the first or primary animals. While this name reflects the egocentricity of man, it is of no help in defining the group of mammals included in it.

Primates are mainly arboreal mammals (Fig. 19-1) but terrestrial forms occur in several groups. Examples of these are the ring-tailed lemur, *Lemur catta;* baboons, *Papio* and *Mandrillus;* the gorilla, *Gorilla gorilla;* and man, *Homo sapiens.* Man is the only completely bipedal primate. Most primates are omnivorous, eating a higher percentage of plant than animal material, and usually softer foods such as fruits, buds, and soft bodied insects are favored. The gorilla, the leaf eating monkeys (the Asian langurs, African colobus monkeys, South American howler monkey) and the Madagascaran indrid lemus are almost entirely herbivorous in the wild and the aye aye, the galagos and the tarsiers are largely insectivorous (Napier and Napier 1967:24).

Figure 19–1.
A capuchin monkey, Cebidae, *Cebus.* (Gray 1865:pl. 45)

Most primates are pentadactyl though some fingers are shortened in some Lorisidae. The thumb is reduced in several brachiating forms and completely absent in the South American spider monkeys, *Ateles,* and the African genus *Colobus* (Napier and Napier 1967: 396). In most primates the digits are prehensile and the pollex (thumb) and/or the hallx (big toe) is more or less opposable. Napier and Napier (1967:396-399) discussed the opposability of the thumb and gave the following classification:

Nonopposable thumb: Tarsiidae, Callitrichidae
Pseudo-opposable thumb: Strepsirini, Cebidae
Opposable thumb: Catarrhini

These prehensile and opposable digits yield a hand or foot with great dexterity, the animal is better able to grasp and manipulate objects. A prehensile tail is present in many Cebidae but occurs in no other primate family.

In general the sense of smell in primates becomes less acute as the hand becomes better adapted for manipulation (Napier and Napier 1967:15) but this should not be interpreted as cause and effect. Along with the loss of olfactory sensitivity primates have developed very good vision. Particularly among the diurnal primates vision is very well developed. The field of vision has a large area of overlap between the fields of each eye resulting in precision in depth perception. Only some nocturnal primates (some Lemuridae; the Galagidae, Lorisidae and Tarsiidae; and the cebid genus *Aotus*) have retinas composed entirely of rods (Napier and Napier 1967:19). Other Primates have both rods and cones in the retinas. Most Strepsirini and Tarsiidae are nocturnal and most Platyrrhini and Catarrhini are diurnal.

Flesh of Primates is eaten in various parts of the world, and the skins of some species, e.g., *Colobus*, are valuable. Some species do a great deal of damage to crops and some, primarily baboons, can be dangerous to man (Fiedler et al. 1975:419). Primates are major attractions at zoos and many species are kept as pets. The Rhesus monkey, *Macaca mulatta*, and several other species are widely used as research animals.

Distinguishing Characteristics

The dentition varies but the cheek teeth are bunodont and brachydont. The orbits are directed more or less anteriorly and either a postorbital bar or a postorbital plate is present. In more "advanced" forms the braincase is large and the rostrum is usually shortened.

The limbs are plantigrade and usually pentadactyl. In some forms the pollex or the hallux is reduced or absent. A nail is always present on the pollex and most species have nails (as opposed to claws) on the other digits.

The uterus is either bicornuate (Strepsirhini and Tarsii) or simplex (Platyrrhini and Catarrhini) and the penis is usually pendant. The testes are scrotal. So far as is known, bacula are lacking only in four cebid species, in the tarsiers, and in man (Hershkovitz 1977: 118).

Distribution

Primates are essentially tropical mammals. Three families are confined to Madagascar (one species also occurs on the nearby Comoro Islands) and three are found only in the tropical portions of the Neotropical Region. One family is confined to the tropical portions of the Ethiopian Region, two are confined to the tropical parts of the Oriental Region and three occur in both of these areas. Five species of Cercopithecidae (four *Macaca* and one *Papio*) inhabit various areas in the southern Palearctic (Corbet 1978:234). Hominidae is worldwide in distribution.

Fossil History**

The fossil record for primates is poorly known. During the Paleocene and Eocene in North America and Europe, there existed several small mammals similar to the tree shrews in being intermediate between the

**Based upon Romer (1966) and Thenius (1975a).

Table 19-1
Living Families of Primates.

Family	Common Name	Number of* Genera	Number of* Species	Distribution
Suborder Strepsirhini				
Lemuridae	lemurs	6	13	Madagascar
Indriidae	indriid lemurs	3	4	Madagascar
Daubentoniidae	aye-ayes	1	1	Madagascar
Lorisidae	lorises, potto, angwantibo	4	5	Oriental, Ethiopian
Galagidae	galagos	1	6	Ethiopian
Suborder Haplorhini				
Infraorder Tarsii				
Tarsiidae	tarsiers	1	2	Indonesia, Philippines
Infraorder Platyrrhini				
Callitrichidae	marmosets, tamarins	4	14	Neotropical
Callimiconidae	Goeldi's monkey	1	1	Neotropical
Cebidae	New World monkeys	11	29	Neotropical
Infraorder Catarrhini				
Cercopithecidae	Old World monkeys	12	73	Ethiopian, Oriental, southern Palearctic
Hylobatidae	gibbons	2	7	Oriental
Pongidae	gorilla, chimps, orangutan	3	4	Central Ethiopian, Java and Borneo
Hominidae	man	1	1	Cosmopolitan

*Based upon Chiarelli (1972) except that Colobidae is here combined with Cercopithecidae and the number of species of Callitrichidae follows Hershkovitz (1977).

insectivores and primates. The first known forms that are definitely primates occurred in the Middle Paleocene in North America. These primitive lemuroid primates were abundant in North America and Europe through the Eocene but then disappear from the fossil record. Lemuridae, Indriidae, and Daubentoniidae are known only from the Pleistocene and Recent of Madagascar. Galagidae is known from the Lower Miocene to Recent of Africa and the Lorisidae from the Pliocene and Recent of south Asia.

The tarsiers, a group that in many respects is intermediate between the primitive primates and the more advanced monkeys and apes (Hershkovitz 1977: 11) are known from numerous Paleocene and Eocene genera in Europe and North America. In North America they lingered into the early Miocene at which time all primates seem to have disappeared from this continent. The Tarsiidae are known from the mid- and upper Eocene of Europe and the Recent of southeast Asia.

The stock ancestral to South American monkeys probably reached the Neotropical in the Oligocene. Callitrichidae are known from the upper Oligocene, Pleistocene and Recent, Callimiconidae only from the Recent and Cebiidae from the upper Oligocene, upper Miocene, the Pleistocene and the Recent. All three families are known only from the Neotropical.

The Old World monkeys and apes are first known from the Oligocene of Egypt and fossils are known from the Pliocene in Eurasia. The oldest known great apes are also from the Oligocene of Egypt and additional specimens are known from subsequent periods in other parts of the Old World tropics and subtropics.

Figure 19–2.
The sifakas, an indriid lemur.
(Flower and Lydekker 1891:686)

Figure 19–3.
Faces of a Catarrhine (A) and Platyrrhine (B) monkeys.
(Duncan 1877-83:6)

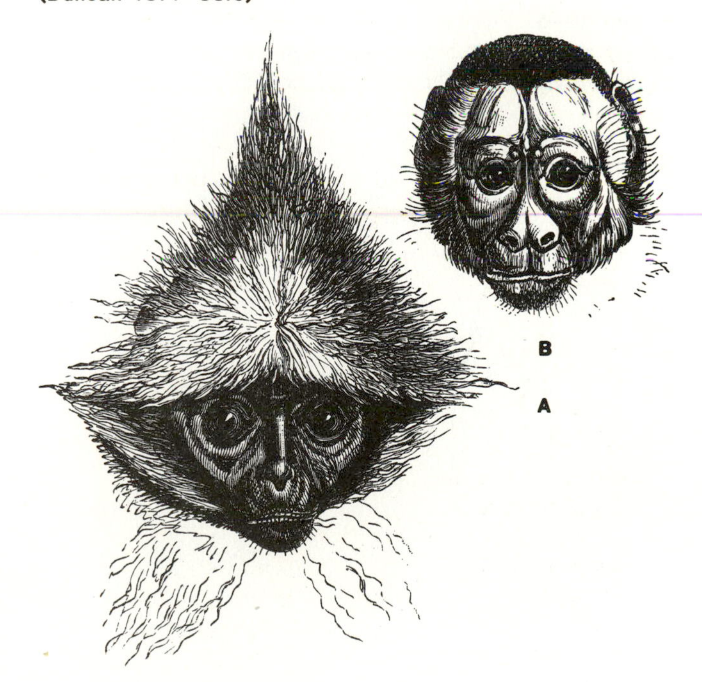

Key to Living Families

1 Postorbital bar present, orbit and temporal fossa broadly confluent; some digits with nails and others with claws; nostrils with crescentic lateral slits* (Fig. 19-2) 2

1′ Postorbital plate present, orbit and temporal fossa completely or mostly separated; all digits with nails or some digits with nails and others with claws; nostrils completely ringed (Fig. 19-3) 7

*The skulls of Tarsiidae could be interpreted by the student as having either a postorbital bar or a postorbital plate. Therefore the key is constructed to allow Tarsiidae to be keyed by either selection in this couplet. However the nostrils of Tarsiidae are consistent only with the condition described in 1′.

2 (1) Teeth number 18 to 20; incisors 1/1, large and similar to those of a rodent with enamel only on anterior surfaces; canines absent (Fig. 19-4); third manual digit much more slender than adjacent digits (Fig. 19-6); hallux with a nail, other digits with claws; tail long and bushy .... **Daubentoniidae** aye aye

2 Teeth number 30 or more; incisors never 1/1, and never resemble those of a rodent; C^1 present, C_1 usually present; third manual digit not appreciably more slender than other digits; distribution of nails and claws varies; tail varies 3

Figure 19–4.
Skull of an aye aye, *Daubentonia madagascariensis.*
(Flower and Lydekker 1891:695)

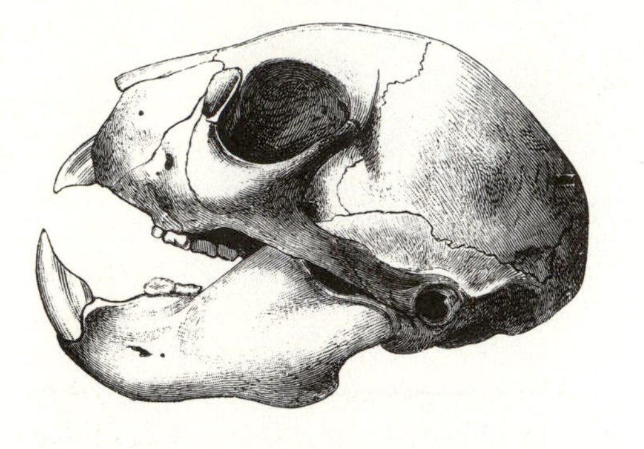

Figure 19–5.
Skeleton of the left manus of an aye aye, *Daubentonia madagascariensis.* (Duncan 1877-83:256)

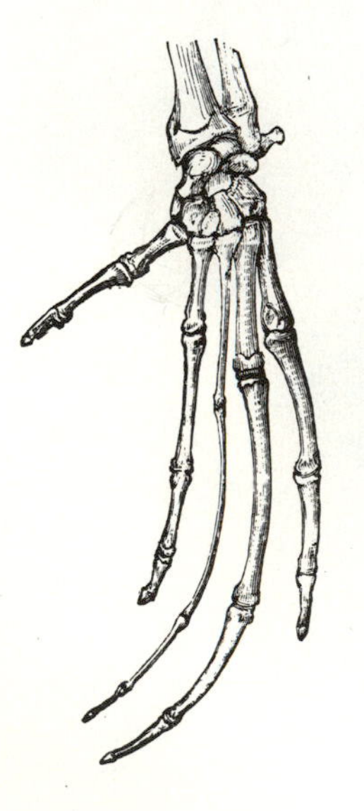

3 (2′) Dental formula 2/2 1/1 2/2 3/3 = 30; P_1 caniniform (Fig. 19-6); basal portion of toes at least partly webbed .... **Indriidae** indriid lemurs

3′ Teeth number 32 or more; P_1 varies; toes never webbed 4

4 (3′) Dental formula 2/1 1/1 3/3 3/3 = 34; I^1 considerably larger than I^2 (Fig. 19-7); rhinarium absent, area around nostrils haired; tail long and naked or only sparsely haired for most of its length (Fig. 19-8) .... **Tarsiidae** tarsiers

4′ Dental formula 0-2/3 1/1 3/3 3/2-3 = 32-36; I^1 and I^2, if present, essentially equal in size; rhinarium present, area around nostrils naked; tail short, or if long, well haired 5

Figure 19–6.
Skull of an indriid lemur, *Indri indri.*
(Duncan 1877-83:219)

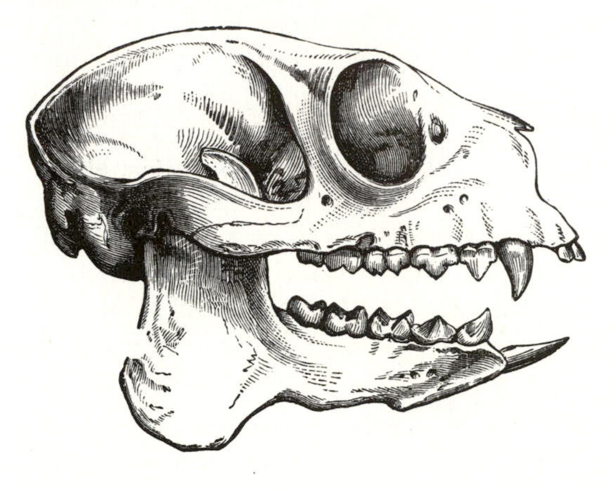

Figure 19–7.
Skull of a tarsier, *Tarsius syrichta.*
(Mary Ann Cramer)

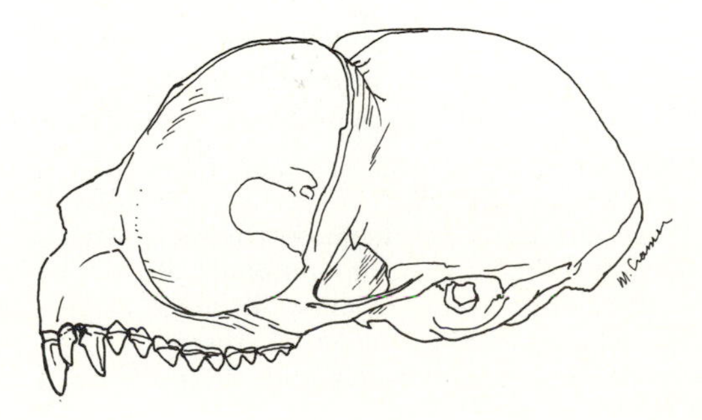

Figure 19–8.
A tarsier, *Tarsius spectrum.*
(Kingsley 1884:490)

5 (4′) Upper incisors 0 or 2; rostrum long; braincase elongate (Fig. 19-9B); ventral surface of pes naked with coarsely ridged pads .... **Lemuridae**
lemurs

5′ Upper incisors 1 or 2; rostrum short, braincase essentially spherical (Fig. 19-10); ventral surface of pes haired at heel (Fig. 19-11) 6

Figure 19–10.
Skull of a galago, *Galago crassicaudatus.*
(Mary Ann Cramer)

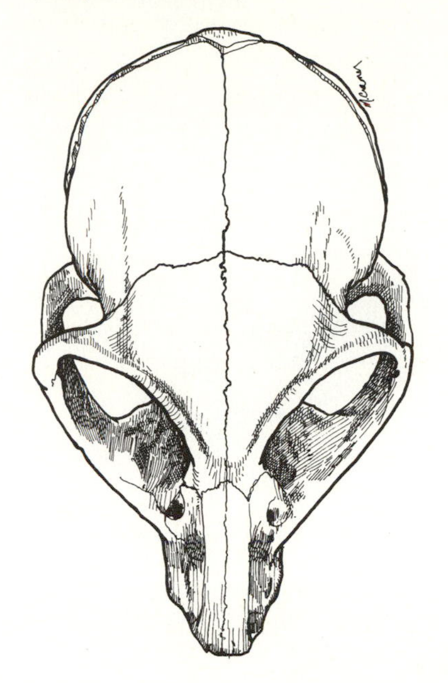

6 (5′) Skull with prominent temporal ridges (Fig. 19-12); tail short or absent (Fig. 19-13) .... **Lorisidae**
lorises, potto, angwantibo

6′ Skull without temporal ridges (Fig. 19-10); tail long and bushy (Fig. 19-14) .... **Galagidae**
galagos

Figure 19–9.
Skull of a ring-tailed lemur, *Lemur catta.*
(Lateral view, Flower and Lydekker 1891:687; dorsal view, Mary Ann Cramer)

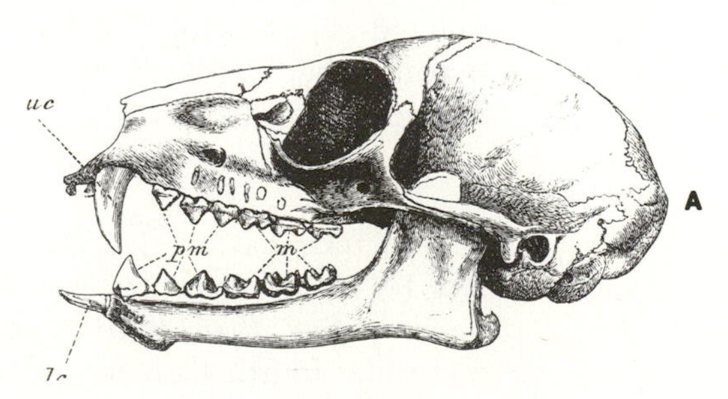

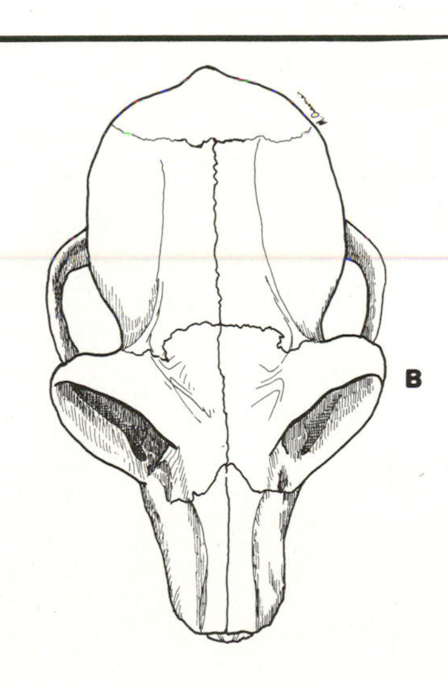

Figure 19–11.
Ventral views of right pes of a thick-tailed galago, *Galago crassicaudatus.*
(Duncan 1877-83:240)

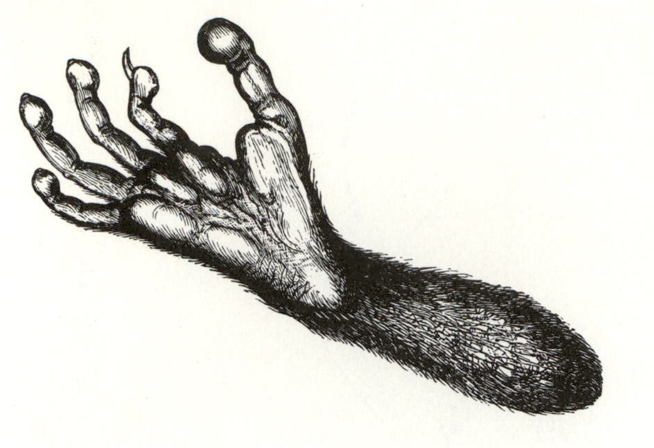

Figure 19–12.
Skull of a slender loris, *Loris tardigradus,* Lorisidae.
(Giebel and Leche 1874: pl. 5)

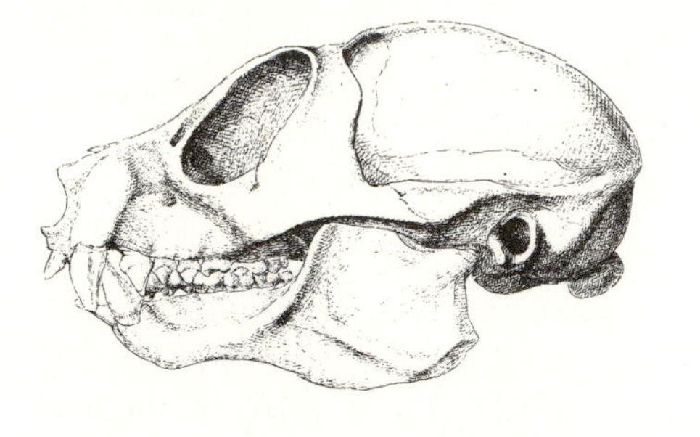

Figure 19–13.
A slow loris, *Nycticebus coucang.*
(Beddard 1902:546)

Figure 19–14.
A thick-tailed galago, *Galago crassicaudatus.*
(Duncan 1877-83:216)

7 (1′) Dental formula 2/1 1/1 3/3 3/3 = 34; orbit and temporal fossa confluent ventrally (Fig. 19-7); two tarsal bones greatly elongated, tips of digits with large circular pads (Fig. 19-18) .... **Tarsiidae**
tarsiers

7′ Dental formula 2/2 1/1 2-3/2-3 2-3/2-3 = 32 or 36; orbit and temporal fossa separated ventrally by postorbital plate (Fig. 19-15); no tarsal bones elongated; tips of digits without enlarged pads 8

8 (7′) Dental formula 2/2 1/1 3/3 2/2 = 32; third postcanine tooth more like second than fourth in shape and size; second incisors pointed (Fig. 19-15); tail long; hallux with a nail, other digits clawed; pollex present but not opposable .... **Callitrichidae**
marmosets and tamarins

8′ Dental formula 2/2 1/1 3/3 3/3 = 36 or 2/2 1/1 2/2 3/3 = 32; if postcanine teeth number five then third postcanine tooth more like fourth than second in shape and size; second incisors chisel-shaped or pointed; tail varies; all digits with nails; pollex absent or if present at least partly opposable Or if claws and pollex as above then dental formula with six postcanine teeth 9

Figure 19–15.
Skull of a marmoset, *Callithrix chrysoleuca.*
(Mary Ann Cramer)

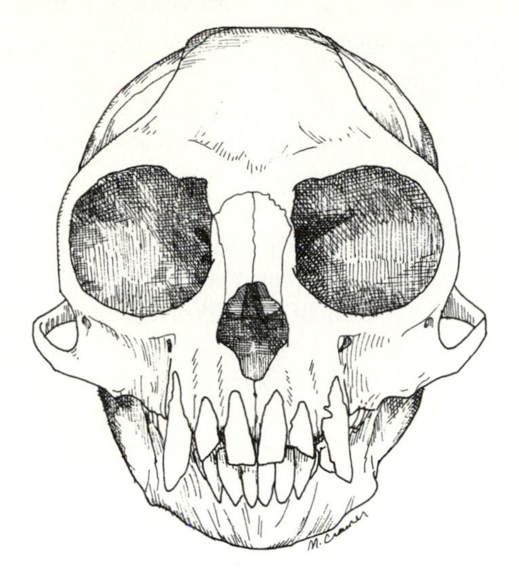

9 (8′)	Dental formula 2/3 1/1 3/3 3/3 = 36; nostrils laterally directed (e.g., Fig. 19-4A); tail present, frequently prehensile 10
9′	Dental formula 2/2 1/1 2/2 3/3 = 32; (Fig. 19-9A); nostrils open forward or down (Fig. 19-4B); tail present or absent, never prehensile 11
10 (9)	Suture between maxilla and premaxilla well defined; hallux with a nail, all other digits clawed; pollex unopposable; tail not prehensile, color black .... **Callimiconidae** Goeldi's monkey
10′	Maxilla and premaxilla fused in adults, suture not distinguishable (Fig. 19-16); all digits with nails; pollex absent or if present at least partly opposable; tail varies (Fig. 19-17), frequently prehensile, color varies .... **Cebidae** New World monkeys

Figure 19–16.
Representative cebids; A, upper and lower tooth rows and skull of the white-throated Capuchin, *Cebus capucinus;* B, skull of a spider monkey, *Ateles* sp.; and C, skull of a howler monkey, *Aloutta* sp.
(Giebel and Leche 1874: pls. 3 and 4)

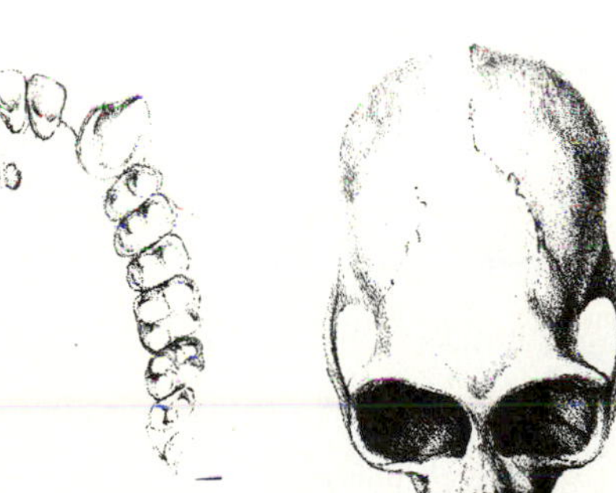

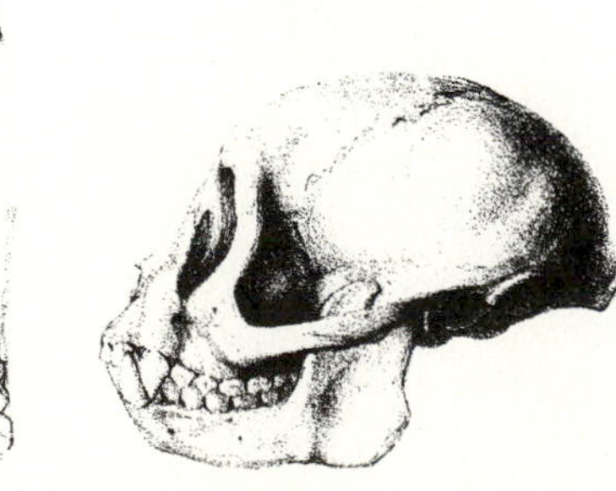

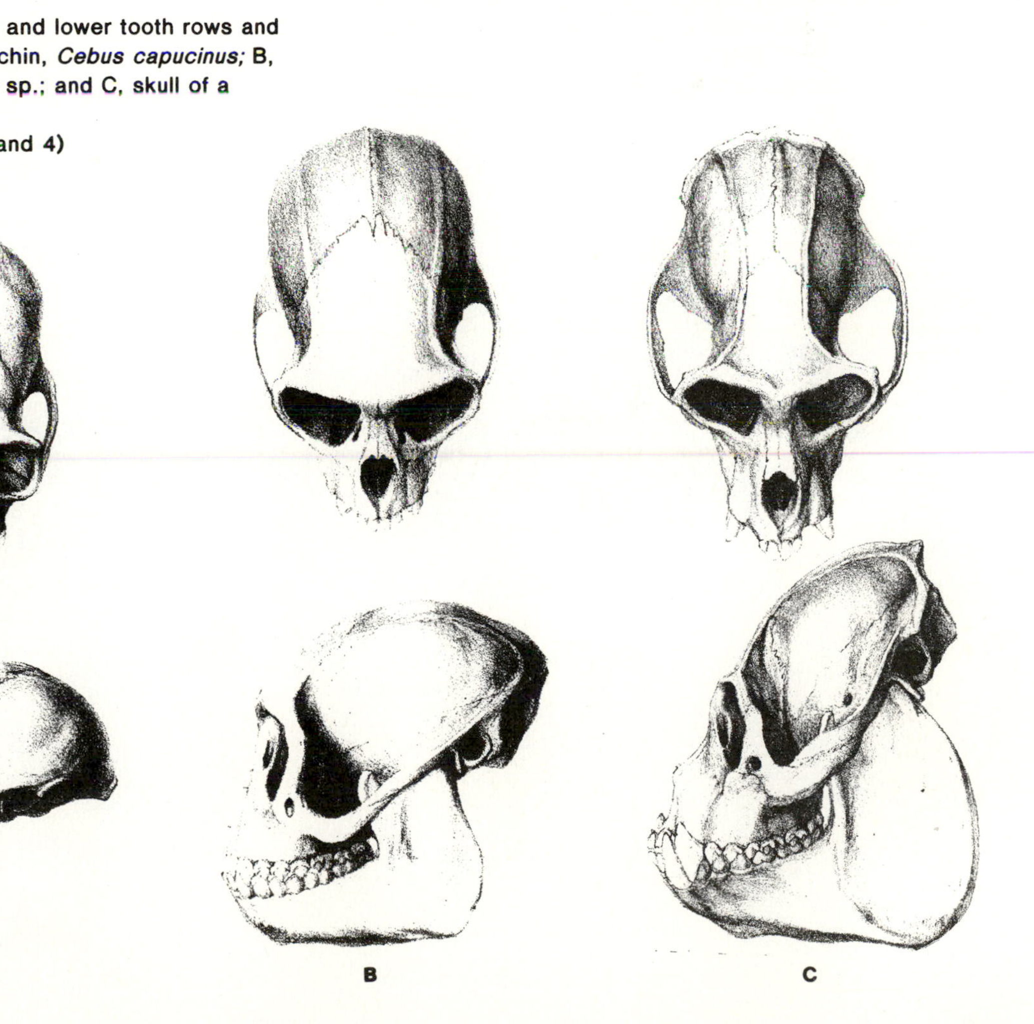

Figure 19–17.
A red ukari, *Cacajao rubicundus*, Cebidae. (Beddard 1902:560)

11 (9′) Greatest length of skull less than 150 mm 12

11′ Greatest length of skull more than 150 mm 13

12 (11) Anterior root of P_1 sloping forward into dentary and exposed for much of its length (Fig. 19-18); wears against upper canine in occlusion; P_1 distinctly different from P_2; tail present (Fig. 19-19) or absent; pollex variable .... **Cercopithecidae** (In part)
most Old World monkeys

12′ Anterior root of P_1 entering dentary vertically or at only a slight anterior angle and unexposed or barely exposed (Fig. 19-20) P_1 similar to P_2; tail absent; pollex reduced .... **Hylobatidae**
gibbons

Figure 19–18.
Skulls of representative cercopithecids. A, a baboon, *Papio ursinus*, showing the distinctive lower premolar; and B, a langur, *Pygathrix nemaeus*, showing the upper tooth row and the more typical skull shape.
(A. Weber 1928:765; B, Flower and Lydekker 1891:725)

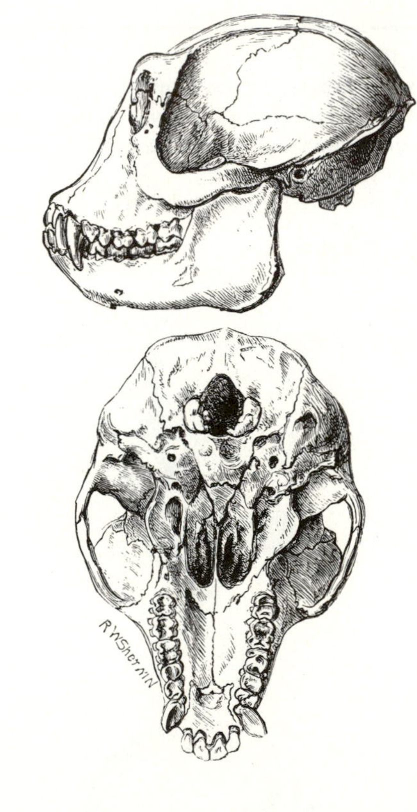

A

B

Figure 19–19.
A Hanuman langur (Cercopithecidae, Colobinae) *Presbytis entellus.*
(Beddard 1902:568)

Figure 19–20.
Dorsal (A) and lateral (B) view of the skull and of the left dentary (C) of a gibbon, *Hylobates syndactylus.*
(A and B, Giebel and Leche 1874: pl. 3; C, Mary Ann Cramer)

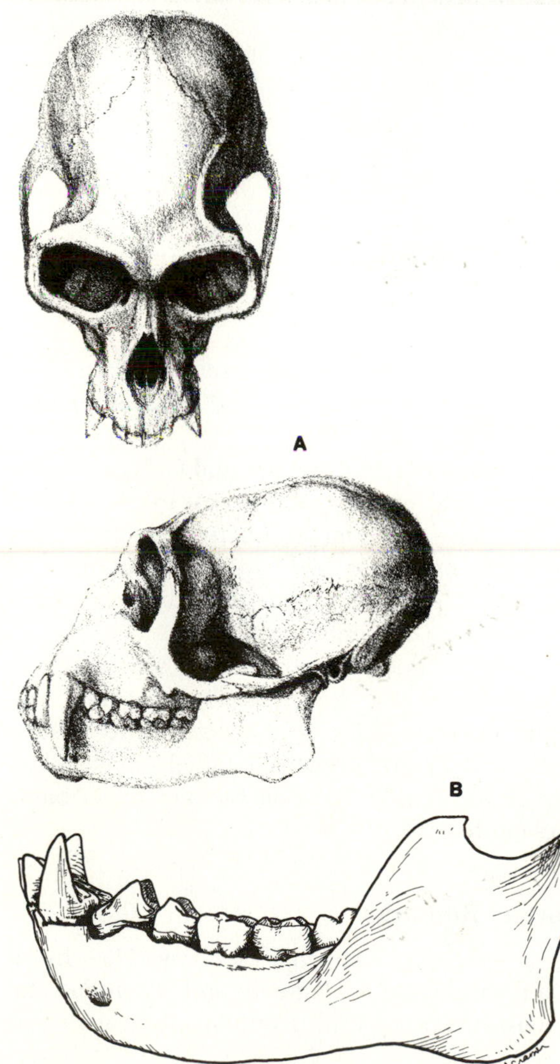

13 (11′) Anterior root of P_1 sloping forward into dentary and exposed for much of its length (Fig. 19-18); longitudinal groove frequently present on anterior surface of C^1 (Fig. 19-21); ischial callosities present .... **Cercopithecidae** (In part)
baboons

13 Anterior root of P_1 entering dentary vertically (Fig. 19-22) or at only a slight anterior angle, unexposed or only barely exposed; canine never grooved; ischial callosities absent 14

14 (13′) Canines usually distinctly larger than adjacent teeth (Fig. 19-23A); rows of cheek teeth parallel (Fig. 19-23A); hallux opposable; forelimb longer than hind limb .... **Pongidae**
gorilla, chimps, and orangutan

14 Canines approximately the same size as adjacent teeth; rows of cheek teeth not parallel, distance between left and right first premolars less than between left and right last molars (Fig. 19-23B); hallux not opposable; hind limb longer than forelimb .... **Hominidae**
man

Figure 19–21.
Skull of a baboon, *Papio ursinus.* Note the longitudinal groove on each canine.
(Giebel and Leche 1874:pl. 4)

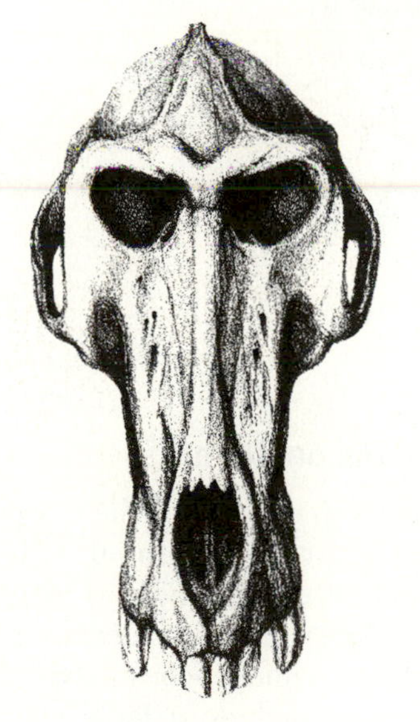

Figure 19–22.
Skulls of the pongids. A, chimpanzee, *Pan troglodytes;* B, gorilla, *Gorilla gorilla;* and C, orangutan, *Pongo pygmaeus.* (A and C, Owen 1866:534, 535; B, Dekeyser 1955:63)

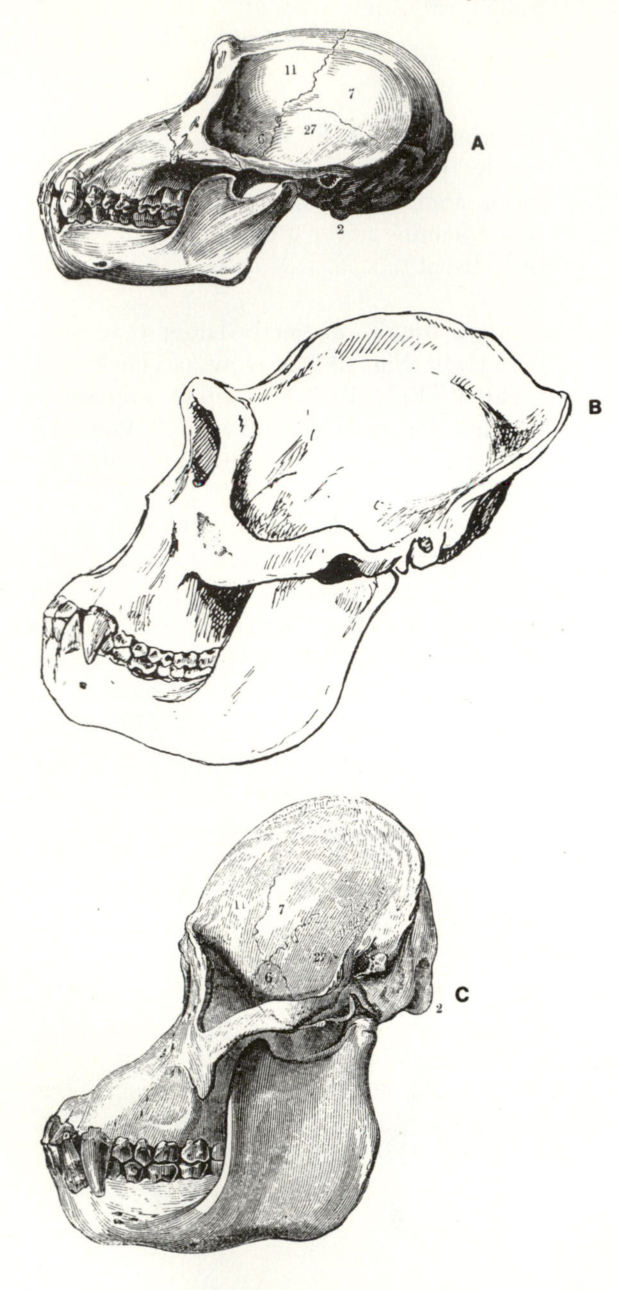

Figure 19–23.
Upper tooth rows of a gorilla (A) Pongidae, *Gorilla gorilla;* and a man (B) Hominidae, *Homo sapiens.*
(A, Duncan 1877-83:20; B, Tomes 1894:481)

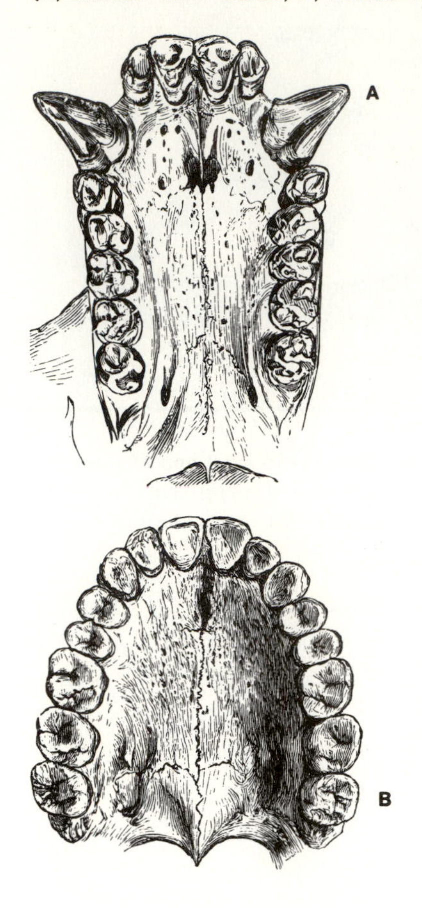

Comments and Suggestions on Identification

Primates as an order are distinctive. Some of the more primitive forms may at first glance be confused with carnivores or rodents, but close examination of general characteristics will reveal their correct placement. Most primate families are rather homogeneous in basic characteristics though they may include a large size range. Some possible sources of confusion are listed below.

The skulls of Daubentoniidae resemble those of rodent. Check the postorbital bar and external characteristics.

Indridae, Lemuridae, Lorisidae and Galagidae have lower incisors and canines resembling the teeth of a comb. Check the tooth number, relative size and shape of rostrum and braincase, and presence or absence of temporal ridges to differentiate these and to distinguish them from Dermoptera.

Some Cebidae and Cercopithecidae are similar in appearance. Check the tooth number, location collected, and characteristics of the nose, third postcanine tooth, and P_1. All prehensile-tailed monkeys are members of the family Cebidae but not all Cebidae have prehensile tails.

Taxonomic Remarks

There is general agreement that the Lemuridae, Indriidae, Daubentoniidae Galagidae and Lorisidae together constitute a natural grouping of primates as

Table 19-2
Classifications of Primates.

Simpson (1945)	Romer (1966)
Order Primates	Order Insectivora
Suborder Prosimii	
Infraorder Lemuriformes	
Superfamily Tupaioidea	
Tupaiidae	Tupaiidae
	Order Primates
Superfamily Lemuroidea	Suborder Lemuroidea
Lemuridae	Lemuridae
Indriidae	
Superfamily Daubentonoidea	
Daubentoniidae	Daubentoniidae
Infraorder Lorisiformes	
Lorisidae	Lorisidae
Infraorder Tarsiiformes	Suborder Tarsioidea
Tarsiidae	Tarsiidae
Suborder Anthropoidea	
Superfamily Ceboidea	Suborder Platyrrhini
Cebidae	Cebidae
Callitrichidae	Callitrichidae
	Suborder Catarrhini
Superfamily Cercopithecoidea	Superfamily Cercopithecoidea
Cercopithecidae	Cercopithecidae
Superfamily Hominoidea	Superfamily Hominoidea
Pongidae	Pongidae
Hominidae	Hominidae

do the Platyrrhini and Catarrhini together. The Tarsiidae have been alternately included in each of these groups and have sometimes been considered to constitute a third group of equal status with these other two. The taxonomic level and name assigned to each of these groups have also varied.

The tree shrews, Tupaiidae, have been included in the Primates by numerous authors (e.g., Simpson 1945; Napier and Napier 1967; Chiarelli 1972; Walker et al. 1975) but Campbell (1974) has reviewed the evidence that Tupaiidae are not closely related to the primates.

The Indriidae are sometimes included in the Lemuridae (e.g. Romer 1966) and the Galagidae are sometimes included in the Lorisidae (e.g. Simpson 1945, Romer 1966). Until Hershkovitz's (1977) recent publication, the Callimiconidae were usually included in the Callitrichidae, but sometimes included in the Cebidae (Wendt 1975a). The Hylobatidae have frequently been included in the Pongidae (e.g. Simpson 1945, Romer 1966). The Colobinae, the leaf-eating monkey subfamily of Cercopithecidae, is sometimes recognized as a distinct family, the Colobidae (e.g. Chiarelli 1972).

The classification of Primates adopted here is that provided by Hershkovitz (1977:9-10). Alternate classifications given by Simpson (1945) and Romer (1966) are listed in Table 19-2. That of Napier and Napier (1967) is identical to that of Simpson (1945) except for recognition of the family Hylobatidae.

Supplementary Readings

Bourne, G. H. (Ed.) 1969. *The chimpanzee.* Vol. I. University Park Press. Baltimore. 466 pp.; 1970, Vol. 2, 417 pp.; 1970, Vol. 3, 402 pp.; 1971, Vol. 4, 407 pp.; 1972, Vol. 5, 273 pp.; 1973, Vol. 6, 406 pp.

Bourne, G. H. 1973. *Nonhuman primates and medical research.* Academic Press. New York. 537 pp.

Carpenter, C. R. 1973. *Behavioral regulators of behavior in primates.* Bucknell Univ. Press., Lewisburg. 303 pp.

Chiarelli, A. B. 1972. *Taxonomic atlas of living primates.* Academic Press, London. 361 pp.

Clutton-Brock, T. (Ed.). 1977. *Primate ecology: studies of feeding and ranging behavior in lemurs, monkeys and apes.* Academic Press, London. 631 pp.

Fiedler, W. and E. Thenius. 1975. The primates, pp. 258-269; Monkeys and apes, pp. 312-325. *In* B. Grzimek (Ed.) *Animal life encyclopedia. Vol. II, Mammals I.* Van Nostrand Reinhold, New York.

Hershkovitz, P. 1977. *Living New World Monkeys (Platyrrhini).* Vol. I. Univ. of Chicago Press, Chicago. 1117 pp.

Holloway, R. L. (Ed.). 1974. *Primate aggression, territoriality and xenophobia.* Academic Press, New York. 514 pp.

Hrdy, S. B. 1977. *The langurs of Abu, female and male strategies of reproduction.* Harvard Univ. Press, Cambridge. 361 pp.

Kortlandt, A. and D. Heinemann. 1975. Chimpanzees, pp. 19-48. *In* B. Grzimek (Ed.) *Animal life encyclopedia. Vol. 11, Mammals II.* Van Nostrand Reinhold, New York.

le Gros Clark, W. E. 1971. *The antecedents of man.,* 3rd Ed. Quadrangle Books, Chicago. 394 pp.

Martin, R. D; G. A. Doyle; and A. C. Walker (Eds.) 1974. *Prosimian biology.* G. Duckworth and Co., London, 983 pp.

Michael, R. P. and J. H. Crook. 1973. *Comparative ecology and behavior of primates.* Academic Press, London. 847 pp.

Napier, J. R. and P. H. Napier. 1967. *A handbook of living primates.* Academic Press, London. 456 pp.

Rumbaugh, D. M. 1977. *Language learning by a chimpanzee, the Lana project.* Academic Press, New York. 312 pp.

Schaller, G. B. 1963. *The mountain gorilla, ecology and behavior.* Univ. of Chicago Press, Chicago. 431 pp.

Sussman, R. W. (Ed.) 1978. *Primate ecology: Problem-oriented field studies.* John Wiley & Sons, New Jersey.

Tuttle, R. H. (Ed.) 1975. *Primate functional morphology and evolution.* Mouton Publ., The Hague. 583 pp.

van Lawick-Goodall, J. 1971. *In the shadow of man.* Houghton Mifflin, Boston. 297 pp.

20 The Anteaters, Sloths and Armadillos

Order Edentata

ORDER EDENTATA

The ordinal name, Edentata, means literally "without teeth." This is misleading since only the members of one of the edentate families lack teeth. The members of the other families have simple, usually homodont teeth. Edentata includes three strikingly different groups of living mammals, the anteaters, tree sloths, and armadillos (Fig. 20-1).

The armadillos, family Dasypodidae, are terrestrial to fossorial mammals possessing a carapace over much of the body. This shell of dermal bony scutes overlaid with epidermal scales (see section on scales in Chapter 5) ranges from a full body armor that completely protects the animals when they roll into a ball to a greatly reduced carapace only loosely attached to the

Figure 20–1.
Representatives of the three living families of Edentata. A, a two-toed sloth, *Choloepus hoffmani;* B, the tamandua, *Tamandua tetradactyla;* and C, the nine-banded armadillo, *Dasypus novemcinctus.*
(A, Sclater and Sclater 1899:57; B and C, Flower and Lydekker 1891:193, 200)

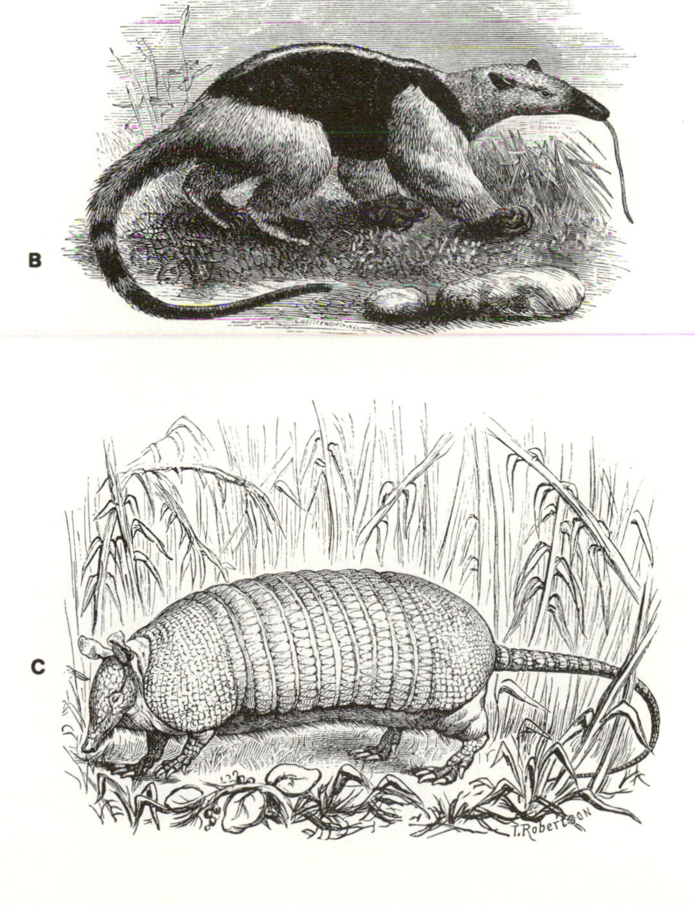

body. Most armadillos are only sparsely haired but the lesser pichiciago, *Chlamyphorus truncatus,* has dense hair over the portions of the body without a shell or to which a shell does not attach, and the rare *Dasypus pilosus* has such dense hair over the trunk that its full body armor is concealed (Moeller 1975). Most armadillos are primarily terrestrial but all are well equipped for digging (Fig. 20-2) for food and burrowing. The pichiciagos, *Chlamyphorus* and *Burmeisteria,* are almost fully fossorial. Armadillos have cylindrical, homodont, evergrowing teeth. Deciduous teeth have been recorded only in the genus *Dasypus.* Most armadillos feed exclusively or primarily upon insects and insect larvae but other invertebrates, small vertebrates, carrion, eggs, berries, fungi and other plant material may be eaten (Davis 1974, Moeller 1975).

The tree sloths, family Bradypodidae, are arboreal animals with long limbs (Fig. 20-1A), completely syndactylous toes, and large curved claws (Fig. 20-3A) used to hang from tree limbs. They are covered by coarse hair that harbors blue-green algae growing in grooves in the hair cuticle. This gives the animal's gray-brown, often mottled, coloration a greenish tinge that aids protective coloration (Moeller 1975). Sloths are strictly vegetarian, eating leaves, buds, flowers, and fruit. The teeth are cylindrical, evergrowing and essentially homodont. In the genus *Choloepus* the anteriormost tooth in each jaw is larger than the others and is somewhat caniniform. Sloths and manatees (Sirenia: Trichechidae) are the only mammals that possess cervical vertebrae that number more or fewer than seven. Both species of three-toed sloths, *Bradypus,* have nine vertebrae in their long, very flexible necks. Some two-toed sloths, *Choloepus,* (and the manatees) have only six cervical vertebrae. Other *Choloepus* have the normal component of seven. The tail is rudimentary in all sloths.

The anteaters, family Myrmecophagidae, include three genera. The giant anteater, *Myrmecophaga tridactyla,* (Fig. 20-4) is a large (TL 165-390 cm, Wt. 30-35 kg) terrestrial anteater. Its tongue is about 60 cm long and can be protruded and withdrawn into the mouth up to 160 times per minute (Moeller 1975). The tamanduas, *Tamandua tetradactyla* and *T. mexicana* (Fig. 20-1B) are medium-sized (TL 108-114 cm, Wt. 3-5 kg) anteaters that are both terrestrial and arboreal. The pygmy or two-toed anteater, *Cyclopes didactylus,* is a small (TL 34-40 cm, Wt. 500 g) anteater that is completely arboreal. Both *Tamandua* and *Cyclopes* have prehensile tails. All of the anteaters have long rostra and very small mouths (Fig. 20-4). They lack teeth and feed primarily on insects. Their large foreclaws (Fig. 20-5) are used to tear open termite mounds,

Figure 20–2.
Lateral and medial views of the right forefoot of the eleven-banded armadillo, *Cabassous unicinctus.* (Pocock 1924C:1005)

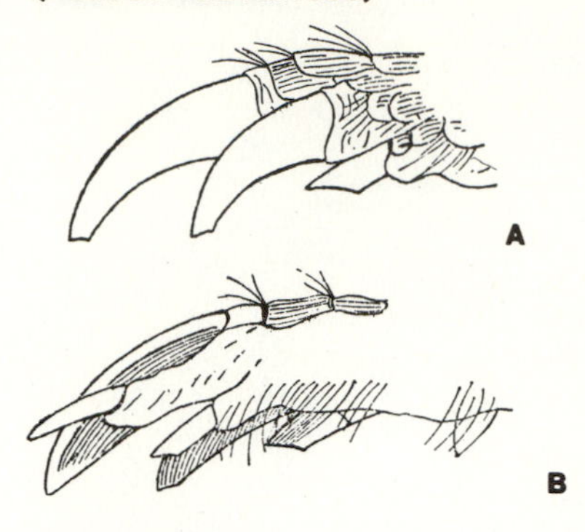

Figure 20–3.
Lateral and ventral views of the right forefoot (A) of the three-toed sloth, *Bradypus tridactylus,* and the ventral view of the right forefoot (B) of a two-toed sloth, *Choloepus.* (Pocock 1924C:1016)

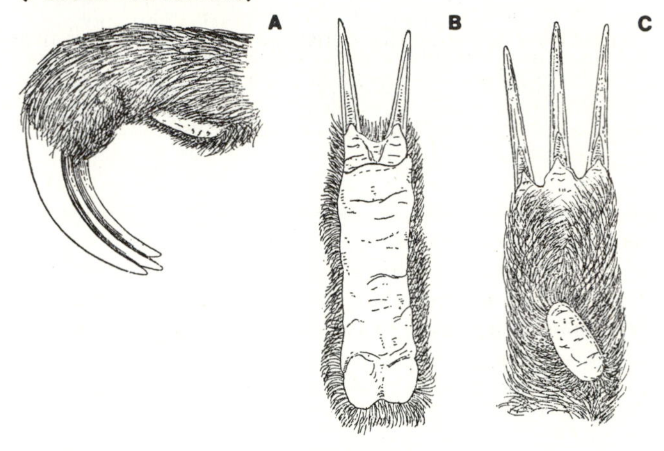

Figure 20–4.
The giant anteater, *Myrmecophaga tridactyla.* (Flower and Lydekker 1891:192)

Figure 20–5.
Forefeet of a tamandua, *Tamandua* sp. (A), and the pygmy anteater, *Cyclopes didactylus* (B). Not to same scale. (Pocock 1924c:1011, 1013)

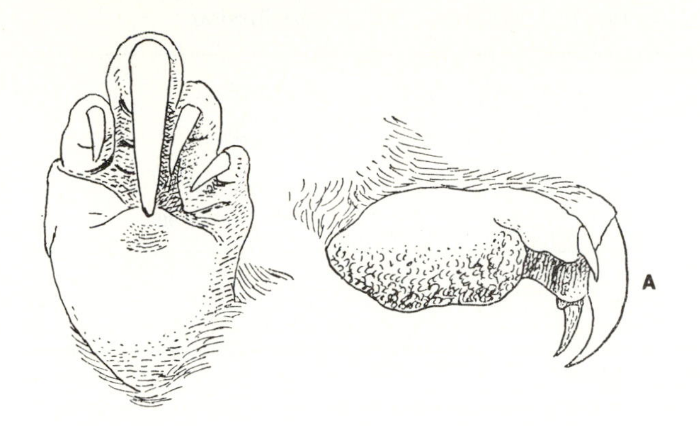

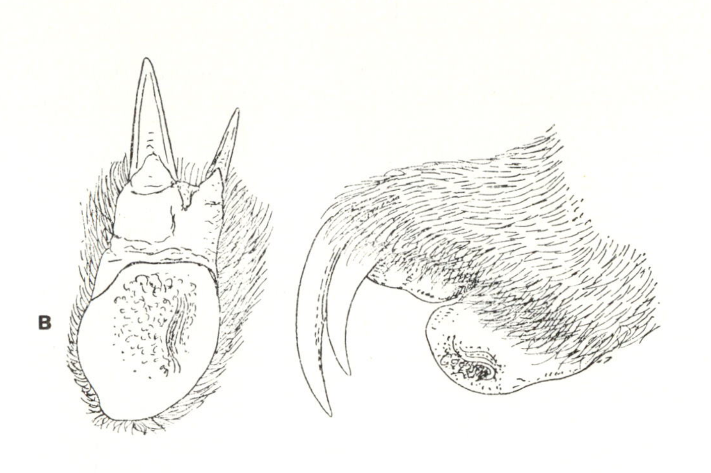

rotting wood and other insect habitats. The insects, their larvae and pupae are then gathered by the long, protrusible, vermiform, mucous coated tongue.

Edentates are of minor economic importance. Some armadillos occasionally consume eggs of ground nesting birds but this habit probably is overstated by those interested in game birds (Davis 1974). They do eat quantities of insects (Davis 1974). Sloths are heavily hunted for meat in Brazil and their pelts are used.

Distinguishing Characteristics

Teeth are completely absent in the Myrmecophagidae. The other two families lack incisors and canines but have cheek teeth that are subcylindrical, lack enamel, and have a single root. In sloths, teeth do not number more than 5/4-5, but in armadillos the tooth number varies from 7/7 to 28/28. The skulls of anteaters and armadillos are essentially conical in shape while those of tree sloths are cuboid. The posterior trunk vertebrae have extra articular surfaces (**xenarthrales**) that provide particularly strong articulations between vertebrate (Fig. 20-6). The uterus is duplex and testes are abdominal. Bacula are absent (Sonntag 1925).

Distribution

Living edentates are confined to the Neotropical Region except for one species. The nine-banded armadillo, *Dasypus novemcinctus,* ranges into the southwestern Nearctic and has been introduced into Florida (Humphrey 1974).

Figure 20–6.
Vertebrae of the giant anteater, *Myrmecophaga tridactyla.* A, side view of twelfth and thirteenth thoracic vertebrae. B, posterior view of second lumbar vertebra. C, anterior surface of third lumbar vertebra. az, anterior zygapophysis; az', az², az³, additional (xenarthrous) anterior articular facets; cc, facet for capitulum of rib; m, metapophysis; pz, posterior zygapophysis; pz', pz², pz³, additional (xenarthrous) posterior articular facets; t, transverse process; tc, facet for articulation of tubercle of rib. (Flower 1885:62–63)

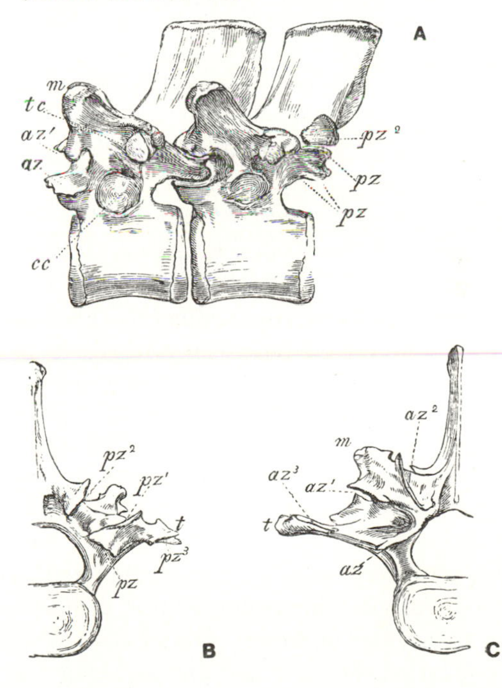

Table 20-1
Living Families of Edentata.

Family	Common Name	Number of* Genera	Number of* Species	Distribution
Myrmecophagidae	anteaters	3	4	Neotropical
Bradypodidae	tree sloths	2	5	Neotropical
Dasypodidae	armadillos	7	16	Neotropical, south Nearctic

*Based upon Moeller (1975).

Fossil History

Neither fossil nor living edentates are known outside the Western Hemisphere. An extinct group, †Palaeadonta, which existed in North America from the late Paleocene to the Oligocene, was considered to be a primitive suborder of Edentata (Simpson 1931, Romer 1966, Vaughn 1978), however Emry (1970) has shown that they are primitive Pholidota rather than Edentata. Thus all known Edentata are in the suborder Xenarthra, characterized by extra points of articulation in posterior trunk vertebrae. Xenarthra are known only from South America until the upper Pliocene. A few genera occur in the West Indies and Central and North America during the Pleistocene and one armadillo occurs in the Nearctic during recent times.

The oldest known xenarthrous specimens are armadilloid scutes from the Paleocene and an armadillo from the early Eocene. Some early armadillos had scutes forming pointed horns on their snouts (Fig. 20-7). †Glyptodonts (Fig. 20-8) appeared in the late Eocene. These animals resembled armadillos in having a dermal bony shell but the shell was one piece rather than a series of plates separated by flexible bands as it is in all Dasyuridae. Several glyptodonts were very large and one species was the size of a present day rhinocerous. Glyptodonts were numerous in South America until the end of the Pleistocene. They also entered North America during the Pleistocene but at the end of this epoch they became extinct on both continents.

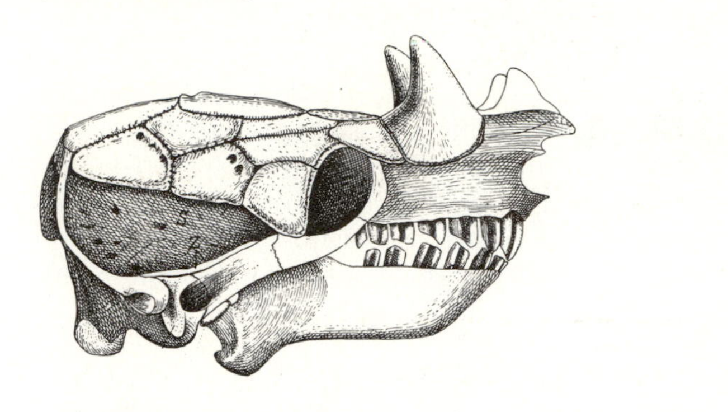

Figure 20–7.
†*Peltephilus*, an armadillo from the lower Miocene bearing horn-like dermal scutes (h) on the rostrum.
(Weber 1928:226)

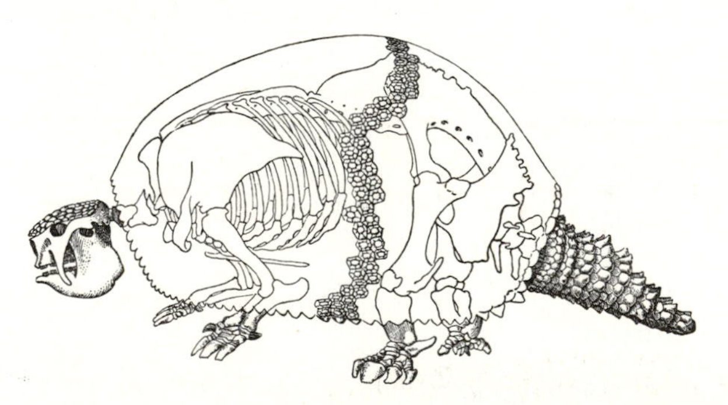

Figure 20–8.
A glyptodont. The shell is shown only in outline except for the tail and a medial band of scutes.
(Weber 1928:227)

Tree sloths, Bradypodidae, are unknown before the recent but their relatives, the ground sloths, first appeared in the Oligocene of South America. Ground sloths flourished during the later Tertiary, reaching the West Indies and North America in the upper Pliocene. Although many ground sloths were the size of living edentates, several became very large. †*Megalonyx* was the size of an ox and †*Megatherium* was larger than a recent elephant. Ground sloths were abundant in the Pleistocene and serveral forms survived the last glacial advance. These survivors were contemporaries of man in the southwestern United States, in Patagonia and on Hispaniola.

Anteaters are poorly represented as fossils and only a few forms are known from the late Miocene, Pliocene and Pleistocene of South and Central America.

Key to Living Families

1 Skull squared, rostrum blunt, not tapered (Fig. 20-9); never more than 3 exposed digits, all syndactylous (Fig. 20-3) .... **Bradypodidae**
tree sloths

1′ Skull conical, rostrum long and tapering (Figs. 20-10, and 20-11); 4 or 5 digits on each foot; (only two on forefoot if pigmy anteater, *Cyclopes didactylus*, Fig. 20-5B); none syndactylous 2

2 Teeth absent (Fig. 20-10); body covered with thick coat of hair; no dermal bone .... **Myrmecophagidae**
anteaters

2′ Teeth present (Fig. 20-11); body sparsely haired in most species; a shell of dermal bone covered with horny epidermal scales is present .... **Dasypodidae**
armadillos

Figure 20–9.
Skulls of A, a two-toed sloth, *Choloepus*, and B, a three-toed sloth, *Bradypus*.
(A, Weber 1928:194; B, Giebel and Leche 1874:pl. 28-5)

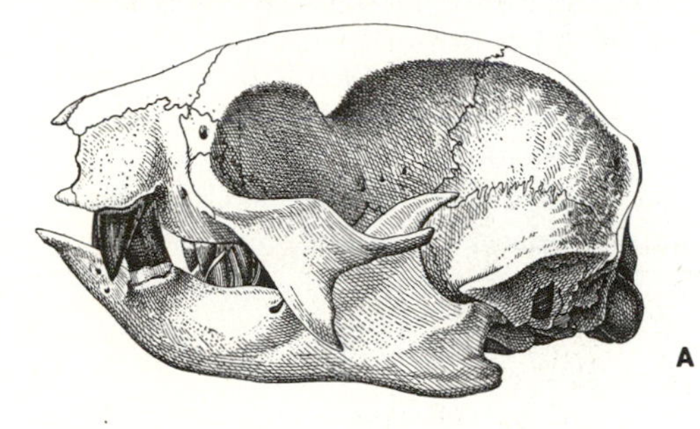

Figure 20–10.
Skulls of A, the tamandua, *Tamandua tetradactyla;* and B, the giant anteater, *Myrmecophaga tridactyla.* The skull of *Cyclopes didactylus* has a proportionally shorter rostrum than these two. Not to same scale.
(A, Giebel and Leche 1874:pl. 29-3; B, Owen 1866:403)

Figure 20–11.
Skulls of A, the giant armadillo, *Priodontes giganteus;* B, the nine-banded armadillo, *Dasypus novemcinctus.* Not to same scale.
(A, Gervias 1855:253; B, Baird 1859; pl. 86)

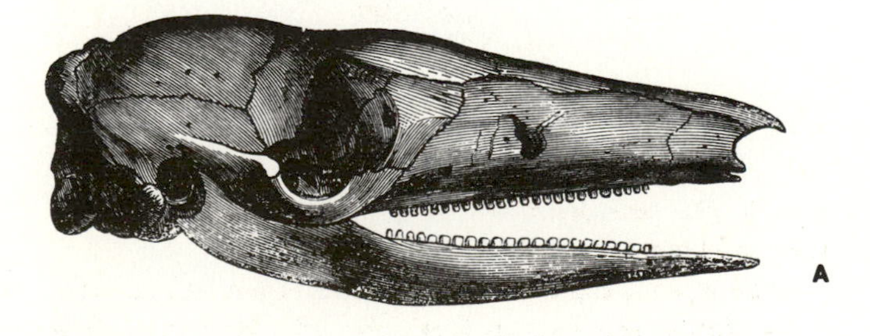

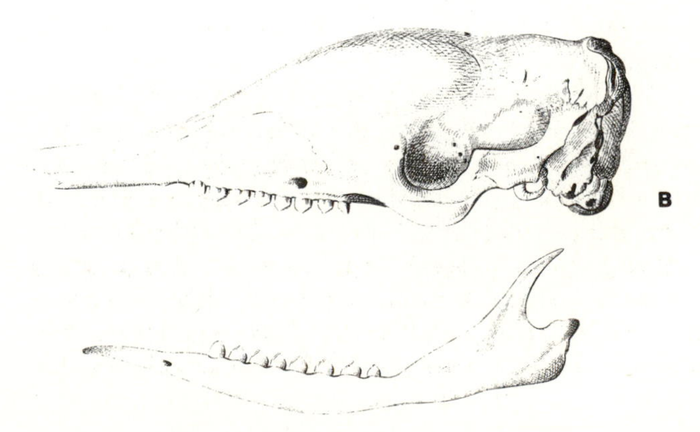

Comments and Suggestions on Identification

The three families of edentates are very distinctive and cannot be confused with each other. Sloths are unique in both external and cranial features. The bony shells of armadillos are unique as skin features. Confusion could arise between skulls of large armadillos and those of aardvarks, but close examination of teeth easily separates these two groups. The toothless skulls of smaller anteaters could be confused with toothless skulls of echidnas and pangolins. Methods of distinguishing these three groups are discussed in the chapter on Monotremata.

Taxonomic Remarks

The order Edentata was formerly considered to include the pangolins (or scaly anteaters) and the aardvarks. Each of these groups is now placed in a distinct order, the Pholidota and Tubulidentata, respectively. Melton (1976:77) has reviewed the history of this classification. Because the name Edentata was long linked with several forms now known to be unrelated some persons have advocated the use of the ordinal name Xenarthra for the New World edentates (e. g. Davis 1974; Hershkovitz 1977). However Simpson (1945:191) argued that Edentata was really Xenarthra plus the †Palaeanodonta. Most others have followed Simpson's usage of the ordinal name Edentata, but many have felt the need to explain what Edentata *did not* include. Now that Emry (1970) has shown that †Palaeanodonta do not belong in the same order with the Xenarthra, Xenarthra becomes a logical ordinal name for this group. The classification of living forms used in this chapter is based upon Moeller (1975).

Supplementary Readings

Goffart, M. 1971. *Function and form in the sloth.* Pergamon Press, Oxford. 225 pp.

Humphrey, S. R. 1974. Zoogeography of the nine banded armadillo (*Dasypus novemcinctus*) in the United States. *BioScience* 24:457-462.

Long, A., R. M. Hanson, and P. L. Martin, 1974. Extinction of the Shasta ground sloth. *Geol. Soc. Amer. Bull.* 85:1843-1848.

Moeller, W. 1975. Edentates. pp. 149-153 *In* B. Grzimek (Ed.) *Animal life encyclopedia, Vol. 11, Mammals II.* Van Nostrand Reinhold, New York.

Wetzel, R. M. 1975. The species of *Tamandua* Gray (Edentata, Myrmecophagidae). *Proc. Biol. Soc. Wash.* 88 (11):95-112.

21 The Pangolins

Order Pholidota

ORDER PHOLIDOTA

The ordinal name, Pholidota, means literally "scaly ones" and points out the major diagnostic characteristic of this group. The pangolins or scaly anteaters are covered with large epidermal scales (Fig. 21-1) that usually are described as being formed by agglutinated hairs. However, Rahm (1975a:183-184) states that

> The scales are not "glued hairs," as was once believed; they are two-sided symmetrical elevations of the skin which are flattened from the back toward the stomach, and which face the back of the animal. The epidermis, as it grows into horn material, leads to scale formation. The horn scales, which are lost through wear, are constantly replaced by the skin base.

Figure 21–1.
An arboreal pangolin, *Manis tricuspidus*. (Flower and Lydekker 1891:206)

> Throughout its life, the pangolin always has the same number of scales. The pattern, quantity, shape, and size of the scales differ from species to species, and can also differ slightly within a species, depending on the part of the body they cover.

Some pangolin species (Fig. 21-1) have a long prehensile tail and are arboreal, others have a short, rather blunt tail and are fully terrestrial (Fig. 21-2). Pangolins are insectivorous, gathering ants or termites with their long, vermiform, sticky tongue. In the four African species, the xiphisternum is greatly elongated, passing between the peritoneum and lower abdominal wall ventral and posterior to the abdominal cavity. It then curves back dorsally to the region of the kidneys where it forms a support for the base of the greatly elongated tongue (Emry 1970:460). The tongue itself, in these species, extends well back into the chest cavity when retracted (Rahm 1975a:184). Teeth are

Figure 21–2.
A terrestrial pangolin, *Manis temmincki*. (Sclater 1901:217)

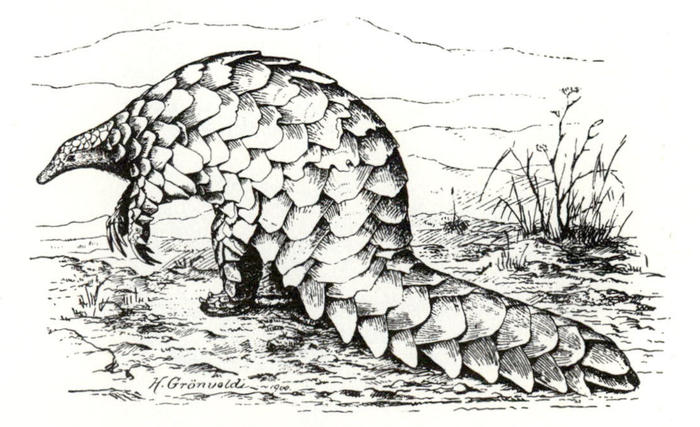

Table 21-1

Living Families of Pholidota.

Family	Common Name	Number of* Genera	Number of* Species	Distribution
Manidae	pangolins	1	7	Ethiopian and Oriental

*Based upon Rahm (1975a). Pocock (1924b) split essentially the same species into six genera. Simpson (1945) reunited these into one genus. Emry (1970) listed only one genus, but stated his belief that the appropriate number of genera lies somewhere between Pocock's six and Simpson's one.

absent and insects are crushed in the stomach where a horny layered epithelium with horny teeth replaces the usual mucous membrane lining.

Distinguishing Characteristics

Teeth are completely absent. The skull is robust and conical and the mandible is slender. The zygomatic arch is incomplete (Fig. 21-3). The limbs are planti-grade, pentadactyl, and possess large claws. The tail is prehensile in arboreal species. The top of the head, the top and sides of the neck, body and limbs, and all parts of the tail are covered with large leaf-shaped epidermal scales. The undersides of the head, neck, and body are scaleless and hairy. The arboreal forms have a scaleless spot at the tip of the tail. The uterus is bicornuate and the testes are inguinal but never scro-tal. Bacula have not been reported.

Figure 21–3.
Skull of a pangolin, *Manis* sp.
(Giebel 1859:317)

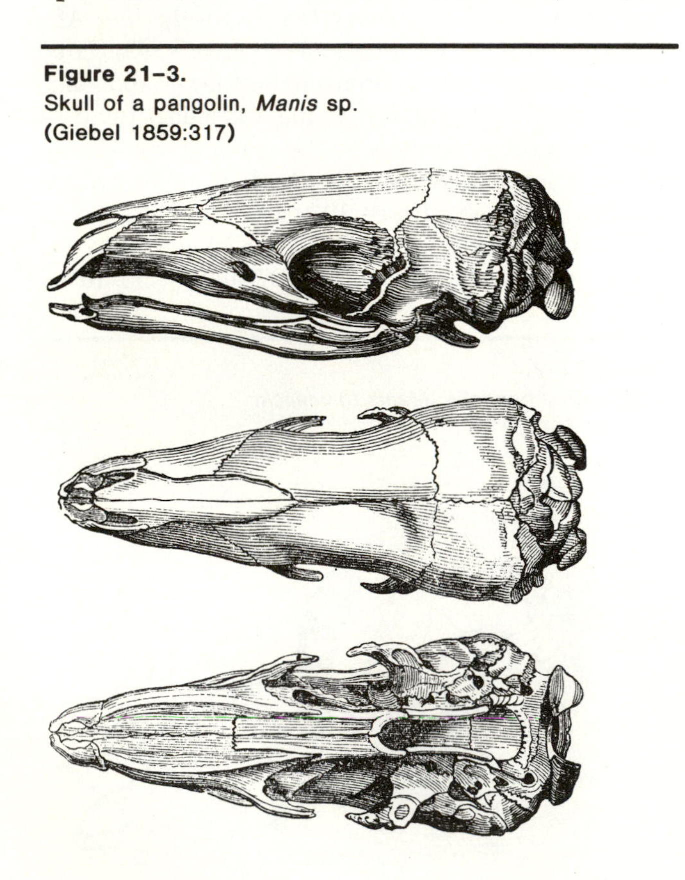

Distribution

Living pholidotes are confined to the Ethiopian and Oriental Regions. Both arboreal and terrestrial forms are found in both regions.

Fossil History

Emry (1970) has described an Oligocene pangolin from North America. This is sufficiently like *Manis* to be placed in the family Manidae. This family is also known from the Oligocene, Miocene and Pleistocene of Eur-ope and the Pleistocene of southeast Asia. Emry (1970) has also shown that the †Palaeanodonta, a poorly known group of North American Eocene and Oligo-cene mammals, are in fact early pangolins rather than early edentates as was previously supposed (Simpson 1931). Emry (1970:507) lists the palaeanodonts as two completely extinct families, †Metacheiromyidae and †Epoicotheriidae, in the order Pholidota.

Comments and Suggestions on Identification

Pangolin skins are unique and cannot be confused with those of any other mammals. The toothless skulls could be confused with those of echidnas or South American anteaters. See chapter 14, Monotremes, for comments on distinguishing these three groups.

Taxonomic Remarks

Pangolins were long included with the Xenarthra in the order Edentata, but are now recognized as a dis-tinct order. Melton (1976) has reviewed the history of this classification. Emry (1970) has reviewed the clas-sification of fossil forms.

Supplementary Readings

Emry, R. J. 1970. A North American Oligocene pangolin and other additions to the Pholidota. *Bull. American Mus. Nat. Hist.* 142:457-510.

Pocock, R. I. 1924. The external characters of the pangolins (Manidae). *Proc. Zool. Soc. London.* 1924:707-723.

Rahm, U. 1975. Pangolins. pp. 182-188 *In* B. Grzimek (Ed.) *Animal life encyclopedia, Vol. 11, Mammals II.* Van Nostrand Reinhold, New York.

22 The Whales

Order Mysticeta
Order Odontoceta

Cetacea, derived from the Greek word meaning whale, is the name often used for an order of mammals with two well defined suborders, the Mysticeti or baleen whales and the Odontoceti or toothed whales. There is no disagreement about what species compose Mysticeti and which Odontoceti but the relationship of the two groups to each other is unclear. The two groups are very similar in many ways but there are striking and basic differences that have prompted several authors to ascribe the similarities to convergent evolution.

The Mysticeta and Odontoceta are known to have been distinct at least since the late Eocene. Kleinenberg (1958) ranked them as separate orders as have many other authors. We follow Rice (1967, 1977) in listing them as separate orders. Both orders are fully aquatic with fusiform bodies and tails flattened dorsoventrally into a pair of large flukes. These flukes have no skeleton other than rudimentary caudal vertebrae and are supported by fibrous connective tissue (Fig. 22-1). A dorsal fin supported by similar connective tissue is present in all but a few species in each order but the height of this fin varies from 22.5% of the body length in the killer whale, *Orcinus orca,* to 1.5% in the sperm whale, *Physeter macrocephalus* (Fig. 22-1B) and 1% in the blue whale, *Balaenoptera musculus,* (Morzer Bruyns 1971:189-190).

Posterior limbs are absent externally but remnants of the pelvic girdle and femur may be present internally (Fig. 22-1). The anterior limb is enclosed in the body contour to the wrist and exposed manus is termed a flipper. The humerus, radius and ulna are greatly shortened and phalanges of at least digits two and three greatly exceed the usual mammalian number

Figure 22–1.
Skeletons and body outlines of representatives of the two orders of whales. Mysticeti (A) *Balaena glacialis,* the right whale and Odontoceta (B) the sperm whale, *Physeter macrocephalus.*
(Lydekker 1909:13, 23)

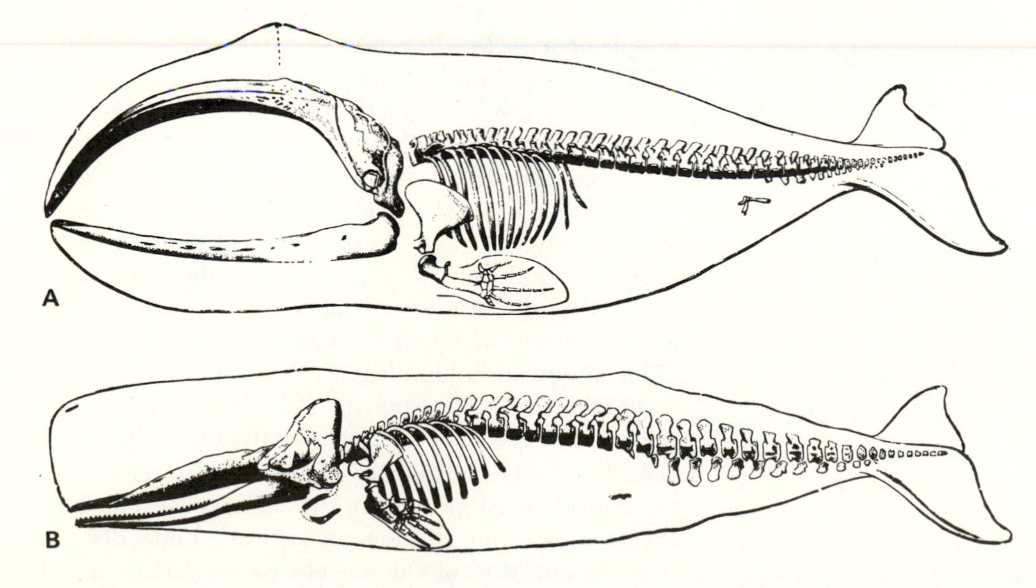

Figure 22-2.
Skeleton of the right forelimb of a longfin pilot whale, *Globicephala melaena.* H, humerus; U, ulna; R, radius; c, l, s, td, and u, carpal bones; digits one through five are indicated by Roman numerals.
(Flower 1885: 302)

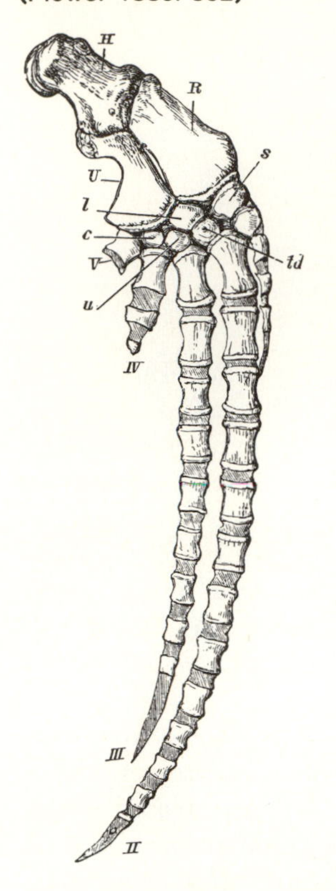

Figure 22-3.
Skulls of (A), the right whale, *Balaena glacialis,* Mysticeta, and (B), the saddleback dolphin, *Delphinus delphis,* Odontoceta. C, occipital condyle; Fr, frontal; Ju, jugal; Mx, maxilla; n, external nare; Na, nasal; Oe, exoccipital; Os, supraoccipital; Pa, parietal; Pal, palatine; Pt, pterygoid; Px, premaxilla; Sq, squamosal; Ty, tympanic bulla.
(Weber 1928: 358-359)

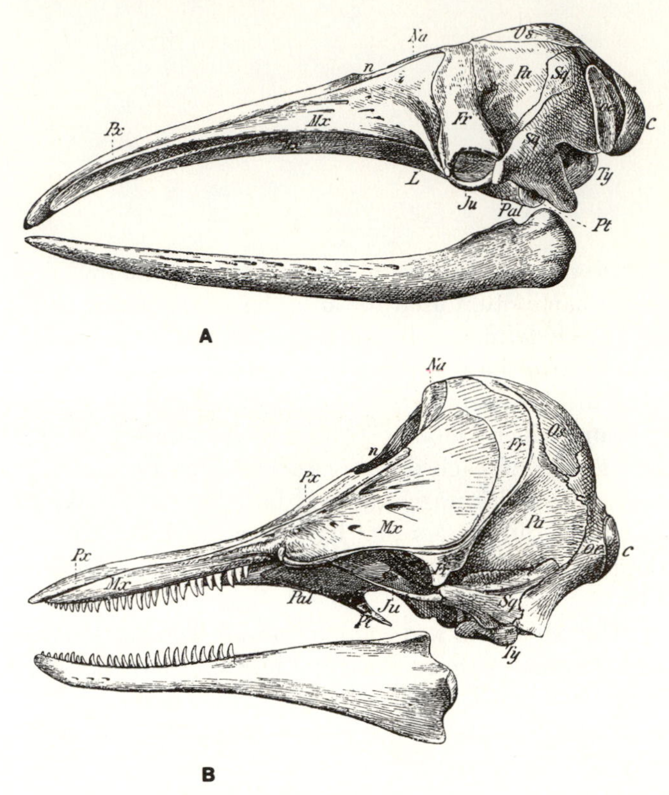

(Fig. 22-2). The nostrils, termed blowholes, are located high on the dorsal surface of the head (Fig. 22-1) and the external nares are located at the proximal end of the rostrum of the skull (Fig. 22-3). To allow for these posteriorly displaced nasal openings, the nasal, maxilla and frontal bones are telescoped and overlap the parietal (Figs. 22-3 and 22-4).

The neck is very short for the length of the animal. Seven cervical vertebrae are present but they may be very thin and fused together (Fig. 22-5). The necks of most whales are virtually inflexible, the white whale or beluga, *Delphinapterus leucas,* is a notable exception.

The skin is essentially hairless and a thick layer of subcutaneous fat provides insulation. The eyes are small and ear pinae are absent. The uterus is bipartite and the urethra and vagina open separately to the exterior. The penis is completely retractile into the body contour and the testes are permanently abdominal.

As air breathing, fully marine animals, the whales have numerous anatomical, physiological and behavioral specializations that allow them to dive to great depths and to remain under water for long periods, to move rapidly through water, to travel great distances and to communicate and reproduce in the vast oceans. For detailed discussions of these mechanisms consult Slijper (1962, 1976), Slijper and Heinemann (1975), Coffey (1977), or other works listed in the Supplementary Readings section.

The oldest fossil whales are the Archaeoceti (Fig. 22-6). They are known from the lower Eocene to the lower Miocene, from Europe, Africa, North America, Australia, and New Zealand (Romer 1966:392-393). The Archaeoceti differ from the Mysticeta and Odontoceta in several ways but the oldest of them are well adapted as fully aquatic mammals, and thus must have been derived from terrestrial ancestral stock in the lower Tertiary or upper Cretaceous (Thenius 1975b: 475). The relationship of Archaeoceti to the Mysticeta and Odontoceta is unknown. If all three groups arose from a common cetacean ancestor the three should be considered suborders of the single order Cetacea. If, as some propose (Slijper 1962), the Archaeoceti and Mysticeta arose from a common ancestor quite distinct from the ancestors of Odontoceta, the Archaeoceta and

Figure 22-4.
Skull of an odontocete, the longfin pilot whale, *Globicephala melaena.* A, ventral view and B, longitudinal section. an, anterior nares; As, alisphenoid; BS, basisphenoid; BO, basi-occipital; Cd, condyle; cp, coronoid process; ExO, exoccipital; Fr, frontal; gf, glenoid fossa of squamosal; IP, interparietal; Ma, jugal; ME, ossified portion of mesethmoid; Mx, maxilla; Na, nasal; OS, orbitosphenoid; Pa, parietal; Per, periotic; Pl, palatine; pn, posterior nares; PS, presphenoid; Pt, pterygoid; SO, supraoccipital; Sq, squamosal; tg, deep groove in squamosal for external auditory meatus; Ty, tympanic; Vo, vomer; ZM, zygomatic process of jugal.
(Flower 1885: 211 and 214)

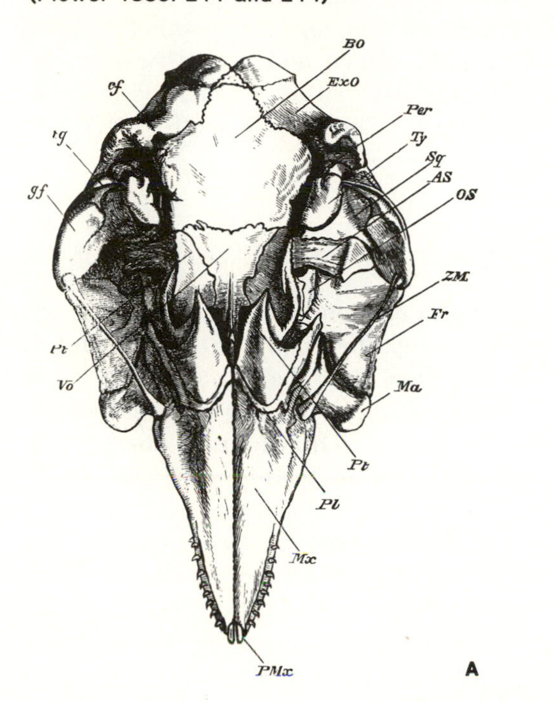

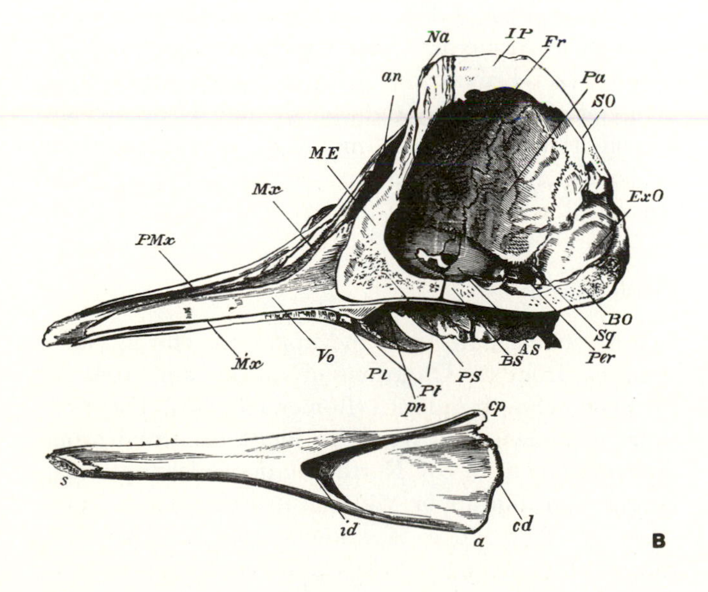

Figure 22-5.
Fused cervical vertebrae of a bowhead whale, *Balaena mysticetus.* a, articular surface for occipital condyle; 1-7, the seven cervical vertebrae.
(Flower 1885: 44)

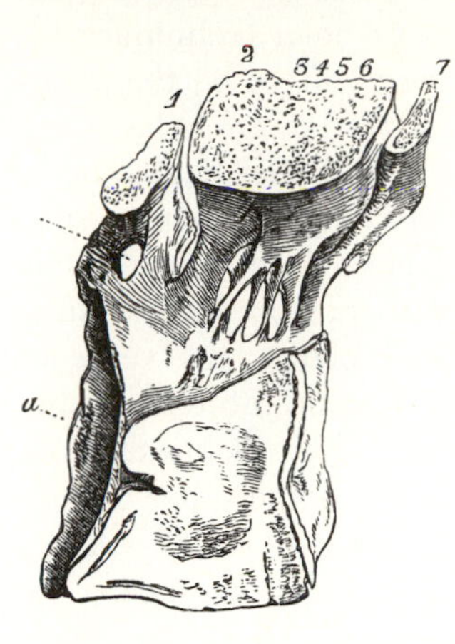

Figure 22-6.
Skull of †*Zygorhiza kochii,* Archaeoceti, a whale from the upper Eocene of Alabama.
(Kellogg 1936:108)

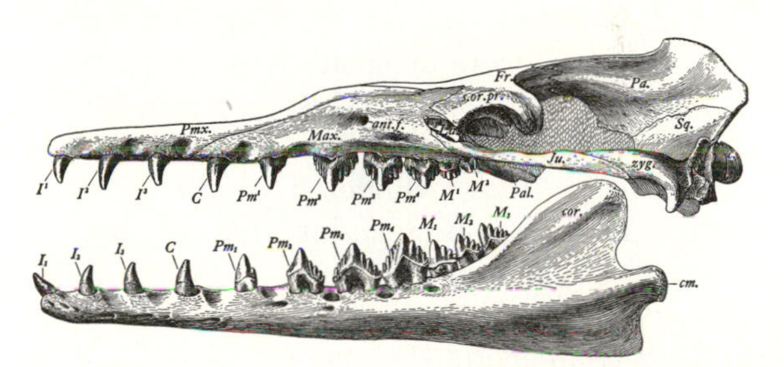

Mysticeta should be regarded as a single order and Odontoceta as another. All authors agree that the Archaeoceti are not themselves ancestral to either of the modern whale groups. Vaughn (1978:200) lists Archaeoceti as a distinct order.

Whales were of major economic importance in the eighteenth and nineteenth centuries but their commercial importance has declined over the last several decades. Whales were hunted by coastal peoples in many parts of the world and a commerical industry developed in Europe, North America, Japan, USSR, and some parts of South America. Whale flesh was eaten in many parts of the world and is still an important food source in Japan and in Arctic areas. Whale oil was an important source of illumination until the development of kerosene and, eventually electricity, replaced it.

The image of a whale was carved on a rock wall in northern Norway about 2200 BC. Ever since then men have hunted whales. Modern whaling began with the invention of the harpoon gun that enabled whalers to catch the fast swimming whales such as the fin and blue and later factory ships that allowed complete processing of whales at sea (Japan Whaling Association 1978). These developments together with faster and bigger ships and advanced techniques for locating whales has had a serious effect on the populations of the "great" whales (the sperm whale and most baleen whales). Tuna fishermen net huge schools of fish and the dolphins that accompany them, resulting in the death of many of these small whales. The great whales are becoming very rare and some, such as the right and blue whales, are near the verge of extinction. The United States has prohibited the importation of all whale products and has instituted controls for the tuna fishing industry. Some other nations have joined in international agreements protecting cetaceans. But the whaling practices of Japan, the USSR and some other nations are still being strongly criticized and more effective international conservation practices still are needed (Morzer Bruyns 1971: 166-170; Schevill 1974; Slijper and Heinemann 1975: 470-472; Coffey 1977:35-37). Matthews (1968) gave an excellent and profusely illustrated account of the history of whales and whaling.

The whales have frequently been described as the most intelligent of nonprimate mammals and some have identified them as more intelligent than the anthropoid apes. They have a large brain in proportion to their body size and they demonstrate complex patterns of behavior and communication. Stories of dolphins aiding man by helping with fishing, driving away attackers (e.g. sharks) and saving people who were drowning date from Greek mythology and continue into the present (Slijper 1976; Coffey 1977; Tin Thein 1977).

ORDER MYSTICETA

The ordinal name, Mysticeta, is derived from the Greek words for "mystic whale" (Jaeger 1955) and it is not difficult to realize why these marine giants could seem mysterious to men in small boats who never sailed out of sight of land. The Mysticeta includes the blue whale, *Balaenoptera musculus*, which at 27.5 to 33.5 meters and 150 tons (Morzer Bryns 1971:178) is not only the largest living creature but is also the largest ever to have lived. It weighs as much as 25 elephants or 1600 men or four of the largest dinosaurs that ever lived (Slijper and Heinemann 1975:477). The smallest mysticete, the pygmy right whale, *Caperea marginata*, is about six meters long and weighs about five tons (Morzer Bryns 1971:161).

The taking of right, gray, blue, and humpback whales is now prohibited by the International Whaling Commission, but not all whaling is conducted by countries that are members of the Commission. The remaining mysticetes, together with one odontocete, the sperm whale, *Physeter macrocephalus*, are the mainstay of the whaling industry.

Distinguishing Characteristics

Baleen whales lack teeth in both upper and lower jaws. From 130 to 400 baleen plates are suspended from each side of the upper jaw (Fig. 22-7). Each plate is composed of longitudinal strands of horny epithelial material embedded in a less resistant matrix. At the ventral edge the matrix is worn away and there is a fringe of the tougher strands (Fig. 22-8). The plates are arranged so that the fringes of adjacent plates overlap to produce a continuous strainer-like network. The baleen whales feed by taking in huge mouthfuls of sea water. Then they close the mouth and use their large tongues to force the water out between the baleen plates. The fringes of fibers strain small organisms from the water and the food is swallowed.

The skull is bilaterally symetrical and the nasal bones extend anteriorly over the nasal passage (Fig. 22-3A). The two nasal passages penetrate to the surface as separate, adjacent blowholes.

Distribution

Mysticetes are found in all oceans but they are more common in Arctic and Antarctic than in tropical waters (Morzer Bruyns 1971). Little is known about the behavior of the pygmy right whale and Byrde's whale. All other mysticetes are migratory, moving between cooler waters in the summer and warmer water in the winter.

Fossil History

The Mysticeta date from the upper Eocene in North America, from the Oligocene of Europe and from the Miocene in South America (Romer 1966:393). Eschrichtiidae is known only from the Pleistocene of Europe and the Recent of the North Pacific. Balaenopteridae ranges from the upper Miocene in Europe and North America. Balaenidae is known from the Miocene through Pleistocene of South America, the Pliocene of Australia and the Pliocene through Pleistocene of North America (Romer 1966).

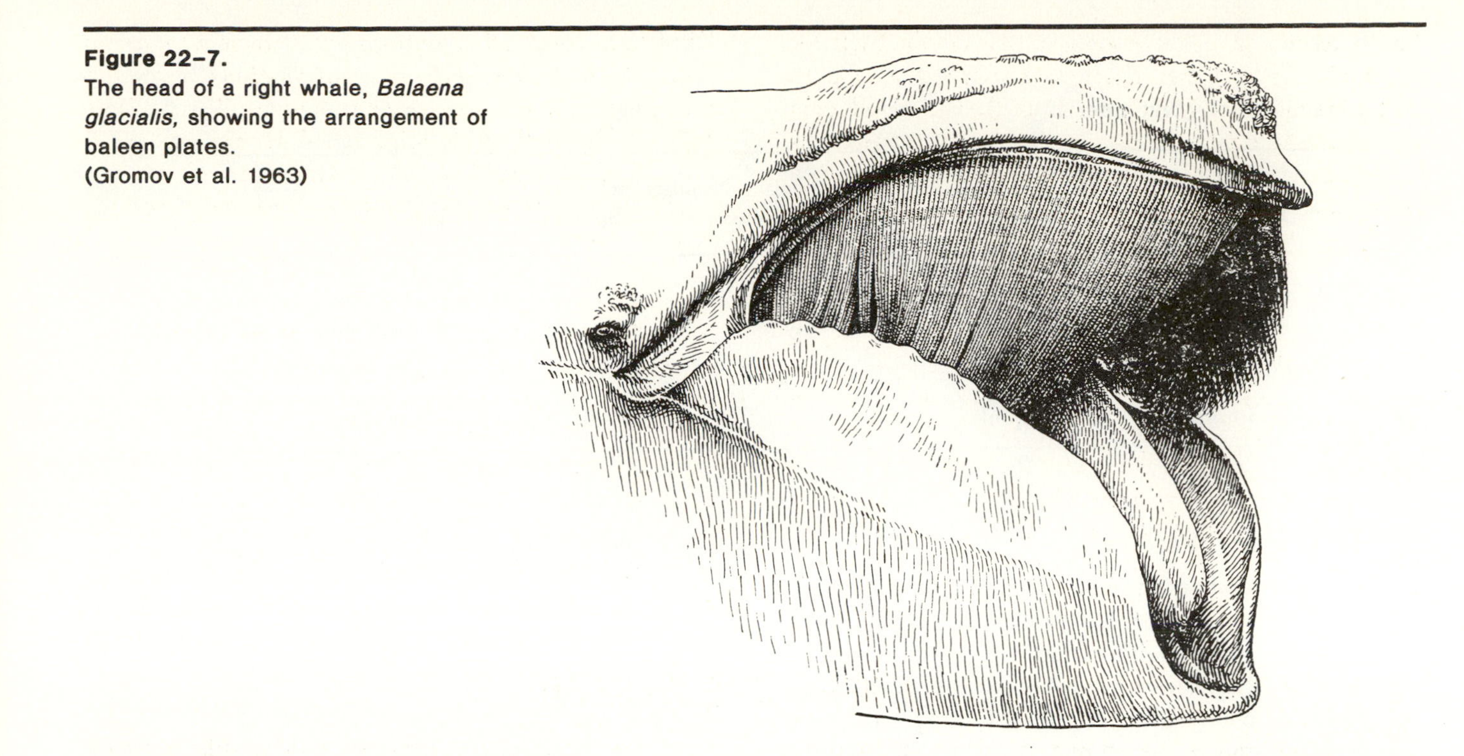

Figure 22–7.
The head of a right whale, *Balaena glacialis,* showing the arrangement of baleen plates.
(Gromov et al. 1963)

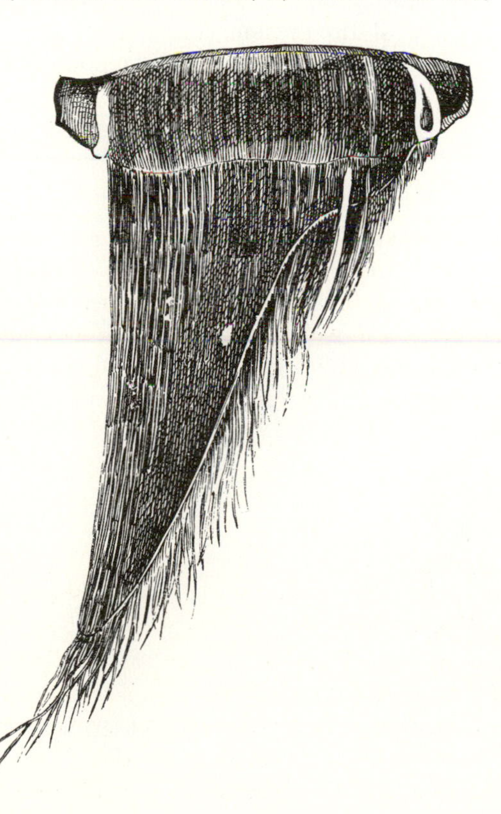

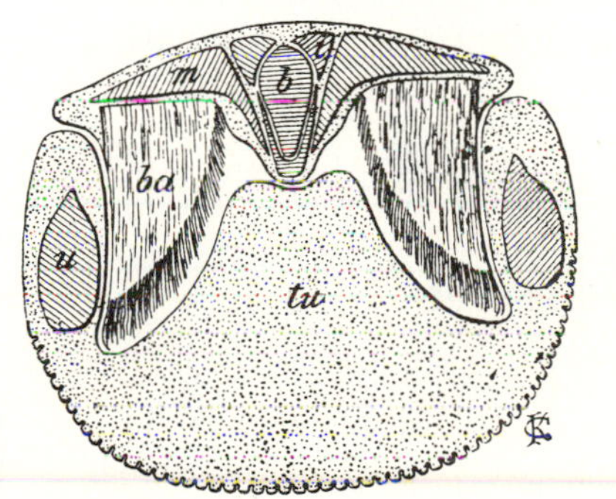

Figure 22–8.
Baleen. A, single sheet of baleen. B, a diagramatic cross section of the head of a rorqual showing the tongue (tu) in position between the two rows of baleen (ba).
(A, Duncan 1877-83:363; B, Weber 1928: 380)

Table 22-1

Living Families of Mysticeta.

Family	Common Name	Number of* Genera	Species	Distribution
Balaenidae	right whales	2	3	All oceans except tropical and south Polar
Eschrichtiidae	gray whale	1	1	North Pacific coasts
Balaenopteridae	rorquals	2	6	All oceans

*Based upon Rice 1977.

Key to Living Families

1 Posterior border of nasals and premaxillae anterior to supraorbital process of frontals; rostrum long, slender and very highly arched (Fig. 22-9); throat not grooved (Fig. 22-10); central baleen plates considerably longer than anterior and posterior plates .... **Balaenidae**
right whales

1′ Nasals and nasal process of premaxillae extending posteriorly beyond anterior border of supraorbital process of frontal (Figs. 22-11 and 22-13); throat grooved (Figs. 22-12 and 22-14); baleen plates all approximately the same length 2

2 (1′) Nasals large, frontals broadly exposed on vertex (Fig. 22-11); rostrum arched (Fig. 22-11B); mandibles not conspicuously bowed outwards; throat with only a few short grooves (Fig. 22-12); dorsal fin absent (Fig. 22-12) .... **Eschrichtiidae**
gray whale

2′ Nasals reduced, frontals scarcely or not at all exposed on vertex (Fig. 22-13); rostrum not arched (Fig. 22-13); mandibles conspicuously bowed outwards; numerous parallel grooves covering entire throat and chest region (Fig. 22-14); dorsal fin present (Fig. 22-14) .... **Balaenopteridae**
rorquals

Figure 22-9.
Skull of a right whale, *Balaena glacialis.*
(Cabrera 1914: 402)

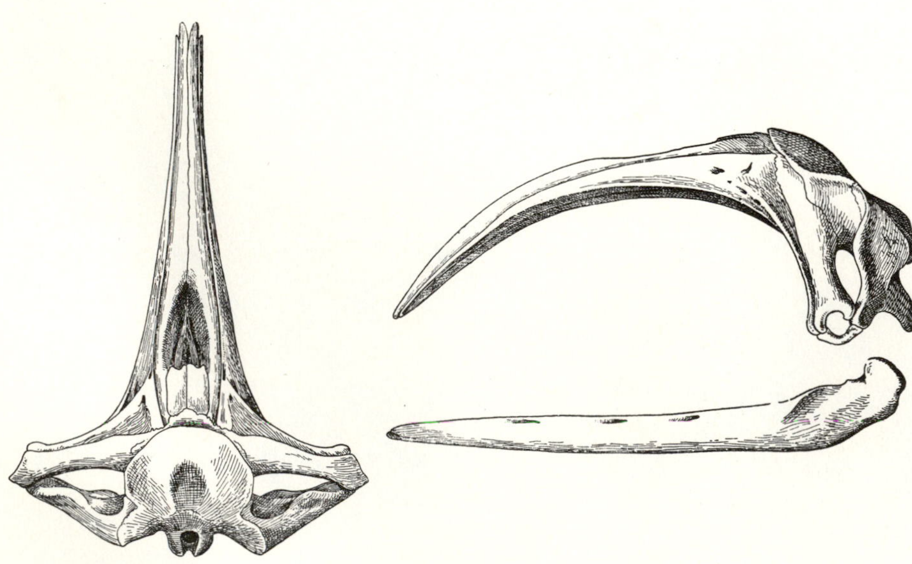

Figure 22–10.
A right whale, *Balaena glacialis.*
(Cabrera 1914: 399)

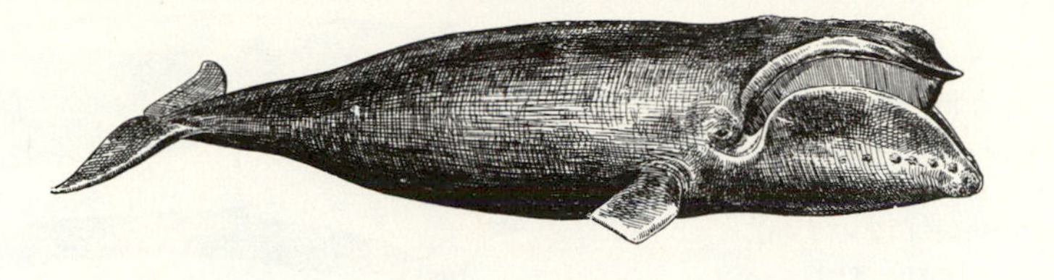

Figure 22–11.
Skull of a gray whale, *Eschrichtius robustus,* A, dorsal view, B, lateral view.
(Mary Ann Cramer)

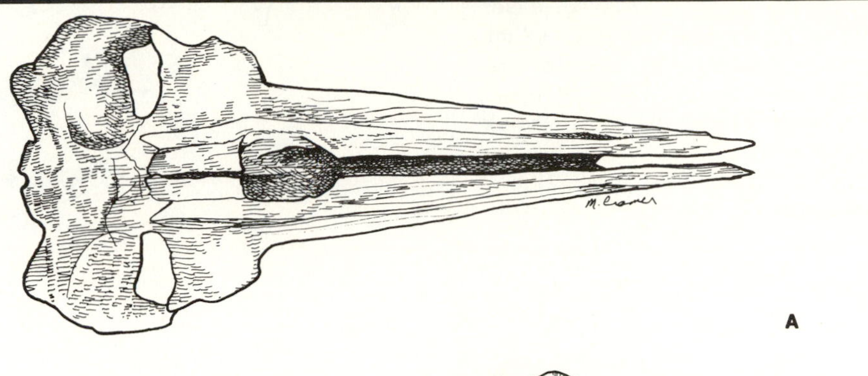

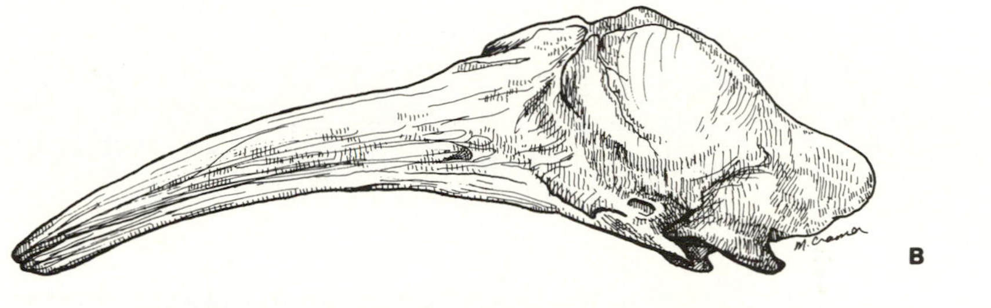

Figure 22–12.
A gray whale *Eschrichtius robustus.*
(Mary Ann Cramer from photograph in Rice and Wolman 1971)

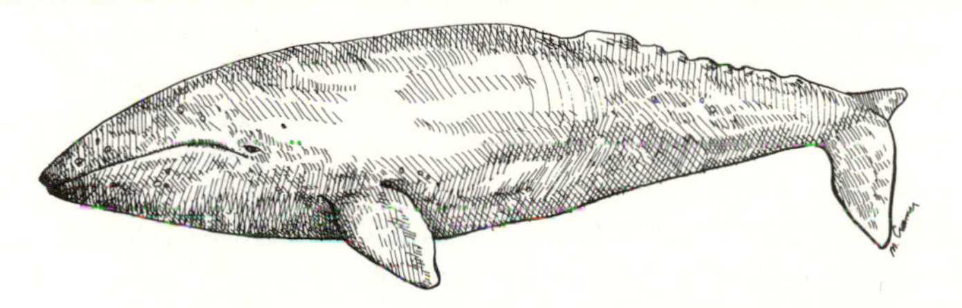

Figure 22–13.
Skull of a fin whale, *Balaenoptera physalus.* A, dorsal view, B, lateral view.
(Tomlin 1962: 47)

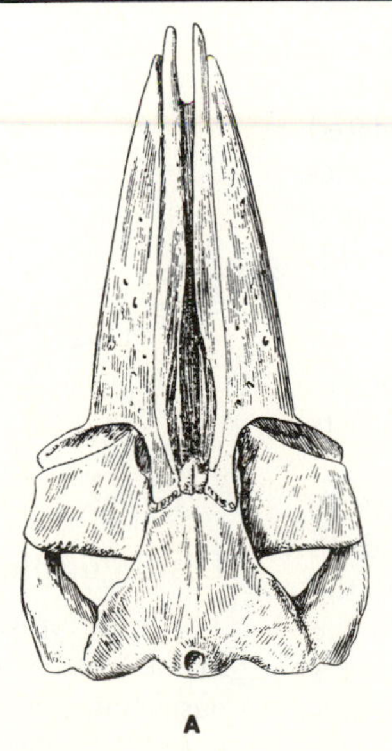

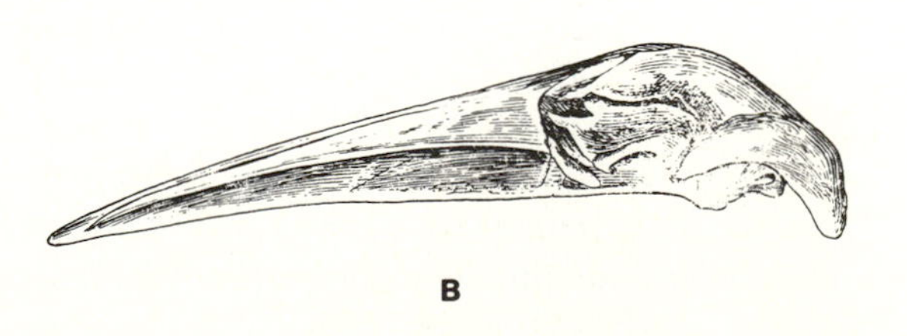

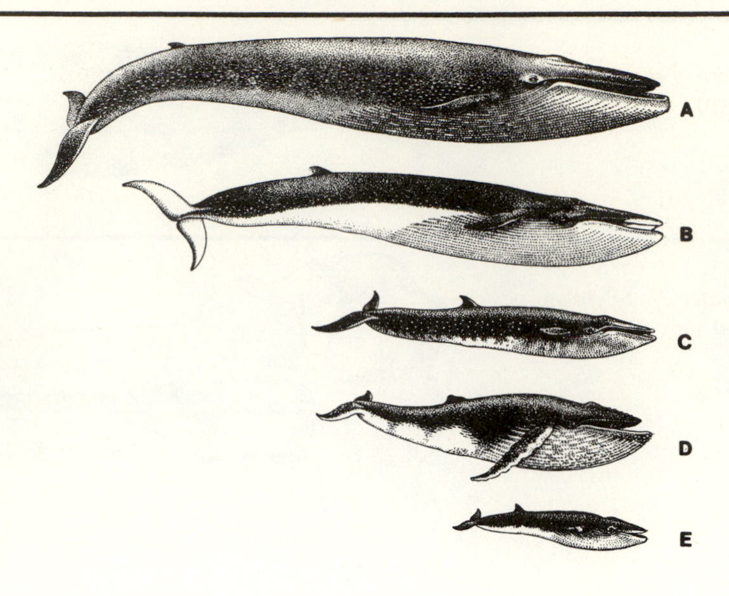

Figure 22–14.
The size range in rorquals, Balaenopteridae. A, right whale, *Balaenoptera musculus* (27-33.2 m); B, fin whale, *B. physalus* (20-25 m); C, sei whale, *B. borealis* (12-18 m); D, humpback whale, *Megaptera novaeangliae* (+/− 17 m); E, lesser rorqual, *B. acutorostrata* (7.6-9.2 m). The only species of Balaenopteridae not pictured, Bryde's whale, *B. edeni,* (12-15 m) is about the same size as the sei whale.
(Figures, Gromov et al. 1963: 727; measurements, Morzer Bruyns 1971: 171-180)

Comments and Suggestions on Identification

Skulls of the three mysticete families can easily be distinguished by the relative arch of the skull (see Figs. 22-9, 22-11, and 22-13). Externally the gray whale could be confused with a small rorqual but it has no dorsal fin and only a few throat grooves. A humpback whale could be confused with a right whale but the humpback has throat grooves and a dorsal fin, both characters lacking in the right whales.

ORDER ODONTOCETA

The ordinal name, Odontoceta, means literally "toothed whale" (Jaeger 1955) and the presence of teeth is the factor that most prominently distinguishes this group from the Mysticeta. The teeth are usually conical, unicuspid and homodont. In some forms teeth are lost entirely in the upper or lower jaw or are greatly reduced in number. But in most forms teeth are numerous (Fig. 22-15) and can range up to a total of 220 in the LaPlata dolphin, *Pontoporia blainvillei* (Morzer Bruyns 1971). The toothed whales specialize in limited food sources; however, the individual related groups hunt different varieties of fish or invertebrates and consequently these groups have developed several very different adaptations to their diets (Slijper and Heinemann 1975:493). Sight is not well developed and olfaction is probably nonfunctional, but hearing is extremely well developed. Most, if not all, odontocetes communicate by sound and sense objects in their environments by sonar.

The sperm whale, *Physeter macrocephalus,* at 20 meters and 60 tons for males and 12 meters and 18 tons for females, is the largest odontocete and the only toothed whale included as a major target by the whaling industry. Dolphins are popular zoo animals and marine exhibits featuring performing delphinids are now common.

Figure 22–15.
Lower jaws of four odontocetes. A, *Ziphius* and B, *Mesoplodon,* both Ziphiidae; C, *Physeter macrocephalus,* Physeteridae; and D, *Delphinus delphis,* Delphinidae. (Gromov et al. 1963: 646)

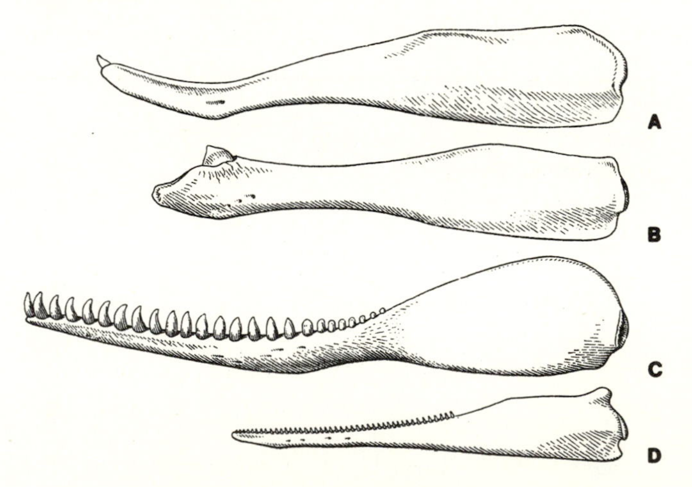

Distinguishing Characteristics

Teeth are present. The skull is bilaterally asymetrical in the area of the external nares (Fig. 22-16) and the nasal bones do not project anteriorly over the external nares (Fig. 22-3 and 22-4). Externally the nostrils are united into a single blowhole (Fig. 22-17).

Figure 22–16.
Dorsal view of the skull of an odontocete whale (*Orcinus*) showing the asymmetry in the region of the external nares. (Gromov et al. 1963)

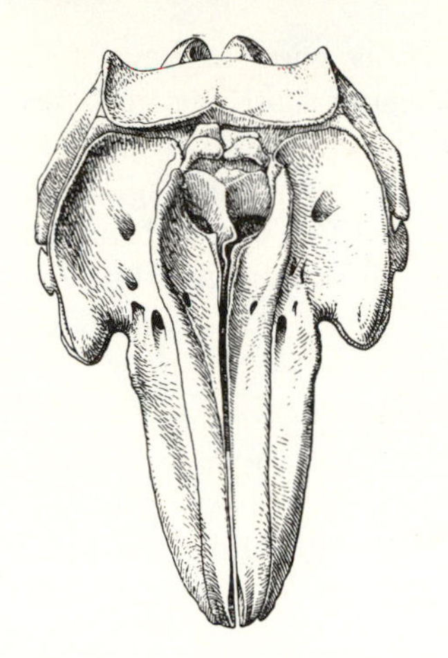

Figure 22–17.
Dorsal view of the bottlenose dolphin, *Tursiops truncatus*, showing the single blowhole.
(Cabrera 1914: 351)

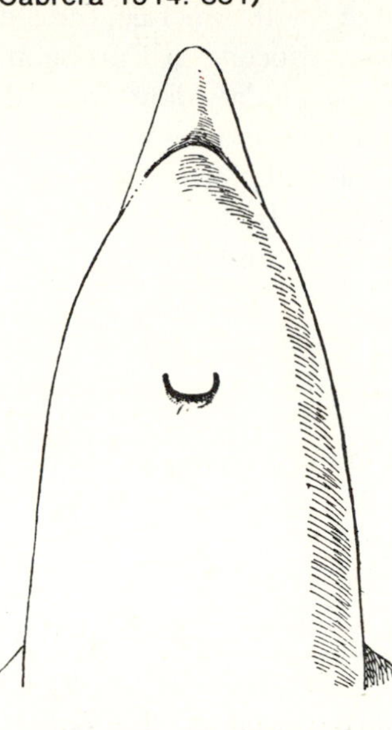

Distribution

Most odontocetes are marine but one family, Platanistidae, primarily inhabits fresh water; certain species in other families are common in fresh water; and members of most families are occasionally found in major rivers.

Odontocetes occur on all oceans and seas but many species never venture far from coastlines. The sperm whale and many other species of Odontoceta are known to be migratory.

Table 22-2

Living Families of Odontoceta.

Family	Common Name	Number of* Genera	Species	Distribution
Platanistidae	freshwater dolphins	4	5	Oriental and Neotropical
Delphinidae	dolphins	16	31	All oceans and seas
Phocoenidae	porpoises	3	6	All oceans except Arctic and Antarctic
Monodontidae	narwhal and beluga	2	2	Arctic Ocean, adjacent seas and large rivers
Physeteridae	sperm whales	2	3	All oceans
Ziphiidae	beaked whales	6	18	All oceans

*Based upon Rice (1977) except that Phocoenidae is here recognized as a family distinct from Delphinidae.

Fossil History*

The oldest known odontocete fossils are from the upper Eocene of North America. The oldest Platanistidae are from the lower Miocene of South America, the Delphinidae date from the lower Miocene in Europe and North America and the Phocaenidae date from the upper Miocene. The Monodontidae are unknown before the Pleistocene of Europe and North America but the Physeteridae and Ziphiidae both date from the lower Miocene of South America and Europe.

Key to Living Families

1 Teeth absent in both upper and lower jaws *or* the upper jaw with one (rarely two) very long spiraling tusk (Fig. 22-18) 2

*Romer 1966:393.

1′ Teeth absent in both upper and lower jaws *or,* if present, not a long spiraling tusk; teeth present in the lower jaw 3

2 (1′) Rostrum broad, nearly as wide as long; two teeth present in upper jaw, right one in males usually unerupted, both may be unerupted in females. Left tooth in males a long spiraling tusk (Figs. 22-18 and 22-19A); teeth absent in lower jaw .... **Monodontidae** (in part) narwhal

2′ Rostrum narrow, much longer than wide (Fig. 22-20); teeth absent in upper jaw, unerupted teeth may be present in lower jaw .... **Ziphiidae** (in part) those without teeth

Figure 22–18.
Skull of a male narwhal, *Monodon monoceros.* The top of the rostrum has been removed to show the root of the large left tusk and the small, unerupted, right tusk. (Flower and Lydekker 1891: 261)

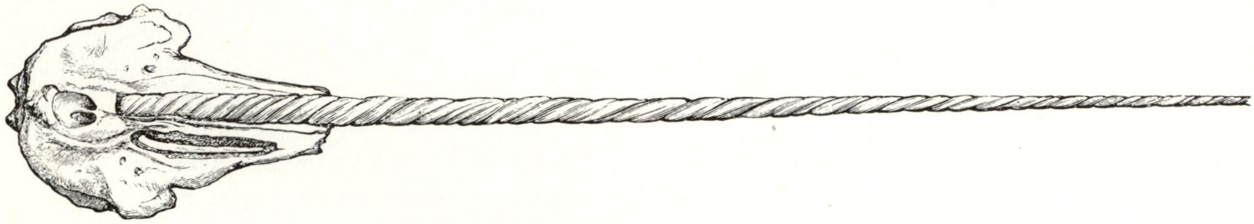

Figure 22–19.
The Monodontidae. A, a male narwhal, *Monodon monoceros* and B, a beluga, *Delphinapterus leucas.*
(Gromov et al. 1963: 680)

Figure 22–20.
Skulls of some beaked whales, Ziphiidae. A, *Mesoplodon bidens;* B, *Berardius bairdii;* C, *Hyperoodon ampullatus;* and D, *Ziphius cavirostris.* (Dorsal views, Gromov et al. 1963: 657-663; lateral view, Cabrera 1914: 379)

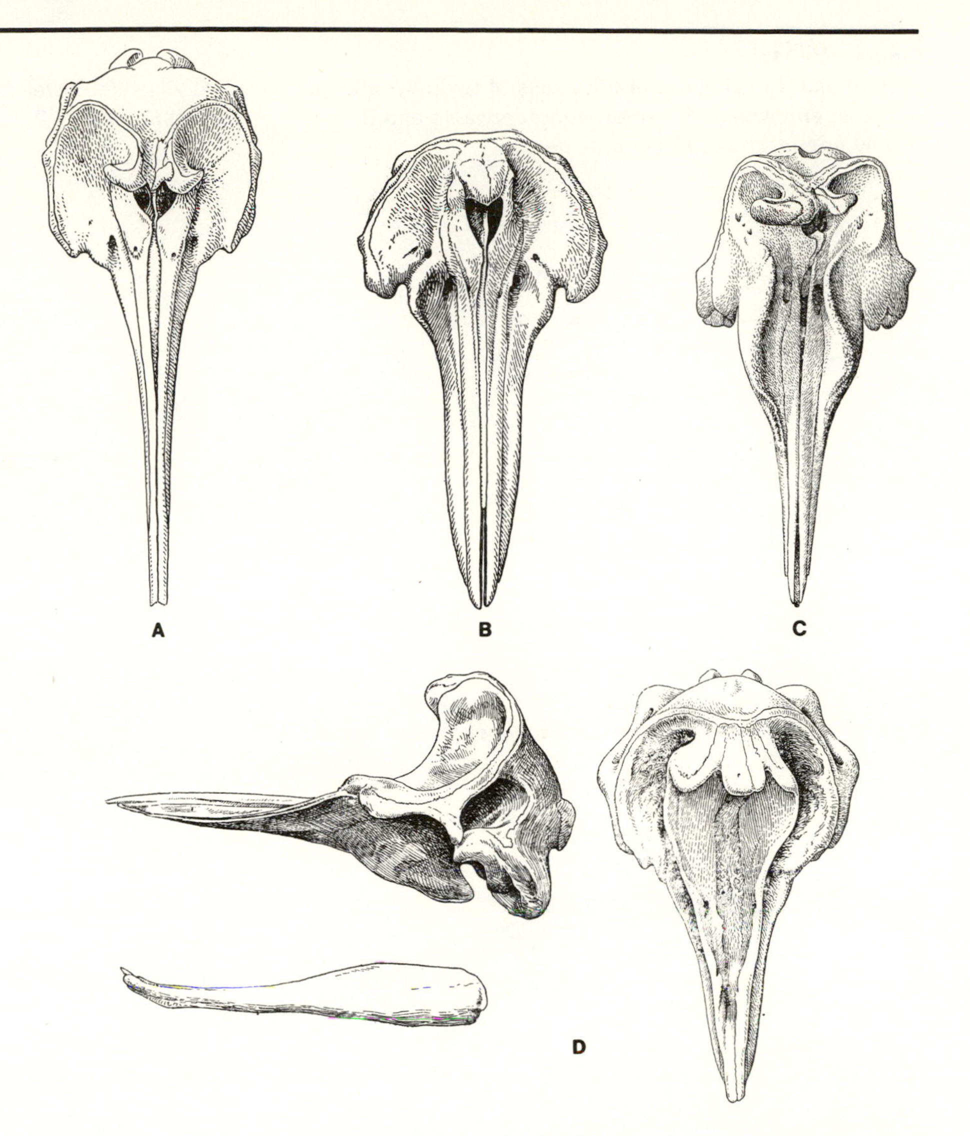

3 (1′) Size very large, total length of skull over two meters, total length of animal 12 to 20 meters; 16 to 30 large conical teeth present in each dentary; no teeth present in upper jaws (rarely a few rudimentary teeth present) (Fig. 22-21); mandibular symphysis at least 35% of mandible length (Fig. 22-21A); head very large and blunt anteriorly (Fig. 22-22A) .... **Physeteridae** (in part) sperm whale

3′ Size large to small, skull less than two meters, (rarely more than one meter), total length of animal less than 12 meters; teeth present in both upper and lower jaw, or if absent in upper jaw, lower teeth do not number more than 15; mandibular symphysis less than 35% of mandible length or if more than 35% many well developed teeth present in upper jaws; head shape varies (Fig. 22-23) 4

4 (3′) Fewer than eight pairs of teeth present in upper jaws or if more than eight pair in upper jaws then anterior one or two pairs of lower teeth much larger than upper teeth or posterior lower teeth 5

4′ At least eight pairs of well developed teeth present in both upper and lower jaws. Anterior lower teeth not appreciably larger than the others 7

Figure 22–21.
Lateral and dorsal views of the skulls of two physeterids. A, the sperm whale, *Physeter macrocephalus* and B, a pygmy sperm whale, *Kogia breviceps*.
(A, *Physeter* lower jaw, Geibel 1859: 506; B, Bobrinskii et al. 1965: 197)

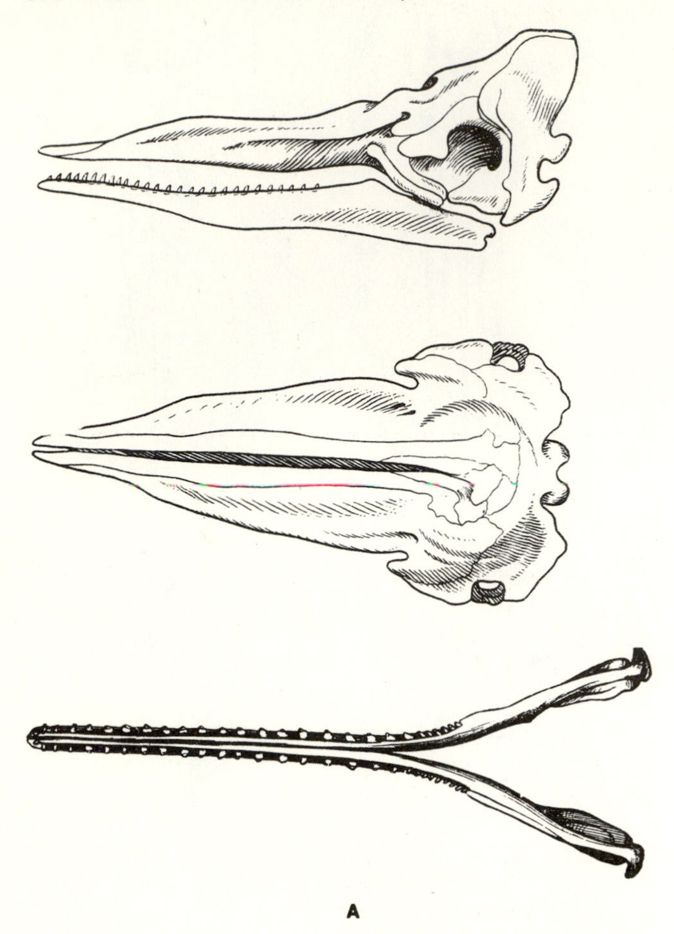

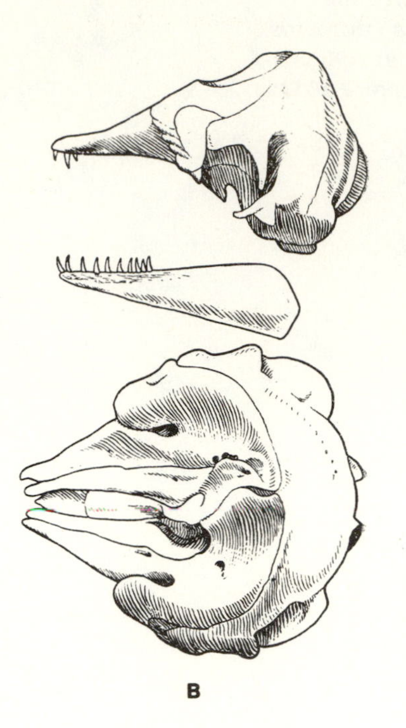

Figure 22–22.
Two physeterids. A, the sperm whale, *Physeter macrocephalus* and B, a pygmy sperm whale, *Kogia breviceps*.
(Gromov et al. 1963: 667)

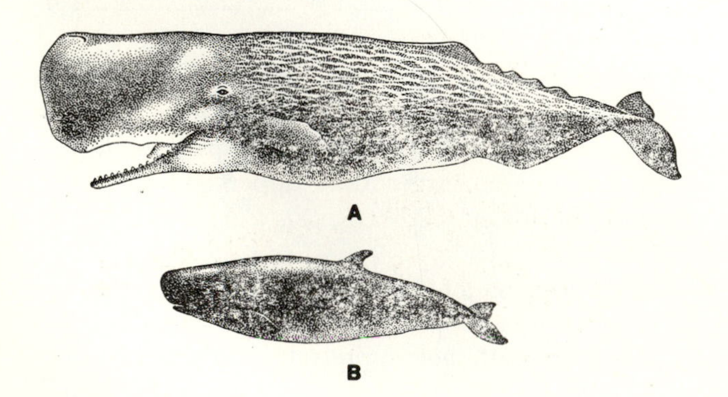

5 (4) 1 or 2 teeth in each dentary, usually large (Fig. 22-15A and B), if more than 2 present, posterior teeth considerably smaller than anterior teeth; rostrum long and narrow (Fig. 22-20); two deep grooves present on throat, converging anteriorly (Fig. 22-24); size large, 3.6 to 12 m total length; head with distinct beak (Figs. 22-23A-D or 22-24) .... **Ziphiidae** (in part)
beaked whales with teeth

5′ 3 or more teeth in each dentary, posterior teeth not appreciably smaller than anterior teeth (e.g. Fig. 22-21B); rostrum short and broad (Figs. 22-21B and 22-25); no grooves on throat; size medium, 3 to 4 m total length; head blunt (Figs. 22-22B or 22-23G) 6

Figure 22–23.
Some examples of head shape in the Ziphiidae (A-D), Delphinidae (E-I) and Phocoenidae (I-J). A, *Mesoplodon steinegeri;* B, *Ziphius;* C, *Berardius;* D, *Hyperoodon;* E, *Delphinus, Stenella* or *Tursiops;* F, *Lagenorhynchus* or *Lissodelphis;* G, *Grampus;* H, *Globicephalus;* I, *Orcinus, Pseudorca, Phocaena,* or *Phocaenoides;* and J, *Neophocaena.*
(Gromov et al. 1963: 654 and 672)

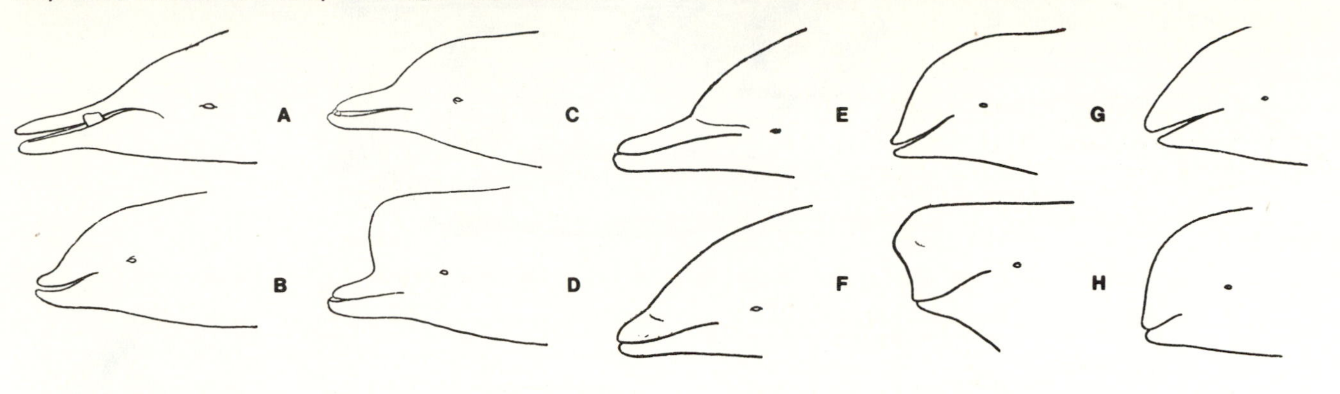

Figure 22–24.
Dorsal and ventral views of the head and a lateral view of a whole beaked whale, *Ziphius cavirostris.* d, grooves on the throat.
(Cabrera 1914: 357-380)

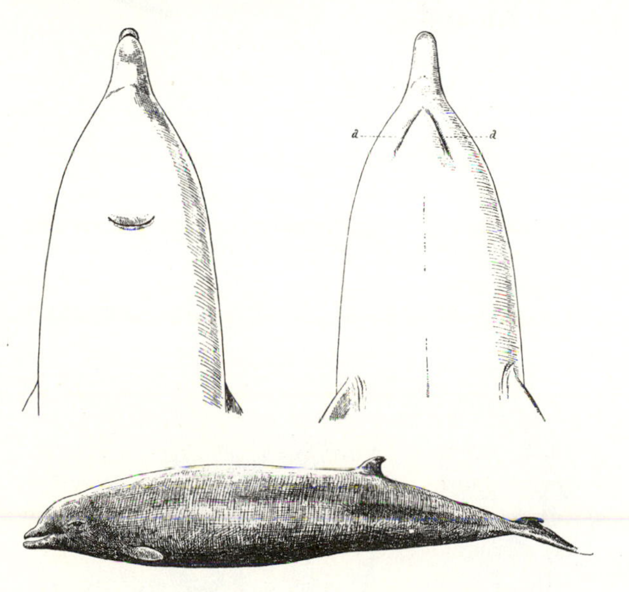

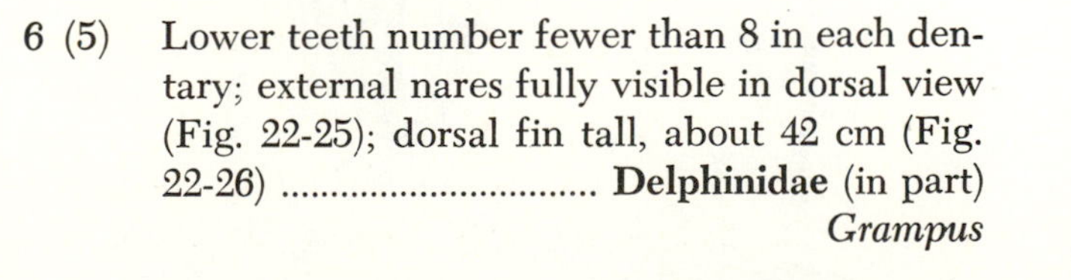

6 (5) Lower teeth number fewer than 8 in each dentary; external nares fully visible in dorsal view (Fig. 22-25); dorsal fin tall, about 42 cm (Fig. 22-26) .... **Delphinidae** (in part)
Grampus

6′ Lower teeth number more than 8 in each dentary; external nares barely visible in dorsal view (Fig. 22-21B); dorsal fin short, about 24 cm (Fig. 22-22B) .... **Physteridae** (in part)
pygmy sperm whales

Figure 22–25.
Skull of a gray grampus, *Grampus griseus.*
(Gromov et al. 1963: 700)

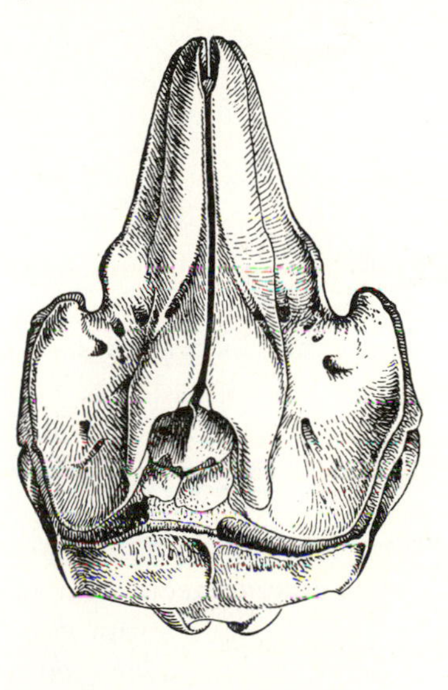

Figure 22–26.
The gray grampus, *Grampus griseus.*
(Cabrera 1914: 374)

Figure 22–27.
Skulls of two platanistids. A, dorsal and lateral views of a Ganges dolphin, *Platanista gangetica.* Note the hood-like bones shielding the blowhole. These are unique to this genus. B, lateral view of the Amazon dolphin, *Inia geoffrensis.*
(A, Duncan 1877-83: 248; B, Geibel 1859: 498)

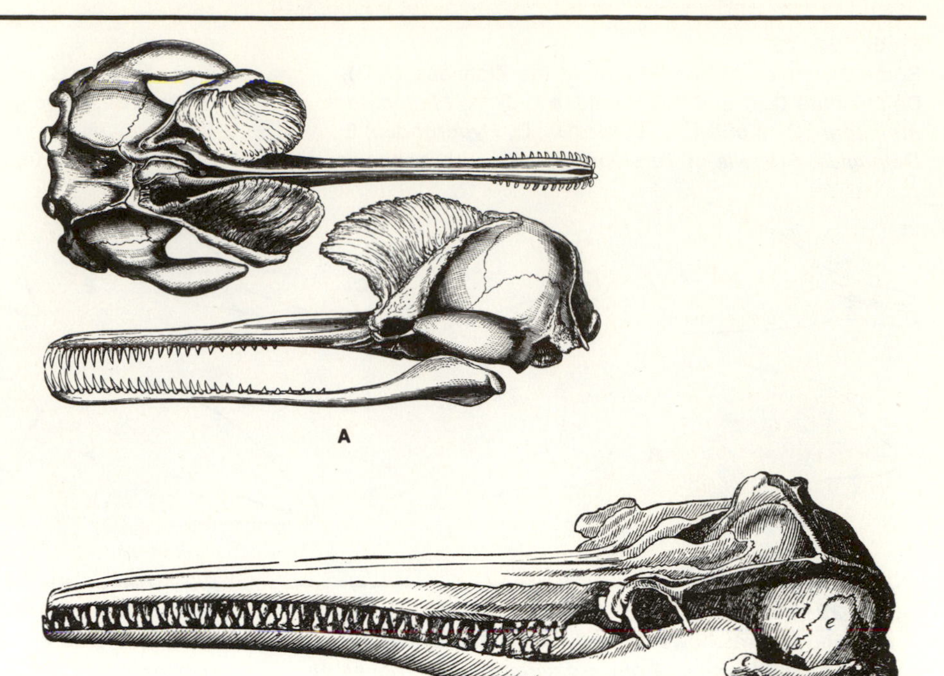

7 (5′) Mandibular symphysis greater than 40% of mandible length; tooth rows parallel for most of their length; teeth number from 25 to 60 in each quadrant; rostrum very long (Fig. 22-27) and very narrow, depth and breadth of rostrum similar .... **Platanistidae**
freshwater dolphins

7′ Mandibular symphysis less than 40% of mandible length; tooth rows diverge, teeth number from 5 to 52 in each quadrant; rostrum variable but never as above, depth always considerably less than breadth 8

8 (7′) Premaxillae with prominent bosses (bumps) immediately in front of external nares (Fig. 22-28); teeth laterally compressed and spadelike (Fig. 22-29); head blunt (Figs. 22-23I and J; 22-30) .... **Phocoenidae**
porpoises

8′ Premaxillae flat or concave immediately in front of narial openings; teeth generally conical, never laterally compressed and spadelike, head shape varies (Fig. 22-23E to I) 9

Figure 22–28.
Skulls of representative Phocaenidae. A, dorsal view of the harbor porpoise, *Phocoena phocoena;* B, dorsal view of the finless porpoise, *Neophocaena phocanoides;* and C, lateral view of a phocaenid skull showing the bosses anterior to the nares.
(A and B, Gromov et al. 1963: 710; C, Gervias 1855: 327)

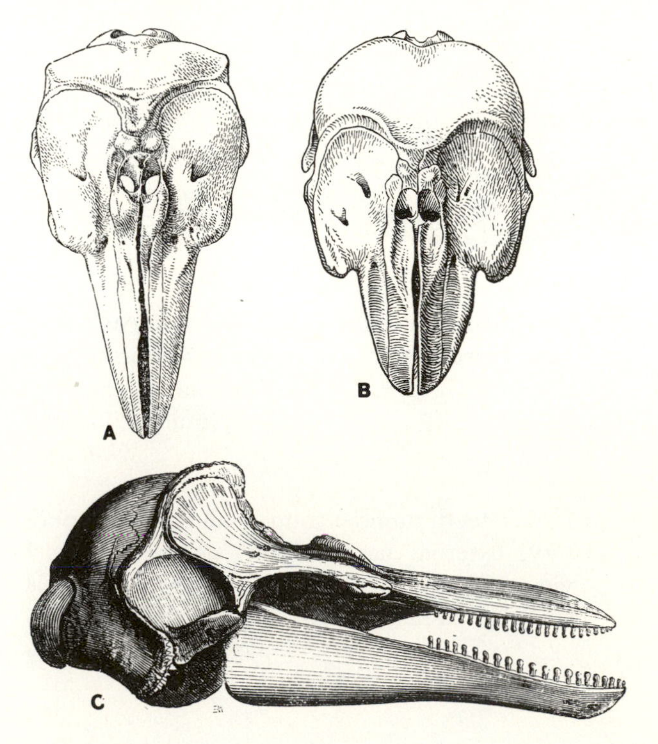

9 (8′) When skull is viewed in profile top of rostrum flat or slightly convex between anterior edge of nares and distal tip of premaxillae (Fig. 22-31); teeth 10/8; dorsal fin absent, color white (Fig. 22-19B) .... **Monodontidae** (in part) beluga

9′ When skull is viewed in profile top of rostrum distinctly concave between anterior edge of nares and distal tip of premaxillae (Fig. 22-32); teeth vary in number, usually exceeding 10/8; dorsal fin usually present, color varies (Fig. 22-33) .... **Delphinidae** (in part) all genera except *Grampus*

Figure 22–29.
The laterally compressed teeth of a porpoise, Phocoenidae.
(Flower and Lydekker 1891: 263)

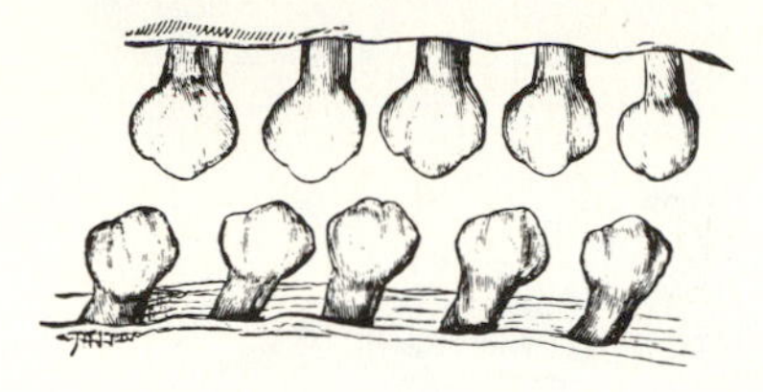

Figure 22–30.
A Phocaenidae, *Phocaena phocaena*.
(Cabrera 1914: 363)

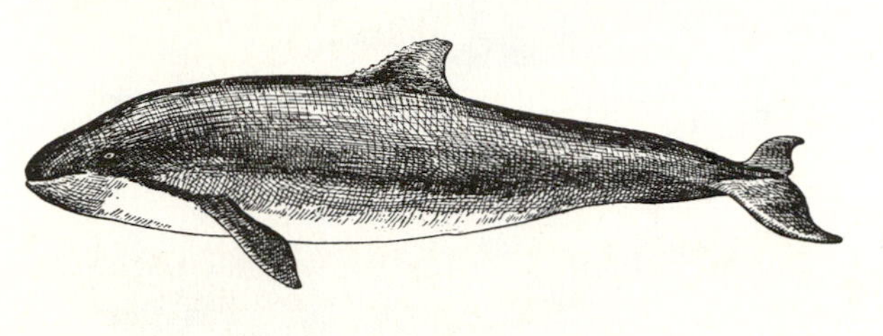

Figure 22–31.
Skull of a white whale, *Delphinapterus leucas*.
(Mary Ann Cramer from Tomlin 1962)

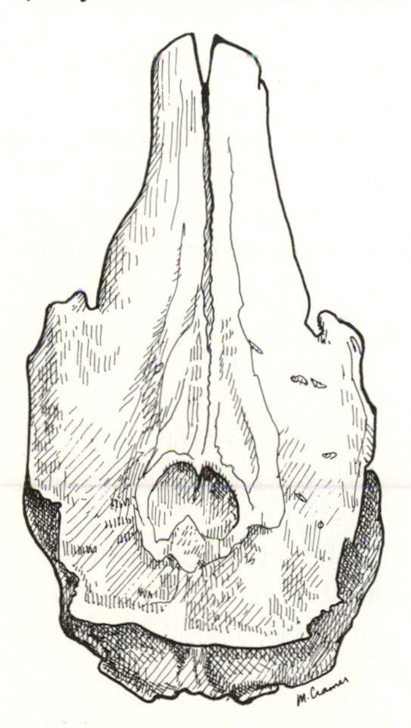

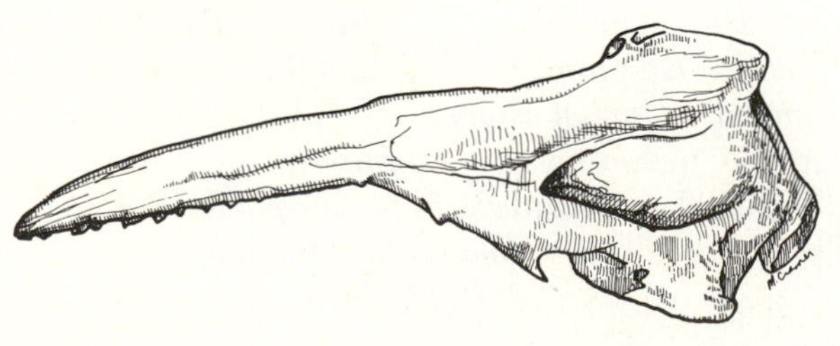

Figure 22–32.
Skulls of two of the many species of Delphinidae. A, the striped dolphin, *Stenella coeruleoalba* and B, the false killer whale, *Pseudorca crassidens*. Not to same scale.
(Tomlin 1962: 157 and 187)

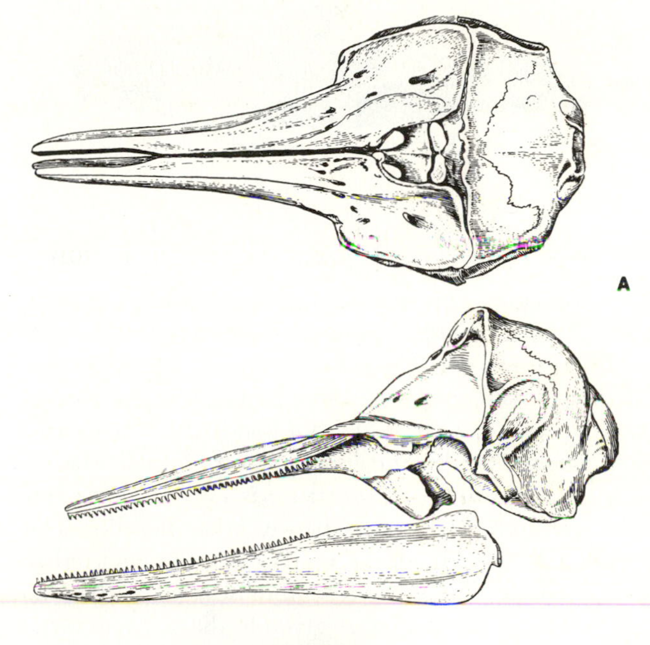

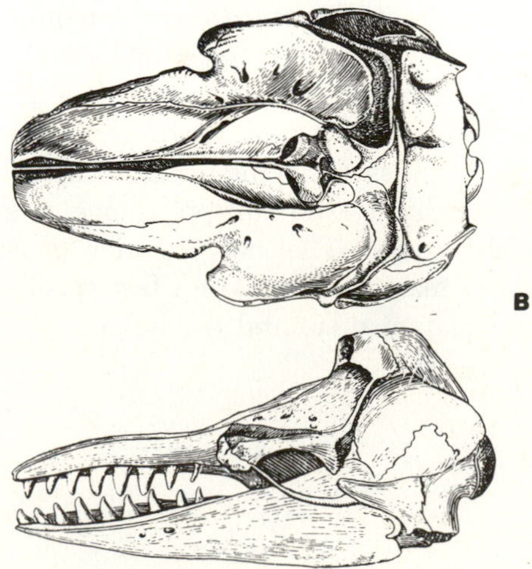

Figure 22–33.
Representative Delphinidae. A, the killer whale *Orcinus orca;* B, the common dolphin, *Delphinus delphis;* C, the bottlenose dolphin, *Tursiops truncatus;* D, the striped dolphin; *Stenella caeruleoalba,* and E, the common pilot whale, *Globicephala melaena.* Not all to same scale. (A-C and E, Cabrera 1914: 356–371; D, Tomlin 1962: 157)

Comments and Suggestions on Identification

The species of the families Monodontidae, Physeteridae and Ziphiidae are each distinctive and, with practice, easily recognizable to family by either external or cranial characters. The Platanistidae resemble some Delphinidae externally but generally have longer, more slender beaks and lower, more broadly based dorsal fins. Cranially the rostrum is more slender and their teeth more numerous than in other families. The members of the Phocoenidae all resemble each other but externally they also resemble several species of Delphinidae. The characters given in the key will identify the skulls, but external identification will require learning to recognize the various species individually.

Many authors have written keys based upon external characteristics of the whales; however, unless you are aboard a fishing vessel or come upon a beached whale, you will rarely have external appearance as a guideline for identification. The only species commonly to be found in marine "zoos" are a few species of Delphinidae and Platanistidae and the beluga.

Taxonomic Remarks

Several classifications have been proposed for the Odontoceta. We have followed the arrangement of families given by Rice (1977) except that we followed Rice (1967) and Slijper (1976) in separating the Phocoenidae from the Delphinidae. The genera *Steno* and *Sousa* were recognized as a distinct family, Stenidae, by Fraser and Purves (1960). Rice (1967) and Slijper and Heinemann (1975) included the genus *Sotalia* in Stenidae. The genus *Kogia* is sometimes split from the Physeteridae and placed in a distinct family, Kogiidae (Nishiwaki 1963, Coffey 1977).

Nishiwaki (1963) revised the Delphinidae and split what we here include as Delphinidae into four families. Imposing Nishiwaki's (1963) classification on Rice's (1977) list yields the following:

Delphinidae, including *Steno, Sousa, Sotalia, Tursiops, Stenella, Delphinus, Lagenodelphis, Lagenorhynchus, Cephalorhynchus,* and *Lissodelphis;*
Grampidae, including only *Grampus;*
Globicephalidae, including *Peponocephala, Feresa, Pseudorca, Globicephala,* and *Orcinus;*

Orcaelidae, including only *Orcaella;*
Phocaenidae, including the genera *Phocoena, Neophocaena* and *Phocoenoides.*

The Platanistidae includes three distinct groups sometimes given full family rank. Hediger (1975) gave the following classification:
Platanistidae, including only *Platanista;*
Iniidae, including *Inia* and *Lipotes;*
Stenodelphidae, including only *Stenodelphis* (=*Pontoporia*).

Supplementary Readings

Andersen, H. T. (Ed.). 1961. *The biology of marine mammals.* Academic Press, New York. 508 pp.

Berzin, A. A. 1971. *The sperm whale.* English Translation by E. Hoz and Z. Blake. 1972. Israel Program for Scientific Translations. 394 pp.

Coffey, D. J. 1977. *Dolphins, whales and porpoises: an encyclopedia of sea mammals.* Macmillan Publishing Co., New York. 223 pp.

Friends of the Earth. 1978. *The whale manual.* Friends of the Earth Books, San Francisco. 153 pp.

Harrison, R. J. (Ed.) 1972. *Functional anatomy of marine mammals,* Vol. I. Academic Press. 441 pp.; 1974. Vol. II. 366 pp.

Kleinenberg, S. E., A. V. Yublokov, B. M. Bel'kovich, and M. N. Tarasevich. 1964. *Beluga (Delphinapterus leucas): investigation of the species.* English translations by O. Theodor (Ed.) 1969. Israel Program for Scientific Translations. 376 pp.

Matthews, L. H. (Ed.) 1968. *The whale.* Simon and Schuster. New York. 287 pp.

Morzer-Bruyns, W. F. J. 1971. *Field guide of whales and dolphins.* Vitgeverij tor/n.v. vitgeverij, v.h.c.a. mees, Amsterdam, 258 pp.

Pilleri, G. (Ed.) 1969. *Investigations on Cetacea* Vol. I. Berne. 219 pp.; 1970, Vol. II, 296 pp.; 1971, Vol. III, 377 pp.; 1972, Vol. IV, 299 pp.; 1973-1974, Vol. V, 365 pp.; 1976, Vol. VI, 151 pp.; 1976, Vol. VII, 249 pp.; 1977, Vol. VIII, 383 pp.

Rice, D. W. and A. A. Wolman. 1971. *The Life History and Ecology of the Gray Whale, (Eschrichtius robustus.)* Spec. Publ. No. 3, Amer. Soc. Mammalogists. 142 pp.

Ridgeway, S. H. 1972. *Mammals of the sea: biology and medicine.* Charles C. Thomas Publ., Springfield, Ill. 812 pp.

Scheffer, V. B. 1969. *The year of the whale.* Chas. Scribner's Sons, New York. 214 pp.

Scheffer, V. B. 1976. *A natural history of marine mammals.* Chas. Scribner's Sons, New York. 157 pp.

Schevill, W. E. 1974. *The whale problem: a status report.* Harvard Univ. Press, Cambridge, Mass. 419 pp.

Slijper, E. J. 1962. *Whales.* Hutchinson and Co., London. 475 pp.

Slijper, E. J. 1976. *Whales and dolphins.* Univ. of Mich. Press, Ann Arbor. 170 pp.

Slijper, E. J., D. Heinemann, H. Hediger, D. R. Martinez, and E. Klinghammer. 1975. Sections on Whales, pp. 457-524. *In* B. Grzimek (Ed.) *Animal life encyclopedia. Vol. 11, Mammals II.* Van Nostrand Reinhold, New York.

Small, G. L. 1971. *The blue whale.* Columbia Univ. Press, New York. 248 pp.

Wood, F. G. 1973. *Marine mammals and man. The navy's porpoises and sea lions.* Robert B. Luce, Inc., Washington. 264 pp.

Key to Living Families

23 The Carnivores

Order Carnivora

ORDER CARNIVORA

Carnivora means flesh eater (Jaeger 1955) and this is an at least partially accurate description of most members of this order. The polar bear (Fig. 23-1), the leopard seal and the lion, are examples of Carnivora that live almost exclusively on the flesh of other vertebrates. Most otarids and phocids feed primarily on fish and several members of other families, e.g, the raccoon dog, the brown bear, the mink, the otters, and the fishing and flat-headed cats, include a significant amount of fish in their diets. The walrus and sea otter feed exclusively, or almost exclusively, on marine invertebrates, and many carnivores, e.g., the sloth bear, the banded mongoose, the meerkat and the aardwolf, are primarily insectivorous. A few members of the order Carnivora, e.g., the spectacled bear, the giant panda and the red panda, are almost exclusively vegetarian.

Regardless of their ordinal name, most Carnivora eat significant quantities of both plant and animal matter. The dietary habits of particular species may change drastically from season to season as various kinds of foods change in abundance and ease of gathering. Berries, nuts, and other fruits, eggs, carrion and invertebrates are eaten by almost all Carnivora. Guggisberg (1960) reported lions eating termites, grass and fruits, Doutt (1967) reported polar bears eating grass and berries, and Ewer (1973:168) mentioned giant pandas eating meat in captivity. The subject of carnivore diets and feeding behavior was discussed by Ewer (1973:139-229) and the above is based primarily upon her review.

Carnivores exhibit a great range in size from the American least weasel with a head and body length of 135-185 mm and a weight of 35 to 70 grams, to the huge brown bear of Kodiak Island, Alaska, which reaches a length of over three meters and a weight up to 780 kg. and the adult male southern elephant seal which can exceed 6 meters in length and can weigh over 3600 kg (Coffey 1977:187).

Figure 23–1.
A polar bear, *Thalarctos maritimus,* one of the most carnivorous members of the order Carnivora.
(Zernov and Pavlovskii 1953: 113)

Carnivores occur in nearly all possible habitats. They range from the ice floes of the Arctic Ocean to the Antarctic continent and from the driest deserts and highest mountains into the oceans. Most are essentially terrestrial but many are variously adapted for arboreal, fossorial, or aquatic existences. Arboreal adaptations range from infrequent tree climbing to the nearly continuous arboreal existence of many procyonids and viverrids. Many carnivores dig for part of their food and several excavate dens, but the badgers in the Mustelidae are probably the most fossorial. Nearly all carnivores can, and most readily do, swim. Several mustelids, mainly the otters, are well adapted for a semiaquatic existance but the families Otariidae and Phocidae are composed of mammals better adapted to life in the water than on land. Only the cetaceans and sirenians are better adapted to aquatic life than the seals and sea lions.

Figure 23–2.
A variety of domestic dogs depicted in ancient Egyptian tombs.
(Geibel 1859: 159)

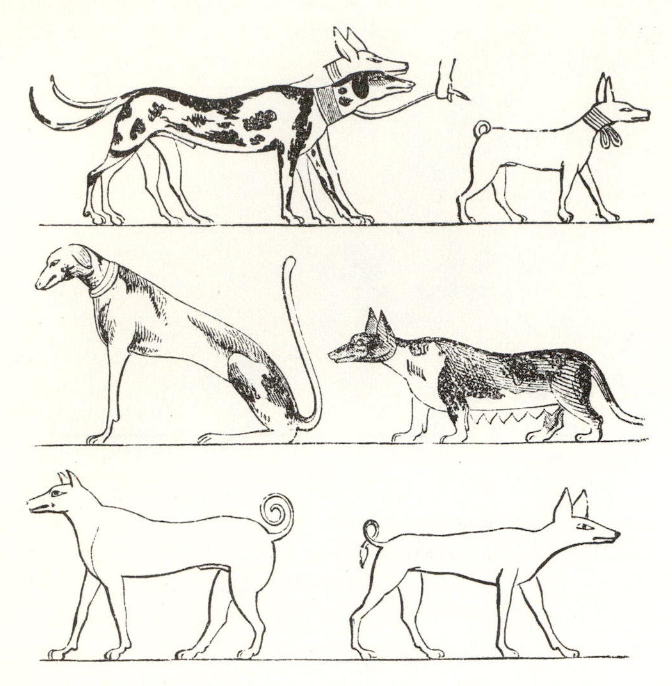

Economically carnivores form one of the most important groups of mammals. One carnivore, the dog, *Canis familiaris,* was the first of man's domesticated animals (Fig. 23-2). The dog and another carnivore, the domestic cat, *Felis catus,* are the most popular nonhuman companions of modern man. The larger species of wild carnivores are important game animals in many parts of the world and many smaller species (e.g. coyote) are considered to be "vermin", although they are important predators of many rodents and other small animals. One family, Mustelidae, includes many of the world's most important fur bearers, (e.g., mink, ermine, sable, sea otter) and most other families include furbearers of commercial importance. Some forms of carnivores (e.g., polar bear, red wolf, tiger) are presently listed as rare or endangered (see Chapter 34 for reference) due to uncontrolled or inadequately controlled sport hunters, "pest" controllers and pelt seekers. Many species (e.g., wolf, black bear, and puma) have declined in number with the increase in human populations while other species (e.g., coyotes, raccoons and some mongooses) have thrived in company with humans.

Figure 23–3.
Labial view of the dentition of a lion. In this species the third upper postcanine tooth is the last upper premolar and the third lower postcanine is the first lower molar. Together these teeth are the carnassial pair.
(Duncan 1877-83: 13)

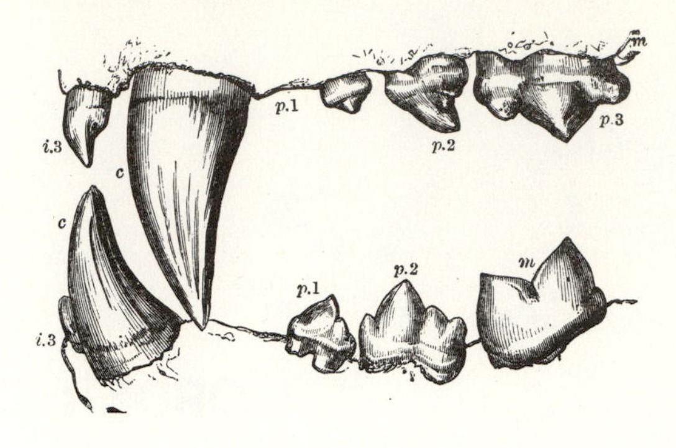

Distinguishing Characteristics

The canine teeth are usually large and conical (Fig. 23-3). Three lower incisors are present in all fissipeds* except the sea otter, *Enhydra lutris,* which has only two. The number of lower incisors in Otariidae and Phocidae varies from 0 to 2. The postcanine teeth of fissipeds are usually secodont and in many species the last upper premolar and the first lower molar are elongated and bladelike (Fig. 23-3). These are termed the carnassial pair and serve for shearing flesh. The premolars and molars anterior and posterior to the carnassials are variously developed, reduced or lost (Fig. 23-4).

In other fissipeds the postcanine teeth may be particularly brachydont (e.g., the sea otter, Fig. 23-5) or may be greatly reduced (e.g., the aardwolf, Fig. 23-21). In the pinnipeds the postcanine teeth are always essentially homodont and frequently unicuspid (Figs. 23-6, 23-9, and 23-11). Carnassials are never present in the pinnipeds.

The zygomatic arches of Carnivora are well developed and a sagittal crest is frequently present. The auditory bullae are usually fully ossified and are frequently large.

The feet of fissipeds are plantigrade or digitigrade and have four or five toes. Each toe ends with a large, sharp, curved claw. The pinnipeds have their knees and elbows included within the contour of the body,

*The term fissiped as used in this chapter refers to all Carnivora other than Phocidae and Otariidae, the pinnipeds. Neither of these terms is intended to have any taxonomic connotation.

Figure 23–4.
The maxillary dentition of representative fissiped carnivores showing varying development of premolars and molars. The vertical line runs through the upper carnassial, the last upper premolar. A, a dog, Canidae; B, a bear, Ursidae; C, a marten and D, a badger, both Mustelidae; E, a mongoose, Viverridae; F, a hyaena, Hyaenidae; and G, a lion, Felidae.
(Weber 1928: 313)

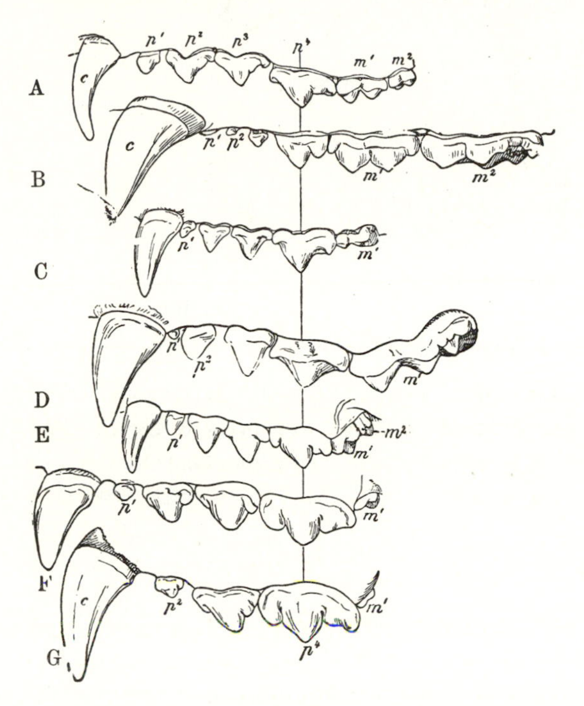

the metatarsals and metacarpals are elongate and the five digits on each foot are fully webbed (Figs. 23-7, 23-10, and 23-12). These feet are termed flippers.

The stomach is simple and the caecum is usually small or absent. The uterus is bicornuate and the testes may be abdominal or scrotal. A baculum is present and is very well developed in many species (Fig. 9-7).

Distribution

Carnivores range from Arctic ice floes through all continents to the Antarctic continent. Though most oceanic islands lack fissipeds, pinnipeds are found on many including Hawaii and New Zealand. Until the first human immigrants brought dingos with them into the Australian region, there were no Carnivora on that side of Wallace's Line. Modern man has introduced other fissipeds such as the domestic dog and cat, the red fox, and certain mongooses to Australia and to New Zealand and many other areas where no terrestrial carnivores had existed before.

Most fissipeds are confined to continental land masses and fresh waters. The polar bear, *Thalarctos maritimus*, the sea otter, *Enhydra lutris*, and the sea "cat," *Lutra felina*, are marine. Conversely, most pinnipeds are marine with landlocked forms only in the saline Caspian Sea and in the fresh water of Lake Baikal in the USSR.

Fossil History*

The Carnivora arose from some primitive insectivore during the mid-Paleocene. In the late Eocene the two living suborders arose from the †Miacidae. This primitive family was itself extinct before the beginning of the Oligocene. The Feliformia differentiated into the Felidae and Viverridae in the late Eocene and the Hyaenidae arose from the Viverridae in the mid-Miocene. The oldest Canidae are known from the Upper Eocene, and the oldest Mustelidae date from the same period. The Ursidae and Procyonidae date from the lower Oligocene and the †Enaliarctidae (Mitchell and Tedford 1973, Tedford 1976) and Otariidae are first known from the lower Miocene. The earliest records of Phocidae are from the middle Miocene. Tedford (1976) placed †Enaliarctidae and Otariidae as more closely related to Ursidae than to Phocidae and Phocidae more closely related to Mustelidae than to the other pinnipeds.

†Creodonta is an extinct group of carnivorous mammals that ranged from the late Cretaceous to the Miocene. While they are sometimes included as a part of the order Carnivora, they are probably not closely related to the †Miacoidea or to living families.

Key to Living Families

1 Lower incisors 0, 1, or 2; cheek teeth essentially homodont and usually unicuspid; carnassials never present; limbs modifed as flippers; tail short or absent 2

1′ Lower incisors 3 (2 in one species); cheek teeth usually heterodont and multicuspid; dentition frequently secodont and carnassials usually well developed; limbs not modified as flippers; tail varies; *if* lower incisors number 2, cheek teeth very broad and flat crowned (Fig. 23-5) and tail long (Fig. 23-29D) 4

*Based upon Romer 1966, except as otherwise noted.

Figure 23-5.
Skull of the sea otter, *Enhydra lutris.* (Novikov 1956: 221)

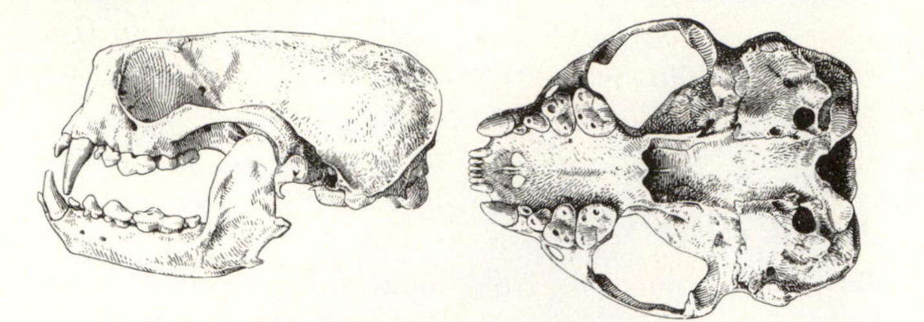

2 Teeth number 24 or fewer; lower incisors absent; upper canines very long, tusklike, (Fig. 23-6); extending well beyond the lips (Fig. 23-7) .... **Otariidae** (in part) walrus

2′ Teeth number 26 or more; lower incisors present; upper canines not tusklike, normally concealed behind lips 3

Table 23-1

Living Families of Carnivora.

Family	Common Name	Number of* Genera	Species	Distribution
Suborder Caniformia				
Canidae	dogs, wolves, foxes, jackals	13	35	Holarctic, Neotropical, Ethiopian, Oriental Introduced worldwide
Ursidae	bears	6	7	Holarctic, Oriental, NW Neotropical
Otariidae	walrus and sea lions or eared seals.	8	14	Colder coastlines of Arctic, Pacific, Atlantic and Indian Oceans
Ailuropodidae	giant panda	1	1	SW China
Procyonidae	red panda, raccoons, and allies	8	16	Nearctic, Neotropical, and SW China
Mustelidae	badgers, skunks, otters, weasels and allies	27	67	Holarctic, Neotropical Ethiopian, Oriental
Phocidae	earless seals	10	19	All oceans and seas, Lake Baikal
Suborder Feliformia				
Viverridae	civets, genets, mongooses and allies	36	72	Ethiopian, Oriental, southern Palaearctic
Hyaenidae	hyaenas, aardwolf	3	4	Ethiopian, S central Palearctic
Felidae	cats	19	37	Holarctic Neotropical Ethiopian, Oriental Introduced worldwide

*Numbers of pinniped genera and species from Rice (1977), numbers of fissiped genera and species from Ewer (1973). The number of genera of Felidae recognized by Ewer is much larger than the two to four genera recognized by most authors. Van Gelder (1978:7-8) included all species of Canidae in only seven genera, one (*Canis*) containing 29 species and the other six monotypic.

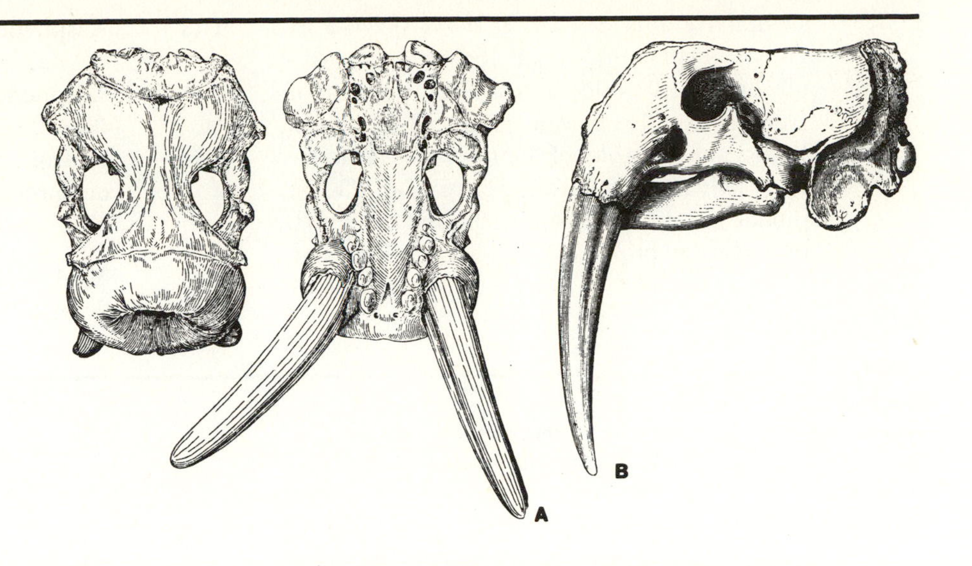

Figure 23–6.
Skull of a walrus, *Odobenus rosmarus.*
(A, Gromov et al. 1963: 916; B, Duncan 1877-83: 212)

Figure 23–7.
A walrus, *Odobenus rosmarus.*
(Gromov et al. 1963: 918)

Figure 23–8.
Section of an otarid skull showing the alisphenoid canal (arrow).
(Gromov et al. 1963: 899)

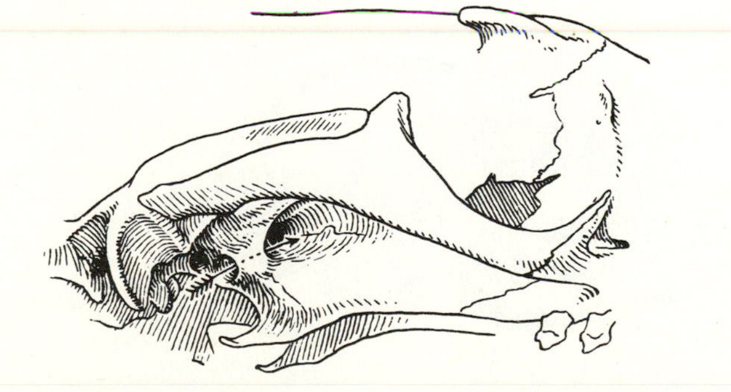

Figure 23–9.
Skulls of two members of the Otaridae; A, dorsal and ventral views of the California sea lion, *Zalophus californianus;* B, lateral view of the Antipodean fur seal, *Arctocephalus forsteri.*
(A, Gromov et al. 1963: 909; B, Flower and Lydekker 1891: 594)

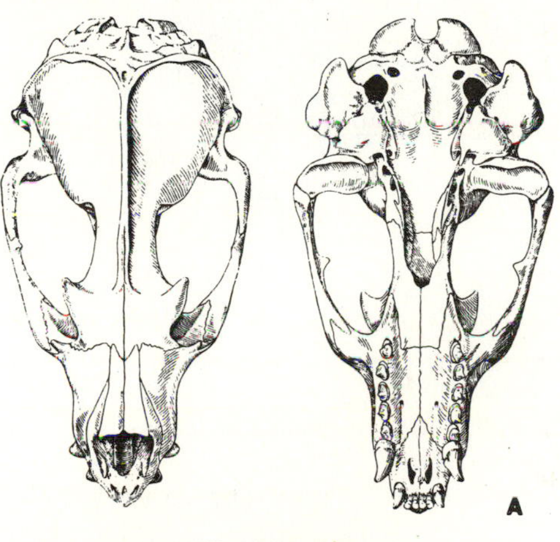

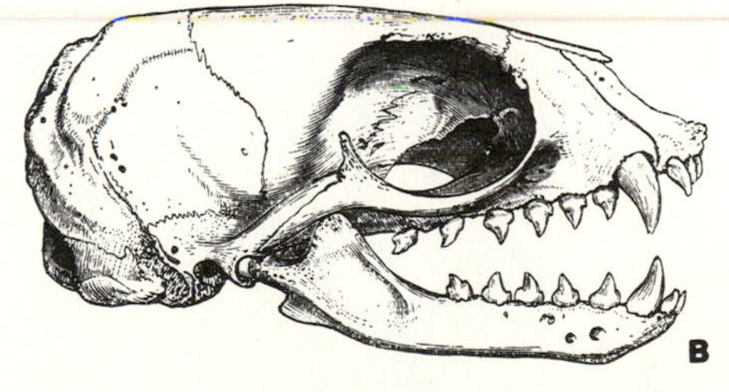

3 (2′) Dental formula 3/2 1/1 4/4 1-3/1 = 34-36; alisphenoid canal present (Fig. 23-8); teeth usually unicuspid; postorbital process of frontal usually well developed (Fig. 23-9); hind limbs capable of being rotated under body (Fig. 23-10); small pinnae present; pelt is usually uniform in color and underfur may be present .... **Otariidae** (in part)
sea lions or eared seals

3′ Dental formula 2-3/1-2 1/1 4/4 0-2/0-2 = 26-36; alisphenoid canal absent; teeth usually not unicuspid; postorbital process of frontal usually absent (Fig. 23-11); hind limbs not capable of being rotated under body, always point posteriorly (Fig. 23-12); pinnae absent; pelt usually spotted or ringed, underfur never present in adults .... **Phocidae**
earless seals

4 (1′) Alisphenoid canal present (Fig. 23-8) 5

4′ Alisphenoid canal absent 8

5 (4) Mastoid process approximately as large as or larger than paroccipital process (Figs. 23-13 and 23-15 6

5′ Mastoid process absent or much smaller than paroccipital process 7

Figure 23–10.
Two members of the Otaridae; A, the northern sea lion, *Eumetopias jubatus;* B, the northern fur seal, *Callorhinus ursinus.*
(Gromov et al. 1963: 905 and 913)

Figure 23–11.
Skull of the harbour seal, *Phoca vitulina.* A, dorsal and ventral views; B, lateral view.
(A, Gromov et al. 1963: 933; B, Flower and Lydekker 1891: 601)

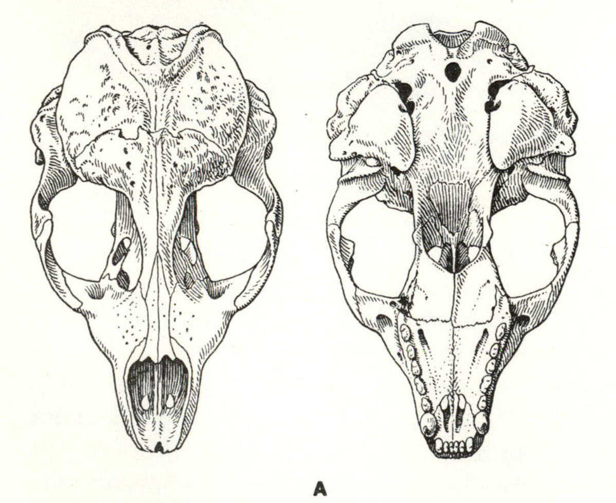

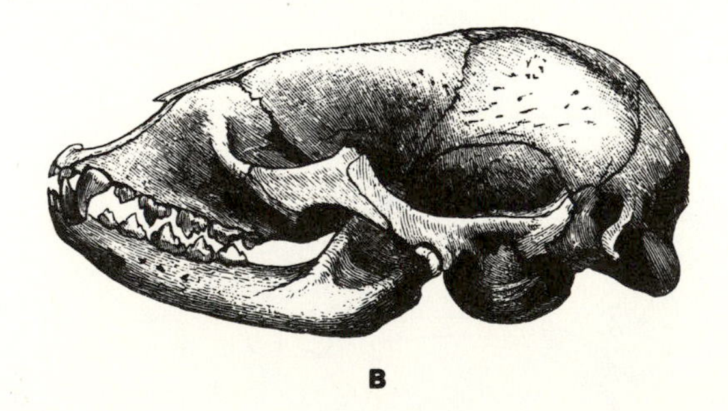

Figure 23–12.
Representative hair seals, Phocidae. A, the ringed seal, *Phoca hispida;* B, the hooded seal, *Cystophora cristata.* (Gromov et al. 1963: 939–962)

Figure 23–13.
The skull of the red or lesser panda, *Ailurus fulgens.* (Flower and Lydekker 1891: 564)

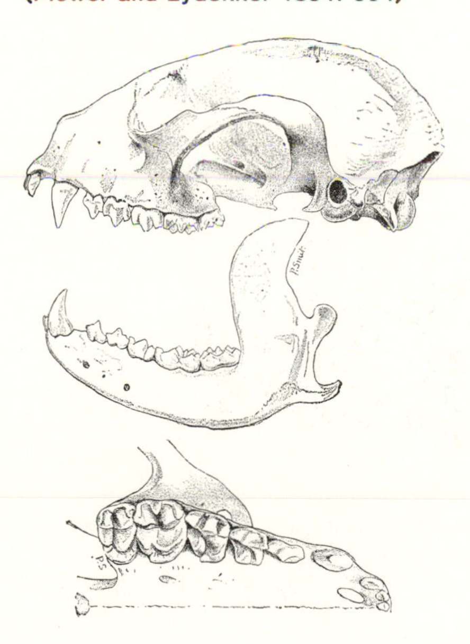

Figure 23–14.
The lesser or red panda, *Ailurus fulgens.* (Hsia et al. 1964: 72)

Figure 23–15.
Skull of the brown bear, *Ursus arctos.* (Novikov 1956: 93)

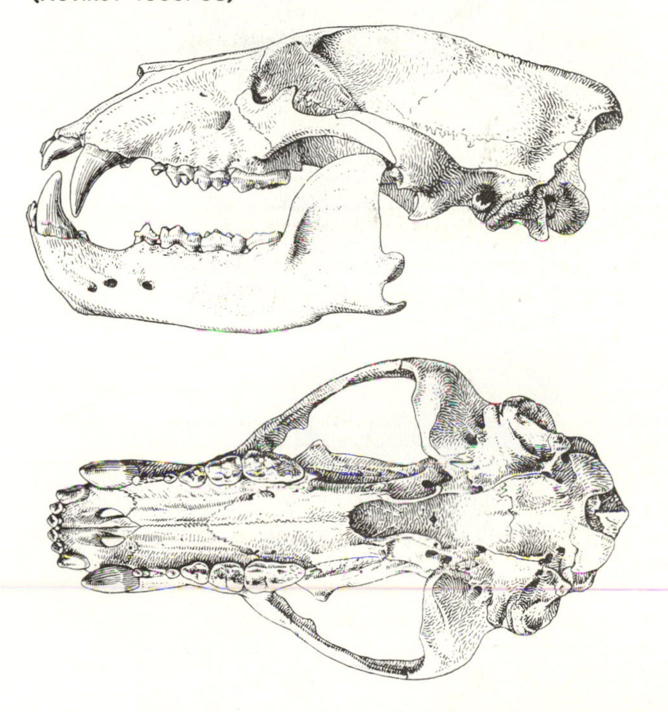

6 (5) Greatest length of skull less than 130 mm; dental formula 3/3 1/1 3/4 2/2 = 38 (Fig. 23-13); Tail long (Fig. 23-14) .... **Procyonidae** (in part)
lesser or red panda

6′ Greatest length of skull more than 130 mm; dental formula variable, usually 2-3/3 1/1 4/4 2/3 = 40-42, but first 3 premolars small and one or more of these sometimes absent (Fig. 23-15); tail short (Figs. 23-1 and 23-16) .... **Ursidae**
bears

Figure 23–16.
Representative bears. A, the Asiatic black bear, *Selenarctos thibetanus;* B, the sloth bear, *Melursus ursinus.*
(A, Gromov et al. 1963: 780; B, Finn 1929: 62)

Figure 23–17.
Skull of a viverrid, the oyan, *Poiana richardsoni.*
(Allen 1924: 145)

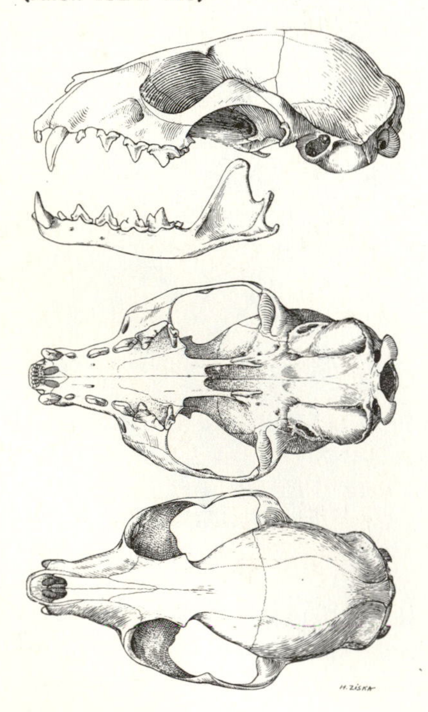

7 (5′) Dental formula 3/3 1/1 3-4/3-4 1-2/1-2 = 36-40; bulla constricted externally (Fig. 23-17); and divided by a septum internally; not dog-like in general appearance (Fig. 23-18) .... **Viverridae** (in part)
most viverrids

7′ Dental formula usually 3/3 1/1 4/4 2/3 = 42, in two genera molars may be 1-2/2 and in another genus 3-4/4-5; bullae not constricted or partitioned (Fig. 23-19); dog-like in general appearance (Fig. 23-20) .... **Canidae**
dogs

8 (4′) Postcanine teeth greatly reduced, widely spaced, number variable; canines well developed; postorbital process of frontal well developed (Fig. 23-21) .... **Hyaenidae** (in part)
aardwolf

8′ Postcanine teeth not as above, or if postcanine teeth reduced and widely spaced (Fig. 23-22), then canines small and postorbital process absent 9

Figure 23–18.
Representative viverrids. A, the African civet, *Civettictis civetta;* B, the oyan, *Poiana richardsoni;* and C, the banded mongoose, *Mungos mungo.*
(Dekeyser 1955: 263-272)

9 (8′) Dental formula 3/3 1/1 2-3/2 1/1 = 28-30; last upper molar tiny, not medially constricted (Fig. 23-23; cat-like in general appearance (Fig. 23-24) .... **Felidae**
cats

9′ Dental formula variable but teeth usually number 32 or more, if fewer than 32, last upper molar is medially constricted into distinct labial and lingual lobes (Fig. 23-28A and B); usually not cat-like in general appearance 10

Figure 23–19.
Skulls of representative canids. A, the wolf, *Canis lupus;* B, a red fox, *Vulpes vulpes.*
(A, Stroganov 1962: 23; B, Gromov et al. 1963: 760)

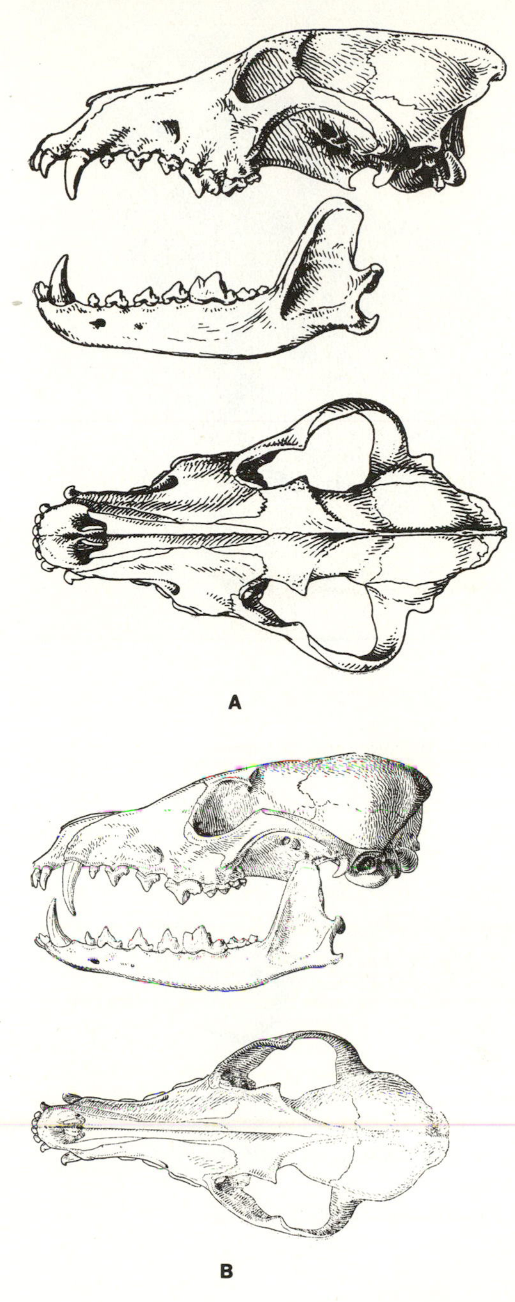

10 (9′) Molars 0-1/1-2; total number of teeth 28-38, usually 34 11

10′ Molars 2/1-3, total number of teeth 36-42, usually 40 13

Figure 23–20.
Representative canids. A, the raccoon dog, *Nyctereutes procyonoides;* B, the red fox, *Vulpes vulpes.*
(A, Stroganov 1962: 67; B, Gromov et al. 1963: 759)

Figure 23–21.
Skull of an aardwolf, *Proteles cristatus,* an insectivorous hyaenid.
(Flower and Lydekker 1891: 540)

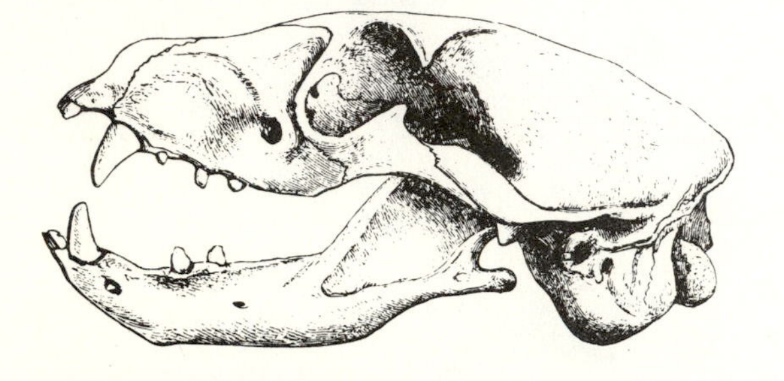

Figure 23–22.
Skull of the Falanouc, *Eupleres goudoti,* a viverrid that feeds upon invertebrates.
(Flower and Lydekker 1891: 539)

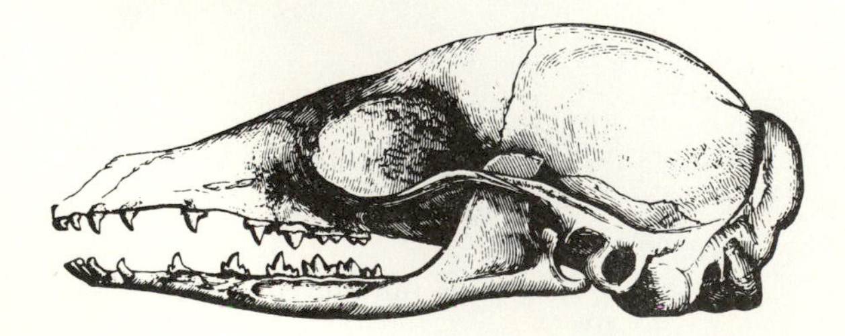

Figure 23–23.
Skulls of representative cats. A, the lynx, *Lynx lynx;* and B, the cheetah, *Acinonyx jubatus.*
(Gromov et al. 1963: 875, 893)

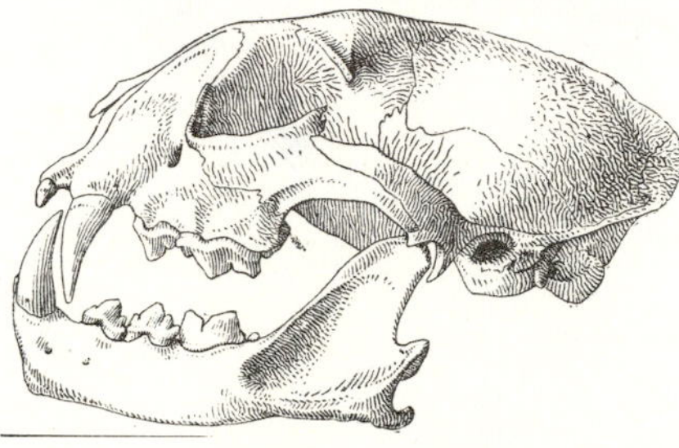

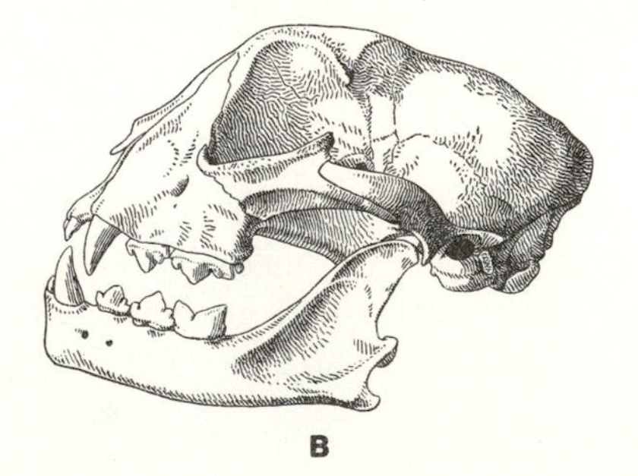

11 (10′) Dental formula 3/3 1/1 4/3 0-1/1 = 32-34; M^1 absent or small and located lingually to posterior portion of P^4 (Fig. 23-25); carnassials well developed and massive; head and shoulders large and powerful, hind quarters smaller, back slopes posteriorly (Fig. 23-26) .... **Hyaenidae** (in part) hyenas

11′ Dental formula variable but never as above; M^1 always directly posterior to P^4 (Fig. 23-28); carnassials variable; head and body usually long and slender, forequarters not noticably larger than hind quarters, back does not slope 12

Figure 23-24.
Representative cats. A, the lynx, *Lynx lynx;* B, the leopard, *Panthera pardus.*
(Gromov et al. 1963: 874, 887)

A

B

Figure 23-25.
Skull of the striped hyena, *Hyaena hyaena.* The small upper molar is absent in the genus *Crocuta.*
(Novikov 1956: 110)

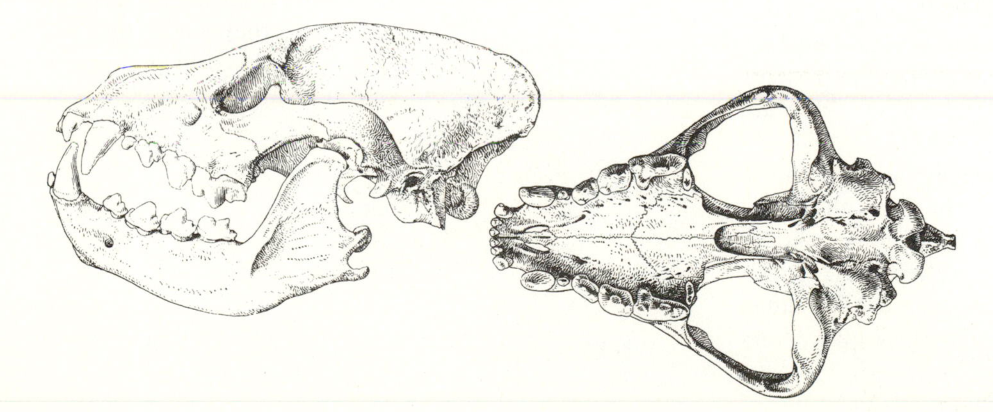

Figure 23-26.
The spotted hyaena, *Crocuta crocuta* (left) and the striped hyaena, *Hyaena hyaena* (right).
(Dekeyser 1955: 275-276)

Figure 23-27.
Skull of a kinkajou, *Potos flavus.*
(Duncan 1877-83: 179)

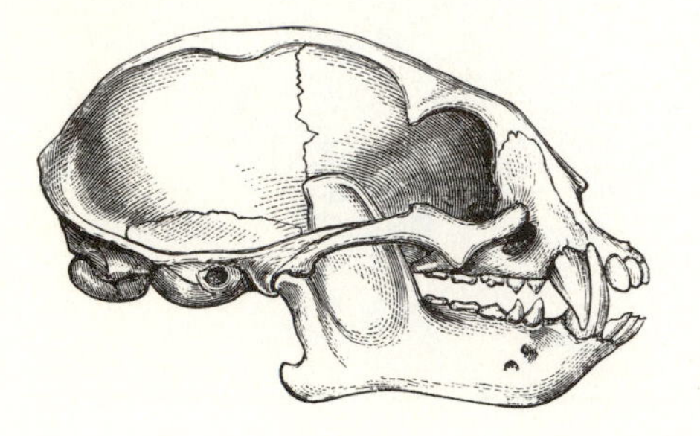

12 (11′) Canines with two or more longitudinal grooves on labial and lingual surfaces (Fig. 23-27); postcanine teeth 5/5, last upper cheek tooth essentially circular in occlusal outline; tail prehensile .... **Procyonidae** (in part)
kinkajou

12′ Canines ungrooved; postcanine tooth number variable, last upper cheek tooth dumbbell-shaped or squarish in occlusal outline (Fig. 23-28); tail never prehensile (Fig. 23-29) .... **Mustelidae**
weasels, badgers, skunks, otters

13 (10′) Mastoid process absent or much smaller than paroccipital process (Fig. 23-22) .... **Viverridae** (in part)
Madagascar viverrids

13′ Mastoid process approximately equal in size to paroccipital process (Fig. 23-31) 14

14 (13′) Size large, condylobasal length over 150 mm; tail short; color black and white (Fig. 23-30) .... **Ailuropodidae**
giant panda

14′ Size small, condylobasal length less than 150 mm, (Fig. 23-31); tail usually long (Fig. 23-32); color never as above .... **Procyonidae** (in part)
New World procyonids

Figure 23–28.
Skulls of representative mustelids. A, the pine marten, *Martes martes;* B, the river otter, *Lutra lutra;* C, the European badger, *Meles meles.*
(Stroganov 1962: 158, 179, 215)

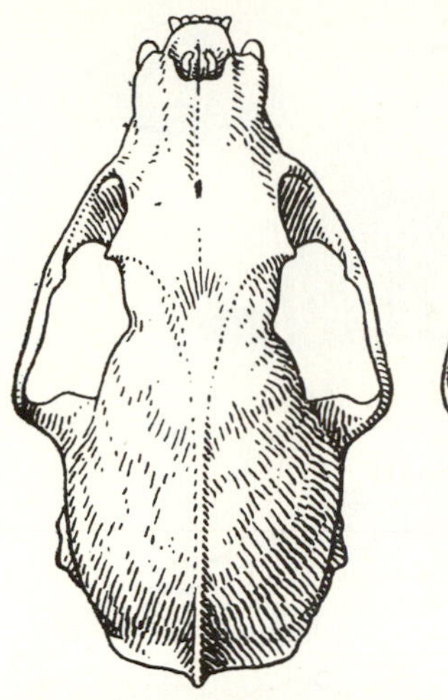

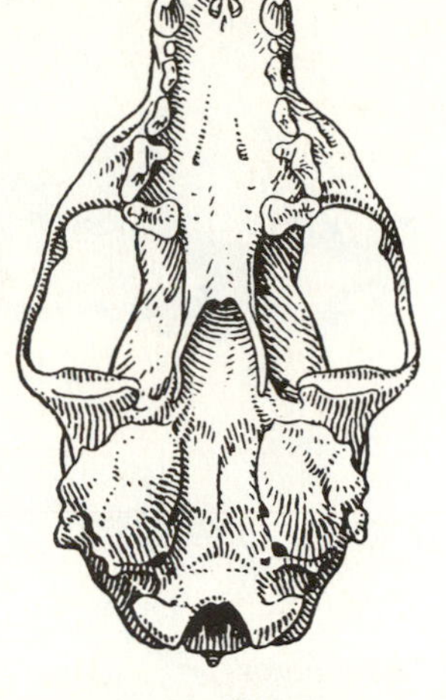

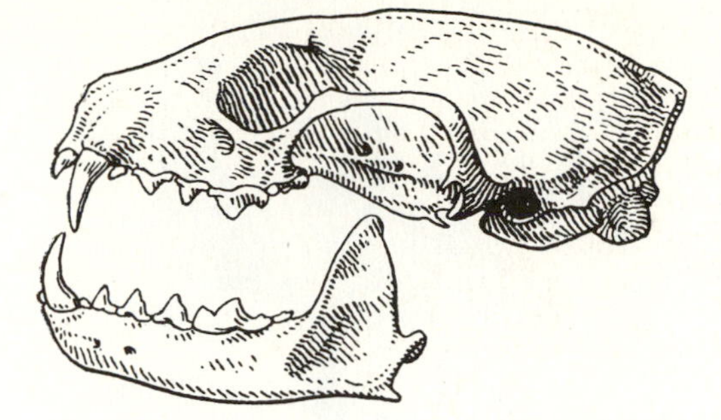

A

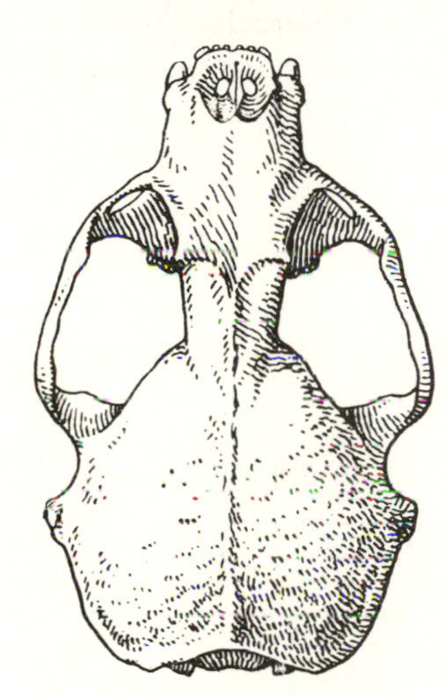

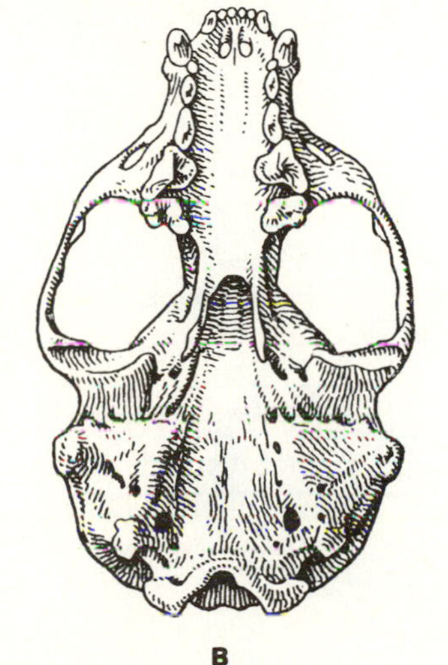

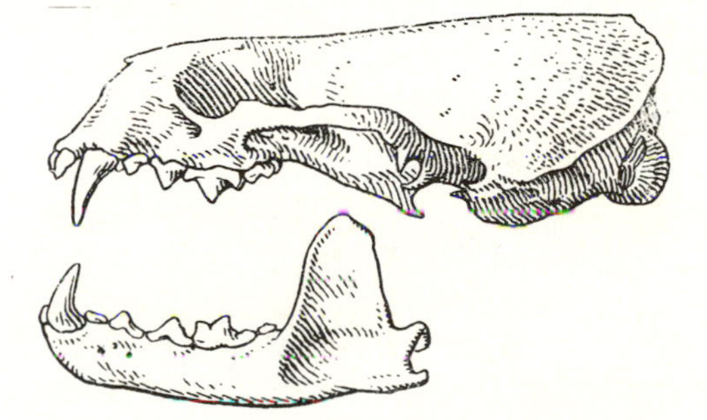

B

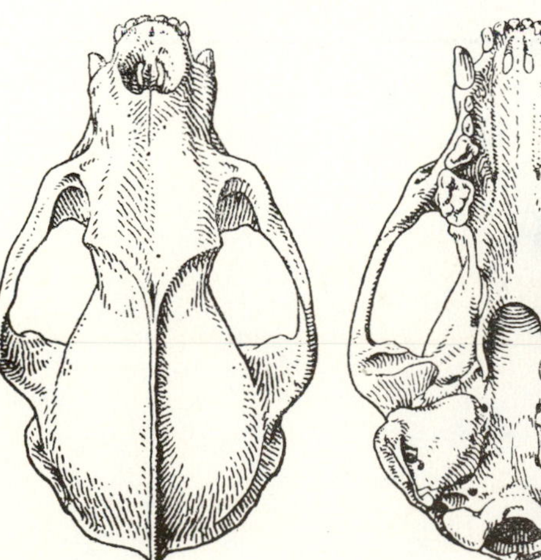

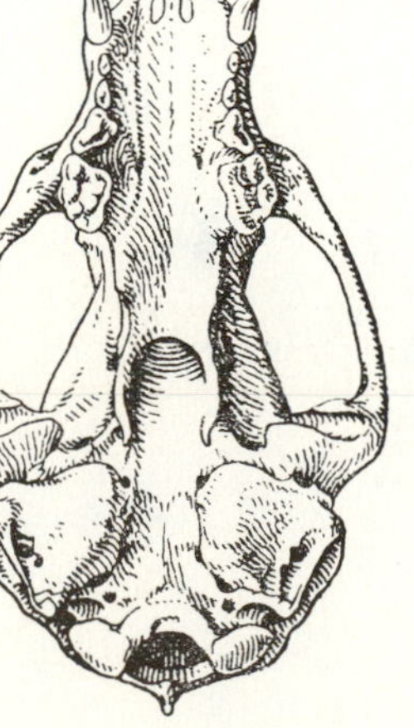

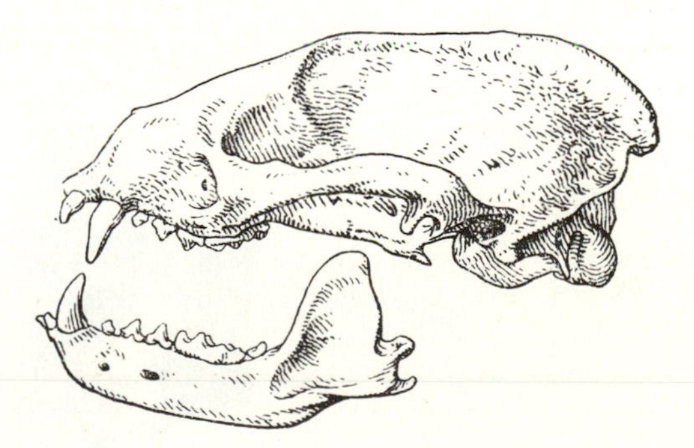

C

Figure 23–29.
Representative mustelids. A, the pine marten, *Martes martes;* B, the honey badger, *Melivora capensis;* and C, the sea otter, *Enhydra lutris;* and D, the spotted neck otter, *Lutra maculicollis.*
(A & C, Gromov et. al. 1963: 830-852; B, Dekeyser 1955; 259)

Figure 23–30.
The giant panda, *Ailuropoda melanoleuca.*
(Hsia et al. 1964: 73)

Figure 23–31.
Skull of the raccoon, *Procyon lotor*.
(Novikov 1956: 114)

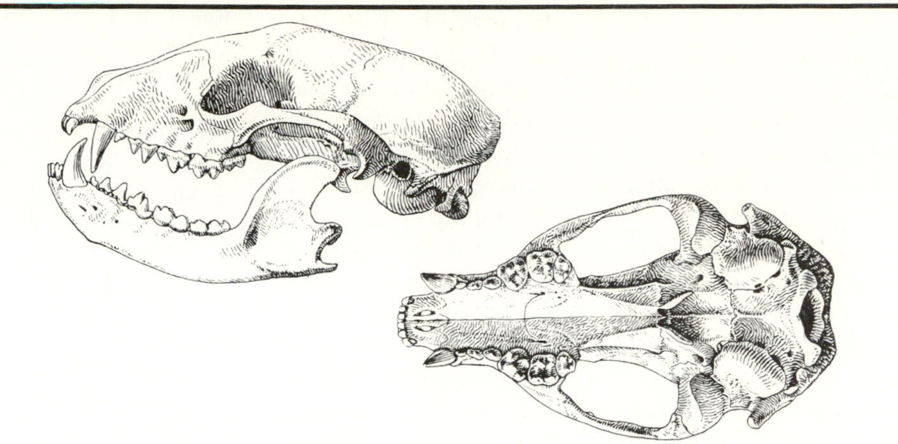

Figure 23–32.
A raccoon, *Procyon lotor*.
(Davis 1964: 33)

Comments and Suggestions on Identification

Most of the carnivores are well known and easily recognized. The dogs, bears, giant pandas, hyaenas and cats each have a basic body form that is consistant throughout the family. Otariids and phocids are easily distinguished from the other mammals by general body form and from each other on the structure of the hind limbs and the relative lengths of the necks. Mustelidae, Procyonidae and Viverridae each contain considerable variation in body shape but with practice each of these groups is easily recognizable.

The skulls of most families of Carnivora will also, with practice, "look right." But caution, remember that even though the basic structure of a family may remain relatively consistant the size can vary considerably—for example the skulls of a house cat and lion are quite similar in shape but the student used to thinking of Felidae as being the size of a tabby will often glance at the lion skull and identify it as a bear without bothering to check the details.

Taxonomic Remarks

Simpson (1945) reviewed the history of the classification of Carnivora and pointed out that the order as he recognized it and as it is listed above constituted the most completely "natural" order listed by Linnaeus (1758), the order Ferae. Linnaeus (1758), Simpson (1945), Romer (1966), and many others have included the pinnipeds in the order Carnivora, usually as a suborder, Pinnipedia, with the other terrestrial carnivores grouped in the suborder Fissipedia. Even though Simpson (1945) has been generally recognized as the standard reference for the systematics of the higher taxa of mammals, most authors over the last few decades have listed the Pinnipedia as an order distinct from the Carnivora (Ellerman and Morrison-Scott 1951; Miller and Kellog 1955; Hall and Kelson 1959; Walker et al. 1964, 1968, 1975; Stains 1967; and Ewer 1973). Many authors have argued that the otariids and phocids had separate origins (Simpson 1945:233). Most recently Mitchell and Tedford (1973) and Tedford (1976) have shown that the otariids are more closely related to the ursids than to the phocids and that the phocids are more closely related to the mustelids than to the otariids. Thus, not only do the pinnipeds not belong in a distinct order, they also do not belong in a unique suborder. We have above followed the arrangement of higher taxa given by Tedford (1976: 372).

The giant and red pandas have variously both been placed in Procyonidae, both in a distinct family, each in a distinct family, or in Ursidae and Procyonidae respectively. (The only arrangement we have not seen is inclusion of the red panda in the Ursidae.) Recently many authors (Stains 1967; Gunderson 1976; Vaughn 1972, 1978) have followed Davis (1964) who did a detailed anatomical study of the giant panda and concluded it was a bear. But Ewer (1973:393) said that Davis (1964) simply brushed aside the features that did not support his preconceived conclusion that the panda was a bear. In spite of Davis (1964), many authors list the giant panda as a procyonid (Simpson 1945; Morris and Morris 1966; Morris 1965; Walker et al. 1968; Matthews 1971; Ewer 1973; Kowalski 1976) and others (e.g. Sarich 1973) listed it as a bear on data of their own. Heine (1973) and Pogloyen-Neuwall (1975) listed both pandas in the family Ailuridae and Chu (1974) placed the giant panda by itself in the family Ailuropodidae. We have left the red panda in the family Procyonidae but have followed Chu (1974) and others before him in placing the giant panda in its own family. We have chosen this arrangement not out of any deep seated conviction that the giant panda is appropriately unique but because this allows us to define the Ursidae and Procyonidae without the giant panda and allows the student to easily add it to which ever of the families he prefers.

Tedford (1976) included the walrus in the family Otariidae but most authors (Simpson 1945; Morris 1965; Romer 1966; Stains 1967; Pedersen 1975; Vaughn 1978) have placed it in a distinct family, Odobenidae. Rice (1977) observed that "Mitchell (1975) regarded the walruses as a subfamily of the Otariidae, whereas Repenning (1975) maintained them as a separate family. Pending consensus I [Rice] follow most previous authors in listing them as a separate family." We have followed Mitchell (1975) and Tedford (1976) in including the walrus in Otariidae.

The aardwolf is now usually considered to be an aberrant member of the Hyaenidae, but it has frequently been listed as an aberrant Viverridae or as a monotypic family, Protelidae (e.g. Ducker 1975). Simpson (1945:229) discussed the history of its classification. We include it in Hyaenidae.

There is considerable disagreement among authors as to the appropriate number of genera of Phocidae (Rice 1977:4), Canidae (Stains 1975; Van Gelder 1978), Ursidae, Viverridae and Felidae (Ewer 1973) but there is little disagreement about family ranking of carnivores other than those disputes listed in the preceeding paragraphs.

Supplementary Readings

Barabash-Nikiforov, I. I. et al. 1947. *The sea otter.* English Translation. 1962. Israel Program for Scientific Translations, Jerusalem. 227 pp.

Bueler, L. E. 1973. *Wild dogs of the world.* Stein and Dry Publishers, New York. 274 pp.

Eaton, R. L. 1974. *The cheetah.* Van Nostrand Reinhold, New York. 178 pp.

Eaton, R. L. (Ed.) 1973. *The world's cats. Vol. I. Ecology and conservation.* World Wildlife Safari, Winston, Oregon. 349 pp.

Eaton, R. L. 1974. *The world's cats. Vol. II. Biology, behavior and management of reproduction.* Feline Research Group, Seattle, 260 pp.

Eaton, R. L. 1976. *The world's cats. Vol. III, No. 1. Contribution to status, management and conservation.* University of Washington, Seattle. 95 pp.

Eaton, R. L. 1976. *The world's cats. Vol. III, No. 2. Contribution to biology, ecology, behavior and evolution.* University of Washington, Seattle. 179 pp.

Ewer, R. F. 1973. *The carnivores.* Cornell Univ. Press, Ithaca. 494 pp.

Fox, M. W. (Ed.). 1975. *The wild canids, their systematics, behavioral ecology and evolution.* Van Nostrand Reinhold, New York, 508 pp.

Hall, E. R. 1951. *American weasels. Univ. Kansas Publ., Mus. Nat. Hist.* 4:1-466.

Harris, C. J. 1968. *Otters. A study of the recent Lutrinae.* Weidenfeld and Nicolson, London. 397 pp.

Hinton, H. E., and A.M.S. Dunn. 1967. *Mongooses.* Oliver & Boyd, Edinburgh. 144 pp.

Howson, R. J. *et al.* (Eds.) 1968. *The behavior and physiology of pinnipeds.* Appleton-Century-Crofts, New York. 411 pp.

King, J. E. 1964. *Seals of the world.* British Museum (Nat. Hist.), London. 154 pp.

Kruuk, H. 1972. *The spotted hyaena: A study of predation and social behavior.* University of Chicago Press, Chicago. 335 pp.

Kruuk, H. and W. A. Sands. 1972. The aardwolf (*Proteles cristatus* Sparrman, 1783) as predator of termites. *E. Africa Wildl. J.* 10:211-227.

Morris, R., and D. Morris. 1966. *Men and pandas.* McGraw Hill Book Co., New York. 223 pp.

Pelton, M. R., J. W. Lentfer, and G. E. Folk. 1976. *Bears—their biology and management.* ICUN Publications, New Series No. 40. Morges, Switzerland. 467 pp.

Peterson, R. S. and G. A. Bartholomew. 1967. *The natural history of the California sea lion.* Spec. Pub. No. 1. Amer. Soc. Mammalogists. 79 pp.

Ronald, K., and A. W. Mansfield (Eds.) 1975. Biology of the seal. *Rapports et Proces—verbaux des Reumons* 169:1-557.

Ronald, K., L. M. Hanly, P. J. Healey, and L. J. Selley. 1976. *An annotated bibliography on the Pinnipedia.* International Council for the Exploration of the Sea, Charlottenlund, Denmark. 785 pp.

Roseveor, D. R. 1974. *The carnivores of West Africa.* British Museum (Nat. Hist.) London. 548 pp.

Schaller, G. B. 1972. *The Serengeti lion: a study of predator-prey relations.* Univ. of Chicago Press, Chicago. 480 pp.

van Lawick-Goodall, H. and J. van Lawick-Goodall. 1971. *Innocent killers.* Houghton Mifflin, Boston. 222 pp.

Verts, B. J. 1967. *Biology of the striped skunk.* Univ. Illinois Press, Urbana. 218 pp.

24 The Rabbits, Hares and Pikas

Order Lagomorpha

ORDER LAGOMORPHA

Lagomorph means literally "hare-shaped." While this name does not identify a diagnostic feature of the order, it does point out the similarity that exists among living species. Lagomorphs are terrestrial mammals, although some species burrow and may be considered semifossorial. Members of the living families have tails that are very short and hind feet that are at least somewhat larger than the forefeet. In rabbits and hares, Leporidae, the hind feet are considerably larger than the forefeet, the external ears are very long, and the tail is short but evident externally (Fig. 24-1). Pikas, Ochotonidae, are smaller than most leporids, the hind feet are only slightly larger than the forefeet, the ears are relatively short and rounded, and the tail is absent externally (Fig. 24-2).

The term "rabbit" is applied to those leporids that have altricial young, that is, born naked, blind, and helpless. The young of "hares" are precocial, that is, born furred, sighted, and capable of moving about on their own. In North America the domesticated rabbit, *Oryctolagus cuniculus,* and the various species of cottontails, *Sylvilagus,* are "rabbits," while the jackrabbits and northern hares, *Lepus,* are "hares."

Lagomorphs are almost totally herbivorous and feed on a wide variety of forbs, grasses, and to some extent, shrubs. Reingestion of caecotrophic feces is known to occur in most species of lagomorphs and enables them to assimilate more plant nutrients and certain B vitamins that are produced by bacteria in the caecum (Hansen and Flinders 1969).

The social structure in lagomorphs ranges from a dispersed system in many hares (*Lepus*) to that of dominance hierarchies in the European rabbit, *Oryctolagus cuniculus* (Eisenberg 1966). Pikas (*Ochotona*) may establish territories based on the defense of accumulated hay piles although these territories are not

Figure 24–1.
The varying hare, *Lepus timidus.*
(Flower and Lydekker 1891: 493)

Figure 24–2.
A pika, *Ochotona*
(Hsia et al. 1964: 24)

Table 24-1
Living Families of Lagomorpha.

Family	Common Name	Number of* Genera	Number of* Species	Distribution
Ochotonidae	pikas	1	*ca.* 14	Western Nearctic mountains, Palearctic mountains and steppes
Leporidae	rabbits and hares	8	*ca.* 49	Worldwide, except SE Oriental and Madagascar. Introduced in many areas of world.

*Layne (1967)

generally defended during the reproductive season (Lutton 1975; Kawamichi 1976). Most species of lagomorphs spend their entire lives above ground but there are notable exceptions. Pikas (*Ochotona*) establish passageways among rock piles and European rabbits (*Oryctolagus cuniculus*) construct extensive underground burrow systems or warrens. Certain other species of rabbits of the genera *Caprolagus, Poelagus,* and *Romerolagus* are also known to construct burrows (Walker et al. 1975).

Wild lagomorphs are of some economic importance as sources of fur and meat and as game animals. Many species are considered vermin. This is particularly true in areas such as Australia where *Oryctolagus* was introduced by man, greatly multiplied and damaged native ecosystems. The domestic rabbit is raised for meat, fur, as a pet, and for laboratory research.

Figure 24–3.
Upper incisors of a lagomorph.
(Guryev 1964)

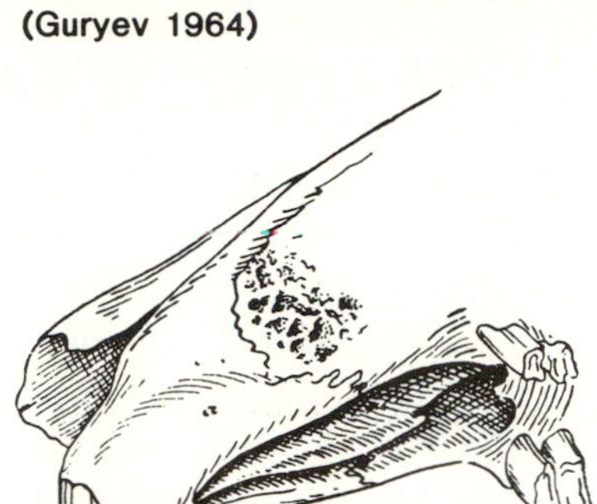

Distinguishing Characteristics

The dental formula is 2/1 0/0 3/2 2-3/3 = 26-28. The first incisors are large and "rodentlike." The second upper incisors are small, peglike teeth located directly behind the first incisors (Fig. 24-3). The cheek teeth are hypsodont, rootless, and evergrowing. Each maxilla is perforated on the side of the rostrum by a single large opening in the Ochotonidae or by numerous small openings separated by a lattice of bone in the Leporidae (Fig. 24-4).

The forefeet are digitigrade and the hind feet plantigrade. The tail is short to absent. There is no baculum and the testes descend seasonally into a scrotum located anterior to the penis. The uterus is duplex.

Figure 24–4.
Skull of varying hare, *Lepus timidus*. Note the fenestration of the rostrum.
(Flower and Lydekker 1891: 492)

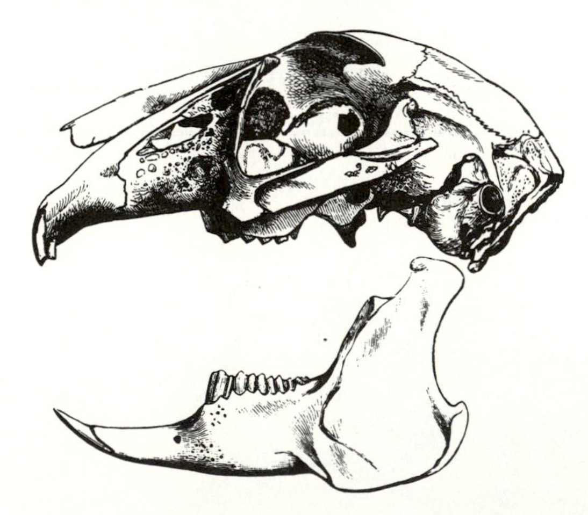

Distribution

Lagomorphs are native to the Holarctic, Ethiopian (except Madagascar), northern Neotropical, and northern and western Oriental Regions. They have been introduced by man into Australia, New Zealand, Java, the southern Neotropical Region, and many oceanic islands.

Fossil History

The oldest known fossil lagomorph, †*Eurymylus,* member of the extinct family †Eurymylidae (upper Paleocene to Oligocene) is from the late Paleocene of Mongolia. No other fossils are known until the late Eocene when both living families are found in the Nearctic and Holarctic. While a group of leporids thrived in the Nearctic until the early Pleistocene (Blancan), all recent leporids seem to be derived from late Tertiary Palearctic forms (Dawson 1958).

Key to Living Families

1 Cutting edge of I^1 with V-shaped notch (Fig. 24-5A); dental formula 2/1 0/0 3/2 2/3 = 26; supraorbital process of frontal absent (Fig. 24-6); external tail absent; ears no longer than wide .... **Ochotonidae** pikas

1′ Cutting edge of each first upper incisor straight (Fig. 24-5B; dental formula 2/1 0/0 3/2 2/3 = 28 (2/1 0/0 3/2 3/3 in *Pentalagus*); supraorbital process of frontal present (Fig. 24-7); short external tail present; ears longer than wide .... **Leporidae** rabbits and hares

Figure 24–5.
Anterior view of the first upper incisors of an ochotonid (A) and leporid (B).
(Gromov et al. 1963)

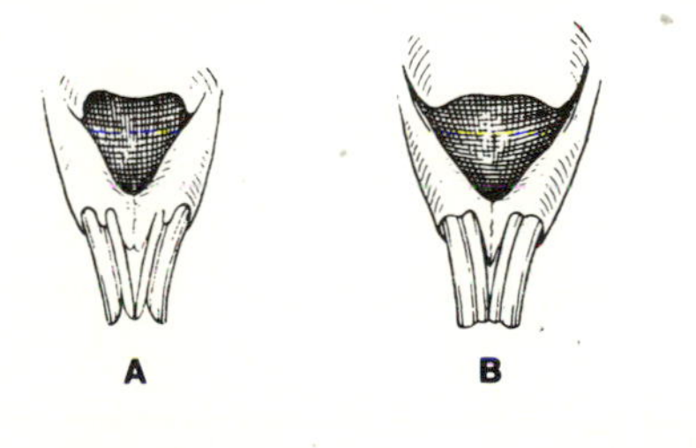

Figure 24–6.
Skulls of (A) a pika, *Ochotona*, and (B) a domestic rabbit, *Oryctolagus cuniculus*.
(Gromov et al. 1963)

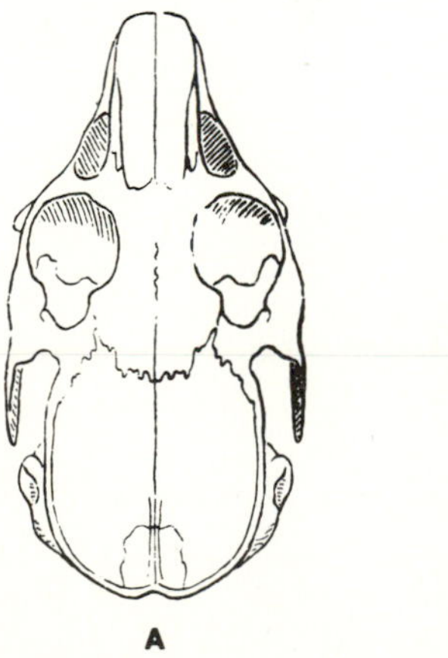

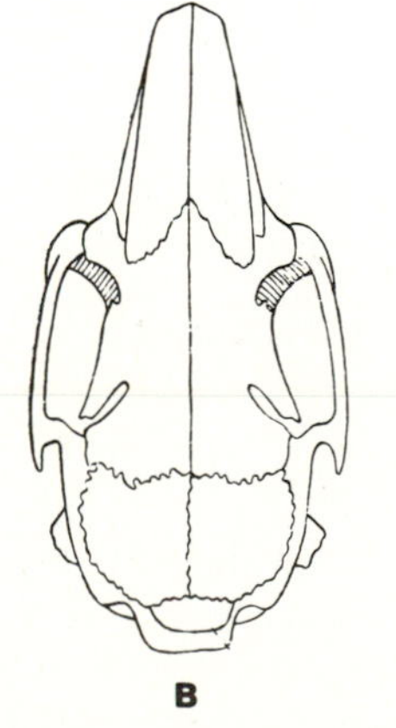

Comments and Suggestions on Identification

All rabbits and hares have a similar basic appearance, but unfortunately, some rodents (such as the springhaas) also superficially resemble lagomorphs. Pikas could be confused with several rodents. Be sure to check for the characteristic second upper incisor.

Taxonomic Remarks

Lagomorphs were long considered to be a suborder, Duplicidentata, of Rodentia, but aside from dental similarities there are few characters that link the two groups. Lagomorpha is now almost universally recognized as an order distinct from, and only distantly related to, the Rodentia (see Gidley 1912).

The genus *Brachylagus* was synomized with the genus *Sylvilagus* by Hall and Kelson (1959). Layne (1967) followed this arrangement but mentioned that not all workers agreed with this interpretation.

Supplementary Readings

Bear, G. D., and R. M. Hansen. 1966. Food habits, growth and reproduction of white-tailed jackrabbits in southern Colorado. *Colorado State Univ., Agricultural Exper. Sta., Tech Bull.* 90:1-59.

Broadbrooks, H. E. 1965. Ecology and distribution of the pikas of Washington and Alaska. *Amer. Midl. Natur.* 73:299-335.

Dawson, M. R. 1958. Later Tertiary Leporidae of North America. *Univ. Kansas Paleont. Contrib. Vertebrata* 6:1-75 + 2 pls.

Flux, J. E. C. 1970. Life history of the mountain hare (*Lepus timidus scoticus*) in northeast Scotland. *J. Zool.* 161:75-123.

Hansen, R. M., and J. T. Flinders. 1969. Food habits of North American hares. *Colorado State Univ., Science Ser., Range Sci. Dept.* 1:1-18.

Ingersoll, J. M. 1964. The Australian rabbit. *Amer. Scientist* 52:265-273.

Kawamichi, T. 1976. Hay territory and dominance rank of pikas (*Ochotona princeps*). *J. Mammal.* 57:133-148.

Keith, L. B., and L. A. Windberg. 1978. A demographic analysis of the showshoe hare cycle. *Wildlife Monog.* 58:1-70.

Marsden, H. M., and N. R. Holler. 1964. Social behavior in confined populations of the cottontail and the swamp rabbit. *Wildlife Monog.* 13:1-39.

Vorhies, C. T., and W. P. Taylor. 1933. The life histories and ecology of jack rabbits, *Lepus alleni and Lepus californicus* ssp., in relation to grazing in Arizona. *Tech. Bull., Agric. Exper. Station, Univ. Arizona* 49:471-587.

Watson, J. S. 1954. Reingestion in the wild rabbit *Oryctolagus cuniculus* (L.). *Proc. Zool. Soc. London* 124:615-624.

Key to Living Families of the World and North America

25 The Rodents

Order Rodentia

ORDER RODENTIA

The order Rodentia, the "gnawing" mammals, contains over forty percent of all species in the class Mammalia. A consistent diagnostic character for all rodents is an upper and lower pair of arc-shaped, chisel-edged incisors.

Rodents have adapted to most habitats and include terrestrial, fossorial, saltatorial (Fig. 25-1), arboreal, gliding, and semiaquatic forms. They range in size from the smallest mice (e.g., *Micromys, Baiomys* and some *Mus*) which weigh only a few grams up to the largest living rodent, the capybara, *Hydrochoerus hydrochaeris* (Fig. 25-48B) that is pig-sized and weighs up to 50 kg.

Most rodents are primarily omnivorous, feeding on bark, grass, seeds, other vegetation, insects, and other animal matter. Some, such as the grasshopper mice, *Onychomys*, feed on small vertebrates during certain months of the year (Flake 1973).

Many species of rodents are very important economically. On the negative side, *Rattus norvegicus, Mus musculus,* and other species commensal with man damage grains stored in unprotected granaries. Under certain conditions rodents may damage field crops, debark trees in orchards or forests, and establish burrows in areas where man does not want them (e.g., lawns, dikes). Most of these depredations can be controlled by agricultural management practices or selective trapping.

Rodents also have many beneficial attributes. The burrowing activities of fossorial species aerate the soil and bring mineral nutrients into the topsoil. Insects are eaten by many species. Beavers, muskrats, and ranch stocks of chinchillas are utilized in the fur industry. Medical and zoological research is highly dependent on laboratory-raised stocks of the Norway rat, *Rattus norvegicus;* house mouse, *Mus musculus;* golden hamster, *Mesocricetus auratus;* guinea pig, *Cavia porcellus;* and, more recently, Mongolian gerbil, *Meriones unguiculatus.* Many people in various parts of the world utilize rodents for food, especially sciurids and the larger hystricomorphs.

Figure 25–1.
The springhaas, *Pedetes capensis*, one of several species of saltatorial rodents.
(Mary Ann Cramer after photograph in Walker et al. 1975)

Rodents, like many other mammals, may live in colonial or noncolonial units. According to Eisenberg (1966), diurnal forms usually inhabit open, savanna or steppe habitat and tend to be colonial and utilize primarily grasses for food. In contrast, an economy based on food storage, as in the Heteromyidae, may promote dispersed social structure.

Distinguishing Characteristics

A single arc-shaped incisor is present in each jaw quadrant (Fig. 25-2). The distal end of each incisor is chisel-edged, and the sharpness is maintained by wear

Figure 25–2.
Components of the cranium (A-C) and dentary (D-E) of a cricetid rodent, *Phyllotis* sp. A, dorsal view; B, ventral view; C–D, lateral views; E, medial view. See explanation of figure for identification of numbered and lettered portions.
(Modified from Hershkovitz 1966: 110–115)

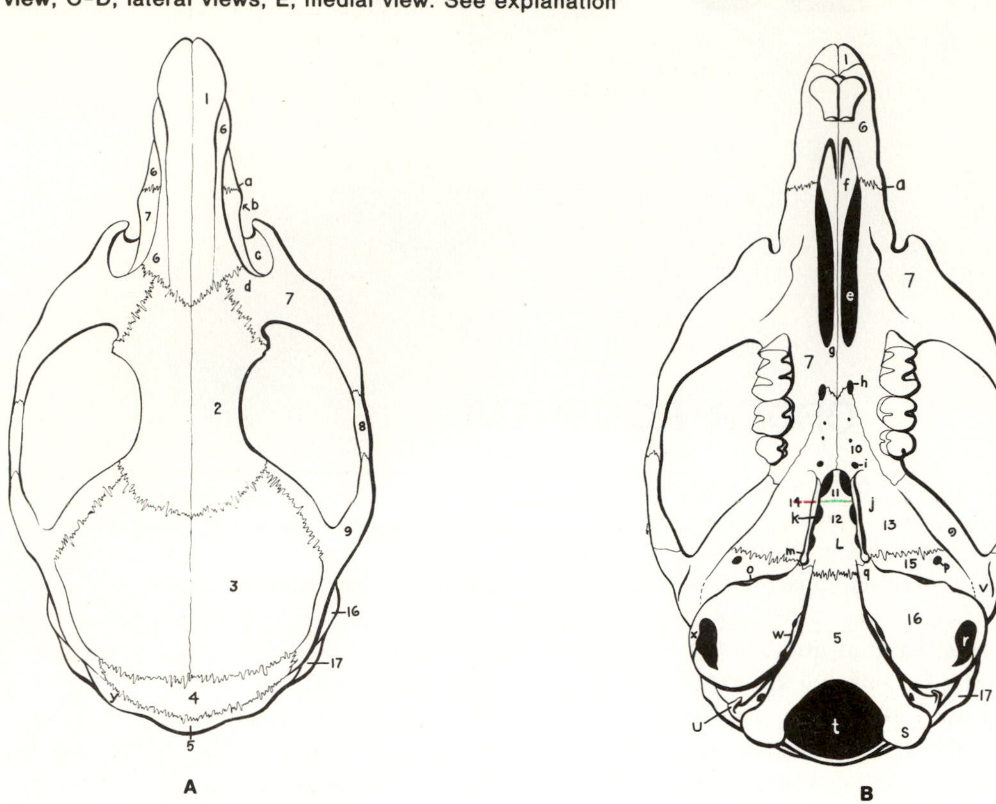

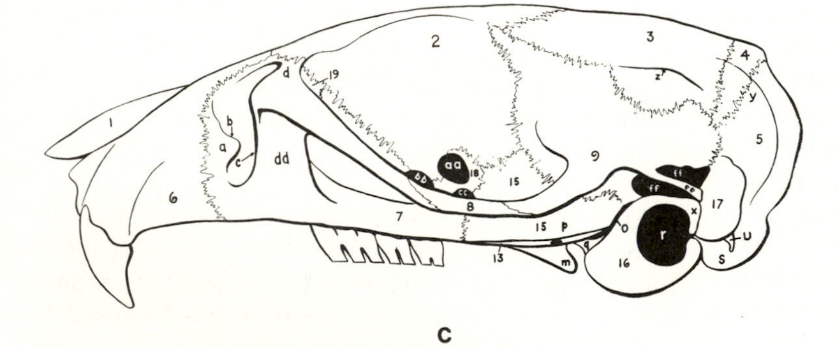

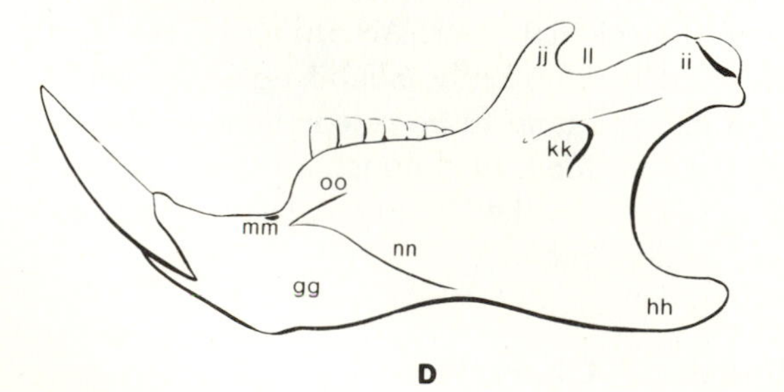

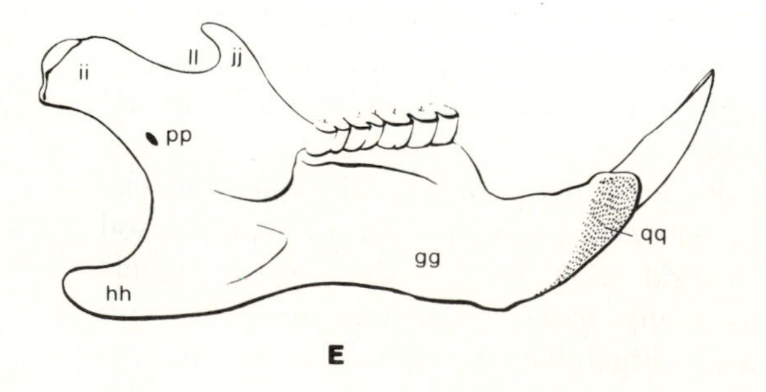

EXPLANATION OF FIGURE 25-2

1. Nasal
2. Frontal
3. Parietal
4. Interparietal
5. Occipital
6. Premaxillary
7. Maxillary
8. Jugal
9. Squamosal or temporal
10. Palatine
11. Presphenoid
12. Basisphenoid
13. Parapterygoid plate (often fused to basisphenoid)
14. Pterygoid process (often fused to basisphenoid)
15. Alisphenoid (more or less fenestrated)
16. Auditory, or tympanic, bulla
17. Mastoidal, or periotic, capsule, or petrosal
18. Orbitosphenoid
19. Lachrymal

a. Capsular projection for upper incisor
b. Preorbital foramen
c. Antorbital, or infraorbital, foramen
d. Antorbital bridge (of maxillary)
e. Incisive, or anterior palatal, foramen
f. Palatal process of premaxillary
g. Palatal process of maxillary
h. Anterior palatal pit (and foramen)
i. Posterolateral palatal pit (and foramen)
j. Parapterygoid fossa
k. Sphenopalatine vacuity
l. Mesopterygoid fossa (between pterygoid processes)
m. Hamular process of pterygoid
o. Petrotympanic fissure
p. Foramen ovale
q. Auditory bullar tube (eustachian canal)
r. Auditory meatus
s. Occipital condyle
t. Foramen magnum
u. Paroccipital process
v. Glenoid fossa
w. Carotid canal and fissures
x. Mastoidal process
y. Lambdoidal crest
z. Temporal ridge

aa. Optic foramen
bb. Sphenopalatine foramen
cc. Anterior lacerated foramen
dd. Zygomatic plate (of maxillary)
ee. Hamular process of squamosal
ff. Temporal vacuity (dorsal and ventral)
gg. Ramus
hh. Angular process
ii. Condyloid process
jj. Coronoid process
kk. Capsular projection
ll. Sigmoid notch
mm. Mental foramen
nn. Inferior masseteric ridge
oo. Superior masseteric ridge
pp. Mandibular foramen
qq. Symphysis

on the hard enamel (anterior surface) and the softer dentine (bulk of tooth). Canines and most premolars are absent resulting in a wide diastema between the incisors and the remaining cheek teeth. The usual complement of cheek teeth is twelve to sixteen, but the range is from four (*Mayermys*) to twenty-eight (*Heliophobius*).

Most rodents are small (less than 300 mm head and body length) but a few approach or exceed one meter in length. With few exceptions (e.g., *Spalax*) an external tail is present, but its form and length vary greatly. All rodents except *Heterocephalus* are well haired over most of their body. Most are quadrupedal but some arid-land species have greatly enlarged hind limbs and are capable of ricochetal locomotion. Primitively, the digits are 5/5 but they may be reduced to 4/3.

The uterus is duplex and the testes may be abdominal, inguinal, or scrotal. A baculum is usually present.

Important Taxonomic Characteristics

Cranial and dental characters are most often used in rodent classification. The cheek teeth show a great diversity in shape and occlusal pattern and may have cusps situated on transverse ridges or **laminae** (Figs. 25-27 and 25-28) or isolated **enamel islands** (Fig. 25-20). An **inner fold** on a cheek tooth is an elongated enamel island (Fig. 25-20). A **re-entrant fold** is an invagination along the side of a cheek tooth (Fig. 25-20). A re-entrant fold that has sharp angles is termed a **re-entrant angle** (Fig. 25-14A).

The zygomatic region of the skull and the infraorbital foramen are variously modified for passage and attachment of different branches of the masseter muscle. In the sewellel, Aplodontidae, (Fig. 25-3A) the masseter originates from the lower edge of the zygomatic arch and does not pass through the infraorbital foramen. In advanced sciurids (Fig. 25-3B) the middle masseter originates high on the zygomatic plate (defined below) at the level of the orbit with no or slight muscle transmission through the infraorbital foramen. In myomorph rodents (Fig. 25-3C) the deep masseter passes through a V-shaped, round, or oval infraorbital foramen. In hystricomorph rodents (Fig. 25-3D) the deep masseter is tremendously enlarged and passes through a correspondingly enlarged infraorbital foramen.

The zygomatic process of the maxilla, the **maxillary process,** is variously modified. When the infraorbital foramen is enlarged (myomorphs and hystricomorphs), the process is divided into an **upper maxillary process** and a **lower maxillary process** (Fig. 25-21A). In many myomorphs (particularly Muridae and Cricetidae) the lower maxillary process may take on a broad, flattened shape, forming a **zygomatic plate** or **lower maxillary process** (Fig. 25-21A).

Distribution

The distribution of rodents is almost worldwide. They are native to most land areas except some Arctic and oceanic islands, New Zealand, and Antarctica. Some members of the Muridae and other families have been introduced into all parts of the world.

Figure 25–3.
Diagrammatic representation of superficial, middle, and deep portions of the masseter muscle in various rodents: A, **aplodontid**, masseter originates from lower edge of zygomatic arch; B, **advanced sciuromorph**, middle masseter (masseter lateralis) originates from outer side of skull in front of orbit, infraorbital foramen small, not transmitting any muscle; C, **myomorph**, deep portion of masseter (masseter medialis) pushes up through orbit and passes through a V-shaped, oval, or round infraorbital foramen; D, **hystricomorph**, masseter medialis enormously developed with portion passing through a greatly enlarged infraorbital foramen. **ms**, masseter superficialis; **ml**, masseter lateralis; **mma**, masseter medialis anterior. (M.A. Cramer)

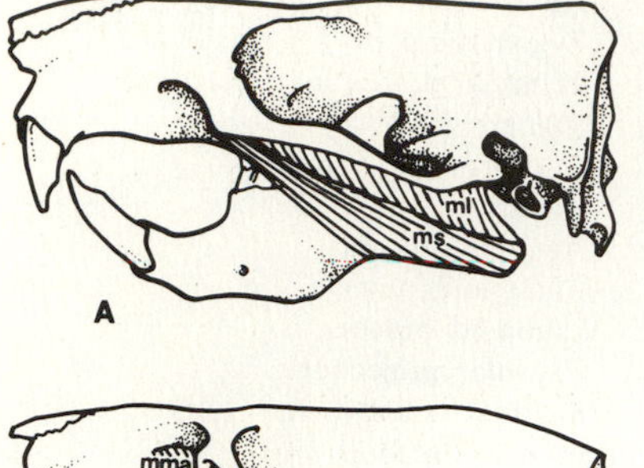

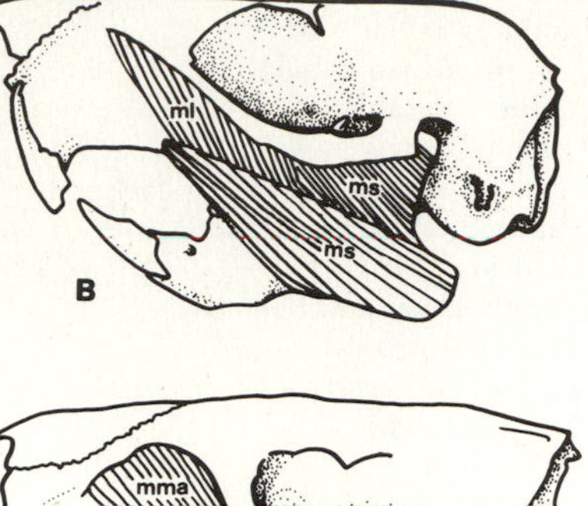

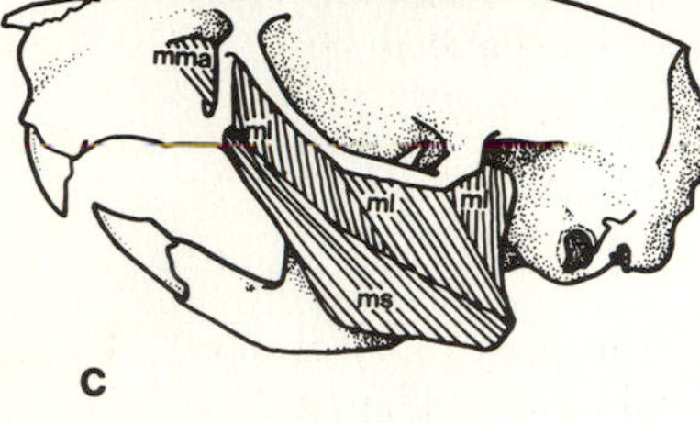

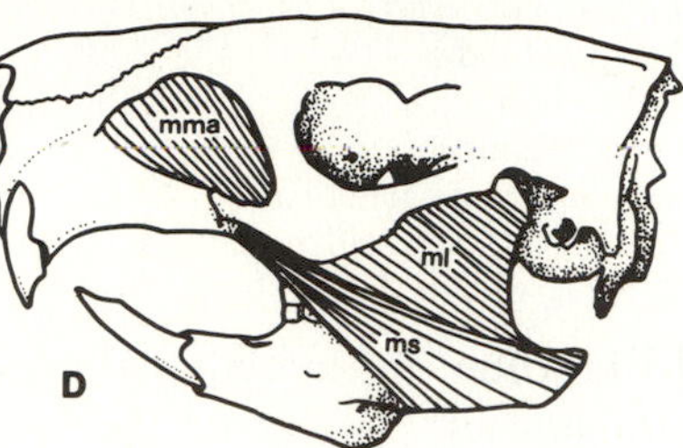

Table 25-1
Living Families of Rodentia.

Family	Common Name	Number of* Genera	Number of* Species	Distribution
Suborder Sciuromorpha				
Aplodontidae	sewellel or mountain "beaver"	1	1	Western Nearctic (California to British Columbia)
Sciuridae	squirrels and marmots	51	*ca.* 261	Worldwide, except southern Neotropical, Madagascar and Australian
Geomyidae	pocket gophers	8	*ca.* 40	Southern and western Nearctic, NW Neotropical
Heteromyidae	kangaroo rats, pocket mice and allies	5	*ca.* 75	Western Nearctic and northern Neotropical
Castoridae	beavers	1	2	Holarctic
Anomaluridae	scaly-tailed squirrels	4	*ca.* 12	West and central Ethiopian
Pedetidae	springhaas	1	2	Southern and eastern Ethiopian
Suborder Myomorpha				
Cricetidae	New World rats and mice, hamsters, voles, lemmings, gerbils, and allies	97	*ca.* 567	Worldwide, except Austro-Malayan region, Antarctica, and some islands
Spalacidae	mole rats	1	3	Palearctic (eastern Mediterranean region and southeastern Europe)

*Based on Anderson (1967).

Table 25-1 *(Continued)*

Family	Common Name	Number of* Genera	Number of* Species	Distribution
Rhizomyidae	bamboo rats	3	18	Eastern Oriental, east-central Ethiopian
Muridae	Old World rats and mice	98	*ca.* 457	Worldwide, except Antarctica, Nearctic and Neotropical (introduced in the latter two areas)
Gliridae	dormice	7	23	Western Palearctic, Ethiopian
Platacanthomyidae	spiny dormice	2	2	Oriental
Seleviniidae	dzhalman	1	1	Central Asia
Zapodidae	jumping mice, birch mice	4	11	Holarctic (except southern fringe and most of Europe)
Dipodidae	jerboas	10	27	Central Palearctic, northern Ethiopian
Suborder Hystricomorpha				
Hystricidae	Old World porcupines	4	15	Ethiopian, southern Palearctic, Oriental
Erethizontidae	New World porcupines	4	8	Nearctic (except much of Mexico and SE USA), northern Neotropical (tropical mainland)
Caviidae	cavies, Patagonian "hares"	5	12	Neotropical (except most of Chile and NE South America)
Hydrochoeridae	capybaras	1	2	Neotropical (tropical mainland north to Panama)
Dinomyidae	pacarana	1	1	Neotropical (Peru, Colombia, Ecuador, Bolivia, Brazil)
Dasyproctidae	agoutis and pacas	4	*ca.* 11	Northern and central Neotropical
Chinchillidae	chinchillas and viscachas	3	6	Southern Neotropical
Capromyidae	hutias	3	11	West Indies
Myocastoridae	nutria or coypu	1	1	Southern Neotropical (introduced elsewhere)
Octodontidae	octodonts	5	8	Neotropical (central Andes region) mostly Chilean
Ctenomyidae	tuco tucos	1	*ca.* 26	Southern Neotropical
Abrocomidae	chinchilla rats	1	2	Neotropical (central Andes region)
Echimyidae	spiny rats	14	*ca.* 43	Northern Neotropical, except northernmost part
Thryonomyidae	cane rats	1	6	West and central Africa
Petromyidae	dassie rat	1	1	Southwestern Ethiopian
Bathyergidae	mole rats	5	*ca.* 22	Ethiopian
Ctenodactylidae	gundis	4	8	Northern Africa

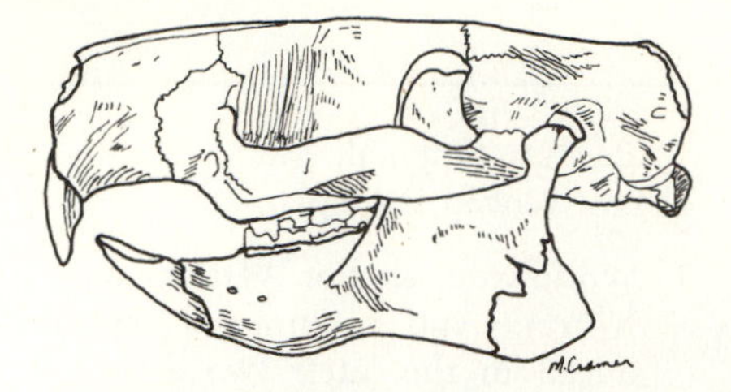

Figure 25–4.
Skull of primitive Eocene rodent, *†Ischyrotomus petersoni.*
(Redrawn from Matthew 1910: 55)

Fossil Record

The oldest known rodents are the extinct †Paramyidae from the late Paleocene of North America and the Eocene of North America and Eurasia. The †Paramyidae (Fig. 25-4) seem to have been ancestral to most rodents and some of their characters are retained in living members of the Aplodontidae (structure of the infraorbital foramen) and Sciuridae (dentition).

The South America hystricomorph rodents first appear in the fossil record in the Oligocene. The largest known rodent was *†Eumegamys,* a Pliocene South American relative of the pacarana, *Dinomys branchii,* Dinomyidae. *†Eumegamys* had a skull length of approximately two feet and was presumably the size of a hippopotamus (Romer 1966). Sciurids date from the lower to middle Oligocene of North America and possess many dental characters of the paramyids. Castorids appeared in the Oligocene and remained in the northern hemisphere throughout their history. The giant beaver, *†Castoroides,* which reached a length of seven feet, was widespread in the upper Mississippi Valley during the Pleistocene (Simpson 1930b).

Myomorph rodents include members of the Cricetidae, Muridae, Zapodidae, and Dipodidae. Four additional families, Seleviniidae, Spalacidae, Gliridae, and Rhizomyidae, are frequently grouped with the myomorphs. The earliest myomorph is probably the extinct *†Simimys,* which dates from the late Eocene of North America. Typical Cricetidae appeared in the early Oligocene of Europe and predate the Muridae, the earliest fossil record for which is from the early Pliocene. The fossil record of the Zapodidae predates that of the Dipodidae (Dawson 1967).

In most other families of rodents the fossil record is very incomplete and the specimens known do not shed much light on their relationships to other rodent groups. The Ctenodactlylidae, which date from the Oligocene, and the Pedetidae and Bathyergidae, which date from the Miocene, pose particularly acute problems in rodent classification (see classifications of Anderson 1967 and Wood 1955, at the end of this chapter).

Keys to Living Families

Since the Rodentia is a very large and complex order, we depart slightly from the usual format and present two family keys. The first is a key to families of the world. The second is a shortened key designed to identify to family all species that occur in North America exclusive of the West Indies.

Key to Living Families of the World

1 Infraorbital canal passes through side of rostrum and emerges anterior to zygomatic plate with the infraorbital foramen visible when the skull is viewed laterally (Fig. 25-5) 2

1′ Infraorbital canal and its foramen variable, but if canal emerges anterior to zygomatic plate, then the foramen is not visible when the skull is viewed laterally 3

2 (1) Infraorbital foramen large, the vacuity extending transversely through the rostrum; zygoma weak and threadlike; tail long, and usually well haired (Figs. 25-6) .... **Heteromyidae**
kangaroo rats, pocket mice, and allies

2′ Infraorbital foramen small, the vacuity never extending transversely through rostrum; zygoma robust, never threadlike; tail short, with sparsely distributed tactile hairs (Figs. 25-5, 25-7) .... **Geomyidae**
pocket gophers

Figure 25–5.
Skull of *Geomys pinetis* showing infraorbital foramen on side of rostrum.
(Redrawn from Tullberg 1899: pl. 23)

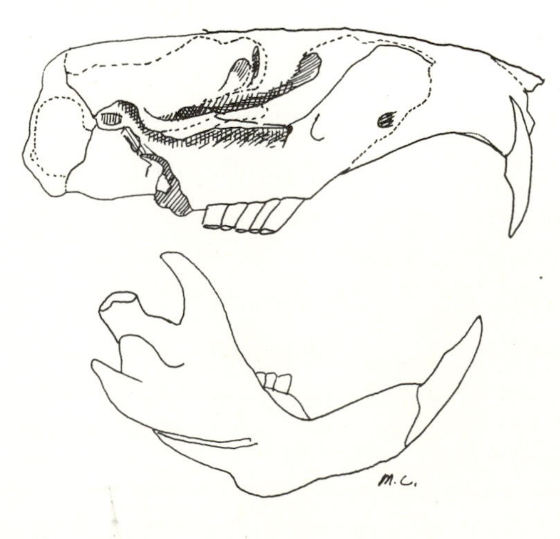

Figure 25–6.
Representative members of the Heteromyidae: A, Merriam's kangaroo rat, *Dipodomys merriami*; B, plains pocket mouse, *Perognathus flavescens apache*. (Mary Ann Cramer; A, after photograph in Walker et al. 1975)

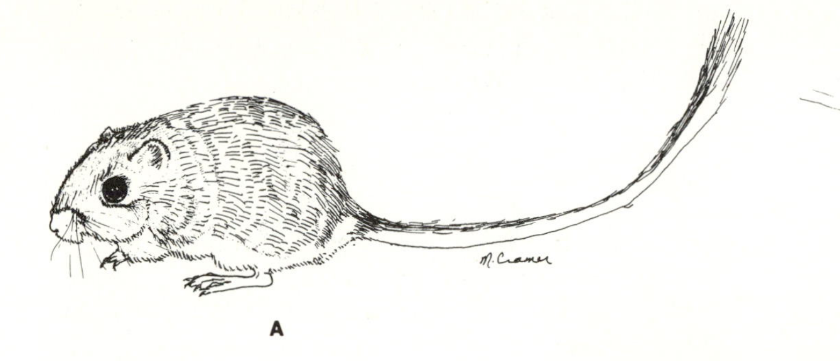

Figure 25–7.
Southeastern pocket gopher, *Geomys pinetis*.
(Merriam 1895: Frontispiece)

Figure 25–8.
Ventral view of sciurognath-type mandible in A, *Castor canadensis*; B, *Marmota marmota*.
(Redrawn A, Landry 1957; B, Weber 1928)

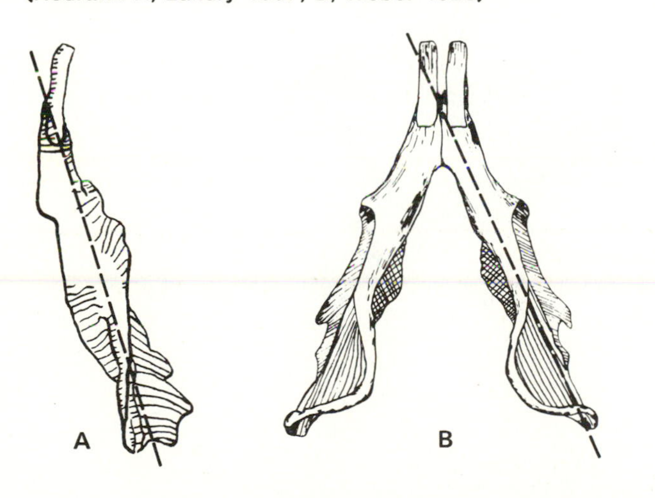

Figure 25–9.
Ventral view of hystricognath-type mandible in A, *Hystrix subcristata*; B, *Bathyergus maritimus*.
(Redrawn A, Landry 1957; B, Weber 1928)

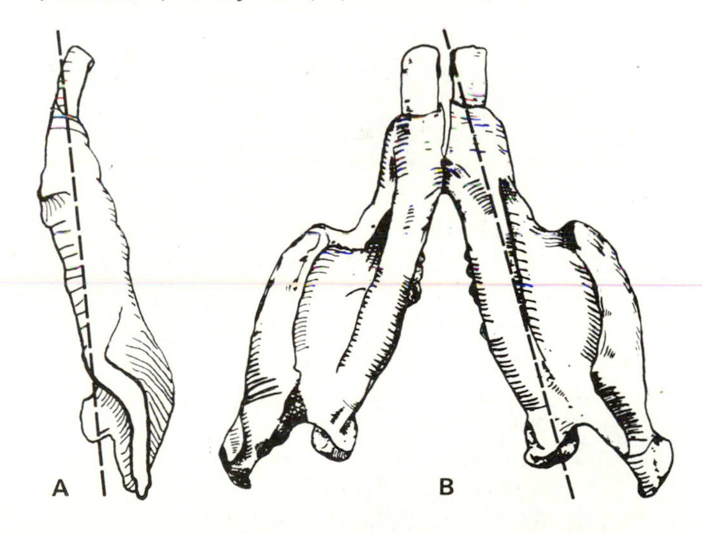

3 (1′) From ventral view, angular process of mandible in line with (Fig. 25-8A) or medial to (Fig. 25-8B) the lateral border of incisive alveolus [sciurognath-type mandible] 4

3′ From ventral view, angular process of mandible lateral to (Fig. 25-9A, B) lateral border of incisive alveolus [hystricognath-type mandible] 29

4 (3) Postorbital process of frontal present and sharply pointed (Figs. 25-10 and 25-11) .... **Sciuridae**
squirrels and marmots

4′ Postorbital process of frontal absent or, if present, small and blunt (Fig. 25-12) 5

Figure 25–10.
Dorsal view of cranium showing sharp-pointed postorbital processes in *Spermophilus fulvus.*
(Vinogradov and Argiropulo 1941: 29)

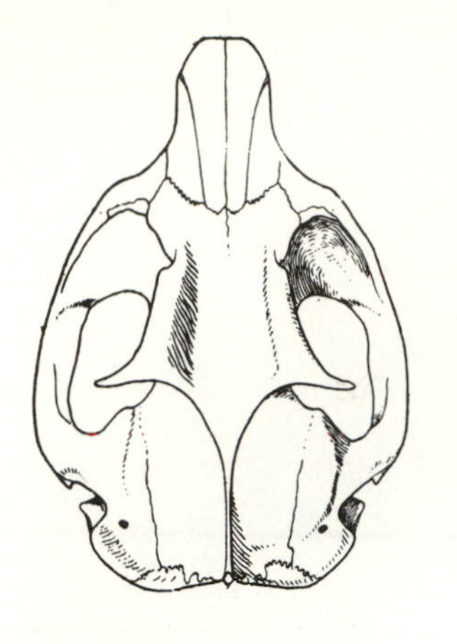

Figure 25–12.
Blunt postorbital processes in the European beaver, *Castor fiber.*
(Redrawn from Tullberg 1899)

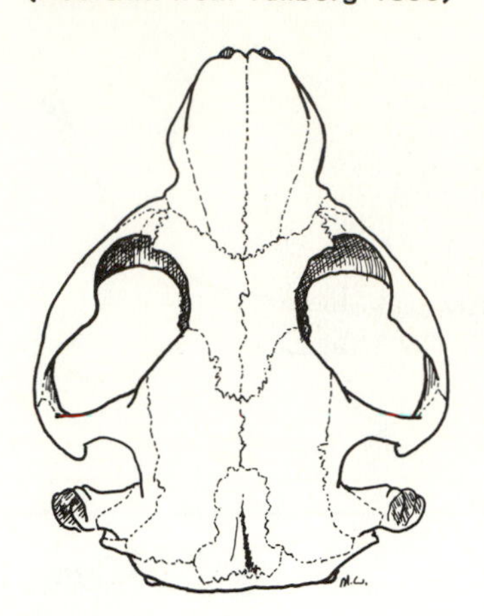

Figure 25–11.
Representative members of the Sciuridae: A, "flying" squirrel, *Hyloptes alboniger;* B, marmot, *Marmota sibirica*; C, striped squirrel, *Tamiops macclellandi.*
(Hsia et al. 1964: 31, 37, 33)

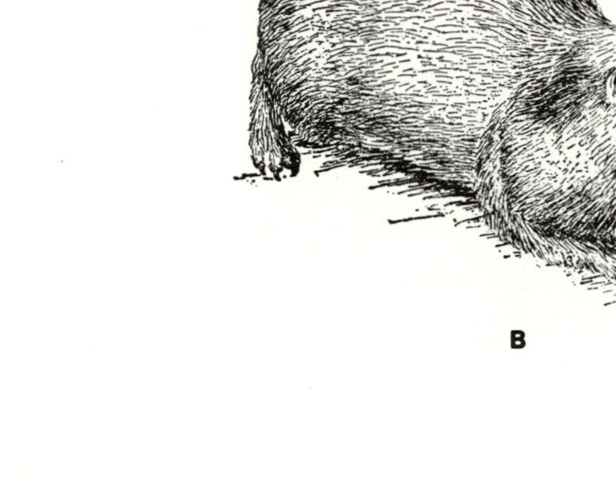

Figure 25–13.
European beaver, *Castor fiber,* showing dorsoventrally flattened tail.
(Hsia et al. 1964: 39)

5 (4′) Postorbital process of frontal small and blunt (Fig. 25-12); tail flattened dorso-ventrally (Fig. 25-13) .... **Castoridae**
beavers

5′ Postorbital process of frontal absent; tail not flattened dorso-ventrally 6

6 (5′) Each dentary with three or fewer cheek teeth 7

6′ Each dentary with four or more cheek teeth 25

7 (6) Occlusal pattern of cheek teeth prismatic, enamel enclosing acute triangles of dentine in all or some of the teeth (e.g., Fig. 25-14), teeth high crowned .... **Cricetidae** (in part)
Microtinae
voles and lemmings

7′ Occlusal pattern of cheek teeth not prismatic, if prismatic in appearance, then no acute triangles formed in any of the teeth *or* cheek teeth low crowned 8

8 (7′) Occipital crest at level of zygomatic process of squamosal (Fig. 25-16) .... **Spalacidae**
mole rats

8′ Occipital crest posterior to level of zygomatic process of squamosal (Fig. 25-17) 9

Figure 25–14.
Prismatic upper cheek teeth in A, *Dicrostonyx;* B, *Myospalax myospalax.* Anterior at top.
(Vinogradov and Argiropulo 1941)

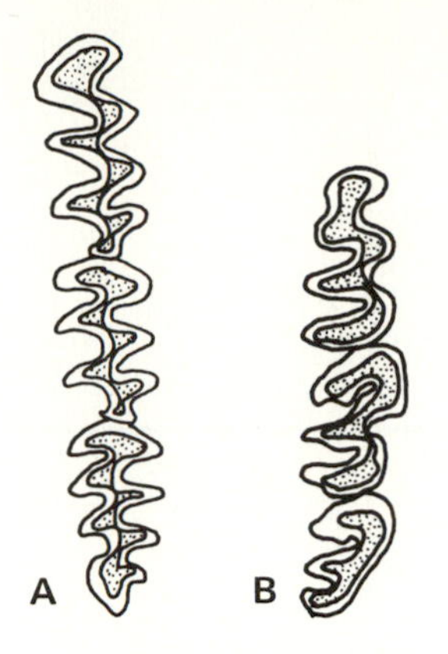

Figure 25–15.
Representative examples of microtine rodents: A, muskrat, *Ondatra zibethicus;* B, Brandt's vole, *Microtus brandti.*
(Hsia et al. 1964: 62, 64)

A

B

Figure 25–16.
Cranium of *Spalax microphthalmus* showing tilted occipital region.
(Vinogradov and Argiropulo 1941: 51)

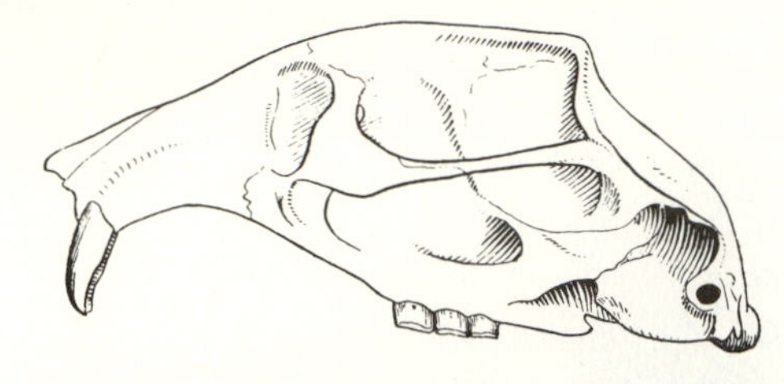

Figure 25–17.
Skull of a mole rat, *Tachyoryctes* sp.
(G.A. Moore)

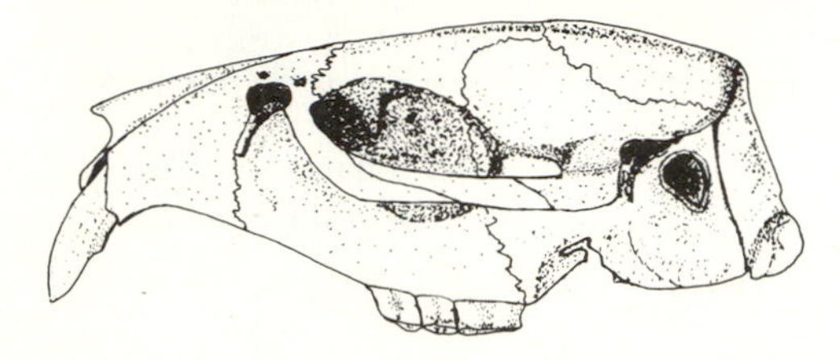

9 (8′) Lower margin of infraorbitaal foramen, in frontal view, even with or above anterior tip of nasal bones (Fig. 25-18) .... **Rhizomyidae** (in part)
Rhizomys and *Cannomys*
bamboo rats

9′ Lower margin of infraorbital foramen, in frontal view, below anterior tip of nasal bones 10

10 (9′) First two cheek teeth each with deep single inner fold that divides occlusal surface (Fig. 25-20), with upper incisors that project beyond tip of nasal bones .... **Rhizomyidae** (in part)
Tachyoryctes
bamboo rats

10′ First two cheek teeth with variable occlusal surfaces, but if single fold present, then incisors do not project anteriorly beyond tip of nasal bones 11

Figure 25–18.
Frontal view of *Rhizomys* skull showing infraorbital foramina.
(Mary Ann Cramer)

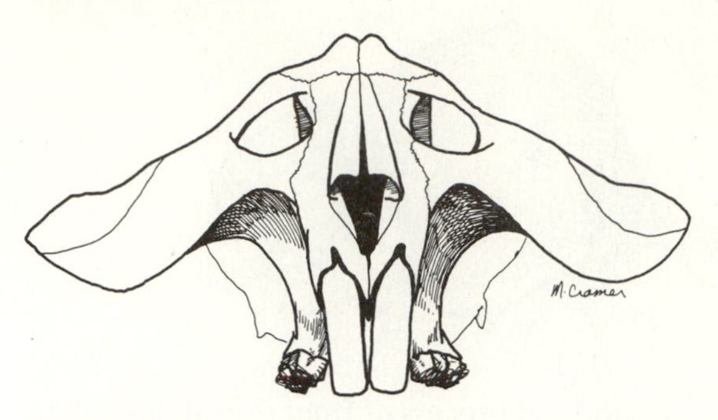

Figure 25–19.
Bamboo rat, *Rhizomys sinensis.*
(Hsia et al. 1964: 44)

Figure 25–20.
Left upper cheek teeth (A) in *Tachyoryctes ruandae* and mole rat (B), *Tachyoryctes splendens.*
(A, Stehlin and Schaub 1951; B, Mary Ann Cramer from photograph in Walker et al. 1975)

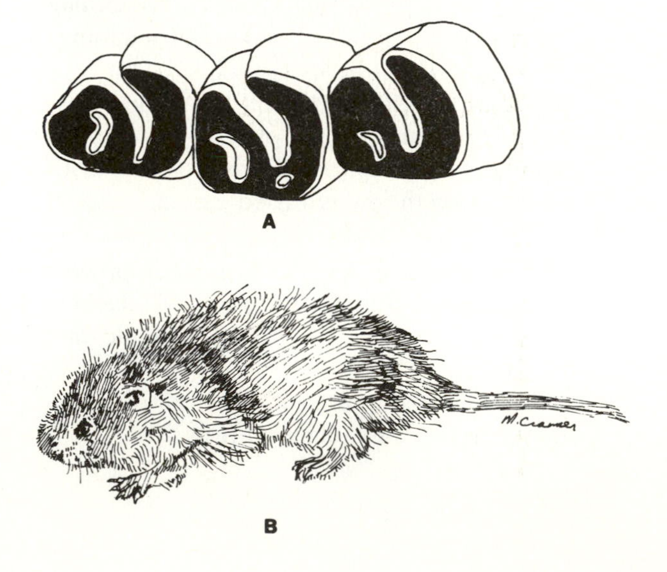

11 (10′) Zygomatic arch tilted upward with a portion situated above lower border of infraorbital canal (Fig. 25-21); lower maxillary process expanded into a zygomatic plate (Fig. 25-21A) that forms most of lateral wall of infraorbital canal (Fig. 25-21) 12

11′ Zygomatic arch horizontal or nearly so with no portion situated above lower border of infraorbital canal (Fig. 25-22); lower and upper maxillary processes together form lateral wall of infraorbital canal (Fig. 25-22) 22

Figure 25–21.
Lower maxillary process elevated above lower border of infraorbital foramen in A, *Mus;* B, *Neotoma cinerea;* C, *Alticola argentata.* **ump,** upper maxillary process; **lmp,** zygomatic plate or lower maxillary process; **zps,** zygomatic process of squamosal.
(A and C, Vinogradov and Argiropulo 1941; B, redrawn from Tullberg 1899: pl. 15-25)

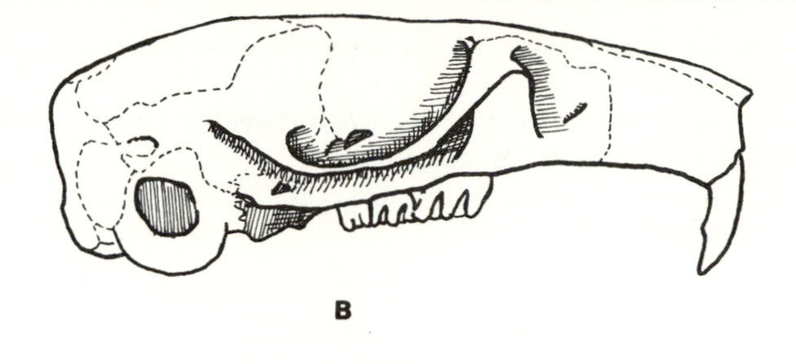

B

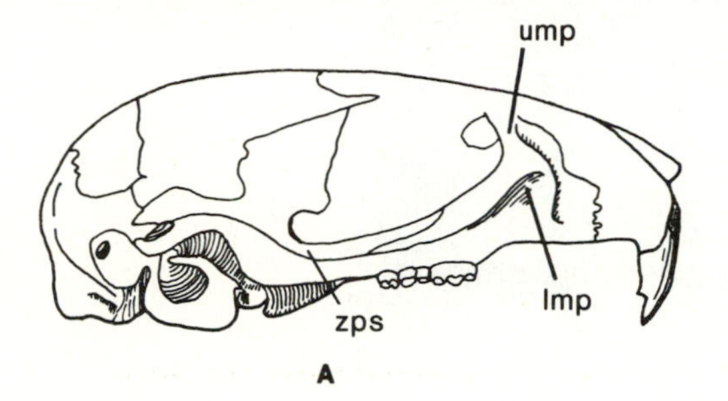

A

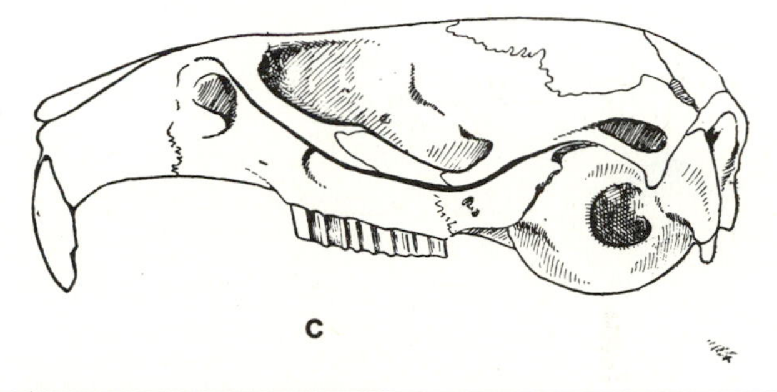

C

Figure 25–22.
Lower maxillary process below infraorbital foramen in A, *Glis glis;* B, *Felovia* sp.; C, *Allactaga severtzovi.*
(A and B, redrawn from Tullberg 1899: pl. 11, 9; C, Vinogradov and Argiropulo 1941: 33)

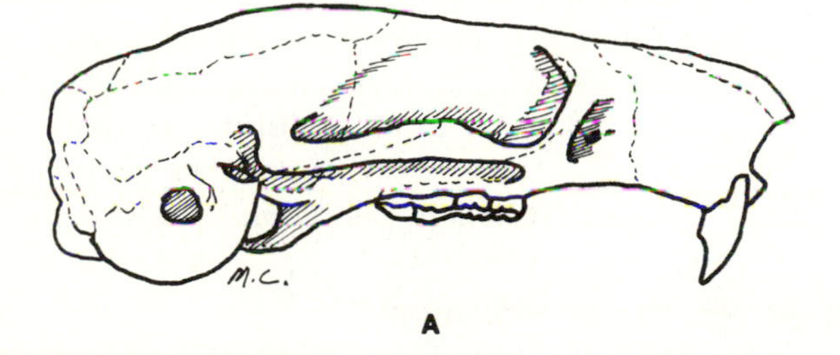

A

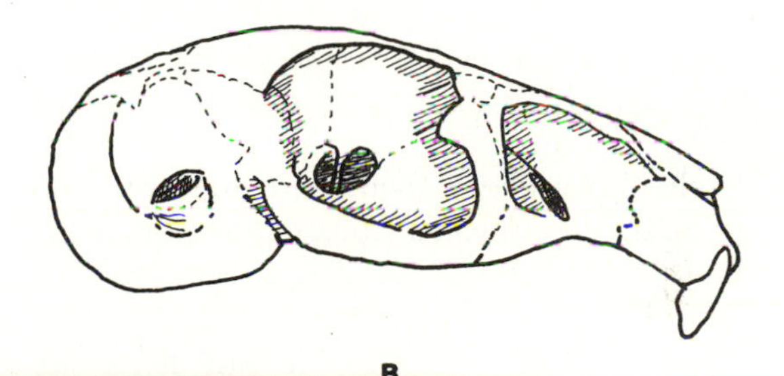

B

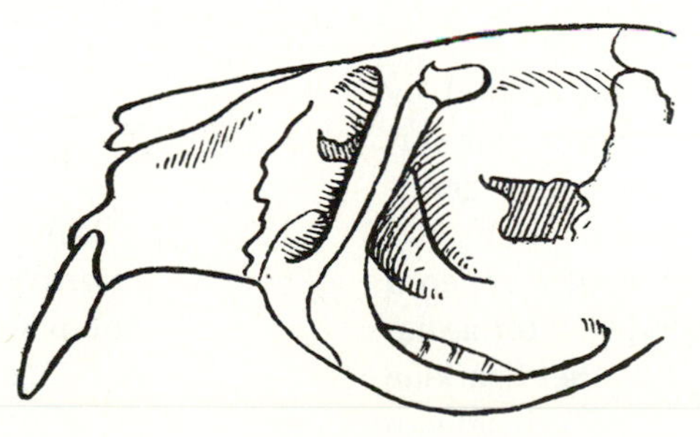

C

Figure 25–23.
Basin-bearing and basin-shaped cheek teeth in A, *Hydromys chrysogaster;* B, *Selevinia betpakdalensis.* Anterior at left, not to same scale.
(A, Stehlin and Schaub 1951; B, Afanasev et al. 1953)

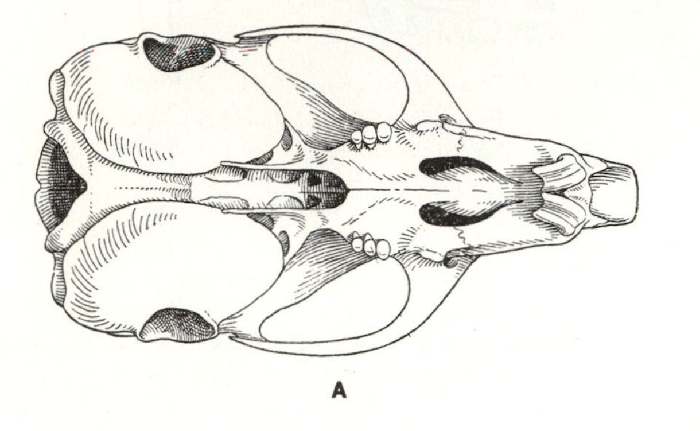

Figure 25–24.
A, ventral view of cranium of *Selevinia* showing grooved incisors and small teeth; B, selevinid.
(A, Gromov et al. 1963; B, Mary Ann Cramer from photograph in Walker et al. 1975)

12 (11) Two or fewer cheek teeth present in each maxilla .... **Muridae** (in part)
some Murinae
shrew rats and water rats

12′ Three or more cheek teeth present in each maxilla 13

13 (12′) Each upper cheek tooth with one, two, or three shallow, basin-shaped depressions (e.g., Fig. 25-23A, B) 14

13′ Upper cheek teeth variable but never as above 15

14 (13) Angular process of mandible perforated by foramen; single deep groove on anterior surface of each upper incisor; size small, greatest length of skull less than 25 mm; tail shorter than head and body length, covered with short hairs; form not modified for aquatic life (Fig. 25-24) .... **Seleviniidae**
dzhalman

14′ Angular process of mandible without foramen; anterior surface of upper incisor without groove, (if grooved, tail naked); size medium, greatest length of skull more than 25 mm; tail longer than head and body length; form frequently modified for aquatic habits (Fig. 25-25) .... **Muridae** (in part)
a few murids
e.g., *Anisomys, Chrotomys, Leptomys*

15 (13′) Occlusal surfaces of cheek teeth consisting of series of transverse ridges and depressions (4 to 9 on last cheek tooth) (Fig. 25-26) .... **Muridae** (in part)
Otomyinae
swamp or groove-toothed rats

15′ Occlusal surfaces of cheek teeth variable; transverse ridges and depressions, if present, number fewer than 4 on last cheek tooth 16

Figure 25–25.
Australian water rat, *Hydromys chrysogaster,* one of the most aquatic-adapted members of the Muridae.
(Brazenor 1950: 64)

16 (15′) Cheek teeth with evidence of cusps (see Fig. Figs. 25-27 and 25-28) 17

16′ Cheek teeth with no evidence of cusps, may possess deep re-entrant folds, transverse non-cuspidate laminae, or sharp re-entrant angles 18

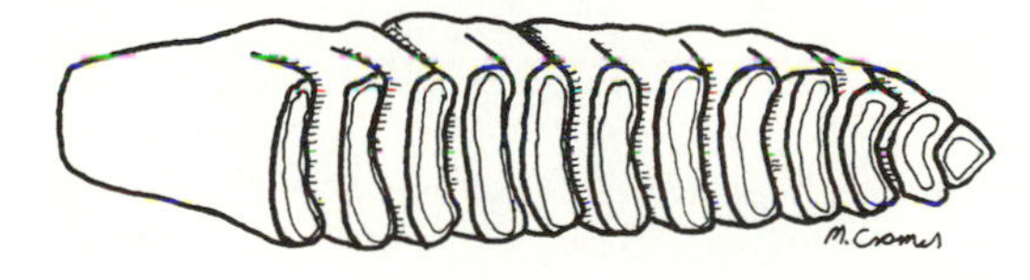

Figure 25–26.
Right upper cheek teeth in *Otomys irroratus.* Anterior at left.
(Redrawn from Tullberg 1899: pl. 29)

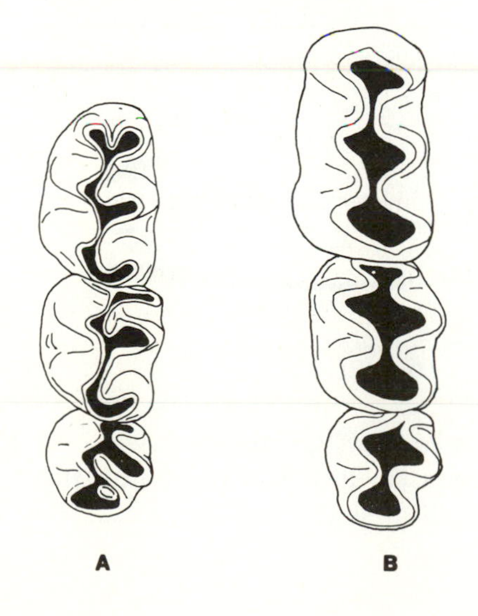

Figure 25–27.
Right upper cheek teeth in A, *Cricetulus migratorius;* B, *Phodopus roborowskii.* Anterior at top, not to same scale.
(Stehlin and Schaub 1951)

17 (16) Maximum of two cusps on each transverse lamina of anterior cheek tooth (Fig. 25-27); laminae that bear cusps interconnected (Fig. 25-27); size and body form variable (Fig. 25-29) .... **Cricetidae** (in part)
most New World rats and mice

17′ Three cusps present on at least one transverse lamina of anterior cheek tooth (Fig. 25-28); laminae that bear cusps separated by valleys or pressed together, but not interconnected (Fig. 25-28); size and body form variable (Fig. 25-30) .... **Muridae** (in part)
most Old World rats and mice

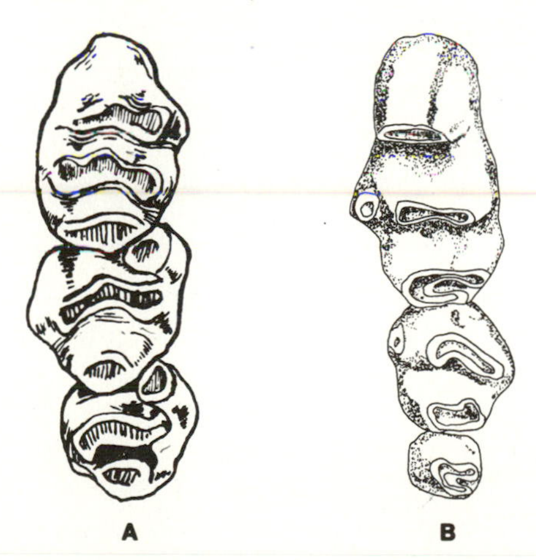

Figure 25–28.
Left upper cheek teeth in A, *Rattus;* B, *Steatomys pratensis.* Anterior at top, not to same scale.
(A, redrawn from Sokolov 1959; B, redrawn from Stehlin and Schaub 1951)

Figure 25–29.
Representative members of the Cricetidae. A, giant crested rat, *Lophiomys imhausi;* B, zokor, *Myospalax myospalax;* C, giant pouched rat, *Cricetomys gambianus.* Not to same scale. A, Lophiomyinae; B and D, Cricetinae; C, Cricetominae. (A, Flower and Lydekker 1891: 460, after Milne-Edwards; B, Hsia et al. 1964: 59; C, Dekeyser 1955: 213; D, Gromov et al. 1963: 481)

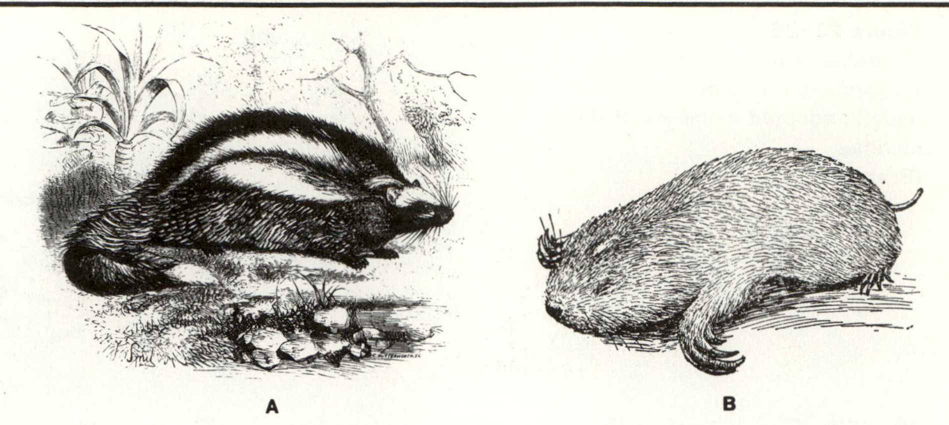

Figure 25–30.
Representative species of the Muridae: A, Australian allied rat, *Rattus assimilis;* B, striped field mouse, *Apodemus agrarius;* C, gray tree mouse, *Dendromus melanotis;* D, hopping mouse, *Notomys mitchelli.* Not to same scale. (A, Sclater 1901: 32; B, Hsia et al. 1964: 47; C and D, Brazenor 1950: 65,70)

18 (16′) Auditory bulla greatly enlarged, its greatest dimension more than one-fourth the greatest length of skull (Fig. 25-31) .... **Cricetidae** (in part) Gerbillinae gerbils

18′ Auditory bulla not greatly enlarged, its greatest dimension less than one-fourth the greatest length of skull 19

Figure 25-29—*Continued.*

Figure 25-31.
Representative gerbilids: A, skull showing enlarged auditory bullae in *Meriones* sp.; B, the great gerbil, *Rhombomys opimus*.
(A, Vinogradov and Argiropulo 1941: 67; B, Hsia et al. 1964: 58)

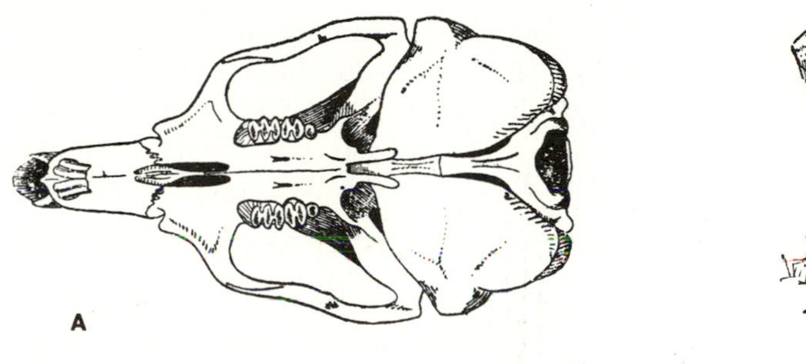

Figure 25-32.
Crania of spiny dormouse, *Platacanthomys lasiurus*, in lateral (A) and ventral (B) views.
(A, Mary Ann Cramer; B, G.A. Moore)

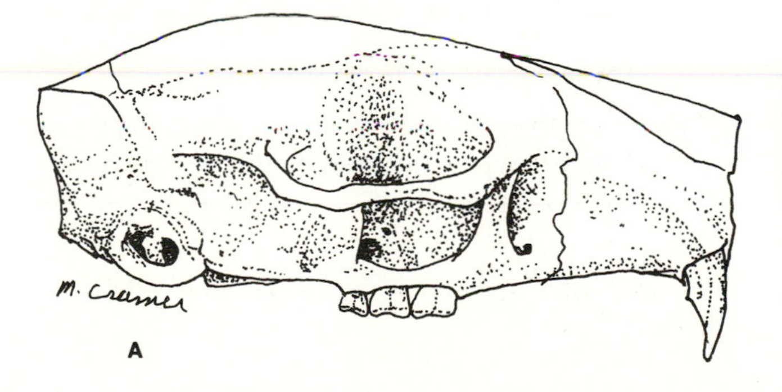

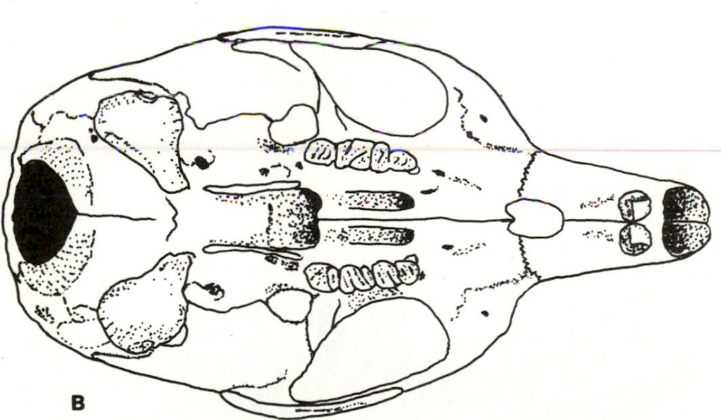

19 (18′) Zygomatic plate narrow, its greatest width equal to or less than the length of first upper tooth (Fig. 25-32A); palatine bones with large foramina medial to first upper cheek tooth (Fig. 25-32B) .... **Platacanthomyidae**
spiny dormice

19′ Zygomatic plate of various types, but greatest width always more than the length of first upper cheek tooth; palatine bones without large foramina medial to first upper cheek tooth 20

Figure 25–33.
Pygmy dormouse, *Typhomys cinereus.*
(Hsia et al. 1964: 43)

20 (19′) Length of second upper cheek tooth 80 to 100 percent length of first upper cheek tooth .... **Cricetidae** (in part)
Nesomyinae, Malagasy rats, and some Cricetinae, New World rats and mice

20′ Length of second upper cheek tooth less than 80 percent length of first upper cheek tooth 21

21 (20′) Occlusal pattern of cheek teeth simple, consisting of two or three noncuspidate laminae on each tooth (e.g., Fig. 25-34); specimen from the Old World (Fig. 25-35) .... **Muridae** (in part)
a few Murinae, Old World rats and mice

21′ Occlusal pattern of cheek teeth variable, but if occlusal pattern that of simple laminae; specimen from the New World .... **Cricetidae** (in part)
some New World rats and mice

Figure 25–34.
Left upper cheek teeth of *Bandicota indica*. Anterior at left.
(Stehlin and Schaub 1951)

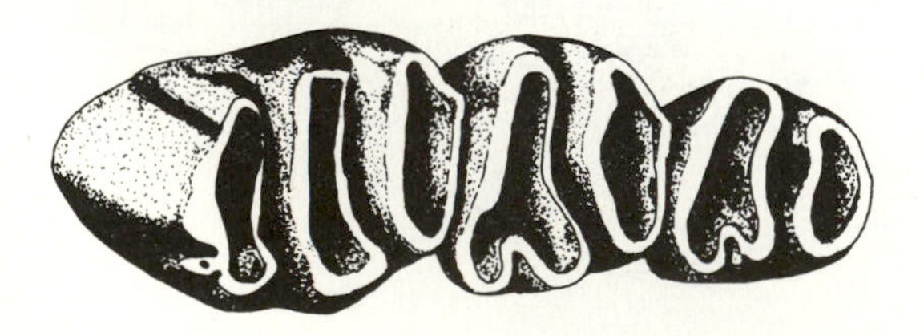

Figure 25–35.
Bandicoot rat, *Bandicota indica*, a murid rodent with noncuspidate laminae.
(Finn 1929: 151)

22 (11′) Foramina (1 or 2) in angular region of mandible (e.g., Fig. 25-36); bullae moderately to greatly inflated (Fig. 25-36); body form modified for saltation (Fig. 25-37); head and body length less than 300 mm .... **Dipodidae**
jerboas

22′ No foramen in angular region of mandible; bullae and body form various, but if modified for saltation then head and body length greater than 300 mm 23

23 (22′) Cheek teeth evergrowing; each lower cheek tooth with a single deep fold on labial and on lingual surface, resulting in 8-shaped pattern (Fig. 25-38) .... **Ctenodactylidae** (in part)
some gundis

23′ Cheek teeth rooted; lower cheek teeth of various types, but never with 8-shaped pattern 24

Figure 25–36.
Skulls of A, small five-toed jerboa, *Allactaga elater* and B, pygmy jerboa, *Salpingotus crassicauda*.
(Gromov et al. 1963: 260)

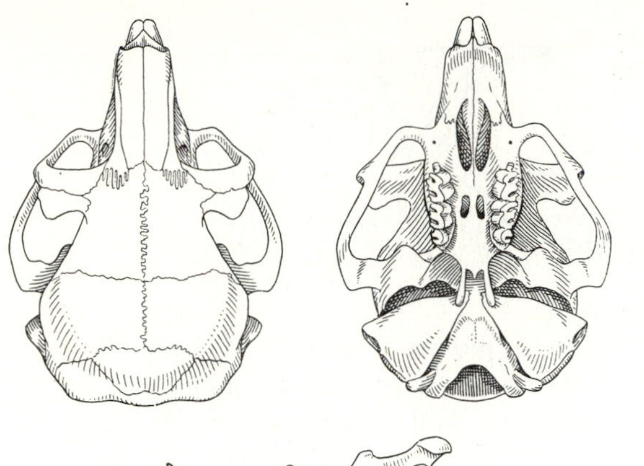

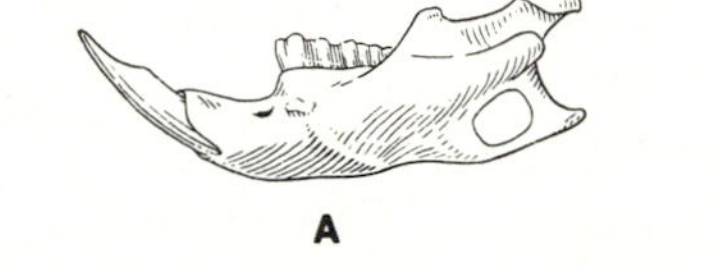

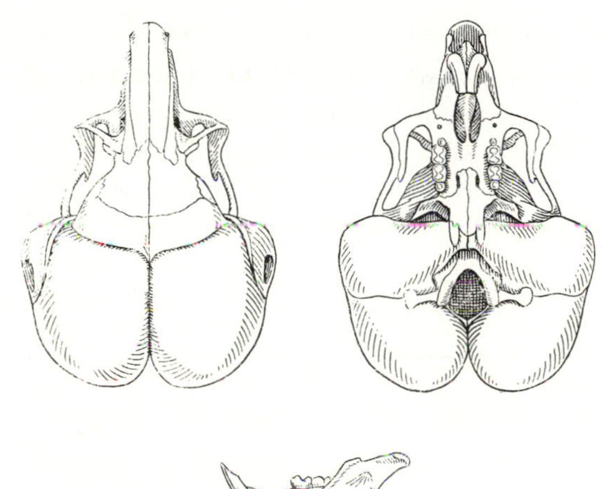

Figure 25–38.
Lower cheek teeth A, and skull B, of gundi, *Ctenodactylus gundi*.
(Redrawn from Tullberg 1899: pls. 27 and 9)

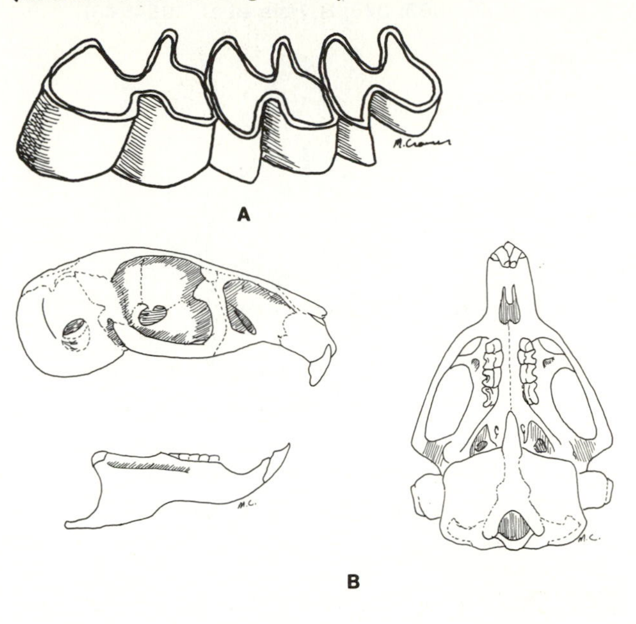

24 (23′) Anterior surfaces of upper incisors smooth or with one groove (Fig. 25-39); size small, greatest length of skull less than 30 mm; pelage fine to hispid, no white at base of hair .... **Zapodidae**
jumping mice and birch mice

24′ Two faint grooves in anterior surface of each upper incisor; size medium, greatest length of skull more than 30 mm; dorsal pelage almost spiny (stiff guard hairs), hair white at base .... **Muridae** (in part)
the link rat, *Deomys*

Figure 25–37.
Body form in some representatives of the Dipodidae. A, thick-tailed pygmy jerboa, *Salpingotus crassicauda*; B, three-toed jerboa, *Dipus sagitta*, in mid hop; C, Mongolian five-toed jerboa, *Allactaga sibirica*.
(A, Gromov et al. 1963: 385; B and C, Hsia et al. 1964: 42)

Figure 25-39.
Skull A, and bipedal posture B, of jumping mice. A, birch mouse, *Sicista subtilis*; B, jumping mouse, *Zapus setchuanus*.
(A, Gromov et al. 1963: 376; B, Hsia et al. 1964: 40)

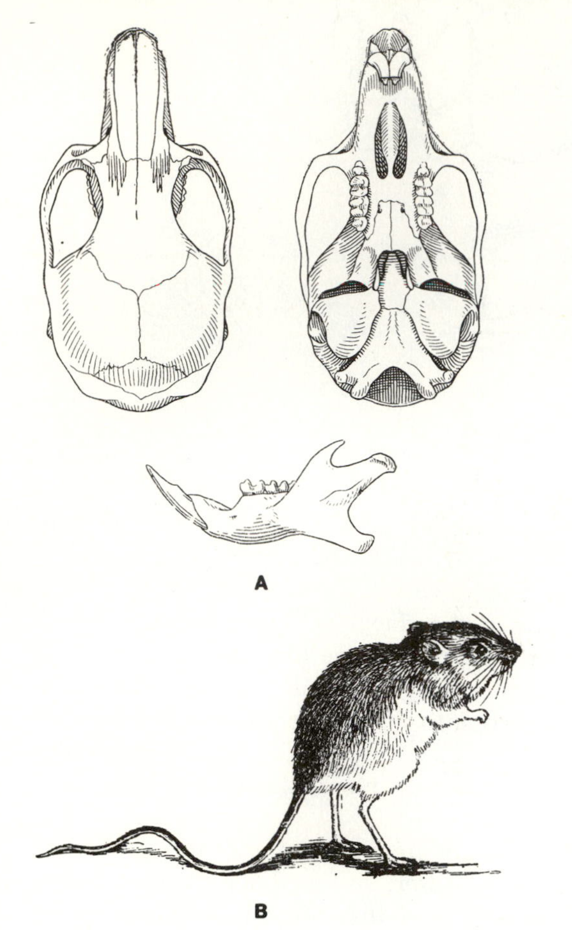

25 (6′) Infraorbital foramen large, maximum dimension greater than 10 mm 26

25′ Infraorbital foramen small, maximum dimension less than 10 mm 28

26 (25) Lower maxillary process projects well forward to a point even with, or anterior to cutting edge of upper incisor (Fig. 25-40); cheek teeth evergrowing; size large (head and body length > 300 mm), form modified for saltation (Fig. 25-1) .... **Pedetidae**
springhaas

26′ Lower maxillary process not as above; cheek teeth rooted; size variable, but head and body length < 300 mm and body form not modified for saltation 27

Figure 25-40.
Anterior portion of the springhaas, *Pedetes capensis*, cranium showing anterior projection of lower maxillary process.
(R.E. Martin)

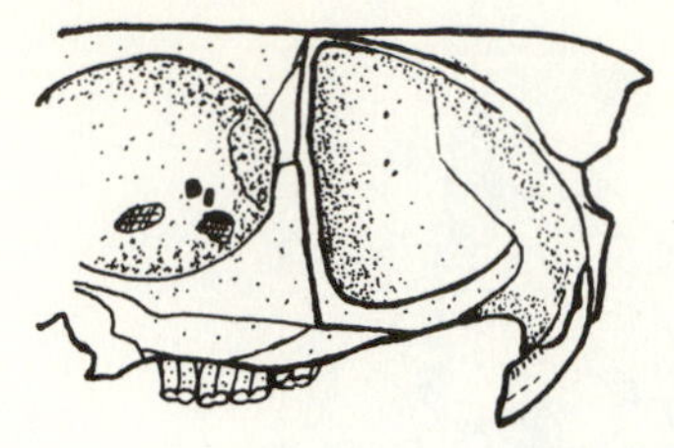

27 (26′) Lower molars 2 and 3 (M_2 and M_3) with one or two deep folds on labial and lingual surfaces, and sometimes resulting in E-shaped pattern (Fig. 25-41); hind foot with four digits, no plate-like scales on ventral surfaces of tail .... **Ctenodactylidae** (in part)
some gundis

27′ Lower molars 2 and 3 (M_2 and M_3) with basin-shaped depressions or enamel islands (Fig. 25-42) and never having E-shaped pattern; hind foot with five digits, plate-like scales on ventral surface of tail (Fig. 25-43) .... **Anomaluridae**
scaly-tailed squirrels

Figure 25-41.
Right lower cheek teeth in a gundi, *Pectinator spekeii*. Anterior at left.
(J. Blefeld)

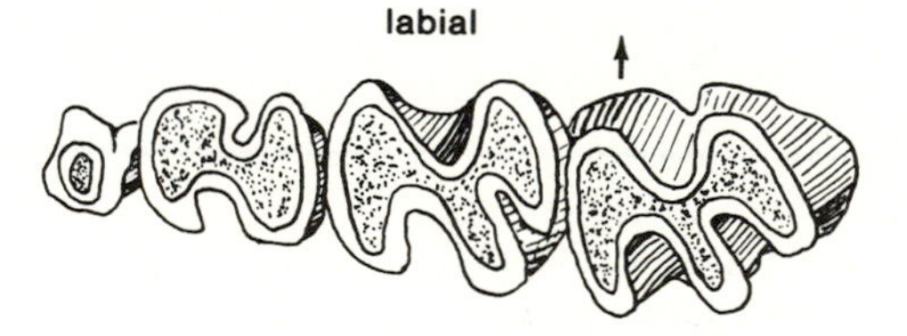

Figure 25–42.
Skull of scaly-tailed squirrel, *Anomalurus peli*.
(Redrawn from Tullberg 1899: pl. 9).

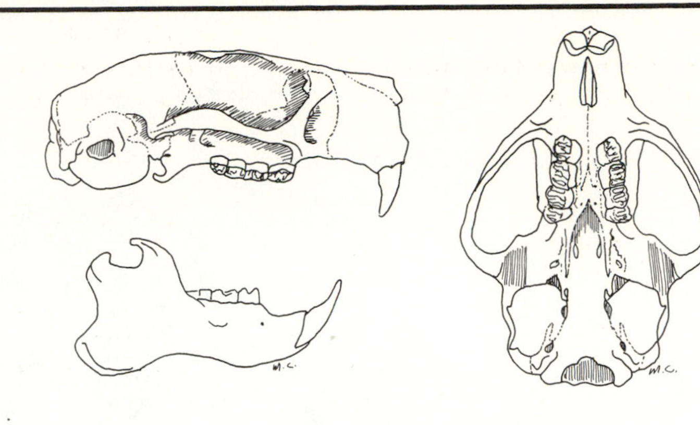

Figure 25–43.
A, scaly-tailed squirrels, *Anomalurus fulgens*, and B, ventral view of base of tail.
(A, Flower and Lydekker 1891: 449 after Alston; B, Dekeyser 1955: 187)

A

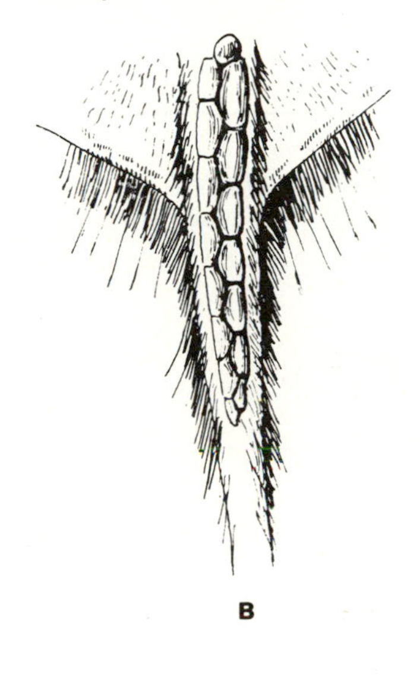

B

28 (25′) Size large, greatest length of skull more than 50 mm; head and body length greater than 300 mm; portion of angular process of dentary projects inward, medial to mandibular condyle (Fig. 25-44); no large foramen in angular process of dentary; tail short, less than 15% of head and body length, well haired (Fig. 25-44) .... **Aplodontidae**
sewellel or mountain "beaver"

28′ Size small, greatest length of skull less than 50 mm; head and body length less than 220 mm; angular process of dentary not turned inward, always projecting lateral to mandibular condyle; large foramen in angular process of dentary (Fig. 25-45A); tail long (more than 50% of head and body length), well haired (Fig. 25-45B) .... **Gliridae**
dormice

29 (3′) Infraorbital foramen smaller than foramen magnum, body haired or nearly hairless (Fig. 25-46) .... **Bathyergidae**
"mole" rats

29′ Infraorbital foramen equal to or larger than foramen magnum; body always well haired 30

Figure 25–44.
Ventral view of mandible (A) of sewellel (B), *Aplodontia rufa.*
(Mary Ann Cramer)

Figure 25–45.
Skull (A) of tree dormouse (B), *Dryomys nitedula.*
(A, Gromov et al. 1963: 364; B, Mary Ann Cramer from photograph in Walker et al. 1975)

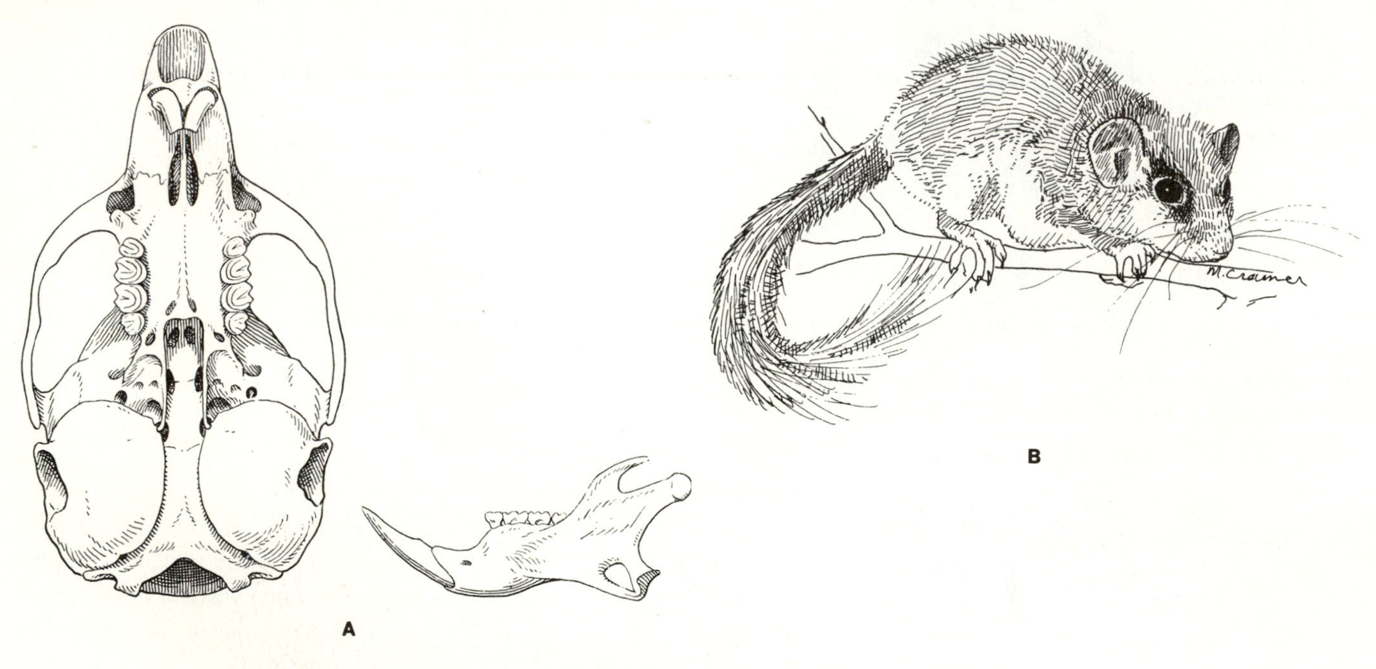

30 (29′) Prominent ridge and groove extending along labial side of dentary, parallel to cheek teeth (e.g., Fig. 25-47) 31

30′ No prominent ridge along labial side of dentary 32

31 (30) Size larger, greatest length of skull more than 175 mm (Fig. 25-48); length of last upper molar greater than combined lengths of preceding cheek teeth in each upper tooth row .... **Hydrochoeridae**
capybaras

31′ Size smaller, greatest length of skull less than 175 mm; length of last upper molar less than combined lengths of preceding cheek teeth in each upper tooth row (Fig. 25-49) .... **Caviidae**
cavies, Patagonian "hares"

Figure 25–46.
Representative bathyergids: A, skull of mole rat, *Bathyergus maritimus*; B, naked mole rat, *Heterocephalus glaber;* C, blesmol, *Georychus capensis.* (A, redrawn from Tullberg 1899: pl. 2; B, Weber 1928: 239; C, Sclater 1901: 75)

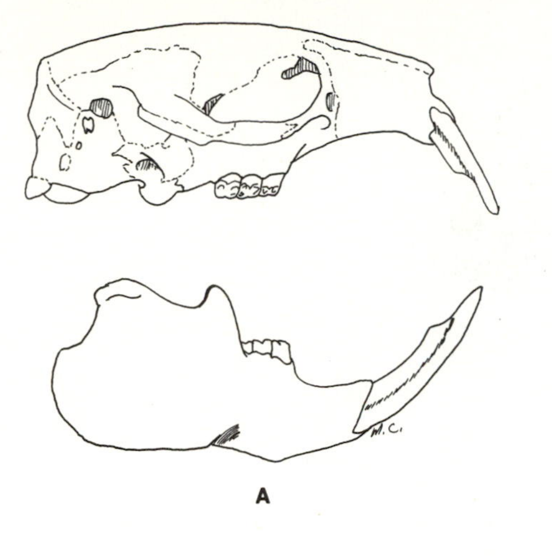

A

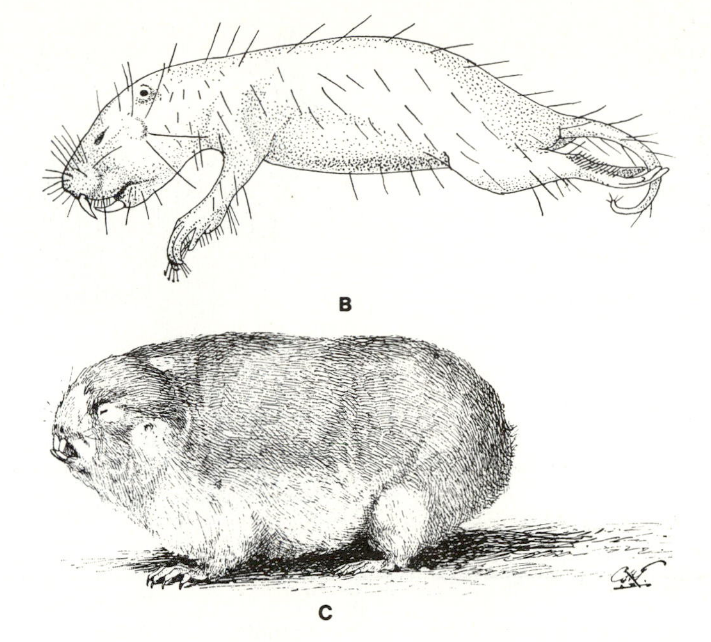

B

C

Figure 25–47.
Right dentary of *Dolichotis salinicola* showing deep groove and ridge. (J. Blefeld)

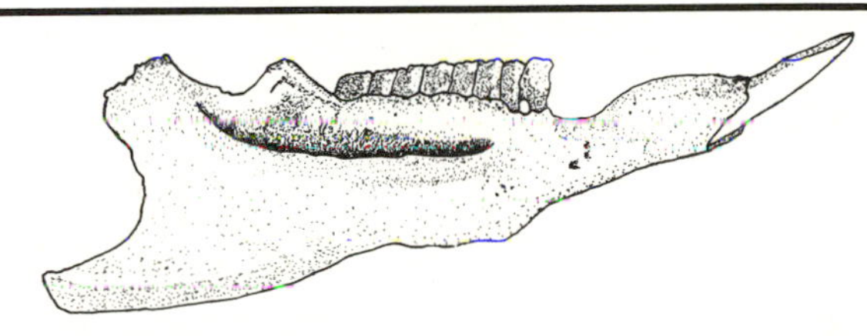

Figure 25–48.
Skull (A) of capybara (B), *Hydrochoerus hydrochaeris.* (A, Flower and Lydekker 1891: 481; B, Beddard 1902: 491)

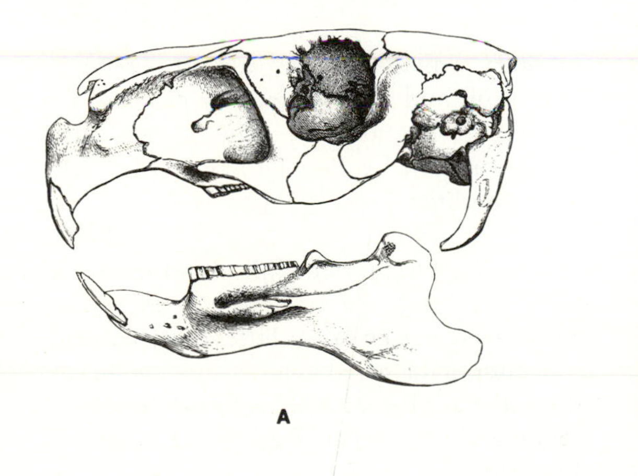

A

B

Figure 25-49.
Representative species of Caviidae: A, skull of *Cavia spixi*; B, Patagonian cavy, *Dolichotis patagona*.
(A, Mary Ann Cramer; B, Beddard 1902: 492)

Figure 25-50.
Bullar region of *Chinchilla* cranium (A) and plains viscacha (B), *Lagostomus maximus*.
(A, G.A. Moore; B, Beddard 1902: 497)

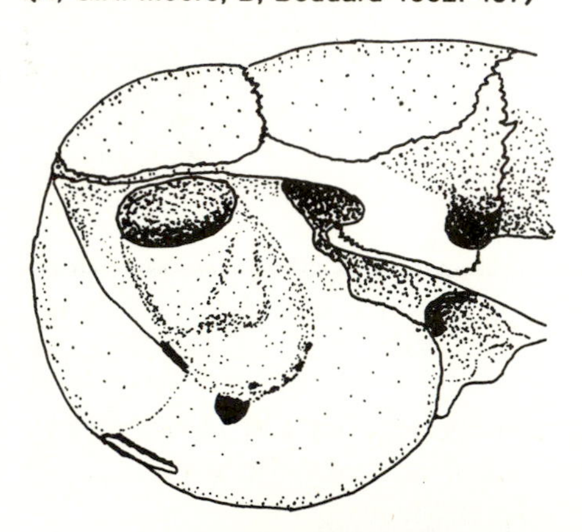

Figure 25-51.
Bullar region of *Trichys lipura* cranium.
(G.A. Moore)

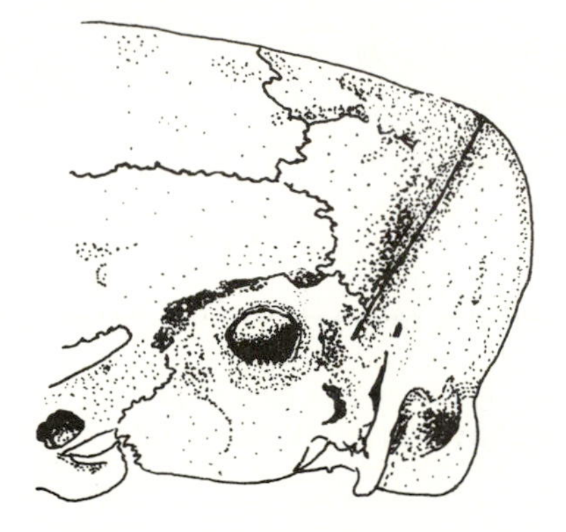

32 (30′) Auditory canal emerges almost vertically from skull and most of interior of canal not evident in lateral view (Fig. 25-50); additional foramen leading into tympanic bulla below auditory canal (Fig. 25-50) .... **Chinchillidae**
chinchillas and viscachas

32′ Auditory canal emerges laterally from skull and interior of canal readily evident in lateral view *or*, if not, then no foramen present below auditory canal (Fig. 25-51) 33

Figure 25–52.
Crania of A, *Abrocoma bennetti* and B, *Dasyprocta* sp., showing exposed lacrimal canals, not to same scale. (A, Mary Ann Cramer; B, J. Blefeld)

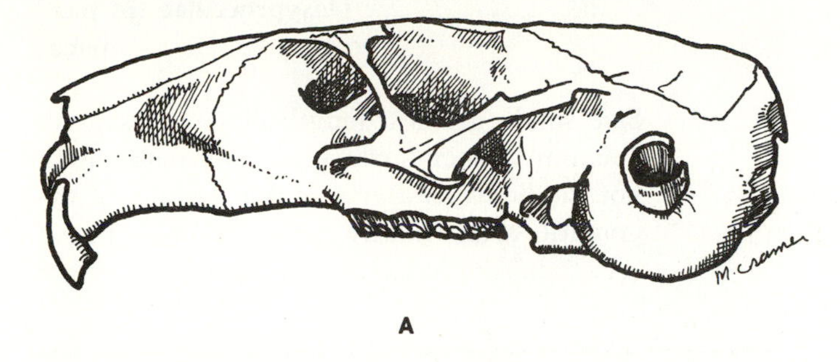

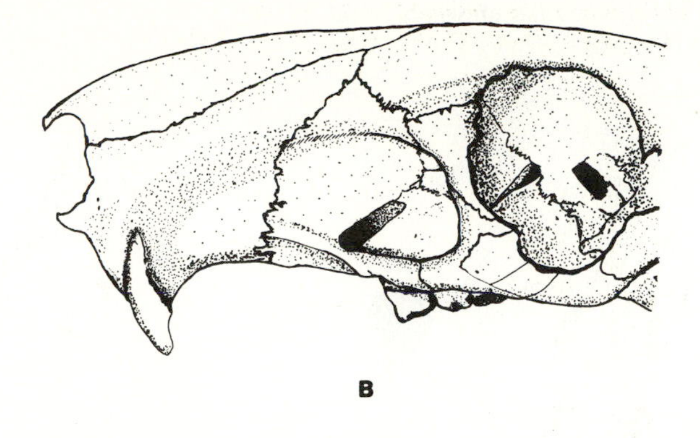

Figure 25–53.
Agouti, *Dasyprocta aguti.* (Beddard 1902: 495)

Figure 25–54.
"Chinchilla" rat, *Abrocoma bennetti.* (Mary Ann Cramer from photograph by R.E. Martin)

33 (32′) Portion of lacrimal canal exposed on side of rostrum (e.g., Fig. 25-52) 34

33′ No part of lacrimal canal exposed on side of rostrum 35

34 (33) Size large, greatest length of skull more than 70 mm; auditory bullae small, never visible from dorsal aspect of skull; cheek teeth with enamel islands; hind limbs elongated, digitigrade, pollex vestigal (Fig. 25-53) .... **Dasyproctidae** (in part) agoutis

34′ Size small, greatest length of skull less than 70 mm; auditory bullae large, visible from dorsal aspect of skull; each cheek tooth with single re-entrant fold on each side, never forming enamel islands; hind limbs not elongated, plantigrade, pollex absent (Fig. 25-54 .... **Abrocomidae** chinchilla "rats"

35 (33′) Additional canal or groove present along ventral, medial wall of infraorbital canal (e.g., Fig. 25-55) 36

35′ No canal or groove present along ventral, medial wall of infraorbital canal 40

36 (35) Upper incisor with three grooves on anterior surface (Fig. 25-56) .... **Thryonomyidae** cane rats

36′ Upper incisor without grooves 37

Figure 25–55.
Crania of A, *Octodon degus* showing groove and B, *Aconaemys fuscus* showing canal.
(G.A. Moore)

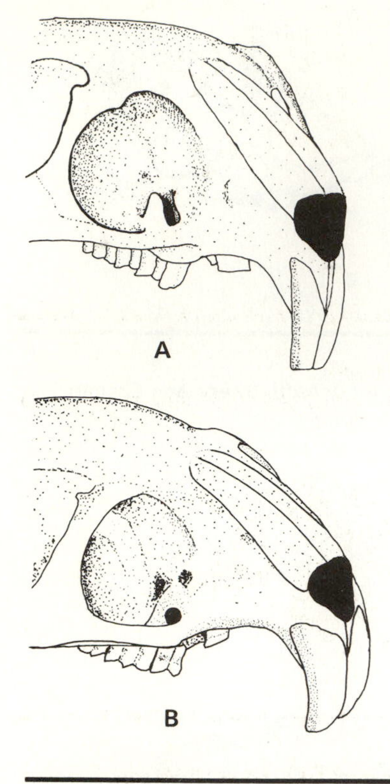

Figure 25–56.
Skull (A) of cane rat (B), *Thryonomys swinderianus*.
(Redrawn from Tullberg 1899: pl. 6; B, Dekeyser 1955: 174)

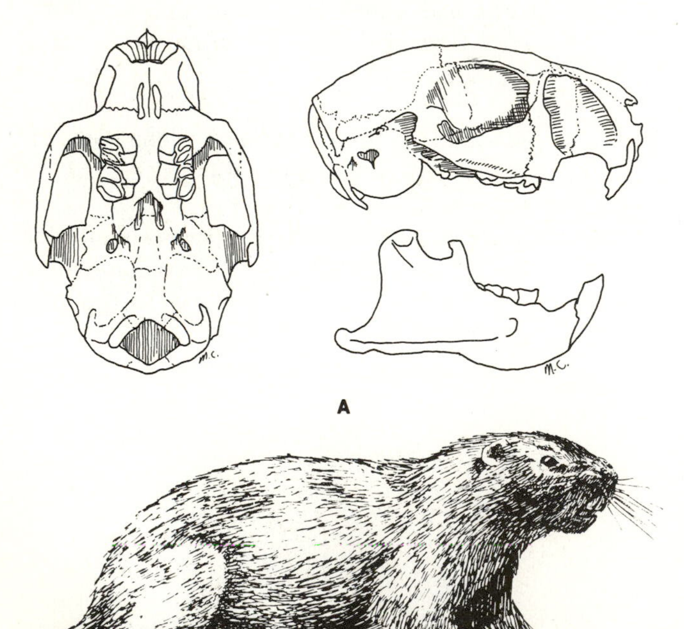

37 (36′) Size large, greatest length of skull more than 100 mm; inner surface of zygomatic arch with deep depression; outer surface of zygomatic arch pitted (Fig. 25-57) .... **Dasyproctidae** (in part)
pacas

37′ Size small, greatest length of skull less than 100 mm; inner surface of zygomatic arch not modified as above; outer surface of zygomatic arch usually smooth, never pitted 38

Figure 25–57.
Ventral view of *Aguti paca* cranium.
(Redrawn from Tullberg 1899: pl. 5)

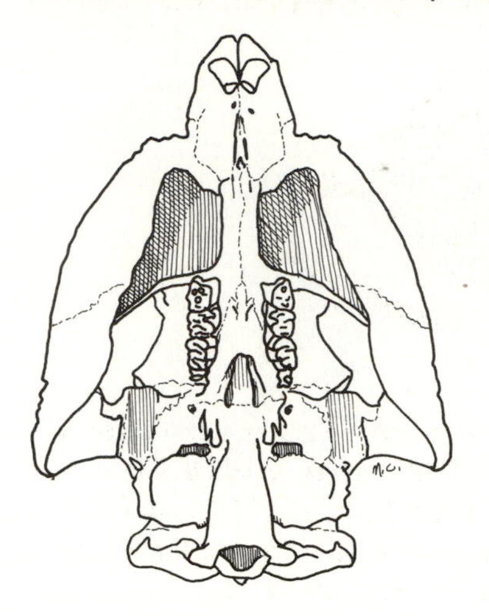

38 (37′) Palate long, posterior border behind last upper cheek tooth; cheek teeth terraced, with marked elevation on lingual side of upper cheek teeth (Fig. 25-58) and labial side of lower cheek teeth .... **Petromyidae**
dassie rat

38′ Palate short, posterior border forward of last upper cheek tooth; upper cheek tooth with occlusal surface almost flat and with no marked elevated areas 39

Figure 25–58.
Terraced cheek teeth (A) of the dassie rat (B), *Petromys typicus.*
(A, Hershkovitz 1962; B, Mary Ann Cramer)

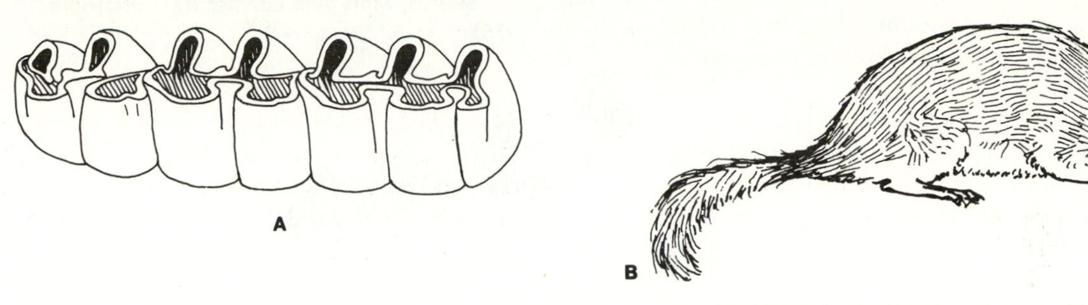

Figure 25–59.
Bridge's octodont, *Octodon bridgesi.*
(Mary Ann Cramer from photograph by R.E. Martin)

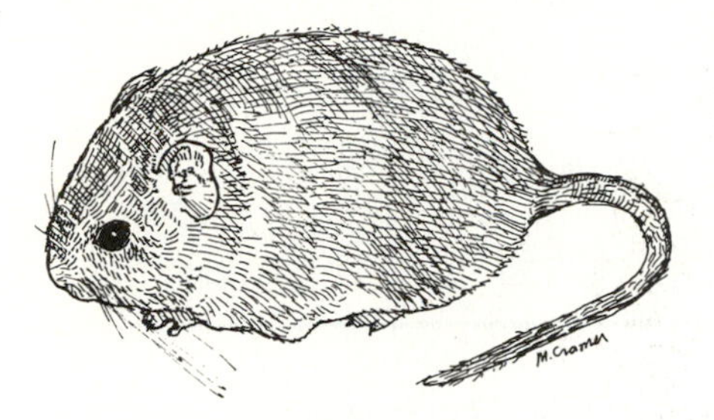

Figure 25–60.
Spiny rat, *Proechimys setosus.*
(Mary Ann Cramer, from photograph in Walker et al. 1975)

39 (38′) Cheek teeth evergrowing, with single main fold along labial border of each cheek tooth (similar to those in Fig. 25-62); and lacking enamel islands in lake of dentine; pelage without spines (Fig. 25-59) .... **Octodontidae** (in part)
degus and other octodonts except *Spalacopus*

39′ Cheek teeth rooted, with two main folds along labial border of each cheek tooth or, if not readily evident, occlusal surface with one or more enamel islands surrounded by dentine; pelage with weak to moderately developed spines (Fig. 25-60) .... **Echimyidae** (in part)
spiny rats of the genera *Cercomys, Eurygomatomys, Hoplomys* and some *Proechimys*

40 (35′) Paroccipital process long, its greatest length more than 15 mm (Fig. 25-61A); hind foot webbed .... **Myocastoridae**
nutria or coypu

40′ Paroccipital process short, its greatest length less than 15 mm; hind foot not webbed 41

Figure 25–61.
Skull (A) of the nutria (B), *Myocastor coypus.*
(A, Gromov et al. 1963: 258; B, Hsia et al. 1964: 40)

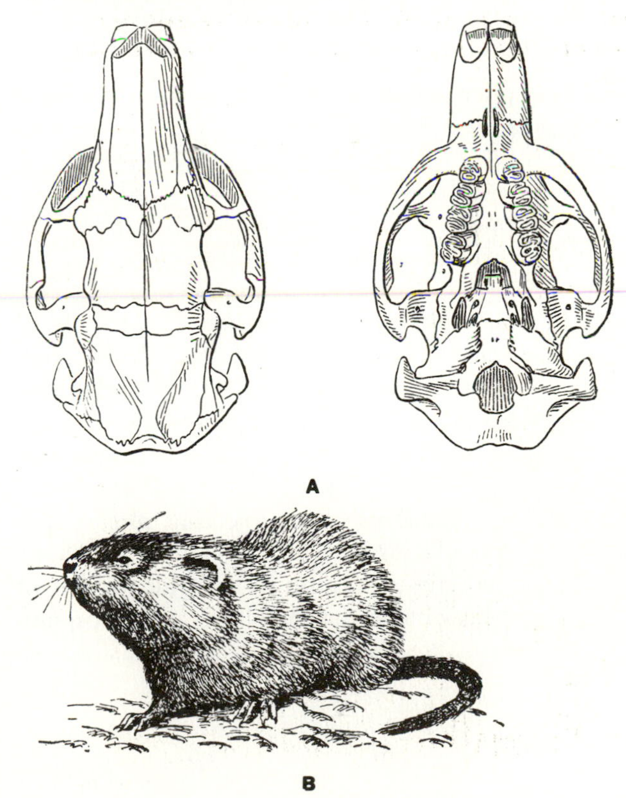

41 (40′) Upper cheek teeth with 8-shaped pattern (Fig. 25-62A); bone enclosing origin of upper incisor forming a prominent ridge (alveolar process) in the orbit (Fig. 25-62B) .... **Octodontidae** (in part)
an octodont, *Scalacopus*

41′ Upper cheek teeth with various patterns, never 8-shaped; usually no prominent ridge in orbit for origin of incisor, but if present, the upper cheek teeth are kidney-shaped 42

Figure 25-62.
Upper left cheek teeth (A) and skull (B) of a coruro, *Spalacopus cyanus*. Anterior end of tooth row at left.
(A, Reig 1970: 596; B, G.A. Moore)

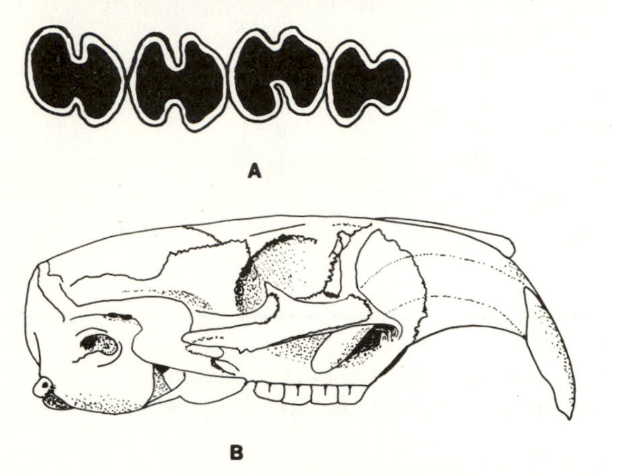

42 (41′) Upper cheek teeth with a kidney-shaped pattern (Fig. 25-63) .... **Ctenomyidae**
tuco tucos

42′ Upper cheek teeth of variable pattern, but never kidney shaped 43

43 (42′) Paroccipital process curving close to ventral surface of auditory bulla (Fig. 25-64); pelage with weak to moderately developed spines (Fig. 25-60) .... **Echimyidae** (in part)
spiny rats

43′ Paroccipital process projecting almost perpendicular to long axis of skull (Fig. 25-66), never curving close to ventral surface of auditory bulla; pelage with or without spines 44

Figure 25-63.
Left upper cheek teeth in *Ctenomys latro* (=*C. mendocinus tucumanus*); B, tuco tuco, *Ctenomys* sp.
(A, Reig 1970: 599; B, Mary Ann Cramer from photograph in Walker 1975)

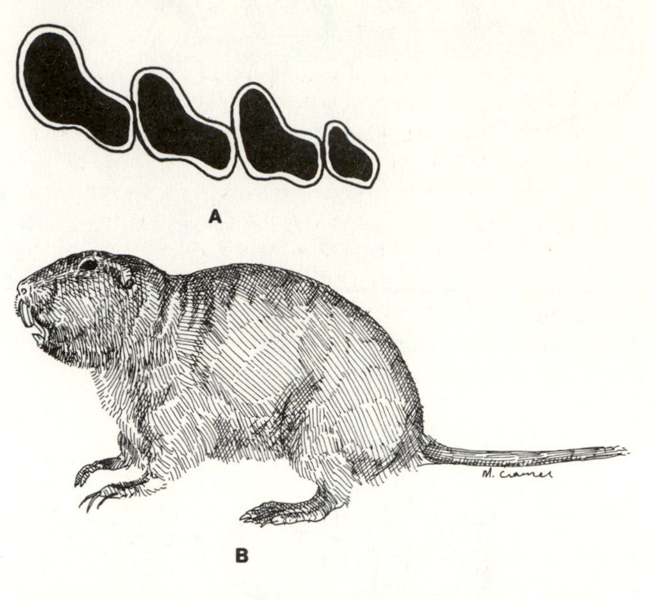

Figure 25-64.
Bullar region in cranium of A, *Hoplomys gymnurus* and B, *Dactylomys bolivianus*.
(G.A. Moore)

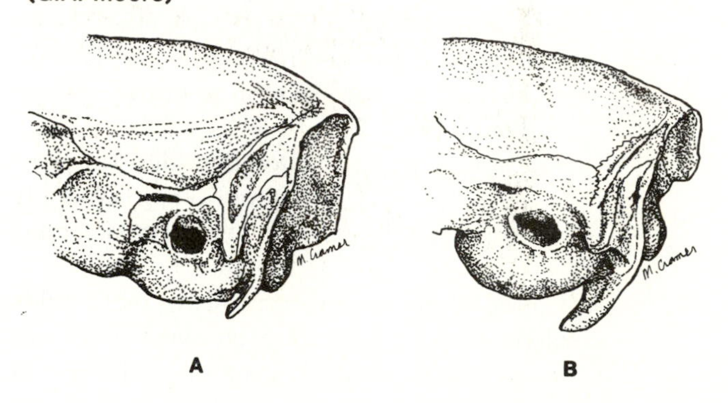

44 (43′) Palate not constricted at anterior end of tooth row, its greatest breadth at first cheek tooth equal to, or more than, greatest breadth at last cheek tooth (Fig. 25-65) .... **Hystricidae**
Old World porcupines

44′ Palate constricted at anterior end of tooth row, its greatest breadth at first cheek tooth less than greatest breadth at last cheek tooth 45

Figure 25–65.
Skull (A) of Old-world porcupine (B), *Hystrix lecura*.
(A, Gromov et al. 1963: 256; B, Finn 1929: 165)

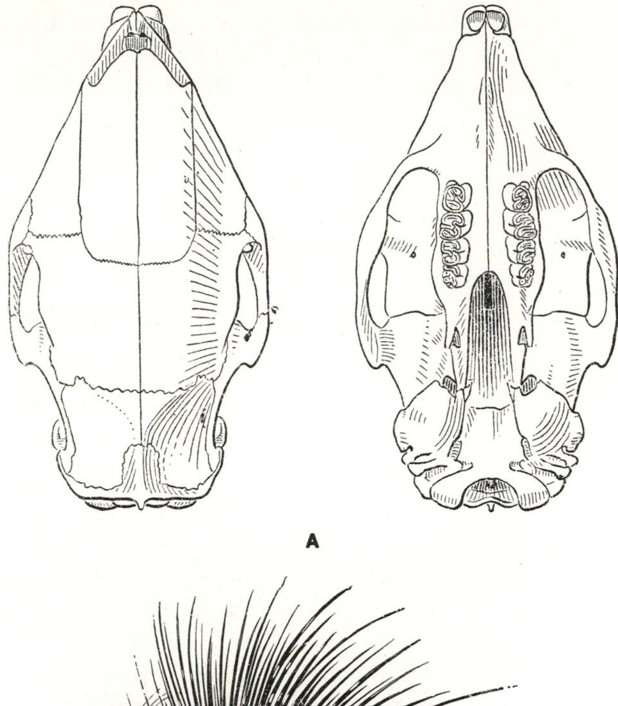

Figure 25–66.
Bullar region in cranium of A, *Capromys pilorides* and B, *Erethizon dorsatum*.
(A, Mary Ann Cramer; B, G.A. Moore)

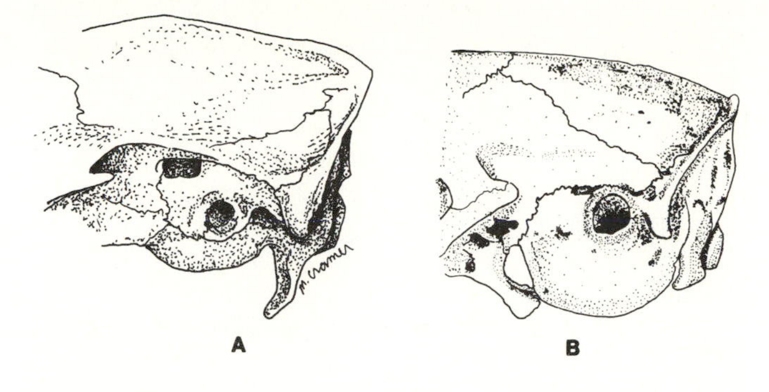

Figure 25–67.
Right lower cheek teeth (A) of pacarana (B), *Dinomys branickii*.
(A, J. Blefeld; B, Duncan 1877–83: 142, after Peters)

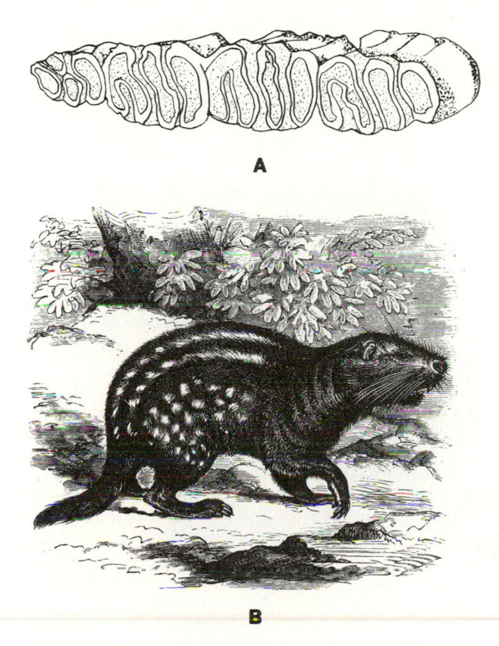

45 (44′) Paroccipital process long, its distal end extending below most ventral point on auditory bulla (Fig. 25-66A) .... **Capromyidae**
hutias

45′ Paroccipital process short, its distal end never extending below auditory bulla (Fig. 25-66B) 46

46 (45) Coronoid process of mandible absent or vestigial; lower cheek teeth with a series of transverse laminae (Fig. 25-67); pelage with spots; spines absent (Fig. 25-67) .... **Dinomyidae**
pacarana

46′ Coronoid process of mandible prominent (Fig. 25-68A); each lower cheek tooth with one lingual re-entrant fold, never laminated; pelage lacks spots; spines present (Fig. 25-68B) .... **Erethizontidae**
New World porcupines

Figure 25–68.
A, skull of *Coendou* sp.; B, tree porcupine, *Echinoprocta rufescens*.
(A, redrawn from Tullberg 1899: pl. 7; Flower and Lydekker 1891: 485)

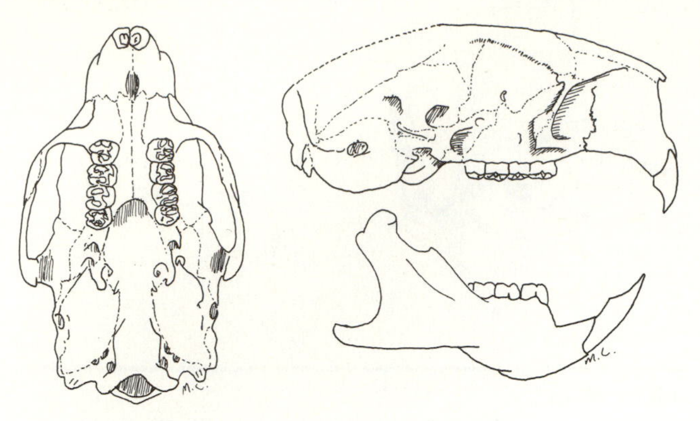

A

B

Key to Living Families of North America*

1 Infraorbital foramen equal to or larger than foramen magnum 2

1′ Infraorbital foramen smaller than foramen magnum 6

2 (1) Greatest length of paroccipital process more than 15 mm (Fig. 25-48A, and 25-61A) 3

2′ Greatest length of paroccipital process less than 15 mm 4

3 (2) Anterior face of each upper incisor with single longitudinal groove; tail very short (less than 10% of head and body length); pes with three digits, not webbed (Fig. 25-48B) .... **Hydrochoeridae**
capybaras

3′ Upper incisors without grooves; tail long (50-80% of head and body length); pes with five digits, webbed (Fig. 25-61B) .... **Myocastoridae**
nutria or coypu
(introduced)

4 (2′) Paroccipital process rudimentary, less than 5 mm in length (Fig. 25-68A); incisors project beyond nasals, visible in dorsal view of skull .... **Erethizontidae**
New World porcupines

4′ Paroccipital process never rudimentary, its length always greater than 5 mm; incisors project beyond nasals and not visible in dorsal view 5

5 (4′) Portion of lacrimal canal exposed on side of rostrum (Fig. 25-52B); claws thick and hooflike (Fig. 25-53) .... **Dasyproctidae**
agoutis and pacas

5′ No part of lacrimal canal exposed on side of rostrum; claws not hooflike .... **Echimyidae**
spiny rats

6 (1′) Infraorbital canal passing through side of rostrum, emerging anterior to zygomatic plate with the infraorbital foramen visible when the skull is viewed laterally (Fig. 25-5) 7

6′ Infraorbital canal and its foramen variable, but if canal emerges anterior to zygomatic plate, the foramen is not visible when the skull is viewed laterally 8

7 (6) Infraorbital foramen large, completely perforating rostrum transversely; tail long, usually well haired (Fig. 25-6) .... **Heteromyidae**
kangaroo rats, pocket mice, and allies

*All of the North American Continent south to Panama but excluding islands of the West Indies.

7′ Infraorbital foramen small, never completely perforating rostrum transversely; tail short and covered with short tactile hairs (Fig. 25-7) .... **Geomyidae**
pocket gophers

8 (6′) Postorbital processes of frontal present and sharply pointed (Fig. 25-10) .... **Sciuridae**
squirrels and marmots

8′ Postorbital processes of frontal absent, or, if present, blunt (Fig. 25-12) 9

9 (8′) Postorbital process of frontal present, blunt and small (Fig. 25-12); tail flattened dorsoventrally (Fig. 25-13) .... **Castoridae**
beaver

9′ Postorbital process of frontal absent; tail not flattened dorsoventrally 10

10 (9′) Infraorbital foramen a long vertical slit 11

10′ Infraorbital foramen round or oval 12

11 (10) Cheek teeth with cusps (or their evidence) arranged in three longitudinal rows (Fig. 25-28A) .... **Muridae**
house mouse, black and Norway rats (introduced)

11′ Cheek teeth with cusps (or their evidence) arranged in two longitudinal rows or with occlusal surfaces having acute triangles or re-entrant folds (Fig. 25-27A) .... **Cricetidae**
voles, deer mice, etc.

12 (10′) Cheek teeth 5/4; greatest length of skull more than 50 mm; size large (head and body length more than 300 mm), tail short, 5-10% of head and body length (Fig. 25-44B) .... **Aplodontidae**
sewellel or mountain "beaver"

12′ Cheek teeth 3/3 or 4/3 (Fig. 25-39A); greatest length of skull less than 50 mm; size small (head and body length less than 200 mm); tail long, more than or equal to head and body length (Fig. 25-39B) .... **Zapodidae**
jumping mice

Comments and Suggestions on Identification

All rodents possess a single pair of evergrowing, arc-shaped upper incisors. This character is unique to the order, although incisors that superficially resemble those of rodents are found in some other mammals:

1. Wombats, Vombatidae, diprotodont marsupials. Check angle of ramus and mandibular fossa.
2. The aye-aye, *Daubentonia,* a primate. Check postorbital bar, position of foramen magnum, and structure of manus.
3. Rabbits, hares, pikas (Lagomorpha) and hyraxes (Hyracoidea). Check the number and shape of upper and lower incisors.

Taxonomic Remarks

Rodents were originally included with lagomorphs in an order called Glires. Gidley (1912) showed that this arrangement was based on superficial evidence and established the orders Rodentia and Lagomorpha. Most mammalogists have accepted his arrangement.

Ellerman (1940-41) recognized twenty-three families of rodents in a monumental work that relied heavily on morphological similarities. We have adopted Anderson (1967) list of families and recognized thirty-three families in the order, but have not included the recently extinct †Heptaxodontidae. The genus *Cricetomys* will key out with the Muridae, although we follow Petter (1964) and Rosevear (1969), in considering this form to be a member of the Cricetidae, subfamily Cricetomyinae. Eisentraut (1976:121) stated that *Phloeomys cumingi* should be placed in a separate family (Phloeomyidae) rather than in a subfamily (Phloeomyinae) of the Muridae. Since Eisentraut's decision was apparently based solely on the structure of the soft palate, we have not adopted his arrangement in this manual. Some taxonomists (e.g., Hershkovitz 1962), included the cricetids and murids in a single family, the Muridae.

Brandt (1855) first proposed that the rodents be classified into three groups: Sciuromorpha, Myomorpha, and Hystricomorpha. Tullberg (1899) divided the order into two groups: Sciurognathi, with subdivisions Sciuromorphi and Myomorphi, and Hystricognathi, with subdivisions Hystricomorphi and Bathergomorphi. Simpson (1945) utilized three suborders in his classification of mammals and relied primarily on morphology of jaw musculature and associated structures. But some forms (e.g., Bathyergidae, Pedetidae) were not easily associated with any of the three groups. The three-suborder arrangement for rodents has, for convenience, been widely used and is adopted in this manual.

Miller and Gidley (1918) arranged the rodents into several superfamilies although their Dipodoidea was an assemblage of miscellaneous forms. Wood (1954, 1955) and Lavocat (1962) established other classification schemes. Stehlin and Schaub (1951) proposed another system based on tooth morphology that was utilized by Grassé and Dekeyser (1955).

Table 25-2 compares a modified version (Anderson 1967) of Simpson's (1945) classification with that of Wood (1955). Notice that there is considerable agreement at the superfamily level, but the classification of Wood (1955) hypothesized a separate origin for the New World hystricomorphs (the caviomorphs). Recent evidence of continental drift (see Rowlands and Weir 1974, Landry 1977), indicates that the New and Old World hystricomorphs had a common origin.

Table 25-2

Classifications of Rodentia.

Anderson (1967)	*Wood (1955)*
"Sciuromorpha"	Suborder Sciuromorpha
Superfamily Aplodontoidea	Superfamily Aplodontoidea
Aplodontidae	Aplodontidae
Superfamily Sciuroidea	Superfamily Sciuroidea
Sciuridae	Sciuridae
Superfamily Geomyoidea	
Geomyidae	
Heteromyidae	
	CF. Sciuromorpha, *Inc. Sed.*
	Superfamily Ctenodactyloidea
	Ctenodactylidae
	Suborder Theridomyomorpha
Superfamily Anomaluroidea	Superfamily Anomaluroidea
Anomaluridae	Anomaluridae
	CF. Sciuromorpha or Theridomyomorpha, *Inc. Sed.*
Pedetidae	Pedetidae
	Suborder Castorimorpha
Superfamily Castoroidea	Superfamily Castoroidea
Castoridae	Castoridae
"Myomorpha"	Suborder Myomorpha
Superfamily Muroidea	Superfamily Muroidea
Cricetidae	Cricetidae
Muridae	Muridae
	Muroidea, *Inc. Sed.*
Spalacidae	Spalacidae
Rhizomyidae	Rhizomyidae
	Superfamily Geomyoidea
	Heteromyidae
	Geomyidae
Superfamily Dipodoidea	Superfamily Dipodoidea
Zapodidae	Zapodidae
Dipodidae	Dipodidae
	CF. Myomorpha, *Inc. Sed.*
Superfamily Gliroidea	Superfamily Gliroidea
Gliridae	Gliridae [includes Platacanthomyidae]
Platacanthomyidae	
Seleviniidae	Seleveniidae

Table 25-2 *(Continued)*

“Hystricomorpha”	Suborder Hystricomorpha
Superfamily Hystricoidea	Superfamily Hystricoidea
Hystricidae	Hystricidae
	Superfamily Thryonomyoidea
	Thryonomyidae
	Petromuridae [=Petromyidae]
	Suborder Caviomorpha
Superfamily Cavoidea	Superfamily Cavoidea
Caviidae	Caviidae
Hydrochoeridae	Hydrochoeridae
Dinomyidae	Dinomyidae
Heptaxodontidae [extinct in recent times]	Heptaxodontidae [extinct in recent times]
Dasyproctidae	Dasyproctidae
	Cuniculidae [combined with Dasyproctidae in this Manual]
Superfamily Chinchilloidea	Superfamily Chinchilloidea
Chincillidae	Chinchillidae
	Capromyidae [includes Myocastoridae of Anderson]
Superfamily Octodontoidea	Superfamily Octodontoidea
Capromyidae	
Myocastoridae	
Octodontidae	Octodontidae
Ctenomyidae	Ctenomyidae
Abrocomidae	Abrocomidae
Echimyidae	Echimyidae
Thryonomyidae	
Petromyidae	
Superfamily Erethizontoidea	Superfamily Erethizontoidea
Erethizontidae	Erethizontidae
	Suborder Bathyergomorpha
Superfamily Bathyergoidea	Superfamily Bathyergoidea
Bathyergidae	Bathyerigidae
Superfamily Ctenodactyloidea	
Ctenodactylidae	

Supplementary Readings

Barnett, S. A. 1963. *The rat: a study in behaviour.* Aldine Publ. Co., Chicago. 288 pp.

Delany, M. J. 1975. *The rodents of Uganda.* Publication No. 764, British Museum (Nat. Hist.), London. 165 pp.

Eisenberg, J. F. 1963. The behavior of heteromyid rodents. *Univ. Calif. Publ. Zool.* 69:1-100, 13 pls.

Eisenberg, J. F. 1967. A comparative study in rodent ethology with emphasis on evolution of social behavior. *Proc. U.S. Nat. Mus.* 122:1-51.

Ellerman, J. R. 1940. *The families and genera of living rodents.* Vol. 1. British Museum (Nat. Hist.), London. 689 pp.; 1941, Vol. 2, 690 pp.; 1949, Vol. 3, 210 pp.

Elton, C. 1942. *Voles, mice and lemmings.* Oxford Univ. Press, Oxford. 496 pp.

Genoways, H. H. 1973. Systematics and evolutionary relationships of spiny pocket mice, genus *Liomys. Spec. Publ., The Museum, Texas Tech Univ.* 5:1-368.

Green, E. L. (Ed.). 1966. *Biology of the laboratory mouse,* 2nd ed. McGraw-Hill Book Co., New York. 706 pp.

Hershkovitz, P. 1966. Evolution of Neotropical cricetine rodents (Muridae) with special reference to the phyllogine group. *Fieldiana: Zoology* 46:1-524.

Hoffman, R. A., P. F. Robinson, and H. Magalhaes (Eds.). 1968. *The golden hamster: its biology and use in medical research.* Iowa State Univ. Press, Ames. 545 pp.

King, J .A. 1955. Social behavior, social organization, and population dynamics in a black-tailed prairie dog town in the Black Hills of South Dakota. *Cont. Lab. Vert. Biol., Univ. Michigan* 67:1-123.

King, J. A. (Ed.). 1968. *Biology of Peromyscus (Rodentia).* Spec. Publ. No. 2, Amer. Soc. Mammalogists. 593 pp.

Landry, S. O. 1957. The interrelationships of the New and Old World hystricomorph rodents. *Univ. Calif. Pub. Zool.* 56:1-118.

Rosevear, D. R. 1969. *The rodents of West Africa.* British Museum (Nat. Hist.), London. 604 pp., 11 pls.

Rowlands, I. W., and B. J. Weir (Eds.). 1974. The biology of hystricomorph rodents. *Symp. Zool. Soc. London* 34:1-482.

Taylor, W. P. 1935. Ecology and life history of the porcupine (*Erethizon epixanthum*) as related to the forests of Arizona and the southwestern United States. *Univ. Arizona Bull. (Biol. Sci.)* 3:1-177.

Thiessen, D., and P. Yahr. 1977. *The gerbil in behavioral investigations.* Univ. Texas Press, Austin. 244 pp.

26 The Aardvark and the Subungulates

Order Tubulidentata
Order Proboscidea
Order Hyracoidea
Order Sirenia

Based upon external appearance, habitat, diet, and nearly every other easily discernable feature the aardvark, elephants, hyraxes and sirenians would have to be regarded as one of the most unlikely assemblages of mammals possible. However fossil evidence indicates that the latter three of these shared a common ancestor in the early Cenozoic of Africa (Romer 1966: 247) and these three, together with two totally extinct orders, the †Desmostylia [Simpson (1945:136) included in order Sirenia] and †Embrithopoda (Romer 1966: 386-387) are frequently grouped as the subungulates.

The aardvark is now considered to be the most primitive living protoungulate (Wendt 1975b:466, Melton 1976:81). It is not a subungulate as defined by Romer (1966:247) but is probably more closely related to these orders than to any other living order of mammals and thus is included with them in this chapter.

ORDER TUBULIDENTATA

Tubulidentata is the smallest living order of mammals. It contains a single living species, the aardvark, *Orycteropus afer*. The ordinal name refers to the unique structure of the teeth described below. The common name is an Afrikaans term meaning "earth hog," which is appropriate since aardvarks are somewhat piglike in appearance (Fig. 26-1) and are semifossorial. The digits terminate in structures that are intermediate between claws and hoofs (Fig. 26-2). These serve the animal well for burrowing and for tearing open termite mounds. Aardvarks have long, extensible tongues

Figure 26–1.
The aardvark, *Orycteropus afer*.
(Dekeyser 1955: 292)

Figure 26–2.
Lateral (A) and ventral (B) views of the forefoot of an aardvark, *Orycteropus afer*.
(Pocock 1924a: 701)

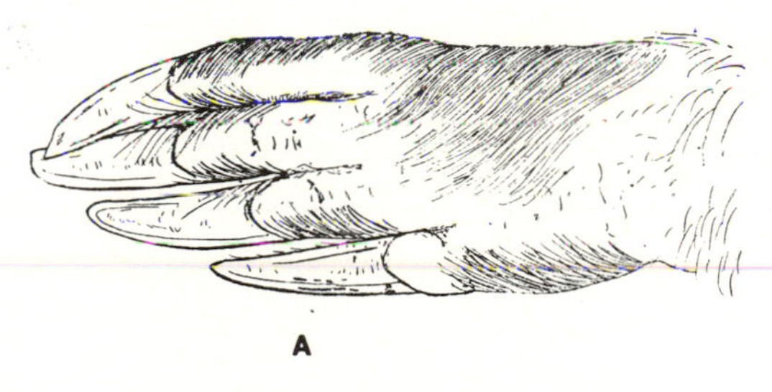

A

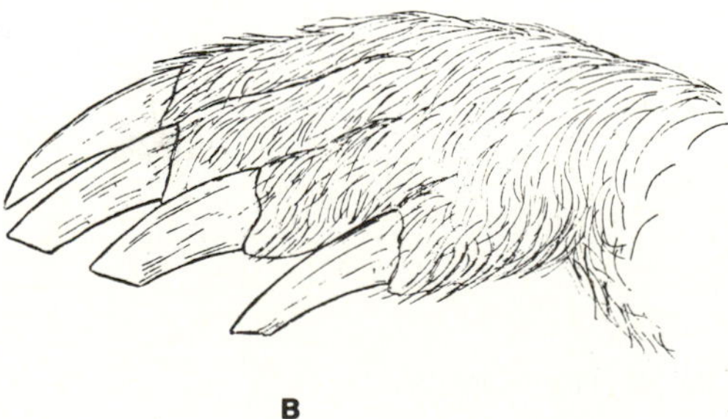

B

that are used to feed upon termites, ants, and other insects (Melton 1976:85) but they are known to feed upon plant materials (Melton 1976, Rham 1975b:476). The flesh and hides of aardvarks are used as meat and leather in some areas (Melton 1976:86). Abandoned aardvark burrows are important to the survival of many other animals that use them as dens and as refuges from bush fires (Rahm 1975b:475).

Distinguishing Characteristics

The skull is elongate and conical in shape (Fig. 26-3). Incisors and canines are absent and cheek teeth usually number 5/5. The cheek teeth are oval or 8-shaped, flat-topped, columnar structures (Fig. 26-4) that lack enamel and that are composed of numerous hexagonal prisms of dentine surrounding tubular pulp cavities (Fig. 26-4).

The limbs are digitigrade with four manual (Fig. 26-2) and five pedal digits. The snout is elongate and pig-like, and the ears are much longer than wide. The tail is long and tapers gradually from its very thick base. The skin is thick with sparse, bristlelike hairs. Subcutaneous fat is lacking (Rahm 1975b:475).

The caecum is large, especially for an insectivorous mammal (Melton 1976:76). The testes are never scrotal and a baculum has not been reported. The uterus is duplex.

Distribution

Aardvarks range throughout Africa south of the Sahara except for areas of dense forest.

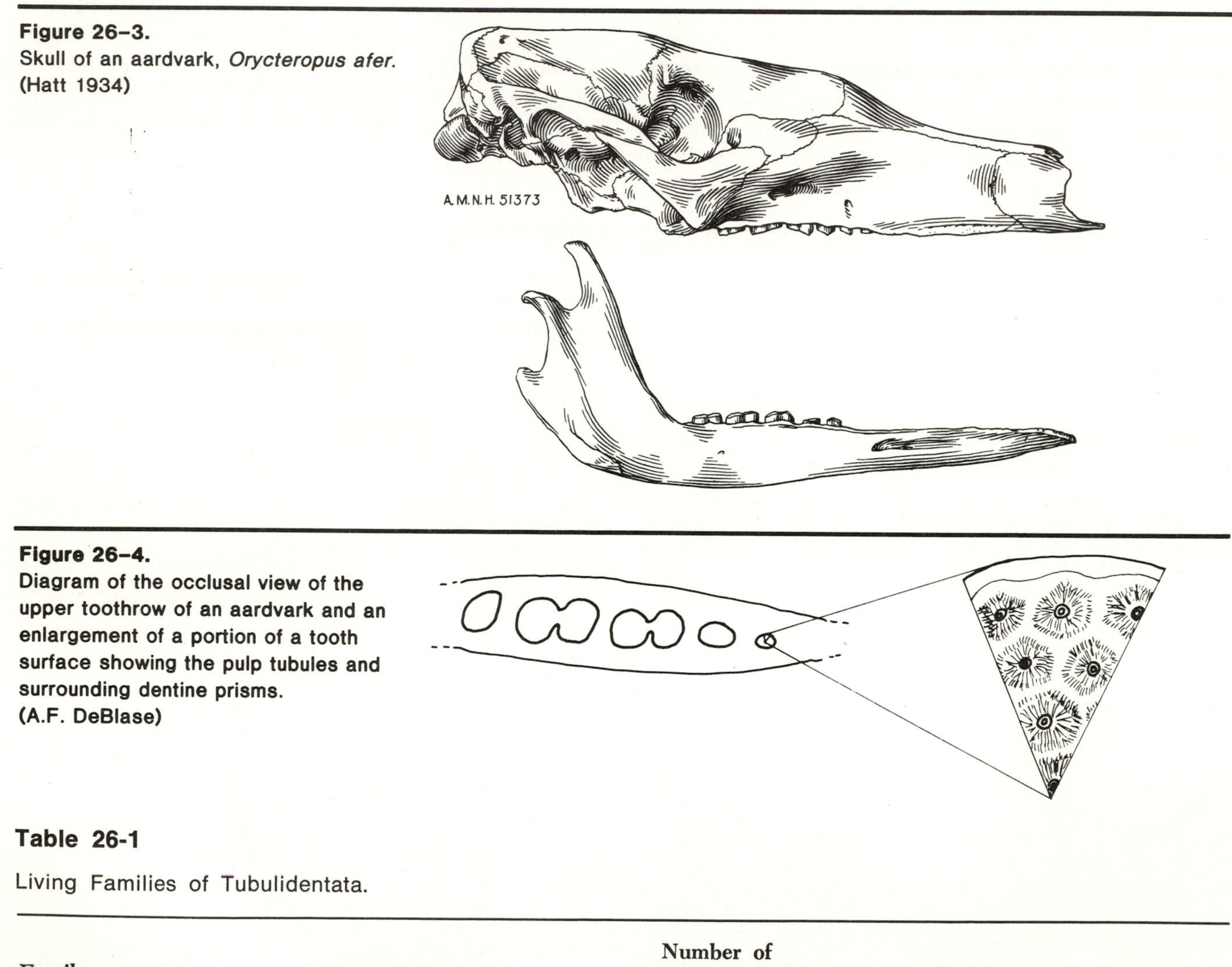

Figure 26–3.
Skull of an aardvark, *Orycteropus afer*.
(Hatt 1934)

Figure 26–4.
Diagram of the occlusal view of the upper toothrow of an aardvark and an enlargement of a portion of a tooth surface showing the pulp tubules and surrounding dentine prisms.
(A.F. DeBlase)

Table 26-1

Living Families of Tubulidentata.

Family	Common Name	Number of Genera	Number of Species	Distribution
Orycteropodidae	aardvark	1	1	Ethiopian

Fossil History

Fossil aardvarks are known from the Miocene in Africa (†*Myorycteropus*) and the Pliocene of the southern Palearctic (*Orycteropus*). Forms known only by fragmentary fossils from the Oligocene in Europe and from the lower Eocene in North America have been tentatively included in this order (Melton 1976:76).

Comments and Suggestions on Identification

Externally aardvarks are unique and easily identified. The skull is also unique and on the basis of size and shape alone could be confused only with that of a giant armadillo. The armadillo, however, has numerous tiny teeth that are very different in structure from those of an aardvark.

Taxonomic Remarks

The aardvark was long considered to be a member of the order Edentata but this association was based entirely upon convergent features related to an insectivorous diet. Tubulidentata is now generally recognized as a distinct order closely related to the †Condylarthra.

ORDER PROBOSCIDEA

The ordinal name, Proboscidea, refers to the elephant's most conspicuous structure—its long prehensile proboscis or trunk. This trunk and other unique external features of elephants are distinctive and well known (Fig. 26-5).

The two extant species of elephants are the largest living land animals. The African elephant, *Loxodonta africana,* is the larger of the two, with large males measuring up to 4 meters at the shoulder and weighing over 7000 kg (Grzimek 1975:500). The Asiatic elephant, *Elephas maximus,* is somewhat smaller, rarely reaching 3 meters at the shoulder and weighing up to 5000 kg (Altevogt and Kurt 1975:486). The African elephant has much larger ears, a flatter forehead, and a somewhat concave profile of the back (Fig. 26-5), while the Asiatic elephant has smaller ears, a more domed forehead, and a convex dorsal profile.

Elephants are grazing mammals that usually live in herds. The Asiatic elephant has been domesticated for centuries and is used as a beast of burden throughout its range in Asia. The African elephant has been domesticated less frequently but was used by Hannibal in early historic times and is presently used in some parts of Africa (Grzimek 1975:512). Both species have been extensively hunted for their ivory tusks and as trophies and are endangered in parts of their range because of hunting or habitat destruction due to agriculture or lumbering.

Distinguishing Characteristics

The incisors, numbering 1/0, are long, evergrowing tusks of solid dentine. Canines are absent. The cheek teeth, consisting of the second, third, and fourth deciduous premolars and the first, second, and third molars, are hypsodont and lophodont (Fig. 26-6). The teeth are replaced from the back of the jaw and worn teeth are shed from the front of each tooth row (Fig.

Figure 26–5.
An African elephant, *Loxodonta africana.*
(Gervias 1855: 133)

Figure 26–6.
Cheek teeth of the Asiatic elephant, *Elephas maximus* (A), and African elephant, *Loxodonta africana* (B). (Flower and Lydekker 1891: 424, 425)

Table 26-2

Living Families of Proboscidea.

Family	Common Name	Number of Genera	Number of Species	Distribution
Elephantidae	elephants	2	2	Ethiopian and Oriental

3-2). Only one or parts of two cheek teeth in a jaw quadrant are functional at any one time (Altevogt 1975:479).

The limbs are graviportal with five digits on each foot. Each digit terminates in a hooflike structure. The upper lip and nose are fused and elongated to form a long, prehensile proboscis with the nostrils at the distal end. The skin is thick and covered with sparse, bristlelike hairs.

The caecum is large. Testes are permanently abdominal and a baculum is absent.

Distribution

The African elephant ranges throughout sub-Saharan Africa and in Roman times also occurred in the Atlas mountains of north Africa (Grzimek 1975:501). The Asiatic elephant ranges through most of the continental portion of the Oriental Region (except central India) and is found on the islands of Ceylon, Sumatra, and Borneo (Altevogt and Kurt 1975:491). Both species have now been eliminated from large areas of this range because of human competition for the land and due to hunting.

Fossil History*

The two living species are the remnants of a once large and widely distributed order. Proboscideans apparently arose in Africa in the Eocene. By the late Cenozoic they had spread through Eurasia and North America and had even reached South America. Several "aberrant" forms developed in various parts of the world. †*Dinotherium,* a proboscidean with no upper incisors but two long, downward curving, lower incisors, was common in Eurasia during the Miocene and Pliocene and survived in Africa into the Pleistocene. †*Amebelodon,* from the Pliocene of North America, is one of the several shovel-tusked proboscideans. These animals had long, dorsoventrally flattened lower jaws and incisors, which together formed a structure resembling a scoop shovel. The two main fossil groups through most of the Holarctic are the mastodons (†*Mastodon*) and mammoths (†*Mammonteus*). The mastodons were browsing mammals with large cuspidate teeth. The mammoths were grazing mammals with lophodont teeth. Both groups existed in the Holarctic well into the Pleistocene and were hunted by primitive man. In addition to fossil bones and teeth, complete cadavers of mammoths have been found in the frozen ground of Siberia and fragments have been found in Alaska.

*Based upon Romer (1966) and Thenius (1975c).

Comments and Suggestions on Identification

Both externally and cranially proboscideans are distinctive and, once seen, are impossible to confuse with any other mammal.

ORDER HYRACOIDEA

The hyraxes are a group of mammals unknown to most peoples in the areas of the world where they do not exist. Hyraxes (also known as "dassies" and referred to in the Scriptures as "conies") are rabbit-sized animals that look rather like rodents (Fig. 26-7). These herbivorous mammals have a unique foot structure (see below) that provides a firm grip on the rocks and trees in which they live. The terrestrial forms, in the genera *Heterohyrax* and *Procavia,* live in colonies of six to fifty individuals in areas of jumbled boulders and rock outcrops. The arboreal species, in the genus *Dendrohyrax,* have none of the limb modifications usually associated with arboreal mammals (*e.g.,* opposable digits, sharp claws), but have great agility and remarkably adhesive pads on the feet.

Figure 26–7.
The rock hyrax, *Procavia capensis.*
(Flower and Lydekker 1891: 415)

Distinguishing Characteristics

The dental formula is 1/2 0/0 4/4 3/3 = 34. The long, evergrowing, upper incisors are triangular in cross section and have pointed tips (Fig. 26-8), while the lower incisors are chisel-shaped and usually tricuspid. The cheek teeth are somewhat lopodont and are separated from the incisors by wide diastemas. The well-developed postorbital processes usually form postorbital bars. The interparietal is well developed. The large jugals contribute to the formation of the mandibular fossae.

The limbs are plantigrade. The four manual digits are syndactylous except for their terminal phalanges. The pes has three digits. All digits terminate in short, flat, hooflike nails, except the second pedal digit which has a long, curved, clawlike nail used for grooming (Rahm 1975c:515). The soles have large, soft, elastic pads that are kept moist by numerous glands. The tail is very short.

The testes are abdominal and a baculum has not been reported. The uterus is duplex.

Figure 26–8.
Skull of a tree hyrax, *Dendrohyrax dorsalis.*
(Flower and Lydekker 1891: 416)

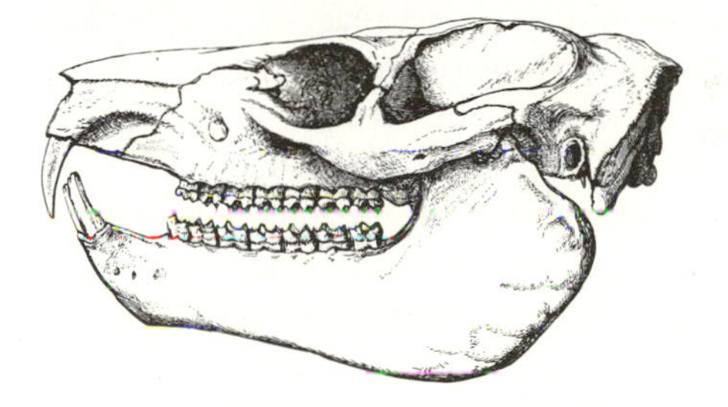

Distribution

Hyraxes range throughout the Ethiopian Region except for Madagascar and the dense forests of the Congo basin. They occur in the Palearctic Region only along the Nile valley and the eastern Mediterranean coast from the Nile to Lebanon.

Table 26-3

Living Families of Hyracoidea.

Family	Common Name	Number of* Genera	Number of* Species	Distribution
Procaviidae	hyraxes	3	11	Ethiopian and south-central Palearctic (see above)

*Based upon Rahm (1975c:514).

Fossil History*

The extinct family, †Geniohyidae, is first known from the Oligocene in Africa. These were large, long snouted animals with limbs resembling those of perissodactyls. Tapir-sized members of this family are known from the Pleistocene of east Asia. Procaviidae is also first known from the Oligocene of Africa. The considerable variety of forms known from lower Oligocene (Fayum) beds of Egypt range in size from that of extant forms to the size of a lion. Except for one form, †*Pliohyrax,* known from the Pliocene in Greece and France, fossil Procaviidae are known only from Africa. The order appears to be African in origin.

*Based upon Romer (1966) and Thenius (1975d:513-514).

Comments and Suggestions on Identification

Hyraxes superficially resemble rodents or large pikas, but are quickly identified by their unique foot structure. The skulls may also, at first glance, look very much like those of rodents. However, the triangular upper incisors and the presence of two lower incisors per side are diagnostic.

ORDER SIRENIA

Sirenians are fully aquatic mammals that lack external hind limbs and have forelimbs modified as flippers (Fig. 26-9). Unlike most whales (the only other fully aquatic mammals), sirenians have a short but flexible neck. (The beluga, Odontoceta: Monodontidae, has a similarly flexible neck.) The mammae are pectoral and the female has been said to float on her back as she clasps her young to her breast to suckle. Presumably the humanoid appearance of these ungainly animals as they floated and nursed their young caused sailors, who must have been long at sea, to give rise to the legends of mermaids. The ordinal name refers to the sea nymph, Siren, who in Greek mythology lured mariners to destruction. Columbus recorded seeing three "sirens" (manatees) in an inlet on the island of Hispaniola and he noted that they were not nearly as beautiful as those described by Horace (Kurt and Wendt 1975:523).

Sirenians feed on aquatic vegetation and in some parts of their range play an important role in keeping navigation channels free of excess vegetation (Kurt and Wendt 1975:526). They are hunted for meat, hides, and oil in various parts of their range.

Living sirenians are tropical or subtropical but one species, the large Steller's sea cow, *Hydrodamalis gigas,* lived in arctic waters. It was discovered on Bering Island and an adjacent island in the Bering Sea in 1742. For twenty-seven years this species served as

Figure 26–9.
The dugong, *Dugong dugong* (left) and the northern manatee, *Trichechus manatus* (right). (Sclater and Sclater 1899: 202-203)

a source of food, oil and boat construction materials to mariners sailing in the area, but by 1768 the last of the population of 1500 to 2000 was exterminated (Kurt and Wendt 1975:532-533). Steller's sea cow is a member of the family Dugongidae. It differs from living representatives of this family in completely lacking teeth and in its much larger size (up to 8 m long and 4000 kg in weight) (Kurt and Wendt 1975:525). This extinct form is not included in the description or key below.

Distinguishing Characteristics

The external nares are situated high on the skull posterior to the anterior margins of the orbits. The nasal bones are rudimentary or absent. Incisors are absent except in the dugong which has I 1/0. Canines are absent. Cheek teeth in the Trichechidae are lost from the front and replaced from the rear as in Proboscidea; in the dugong cheek teeth are quickly worn away and replaced by horny plates in adults (Kurt and Wendt 1975:524).

The vestigial pelvic limbs are absent externally. The pectoral limbs are paddlelike with their five digits indistinguishable externally. The tail is a horizontally flattened fluke that may (Dugongidae) or may not (Trichechidae) be cleft (Fig. 26-9). The dugong has the normal mammalian component of seven cervical vertebrae but the species of Trichechidae have only six. (Variation from the basic number of seven cervical vertebrae occurs among mammals only here and in the Bradypodidae.)

The nostrils are on the upper surface of the snout. The neck is short but somewhat flexible and the body is generally fusiform. Eyes are small and pinnae are absent. The lips are large and highly mobile. Numerous, stiff, vibrissaelike hairs are present on the upper lip but the body is otherwise nearly naked.

The uterus is bicornuate. Testes are permanently abdominal and a baculum has not been reported.

Distribution

In the Western Hemisphere manatees (Trichechidae) range along the coast from North Carolina to southern Brazil and throughout the West Indies. They are found in rivers in Florida and through the Amazon and Orinoco drainages of South America. In the Eastern Hemisphere they range along the western coast of Africa from 10°N to 10°S, in Lake Tchad, and throughout the drainages of the Congo, Niger, and several other west and central African rivers (Jones and Johnson 1967:370, Kurt and Wendt 1975:524). The dugong, (Dugongidae) occurs in the Red Sea and ranges through coastal waters of the Indian Ocean from Mozambique and Madagascar to northern Australia. It inhabits all of the Indonesian region and extends east through the Solomons, north through the Philippines and along the Chinese coast almost to Japan (Jones and Johnson 1967:370).

The ranges given above have been greatly reduced and the various sirenian species now are found only in scattered portions of their former ranges.

Fossil History**

The oldest known Sirenia are Dugongidae from the middle Eocene of Egypt and Jamaica. They were already aquatic animals but their pelvic girdles were fully developed and their hind limbs were rudimentary but still evident externally. In the upper Tertiary, Dugongidae ranged from the Mediterranean through the Atlantic and Caribbean to the eastern Pacific. The oldest known Trichechidae is from the Miocene in South America. Manatees are known from the Miocene and Recent in South America, the Pleistocene and Recent in North America, and only the Recent in West Africa. Trichechidae may have arisen from early Dugongidae.

**Based upon Thenius (1975e:525).

Table 26-4

Living Families of Sirenia.

Family	Common Name	Number of* Genera	Species	Distribution
Trichechidae	manatees	1	3	See "Distribution" Section
Dugongidae	dugong or sea cow	1	1	See "Distribution" Section

*Based upon Kurt and Wendt (1975).

Key to Living Families

1 Incisors absent; cheek teeth with two cuspidate, transverse crests (Fig. 26-10); small nasal bones present; tail spatulate (Fig. 26-9 right); upper lip deeply cleft (bilobed) (Fig. 26-11) .... **Trichechidae**
manatees

1′ Upper incisors present (Fig. 26-12); cheek teeth simple or absent; nasal bones absent; tail cleft (Fig. 26-9 left); upper lip only partly cleft .... **Dugongidae**
dugong

Figure 26–10.
Skull of the West African manatee, *Trichechus senegalensis.*
(Flower and Lydekker 1891: 217)

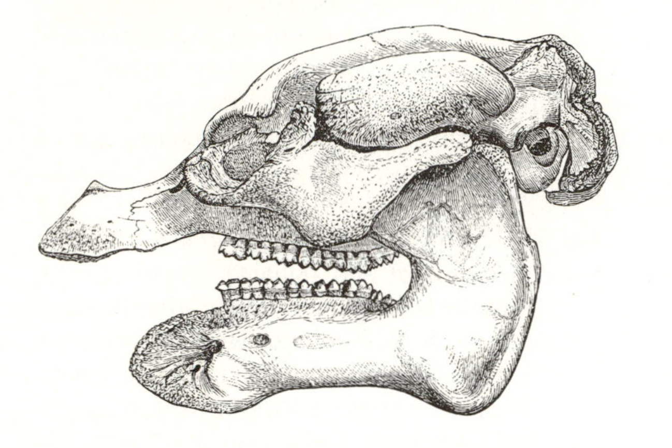

Figure 26–11.
Anterior view of a manatee showing bilobed upper lip.
(Flower and Lydekker 1891: 217)

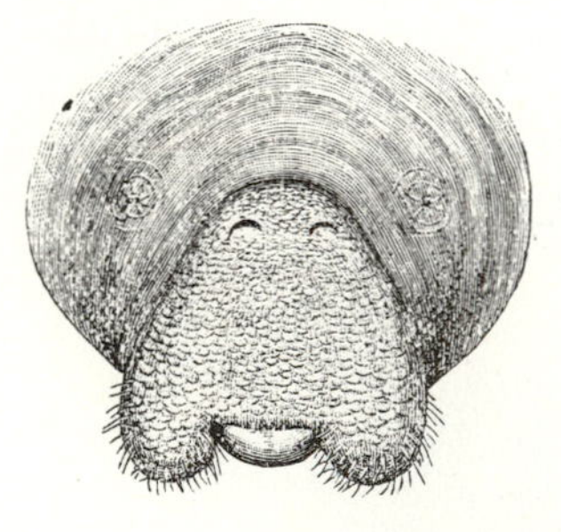

Comments and Suggestions on Identification

Both the external appearance and the skulls of sirenians are distinctive and not easily confused with those of any other mammals. Dental characters and the shapes of the skulls and tails serve to differentiate the two families.

Figure 26–12.
Skull of a dugong, *Dugong dugong.*
(Giebel 1859: 491)

Taxonomic Remarks

The recently extinct Steller's sea cow, *Hydrodamalis* (=*Rhytina*) *gigas,* is sometimes considered to constitute a monotypic family, Hydrodamalidae.

Supplementary Readings

Altevogt, R., F. Kurt, and B. Grzimek. 1975. Elephants, pp. 478-512, *In* B. Grzimek (Ed.), *Animal life encyclopedia. Vol. 12, Mammals III.* Van Nostrand Reinhold, New York.

Bertram, G. C. L. and C. K. Ricardo Bertram. 1973. The modern Sirenia: Their distribution and status. *Biol. J. Linn. Soc.* 5:297-338.

Brandt, J. F. 1849. *Contributions to sirenology, being principally an illustrated natural history of Rhytina.* English translation 1974. Nolit Publishing, Belgrade.

Kurt, F. and H. Wendt. 1975. Sirenians, pp. 523-533, *In* B. Grzimek (Ed.), *Animal life encyclopedia. Vol. 12, Mammals III.* Van Nostrand Reinhold, New York.

Laws, R. M., I. S. C. Parker, and R. C. B. Johnstone. 1975. *Elephants and their habits.* Clarendon Press, Oxford. 376 pp.

Maglio, N. J. 1973. Origin and evolution of the Elephantidae. *Trans. Amer. Phil. Soc.* 63:1-149.

Melton, D. A. 1976. The biology of aardvark (Tubulidentata-Orycteropodidae). *Mammal Rev.* 6:75-88.

Rahm, U. 1975. Aardvarks, pp. 473-477, *In* B. Grzimek (Ed.), *Animal life encyclopedia, Vol. 12, Mammals III.* Van Nostrand Reinhold, New York.

Rahm, U. 1975. Hyraxes, pp. 513-522, *In* B. Grzimek (Ed.), *Animal life encyclopedia, Vol. 12, Mammals III.* Van Nostrand Reinhold, New York.

Dhrlftivk, F. 1973. *The Tsavo story.* Collins and Harvill Press, London. 288 pp.

Sikes, S. K. 1971. *The natural history of the African elephant.* American Elsevier Publishing Co., New York, 397 pp.

27 The Perissodactyls

Order Perissodactyla

ORDER PERISSODACTYLA

The ordinal name, Perissodactyla, which means literally "odd-fingered ones," points out the major distinctive feature of the order and the one feature that unites the three very different extant families. The horses (including zebras and asses, Fig. 27-1), rhinoceroses, and tapirs all have a mesaxonic limb structure in which a large central digit carries the bulk of the weight of the animal, while smaller lateral digits may or may not be present. Hoofs are present on all exposed digits.

All living perissodactyls are herbivorous. Horses and some rhinos are primarily grazing animals inhabiting plains and savannas, while tapirs and other rhinos are generally browsing, forest-dwelling mammals.

All extant odd-toed ungulates have been used as a source of hides and meat, but only a few species are of major economic importance. The unique horns of rhinos (see chapter 5) are considered in the Orient to be both a powerful aphrodisiac and a neutralizer of poisons. Rhinos in all parts of their range are slaughtered by poachers to obtain these valuable horns, and today most of the five living species are considered endangered primarily because of this poaching. The domestic horse, *Equus caballus,* and the domestic donkey, *Equus asinus,* are the major beasts-of-burden over most of the world. The third important equine beast-of-burden is a sterile hybrid produced by crossing a

Figure 27–1.
The African wild ass, *Equus asinus.*
(Kingsley 1884: 256)

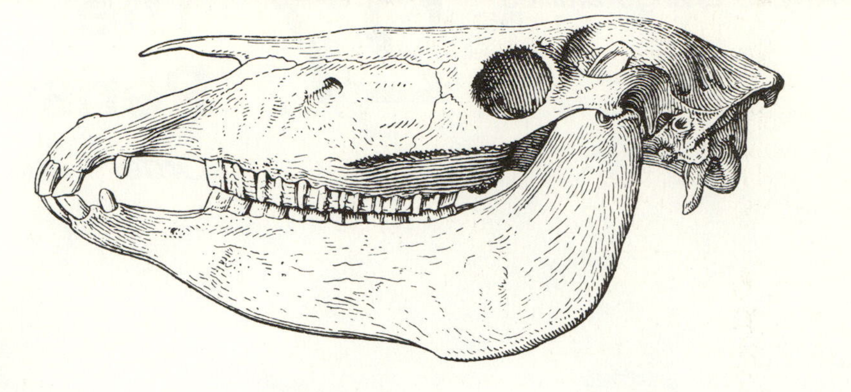

Figure 27–2.
Skull of a male Przewalski's horse, *Equus przewalskii.* The canine teeth are usually lacking in females and in males gelded before the adult dentition develops. (Sokolov 1959)

horse and donkey. The result of a cross between a male donkey and female horse is termed a mule, while the result of a cross between a male horse and female donkey is termed a hinny.

Distinguishing Characteristics

The limbs are unguligrade and mesaxonic. A large central digit carries the main axis of weight and smaller lateral digits may be present. Manual digits number one (Equidae), three (Rhinocerotidae), or four (Tapiridae). Pedal digits number one (Equidae) or three (Rhinocerotidae and Tapiridae). The skull is usually elongated. Canine teeth, if present, are small and the molars, and usually most premolars as well, have a more or less complex pattern of lophs and ridges.

Equids have a dental formula of 3/3 0-1/0-1 3-4/3 3/3 = 36-42. The cheek teeth are hypsodont and have a complex grinding surface of folded enamel ridges. Canines, when present, are small and located in the wide diastema between the incisors and premolars (Fig. 27-2). The orbit is completely separated from the temporal fossa by a postorbital plate. The body is fully haired.

Rhinos have a dental formula of 0-2/0-1 0/0-1 3-4/3-4 3/3 = 24-34. The cheek teeth are basically lophodont. The orbit and temporal fossa are confluent (Fig. 27-3). One or two unique "horns," composed of agglutinated "hairs," are present on the rostrum (Fig. 27-4) (see chapter 5 for a detailed description of rhino "horn"). The body is covered with thick skin that is sparsely-haired in most species; but the Sumatran rhinoceros, *Dicerorhinus sumatrensis,* is relatively well haired. The Palearctic Pleistocene wooly rhinoceros, †*Coelodonta antiquitatis,* had a thick coat of long hair (Thenius 1972:35).

Tapirs have a dental formula of 3/3 1/1 4/4 3/3 = 44. A diastema is present between the canines and premolars. The cheek teeth have transverse ridges. The orbit and temporal fossa are confluent (Fig. 27-5). The upper lip and nostrils are elongated to form a short proboscis. In three of the four living species the body is covered with a short, sleek coat of hair, but on the mountain tapir, *Tapirus pinchaque,* the coat is soft and wooly (Fradrich 1972:19).

All three families have a simple stomach and large caecum. The uterus is bicornuate. Testes are scrotal in Equidae and abdominal in Rhinocerotidae and Tapiridae. A baculum is never present.

Figure 27–3.
Skull of a Javan rhinoceros, *Rhinoceros sondaicus.* (Gray 1869: 306)

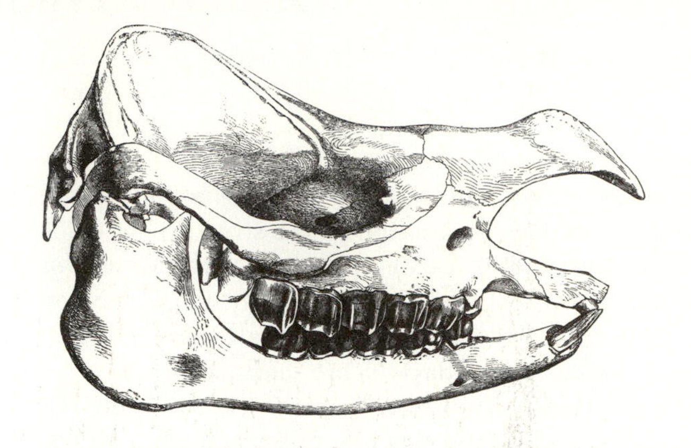

Figure 27–4.
A black rhinoceros, *Diceros bicornis.* (Dekeyser 1955: 315)

Figure 27-5.
A South American tapir, *Tapirus terrestris*, and a skull of the same species.
(Tapir, Beddard 1902: 251; skull, Gray 1869: 257)

Distribution

Indigenous equids are found today only in southern and eastern Africa and in the arid regions of southwestern and central Asia (Volf 1972). However, the domesticated forms have been introduced to all parts of the world and feral populations exist in many areas. Rhinocerotids presently are found in central and eastern sub-Saharan Africa and in the Oriental Region from eastern India to Borneo. Through much of this range they are now extinct or very rare and endangered (Lang 1972). Tapirids presently are found in the tropical portions of the mainland Neotropical, on the Malayan Peninsula, and on Sumatra (Frädrich 1972).

Fossil History

The six genera and fifteen species of living Perissodactyla are but a small remnant of the approximately 150 genera known as fossils. The oldest fossils, including two of the three living families, are known from the lower Eocene. During the Eocene and Oligocene, perissodactyls were numerous and fourteen families are recognized. By the end of the Oligocene, however, most of these families had become extinct. Only four survived, the three living families and the †Chalicotheriidae. The chalicotheres, which became extinct in the Pleistocene, were unique animals. They resembled horses in size and appearance but had rather long forelegs and necks. Their most striking characteristic, however, was the presence of large, stout claws on the ends of the digits. The function of these claws on an animal, which in all other respects is a good ungulate, remains a mystery. Some authors suggest that they were used to pull down branches to browse on leaves, while others believe that they were used to dig for roots and tubers.

All three living families of perissodactyls were present in the Holarctic throughout much of the Cenozoic. However, with the exception of tapirs in the Neotropical, all families had disappeared from the Western Hemisphere by the early Holocene. The largest land mammal ever known to have lived was a giant, hornless rhinoceros, †*Baluchithrium*, from the Oligocene and Miocene of Asia. Its skull was over 1.2 meters in length and its height at the shoulder was approximately 5.5 meters. While rhinos are essentially restricted to tropical regions today, the extinct woolly rhinoceroses ranged along the Pleistocene ice fronts of Eurasia where they were hunted by Stone Age man.

Table 27-1

Living Families of Perissodactyla.

Family	Common Name	Number of* Genera	Number of* Species	Distribution
Equidae	horses, asses, zebras	1	6	Ethiopian, Palearctic
Tapiridae	tapirs	1	4	Neotropical, Oriental
Rhinocerotidae	rhinoceroses	4	5	Ethiopian, Oriental

*Classification based upon Volf (1972:539-579), Equidae; Frädrich (1972:17-33), Tapiridae; and Lang (1972:36-70), Rhinocerotidae.

The evolution of the equids is better documented by fossils than that of any other group. Primitive four-toed, browsing animals are known from the Eocene and species representing an almost continuous development of limb (Fig. 27-6) and tooth modifications are known from the later Cenozoic epochs.

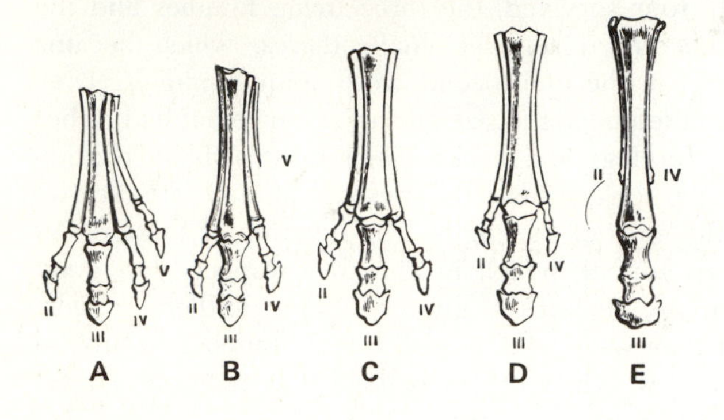

Figure 27–6.
Representative stages in the development of the equid limb showing an increase in the size of digit III and a reduction and elimination of the other digits. A, †*Pachynolophus* (Eocene); B, †*Anchitherium* (Early Miocene); C, †*Anchitherium* (Late Miocene); D, †*Hipparion* (Pliocene); E, *Equus* (Pleistocene).
(Flower and Lydekker 1891: 377)

Key to Living Families

1 Orbit and temporal fossa separated by a post-orbital plate (Fig. 27-2); dental formula 3/3 0-1/0-1 3-4/3 3/3 = 36-42; digits 1/1 (Fig. 27-1) .... **Equidae**
horses, zebras, and asses

1′ Orbit and temporal fossa confluent; teeth number fewer than 36 or more than 42; digits 3-4/3 2

2 (1′) Dental formula 3/3 1/1 4/4 3/3 = 42; upper canines well developed (Fig. 27-5); short muscular proboscis present (Fig. 27-5); no horns .... **Tapiridae**
tapirs

2′ Dental formula variable, teeth number 34 or fewer; upper canines absent (Fig. 27-3); proboscis absent; 1 or 2 horns on rostrum (Fig. 27-4) .... **Rhinocerotidae**
rhinoceroses

Comments and Suggestions on Identification

Externally the three families are distinctive and cannot easily be confused with each other or with any other mammals. The general appearance of the skulls of each of the three also is distinctive and easily recognized with a little practice.

Supplementary Readings

Geist, V. and F. Walther (Eds.). 1974. *The behavior of ungulates and its relation to management.* 2 Vols. New Series No. 24, IUCN Publications. Morges, Switzerland. 941 pp.

Goodall, D. M. 1965. *Horses of the world.* Macmillan Co., New York. 272 pp.

Groves, C. P. 1974. *Horses, asses and zebras in the wild.* David & Charles, London. 192 pp.

Heller, E. 1913. The white rhinoceros. *Smithsonian Misc. Coll.* 61(1):1-77.

MacClintock, D. 1976. *A natural history of zebras.* Chas. Scribner's Sons, New York. 134 pp.

Mochi, U., and T. D. Carter. 1971. *Hoofed mammals of the world.* Chas. Scribner's Sons, New York. 268 pp.

Mochi, U., and D. MacClintock. 1976. *A natural history of zebras.* Chas. Scribner's Sons, New York. 134 pp.

*Radinski, L. B. 1966. The adaptive relationships of the phenacodontoid condylarths and the origin of the Perissodactyla. *Evolution* 20:408-417.

Schenzel, R., and L. Schenzel-Hulliger. 1969. *Ecology and behavior of black rhinoceros* (*Diceros bicornis* L.) Verlag Paul Parey, Hamburg. 101 pp.

Simpson, G. G. 1951. *Horses.* Oxford Univ. Press, New York. 247 pp.

*Included in Jones, Anderson and Hoffman (1976).

28 The Artiodactyls
Order Artiodactyla

ORDER ARTIODACTYLA

The ordinal name, Artiodactyla, meaning literally "even-fingered ones," points out the major feature uniting the families of artiodactyls and distinguishing these ungulates from the perissodactyls. The limbs are paraxonic with the plane of symmetry passing between the third and fourth digits of each foot which are about equal in size and equally share the main bulk of the weight of the animal. The second and fifth digits, when present, are distinctly smaller than the third and fourth and frequently do not touch the ground except on a very soft substrate.

Figure 28–1.
The pronghorn, *Antilocapra americana*, Antilocapridae. (Kingsley 1884: 333)

Most living artiodactyls are herbivorous, but some of the more primitive forms (e.g., the hogs) are omnivorous. The types of plant material eaten by artiodactyls varies from lichens on the Arctic tundra to fruits and tubers in tropical forests.

Wild artiodactyls are important as sources of meat and hides in many cultures and as game animals in others. By far the largest number of domestic animals comes from this order. Domesticated artiodacyls provide meat, hides, milk, and beasts-of-burden in nearly all parts of the world. These domestics include the hog, *Sus scrofa;* llama, *Llama g. glama;* alpaca, *Llama glama pacos;* bactrian (two humped) camel, *Camelus bactrianus;* dromedary (one humped camel), *Camelus dromedarius;* reindeer, *Rangifer tarandus;* water buffalo, *Bubalus bubalis;* cattle, *Bos taurus, B. frontalis,* and *B. indicus;* yak, *Bos grunniens;* goat, *Capra hircus;* and sheep, *Ovis aries* (Zeuner 1963).

Distinguishing Characteristics

The third and fourth digits of each foot are subequal in size, with the main axis of weight passing between them. The second and fifth digits are reduced in size or are absent (Fig. 28-2). The first digit is absent. Metapodials of the third and fourth digits may (Tylopoda, Ruminantia) or may not (Suiformes) be fused into a **cannon bone.** All digits terminate in hoofs (modified in Camelidae and Hippopotamidae).

Cheek teeth range from bunodont (Suiformes) to selenodont (Tylopoda and Ruminantia) and from brachydont to hypsodont. The full primitive placental dental formula is present in some forms (most Suidae). The upper incisors are lost (Ruminantia) or reduced in number (Tylopoda) in most species. Upper canines are frequently lost (most Ruminantia) or reduced in

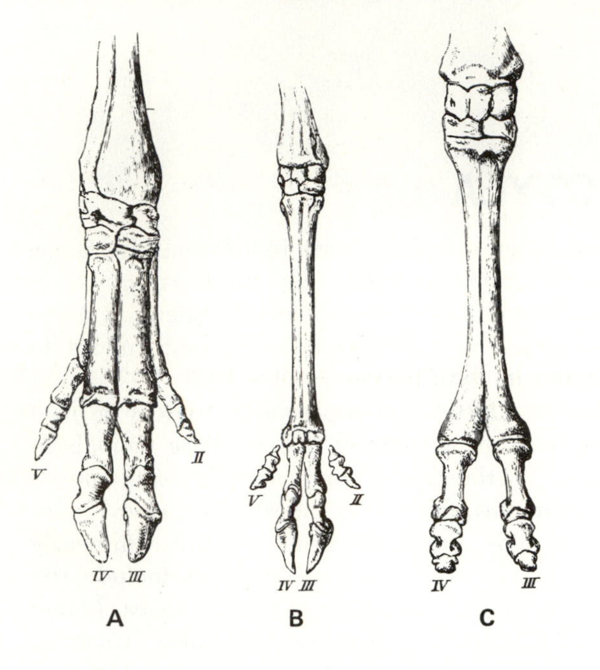

Figure 28–2.
Bones of the right forelegs of three representative artiodactyls: A, hog, *Sus scrofa*; B, red deer, *Cervus elephus*; C, camel, *Camelus bactrianus*. Note that in the Suiformes (A) the metacarpals are each distinct, in the Tylopoda (C) the metacarpals of digits III and IV are fused for most of their length but separated at their distal end, and in the Ruminantia (B) the two metacarpals are fused for their entire length. Fused metacarpals or metatarsals are termed a **cannon bone.**
(Flower and Lydekker 1891: 276)

size (Tylopoda, some Ruminantia). A complete postorbital bar may (Hippopotamidae, Tylopoda, Ruminantia) or may not (Suidae and Tayassuidae) be present.

In many Ruminantia an **antorbital pit** (in which the antorbital gland is located) is conspicuous as a depression on the side of the rostrum just anterior to the orbit (Fig. 28-3). A hole or fenestra is present on the rostrum of many ruminants at the point where the frontal, nasal, lacrimal, and maxillary bones meet (Fig. 28-3). Most male Ruminantia have horns or antlers and these are present in females of some species. The structure of these is different in each of the four families, Cervidae, Giraffidae, Antilocapridae, and Bovidae. See chapter 5 for the anatomy and development of these structures. Males of the Tragulidae (Fig. 28-4) and of the two genera of Cervidae that lack antlers, *Moschus* and *Hydropotes,* are equipped with very long upper canines. Two genera of Cervidae, *Muntiacus* and *Elaphodus,* possess both antlers and long upper canines.

The stomach of nonruminant artiodactyls is a relatively simple, two-chambered structure in Suidae and a three-chambered structure in Tayassuidae and Hip-

Figure 28–4.
The African water chevrotain, *Hyemoschus aquaticus,* Tragulidae.
(Flower and Lydekker 1891: 306)

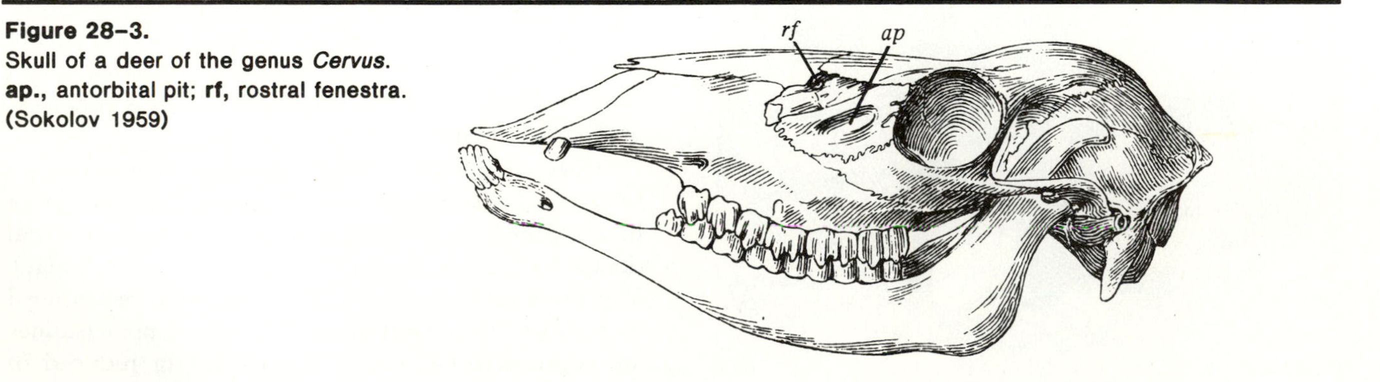

Figure 28–3.
Skull of a deer of the genus *Cervus.*
ap., antorbital pit; **rf,** rostral fenestra.
(Sokolov 1959)

Figure 28–5.
A generalized four-chambered stomach of a ruminant. a, esophagus; b, first chamber or rumen; c, second chamber or reticulum. d, third chamber or omasum; e, fourth chamber or abomasum; f, duodenum.
(Flower and Lydekker 1891: 312)

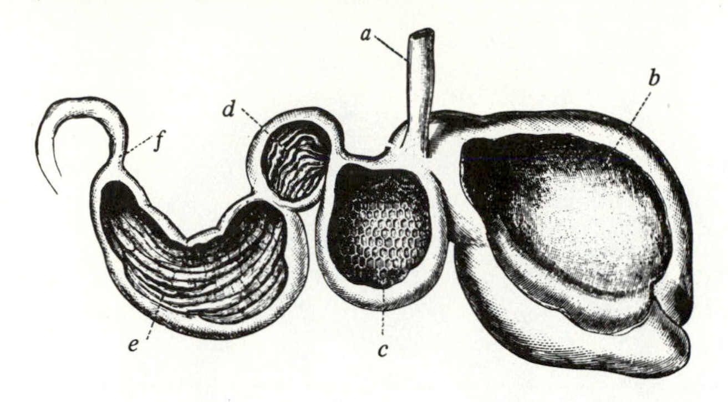

popotamidae. Among those Artiodactyls that ruminate the stomach is a three-chambered structure in Camelidae and Tragulidae* and a complex, four-chambered structure (Fig. 28-5) in the remaining four families. The caecum is absent and the uterus bicornuate. Testes are scrotal in some species and a baculum is never present.

*In Tragulidae the stomach is four chambered but the omasum is rudimentary.

Distribution

Artiodacyls are native to all parts of the world's land surface except Antarctica, oceanic islands, and the Australian Region. (One species of Suidae occurs on Celebes, barely penetrating the Australian Region.) Domestic herds and feral populations of domesticated artiodactyls now occur over much of the Australian Region and on many oceanic islands.

Fossil History***

The first artiodactyls are known from the early Eocene and twenty families are known from that epoch. But this diversity was exceeded by the primitive perissodactyls. By the late Cenozoic, perissodactyls had declined drastically; and, while the number of artiodactyl families decreased, the number of species increased. Artiodactyls then became the dominant large terrestrial herbivores. The Suidae arose during the Oligocene and the Tayassuidae in the late Eocene. Both continue into the Recent. The Hippopotamidae originated in the Pliocene from a large, now extinct family, the †Anthracotheriidae. The Camelidae arose during the late Eocene as one of the eight tylopod families. Six of these were extinct by the mid-Miocene and a seventh by the mid-Pliocene. Only the two geo-

***Based upon Romer 1966.

Table 28-1

Living Families of Artiodactyla.

Family	Common Name	Number of** Genera	Species	Distribution
Suborder Suiformes				
Suidae	hogs	5	8	Ethiopian, Palearctic, Oriental, Celebes in Australian
Tayassuidae	peccaries	2	3	Neotropical, SW Nearctic
Hippopotamidae	hippos	2	2	Ethiopian
Suborder Tylopoda				
Camelidae	camels and llamas and allies	2	4	Western Neotropical, southern Palearctic
Suborder Ruminantia				
Tragulidae	chevrotains	2	4	Portions of Ethiopian and Oriental
Cervidae	deer	16	*ca.* 37	Holarctic, Neotropical, Oriental
Giraffidae	giraffe and okapi	2	2	Ethiopian
Antilocapridae	pronghorn	1	1	Nearctic
Bovidae	cattle, antelope, sheep, goats, and allies	44	*ca.* 111	Holarctic, Ethiopian, Oriental

**Based upon Koopman (1967), and Wetzel (1977).

grahpically widely separated groups of the living family remain. The oldest Tragulidae are known from the early Miocene. They were never very numerous and by the end of the Pliocene were further reduced. The Cervidae and Giraffidae both split from the extinct family †Palaeomerycidae in the early Miocene. The Cervidae have steadily increased in numbers and diversity, while the Giraffidae declined during the Pliocene and Pleistocene. The Antilocapridae originated during the Miocene and became numerous in the Nearctic during the Pliocene and late Miocene. During the Pleistocene they declined to the single extant species. The Bovidae originated in the early Miocene and have increased in number and diversity to the point that they now are the dominant family of Artiodactyla.

Key to Living Families

1 Postorbital bar absent or incomplete; upper incisors present 2

1′ Postorbital bar complete;* upper incisors present or absent 3

2 (1′) Dental formula 2/3 1/1 3/3 3/3 = 38; upper canines straight (Fig. 28-6); three pedal digits .... **Tayassuidae** peccaries

2′ Dental formula 3/3 1/1 4/4 3/3 = 44, 1/3 1/1 3/2 3/3 = 34, or 2/3 1/1 2/2 3/3 = 34; upper canines curve either outward, upward, or both (Figs. 28-7, and 28-8); four pedal digits .... **Suidae** hogs

3 (1′) Lower canine alveoli anterior to alveoli of lower incisors (Fig. 28-9); lower canines larger than upper canines; skull very massive (Fig. 28-9); rostrum broader distally than proximally .... **Hippopotamidae** hippos

3′ Lower canine alveoli posterior to alveoli of lower incisors (Figs. 28-10); incisors rooted; lower canines, if present, smaller or equal to upper canines; skull not particularly massive; snout narrower distally than proximally 4

*Incomplete in some specimens of hippos.

Figure 28–6.
The collared peccary, *Tayassu tajacu*, Tayassuidae, and a skull from the same species.
(Peccary, Flower and Lydekker 1891: 290; skull, Mary Ann Cramer)

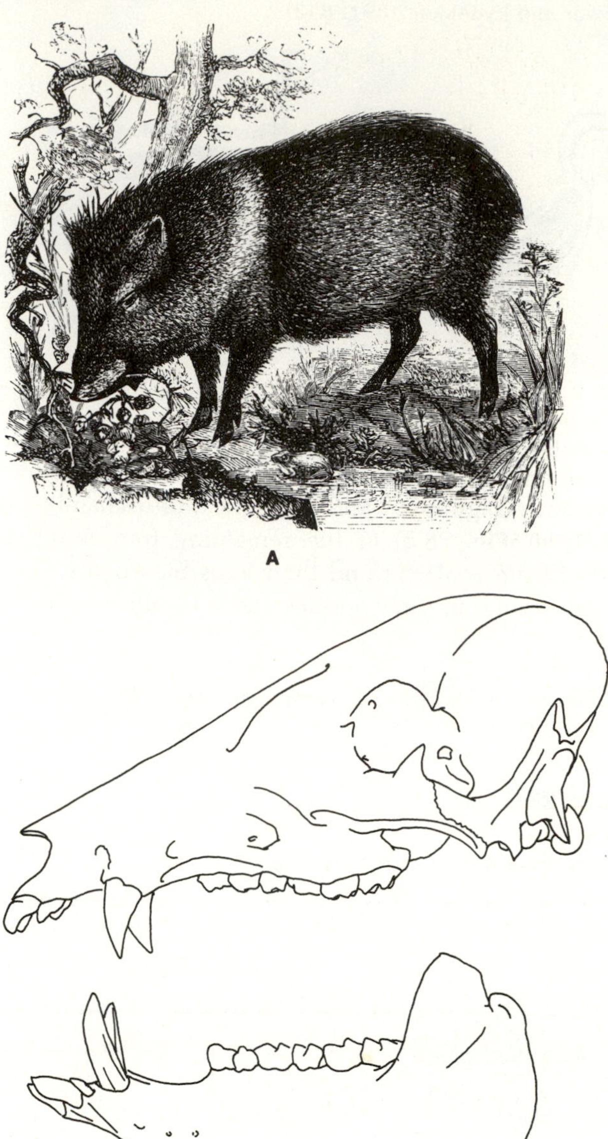

4 (1′) Upper incisors present (Fig. 28-10); hoofs nail-like; large pads present on bottom of feet posterior to hoofs .... **Camelidae** camels and llamas and allies

4′ Upper incisors absent; well-developed hoofs present; pads absent, only hoofs touch ground 5

Figure 28–7.
The wild boar, *Sus scrofa*, and its skull. The domestic hog is the same species but its skull has a concave dorsal profile.
(Boar, Hsia et al. 1964: 91; skull, Gromov et al. 1963)

Figure 28–8.
The babirusa, *Babyrousa babyrussa*, [yes, each of the three names is spelled differently!] a unique hog confined to Celebes. The sockets of the upper canines turn up alongside the rostrum so that the teeth grow dorsally through the skin. The upper and lower canines do not come in contact.
(Flower and Lydekker 1891: 287)

Figure 28–9.
A hippopotamus, *Hippopotamus amphibius*, and a hippo skull.
(Hippo, Sclater and Sclater 1899: 91; skull, Sclater 1900: 266)

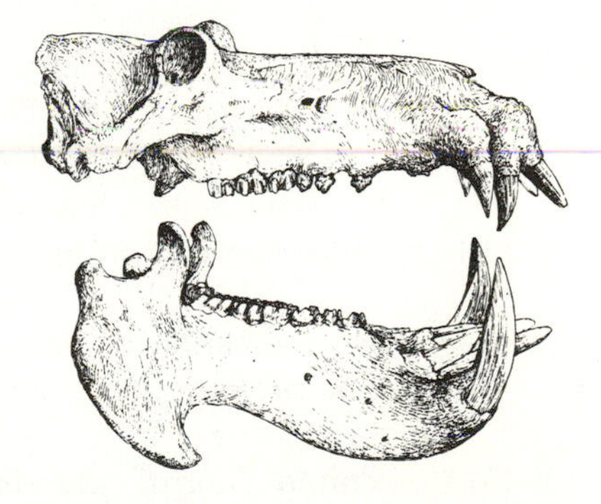

Figure 28–10.
A Bactrian camel, *Camelus bactrianus*, and the skull of a Bactrian camel.
(Camel, Hsia et al. 1964: 91; skull, Sokolov 1959)

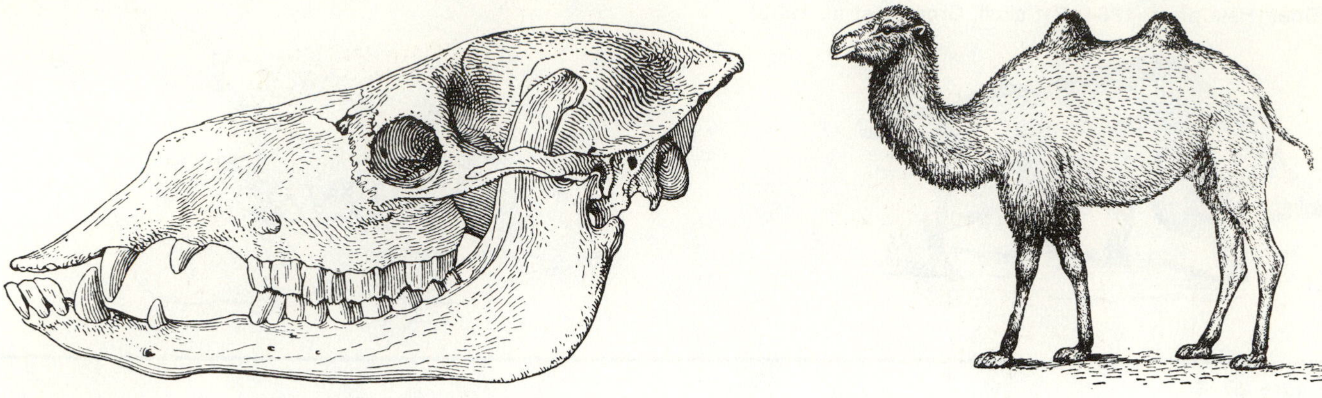

Figure 28–11.
Skull of the giraffe, *Giraffa camelopardalis*. Note that the paired "horns" are distinct bones and are separated from the frontals by sutures.
(Dorsal view, Owen 1866; lateral view, Giebel 1859: 370)

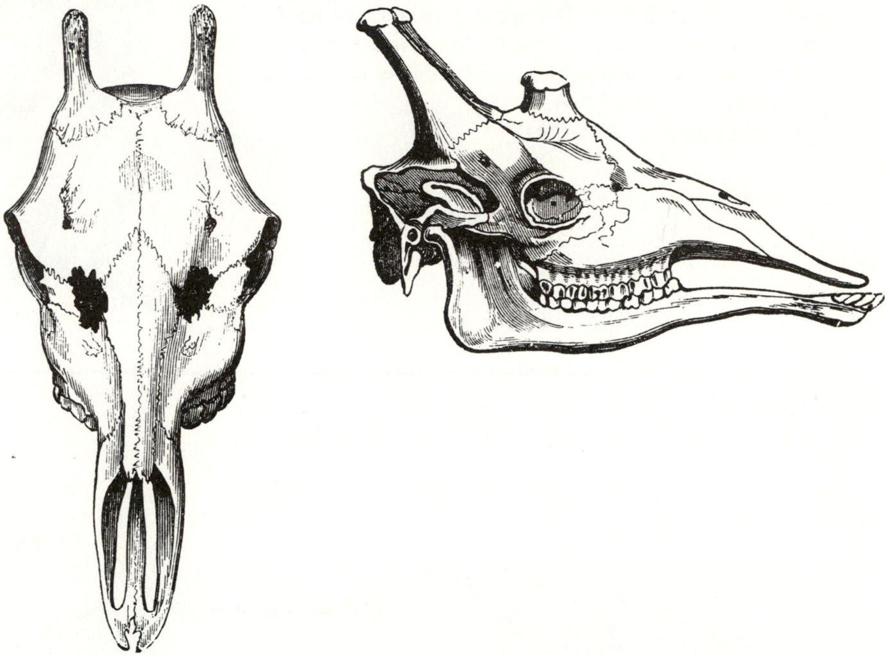

5 (4′) Horn cores or antler pedicels present 6

5′ No indication of horn cores or antler pedicels present 9

6 (5) Paired "horns" are distinct bones situated over suture of frontal and parietal bones (Fig. 28-11); a third "horn" sometimes present on midline of rostrum .... **Giraffidae** giraffe and okapi

6′ Paired horns or antlers are projections of frontal bone; third medial "horn" never present 7

7 (6′) Antlers or antler pedicels present (Fig. 28-12); antorbital pit and rostral fenestration both present (e.g. Fig. 28-3) .... **Cervidae** (in part) deer with antlers

7′ Horns or horn cores present (Fig. 28-13 and 28-14); antorbital pit and rostral fenestration variable, one or both frequently absent 8

Figure 28–12.
Representative cervids. A, Pere David's deer, *Elaphurus davidianus*; B, Sika deer, *Cervus nippon*; and C, tufted deer, *Elaphodus cephalophus*.
(Hsia et al. 1964: 94-96)

8 (7′) Horn core with sharp anterior edge; one or two large foramina present in frontal at anteromedial base of horn cores (Fig. 28-13A); horn forked, having small anterior projection (Fig. 28-1); digits 2/2 .... **Antilocapridae** (in part)
most pronghorns

8′ Horn core with rounded anterior edge, or if anterior edge sharp, no foramina in frontal at base of horn core; horn not forked (Fig. 28-14); digits 2/2 or 4/4 .... **Bovidae** (in part)
horned bovids

Figure 28–13.
The skull of pronghorn, *Antilocapra americana*. A, foramen at base of horn core; B, rostral fenestra; C, outline cross section of horn core.
(G.A. Moore)

Figure 28–14.
Representative bovids, Bovidae. A, the gaur, *Bos gaurus*; B, the takin, *Budorcas taxicolor;* C, the ibex, *Capra ibex*; D, the Palearctic wild sheep, *Ovis ammon*. Not to same scale.
(Hsia et al. 1964: 98–104)

A

B

C

D

Figure 28–15.
Skull of a female gazelle, *Gazella*.
(Sokolov 1959)

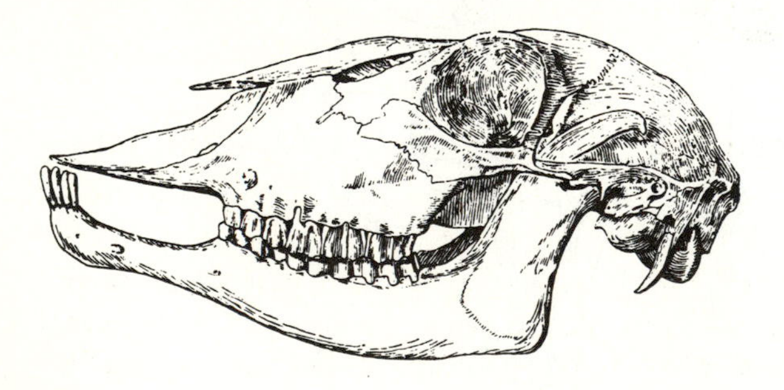

9 (5′) Either antorbital pit or rostral fenestra or both absent; upper canine teeth absent (Fig. 28-15) 10

9′ Both antorbital pit and rostral fenestra present, or if one absent, upper canine teeth present (Figs. 28-16, and 28-17) 12

10 (9) Antorbital pit absent; rostral fenestra narrow dorsoventrally and elongate anteroposteriorly (Fig. 28-13B) .... **Antilocapridae** (in part) hornless female pronghorns

10′ Antorbital pit present or absent; rostral fenestra, if present, not as above 11

Figure 28–16.
The musk deer, *Moschus moschiferus*, Cervidae, one of the small deer that lacks antlers and possesses long upper canines.
(Gromov et al. 1963)

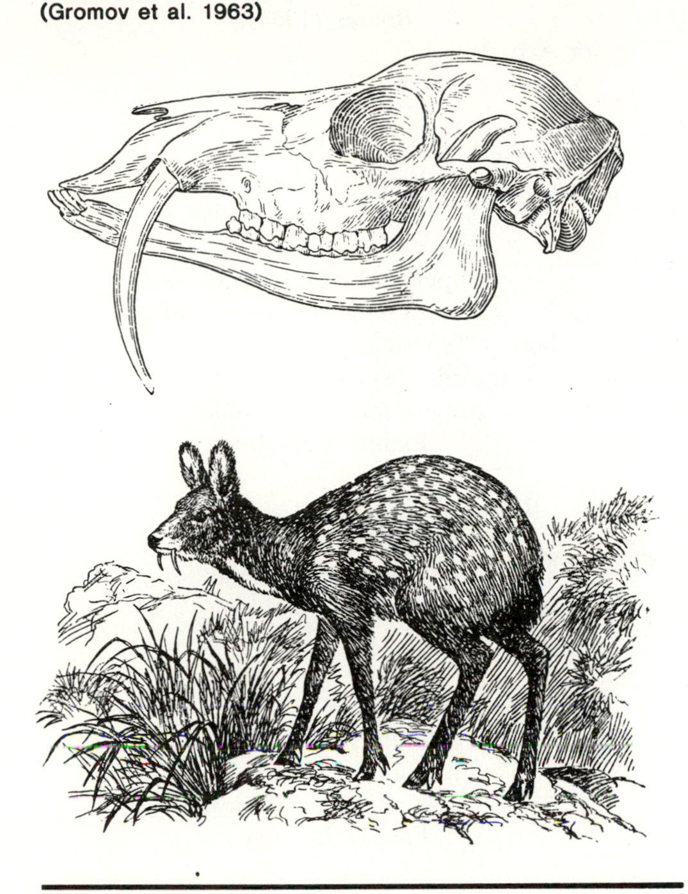

Figure 28–17.
Skull of a female Sika deer, *Cervus nippon*.
(Gromov et al. 1963)

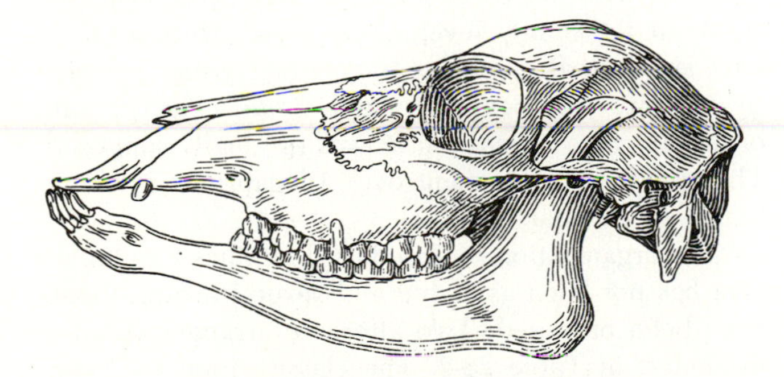

11 (10′) Antorbital pit absent; rostral fenestra very small, less than 4 mm in length; greatest length of skull 100 mm or less .... **Tragulidae** (in part)
some female chevrotains

11′ Antorbital pit present or absent, if absent, greatest length of skull more than 100 mm; rostral fenestra usually more than 4 mm in length .... **Bovidae** (in part)
most hornless female bovids

12 (9′) Antorbital pit absent; upper canines present; condylobasal length less than 100 mm .... **Tragulidae** (in part)
most chevrotains

12′ Antorbital pit present; upper canines present or absent; condylobasal length usually greater than 100 mm 13

13 (12) Orifice of lacrimal canal double (Fig. 28-18B); upper canines present or absent .... **Cervidae** (in part)
most female deer

13′ Orifice of lacrimal canal single (Fig. 28-18A); upper canines absent .... **Bovidae** (in part)
some female antelope

Figure 28–18.
Orbital regions of the skulls of a bovid (A) and a cervid (B). Cervids have a double lacrimal orifice in the orbit while most bovids have a single orifice.
(Gromov et al. 1963: 982, 983)

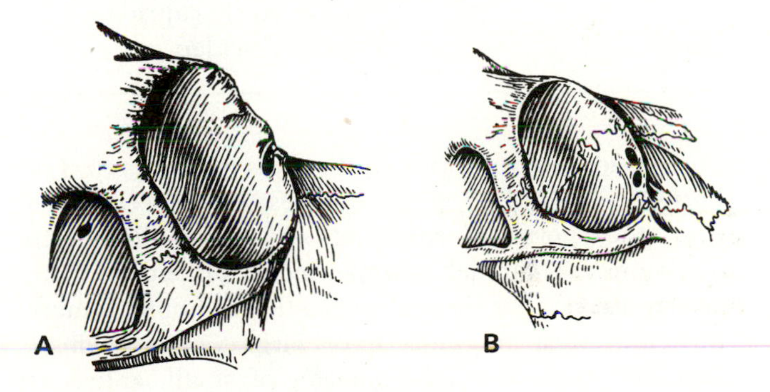

Comments and Suggestions on Identification

The structure of the limbs readily distingushes artiodactyls from all other orders. Cranially, only skulls of perissodactyls could be confused with those of artiodactyls. The massive skull size and enormous lower canines serve to identify hippo skulls. Suid skulls may be distinguished from those of tayassuids by the curvature of the upper canines of the former. The rudimentary upper incisors of camelids distinguish them from other families and the unique "horns" of the giraffe and okapi similarly identify this family.

Table 28-2

Classifications of Artiodactyla.

Simpson (1945)	*Romer (1966)*
Order Artiodactyla	Order Artiodactyla
Suborder Suiformes	Suborder Suina
Infraorder Suina	
Superfamily Suoidea	Superfamily Suoidea
Family Suidae	Family Suidae
Family Tayassuidae	Family Tayassuidae
Infraorder Ancodonta	
Superfamily Anthracotherioidea	Superfamily Hippopotamoidea
Family Hippopotamidae	Family Hippopotamidae
Suborder Tylopoda	Suborder Ruminata
	Infraorder Tylopoda
	Superfamily Cameloidea
Family Camelidae	Family Camelidae
Suborder Ruminantia	
Infraorder Tragulina	Infraorder Pecora
Superfamily Traguloidea	Superfamily Traguloidea
Family Tragulidae	Family Tragulidae
Infraorder Pecora	
Superfamily Cervoidea	Superfamily Cervoidea
Family Cervidae	Family Cervidae
Superfamily Giraffoidea	
Family Giraffidae	Family Giraffidae
Superfamily Bovoidea	Superfamily Bovoidea
Family Antilocapridae	Family Antilocapridae
Family Bovidae	Family Bovidae

Cervids, antilocaprids, and bovids having antlers or horns are identified easily by these structures. However, the "hornless" ruminants are more difficult to distinguish. Their lack of upper incisors identifies them as artiodactyls but their identification to family is more difficult. Most tragulids have large upper canines; however, there are some species of small, antlerless deer that also have long upper canines. The Tragulidae have a tiny rostral fenestration and no antorbital pit. The long-canined, antlerless deer also have a small rostral fenestration but do have an antorbital pit. Unfortunately this pit is very shallow and frequently difficult to distinguish. Among the larger "hornless" ruminants, the cervids sometimes possess small upper cannines, while the antilocaprids and bovids never do. The cervids always have a rostral fenestra and an antorbital pit. The antorbital pit is always lacking in the Antilocapridae. However, the Bovidae are variable and may have both the fenestra and the pit, one of these structures, or neither of them.

Taxonomic Remarks

The classification of artiodactyls has been relatively stable at the family level. O'Gara and Matson (1975) have shown that several of the characteristics believed to separate Antilocapridae and Bovidae are invalid or at least not as clear cut as was originally supposed. They proposed placement of Antilocaprinae as a subfamily of Bovidae.

The organization of artiodactyl families into higher taxa has not been as stable and several arrangements have been proposed. Two alternate arrangements are presented in Table 28-2. The classification by Simpson (1945), as adapted by Koopman (1967), was used in Table 28-1. The differences in the use of the terms Ruminantia (or Ruminata) and Pecora are the most important differences between these two systems.

Supplementary Readings

Clark, R. L. 1964. *The great arc of the wild sheep.* Univ. of Oklahoma Press, Norman. 247 pp.

Dagg, A. I., and J. B. Foster. 1976. *The giraffe, its biology, behavior and ecology.* Van Nostrand Reinhold, New York. 210 pp.

Geist, V., and F. Walther (Eds.). 1974. *The behavior of ungulates and its relation to management,* Vol. I & II. IUCN Publications, New Series No. 24. Morges, Switzerland. 941 pp.

Hoffmann, R. R. 1973. *The ruminant stomach.* East African Monographs in Biology, Vol. II. East African Literature Bureau. Nairobi. 354 pp.

Kitchen, D. W. 1974. Social behavior and ecology of the pronghorn. *Wildlife Monographs* 38:1-96.

Leuthold, W. 1977. *African ungulates, zoophysiology and ecology,* 8. Springer Verlag, Berlin. 307 pp.

Mochi, U., and Carter, T. D. 1971. *Hoofed mammals of the world.* Chas. Scribner's Sons, New York. 268 pp.

McHugh, T. 1972. *The time of the buffalo.* Alfred A. Knopf, New York. 339 pp.

Prior, R. 1968. *The roe deer of Cranborne Chase, an ecological survey.* Oxford Univ. Press. London. 222 pp.

Schaller, G. B. 1977. *Mountain monarchs, wild sheep and goats of the Himalaya.* Univ. Chicago Press, Chicago. 425 pp.

Sinclair, A. A. E. 1977. *The African buffalo. A study of resource limitation of populations.* Univ. Chicago Press, Chicago. 354 pp.

Talbot, L. M., and M. H. Talbot. 1963. The wildebeest in western Masailand. *Wildlife Monographs* 12:1-88.

Wetzel, R. M. 1977. The Chacoan peccary, *Catagonus wagneri* (Rusconi). *Bull. Carnegie Mus. Nat. Hist.* 3:1-36.

Whitehead, G. K. 1972. *Deer of the world.* Constable, London. 194 pp.

29 Statistical Analysis and Representation of Data

One of the most remarkable developments in recent years has been a tremendous increase in the availability of calculators, computers, terminals, and other devices as aids to analyses of numerical data. Still, many persons avoid learning methods for handling and interpreting numbers because the terminology looks strange or they presume that they do not have an adequate mathematical background. In most instances, these fears are unwarranted because one can learn many techniques for analyses quickly and easily.

The purpose of this chapter is to provide an overview of **statistics,** the scientific analysis of numerical data. Learning how to collect, organize, analyze, and interpret numerical data is vitally important if one is to be an effective scientist and researcher. As biologists, we are increasingly involved with the collection and analysis of large quantities of data on environmental, morphological, and physiological variables. Statistics provides techniques for objectively identifying sources of variation, for comparing data sets, and for establishing confidence limits for various estimates.

Students are stronglv urged to enroll in a statistics course or courses sometime during their tenure at college. There are many statistics books on the market but the following deserve special comment. Steel and Torrie (1960), while technically good, is oriented toward applied and experimental agriculture. Three books that directly concern biologists include the introductory account by Simpson *et al.,* (1960) and the more extensive treatments by Sokal and Rohlf (1969) and Zar (1974).

Data and Sampling Units

Like most scientific disciplines, statistics does have special terminology to aid standardization in publications and clarity in communication between workers in the field. In many instances, these words are familiar to you but may be defined more rigorously.

Data (singular **datum**) are observations or measurements taken on a sampling unit. These data or observations are taken from objects or entities termed **sampling units.** Sampling units could be individuals, leaves, soil samples, or other objects.

A **variable** or **character** is the property being measured or observed on the sampling unit. For example, a mammalogist interested in the variable "body weight in grams" weighed ten rats (the sampling units) producing the following set of data: 105, 110, 90, 85, 120, 115, 110, 100, 90, 95.

Many people use the term statistics to refer to the observations or data. Statisticians restrict the usage of this term to apply only to the discipline *or* a computed quantity such as the mean (arithmetic average).

29-A. In the following list, circle all words or statements representing variables and underline all those representing sampling units:

Rat 1421	Weight	Female 201
Row 114	Temperature	Glucose Level
Ear Length	Sex	Quadrat 1102

Kinds of Variables

Sokol and Rohlf (1969:11-12) stated that variables may be classified into those resulting from measurements (continuous or discrete), counts based on attributes, and those based on relative rankings. **Discontinuous variables** (meristic or discrete) have fixed values with no intermediate values possible, e.g., 1, 2, or 3 toes present, but never 2.4 toes present. **Continuous variables** can assume, theoretically, an infinite number of values between any two fixed points, e.g., between points 1 and 2 there exist 1.1, 1.112, 1.0113, etc., depending on the accuracy of the measuring instrument and the patience of the data recorder. An **attribute** is a qualitative measure or property, e.g., black, brown, or blue eyes; male or female; dead or alive. When attributes are combined with frequencies into tables of numbers they are referred to as **enumeration data.** **Ranked variables** can be placed on an ordinal scale but differences in ranks remain relative rather than absolute. For example, the widths of body stripes on a skunk might be ranked from 1 to 5, but a stripe in a class "1" may not be five times narrower than a class "5" stripe. The letter grades that you receive in a course are ranked values derived from measurement values (e.g., percentages).

29-B Using the letters in parentheses, classify the following data as continuous (C), discrete (D), or rank (R) variables:

__ 90.1°C	__ Code 3
__ density "3"	__ 23.02 grams
__ 14 (stems)	__ 6 (bites)
__ I Class	__ male 6 Largest
__ 114.22 mm	__ 242 Kilos
__ 0.01 grams	__ Incisors 3/3

Accuracy and Precision

The researcher must strive for accuracy and precision in measurements. **Accuracy** is the nearness of a measured value to its true value. **Precision** is the nearness of values of successive measurements of the same character of the same specimen. Both accuracy and precision depend on the skill of the person making the measurement. Accuracy, however, can also be lost by a measuring device that is incapable of measuring to the researcher's goal of accuracy; e.g., vernier calipers accurate to 0.1 mm cannot produce results accurate to 0.01 mm. But an improperly calibrated balance could produce precise successive measurements even though these measurements may be inaccurate (e.g., the balance may show that animals weight 0.5 grams less than their true weights).

29-C. To the right of each numerical observation listed, record the range of accuracy implied by the value. For example, for the value 0.1, the range of accuracy implied is 0.050-0.149.

5	1000
0.01	100.1
100	10
12.3	0.001
10.5	2.06

Populations, Samples, Sampling

In statistics a **population** consists of all values of a particular variable within a specified space or time. For practical and logistical reasons it is usually impossible to record observations on all individuals in the population. Instead, the biologist works with randomly selected individuals from the population. A set of such individual observations is called a **sample.** The biologist then makes inferences about the entire population based on the available samples. Thus, **sample statistics** (usually symbolized by Latin letters) are estimators of **population parameters** (usually symbolized by Greek letters).

A population or **universe** may refer to all possible individuals or objects (units) or may be restricted by a space or time limitation. For example, a population could consist of all human inhabitants of the world or be all males 15 years of age in New York City. It would be extremely difficult to devise a sampling scheme to estimate the population parameters of the entire world population. Most scientists and statisticians make estimates of populations that have been more narrowly defined.

Many statistical procedures assume that samples will be obtained in a **random** fashion, such that each member of a population has an equal and independent chance of being selected. Randomness in sampling might be achieved by assigning all individuals in a population a number and then drawing numbers (representing the individuals) out of a hat. Better still, a **Table of Random Digits** (Table 29-1 in this text; Table 0 in Rohlf and Sokal, 1969; Table D.45 in Zar, 1974) should be consulted for selecting samples. Suppose that we wish to sample plants in 15 of 64 possible quadrants in an 8 × 8 grid. Since each quadrat can be represented by an X and Y coordinate, then pairs of numbers can be selected, in order, from the table of random numbers. Prior to consulting the table, determine the direction that you will follow in the table (e.g., across page, top to bottom, etc.). Then select a starting point in the table, proceed in the predetermined direction and write down pairs of unique numbers between 1 and 8. Fifteen of these pairs will then indicate the quadrants that should be sampled.

The number of sampling units or the **sample size** (**N**) that is used to estimate the population parameters is a matter of prime importance in any statistical analysis. In many instances, the biologist has no control over the sample size available for study. This frequently occurs in taxonomic studies when the scientist must work with existing collections of specimens (see Simpson et al., 1960:102-107). Ideally, the investigator should work with a sample size sufficient for

Table 29-1

Two thousand and five hundred random digits. (Brower and Zar (1974:4).

72965	92280	85318	98478	05200	26558	04697	63195	41679	24133
25182	09959	91375	97794	50193	25930	47938	95633	22271	15628
78812	39100	81576	84683	47466	04204	86339	31919	83404	48293
87264	75327	92529	25409	52589	20914	58768	46171	32657	89750
21571	57796	67813	88705	52576	51712	12407	00644	81748	04204
98532	11191	63198	79306	04193	00859	83906	30625	67175	37774
38981	76006	33931	22225	00014	37716	67499	90402	08962	88602
11305	19964	22932	62300	64508	32996	05699	06536	22619	89725
96753	89989	67869	65743	65353	55722	91650	77833	05353	05950
28316	27206	32507	96140	83430	75357	57822	75247	93486	20481
24390	09214	19493	94975	71393	54675	51712	00581	11187	73464
23995	32726	41075	32118	63946	62464	60599	81670	73097	78553
41920	60706	55864	70343	61238	06810	53263	07815	56588	29384
78281	15410	26154	70445	27828	38282	29051	13433	84405	82969
92910	17017	92704	25210	63833	04909	02571	58402	62649	86771
29265	89779	95437	51929	75534	70858	54623	99661	87146	16775
60422	65242	57037	95091	25582	76743	95890	09033	08368	62677
42748	43783	94238	97764	64110	68935	21057	14994	94235	53722
39611	11320	52913	20490	84147	59510	45967	93742	71756	09298
74011	92403	54878	91689	20402	20287	05402	16617	86101	28192
49056	17282	52320	73306	91759	85329	88229	62615	25802	28655
06572	13935	69948	12322	84900	85760	67583	36717	75897	39169
32726	45220	41600	61236	55701	08181	26259	49841	88968	83197
13800	03061	28494	09432	95359	92550	11251	76533	51923	34450
09838	95794	39792	06406	81584	49541	20520	91941	43448	91692
86499	23583	61444	72616	78692	50822	10283	23499	17883	21908
19618	23145	32406	91793	50163	72615	61939	18183	20368	51482
04145	26409	44737	98157	14158	94981	66518	84956	65372	00578
44083	35657	49215	93131	41815	34454	46347	02783	27988	86461
13883	40605	76333	56473	27866	16074	00939	05149	14090	70080
08697	34971	19204	70701	56065	23839	45794	62036	07594	36604
86447	56887	61107	63246	88350	51579	95387	03708	16441	64848
37914	39110	60363	95348	96498	17447	18058	36020	57301	50492
08771	12569	06379	51277	88233	45879	89353	82759	16691	20680
65529	84747	61160	19575	98709	23055	37992	82397	62884	63738
53783	03060	00563	21869	41559	85468	37401	81331	62733	10999
40881	01466	66439	92600	95878	43878	76006	93166	20603	76173
81424	81842	17993	63784	39351	41580	89006	47888	92753	45323
47362	92940	89774	05283	49461	21521	72572	37403	90574	22562
79898	44180	49706	58783	47012	90892	89032	56904	56473	38246
98433	36491	48288	53653	77220	82969	70063	58551	20025	83414
79849	94549	69691	11789	43233	46831	08737	25992	11296	69195
26004	14598	80743	25043	45287	35345	46914	71487	10345	48236
46218	40835	82386	91946	14266	77484	02759	92164	77842	21600
49618	10730	47690	44746	09566	36769	39108	47001	62935	10227
66259	25266	88651	56018	68181	45119	91387	37257	83610	53138
65170	81485	14727	22898	63815	17317	68293	06449	91890	49994
82679	72969	04512	11079	95969	87389	46263	96780	78124	04120
37900	90316	47434	60701	89649	51773	26139	39231	72264	17654
27111	31679	71539	61375	58691	20215	91170	44290	91396	90173

the chosen statistical technique and aims of the investigator. Sample sizes numbering fewer than 6 or 8 individuals are rarely useful for statistical purposes since the investigator can place little confidence in the statistics derived from such small numbers. Most statistics textbooks give guidelines for determining adequate sample sizes and discussions of the value of larger sample sizes. A procedure for estimating adequate sample sizes is described in Sokal and Rohlf (1969:246-249).

29-D. Examine a table of random digits (e.g., Table 29-1). Select numbers to take 20 random samples from each of the following situations:

1. Location of vegetation plots alongside 100-meter transect divided into 100 one-meter segments
2. Locations for taking ten soil samples in 100 x 100 meter grid divided into 100 10 x 10 meter quadrats

What was similar about the methods used to select the locations? What was different?

Frequency Distributions

Inspection of the distribution of a variable is an important step in the analysis of data since it frequently guides our decisions on what statistical analyses to utilize. A **frequency distribution** is prepared by listing all observed values and then noting how many times each value is observed. For example, we autopsy a sample of 29 female rats and wish to examine the frequencies of various numbers of embryos as an indication of potential litter size. A **frequency table** of the distribution of these data appears below (Table 29-2).

If the shape of a frequency distribution needs to be examined, a graphical technique is employed. For discrete, ranked, or attribute data, a **bar graph** (Fig. 29-1) should be prepared, keeping each vertical bar separate. For continuous data, a **histogram** (Fig. 29-2) is prepared with the vertical bars adjoining one another. Data of continuous variables may also be represented by a **frequency polygon** (Fig. 29-3) formed by a line that connects points. A relative **cumulative frequency polygon** (Fig. 29-4) may be plotted when you wish to examine the contribution of particular values to overall totals.

Table 29-2
Frequency table of numbers of embryos in a sample of the cricetine rodent, *Holochilus sciureus berbicensis.* (Twigg 1965:271)

Number of Embryos	*Frequency*
1	19
2	94
3	135
4	67
5	25
6	10
7	3
8	1

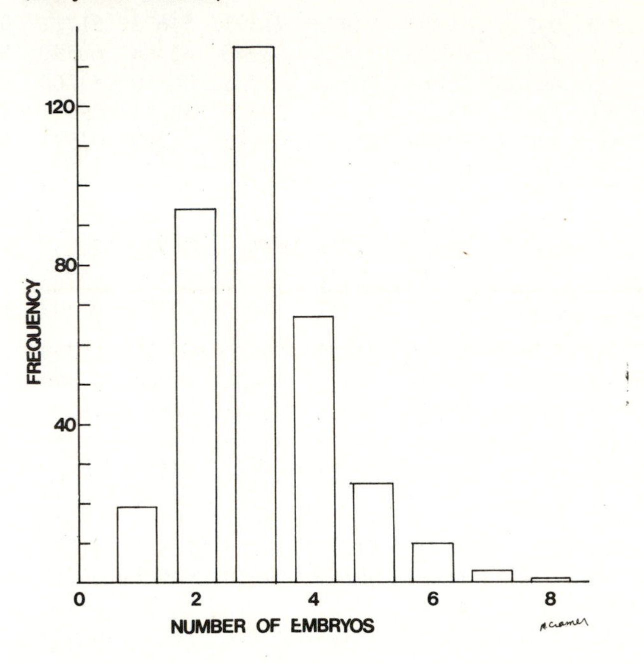

Figure 29–1.
Bar graph of the frequency data in Table 29-2. (Mary Ann Cramer)

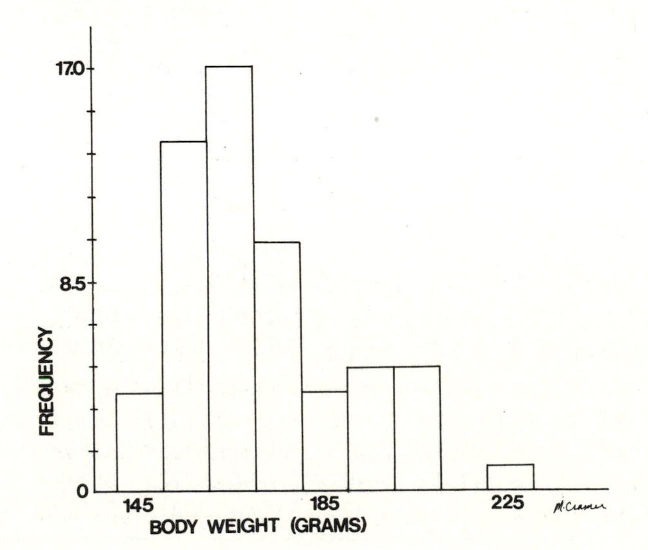

Figure 29–2.
Histogram of body weights of the degu, *Octodon degus.* (Mary Ann Cramer)

Figure 29–3.
Frequency polygon of body weights of the degu, *Octodon degus.*
(Mary Ann Cramer)

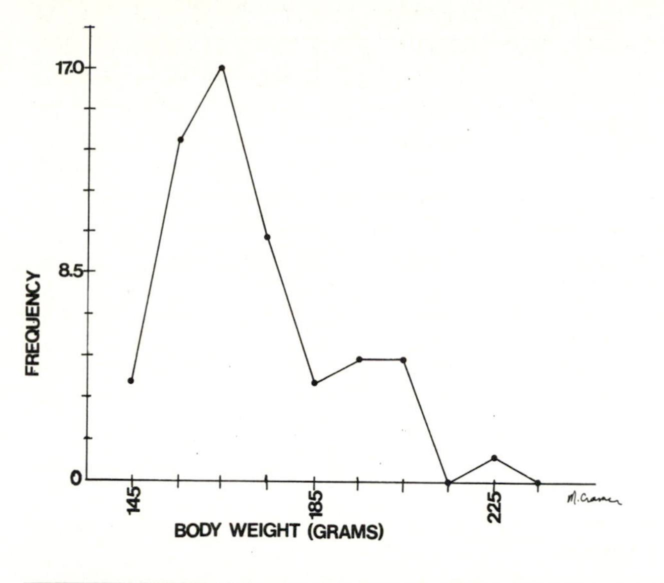

Figure 29–4.
Cumulative frequency polygon of body weights of the degu, *Octodon degus.*
(Mary Ann Cramer)

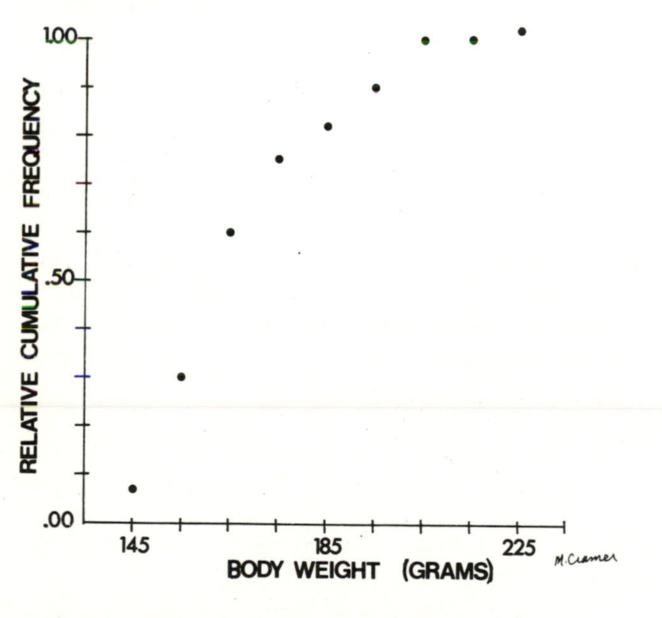

Central Tendency and Dispersion

In analyses of data, we generally wish to know something about **central tendency,** the localization of values near a central point, and **dispersion,** the scatter of these values from the central region. The **sample mean** ($\overline{X}$), the average of a set of numbers (N), is one of the best estimates of central tendency and the best and most consistent estimator of the **population mean** (μ).

$$\overline{X} = \sum_{i=1}^{N} Xi/N, \text{ where } \Sigma = \text{summation} \qquad \text{(formula 29-1)}$$

Other measures of central tendency will be discussed in the section on "Basic Statistics."

The **range,** the difference between the highest and lowest values, is a measure of dispersion or variability familiar to most persons. For statistical purposes, it is a crude measure since it frequently underestimates the range of the population.

The sum of all deviations from the mean is equal to zero.

$$\sum_{i=1}^{N}(X_i - \overline{X}) = 0 \qquad (29\text{-}2)$$

The **population variance** (σ^2, sigma squared) is defined as the mean sum of squares of the deviations from the population mean (μ).

$$\sigma^2 = \frac{\Sigma(X_i - \mu)^2}{N} \qquad (29\text{-}3)$$

The population (parametric) standard deviation (σ), the square root of the variance, is a useful measure since it is expressed in the same scale as the population values.

The **sample variance** (s^2) is the best estimate of the population variance and is distinctly superior to the mean deviation for hypothesis testing (Zar 1974). Computational methods for calculating s^2 and its derivatives will be described in the section on "Basic Statistics."

Basic Statistics

Descriptive statistics provide a numerical summary on the properties of an observed frequency distribution. Measurements made for computing these statistics are generally recorded in tabular form (Table 29-3). Standard descriptive statistics include **sample size** (N), **degrees of freedom** (generally N–1), **range,** arithmetic **mean** ($\overline{X}$ or $\overline{Y}$), **variance** (s^2) **standard deviation** (s), **standard error of the mean** ($s_{\overline{X}}$), and the **coefficient of variation** (CV).

The symbol Σ indicates that a set of observations must be summed. By reference to Table 29-3 the mean for the character "condylobasal length" can be computed as follows:

$$\overline{X} = \frac{\Sigma X_i}{N} = \frac{45.4 + 48.7 + \ldots + 43.9}{8}$$

$$= \frac{384}{8} = 48.0$$

The range of values for character 1 of taxon B is 43.9 to 51.8. Variance is a measure of dispersion of a set of data about the mean.

The variance is expressed in squared units. The standard deviation is the square root of the variance and is expressed in the same units as the original observations (X). From data in Table 29-3 the values for variance and standard deviation are then calculated using computational formulas:

$$s^2 = \frac{\Sigma X^2 - (\Sigma X)^2/N}{N-1} \quad (29\text{-}4)$$

$$s^2 = \frac{(45.4)^2 + \ldots (43.9)^2 - (45.4 + \ldots 43.9)^2/8}{7}$$

$$s^2 = \frac{18478.54 - (384.0)^2/8}{7}$$

$$s^2 = 6.648$$

$$s = \sqrt{\frac{\Sigma X^2 - (\Sigma X)^2/N}{N-1}} \text{ or } s = \sqrt{s^2} \quad (29\text{-}5)$$

$$s = \sqrt{6.648} = 2.578$$

The standard error of the mean ($s_{\bar{x}}$) is the standard deviation of the means for a sample. From the quantities above, $s_{\bar{x}}$ is computed as follows:

$$s_{\bar{x}} = \sqrt{\frac{s^2}{N}} = \sqrt{\frac{6.648}{8}} = 0.912 \quad (29\text{-}6)$$

The coefficient of variation is the standard deviation expressed as a percentage of the mean. This permits comparison of variation in data when the mean or standard deviation values are very different; e.g., the measurements of a horse can be compared with those of a small shrew. The CV is computed on the basis of the above data as follows:

$$CV = \frac{(s)\ (100)}{\bar{X}} = \frac{257.8}{48.0} = 5.37 \quad (29\text{-}7)$$

One disadvantage of the coefficient of variation, as pointed out by Lewontin (1966) and Moriarty (1977), is the inability to perform exact statistical tests to compare CV values. Lewontin suggests transforming the measurements to logarithms (to any base) and then computing descriptive statistics on the characters, (e.g., A and B). Then, the ratio: $s^2 \log A / s^2 \log B$ can be compared with an F-distribution to test the magnitude of the difference (if any). Lande (1977) points out other precautions that must be observed when using coefficients of variation (e.g., should not compare CV's based on discrete data).

Table 29-3
Cranial and bacular measurements and descriptive statistics based on them. Measurements are of a small sample of a single species at a single locality. Refer to text for explanation of symbols.

Identification Number of Individual (X_i)	*Measurements*			
	Condylo-basal Length	*Zygo-matic Breadth*	*Bacular Length*	*Bacular Width*
A01	45.4	25.3	6.3	1.6
A02	48.7	25.8	—	—
A03	51.8	27.3	8.3	1.7
A04	49.3	25.5	8.4	2.0
A05	47.5	25.4	6.3	1.5
A06	47.1	24.6	—	—
A07	50.3	26.4	6.7	1.7
A08	43.9	23.8	7.4	1.7
Sample Statistics				
N	8	8	6	6
Σ	384.0	204.1	43.4	10.2
$\bar{X}$	48.00	25.51	7.23	1.70
s^2	6.648	1.127	0.911	0.028
s	2.578	1.062	0.954	0.167
$s_{\bar{x}}$	0.912	0.375	0.390	0.068
CV	5.37	4.16	13.19	9.82
$\bar{X} \pm t_{.05,\ df}$	48.0 ± 2.2	25.5 ± 2.1	7.2 ± 1.0	1.7 ± 0.2

To compute the confidence limits on these values we utilize the t-distribution (see section on Two-Sample Comparisons). Thus, to compute the 95% confidence limit for the measurement "condylobasal length," we must utilize the following values: degrees of freedom (df), t-statistic from table ($t_{tab.}$), standard error of the mean ($s_{\bar{x}}$), and mean ($\bar{X}$).

$df = N - 1$ or 7 $\qquad t_{tab\ (.05),\ 7df} = 2.365$

$s_{\bar{x}} = 0.912 \qquad \bar{X} = 48.0$

The general formula for obtaining the confidence limits for the mean is the following:

$$\bar{X} - t_{\alpha,\ df} s_{\bar{x}} \leq \mu \leq \bar{X} + t_{\alpha,\ df} s_{\bar{x}} \quad (29\text{-}8)$$

Thus,

$$48.0 - (2.365)\ (0.912) \leq 48.0 \leq 48.0 + (2.365)\ (0.912)$$

or, in abbreviated form,

$$\bar{X} \pm t_{.05,}\ df\ s_{\bar{x}} \ \ or \ \ 48.0 \pm 2.16.$$

29-E. Verify this calculation and then substitute an α-level of .01 to see how this changes the value.

$\overline{X} \pm t_{.01,7}\, s_{\bar{x}}$ *or* 48.0 ± 3.19.

29-F. Compute the mean, standard deviation, standard error of the mean, coefficient of variation, and confidence limits for the measurements of bacular width and bacular length in Table 29-4. Check your answers with the values in the table. Compare the values of coefficient of variation for all four measurements. What do these values indicate?

Descriptive statistics, such as mean, range, standard deviation and standard error, can be presented in the form of Dice-Leraas diagrams or Dice-grams (Fig. 29-5). These diagrams are helpful to see overall patterns of variation but should not be used for extensive testing of differences between means (see section on Multiple Samples and Comparisons).

Probability Distributions

Probability and Binomial Distribution

Probability and probability distributions are important concepts for understanding many statistical procedures. **Probability** is the chance for the occurrence of a particular event given the total number of possible outcomes of all events. For example, when a die (pl. dice) is thrown, one of six possible numbers may appear. Thus, the probability (p), that the number "5" will appear is

$$p_{(5)} = 1/6 \text{ or } 0.167.$$

The probability (k) of any of the other numbers occurring (i.e., 1,2,3,4,6) is

$$1 - 0.167 = 0.833.$$

Probability values always range between 0 and 1.

Figure 29–5.
Variation in measurements of two external characters and one cranial character between populations of the rodent, *Phyllotis andium*. Descriptive statistics are indicated by modified Dice-Leraas diagrams. Each bar shows the mean (vertical line), twice the standard error of the mean (black rectangle), and standard deviation (black plus open rectangles). Sample sizes (*N*) are indicated in the parentheses above the diagrams.
(Pearson 1958: 438)

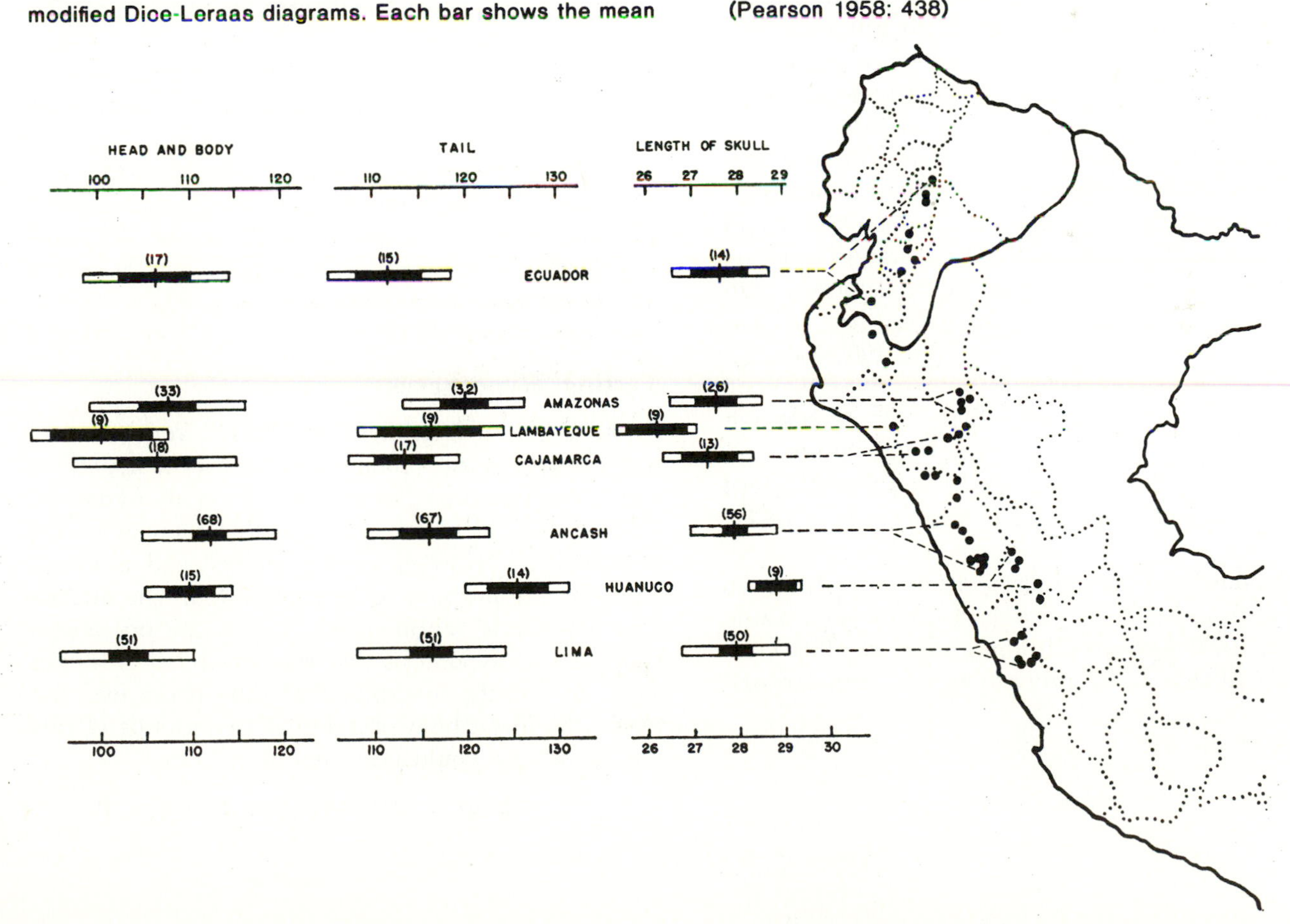

When a pair of dice is thrown, the probability of obtaining a pair of "5's" is the product of the independent probabilities:

$$p_{(5,5)} = (p_1)\ (p_2) = (1/6)\ (1/6) = 1/36 \textit{ or } 0.028.$$

Further details on methods for computing probabilities can be found in textbooks such as Snedecor and Cochran (1967:199-202) and Sokal and Rohlf (1969: 69-71).

The theoretical frequency distribution or **probability distribution** of events that can occur in two classes is known as the **binomial distribution** (Sokal and Rohlf 1969:71-81). Actual proportional data of a given sample size for two classes can then be compared with the theoretical distribution.

29-G. Compute probability values for the following situations:

1. Probability of obtaining two heads in two tosses of a coin: _______*
2. Probability of selecting on one occasion a male from a cage containing five male and 10 female rats: _______*

Normal Distribution

Data that approximate a normal probability density function or bell-shaped **normal distribution** (Fig. 29-6) are necessary for conducting most **parametric** kinds of statistical tests. Many kinds of biological data such as lengths, weights, heights, and rates conform reasonably well to these distributions (Brower and Zar 1974). Data based on counts, frequencies, and percentages generally are not normally distributed and thus **nonparametric** methods of data analysis must be utilized (unless these data can be transformed into approximate normal distributions by the use of logarithms or square roots; Sokal and Rohlf 1969:380-387).

29-H. Examine Figure 29-6. What would happen to the shape of the normal distribution if the mean (μ) was as 10 and the standard deviation (σ) 0.5? if μ was 10 and σ 1.5?

Figure 29–6.
A normal distribution. These data are a hypothetical population of tree heights (X), with a mean, μ, of 11.72 m, and a standard deviation, σ, of 1.16 m. The mean $\pm$ 1 standard deviation includes 68.3% of any normal curve; the mean $\pm$ 2 standard deviations encompasses 95.5%; and $\mu \pm 3\sigma$ includes 99.7%.
(Brower and Zar 1974: 12)

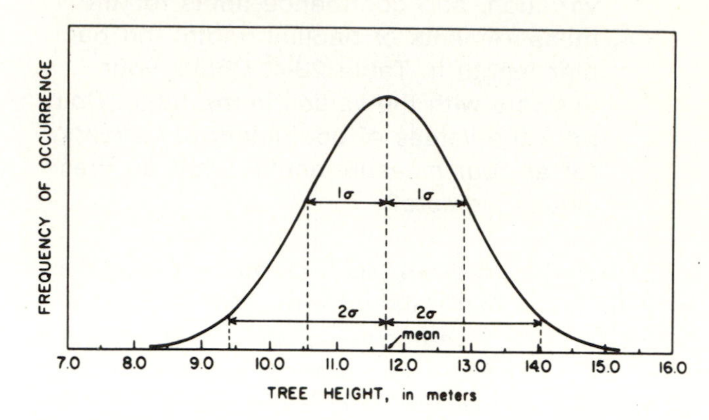

A **standardized normal distribution** is one in which the $\mu = 0$ and $\sigma = 1$. Standard tables (e.g., Table D.9, Zar, 1974; Appendix I, Simpson, *et al.*, 1960) enable one to determine proportions of normal distributions. Thus, for a normal population with a mean (μ) and standard deviation (σ), the expression

$$Z = \frac{X_i = \mu}{\sigma} \qquad (29\text{-}9)$$

yields a **Z-value**, which indicates the number of standard deviations from the mean that an X value is located. These Z-values are termed **normal deviates** or **standard scores** and the calculation is referred to as normalizing or standardizing X_i-values. Since we rarely know the population mean (μ) and the standard deviation (σ), we must use the sample approximations, $\overline{X}$ and s, respectively. However, for small samples these statistics are poor approximations of the population parameters (Zar 1974:87).

Testing Hypotheses

In statistics, an hypothesis is phrased very carefully and consists of two components: the null hypothesis and the alternative hypothesis. The **null hypothesis** (abbreviated H_o) is a statement that there is no difference (e.g., between sample groups) and is formulated for the purpose of being rejected. The **alternative hypothesis** (abbreviated H_A) is the operational statement or hypothesis that the researcher is testing. Thus, to test the assertion that differences exist between the mean body weights of two groups of rats, the hypothesis would be stated as follows:

H_o: Group A = Group B *or* H_o: A = B

H_A: A $\neq$ B

*Answers: 1.) = 0.25; 2.) = 0.33

Once the null hypothesis (H_o) has been formulated, there must be an objective method for determining when to reject this hypothesis. First, let us examine the two types of errors that can be made when hypotheses are tested. A **Type-I error** (symbolized by α) is the *rejection of H_o when it is true* (*or* accepting a difference when there is none). A **Type-II error** (symbolized by β) is the *acceptance of H_o when it is false* (*or* failing to find a difference when there is one).

Since our primary goal in hypothesis testing is to reject the null hypothesis when it is false, we generally wish to keep the probability of a Type-I error minimized to a stated α-level. The larger the value of α, the more probable that the null hypothesis will be rejected falsely (i.e., committing a Type-I error). Remember that all probability values, whether α or β or some other parameter, range from 0 to 1.

The levels of α and β are inversely related to each other and dependent on the sample size (N). Thus, to decrease the possibility of both types of error, N must be increased. The *power* of a statistical test is the probability of rejecting H_o when it is false (*or* the probability of finding a difference when there is one). Stated in another form, the power of a test is

$$1 - \text{probability of a Type-II error } or \ 1 - \beta. \quad (29\text{-}10)$$

Generally, the power of a test increases with an increase in the sample size (N) (Siegel 1956). A **test of significance** evaluates the probability of rejecting the null hypothesis when it is true. The probability of making a Type-I error (α), expressed as a percentage, is termed the **significance level.** For example, if α is .01, then the significance level is 1% for a given test. If we choose a 5% level of significance, then we expect that only 5 of 100 samples examined will result in making a Type-I error (i.e., rejection of a true null hypothesis). In many scientific disciplines, α-levels of .05, .01, and .001, are utilized for hypothesis testing. However, the choice of an appropriate α-level is somewhat arbitrary and will depend on the nature of the investigation and the degree of predictability required. Refer to Sokal and Rohlf (1969:155-166) for further discussion on the selection of appropriate α-levels.

29-I. If the significance level was set at 1% ($\alpha = .01$) rather than 5%, would you be more likely to make a Type-I error at the 1% level if the sample size was smaller? How can you decrease the probability of making Type-I and Type-II errors?

Once the hypothesis has been formulated, an appropriate statistical procedure and test must be utilized. Data obtained from continuous variables are generally analyzed by **parametric statistics** since these variables most nearly follow a normal distribution (see section on "Probability Distributions"). Data from enumeration, discontinuous, and ranked types are generally analyzed by **nonparametric statistics** since no assumptions about the shape of the distributions are required to utilize these procedures. Large samples of enumeration and discontinuous data often have a nearly normal distribution and can be analyzed as if they were continuous. Ranked data, however, can never be analyzed with parametric statistics since ranks are relative and cannot be multiplied or divided.

Two-Sample Comparisons

In biological problems, we frequently wish to know whether or not the means of two sample groups are significantly different from one another (e.g., $H_o : \overline{X}_A = \overline{X}_B$ *vs* $H_A : \overline{X}_A \neq \overline{X}_B$). If the sample values for these two groups (1) follow a normal distribution (or nearly so) and (2) the sample variances are not significantly different from one another, then parametric statistical procedures can generally be used for testing. If not, nonparametric statistics must be utilized.

One of the most common and useful tests for comparing sample means is Student's *t*-test or, simply, the ***t*-test.** The *t*-distribution is like a normal distribution when sample sizes approach infinity but the curve is more flattened for smaller sample sizes. Prior to utilizing the *t*-test or other parametric test for two samples, the **homogeneity of variances** (i.e., equality of variances) between the groups must be tested. For this purpose, we use the following statistic:

$$F_s = \frac{s^2_{\text{larger}}}{s^2_{\text{smaller}}} \quad (29\text{-}11)$$

$$\text{for } df_{\text{larger } s^2}, df_{\text{smaller } s^2}$$

Then, substituting the appropriate variances (s^2) from (Table 29-5) into the formula, we obtain the following:

$$F_s = \frac{5.67}{2.00}, \text{ df } 9, 8 \text{ (numerator, denominator).}$$

Since the $F_{tab\ .05}$ value for 9 (numerator) and 8 (denominator) degrees of freedom is 3.39, we accept the null hypothesis that the variances are equivalent and thus can proceed with testing the equality of the group means.

To calculate a *t*-statistic ($t_{cal.}$), we utilize the same procedures that were employed for obtaining basic statistics, with these exceptions. A **sum of squares (SS)** is calculated for each group utilizing the basic formula for estimating variance (s^2) with the exception of omitting the step where the quantity is divided by N-1. Thus, utilizing the values from Table 29-5, the sums of squares for the two groups are calculated as follows:

$$SS_A = \Sigma X_A^2 - (\Sigma X_A)^2/N \quad (29\text{-}12)$$

$$= 23622.07 - (485.5)^2/10$$

$$= 51.045$$

$$SS_B = \Sigma X_B^2 - (\Sigma X_B)^2/N$$

$$= 20227.37 - (426.5)^2/9$$

$$= 16.009$$

To evaluate the means of the two groups, it is necessary to obtain estimate of the pooled variance (s^2_{pooled}) and from this the standard error of the pooled mean ($s_{\bar{x}_{pooled}}$) as follows:

$$s_p^2 = \frac{SS_A + SS_B}{df_A + df_B} \quad (29\text{-}13)$$

$$= \frac{51.045 + 16.009}{9 + 8}$$

$$= 3.94$$

$$s_{\bar{X}_A - \bar{X}_B} = \sqrt{\frac{s_p^2}{N_A} + \frac{s_p^2}{N_B}} \quad (29\text{-}14)$$

$$= \sqrt{\frac{3.94}{10} + \frac{3.94}{9}}$$

$$= \sqrt{.832}$$

$$= 0.912$$

Then, $s_{\bar{x}_{pooled}}$ is substituted into the formula below to obtain $t_{cal.}$:

$$t_{cal.} = \frac{X_A - X_B}{s_{\bar{X}_A - \bar{X}_B}} \quad (29\text{-}15)$$

$$= \frac{48.6 - 47.4}{0.912}$$

$$= 1.32$$

Inspection of Table 29-4 reveals that the tabulated value of the t-statistic for 17 df and $\alpha = 0.05$ is 2.11. Thus, the null hypothesis of no difference in mean skull lengths of the male and female samples is accepted at the 5% level because $t_{cal.} < t_{tab.}$

Table 29-4
Critical values of student's t (Brower and Zar 1974:10)

DF	$\alpha = 0.10$	$\alpha = 0.05$	$\alpha = 0.02$	$\alpha = 0.01$
1	6.31	12.71	31.82	63.66
2	2.92	4.31	6.96	9.92
3	2.35	3.18	4.54	5.84
4	2.13	2.78	3.75	4.60
5	2.01	2.57	3.36	4.03
6	1.94	2.45	3.14	3.71
7	1.89	2.36	3.00	3.50
8	1.86	2.31	2.90	3.36
9	1.83	2.26	2.82	3.25
10	1.81	2.23	2.76	3.17
11	1.80	2.20	2.72	3.11
12	1.78	2.18	2.68	3.06
13	1.77	2.16	2.65	3.01
14	1.76	2.14	2.62	3.00
15	1.75	2.13	2.60	2.95
16	1.75	2.12	2.58	2.92
17	1.74	2.11	2.57	2.90
18	1.73	2.10	2.55	2.88
19	1.73	2.09	2.54	2.86
20	1.72	2.09	2.53	2.85
22	1.72	2.07	2.51	2.82
24	1.71	2.06	2.49	2.80
26	1.71	2.06	2.48	2.78
28	1.70	2.05	2.47	2.76
30	1.70	2.04	2.46	2.75
35	1.69	2.03	2.44	2.72
40	1.68	2.02	2.42	2.70
45	1.68	2.01	2.41	2.69
50	1.68	2.01	2.40	2.68
60	1.67	2.00	2.39	2.66
70	1.67	1.99	2.38	2.65
80	1.66	1.99	2.37	2.64
90	1.66	1.99	2.37	2.63
100	1.66	1.98	2.36	2.63
120	1.66	1.98	2.36	2.62
150	1.66	1.98	2.35	2.61
200	1.65	1.97	2.35	2.61
300	1.65	1.97	2.34	2.59
500	1.65	1.96	2.33	2.59
∞	1.65	1.96	2.33	2.58

The above values were computed as described by Zar (1974:414). More extensive tables of Student's t are found in Rohlf and Sokal (1969:160-161) and Zar (1974:413-414).

Table 29-5
Skull measurements (condylobasal lengths) of two samples (males and females) of a single species from the same locality.

Males		*Females*	
Specimen No.	*CBL in mm*	*Specimen No.*	*CBL in mm*
068	51.4	059	48.2
064	51.6	051	48.9
067	51.4	073	49.0
056	49.8	009	47.1
048	46.7	062	46.2
071	46.4	061	45.2
053	49.1	064	47.7
065	47.2	057	48.5
0.72	45.6	052	45.7
054	46.3		
N	10	9	
df	9	8	
$\overline{X}$	48.6	47.4	
SS	51.045	16.009	
s^2	5.67	2.00	
s	2.38	1.41	
$s_{\overline{x}}$	0.753	0.472	
$\overline{X} \pm t_{.05}\, s_{\overline{x}}$	48.6 ± 1.70	47.4 ± 1.09	

t_{tab} (.05), 17 df = 2.110

$$t_{cal.} = \frac{\overline{X}_m - \overline{X}_f}{s_{\overline{x} - \overline{x}}} = \frac{48.6 - 47.4}{0.9} = 1.32$$

Since $t_{tab.} > t_{cal.}$, the null hypothesis, H_o : A=B, is accepted

A single-classification **analysis of variance** (ANOVA) can also be used to compare group means (Steel and Torrie 1960; Sokal and Rohlf 1969:218-219).

For nonparametric data, the Mann-Whitney U test (Siegel 1956:116-127; Sokal and Rohlf 1969:392-394) and the Kolmogorov-Smirov two-sample test (Siegel 1956:127-136) are appropriate. The latter test should be applied only to nonparametric data of a continuous variable (e.g., continuous data not meeting the assumptions of a normal distribution).

For nonparametric discrete data, the **chi-square test** for independent samples is frequently very useful. Suppose, for example, that we wish to compare the sex ratios in several litters (pooled) of a species of rodent. The null hypothesis to be tested states that half of the sample will be male and the other half female (H_o : $P = 0.5$). Upon examination of the litters, we discover that 20 are male and 17 are female.

The chi-square (X_s^2) statistic* is computed according to the following formula:

$$X_s^2 = \sum_{i=1}^{r} \sum_{j=1}^{k} \frac{(O_{ij} - E_{ij})^2}{E_{ij}} \qquad (29\text{-}16)$$

where O_{ij} = observed number of cases in the ith row (horizontal) of jth column (vertical) and

E_{ij} = number of cases expected under H_o in ith row and jth column

and $\sum_{i=1}^{r} \sum_{j=1}^{k}$ indicates to sum over all rows (r) and all columns (k), i.e., over all cells.

Since, there are only two cells, there is only one k-category and the expected value is determined by multiplying N times the predicted probability of occurrence (i.e., 0.5). Thus, the expected value for each cell is 18.5 (0.5 × 37). Substituting the observed and expected values into the equation, the X_s^2 statistic is computed as follows:

$$X_s^2 = \frac{(20 - 18.5)^2}{18.5} + \frac{(17 - 18.5)^2}{18.5}$$
$$= (.1216) + (.1216)$$
$$= 0.243$$

In order to compare the calculated value of X^2 with the tabulated value, we must determine the number of degrees of freedom and then utilize Table 29-6. The general formula for determining the degrees of freedom is (r–1) (k–1). Since k = 1 in the present example, the appropriate degrees of freedom is r–1 or 2–1 = 1. Thus, the tabulated value of chi-square for an α-level of .05 is 3.84. Since $X^2_{cal.} < X^2_{tab,}$ we accept the null hypothesis that the sex ratio does not differ significantly from a 1:1 ratio.

There are special formulas for calculating X^2 values when data are arranged in 2 × 2 contingency tables and when the expected frequencies must be calculated from the marginal totals. Refer to Siegel (1956: 42-47, 104-111, and 175-179) or Sokal and Rohlf (1969: 549-620) for additional information.

Multiple Samples and Comparisons

Several statistics can be used to test for the significance of differences between the means of samples. An analysis of variance (*ANOVA*) not only gives an indication of differences between means but provides

*See Sokol and Rohlf (1969:553) for the rationale behind using the symbol X^2 rather than χ^2 for this quantity.

Table 29-6
Critical values of Chi-square. (Brower and Zar 1974:15-16)

DF	$\alpha = 0.10$	$\alpha = 0.05$	$\alpha = 0.025$	$\alpha = 0.01$
1	2.706	3.841	5.024	6.635
2	4.605	5.991	7.378	9.210
3	6.251	7.815	9.348	11.345
4	7.779	9.488	11.143	13.277
5	9.236	11.070	12.833	15.086
6	10.645	12.592	14.449	16.812
7	12.017	14.067	16.013	18.475
8	13.362	15.507	17.535	20.090
9	14.684	16.919	19.023	21.666
10	15.987	18.307	20.483	23.209
11	17.275	19.675	21.920	24.725
12	18.549	21.026	23.337	26.217
13	19.812	22.362	24.736	27.688
14	21.064	23.685	26.119	29.141
15	22.307	24.996	27.488	30.578
16	23.542	26.296	28.845	32.000
17	24.769	27.587	30.191	33.409
18	25.989	28.869	31.526	34.805
19	27.204	30.144	32.852	36.191
20	28.412	31.410	34.170	37.566
21	29.615	32.671	35.479	38.932
22	30.813	33.924	36.781	40.289
23	32.007	35.172	38.076	41.638
24	33.196	36.415	39.364	42.980
25	34.382	37.652	40.646	44.314
26	35.563	38.885	41.923	45.642
27	36.741	40.113	43.195	46.963
28	37.916	41.337	44.461	48.278
29	33.711	39.087	42.557	45.722
30	40.256	43.773	46.979	50.892
31	41.422	44.985	48.232	52.191
32	42.585	46.194	49.480	53.486
33	43.745	47.400	50.725	54.776
34	44.903	48.602	51.966	56.061
35	46.059	49.802	53.203	57.302
36	47.212	50.998	54.437	58.619
37	48.363	52.192	55.668	59.893
38	49.513	53.384	56.896	61.162
39	50.660	54.572	58.120	62.428
40	51.805	55.758	59.342	63.691

The above values were computed as described by Zar (1974:411). More extensive tables of chi-square are found in Rohlf and Sokal (1969:164-167) and Zar (1974:409-410).

a measure of variation within samples. Sokal and Rohlf (1969:173-366), Steel and Torrie (1960:99-160), and Zar (1974) give extensive accounts of the use of ANOVA's.

A *t*-test is inappropriate for making multiple paired comparisons of means (Sokal 1965). When more than two samples are involved an ANOVA can be used to test for overall difference between the means, although significant differences between pairs of means cannot be established. *A posteriori* tests, such as the sum of squares simultaneous test procedures (SS-STP) described in Sokal and Rohlf (1969), permit determination of homogeneous subsets of means within the total collection of means, *e.g.*, between means from different geographic localities. The Student-Newman-Keuls (SNK) test (Sokal and Rohlf 1969) and Duncan's multiple range test (Steel and Torrie 1960) have also been used to test multiple means. Some researchers believe that the last two tests are more useful, since the experimental error rate is not altered. In contrast, the SS-STP procedure generates more possible answers than can be realistically evaluated.

Covariate Analysis

Correlation and regression are techniques of covariate analysis. **Correlation analysis** is an investigation of the degree of association between pairs of variables (e.g., forelimb length versus hind limb length). Correlation analysis estimates the strength of the relationship between variables but implies no cause and effect relationship between the two. **Regression analysis** seeks to estimate the dependence of one variable (Y) on another, independent (X) variable (Fig. 29-7). Such a relationship, expressed mathematically, is generally written as a function (termed the regression equation), such as $Y = fX$, where the magnitude of a given Y is dependent upon the value of a given X. The slope of a regression line is termed the regression coefficient (b).

Regression analysis can be used in studies of differential growth (allometry) of body parts or regions. Differences between regression coefficients can be tested using a *t*-test or analysis of covariance (Steel and Torrie 1960; Sokal and Rohlf 1969).

A correlation coefficient (r) is a measure of interrelation between two variables, independent of the scale of measurement. The most commonly used correlation coefficient is Pearson's product-moment correlation coefficient. Procedures for the calculation of this statistic may be found in Sokal and Rohlf (1969:508-515) and Zar (1974:236-240). In addition, many of the programmable calculators currently on the market have routines to calculate this statistic.

Figure 29–7.
Relationship between mean body weights of three species of *Neotoma* from ten populations and mean annual temperatures. The regression line, its slope (*b*), and the correlation coefficient (*r*) are given. *N. cinerea*, circles; *N. albigula*, squares; *N. lepida*, triangles.
(Brown and Lee, 1969)

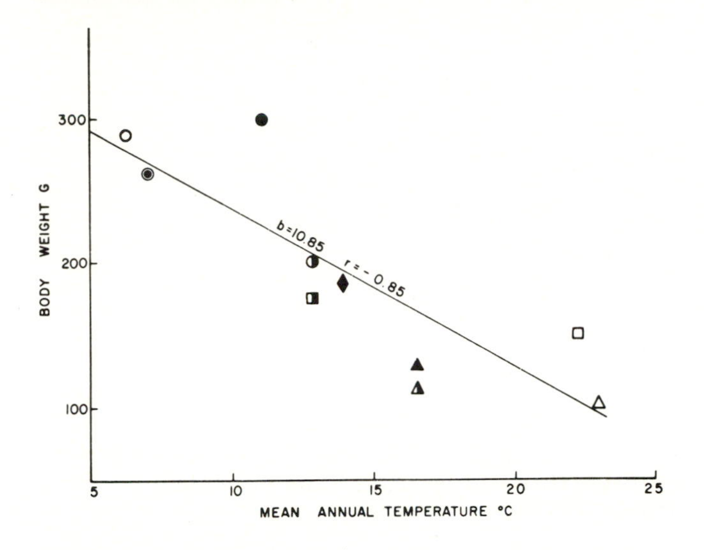

Multivariate Analysis

Multivariate analysis simultaneously considers variation and covariation of two or more variables. Computations involving three or more variables are extremely complex and time consuming. Thus, the wide application of multivariate statistical techniques in biological studies awaited the development of electronic digital computers with their capability for rapidly processing numerous variables and data points.

General references on multivariate analysis include Cooley and Lohnes (1971), Morrison (1967), Seal (1964), and Anderson (1958). A working knowledge of matrix algebra is helpful, though not essential, for using multivariate statistical procedures (Searle 1966 is a useful reference on matrix manipulations). Sneath and Sokal (1973) provide information on multivariate analyses in phenetic classification studies. A useful key for determining what types of multivariate analyses to utilize may be found in Atchley and Bryant (1975: 3-4) and Bryant and Atchley (1975:2-3).

In multivariate statistical analyses, the biologist is interested in one or all of the following: (1) a measure of similarity between groups (e.g., taxa); (2) reduction in the number of variables; and (3) discrimination between groups. Similarity can be measured by correlation, association, or distance coefficients. Phenograms are frequently constructed using the Unweighted Pair Group Method of Analysis (UPGMA) on correlation and average taxonomic distance matrices. Choate (1970), Genoways and Jones (1971), and Johnson and Selander (1970) and Patton, et al. (1975) are examples of phenetic cluster schemes using the UPGMA technique. Seal (1964), in contrast, recommended the use of a distance coefficient that considers relative correlation, such as generalized or Mahalonobis distance (D^2).

Factor analysis is a general term for several multivariate techniques that convert a large number of original variables into a smaller set of new variables. Two techniques that are used in systematics research include principal components analysis and multiple-factor analysis (frequently with rotation to simple structure). The principal components analysis has a sound mathematical basis although interpretation is sometimes difficult. Multiple-factor analysis is less exact mathematically since there are no unique solutions for obtaining communalities (summarization of intercorrelations among variables) or for estimating the number of factors to extract from the many potential factors. Despite these difficulties, factor analysis is an important summarization technique. Genoways and Choate (1972) utilized principal components analysis in a study of geographic variation in Nebraska populations of *Blarina*. In their study the first three principal components accounted for approximately ninety-two per cent of the total variance in nine cranial and three external measurements (Fig. 29-8). Multiple-factor analysis, with rotation to simple structure, was used by Wallace and Bader (1967) in a study of twenty-seven morphometric variables in a single sample of the house mouse, *Mus musculus*. To improve interpretability and understanding of the forces affecting tooth size, the twenty-seven variables were reduced to five factors, of which the first three were identified as width, anterior length, and posterior length factors. Poole (1971, 1974) utilizied factor analysis for modeling natural communities of plants and animals and for measuring the structural similarity of communities composed of the same species.

Discriminant functions were developed by Fisher as a means to distinguish members of closely related taxa. The computations produce differential weights for the various characters. Those characters with the highest weights, loadings, are the most useful "discriminators" for separating two groups or taxa. A stepwise discriminant analysis can be used if more than two reference groups or taxa must be separated. Summed values of the discriminant scores are frequently plotted on a frequency histogram (Y-axis, individuals; X-axis, discriminant scores) to illustrate the separation between taxa (Fig. 11-8). Genoways and Choate (1972) were interested in analyzing the specific relationships

Figure 29–8.
Three-dimensional projection of 83 specimens of *Blarina* onto the first three principal components (the third component is indicated by height). Solid circles, *B. b. brevicauda* reference sample; half solid circles, *B. b. carolinensis* reference sample; open circles test specimens of both taxa collected near zones of contact. (Genoway and Choate, 1972)

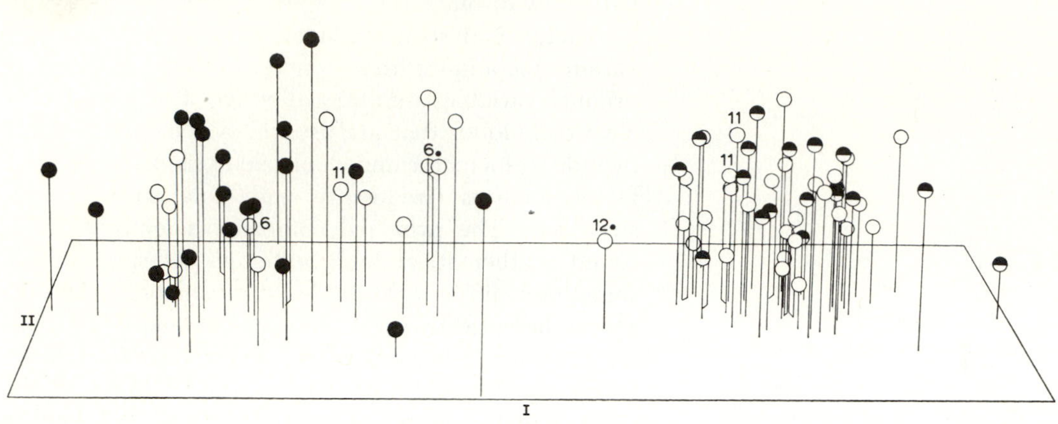

of two previously defined subspecies of short-tailed shrews (*Blarina*) occurring in a contact zone in Nebraska. After collecting specimens of these shrews from the contact zone, they wished to compare the morphology of these specimens with the morphology of reference specimens representing each of the two subspecies. The technique of discriminant function analysis permitted the calculation of discriminant scores for each of the two reference samples. Then, when the discriminant scores for the specimens in the contact zone were compared with the scores of the reference samples, the taxa were easily separated and potential hybrids (or intergrades) spotted (Fig. 11-8). Discriminant analyses were also utilized by Jolicoeur (1959) and Lawrence and Bossert (1967) in studies of canid populations and Robinson and Hoffmann (1975) in studies of geographical and interspecific cranial variation in big-eared ground squirrels (*Spermophilus*).

Supplementary Readings

Bliss, C. I. 1967. *Statistics in biology,* Vol. 1. McGraw-Hill Book Company, New York. 558 pp.

Campbell, R. G. 1974. *Statistics for biologists,* 2nd ed. Cambridge Univ. Press, London. 385 pp.

Rohlf, F. J., and R. R. Sokal. 1969. *Statistical tables.* W. H. Freeman and Co., San Francisco. 253 pp.

Siegel, S. 1956. *Nonparametric statistics for the behavioral sciences.* McGraw-Hill Book Co., New York. 312 pp.

Simpson, G. G., A. Roe, and R. C. Lewontin. 1960. *Quantitative zoology,* rev. ed. Harcourt, Brace and Co., New York. 440 pp.

Sokal, R. R. 1965. Statistical methods in systematics. *Biol. Rev.,* 40:337-391.

Sokal, R. R., and F. J. Rohlf. 1969. *Biometry.* W. H. Freeman and Co., San Francisco. 776 pp.

Steele, R. G. D., and J. H. Torrie. 1960. *Principles and procedures of statistics.* McGraw-Hill Book Co., New York. 481 pp.

Zar, J. H. 1974. *Biostatistical analysis.* Prentice-Hall, Englewood Cliffs, New Jersey. 620 pp.

30 Age Determination

Age is one of the most important parameters in studies of mammals. Knowledge of age structures of populations is necessary for the proper management of populations (Morris 1972; Taber 1969) and for understanding the life history strategies of mammals (Deevey 1947; Cole 1954; Caughley 1966; Wilson 1975). Methods of determining the ages of individuals in a population are reviewed by Gandal (1954), Morris (1972), Spinage (1973), Pucek and Lowe (1975), and Harris (1978); also see the bibliography by Madson (1967), and articles in journals such as *Acta Theriologica,* the *Journal of Mammalogy,* and the *Journal of Wildlife Management.*

Types of Aging Criteria

When samples of mammals are taken from a population, it is usually impossible to assign a **known age** to the specimens unless the date of birth was observed and the individual uniquely marked for later identification. An **absolute age,** can often be determined by counting incremental growth lines in various structures of wildcaught individuals. More commonly, a **relative age** may be assigned to an individual based on comparison with other individuals in the sample. The procedures for establishing absolute and relative ages should be standardized with known age individuals (see Morris 1972). However, if animals of known age live under captive conditions, the morphological changes that occur during development may be different than the changes experienced by individuals living in the wild and thus the results obtained from captive animals may not be accurate for purposes of aging.

A rough estimate of relative age is used by many taxonomists to segregate adult from younger individuals of a sample. In mammalian population biology, the category **adult** generally refers to the larger and potentially breeding members of a population. A **subadult** individual is generally a young of the year (in species with lifespans of 1-5 years) that may or may not be in breeding condition. This individual is typically smaller than an adult, but otherwise is similar to the adult. A **juvenile** individual is smaller than a subadult and often (but not always) has a pelage coloration that is distinct from that of subadults and adults. A **nestling** is a recently born individual that is still confined to a nest. In precocial species (e.g., artiodactyls, hares), the nestling stage is virtually nonexistant and only a juvenile stage is present.

30-A. Assemble a sample of mammals skulls of one species from a single locality and, preferably, a limited interval of dates of collection. Arrange the skulls in the order of their presumed age by examining tooth wear, degree of ossification of bones, etc. (see Chapter 2, The Skull). Can you separate these skulls into adult, subadult, and juvenile categories? If not, could arbitrary groups be established? How could you determine whether or not these groupings were meaningful biologically?

30-B. Examine skins from the same sample of speciments assembled in Exercise 30-A. Does information from pelage morphology and coloration aid in classifying these specimens into age groups? Do these procedures tell you the absolute ages of these specimens?

Use of Statistics and Known-Age Samples

It is important to have a reference sample of known-age individuals so that the efficacy of an aging procedure can be evaluated. Some of the aging techniques discussed will be more accurate than other techniques although they may involve more time and care in processing the samples. The nature of the investigation will determine what level of accuracy you should strive for and whether you can accept a larger error rate to save time for other procedures.

In studies of relative age, the technique of regression analysis is often used to predict ages (see Chapter 29, this book; Sokal and Rohlf 1969). With known-age individuals, age is a nonrandom variable (i.e., it is measured without error) and thus the quantity measured (Y, dependent variable) must be regressed on age (X, the independent variable). With samples of unknown age, age can be estimated by the method of inverse prediction (Sokal and Rohlf 1969:446-448; Dapson and Irland 1972; Dapson 1973) using values obtained from the measured variable (e.g., weight). In such cases, the computation of confidence limits for these regression lines is different from that of a Model I regression (Sokal and Rohlf 1969).

A representative sample of relative and absolute age determination methods will be described below under the type of structure involved.

Relative Growth of Skull, Skeleton, and Body Dimensions and Weight

In early life, growth in mammals is continuous and thus provides a means for separating the youngest individuals from adult members of a population. Some structures cease growth sooner than others. Not all morphological features are useful for aging individuals in a population. Increases in linear dimensions and weight may be useful indicators of age during the earliest portions of a mammal's life (Fig. 30-1) but rapidly lose their usefulness once adult dimensions are reached (Kirtpatrick and Hoffman 1960). These measurements are frequently utilized in live-trapping studies because of their simplicity and because of the frequent lack of other suitable criteria obtainable from live mammals.

Degree of Fusion of Epiphyseal Cartilage

The degree of fusion (determined by X-ray analysis or analysis of skeletons) of the distal epiphyseal cartilages of the radius and humerus has been used in many management studies to age hunter-killed specimens of rabbits and other animals (Sullivan and Haugen 1956). The technique is less accurate than most other methods (Wight and Conaway 1962) but is readily applicable to samples where the investigator has little control over the type and method of preservation employed on the samples.

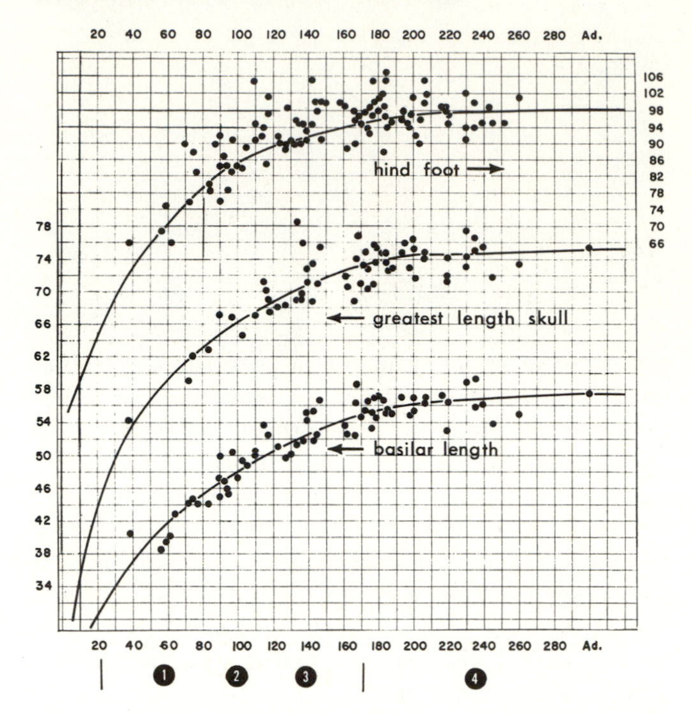

Figure 30–1.
Changes in the length of the hind foot and two cranial measurements in known-age individuals of *Sylvilagus floridanus*.
(Hoffmeister and Zimmerman 1967: 200)

30-C. Assemble a sample of known-age skulls (perhaps obtainable from laboratory colonies, mink ranch, research project). Select several cranial measurements (see Chapter 2) and measure the skulls. Using age as the independent variable (X) and a measurement variable as the dependent variable (Y), perform a regression analysis on these data (see Sokal and Rohlf 1969 or similar reference).

30-D. Examine skeletal elements (pelves, distal elements of limbs) of a sample of known-age material (see 30-C). Is the degree of fusion of the epiphyses a useful indicator of age? At what age (in the species studied) do the epiphyses completely ossify?

Baculum

For male mammals, the weight, length, and volume of the baculum often provides a means for separating a juvenile from an adult individual (Friley 1949; Elder 1951). More precise age determinations are generally not possible using this bone (Morris 1972; Harris 1978).

Figure 30–2.
Relative age classes based on tooth wear in museum specimens of spiny rats, *Proechimys guyannensis.*Tooth rows (A-E, left upper; F-J, left lower) arranged from left to right in order of increasing tooth wear and presumed age. Dotted lines indicate teeth that are not at occlusal level. True chronological age unknown.
(Martin 1970: 4)

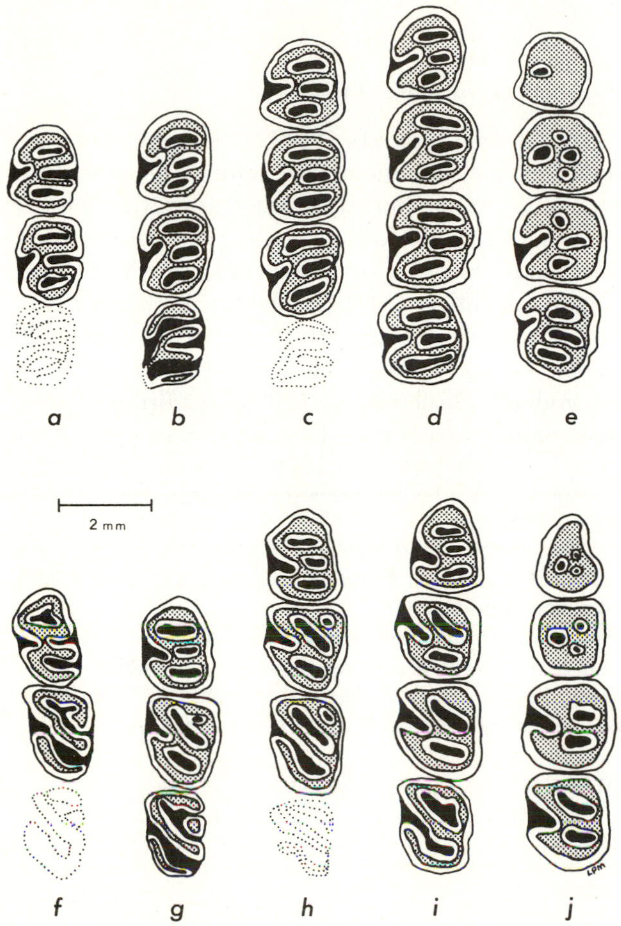

Figure 30–3.
Lateral views of upper molars (M^3) in wart hogs (*Phacochoerus aethiopicus*) of two to 15 years of estimated age. The X (occlusal surface) and Y refer to measurements used to derive ratios (X/Y) useful for aging this species.
(Spinage, C.A., and G.M. Jolly. 1974. Age estimation of warthog. J. Wildlife Manag. 38(2): 229- 233, fig. 1.)

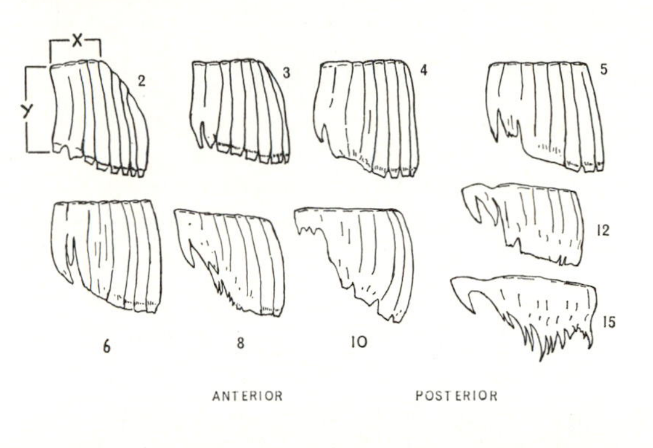

Relative Growth and Morphology of Teeth

Tooth Wear

Examination of wear on teeth has been widely utilized to separate mammals into age groups for taxonomic studies (Fig. 30-2) and for population and management studies (Fig. 30-3). Spinage (1973) noted that the pattern of wear on teeth generally follows a negative exponential curve. Thus, younger animals tend to be classified as older than their true age. Harris (1978) found that 65.5% of the red foxes (*Vulpes vulpes*) he examined could be aged correctly with an error of 1 year or less. Aging was accurate in 93.3% of the cases with foxes up to four years of age. The accuracy of this technique also depends on the type of tooth examined. The incisors are useful for aging many species of canids (Harris 1978) whereas molars are useful for aging cervids (Taber 1969). Generally, this technique is less accurate for aging than methods based on discrete data (e.g., tooth eruption, annulations in tooth or cementum; Morris 1972).

Tooth Eruption

In many rodents, the pattern of tooth eruption is a useful indicator of age during the first few months following birth. In artiodactyls, tooth eruption sequences may be used to group individuals into year or seasonal classes during the first two to three years of life (Taber 1969).

30-E. Examine teeth in the sample of skulls assembled for Exercise 30-C. Can you group these specimens into age categories (or an age sequence) based upon tooth eruption sequences and tooth wear without relying upon the known ages? How do your tooth wear age categories compare with the true ages of the specimens? Are the age categories arbitrary? Perform a nonparametric test (rank correlation) on these data to compare estimated age with true age (Siegel 1956: 202-223; Sokal and Rohlf 1969: 532-538).

30-F. Examine the sample of skulls utilized in Exercise 30-E. Can you arrange these specimens into age groups based on the sequence of tooth eruption (and/or replacement of the deciduous dentition by the permanent dentition)? For what ages of animals do tooth eruption sequences appear to be most useful? What statistical test(s) could be used to check your results?

Relative Growth of Eye Lenses

Weight of Eye Lens

The weight of the eye lens increases with age (Fig. 30-4) and thus many studies of age structures of populations have relied on weighing the lenses removed from mammal specimens (see Lord 1959 and review in Friend 1968; Morris 1972). Several investigtors (Adamczewska-Andrzejewska 1973; Myers, *et al.* 1977) found eye lens weights to be accurate for estimating ages of several species of myomorph rodents. Morris (1972) reported that the technique has been most successfully applied with animals of medium size (e.g., rabbits, hares) during the period of rapid growth prior to the attainment of adult size. Harris (1978), studying red foxes (*Vulpes vulpes*), did not find lens weights to be useful for separating year classes of animals that had attained adult size.

Refer to Friend (1968), Morris (1972), and Harris (1978) for precautions that must be followed to insure accuracy and precision in the measurement of lens weights.

Measurement of Lens Protein

Dapson and Irland's (1972) refined technique found that the amount of the soluble tyrosine (lens protein) increased linearly during approximately the first year of life in old-field mice (*Peromyscus polionotus*) and was superior to any other method known for aging mammals. They also found that the increase in the insoluble tyrosine fraction was a curvilinear function of age during the first 750 days of life in these rodents. Birney et al. (1975)confirmed the accuracy of this technique (Fig. 30-5) for aging cotton

Figure 30–4.
Changes in body weight and lens weight with age in cotton rats (*Sigmodon hispidus*). Heavy lines indicate 95% confidence intervals. (Birney, E.C., R. Jenness, and D.D. Baird. 1975. Eye lens proteins as criteria of age in cotton rats. J. Wildlife Manag., 39(4): 718–728, fig. 1. Copyright 1975 The Wildlife Society. Reprinted with permission.)

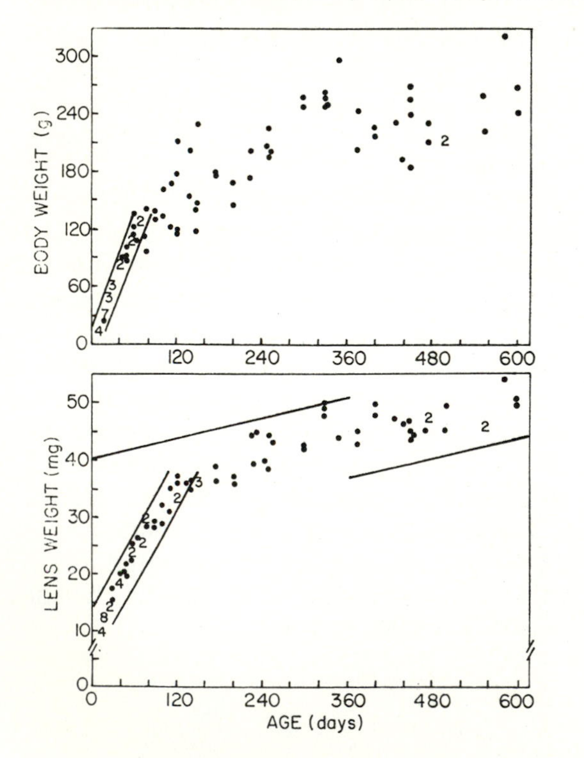

Figure 30–5.
Changes in soluble and insoluble lens protein with age in cotton rats (*Sigmodon hispidus*). Heavy lines indicate 95% confidence intervals.
(Birney, E.C., R. Jenness, and D.D. Baird. 1975. Eye lens proteins as criteria of age in cotton rats. J. Wildlife Manag., 39 (4): 718–728, fig. 2. Copyright 1975 The Wildlife Society. Reprinted with permission)

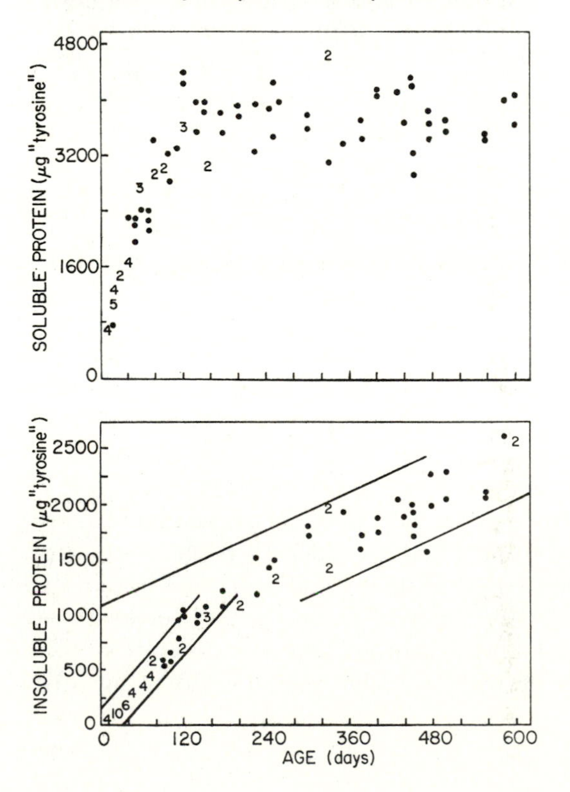

rats (*Sigmondon hispidus*) but cautioned that the most accurate technique is not necessarily the one that should be used if time is limited or the needed specialized equipment is lacking.

30-G. Compare the changes in body weight and lens weight in Figure 30-4 with the changes in soluble and insoluble lens protein in Figure 30-5. How do the estimates differ? At what ages do the techniques produce similar results. Examine papers by Wight and Conaway (1962), Tiemeir and Plenert (1964), Harris (1978) and/or other papers dealing with multiple techniques for aging mammals.

Absolute Growth Lines

The growth of teeth and bones in mammals is not uniform throughout the year. Thus, narrow and wide layers of dentine, cementum, and bone may be laid down in different seasons of the year or in annual increments (Klevezal and Kleinenberg 1967; Morris 1972). Generally, these **growth lines** or **incremental lines** are the only useful criteria for distinguishing year classes of adult individuals in a population other than marking groups of the same age (i.e., cohorts) when they are young or recently born. These growth lines also provide an absolute age for an individual since the units are discrete and not subject to the continuous variation inherent in the criteria for relative growth described previously (but see cautions in Harris 1978).

Teeth: Dentine and Cementum

Growth lines in teeth have been utilized widely for aging the teeth (Fig. 30-6) of marine mammals since publication of the paper by Laws (1952). Subsequently, the technique has been applied successfully to other long-lived species such as moose, bear (Stoneberg and Jonkel 1966), caribou (Miller 1974), and other species. Harris (1978), in studies with red foxes, found that only sections of the tooth cemetum gave good resolution of growth lines that could be demonstrated consistently. Growth lines of dentine were less reliable for aging and he also cautioned that the determined age should be based on sections of at least two teeth from the same individual.

Erickson and Seliger (1969) and Morris (1972) summarized the various methods used to prepare sections of teeth for study. For larger teeth (whales, pinnipeds, artiodactyls), a section is made of a tooth and then ground down to a thin layer using carborundum powder or a grindstone. Smaller teeth can be treated similarly if the teeth are mounted on a cork block (or slide) for easier handling. A second method for making sections involves decalcification and sectioning of the teeth using histological procedures. Formic acid is the most commonly used decalcifying agent although, with caution, a dilute solution of nitric acid is also permissible (Morris 1972). Sections are cut with a microtome. Hematoxylin stain generally gives good resolution of the growth lines (Harris 1978).

Figure 30–6.
Outline drawing of tooth section from a bottlenose dolphin (*Hyperoodon ampullatus*) showing growth lines in the dentine. C, cementum; D, dentine; E, enamel; O, osteodentine; P, prenatal dentine.
(Christensen 1973: 333)

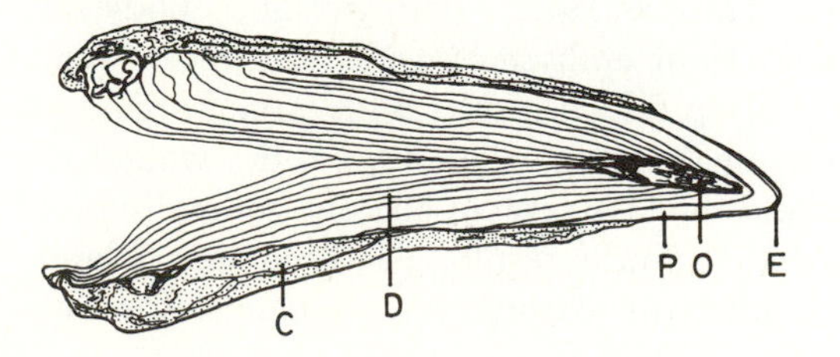

30-H. Select a carnivore or artiodactyl tooth (incisors are frequently utilized since the permanent dentition in these teeth appears early in development) with which you can practice the preparation of thin sections. Cut thin slices of the tooth and mount in resin on a slide. Grind the tooth to a thin section using various abrasive grits and then examine the section under transmitted light of a microscope (students should select different ages of teeth for later comparison among themselves). Can growth lines be detected in the sections (it may be necessary to stain them lightly with hematoxylin; Morris 1972)? If known age material is available, how well do the growth lines compare with the true chronological ages? Is there always only one growth line per year? Two lines per year? Can you correlate growth lines with seasonal or climatic effects in the environments where the animals live?

Periosteal Lines in Bone

Millar and Zwickel (1972) and Franson et al. (1975) successfully utilized sections of mandibles to age, respectively, pikas (*Ochotona princeps*) and mink (*Mustela vison*). Usually, sections of mandibles are prepared by decalcification and sectioned using histological procedures. Morris (1972) recommends that formalin-fixed material be utilized since the process of cleaning the bones may result in the loss of some structural detail. Minor accessory growth lines may be seen among the true annual growth lines in bones (Morris 1972). Periosteal growth lines are generally thicker than growth lines in cementum and are thought by some workers to be easier to interpret for this reason.

Growth Lines in Horns

Seasonal changes in forage quality cause the deposition of the keratinized epithelian layers to be unequal (Taber 1969). Thus, these growth lines (Fig. 30-7), with adjustments, can be used to estimate year classes of members of the family Bovidae (Artiodactyla) that possess horns (Morris 1972). Murie (1944) successfully utilized this technique in a classic study of Dall sheep (*Ovis dalli*) mortality. Caughley (1965, 1966) summarized the use of this technique in the preparation of a life table for the Himilyan thar (*Hemitragus jemlahicus*) that was introduced into New Zealand in 1904. Geist (1966), working with known-age bighorn sheep *(Ovis canadensis)*, found that counts of growth lines were satisfactory for aging males but gave erratic results with females.

30-I. Examine a series of bovid horns (preferably of known age) of the same species. Can growth lines be detected on the horns? Can you differentiate between annual and minor growth lines? If possible, compare horns of species living in temperate climates with those of species living in tropical climates. Does this difference in seasonal change affect the growth lines?

Figure 30–7.
Horn of Himalayan thar (*Hemitragus jemlahicus*) showing minor growth lines and true annual growth lines. (After Caughley 1965)

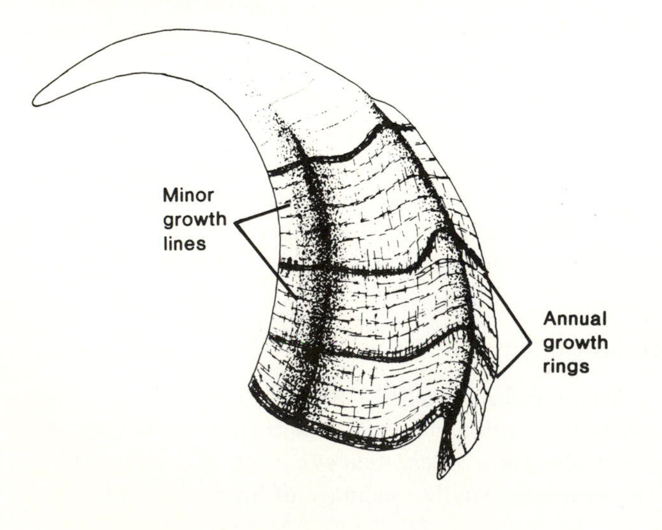

Growth Ridges in Baleen

Growth ridges are formed in the baleen plates of mysticete whales as new material is added at the base of these keratinized structures. Seasonal variations in growth rates result in ridges of different heights that can be related to annual increments of age. Although Morris (1972) indicated that it is difficult to obtain precise results. Further details and a review of the literature can be found in Jonsgard (1969).

Epithelial Ear Plugs of Whales

Each auditory meatus of a mysticete whale is closed below the layer of the blubber. Consequently, the epithelial lining that sloughs off the walls of this canal cannot exit from the body of the animal and instead forms a layered ear plug (Morris 1972). Alternate light (higher fat content) and dark layers apparently represent seasonal feeding patterns. Roe (1967) provided a review of the literature in this field along with data concerning the fin-backed whale (*Balaenoptera physalus*).

Counts of Antler Tines Inaccurate for Aging Individuals

There is a popular misconception among many hunters and laymen that the numbr of tines or "points" possessed by a deer is an accurate index of the age of the animal. Cahalane (1932) reviewed the evidence for this claim and found that there was no correlation of age with the number of tines on a set of antlers.

Limitations of Age Determination Methods

Few studies of age determination of mammals demonstrate equivalent age estimates when different methods are compared (Morris 1972). Body weights, body dimensions, degrees of epihyseal fusion, and tooth wear indices generally produce results with large variances. For aging mammals prior to adulthood, the weight of the eye lens and the lens protein content generally produce the most accurate results (Dapson and Irland 1972). For animals in seasonal environments that live more than one year, the development of annuli in teeth (dentine or cementum) or bones has proved to be a very successful and accurate method for aging mammals (Morris 1972; Harris 1978). For any species studied, tests should be made with known-age animals using a variety of methods to evaluate the accuracy of the procedures and determine whether measurements can be made precisely to minimize measuring error.

Supplementary Readings

*Caughley, G. 1966. Mortality patterns in mammals. *Ecology* 47:906-918.

Friend, M. 1968. The lens technique. *Trans. 33rd. N. Amer. Wildl. Nat. Res. Conf.* 33:279-298.

Gandal, C. P. 1954. Age determination in mammals. *New York Acad. Sci. Trans.* (Ser. 2) 16:312-314.

*Included in Jones, Anderson and Hoffman (1976).

Harris, S. 1978. Age determination in the red fox (*Vulpes vulpes*)—an evaluation of technique efficiency as applied to a sample of suburban foxes. *J. Zool., London* 184:91-117.

Jonsgard, A. 1969. Age determination of marine mammals, pp. 1-30. *In* H. T. Andersen (Ed.). *The biology of marine mammals.* Academic Press, London.

Klevezal, G. A., and S. E. Kleinenberg. 1967. *Age determination of mammals from annual layers in teeth and bones.* (English Translation, Israel Program Scientific Translations, Jerusalem, 1969. 127 pp.)

Madson, R. M. 1967. Age determination of wildlife. A bibliography. *U.S. Dept. Interior, Bibl.* 2:1-111.

Morris, P. 1972. A review of mammalian age determination methods. *Mammal Review* 2:69-104.

Pucek, Z., and V.P.W. Lowe. 1975. Age criteria in small mammals, p. 55-72. In F. B. Golley et al. (Eds.). *Small mammals: their productivity and population dynamics,* I. B. P. Handbook No. 5. Cambridge Univ. Press, Cambridge.

Spinage, C. A. 1973. A review of age determination of mammals by means of teeth, with especial reference to Africa. *E. African Wildl. J.* 11:165-187.

31 Diet Analysis

Studies of mammalian diets are important for understanding niche relationships, competitive processes, predation, and the influences that mammals exert on natural and cultivated ecosystems. Today, mammalian diet studies not only involve quantitative determinations of foods consumed but quantitative estimates of food resource abundance so that the role of diet selectivity and preference can be examined.

For brevity, coverage in this chapter is limited principally to discussions of some techniques for analyzing diets by macroscopic and microscopic means. Some material is also presented on methods for determining resource levels and making ecological interpretations. Further information on many of these topics can be found in Baumgartner and Martin (1939), Scott (1943), Dusi (1949), Adams (1957), Storr (1961), Gebczynska and Myrcha (1966), Sparks and Malechek (1967), Williams (1962), Korschgen (1969), Hansson (1970) Cox (1976), and Westoby et al. (1976).

Collection of Samples

Types of Samples

Most studies of diet in mammals rely on fecal samples obtained from free-living or live-trapped mammals or stomach or intestinal samples from dead mammals. In studies of domestic mammals, a fistula operation may be performed to allow the sampling of rumen contents of live animals at periodic intervals. In collecting fecal samples from free-living mammals, an investigator must be reasonably confident that a sample to be analyzed is derived from a single species. Fecal samples from live-trapped individuals or stomach and/or intestinal content samples from dead specimens are preferred since these allow verification of the specific identity of the individual whose diet is being measured. Samples obtained from stomach contents generally present a less biased picture of the food items eaten by the individual (Hansson 1970). Information on sex and age variables cannot be determined from an isolated pellet or scats collected in the field, but these data are available with the other techniques for gathering samples.

Live traps provide a way to obtain fecal samples without killing the animals. Generally, the baits (e.g., peanut butter, seeds, rolled oats) used in the traps are readily distinguishable from the foods that the animals are feeding on under wild conditions. This technique, most often used with rodents (e.g., Meserve 1976), is potentially adaptable to other species and offers a way to monitor individual and population dietary patterns throughout seasons of the year. The live traps must be thoroughly cleaned before they are reset.

Contents of the cheek pouches of geomyid and heteromyid rodents may provide information on the diet of these mammals although the results of Reichman (1975) suggest that the diet determined by these samples underestimates the total diet.

Analyses of food caches of mammals can provide information on accumulated dietary preferences of a species (Korschgen 1969). But little quantitative information on individual and seasonal dietary trends can be determined by this method.

Number of Samples

The number of samples collected for analyses is determined partly by the nature of the investigation (e.g., local or regional applicability) and the amount of time available for collecting the samples. Hanson and Graybill (1956) utilized a statistical procedure to determine the number of samples required to estimate the food habits of a population. Korschgen (1969) provided additional guidelines for determining adequate sample sizes.

Preservation and Labeling of Samples

Korschgen (1969) and Hansson (1970) discussed methods used to preserve samples for dietary analyses. Freezing of samples produces the least alteration of material but is frequently impractical for field situations. Drying of scats or pellets is a convenient method for most mammals, especially carnivores, if precautions are taken to prevent mold (oven dry at 80-85° C for several hours) and later insect infestations (store in tight container with moth flakes, napthalene).

Stomach or intestinal contents may be frozen (possibly impractical for field situations) or preserved in 5% formalin (1 part stock formaldelhyde in 19 parts of water) for small to medium-sized species and in 10% formalin (1 part stock formaldehyde in 9 parts of water) for larger species. Preservation with 70% alcohol can also be utilized although this preservative will extract chlorophyll and is thus less satisfactory for studies with herbivorous species. Hansson (1970) found that formalin-preserved material was superior to alcohol-preserved material for quantitative determinations although it caused slight overstaining in final slide preparations. To achieve better preservation, the stomachs or intestines of larger species should be injected with the fixative and then immersed into the fluid. For the largest species (e.g., ruminants) a quart or liter sample may be taken from the stomach or intestinal contents and then preserved in a 5% formalin solution (Korschgen 1969).

The sample for dietary analysis should be properly tagged (labeled) with full information on species, sex (if known), date of collection, locality, and the collector's number (see Chapter 33). For samples from dead material. the collector's number is adequate if the remainder of the field data are available in the catalog (see Chapter 33). In live trapping studies, the number of the individual, sex, date of trapping session, and grid should be indicated on the label or package that holds the sample. For fluid-preserved material, always place the label *inside the container* that holds the sample to prevent its accidental loss. Stomachs or intestines may be individually wrapped in cheesecloth and labeled and then placed in a single container of preservative to save space (unless the results of certain biochemical analyses might be altered by this combined storage).

Identification of Food Items

Reference Samples

Properly identifed reference materials are essential for anlyzing dietary samples of mammals (Korschgen 1969; Hansson 1970). For larger mammals that consume other vertebrates as prey, these reference materials should include skulls and skeletons of the prey species and their epidermal elements such as scales (fish), feathers (birds), and hair (mammals). Mammalian prey items can be identified most easily using skull and skeletal elements. Where skeletal elements are fragmentary or inconclusive, it may be necessary to have a reference collection of guard hairs or impressions of the cuticular scales of the guard hairs. To observe pigmentation patterns and medullary characteristics, hair samples can be cleaned in ether (**Caution:** extremely flammable and volatile), dried, and covered with Permount® (or similar mounting medium), and then sealed with a coverslip. Cross sections of hairs can be prepared using the technique of Mathiak (1938).

Cuticular scale patterns provide some of the most useful characters for identifying hair samples (Fig. 4-2). The technique of Williamson (1951) and Korschgen (1969), or that described in Exercise 4-I will enable you to obtain impressions of these cuticular scale patterns.

This technique has proved useful in anlyses of the diets of stoats and weasels (*Mustela,* Day 1966) and wolverines (*Gulo gulo,* Myhre and Myrberget 1975).

31-A. Select samples of guard hairs from several species of prey mammals that occur in a local geographical area (collect hairs from comparable areas of the body since hair shape and pattern may vary between regions of the body). Prepare reference slides using the technique described in Exercise 4-I. Do you see differences in cuticular scale patterns? Would these patterns be useful for identifying species? If not, to what taxonomic level could the prey items be identified?

For invertebrate prey items, appropriate dry and wet specimens of identified taxa should be available for reference. Invertebrate prey items are generally partially digested and broken up in the intestinal tract of the predator. Consequently, Hansson (1970) utilized the following technique to prepare reference slides of these prey animals so that similar-sized fragments would be represented in the reference slides:

1. Place reference animals in a 1% solution of pepsin at pH 1.0-2.0 and incubate at 40°C for about 4 hours.
2. Wings, legs, and other sclerotized structures should be cut into small pieces and placed in a vial of clearing solution.
3. Krantz (1978) discusses several agents that can be used for clearing sclerotized tissues. **Lactophenol** is one

of the most commonly utilized clearing agents for invertebrates. This agent should be prepared with the following ingredients added *in sequence*:

Lactic acid	50 parts
Phenol crystals	25 parts
Distilled water	25 parts

Immersion of material in lactophenol for 24-72 hours should be sufficient for clearing most specimens. Rinse in 3-4 changes of water until cloudiness disappears.

4. Following clearing, the fragments should be spread in a thin layer on a microscope slide. Place a drop(s) of an aqueous mounting medium (Krantz 1978) followed by the animal fragments onto a microscope slide. Then, using a pair of forceps, pick up a coverslip at its rim, apply the opposite edge to the droplet of mounting medium, and then allow the coverslip to fall into place.

   Hoyer's medium is a widely utilized aqueous mounting medium and can be prepared by mixing the following ingredients *in sequence* (make sure each solid ingredient is completely dissolved before adding additional reagents; *do not heat* the mixture):

Distilled water	50 ml
Gum arabic (crystalline)	30 g
*Chloral hydrate	200 g
Glycerine	20 ml

5. Mark the slide with an identifying number and name until a permanent label can be prepared. Allow the newly prepared slide to dry in an oven at 45°C for 48 hours to 1 week.
6. Remove the slide from the oven, allow it to come to room temperature, and then place a ring of **Glyptal® (waterproof paint for electrical circuits; General Electric Company) around the perimeter of the coverslip to prevent movement of water into or out of the mounting medium (infiltration of atmospheric moisture causes a breakdown of the mounting medium and deterioration of the slide preparation).
7. Prepare a permanent label for the slide that indicates the name of the reference taxon, the components preserved, date and locality, and the names of the clearing and mounting media.

31-B. Collect several species of invertebrates (insects, arthropods, etc.), preserve in 70% alcohol, rinse in decreasing concentrations of alcohol followed by a final rinse in water. Then, separate the specimens by taxa and place into individually labeled vials. Fresh dead or dried material and those in an aqueous medium can be passed directly to the water rinse.

*Regulated substance, available only for research and professional purposes.

**Clear nail polish or Permount® (or similar synthetic mounting medium) can also be used but the preparations will not be as permanent.

Use a pepsin solution to dissociate the elements (see above or Hansson 1970). Prepare slides using an aqueous mounting medium. What types of structures appear to be diagnostic for identification? To what taxonomic level can the taxa be identified?

Reference material of plants should include seeds and intact seed pods (generally, items > 0.4 mm in diameter) and tissues taken from the plants for microscopic analysis (Hansson 1970; Korschgen 1969). Often seeds can be identified by using references such as Martin (1946), USDA (1948), Martin and Barkley (1961), and Musil (1963).

Microscope reference slides (or photomicrographs) are necessary for the identification of finely masticated plant materials found in the stomach, intestines, or feces of mammals. These slides are generally prepared using the techniques pioneered by Baumgartner and Martin (1939) and Dusi (1949) and discussed in detail by Hansson (1970). Ideally, plant material for reference slides can be prepared by feeding plant samples to a target species and then making slide preparations from samples (feces, stomach or intestinal contents) obtained from individuals of this species (Hansson 1970). However, this method is tedious (Hansson 1970) and not often utilized. Another alternative is to grind dried plant material to uniform size in a Wiley Mill and then prepare reference slides (Reichman 1975). Lastly, epidermal tissues can be stripped from plant samples, treated with Hertwig's solution (Baumgartner and Martin 1939), and microscope slides prepared for use in identification.

Epidermal tissues of plants are frequently utilized for identification purposes since they possess useful diagnostic features (Fig. 31-1) such as siliceous cells, stomata, spines and hairs (Dusi 1949; Williams 1962). But reference slides should include other material (e.g., portions of flowers with pollen, seed husks and seed endosperms) that may be encountered in the samples to be analyzed. The following technique, adapted, in part, from Baumgartner and Martin (1939), Dusi (1949), Hansson (1970), Reichman (1975), and Meserve (1976) can be utilized to prepare reference slides of a variety of materials:

1. From the reference sample, strip pieces of epidermis from leaves (both sides) and stems and section portions of other plant material (flowers, seeds) and place on a microscope slide (the material may need to be mascerated in Hertwig's solution before it can be stripped; Dusi 1949). Alternatively, the plant material can be dried, and a sample ground in a **Wiley Mill** to uniform particle size (this helps to minimize bias since in examining slides, larger particles

Figure 31–1.
Examples of types of stomata found in different species of fox squirrel (*Sciurus niger*) foods. A,B,C, *Juglans nigra*; D, *Carya alba*; E, *Prunus serotina*; F, *Cornus alternifolia*. (Baumgartner and Martin 1939)

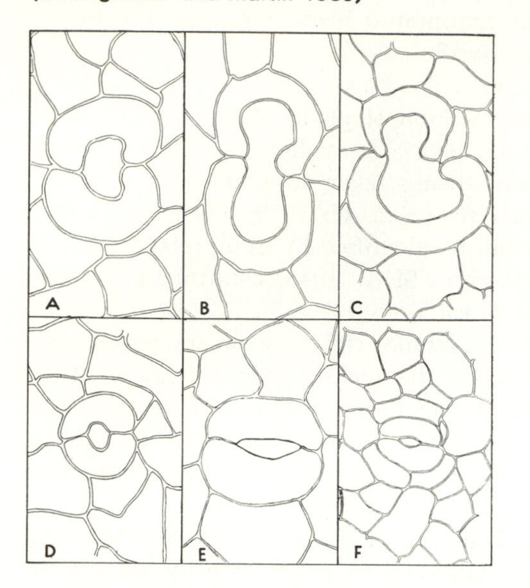

tend to be encountered by the eyes more readily than smaller particles; Westoby et al. 1976).

2. Place a few drops of **Hertwig's solution** (Baumgartner and Martin 1939) on the slide containing the reference material. Allow the solution to boil for a few seconds on a hot plate (100-105° C).

Hertwig's solution (add reagents *in sequence*)	Distilled water	150 ml
	HCl	19 ml
	Glycerin	60 ml
	*Chloral hydrate	270 g

3. Allow the slide to cool quickly in air, rinse *very gently* with tap water to remove most of the Hertwig's solution (be careful not to lose the plant material).
4. The sample can now be stained by any aqueous mixture unless it is determined that staining is not necessary for identification of the material. Hansson (1970) recommended hematoxylin as a general purpose stain since it does not stain suberin (a constituent of some cell walls) and thus provides good contrast with tissues that are stained by this dye. For a hematoxylin stain, 10-15 minutes should be sufficient to produce the desired color (light blue). Overstaining can be corrected with acid alcohol using techniques described in Guyer (1953) or Humason (1967).
5. Rinse the slide *gently* with tap water to remove excess stain and then touch a piece of blotting paper to the slide to draw off excess water.
6. Add one or two drops of Hoyer's medium (see section on preparing slides of invertebrates) or any commercially available water-soluble mounting medium to the slide preparation, arrange the reference material in the desired positions, and then drop a cover slide on top. Mark the slide with a temporary label (I.D. No. and Initials, e.g.) and then place it in an oven (45° C) to dry for 48 hours to 1 week. Remove the slide from the oven, allow it to come to room temperature, and then ring the slide with Glyptal® to make a permanent preparation (Krantz 1978).
7. Prepare a permanent label for the slide listing the name of the reference material, the portions of the plant preserved, the date collected, and the techniques utilized.
8. A photomicrographic key can then be prepared using these slides.

*Regulated substance, available only for research and professional purposes.

31-C. Collect samples of several species of woody and herbaceous plants and grasses from a riparian area. Each student should prepare a reference slide using the technique described above. Exchange and compare slides. What structures appear to be most useful for identification of the taxa? Are species easily identified as shrubs, herbs, or grasses? Are genera easier to separate than species? What structural types of plants are most easily identified? Repeat the procedure using plants from a grassland area.

31-D. Examine reference slides and prepare a diagnostic key (see Chapter 12 for hints on writing keys) to identify a sample of at least 10 taxa. What characteristics seem to be most useful for identification purposes? Would it be easier to learn to identify the taxa on sight rather than key them out?

Analysis of Dietary Samples

Preparing Samples for Study

The contents of cheek pouches and the dried scats of large mammals (other than herbivores) can be examined with little or no special preparation. If only bones are to be examined, the hair in the scats can be separated with a dilute (10-20%) solution of NaOH. The dietary items in these samples can be compared with a reference collection to identify the contents. Weight, volumetric and/or frequency of occurrence determinations can then be made (Korschgen 1969; Hansson 1970).

Microscope slides must generally be prepared for the dietary samples (stomach contents, feces) of herbivorous species (Baumgartner and Martin 1939; Dusi 1949; Hansson 1970) since identifying characteristics may only be apparent under microscopic examination.

Generally, fresh, dried, or preserved material can be prepared using the technique described for making reference slides. Additional recommendations for analytical procedure can be found in Hansson (1970).

Quantitative Determinations

Ecologically, one of the most meaningful statistics in dietary analyses is a measure of the *weight* of food consumed of each species (or taxon). This measure can then be directly related to energy flow and analyzed by parametric statistical procedures (see Chapter 29). A volumetric determination is also meaningful and often utilized with species that do not masticate their food into fine particles. Measurements of volumes are made using a graduated cylinder or by displacement of a known quantity of water in a burette (Inglis and Barstow 1960).

Volumetric determinations are very difficult to accomplish with finely masticated food items and investigators utilize two approaches to obtain estimates of dietary intake. One method estimates the *percent coverage* of a species on a slide (generally estimated to nearest 10-20%) to obtain an index of volume (Keith et al. 1959; Myers and Vaughn 1964; Gebczynska and Myrcha 1966; Meserve 1976). Usually ten randomly selected fields are examined under a microscope (35-125X) and the percent coverage of fragments noted in each field.

Another widely used method (Sparks and Malechek 1967; Free et al. 1970; Hansen and Ueckert 1970; Reichman 1975) is based on determining the *relative frequency of occurrence* of a food item (number of occurrences of a species/total number of occurrences of all species) and then converting these values to *relative densities* using the table in Fracker and Brischle (1944).

As detailed by Hansson (1970) and Westoby et al. (1976), neither of these techniques yield results that are satisfactory for all types of plants and plant communities. Plant taxa that are very rare may be missed completely and those that are abundant in the plant community may be overestimated in the analyses. However, these techniques are the best that are presently available and, as Hansson (1970) noted, may represent the practical limits for further quantitative refinements. Standardization using uniform diets (see next section) is an important component of dietary studies to identify sources of error in the analyses.

Standardization Using Uniform Diets

It is useful to conduct feeding trials (Meserve 1976; Westoby et al. 1976) using plant and/or animal mixtures of known composition to minimize bias in analyses of dietary samples. These mixtures can then be fed to animals and samples collected from the feces or stomachs of these individuals. Then, after preparing microscope slides of the samples (or examining the samples macroscopically in the case of carnivores or other large nonherbivorous species), the dietary compostion can be estimated using the volumetric and relative frequency methods described above. Experimentor bias can be minimized by randomizing the samples so that the estimator does not know the true dietary composition until the completion of the estimates. Westoby et al. (1976) standardized the procedure further by preparing diet mixtures and then grinding them to uniform particle sizes (to eliminate bias in degree of mastication in different species and individuals). Then, slides were made from these artificial samples, analyzed, and the estimates compared with the known values.

31-E. Obtain a series of microscope slides of feces or stomach contents.

Examine 100 fields at 100X magnification. Obtain frequency data by recording the presence or absence of a taxon in each field. Obtain volumetric data by determining the proportion of the field occupied by each taxon. Convert the frequency data to relative frequency by dividing by the total number of fields that contain fragments of any taxon.

How do these two methods compare in ease of application and in results? Are the results comparable? Did this exercise provide information on actual dietary composition? If not, what could be done to test the accuracy of these techniques?

Determination of Resource Levels

Before preferences (and conversely, aversion or disregard) of animals for food items can be determined, the resource levels of the plant and animal community must be measured. For plants, these measurements are straightforward and techniques can be found in most plant ecology and general ecology tests (e.g., Oosting 1956; Brower and Zar 1974; Cox 1976). The measurement of mammalian resource levels (and those of many vertebrates) may involve trapping or line censuses (see Chapter 38). Population levels of nonmammalian vertebrates can be estimated using techniques described in Pettingill (1956) for birds and Tinkle (1977) for reptiles and amphibians.

The measurement of arthropod population levels may be more difficult due to the tremendous variety of habitats that these animals occupy and the seasonal and diurnal variations in their activity. Southwood (1966) and Bram (1978) give methods for estimating population levels in many species of arthropods.

Seed resource levels can be determined by removing seed pods or heads directly from the plants along transects or quadrats selected from the vegetation analyses. Seeds can also be extracted from soil samples using a flotation technique (see Franz et al. 1973; Reichman 1975).

Calculating Indices of Preference and Dietary Overlap

Mammals can and often do exhibit dietary preferences and thus it is important to be able to measure such preferences in the analyses of samples.

Preference Indices

Food preferences can be determined once resource levels have been quantified and analyses completed on the food samples. Indices of preference are based on a comparison of the frequency of an item in the dietary sample compared with that in the resource. In the method detailed by Reichman (1975), values greater than 1.0 indicated preference while those less than 1.0 suggested avoidance or disregard of the resource.

$$\text{Preference} = \frac{\text{relative frequency in dietary sample}}{\text{relative frequency in resource}}$$

(after Reichman, 1975)

The **electivity index** of Ivlev (from Alcoze and Zimmerman 1973) is another approach to calculating dietary preference. This index is calculated using the formula

$$E = \frac{(r_i - p_i)}{(r_i + p_i)}$$

where r_i = % volume of a food item in the diet and p_i = % density of the same food item in the resource. A value of zero indicates randomness in resource selection, a positive index value indicates presumed preference for a dietary item, and a negative value indicates presumed avoidance or indifference to a dietary item.

31-F. Compute dietary preference indices using data in Reichman (1975) or Meserve (1976). Compare results from using the two techniques described above. Do the methods produce similar results? Recompute the indices after changing the proportions of the food items in the resource and/or samples. What differences did these changes produce in the values of the indices?

Dietary Overlap Indices

Mammalian ecologists frequently want to know the degree of **dietary overlap** that exists between two or more species. The degree of dietary overlap measured between species may vary along the sampled resource gradiant (Colwell and Futuyma 1971; Cody 1974) and thus it is important to consider this effect in formulating problems.

Dietary overlap is a subset of a broader concept termed **niche overlap** that may include resource states other than those of diet. For many practical reasons, diet and habitat measures are frequently major components of niche measurements of mammals and other vertebrates. You may recall that most studies of niche relationships actually measure properties of the **realized** (or existing) **niche** of a species rather than the broader and theoretical (niche in absence of all other species) **fundamental niche** of a species (Hutchinson 1957; Colwell and Fuentes 1975). **Niche breadth** is a measure of the magnitude of niche dimensions possessed by a species. Further discussion on niche concepts can be found in Hutchinson (1957), Vandermeer (1972), and Colwell and Fuentes (1975). Insight into the state of analyses of niche overlap, niche breadth, and similar measures can be found in Horn (1966), Levins (1968), Colwell and Futuyama (1971), Cody (1974), Sale (1974), and Hurlbert (1978).

One of the simplest measures of niche overlap is an unweighted proportional method (see Colwell and Futuyma 1971 and review by Hurlbert 1978). In this method, niche overlap (C) is computed according to the following formula:

$$C_{xy} = 1 - 1/2\ (\sum_i |p_{xi} - P_{yi}|),$$

where $p_{xi} = x_i/X$ (i.e., the relative proportion of the total values for species x that occur in category i) and $P_{yi} = y_i/Y$ (i.e., the relative proportion of the total values for species y that occur in category i). To compute these indices for a set of species, a **resource matrix** (Table 31-1) is prepared by listing the species along

Table 31-1
Format for resource matrix to facilitate data recording on proportions in each resource state (i) or niche category for various species. Resource states could be habitat types, soil types, food categories, or other components. See Colwell and Futuyma (1971) or Cox (1976) for further details.

Species (x, y, . . . s)	*Resource States (i)*							

the left margin (rows) and resource or habitat states along the top margin (columns). All resource state values must then be converted to decimal proportions (e.g., 0.21 of total abundance of species A in Oak Forest resource state) prior to making the calculation of niche overlap between two species. Like most indices of niche overlap, no component is included in this formula for differences in resource abundance. Colwell and Futuyma (1971) modified this method by introducing weighting to reduce the effect of the relative abundance of the species and the number of resource states measured although Hurlbert (1978) questioned the appropriateness of their method. Thus, prior to adopting this or other techniques for general use, you should consult the review papers cited above.

Supplementary Readings

Alcoze, T. M. and E. G. Zimmerman. 1973. Food habits and dietary overlap of two heteromyid rodents from the mesquite plains of Texas. *J. Mammal.* 54:900-908.

Baumgartner, L. L. and A. C. Martin. 1939. Plant histology as an aid in squirrel food-habits studies. *J. Wildl. Manag.* 3:266-268.

Colwell, R. K. and D. J. Futuyma. 1971. On the measurement of niche breadth and overlap. *Ecology* 52:567-576.

Cox, G. W. 1976. *Laboratory manual of general ecology,* 3rd ed. Wm. C. Brown Company Publishers, Dubuque, Iowa. 232 pp.

Day, M. G. 1966. Identification of hair and feather remains in the gut and feces of stoats and weasels. *J. Zool.* 148:201-217.

Dusi, J. L. 1949. Methods for determination of food habits by plant microtechnics and histology and their application to cottontail rabbit food habits. *J. Wildl. Manag.* 13:295-298.

Hansson, L. 1970. Methods of morphological diet micro-analysis in rodents. *Oikos* 21:255-266.

Hurlbert, S. H. 1978. The measurement of niche overlap and some relatives. *Ecology* 59:67-77.

Korschgen, L. J. 1969. Procedures for food-habits analyses, pp. 233-250. *In* R. H. Giles, Jr. (Ed.). *Wildlife Management Techniques.* The Wildlife Society, Washington.

Meserve, P. L. 1976. Food relationships of a rodent fauna in a California coastal sage scrub community. *J. Mammal.* 57:300-319.

Oosting, H. J. 1956. *The study of plant communities,* 2nd ed. W. H. Freeman and Co., San Francisco. 440 pp.

Reichman, O. J. 1975. Relation of desert rodent diets to available resources. *J. Mammal.* 56:731-751.

Sparks, D. R. and J. C. Malechek. 1967. Estimating percentage dry weights in diets. *J. Range Manag.* 21:203-208.

Westoby, M., G. R. Rost, and J. A. Weis. 1976. Problems with estimating herbivore diets by microscopically identifying plant fragments from stomachs. *J. Mammal.* 57:167-172.

Williamson, V. H. H. 1951. Determination of hairs by impressions. *J. Mammal.* 32:80-85.

32 Identifying Mammal Sign

Since most mammals are secretive or nocturnal, they are seldom seen by the casual observer. But their presence is often revealed by tracks, burrows, nests, runways, evidence of feeding and other **sign** (Fig. 32-1). This sign is often useful in determining the distribution of mammals and learning something of their habits. Learning to recognize and interpret mammal sign is essential to efficient collecting and field investigation.

Footprints and tail markings can be found in mud along streams and lakes or in any other suitable substrate such as dust, soft soil, sand, or snow. Such tracks can usually be identified and, with experience, are a reliable means for verifying the presence of a particular species in a given locality. The tracks of most larger mammals can be readily identified by their characteristic patterns (Fig. 32-2), but it is sometimes difficult to distinguish closely related species. The tracks of small mammals are difficult to distinguish unless one knows which species can be expected in a particular area. Even then, evidence other than tracks is often necessary for positive identififcation. Olaus Murie (1954) in his *Field Guide to Animal Tracks* described and illustrated footprints and other sign of many North American mammals. This book and several other books by Perkins (1954) and Seton (1958) for U.S. species, and Lawrence and Brown (1967), and Twigg (1975a) for British species may be used to identify tracks in the field or casts or photographs of them in the laboratory.

Figure 32–1.
Nest (A) of the dormouse, *Muscardinus avellanarius* and tracks (B) of the brown hare, *Lepus capensis.*
(A, Ognev 1947: 542; B, Ognev 1940: 131)

Figure 32–2.
Tracks of beaver, *Castor canadensis*. Forefoot, f; hindfoot, h. Bottom figure shows footprints and impression of tail. (Henderson 1960: 54)

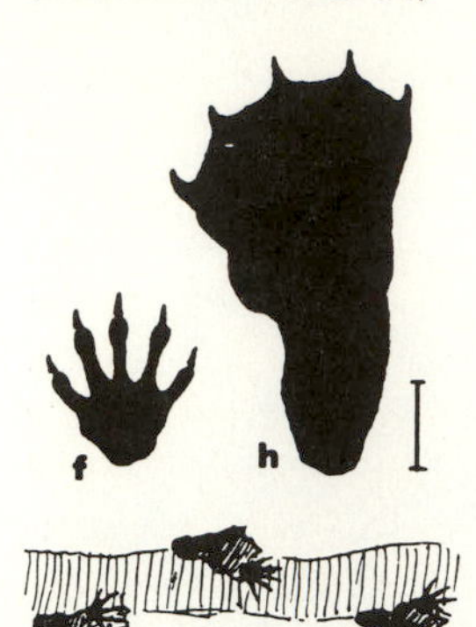

32-A. Use Murie (1954), Stains (1962), or another book on mammal tracks to answer the following questions. Use casts when available.

a. How do the footprints of carnivores (*Felis* and *Canis*) differ? What is the reason for this difference?

b. How, other than by size, can you differentiate between the tracks of a deer and cow? Between a deer and a wild sheep?

c. What similarities do you note between the footprints of an opossum and a raccoon? How are these two distinguished?

d. Why does a hog walking on a relatively hard surface leave the impression of two digits per foot while on softer ground it registers four digits per foot?

Scats

A mammal's fecal material or **scat** is also frequently species distinctive and can yield important information on feeding habits, occurrence, and activity. Scats of small mammals can usually be identified only to the generic level while scats of carnivores and ungulates are often species distinctive (Fig. 32-3). However, the shape and appearance of scat varies with the diet and age of the animal.

Figure 32–3
Scats of some North American artiodactyls. A, peccary; B, mule deer, two examples; C, white-tailed deer; D, mountain sheep; E, domestic sheep; F, mountain goat; G, domestic goat; H, caribou, two examples; I, pronghorn, two examples; J, moose; K, elk, two examples; L, bison, two examples.
(Murie, O. 1954. A field guide to animal tracks. Houghton Mifflin Co., Boston, 374 pp., fig. 132. With permission of the publisher.)

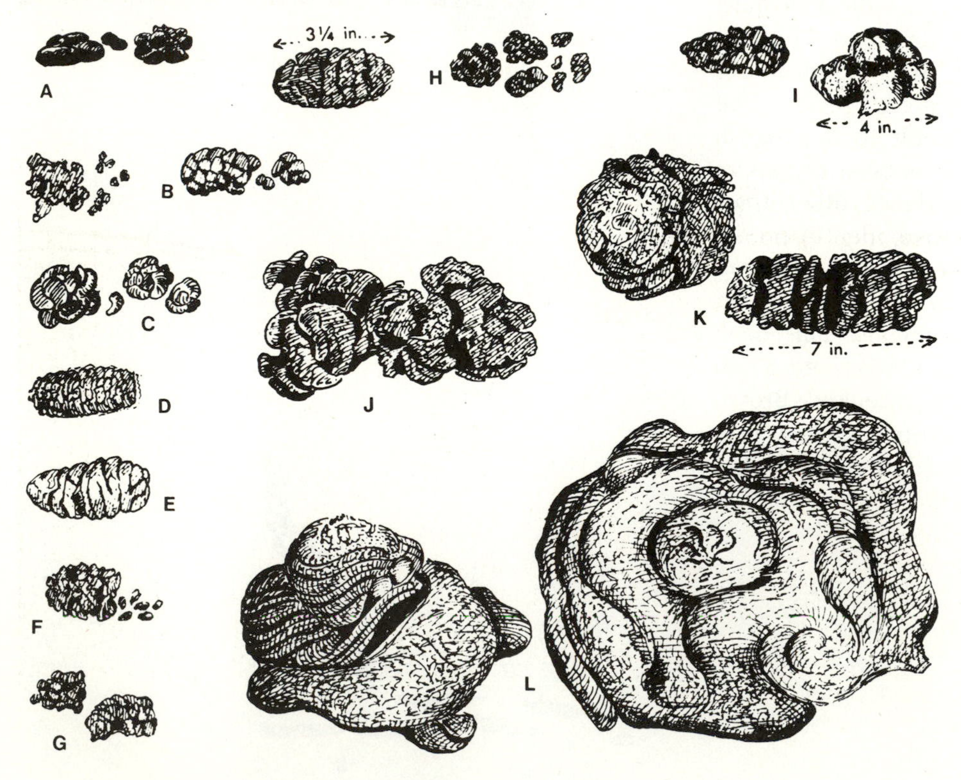

Examinations of scats can help determine the dietary habits of a species. Relative amounts of vegetable and animal matter can be determined by indigestible portions, such as hair, feathers, seed coats, and chitin of insect skeletons. Microscopic analyses of fecal material (and stomach contents) are usually necessary, however, for quantitative determinations of food habits (see Chapter 31).

32-B. Use the illustrations in Murie (1954) or collections of preserved (dried) scats to answer the following questions. How do you differentiate between the droppings of a deer and a sheep? Between those of a deer and a rabbit? Between those of a *Peromyscus* and a *Microtus?*

Trails, Runways, and Burrows

Terrestrial mammals of all sizes may form conspicuous **trails** in their travels to and from feeding sites, water holes, or other areas within their home ranges. Many small rodents and insectivores establish definite **runways** that may be completely open (e.g., ground squirrels, *Spermophilus*), partly covered (e.g., cotton rats, *Sigmodon*), or entirely roofed over by surrounding vegetation (e.g., voles, *Microtus*). Some species which usually do not form distinct runways themselves (e.g., deer mice, *Peromyscus*) will frequently use runways formed by others, but other species (e.g., harvest mice, *Reithrodontomys*) apparently disregard runways completely.

The excavations made by individuals of different species of terrestrial mammals are highly variable. Many mammals, such as the hares, *Lepus*, use only shallow depressions, termed **forms**, to rest in. Some species such as sewells, *Aplodontia rufa*, and northern pocket gophers, *Thomomys talpoides*, dig elaborate networks of tunnels and nests and storage chambers. Several burrows, each with one or more entrances, may be located within an individual's home range.

The location and shape of the burrow entrance, coupled with a knowledge of the species that potentially inhabit an area, can be used to ascertain the animal responsible. The diameter of the hole places a limit on the size of animal occupying it. Some species leave excavated soil in large mounds at the burrow entrance, while others scatter the soil. Individuals of some species dig burrows that enter the ground at slight angles, whereas others dig nearly vertical entrances. Members of some species normally locate burrow entrances at the base of a rock or vegetation; others select open areas. Many species leave their burrows open but some place a soil plug in the entrance. Many terrestrial species do not usually dig their own burrows, but inhabit those abandoned by other species.

Certain fully fossorial mammals such as moles, pocket gophers, and mole rats have elaborate tunnel systems that may be barely subsurface or very deep. Some moles, Talpidae, usually dig tunnels just below the surface leaving serpentine-like ridges visible above ground with radially symmetrical eruptions of excavated soil at intervals along the ridges (Fig. 32-4).

Pocket gophers (Geomyidae) and some other burrowing rodents (e.g., *Spalax, Ellobius, Heterocephalus*) excavate tunnels that are several inches underground. The tunnels themselves are not visible at the surface, but mounds of excavated soil indicate their presence. A short tunnel connecting these mounds with the main underground runway system is generally opened to permit removal of excavated soil, air-condition the burrow system, or to allow the animal to exit and forage aboveground (Fig. 32-5).

32-C. Compare illustrations of mole and pocket gopher tunnel systems. What similarities do you see? What differences?

Figure 32–4.
Ridges and mounds made by a mole (Talpidae) with sectional view of tunnel.
(Silver and Moore 1941: 4)

Figure 32–5.
Runway system of pocket gopher (Geomyidae) viewed from above (A) and in cross-section (B). Nest in use (a), old nest filled with dirt (b) and chamber (c) for storage of roots and other materials.
(Crouch 1933: 8)

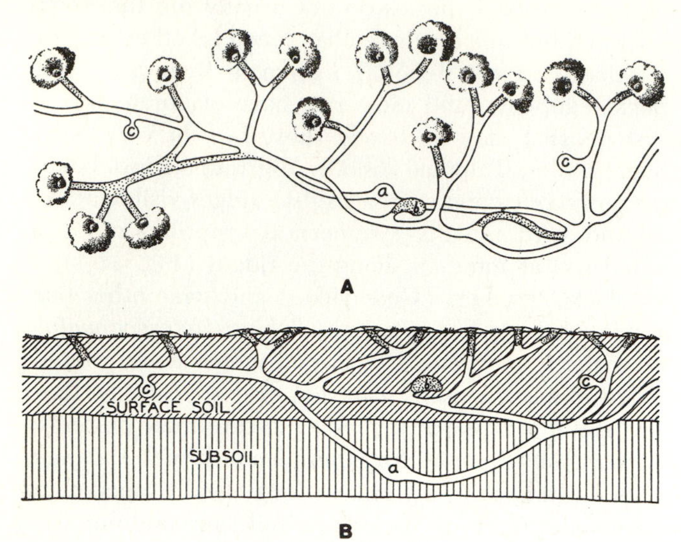

Figure 32–6.
Diagrammatic side views of rock (A), stick (B), and cactus (C) houses (dens) made by the desert woodrat, *Neotoma lepida*. Blind passageway, b; entrance, e; food cache, fs; hollow in tree used as passageway, x. Houses may also be made of combinations of these and other materials.
(Cameron and Rainey 1972: 256, 258, 261)

Nests and Dens

Burrowing species usually include nest chambers in their network of tunnels. Frequently these chambers are lined with dry grass, leaves or some other cushioning and insulating material. Nonburrowing small mammals may construct nests for retreat, resting, and/or the rearing of young. These may be located in small depressions in the ground, in tangled vegetation, in cracks in rocks, or in hollows in trees. Some wood rats or pack rats, *Neotoma*, use twigs, rocks, dried dung, and any other handy material to construct elaborate domed lodges, or dens (Fig. 32-6). Grass and fur-lined nests are constructed in chambers within these dens. Beavers, *Castor* (Fig. 32-7), and muskrats, *Ondatra*, build elaborate houses when appropriate habitat and sufficient construction materials (sticks, limbs, etc.) are available. When these are lacking, they burrow into the side of a pond or into a stream bank.

Arboreal species frequently construct leaf and twig platforms or use cavities in tree trunks. Many tree squirrels inhabit leaf nests during the warm months and move to hollow trees or underground burrows during the winter. Gorillas, orangutans, and chimpanzees (Pongidae), build sleeping platforms in trees for overnight occupancy. The red tree vole, *Arborimus longicaudus*, normally constructs an arboreal nest with Douglas-fir twigs and resin ducts from the needles. Such a nest may grow to three feet in diameter through generations of use.

Figure 32–7.
View of beaver (*Castor canadensis*) pond and associated sign. A, den; B, feeding sign near shore; C, trails leading to logging areas; D, dam.
(Henderson 1960: 55)

Figure 32–8.
Tree that has been clawed by a bear, *Ursus arctos*. (Novikov 1956)

Hollow trees, caves, protected areas under fallen logs, and other similarly protected spots may be used by mammals of different sizes as dens for a temporary rest or for a long period of hibernation.

Bats may roost in caves, buildings, or hollow trees. Some species select crevices in rocks or spaces under tree bark. Many tropical forms conceal themselves among the leaves of palm trees and some (e.g., *Artibeus watsoni, Uroderma bilobatum*) modify leaves to construct their shelters.

Feeding Residues

Mammals leave many signs of their feeding activities. Species that browse on woody vegetation break twigs and branches and leave conspicuously nipped off twigs. Rodents of all sizes chew nuts and large seeds and discard the shells. Shrews and rodents eat the soft bodies of snails and leave the opened shells. Many species strip bark from trees, and beavers cut down large numbers of saplings as food (Fig. 32-7). Hogs, skunks, armadillos, and many other mammals dig and root in the ground for their food and leave evidence of this activity. Some bats (e.g., pallid bats, *Antrozous pallidus*) feed upon large insects and/or vertebrates and leave piles of uneaten feathers, wings and legs under their roosts.

Many mammals, such as squirrels, wood rats, and kangaroo rats, *Dipodomys,* cache quantities of food in or around their nests or dens. Pikas, *Ochotona,* and some mice cut vegetation and allow it to dry. Carnivores may cache their kills by hanging them in trees (e.g., leopard) or by covering them with brush (e.g., tiger) and return to them several times to feed.

32-D. Examine some nuts that have been gnawed by various rodents. How can you tell which were chewed by squirrels and which by mice?

Miscellaneous Sign

Many indications of an animal's presence do not fit into any of the above categories.

Antlers and other bone are frequently well gnawed by rodents, leaving tooth marks as sign of their presence. Apparently, the animals secure minerals and salts from these objects.

During the fall, male deer lose velvet from their antlers as they rub the antlers against rocks, trees, or brush. Shreds of discarded skin and/or worn spots on the rubbing post indicate this activity that may be partially related to sign posting. Some carnivores, such as bears and cats, scrape their claws on trees, leaving identifiable marks (Fig. 32-8).

Miscellaneous strands and tufts of hair may be found clinging to brush, barbed-wire fences, or rocks along a trail where a mammal has brushed against something sharp or rubbed to scratch irritated areas of the body.

Many mammals mark areas within their home range or territory. Zebras and African buffalos may rub a termite nest until the ground around the nest is bare. The European bison may select a particular tree and remove portions of the bark by rubbing it with their horns. Dung heaps and other signposts usually have a territorial function.

32-E. Field exercise. Choose an area with at least two distinct habitats (such as an open field and an adjacent wooded area). Mark off two parallel lines 100 meters long and 2 meters apart. The transect thus marked should extend into both habitats. Carefully examine the area between the lines for mammal sign. Plot the location and nature of all sign on a map or piece of graph paper. Using Murie (1954) or another similar reference, identify the sign as precisely as you can.

How many kinds of mammals can you now say inhabit the area? Does there seem to be a higher concentration of mammals in any one part of the transect? Where would you set traps to catch the mammals whose sign you have identified?

Supplementary Readings

Hamilton, W. J., Jr. 1934. The life history of the rufescent woodchuck *Marmota monax rufescens* Howell. *Ann. Carnegie Mus.* 23:85-178.

Lawrence, M. J. and R. W. Brown. 1967. *Mammals of Britain—Their tracks, trails and signs.* Blanford Press, London. 223 pp.

Leutscher, A. 1960. *Tracks and signs of British animals.* Cleaver-Hume Press, London. 252 pp.

Murie, O. J. 1954. *A field guide to animal tracks.* Houghton-Mifflin Co., Boston. 374 pp.

Perkins, H. M. 1954. *Animal tracks: the standard guide for identification and characteristics.* Stackpole, Harrisburg. 63 pp.

Seton, E. T. 1958. *Animal tracks and hunter signs.* Doubleday and Co., New York. 160 pp.

Stains, H. J. 1962. *Game biology and game management: a laboratory manual.* Burgess Publishing Co., Minneapolis. 143 pp.

Twigg, G. I. 1975. Finding mammals—their signs and remains. *Mammal Review* 5:71-82.

33 Recording Data

Field observations yield valuable data on mammalian distribution, abundance, habitat tolerances, feeding habits, reproductive cycles, behavior, etc. In order to preserve these observations for later reference, it is essential to develop good note-taking practices. Accurate and complete field notes will be valuable references for the observer and for other scientists at later dates. With the passage of time environmental conditions change and often old field notes are the most valuable sources for documenting and analyzing changes.

The format of the field notes should be flexible and permit changes to fit any given situation, but some guidelines are necessary to produce notes with the greatest scientific value and to expedite filing and subsequent use. A suggested method for recording data is given below with various examples of notes and labels. Additional suggestions can be found in Hall (1962) and Mosby (1969a:61-72).

Equipment

Paper and Notebook

Good quality white paper (of neutral or slightly alkaline pH) should be used for notes. A bond paper having a rag content of fifty to one hundred percent is preferable. This paper is durable and will resist water damage and deterioration due to age. Sulphide papers (e.g., ditto, mimeograph, and most light weight notepaper) should *not* be used since they are highly susceptible to age deterioration and water damage. Paper with dimensions of either 5 3/4" x 8 1/2" or 6 1/4" x 8 1/2" is of a convenient size to work with.

We recommend that the paper be placed in a sturdy three-ring binder or notebook. Index dividers may be used to separate different sections of the field notes. The use of a ring binder has advantages and disadvantages. With the binder it is easy to lose a few sheets of paper and thus render your field notes incomplete. This possibility could be avoided by using a prebound volume of blank pages; however, the greatest danger to the field notebook is loss or damage in the field. Thus many fieldworkers use two notebooks. One contains past field notes and is stored in a safe place in the home or office, whereas the other, containing only current notes, is taken into the field. At the end of each segment of fieldwork the notes are removed from the field notebook and inserted into the permanent one. This system, impossible with a prebound volume, frees the bulk of the data from the chance of loss or destruction in the field.

Ink and Pens

A black waterproof ink should be used for writing field notes and data on specimen tags. The ink must also be resistant to alcohol and formalin preservatives, to grease and other animal fluids, and to the ammonia or detergent frequently used in cleaning skeletal material. The following inks were tested and found to be most suitable for writing notes and data on tags: (1) Pelikan Drawing Ink, (2) Higgins Eternal Black Ink, (3) Koh-I-Noor Rapidograph Ink. Three other inks tested, Higgins Engrossing and Higgins India Inks Nos. 4411 and 4415, rapidly lost legibility in ammonia solutions. Ball point, nylon- or felt-tipped pens, or washable fountain pen inks should *never* be used. If permanent ink is not available, a No. 2 or 2 1/2 pencil is the best substitute.

Permanent inks will quickly clog most fountain pens. The Rapidograph made by Koh-I-Noor and similar pens made by Pelikan and Castell are three of the few pens that will work with these heavy inks. Although less convenient (but considerably cheaper), a staff pen and point may be used.

Specimen Labels

Labels for study skins, tanned skins, skulls, and entire skeletons should be made of 100 percent rag stock white paper and should be of sufficient thickness to allow string to be attached without danger of tearing the paper. Labels for fluid preparations (entire animals, organs, etc.) must be made of heavy weight, 100% rag stock.

Field Notes

Some mammalogists and other vertebrate biologists organiz their field notes into three sections: (1) **journal,** (2) **catalog,** and (3) **species accounts.** The catalog, a numbered listing of all specimens preserved, is essential if any capture and preservation of mammals is done. The journal, a field diary of all activities and observations, is highly recommended for all types of field work. The species account section includes detailed observations on species. Special data recording forms (see: "Field Notes: Special Forms") may be filed separately or included with the journal and species accounts sections.

Journal

The field journal is a complete, chronological record of the activities and observations of an investigator. The what, when, where, and how of all fieldwork should be recorded here. Results of the work should be described along with supplemental information on habitat, general impressions of mammal populations, conversations with residents of the area, and any additional information that is potentially helpful.

The name of the investigator and the year should be recorded at the top of each page. Number the pages consecutively beginning with one. Record the exact locality and date of each observation or account. The locality should list country, state, county, and miles (or kilometers) and direction from a permanent map feature (e.g., town, city, mountain). The date should be written out fully (e.g., 14 July 1980). Figure 33-1 is an example of a journal page. Methods of recording location, date, and other data are given below.

Figure 33–1.
Sample field journal page. Original size 6¼ × 9 inches. (K.G. Matocha)

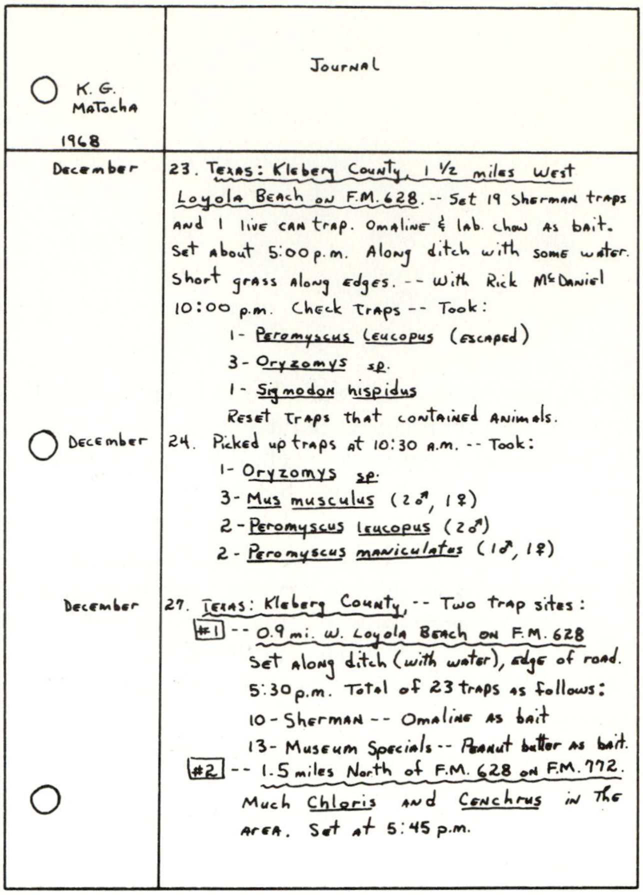

K. G. Matocha 1968	Journal
December	23. Texas: Kleberg County, 1 ½ miles West Loyola Beach on F.M. 628. -- Set 19 Sherman traps and 1 live can trap. Omaline & lab. chow as bait. Set about 5:00 p.m. Along ditch with some water. Short grass along edges. -- with Rick McDaniel 10:00 p.m. Check traps -- Took: 1- Peromyscus leucopus (escaped) 3- Oryzomys sp. 1- Sigmodon hispidus Reset traps that contained animals.
December	24. Picked up traps at 10:30 a.m. -- Took: 1- Oryzomys sp. 3- Mus musculus (2♂, 1♀) 2- Peromyscus leucopus (2♂) 2- Peromyscus maniculatus (1♂, 1♀)
December	27. Texas: Kleberg County, -- Two trap sites: #1 -- 0.9 mi. W. Loyola Beach on F.M. 628 Set along ditch (with water), edge of road. 5:30 p.m. Total of 23 traps as follows: 10 - Sherman -- Omaline as bait 13 - Museum Specials -- Peanut butter as bait. #2 -- 1.5 miles North of F.M. 628 on F.M. 772. Much Chloris and Cenchrus in the area. Set at 5:45 p.m.

Catalog

The catalog (Fig. 33-2) is a record of all specimens that are preserved in any manner. Each specimen is assigned a number that is associated with the name of the collector. If you have never captured and preserved animals previously, your first entry will be designated number one. Throughout your life, *never repeat a number* once it has been used for an animal. Record the following elements of data for each specimen.

Locality

As described in Williams et al. (1977), there are two principal systems for writing locality descriptions: specific to general and general to specific. Various institutions have adopted standard ways for locality descriptions to be written on specimen labels. In this manual, we give examples of the general to specific scheme since it clearly has a distinct advantage for retrieval using automatic data processing techniques. Data for localities in the United States should include *state, county* (or parish), and *mileage* and *direction* from a recorded permanent map feature.

TEXAS: Hockley County; 3 miles N Ropesville.

For locations in countries other than the United States use country, political subdivision (e.g., department,

Figure 33–2.
Sample field catalog page. Original size 6¼ × 9 inches. (K.G. Matocha)

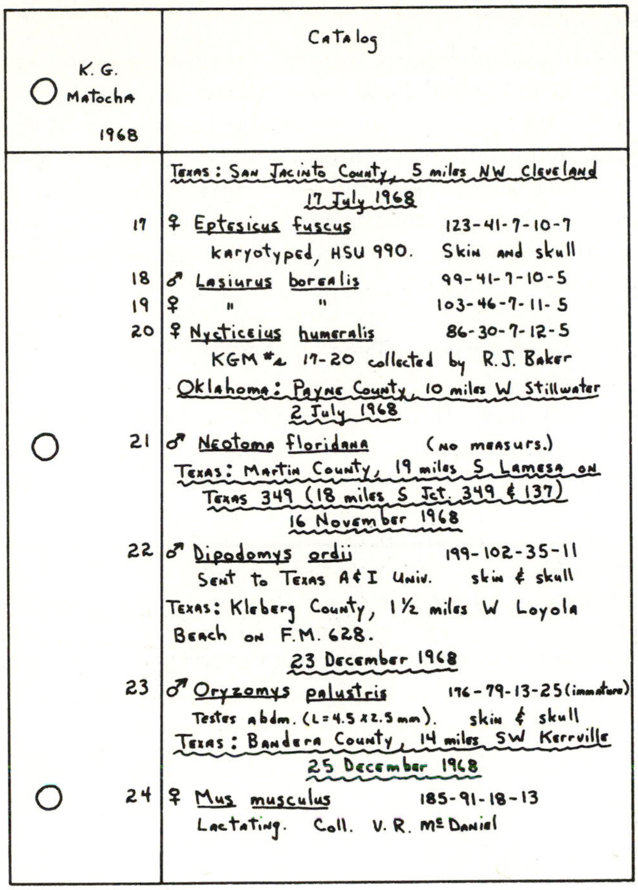

K. G. Matocha 1968

Catalog

Texas: San Jacinto County, 5 miles NW Cleveland
17 July 1968

17 ♀ Eptesicus fuscus 123-41-7-10-7
karyotyped, HSU 990. Skin and skull
18 ♂ Lasiurus borealis 99-41-7-10-5
19 ♀ " " 103-46-7-11-5
20 ♀ Nycticeius humeralis 86-30-7-12-5
KGM #'s 17-20 collected by R.J. Baker

Oklahoma: Payne County, 10 miles W Stillwater
2 July 1968

21 ♂ Neotoma floridana (no measurs.)

Texas: Martin County, 19 miles S Lamesa on Texas 349 (18 miles S Jct. 349 & 137)
16 November 1968

22 ♂ Dipodomys ordii 199-102-35-11
Sent to Texas A&I Univ. skin & skull

Texas: Kleberg County, 1½ miles W Loyola Beach on F.M. 628.
23 December 1968

23 ♂ Oryzomys palustris 176-79-13-25 (immature)
Testes abdm. (L=4.5×2.5 mm). skin & skull

Texas: Bandera County, 14 miles SW Kerrville
25 December 1968

24 ♀ Mus musculus 185-91-18-13
Lactating. Coll. V. R. McDaniel

province) and kilometers (or miles) from a recorded prominent map feature.

IRAN: Fars Province; 15 km N Shiraz.

More precise locality data is valuable and should be recorded when possible. Geographic coordinates (latitude and longitude, accurate to one minute), legal land descriptions, and elevations provide the most accurate and lasting data.

NEW MEXICO: Dōna Ana County; south edge of Red Lake, T 18S, R 1E, SW 1/4 of Section 27
or
NEW MEXICO: Dōna Ana County; south edge of Red Lake, 107° 10′ W, 32° 42′ N

Sometimes the reference landmark may be in one county (or other political subdivision), while the specimen was actually collected in an adjacent county. Always list the county in which the specimen was collected.

OKLAHOMA: in Harper County; 3 miles N Fort Supply
(Note: Fort Supply is a town in Woodward County, Oklahoma.)

It is recommended that only cardinal compass directions (N, S, E, W) be utilized and that N or S precede E or W, where appropriate. A direction given as NE may be difficult to pinpoint (i.e., is the direction *exactly* 45° from the N-S axis?). In addition, one should not use road junctions or railroad intersections as locality descriptors for specimens. If used, they should be listed as additional comments in the journal.

CHILE: Santiago Province; 1 km N, 0.5 km E Cerro Manquehue

Date

The date of collection should be written out completely. Do not write "4/1/80" since this may be interpreted as either "1 April 1980" or "4 January 1980." Likewise, never use only the last two digits (e.g., "80") for the year. Since collections have and should last for hundreds of years, the above date could in the future be interpreted as "1 April 1880" or "1 April 2080." The recommended method for writing the date is "1 June 1981" since placing the numeral before the month eliminates the necessity of a comma. If the specimen is taken alive and dies or is killed at a later date, the date of death and date of collection should both be recorded.

Measurements, Weight, and Sex

Mammal specimens should be measured prior to preparation. The standard measurements for a mammal are always listed in the following order: (1) *total length,* (2) *tail length,* (3) *hind foot length,* (4) *ear length,* and in bats (5) *tragus height,* and (6) *forearm length.* The measurements are taken in millimeters and each measurement is separated by a dash since this punctuation is least likely to be interpreted as a numeral. The weight is recorded in metric units: grams or kilograms.

The sex should be recorded using the symbols ♂ for male and ♀ for female. Use a question mark, ?, if the sex cannot be determined. Indicate immature, juvenile, or subadult if one of these terms is appropriate.

Refer to Chapter 35 for the techniques of measuring, weighing, and sexing mammals.

Reproductive Condition

If the specimen is a male, measure and record the length (exclusive of epididymis) and breadth of the testis. If the species is one in which the testes descend seasonally, record their position. If the specimen is a female, check the uterus for embryos and record your observation. If embryos are present, record their number, location in the uterus, and their head-to-rump length in millimeters. If the female is lactating, make note of this. If any portion of the reproductive tract or embryos is preserved for later examination, attach a label with the collector's initials and catalog number and note the type of preservation (see Table 33-1) in the catalog. Refer to Taber (1969) or Brown and Stoddart (1977) for additional information on ascertaining the reproductive condition of a specimen.

In conjunction with censuses of populations, autopsies may be performed on animals to ascertain their reproductive status and condition. While such data could be recorded in simple tabular form, special forms have been designed for particular studies to facilitate data processing (Fig. 33-3).

Parasites

In epidemiological surveys, extensive records are kept on individual hosts. An ectoparasite survey data sheet utilized by personnel of the Division of Mammals, Smithsonian Institution, is shown in Figure 33-4. One data sheet was completed for each host captured and sampled for ectoparasites. See Chapter 36 for additional information on collecting and preserving ectoparasites.

Portions Preserved (Type of Preservation)

The usual mammal specimen consists of a study skin and skull. If the postcranial skeleton, baculum, embryos, stomach contents, or any other portions of the specimen are preserved, these should be noted in

Figure 33–3.
Original size 5½″ × 8½″. Reproductive data sheet. (R.E. Martin)

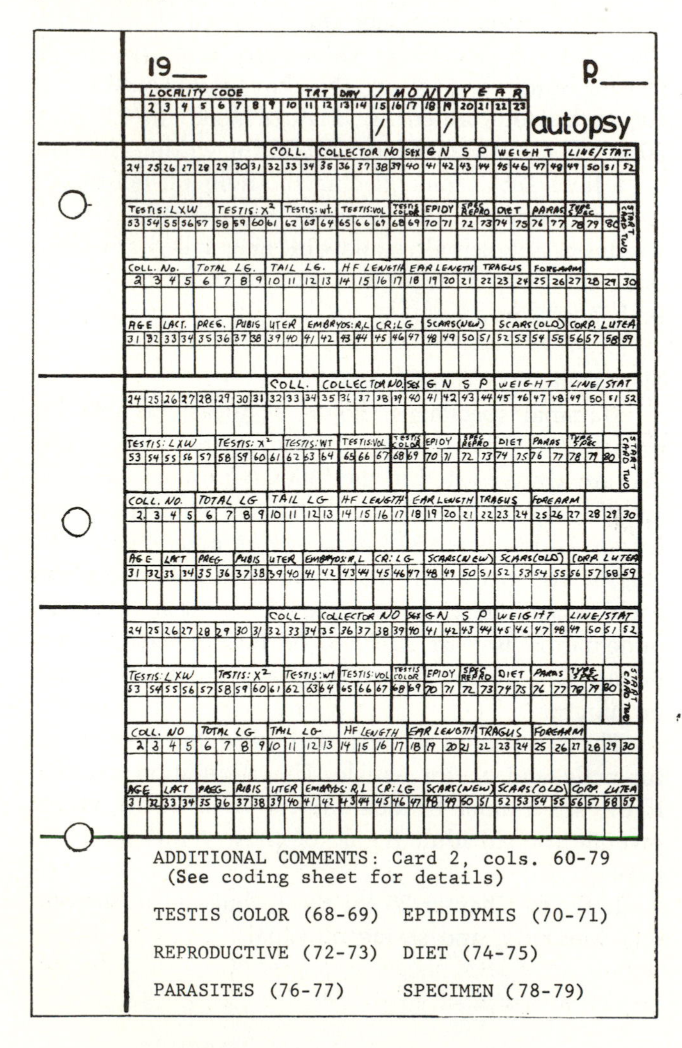

19___ R___

LOCALITY CODE | TAT | DAY / MONTH / YEAR

autopsy

COLL. | COLLECTOR NO | SEX | G N S P | WEIGHT | LINE/STAT.

TESTIS: LXW | TESTIS: X² | TESTIS: wt. | TESTIS: VOL. | TESTIS COLOR | EPIDY | SPEC REPRO | DIET | PARAS | TYPE SPEC | START CARD TWO

COLL. NO. | TOTAL LG. | TAIL LG. | HF LENGTH | EAR LENGTH | TRAGUS | FOREARM

AGE | LACT. | PREG. | PUBIS | UTER | EMBRYOS: R,L | CR:LG | SCARS(NEW) | SCARS(OLD) | CORP. LUTEA

ADDITIONAL COMMENTS: Card 2, cols. 60-79
(See coding sheet for details)

TESTIS COLOR (68-69) EPIDIDYMIS (70-71)

REPRODUCTIVE (72-73) DIET (74-75)

PARASITES (76-77) SPECIMEN (78-79)

Figure 33–4.
Ectoparasite survey data sheet. Original size 5½ × 8½ inches.
(Division of Mammals, Smithsonian Institution)

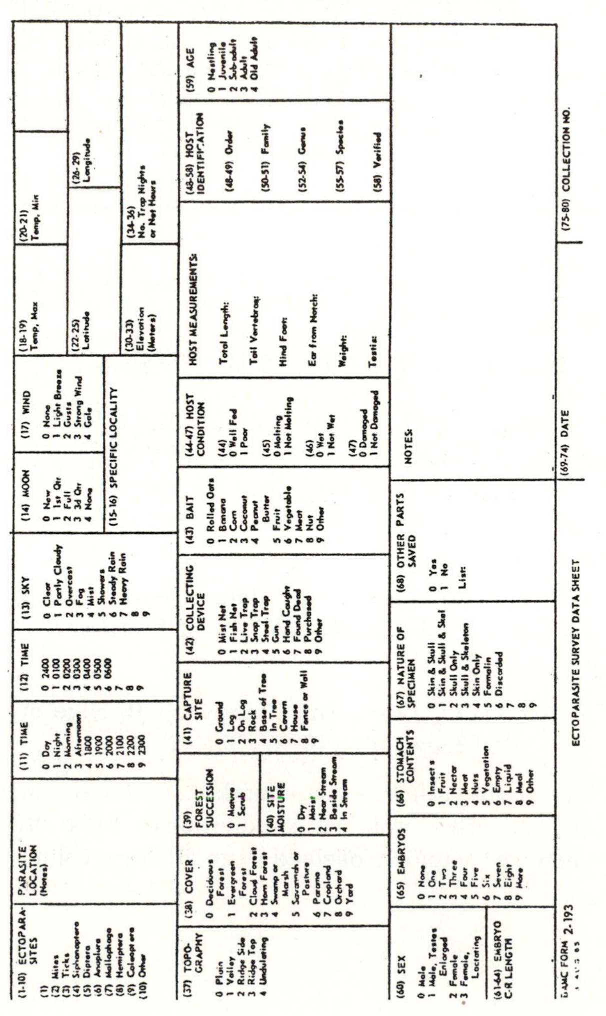

(1-10) ECTOPARASITES: (1) Mites (2) Ticks (3) Siphonaptera (4) Diptera (5) Anoplura (6) Mallophaga (7) Hemiptera (8) Coleoptera (9) Coleoptera (10) Other

PARASITE LOCATION (Notes)

(11) TIME: 0 Day, 1 Night, 2 Morning, 3 Afternoon, 4 1800, 5 1900, 6 2000, 7 2100, 8 2200, 9 2300

(12) TIME: 0 2400, 1 0100, 2 0200, 3 0300, 4 0400, 5 0500, 6 0600, 7, 8, 9

(13) SKY: 0 Clear, 1 Partly Cloudy, 2 Overcast, 3 Fog, 4 Mist, 5 Showers, 6 Steady Rain, 7 Heavy Rain, 8, 9

(14) MOON: 0 New, 1 1st Qr, 2 Full, 3 3d Qr, 4 None

(15-16) SPECIFIC LOCALITY

(17) WIND: 0 None, 1 Light Breeze, 2 Gusts, 3 Strong Wind, 4 Gale

(18-19) Temp, Max

(20-21) Temp, Min

(22-25) Latitude

(26-29) Longitude

(30-33) Elevation (Meters)

(34-36) No. Trap Nights or Net Hours

(37) TOPOGRAPHY: 0 Plain, 1 Valley, 2 Ridge Side, 3 Ridge Top, 4 Undulating

(38) COVER: 0 Deciduous Forest, 1 Evergreen Forest, 2 Cloud Forest, 3 Thorn Forest, 4 Swamp or Marsh, 5 Savannah or Pasture, 6 Paramo, 7 Cropland, 8 Orchard, 9 Yard

(39) FOREST SUCCESSION: 0 Mature, 1 Scrub

(40) SITE MOISTURE: 0 Dry, 1 Moist, 2 Near Stream, 3 Beside Stream, 4 In Stream

(41) CAPTURE SITE: 0 Ground, 1 Log, 2 On Log, 3 Rock, 4 Base of Tree, 5 In Tree, 6 Cavern, 7 House, 8 Fence or Wall, 9

(42) COLLECTING DEVICE: 0 Mist Net, 1 Fish Net, 2 Live Trap, 3 Snap Trap, 4 Steel Trap, 5 Gun, 6 Hand Caught, 7 Found Dead, 8 Purchased, 9 Other

(43) BAIT: 0 Rolled Oats, 1 Banana, 2 Corn, 3 Coconut, 4 Peanut Butter, 5 Fruit, 6 Vegetable, 7 Meat, 8 Nut, 9 Other

(44-47) HOST CONDITION: (44) 0 Well Fed, 1 Poor; (45) 0 Molting, 1 Not Molting; (46) 0 Wet, 1 Not Wet; (47) 0 Damaged, 1 Not Damaged

HOST MEASUREMENTS: Total Length: Tail Vertebrae: Hind Foot: Ear from Notch: Weight: Testis:

(48-58) HOST IDENTIFICATION: (48-49) Order, (50-51) Family, (52-54) Genus, (55-57) Species, (58) Verified

(59) AGE: 0 Nestling, 1 Juvenile, 2 Sub-adult, 3 Adult, 4 Old Adult

(60) SEX: 0 Male, 1 Male, Testes Enlarged, 2 Female, 3 Female, Lactating

(61-64) EMBRYO C-R LENGTH

(65) EMBRYOS: 0 None, 1 One, 2 Two, 3 Three, 4 Four, 5 Five, 6 Six, 7 Seven, 8 Eight, 9 More

(66) STOMACH CONTENTS: 0 Insects, 1 Fruit, 2 Nectar, 3 Meat, 4 Nuts, 5 Vegetation, 6 Empty, 7 Liquid, 8 Meal, 9 Other

(67) NATURE OF SPECIMEN: 0 Skin & Skull, 1 Skin & Skull & Skel, 2 Skull Only, 3 Skull & Skeleton, 4 Skin Only, 5 Formalin, 6 Discarded, 7, 8, 9

(68) OTHER PARTS SAVED: 0 Yes, 1 No, List:

NOTES:

(69-74) DATE

(75-80) COLLECTION NO.

ECTOPARASITE SURVEY DATA SHEET

SAMC FORM 2-193

the catalog. If ecto- and/or endoparasites are preserved, these should be mentioned. If the specimen is preserved in liquid or in any manner other than a standard study skin and skull, note this fact. If either the skin or skull is badly damaged, or if for some reason either is not preserved, record exactly what is included as a specimen.

Standard types of preservation recognized by the NIRM standards can be found in Williams et al. 1979. Explain the nature of the specimen if it does not fit one of the standard categories.

Methods

A brief mention should be made of the method used to secure the specimen. "Snap trapped" or "shot" are examples of adequate catalog descriptions, but more detailed notes should be made in the journal. The abbreviation **"DOR"** is frequently used to designate an animal found dead on the road. If the animal is caught in a baited trap, it is useful to name the bait used.

The six items or elements of data listed above are usually included with each catalog entry. The locality, date, sex, and type of preservation are essential elements that must be included with each entry. At times it may not be possible to record the other data. For example, it is not possible to record tail length or reproductive data for a weathered skull found in the field.

Species Accounts

This section of the field notes can be very useful when many observations are being recorded about a single species. Information on a particular species is called a species account. List the scientific name or common name of the animal at the top of the page. Give the date and location of the observation. As time permits, record observations made on the species, even those that may seem insignificant at the time. These additional observations may prove valuable in later analyses of data. Figure 33-5 is an example of a species account.

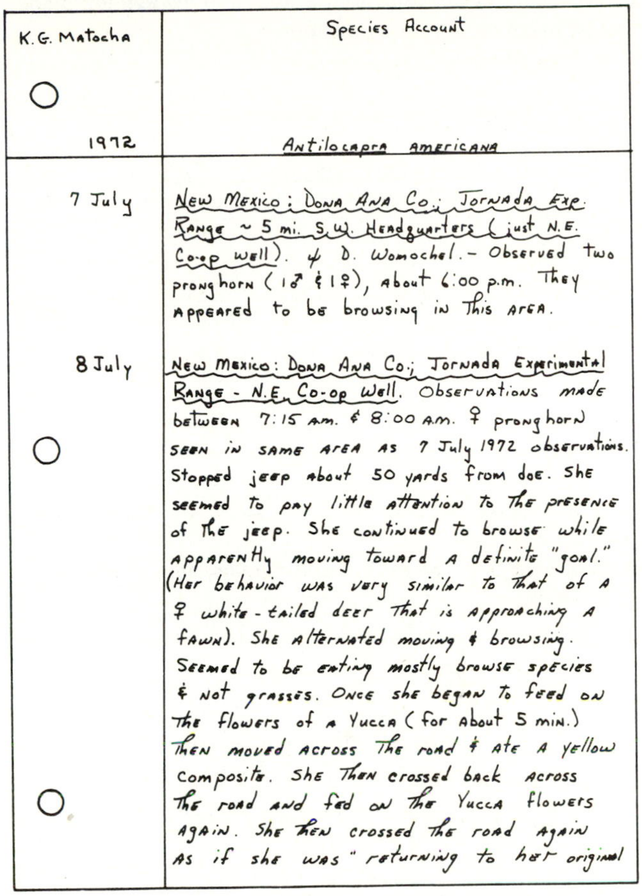

Figure 33-5.
Sample species account page. Original size 6¼ × 9 inches.
(K.G. Matocha)

K.G. Matocha 1972	Species Account *Antilocapra americana*
7 July	New Mexico: Dona Ana Co.; Jornada Exp. Range ~ 5 mi. S.W. Headquarters (just N.E. Co-op well). & D. Womochel.– Observed two pronghorn (1 ♂ & 1 ♀), about 6:00 p.m. They appeared to be browsing in this area.
8 July	New Mexico: Dona Ana Co.; Jornada Experimental Range - N.E. Co-op Well. Observations made between 7:15 am. & 8:00 A.m. ♀ pronghorn seen in same area as 7 July 1972 observations. Stopped jeep about 50 yards from doe. She seemed to pay little attention to the presence of the jeep. She continued to browse while apparently moving toward a definite "goal." (Her behavior was very similar to that of a ♀ white-tailed deer that is approaching a fawn). She alternated moving & browsing. Seemed to be eating mostly browse species & not grasses. Once she began to feed on the flowers of a Yucca (for about 5 min.) then moved across the road & ate a yellow composite. She then crossed back across the road and fed on the Yucca flowers again. She then crossed the road again as if she was "returning to her original

Specimen Labels

For each specimen recorded in the field catalog, a corresponding data tag (or tags) should be attached to the specimen (and all of its separate parts). Depending upon the specimen preparation technique, several different kinds of tags may be required for each mammal. If you are collecting for a museum or university collection, the institution will usually provide you with the necessary specimen tags. If these are not provided, use the types of paper recommended above.

Study Skins

Labels or tags for study skins are usually about 3" x 3/4". All of the data from the catalog entry should be recorded on the tags in permanent ink. Include your field catalog number, initials, and entire surname (Fig. 33-6). The exact arrangement of data on the tag varies from collection to collection. If you are collecting for an institution, use the format they prefer. If you are making your own collection, all data, except perhaps ecological notes, should go on one side of the tag. This will make your specimens easier to work with. While most of the data is entered in permanent ink, the identification (scientific or common name) should be entered in pencil since this is the only portion of the tag that might require changing.

Figure 33-6.
Obverse and reverse of two specimen tags. A, the style used by the University of Kansas Museum of Natural History. B, the style used by the Oklahoma State University Museum. Note that a space is always left blank for the Museum's catalog number.
(K.G. Matocha)

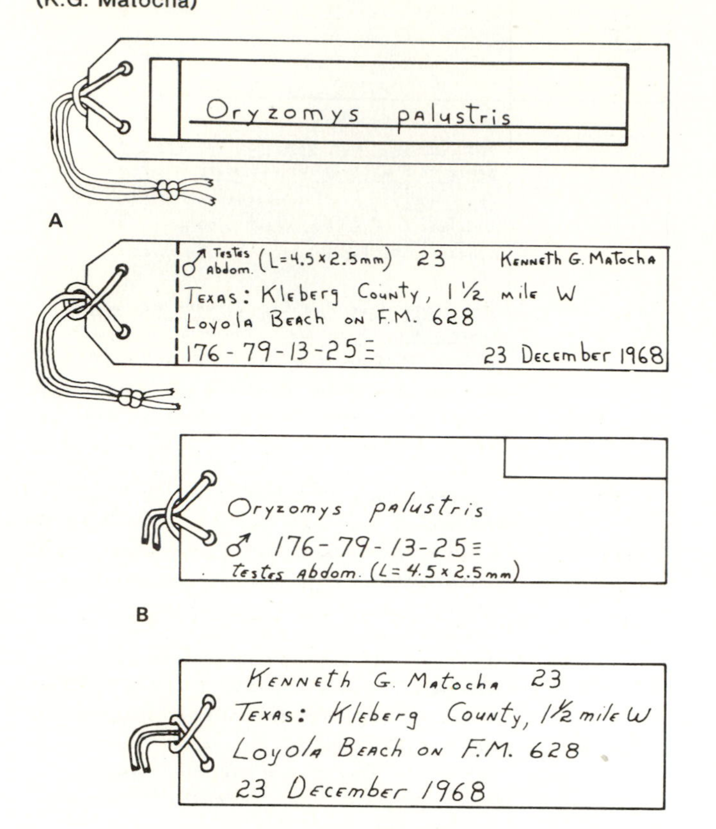

Thread the tag through two holes punched at least 1/8" from one end and tie a knot in the string about 1" from the tag. See Figure 33-6A for the appropriate method of stringing a tag. Use a square knot to securely attach the tag just above the right hind ankle of a study skin. With bat skins the label is sometimes attached above the knee joint to allow for greater visibility of the calcar and to protect the feet (Fig. 35-20).

Use the same type of label for a skull or any other bones that will not require cleaning or for any other portion of the specimen that is not to be tanned or preserved in liquid.

Skulls and Skeletons to be Cleaned

The type of label described above will usually not survive the cleaning processes used for skeletal material. Thus the "skull tag" is usually a small piece of resistant paper that has only the collector's name and field catalog number and the sex of the specimen.

Figure 33-7.
A skull tag. Note that a knot is tied in the string about one inch from the tag.
(K.G. Matocha)

After the skull or skeleton has been cleaned a permanent label, including data from the museum catalog, will be placed with the specimen.

Figure 35-21 illustsrates a skull tag. Again the string should be knotted about one inch from the tag. Attach the tag loosely around the mandible of an uncleaned skull and secure with a square knot (Fig. 33-5). In larger species the string may be attached to the zygomatic arch. Attach the tag to the pelvis of a complete skeleton or to secure locations on each portion of a partly disarticulated skeleton.

Skins to be Tanned

Since few tags can survive the tanning process, either of the two types described above may be attached to a skin to be tanned. Prior to tanning, special code marks are punched into the skins to insure correct reference to the catalog number of the specimen (see Fig. 35-28). A secure point of attachment for the tag is through a nostril or between the eye opening and upper lip. After the specimen has returned from the tanner, a permanent specimen tag bearing complete data will be attached to the skin.

Fluid Preservation

The type of specimen labels described above are usually not substantial enough to survive long immersion in alcohol or formalin. Thus special, heavy duty tags (usually parchment) are required (Fig. 33-8). Complete data from the field catalog is entered on these tags with permanent ink. If a whole animal is preserved tie the tag securely *above the ankle* of the right hind foot as with the specimen tag used on a study skin. Several specimens may then be placed in one container.

If embryos, stomach contents, parasites, or any other portion of the specimen are preserved in fluid, a separate container must be used for each *or* each portion may be securely tied in cheesecloth and a tag attached

Figure 33–8.
The two sides of a tag for a specimen preserved in fluid. (K.G. Matocha)

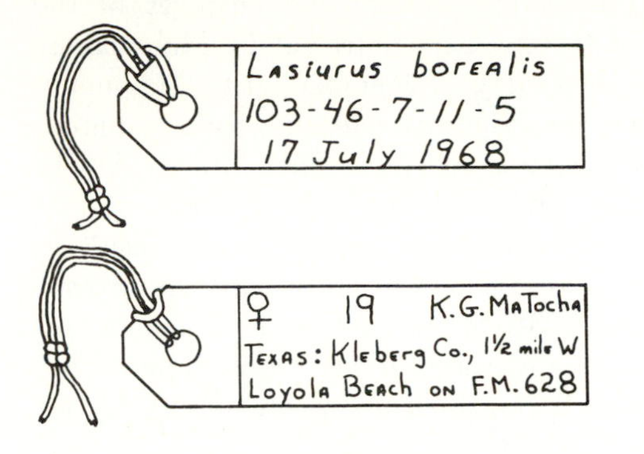

to each cheesecloth package. Several of these packages can then be placed in one container (*NOTE*: This recommendation holds only if mixing of fluids from several specimens would not alter the results of subsequent studies; mixing might be a problem with certain biochemical studies). A label with complete data should be inserted *into* the container. Labels attached to the outsides of the containers all too frequently come off and are lost.

Special Data Forms

In many situations, it is more appropriate and convenient to utilize special forms for recording data in the field and laboratory. Examples of these forms were described in the section on field notes (e.g., Figs. 33-3 and 33-4).

Census Records

In live-trapping studies, data on species, spatial location, condition of animal, and other information must be maintained. For these purposes, a talley sheet (Fig. 33-9) or a more elaborate form is utilized to facilitate subsequent transcription of the records onto tabulating cards, magnetic tape, or other input media.

Recording Measurements

For quantitative analyses, extensive measurements are generally recorded on coding sheets that permit accurate transcription to data processing equipment. Semiautomatic systems for making linear measurements are now in use and will probably be utilized more widely in the future (Anderson 1972, Calaprice and Ford 1969, Sneath and Sokal 1973:452-453).

Figure 33–9.
Data recording sheet for small mammal censusing. Original size 8½ × 11 inches.
(P.L. Meserve)

PERMANENT RECORD ( )   PLOT__________   INCLUSIVE DATE__________

DATE	TOT. NO.	Species	Coordinates	Wt.	Sex	Sexual Condit.	Mat. Stag	No.	Remarks

Recording Behavior

Since behavioral events can occur in quick succession, coding sheets are designed for rapid recording. Even so, instrumentation and keyboards are usually necessary for recording sequences of behavior.

Recently, investigators developed instrumentation for direct recording and transcription of behavioral events under field conditions (Butler and Rowe 1976; Stephenson et al. 1975). With the advent of portable terminals and microprocessors, we can expect to see additional uses of instrumentation for gathering and analyzing data under field and laboratory conditions. Techniques for recording behavior were described in detail by Altmann (1974).

Processing of Data

Edge-Sort Cards

For relatively small quantities of data (generally fewer than 1000 records), *edge sort cards* may be used for uncomplicated analyses of data (e.g., counts, frequencies). Holes in the edges of the cards are notched using a coded scheme appropriate for the individual record (e.g., "food habits" in "Cricetidae"). Mosby (1969a:62-63) provided further details on this system.

Tabulating Cards

Tabulating or "IBM-type" *cards* are more versatile than edge-sort cards. For one, they can be read into a computer via a card reader and analyses conducted without further coding or transcription. *Keypunch machines* for producing these cards are widely available at many colleges, universities, and museums. Neuner (1976:1-5) provided a concise set of instructions for using these machines. For field use, the Porta Punch® has been used for primary recording of data in some studies (Patton and Casner 1970).

Table 33-1
Mandatory and selected optional data categories in mammalogy. Items marked with an asterisk are mandatory under NIRM standards. Refer to Williams et al. (1979) for additional information.

Category	*Category*
*Catalog Number	Latitude and Longitude
*Museum Acronym	Elevation
*Genus	Collector's Name
*Species	Collector's Number
Subspecies	*Sex
*Date Collected	*Type of Preparation
*Continent or Country	Weight
*State or Province	External Measurements
*County, District, or Major Island Group	Reproductive Data
Specific Locality	
*Ocean	
*Sea	
*Bay, Inlet, Strait, Estuary Gulf, or Channel	

Interactive Terminals

Interactive terminals (both hardcopy and video display) are rapidly replacing keypunch machines for entry of data. These devices permit much faster entry with less fatigue to the operator. In addition, terminals are frequently tied into a time-sharing system so that data files can be stored and retrieved very easily.

Data Standards for Museum Specimens

In 1975, the American Society of Mammalogists sponsored a workshop for the National Network for Information Retrieval in Mammalogy (NIRM; see discussion in Williams et al., 1977). An outgrowth of this workshop was the selection of optional and mandatory categories of data for specimen records in museums. Williams et al. 1979 recently drafted a set of "Documentation Standards for Automatic Data Processing in Mammalogy," under the auspices of the Committee on Information Retrieval in Mammalogy of the American Society of Mammalogists. Mammalogists and other investigators working with mammals should strive to record specimen data in accord with these standards. This will not only promote consistency in data recording but hopefully will result in more accurate and useful specimen data for investigators. In Table 33-1 we have listed all mandatory NIRM categories and some of the more common optional categories. Refer to Williams et al. (1979) for a complete listing of these standards.

Supplementary Readings

Cutbill, J. L. (Ed.). 1971. *Data processing in biology and geology.* Systematics Association, Special Vol. 3, Academic Press, London and New York. 346 pp.

Dixon, W. J. (Ed.). 1971. *BMD biomedical computer programs.* Univ. California Publ. Automatic Computation, 2:1-600.

Hall, E. R. 1962. Collecting and preparing study specimens of vertebrates. *Univ. Kansas Mus. Natur. Hist., Misc. Publ.* 30:1-46.

Mosby, H. S. 1969. Making observations and records, pp. 61-72. *In* R. H. Giles, Jr. (Ed.). *Wildlife management techniques,* 3rd ed. The Wildlife Society, Washington.

Neuner, A. M. 1976. *SELGEM workbook.* Assoc. Systematics Collections, Lawrence, Kansas. 51 pp.

Williams, S. L., R. Laubach, and H. H. Genoways. 1977. A guide to the management of Recent mammal collections. *Carnegie Mus. Natur. Hist., Spec. Publ.* 4:1-105.

Williams, S. L., M. J. Smolen, and A. A. Brigida. 1979. *Documentation standards for automatic data processing in mammalogy.* The Museum of Texas Tech Univ., Lubbock. 48 pp.

34 Collecting

The study of mammals frequently requires that they be captured or collected. Many detailed investigations can be carried out only when the mammal is in captivity or has been prepared as a museum specimen. Taxonomists require specimens to document the presence of species in particular localities, to assess geographic and other types of variation, and to prepare reviews and keys on various groups of mammals. Scientists interested in life history studies of mammals collect specimens to assess reproductive patterns or food habits. Population biologists capture mammals alive, mark them for permanent identification, and later try to recapture the marked animals to obtain data on population sizes and structures. Behavioral and cytological studies generally require live mammals for field or laboratory investigations. These are only a few of the many types of research that require the collection or capture of mammals.

Collecting, Conservation, and the Law

In recent years, increasing concern on the part of many individuals for the conservation of flora and fauna and for the protection of rare and endangered species, prompted the passage of many federal and state laws. These and earlier laws regulate the salvage, capture, possession, transport, and sale of certain species of animals and affect, in many instances, the manner in which scientific and educational activities involving these animals can be conducted.

Federal and International Regulations

The most important of recent federal laws are the Endangered Species Act of 1973 (Code Federal Register, CFR 50.17) and the Marine Mammal Protection Acts (CFR 50.18, 50.216). In addition, the Lacey Act, in part, regulates the importation and exportation of nonendangered species and injurious wildlife (CFR 50.10, 50.13, 50.14, 50.16).

Important amendments to the regulations of the Endangered Species Act of 1973 (CFR 50.17) were published February 22, 1977 in the Federal Register (42:35). These published regulations implement the provisions of the Convention on International Trade in Endangered Species of Wild Fauna and Flora, TIAS 8249 to which the U.S. is one of 34 participants as of 1977. In addition to the list of endangered species that the U.S. Fish and Wildlife Service maintains (Federal Register 41:191, Sept. 30, 1976), the Convention defined three categories of species (listed in Appendices I, II, and III of the convention document described below:

Appendix I: All species or other taxa threatened with extinction.

Appendix II: All species or other taxa not threatened with extinction but those in which trade must be restricted to insure their survival.

Appendix III: All species or other taxa identified by a particular country as subject to conservation regulations within that country and requiring the cooperation of other parties to the convention.

For importation (or exportation) of animals and their products into (or out of) the U.S. the provisions of the international convention of the Endangered Species Act of the Marine Mammal Protection Acts, and of other applicable regulations of the Department of Agriculture must be adhered to. Refer to Genoways and Choate (1976), Berger and Phillips (1977) and the Federal Code for specific guidelines and provisions of these regulations.

State Regulations

The complexity, number, and lack of uniformity in state laws regulating scientific collecting have prompted some scientists to urge the adoption of uniform regulations among states for this activity (see McGaugh and Genoways 1976). Until realistic and uniform regulations exist for all states, it is incumbent upon scientists and students to follow the appropriate state and local regulations pertaining to specimen acquisition. Refer to McGaugh and Genoways (1976) for a summary of state laws regarding scientific collecting permits and addresses where copies of the latest regulations for each state can be obtained.

Conservation

Many species (e.g., cave bats) not considered threatened or endangered can also be permanently harmed by indiscriminate collecting. All collecting should be carried out for a valid purpose, and the purpose should dictate the amount and type of collecting. For instance, a project to document the occurrence or distribution of a species in a given area would normally require collecting only a few individuals at a variety of locations, while a study of individual variation within a population might require larger numbers from fewer locations.

Many bat colonies are particularly susceptible to disturbance. The slightest provocation will cause many species to desert roost sites, and such provocation can be extremely harmful if it disturbs a nursery colony or a hibernaculum. Whenever possible, get an expert opinion before disturbing bat colonies, and when collecting in them, create as little disturbance as possible.

Locating a Collecting Area

Obtain permission from landowners or leaseholders before collecting in an area. Failure to do so may subject you to arrest for trespassing. Often the landowners will be able to direct you to favorable collecting sites when you observe this courtesy. A large percentage of the federal land in the United States is available for collecting, with permission, but state game laws and applicable Federal laws must be followed (see McGaugh and Genoways 1976).

After you reach the collecting area, make a visual survey to determine where to concentrate your efforts. Some species may be restricted to particular types of habitat but others have a tolerance for several habitats. After the appropriate habitat area is located, look for "sign" before placing traps (see Chapter 32).

After you finish collecting in an area, always remove all trap markers, wire, plastic, cloth, paper and any other debris from the site.

Records

Keep a careful record of all collecting activities. It is very important to know when, where, and how the specimens were collected. Always keep the data closely associated with live animals housed in cages and with dead animals stored in a freezer. Refer to Chapter 33 for record-keeping techniques.

Methods of Trapping Mammals

Kill Traps

The traps listed below are valuable for most forms of collecting when it is not necessary to acquire live animals. They are frequently more effective and less expensive than live traps.

Snap Traps

Snap traps are by far the most important trap type for general collecting. The snap trap comes in three basic sizes. The largest, the **rattrap** (Fig. 34-1C), and the smallest, the **mousetrap** (Fig. 34-1A), are generally available and familiar to most people. The rattrap is used to catch rats, *Rattus;* wood rats, *Neotoma* chipmunks, *Tamias* and *Eutamias;* the smaller ground squirrels, *Spermophilus;* and similarly sized mammals. Smaller mammals, e.g., most *Peromyscus, Microtus, Blarina,* can be taken in rattraps, but the strength of the springs on these traps frequently causes damage to the specimen. Because of the small distance between the sprung wire bail and the bait pan of mousetraps, the skulls of mammals collected in them are usually crushed. Since the skull is an important part of any museum specimen, mousetraps are very seldom used for scientific collecting. The **Museum Special** (made by Woodstream Corp., Lititz, Pennsylvania) is especially designed for collecting small mammals for scientific purposes (Fig. 34-1B). It is slightly larger than a mousetrap but less powerful than a rattrap. Its design greatly reduces the frequency of skull damage.

To set any of the snap traps, place a small quantity of bait on the pan and adjust the trigger mechanism so that it will be set off with a light touch. Since wood rats are adept at collecting these traps and adding them to their den debris, it is wise to wire the traps to a rock or branch when trapping near wood rat dens.

Figure 34–1.
Examples of snap traps. A, mouse trap; B, museum special; C, rattrap.
(R.E. Martin)

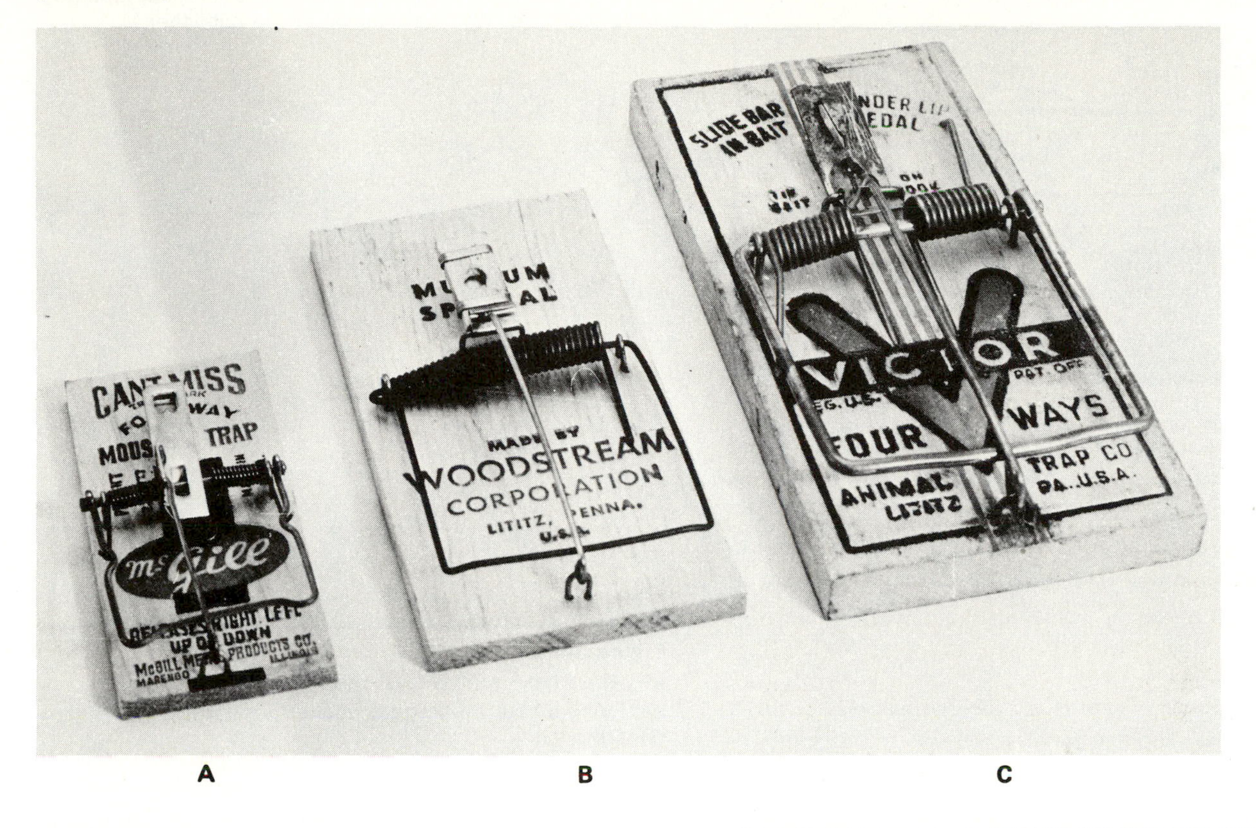

Steel Leg-traps

The type of trap illustrated in Figure 34-2 is used primarily by fur trappers and to some extent for scientific collecting. These traps catch mammals by one leg and generally do not kill. However, they are included under the "kill trap" classification since they usually severely damage the animal and necessitate killing it. These traps cannot be considered humane and should be used only when absolutely necessary. The Conibear trap, discussed below, is a more humane substitute.

Steel traps come in a variety of sizes with the smallest code numbers designating the smallest trap. They are also classified according to the type of springs they possess and to their mode of action. The **long spring trap** (Fig. 34-2) has one or two U-shaped springs extending away from the jaws. The **jump trap** has a flat spring underneath the bait pan. The **coil spring trap** has a pair of spiral springs underneath the bait pan. A stop-loss device that prevents the captured animal from gnawing its leg off may be attached to any of the above traps.

Steel traps may be set under water for beaver (Fig. 34-3) or muskrat, on the ground surface, or partially buried and camouflaged to capture many carnivores. Traps for large carnivores are usually not baited but are placed along trails or runways that the animals frequent. When setting traps for these animals, it is important to reduce the human odor on the traps by wearing clean gloves. A scent bait is frequently used around the site. Steel traps should be tied to heavy weights (e.g., rocks or logs) or stakes or trees to prevent the animal from dragging the trap away. Traps

Figure 34–2.
Example of double long-spring steel trap with components identified.
(Henderson 1960: 52)

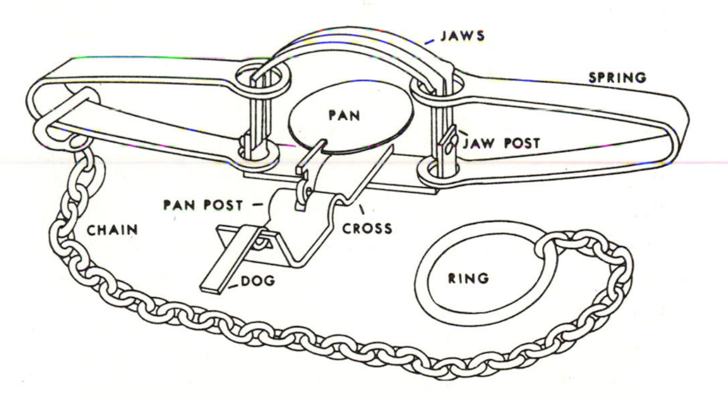

Figure 34–3.
Underwater set for beaver, *Castor canadensis*. A, Stake; B, tunnel and den in earthen bank; C, food cache. (Henderson 1960: 56)

set in water are usually tied to weights or stakes set in deeper water so that the animals will drown as they try to escape. If the animal is alive when the traps are checked, it must be killed quickly and humanely in a manner that will not damage the skin or skull.

Conibear Traps

The **Conibear Trap** (Fig. 34-4) also comes in a variety of sizes and frequently can be substituted for the smaller sized leg-traps. Compared with leg-traps, the Conibear trap is more humane since it generally kills animals very quickly.

Gopher Traps

Several traps designed specifically to catch gophers are manufactured in the United States. A style that is widely available and easily set is shown in Figure 34-5. The traps are set in underground runways, singly (Fig. 34-6A) or in pairs (Fig. 34-6B), and are tied to a stake to prevent them from being dragged deep into the burrow system by the captured animal. The tunnel is left open so that the gopher will push dirt over the trap and in so doing set off the trigger mechanism. These traps must be checked every 15 to 30 minutes to get maximum benefit.

Mole Traps

The **harpoon mole trap** (Fig. 34-7) is the most commonly used mole trap in the USA. It is set on the surface. The mole tunnel is pressed down and the trap pushed into the ground so that it straddles the tunnel. The trigger mechanism is cocked, and the height of the trap adjusted so that the mechanism will be raised and thus set off by the mole as it travels just under the surface of the ground.

Figure 34–4.
Conibear trap, shown set.
(Stains, H.J. 1962. Game biology and game management; a laboratory manual. Burgess Publishing Co., Minneapolis. 143 pp., fig 7e. With permission of the publisher.)

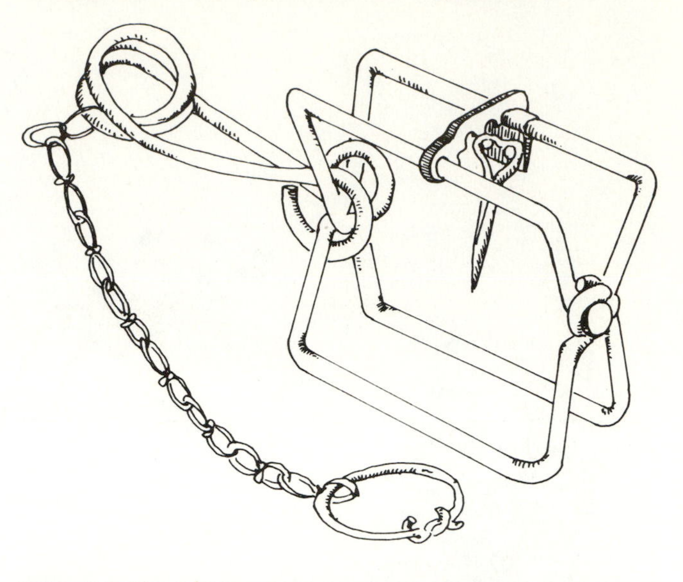

Figure 34–5.
Macabee-type gopher trap shown set. Suitable for capturing all but the largest pocket gophers.
(R.E. Martin)

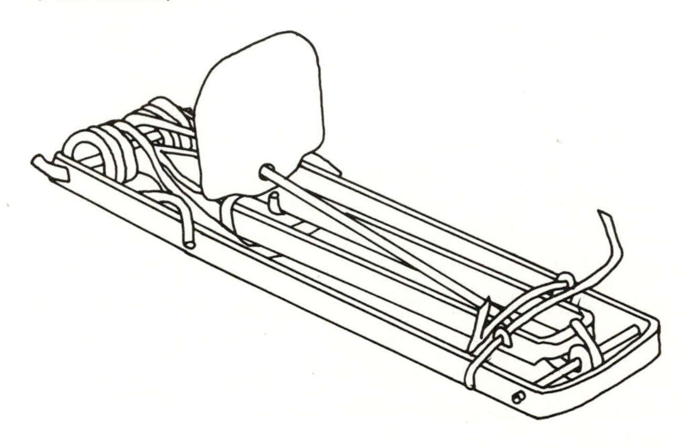

Deadfalls and Snares

These devices are primarily human survival equipment for use in remote areas when it may be necessary to "live off the land." They are rarely used for scientific collecting since they must be checked frequently and often severly damage specimens. Taber and Cowan (1969), Stains (1962), Petrides (1946), and Soper (1944) illustrated various kinds of these devices.

Figure 34–6.
Procedure for making single and dual-trap sets for pocket gophers. A, single set in lateral tunnel; B, dual set, back to back, in main tunnel.
(Crouch 1933: 17)

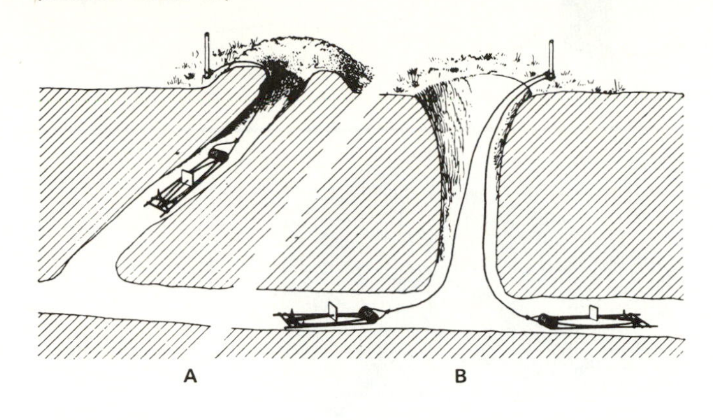

Figure 34–7.
Harpoon mole trap, shown set.
(R.E. Martin)

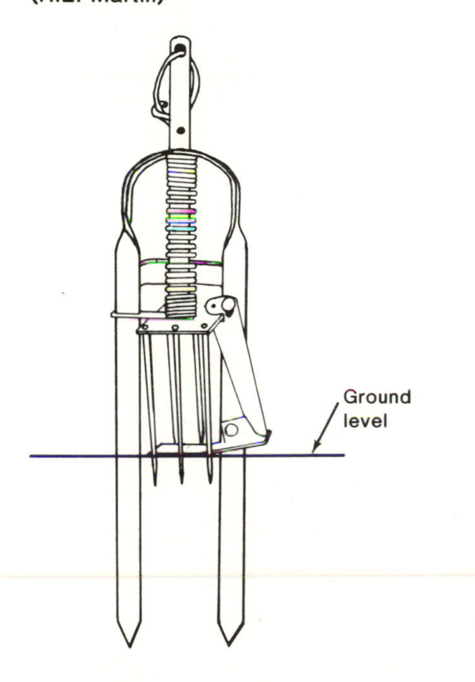

Pitfalls

Tall cans or jars, placed in the ground with their rims even with the surface, are useful in catching small animals, particularly shrews. The containers can be partially filled with water so that the animals drown. Be sure to check pitfalls regularly.

Live Traps

Live traps offer several advantages over kill devices. Animals that are needed alive can be taken back to the laboratory for further investigation or studied or marked at the trap site and then released. In areas where endangered or threatened species may occur, the use of live traps will minimize the chance of harming these forms.

Arrange the traps so that the animals will be protected from extremes of cold and heat, and check them frequently for any captures. In cold weather provide cotton or other insulating material.

Commercial Traps

Several manufacturers produce and market live traps in a range of sizes to capture everything from mice to coyotes. Some firms make types large enough to capture small deer. Welded wire traps (Fig. 34-8) are made by at least three companies, Allcock Manufacturing Co. (Havahart® brand traps), Tomahawk Live Trap Co., and National Trap Co. These traps (made in collapsible or noncollapsible varieties) have either one or two doors and are activated by a small bait pan in the center of the trap. The **Sherman trap** (H. B. Sherman Traps) is made of galvanized sheet iron or aluminum in folding (Fig. 34-9A) or nonfolding (Fig. 34-9B) varieties. The **Longworth trap** (Longworth Scientific Instrument Co., Ltd.) is made of aluminum and composed of two sections: a trap and trigger mechanism and a detachable nest box (Fig. 34-10). This added feature provides greater protection of the captured animal from the elements and makes for easier handling of the live mammal in the

Figure 34–8.
Havahart® trap, one of several varieties of welded-wire live traps.
(Allcock Manufacturing Company)

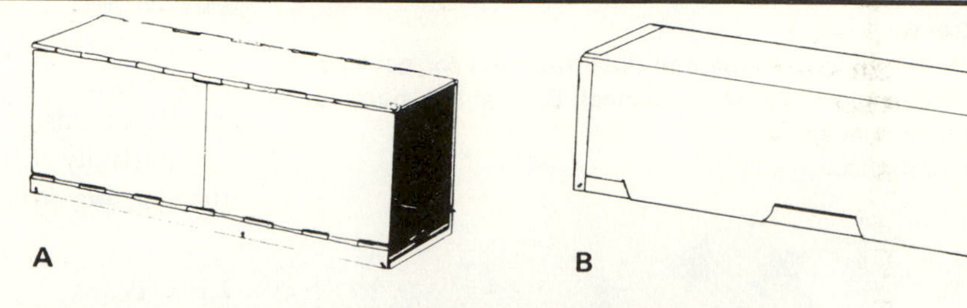

Figure 34–9.
Folding (A) and nonfolding (B) models of Sherman live traps.
(G.A. Moore)

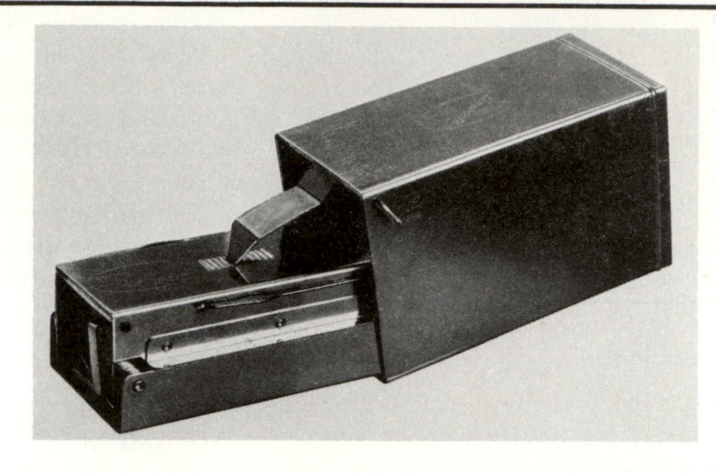

Figure 34–10.
Longworth live trap with trap mechanism and detachable nest box.
(Longworth Scientific Instrument Company)

field. Grant (1970), however, reported that lighter-weight animals are not caught as readily in Longworth traps as are heavier animals.

Homemade Traps

Many types of live traps can be made relatively easily and inexpensively. If a particular trap design does not suit your needs, the plans may be modified so that the trap will capture the species that are desired. The **Gen trap** (Fig. 34-11) was originally designed by Shemanchuk and Bergen (1968) for use in ground squirrel burrows. However, this trap is adaptable to other species by modifying the diameter of the pipe and method of setting (e.g., placing flat on ground for mice). The Gen trap is constructed of economical materials and is easily made and transported. The **Fitch trap** is dependable and widely used for many species of rodents. This trap is constructed of hardware cloth, galvanized metal, wire, and a metal can (see Fitch 1950, for details on original design and Rose 1973, for modified version). Pocket gophers, Geomyidae, are somewhat difficult to live trap but Baker and Williams (1972) reported good results with their design (Fig. 34-12). Earlier, Howard (1952) designed a live trap for pocket gophers that was the standard for many years.

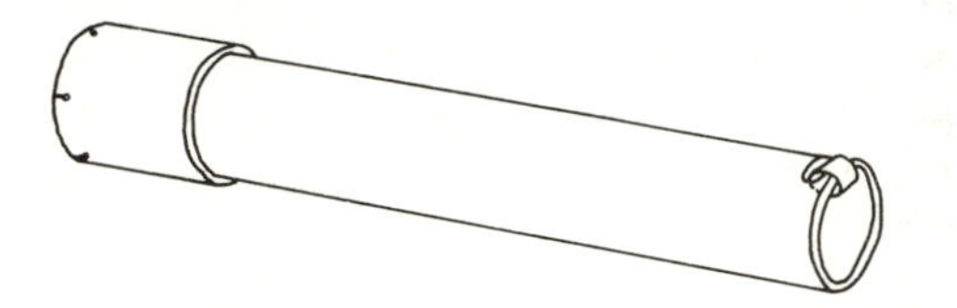

Figure 34–11.
Gen trap, with removable rear closure.
(R.E. Martin)

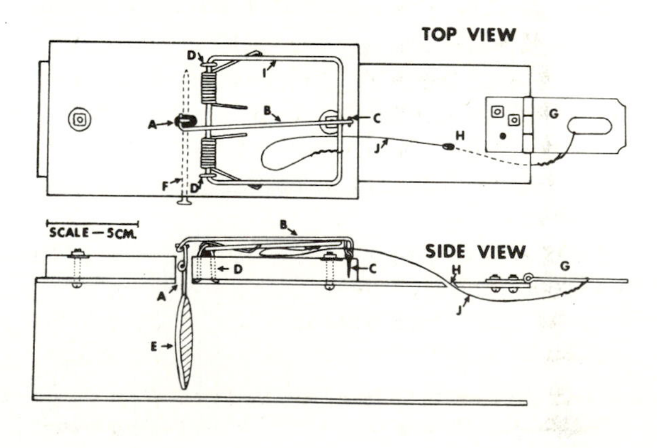

Figure 34–12.
Live trap for pocket gophers.
(Baker and Williams 1972)

Figure 34–13.
Live mole trap.
(Moore 1940: 224)

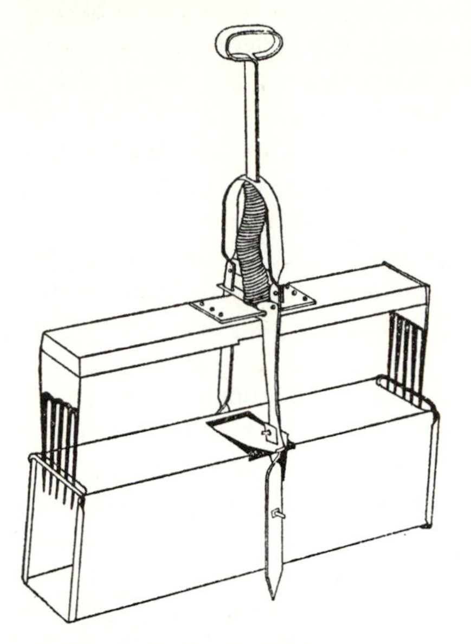

According to Moore (1940), moles can be captured alive using modified kill traps (Fig. 34-13). Additional designs for live traps can be found in Taber and Cowan (1969), LoBue and Darnell (1958), Howard (1953), Gilmore (1943), Garlough et al. (1942), and Burt (1927).

Pitfalls

Tall cans or jars, placed in the ground with their rims even with the surface, are useful in catching small mammals, particularly shrews. Provide an adequate quantity of food in the container and check frequently for captures. Baffles pressed vertically into the ground increase the effectiveness of pitfalls by guiding mammals toward them. Howard and Brock (1961) designed a pitfall trap to capture and preserve small rodents. By omitting the preservative and checking the traps frequently, these traps can be used to capture small mammals alive.

Runways and Corrals

Larger mammals (even rabbits and hares) may be herded along runways into corrals. Care must be taken to avoid injuring the animals during the chase since most will be released. Most states restrict this activity, and usually only qualified professional biologists or game wardens engaged in management or research programs may utilize this technique. Check with game officials before attempting this procedure.

Special Purpose Traps

Beaver may be taken in either Bailey or Hancock traps and muskrat in Snead traps (see Taber and Cowan 1969, and Couch 1942).

Methods of Trap Placement

Snap traps and most types of commercial and homemade live traps can be distributed by any of the three methods listed below. The other types of traps are almost always placed by the "sign method."

Sign Method

With this method of trap placement appropriate fresh mammal sign is located and a trap positioned to catch the animal. While this method generally produces the highest results, it is more time consuming than the methods listed below. To facilitate relocating traps placed in this manner, either each trap must be marked with a conspicuous trap marker, or detailed notes must be kept. The sign method will generally result in a biased sample of the species present since those small mammals which leave little or no conspicuous sign will usually not be collected.

Paceline Method

This method places traps at regular intervals along a straight line. One technique is to select a starting point and mark it with a strip of conspicuously colored cloth or plastic surveyor's tape. Sight toward a distant landmark and place traps at regular intervals as you walk toward that point. Trap markers are usually placed at the beginning and end of each line. If the line is particularly long or if the habitat has dense ground cover, it is prudent to also mark every fifth or tenth trap location. The paceline method generally allows for the placement of a maximum number of traps in a minimum amount of time and also is free of the bias mentioned above. However, it will usually result in a lower percentage of captures than the sign method.

Grid Method

A trap grid is a series of parallel trap lines. Grid trapping is most frequently utilized in life history studies where mammals are marked and recaptured (see Chapters 37 and 38). The size of the grid is dependent on the species to be captured and the type of study. In small rodent studies, for example, the grids are often arranged in twelve rows and twelve columns (144 traps total), with each trap station fifteen to twenty meters from an adjacent station. The traps are usually placed without regard to sign or habitat.

Timing of Trapping Activity

Since most small mammals are crepuscular or nocturnal, traps are usually set just before dusk and checked or collected in early morning. Ants and other small invertebrates will eat the bait if the traps are set too early in the day, and these small animals can ruin trapped specimens if the traps are not checked soon after dawn. It is frequently productive to check traps during the night, remove trapped mammals, and reset the traps. Fleas will leave a specimen as soon as the body cools, so frequent trap checks are necessary if you are interested in collecting these ectoparasites.

To catch diurnal mammals such as squirrels, traps can be set any time during the day. But they should be checked at frequent intervals, particularly in hot weather.

Weather conditions will frequently influence trapping success. A heavy rain may wash the bait off of traps or may set them off. A heavy snowfall can bury traps and make them very difficult to relocate. The activity patterns of some mammals may be influenced by the weather or by the phase of the moon. Kangaroo rats, *Dipodomys,* for example, are more active and therefore more readily collected, when there is no moonlight.

Calculating Trapping Success

Trapping success is given as the percentage of the traps that produce specimens. Thus if 100 traps were set and ten mammals collected, the trapping success would be ten percent. The number of traps used multiplied by the number of nights they were in position is referred to as the number of **trap nights.** Thus 100 traps set for one night equals 100 trap nights, twenty traps set for five nights equals 100 trap nights, and twenty, forty, thirty, and ten traps set on four successive nights still equals 100 trap nights. Trapping success is also frequently given as a percentage of trap nights.

The trap success to be expected varies considerably with the habitat, type of trap, trapping method, population density, and many other variables. In the U.S., a pace line of snap traps in moderately good habitat will usually produce about ten percent success. In the American tropics, trapping success is generally much lower.

See Chapter 38 for techniques that can be used to estimate densities.

Baits and Scents

Baits

Bait may be preferred seasonal food or a substance entirely new to the animal. For rodents the most commonly used bait in temperate North America is a mixture of peanut butter and oatmeal (mixed bird seed is sometimes added). To catch insectivores, it is helpful to add a small quantity of meat, especially fish or bacon, to the mixture. Most carnivores will be attracted by canned dog food or fish products, while rabbits and similar herbivores require apples, carrots, or lettuce. Dead or live animals may also be used to attract mammals to traps.

In many areas, ants can kill mammals in live traps, remove bait, or otherwise be a big problem. A change of bait may be effective in minimizing its loss to insects. Various insecticides (e.g., DDT), formerly were recommended as additives to bait to prevent its loss to insects (especially ants), although the use of these materials is often restricted or not recommended now. The technique of mixing shredded cotton into heated peanut butter to retard the removal of the bait by insects, as suggested by Getz and Prather (1975), offers one solution to avoid the indiscriminate use of insecticides in bait preparations.

Scents

Scent baits are very effective with many mammal species since the odor may arouse sex interest, antagonism, or curiosity. Scents may be superior to food baits for mammals that communicate by odor. In general, valerian (prepared from *Valeriana* sp.) is attractive to carnivores of the family Canidae and catnip (*Nepeta* sp.) attracts members of the family Felidae. Anise (*Pimpinella* sp.) is attractive to the black bear, *Ursus americana;* and beaver castor or castoreum (prepared from musk sacs of both sexes of *Castor*) elicits a positive response from beaver. Refer to Taber and Cowan (1969) for instructions on how to prepare and use various scents.

Minimizing Damage to Trapped Mammals

At times, larger mammals (particularly carnivores) may disturb or injure smaller mammals caught in live traps. Getz and Batzli (1974) designed a cage that is placed over a live trap to minimize this possibility.

In areas where ants may cause damage to specimens (e.g., tropics, deserts), traps must be checked frequently (including nighttime hours) to remove captured mammals. Frequent checking of traps may also minimize damage to kill-trapped mammals resulting from the activities of shrews, carnivorous rodents (e.g., *Onychomys*), and small carnivores.

Traps must be checked frequently during warm weather. Mammals held in live traps may overheat and die and specimen in kill traps may decompose quickly under such conditions.

Adjusting Trapping Methods for Regional and Climatic Differences

Most of the advice contained in this chapter applies to North American temperate conditions. You may have opportunities to collect in other countries and other climatic areas. Often, you can learn important techniques by talking with persons experienced in trapping in these areas. Papers in the Journal of Mammalogy and similar journals are also helpful as are the introductory sections of many regional faunal works.

Other Methods of Capturing or Collecting Mammals

Hunting

Larger species of mammals are generally collected using a rifle or shotgun. Small species, if not trapped or collected in some other manner, can be collected using a shotgun. Specimens collected for scientific purposes must be killed in a manner that causes the least damage to the specimen or to the body parts required for study. Remember that firearms impose an obligation upon the user to handle them safely and in compliance with local and state laws.

Shotguns of 12, 20, or .410 gauge are generally used to collect smaller mammals. Very small species are best collected with .22 caliber rifles or pistols (preferably smoothbore) loaded with birdshot or a .410 shotgun with half-loads of No. 12 shot. Rifles loaded with ball ammunition are generally used on the larger species. A .22 cartridge will handle most species up to the size of a raccoon or badger. Larger calibers are necessary for large carnivores and herbivores. If you anticipate collecting some of the larger mammals, it would be well to seek the advice of a person experienced in firearms and hunting.

Drug Immobilization

Large mammals that must be captured alive are usually immobilized with drugs to prevent harm to the animal and the captor. The drugs most commonly used are nicotine salicylate and succinyl choline chloride. To achieve proper dosages, it is necessary to have a fairly accurate method of estimating weights of the target animals. Taber and Cowan (1969) list approximate dosages for a selected list of mammals. Harthoorn (1965, 1975) provided a comprehensive treatment of this specialized field and Twigg (1975b:89-91) gave a particularly succinct account.

The drugs are administered by means of specially designed syringes attached to arrows or firearm projectiles. The Cap-Chur gun is a commercial form of these devices. In most states drug immobilization can be conducted only by qualified scientists, game management personnel, or zoological garden staffs. Consult local game officials before attempting this procedure.

Corral Traps and Nets

Taber and Cowan (1969:286-290) described and illustrated several types of herd and drift traps for the capture of large mammals. A radio-controlled drop net was used by Ramsey (1968) to capture white-tailed deer (*Odocoileus virginianus*) and axis deer (*Axis axis*).

Den Investigation

Woodrats, *Neotoma,* some species of kangaroo rats, *Dipodomys,* and several other species frequently can be dug out of their dens and then captured by hand, with a net, or shot with firearms. In most states it is illegal to collect game and furbearing species in this manner. Consult game officials before using this procedure.

Hand Capture

Many nocturnal mammals, particularly certain rodents, can be captured by hand or with a net. A light is shined in the animal's eyes causing it to remain motionless while approached and captured. You will feel safer wearing gloves but many species make no attempt to bite. Hand capture can be conducted by walking through fields with a lantern or driving slowly along infrequently traveled roads. Night-lighting for game species is illegal in most states, but the species that can be caught by hand usually will not fall into this protected category.

Bat Collecting

Bats may be collected by shooting, by netting, or by hand. Shooting is difficult since most bats fly at dusk and their flight is erratic. To improve your chances, select a location where the horizon is free

Figure 34–14.
Mistnet used in capturing bats.
(Greenhall and Paradiso 1968)

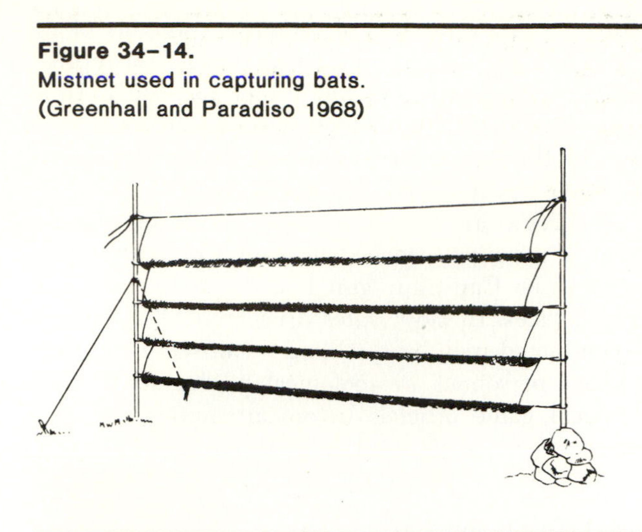

Figure 34–15.
Double-frame Tuttle trap showing details of construction.
(Tuttle 1974: 476)

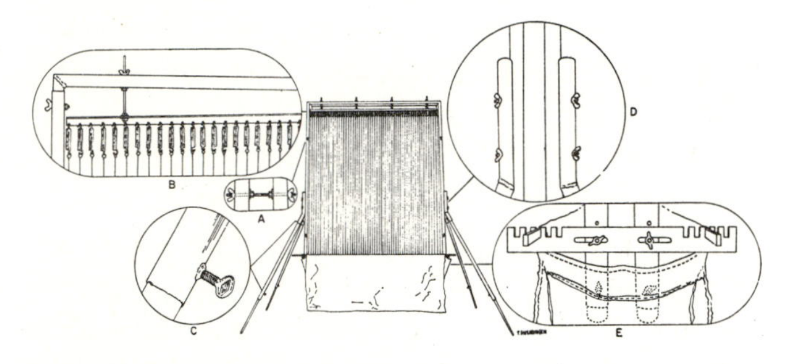

of trees or other obstructions and where fallen bats will not be lost in the vegetation. Use size 8 shot or smaller for North American species. Bats may be caught in **mistnets** (Fig. 34-14), bat traps (Fig. 34-15) or by hand. The nets are strung over water holes (especially in desert regions), along streams, or across narrow clearings in timbered areas. Bats can also be collected by hand from bridge expansion joints, old mine tunnels, caves, attics, belfreys or other sites (see Greenhall and Paradiso 1968).

Many species of bats have had their numbers seriously depleted by indisriminate collecting. Be especially careful about collecting bats from suspected breeding or nursery colonies. The gray bat, *Myotis grisescens,* and the Indiana bat, *Myotis sodalis,* are endangered species and must not be disturbed or captured without official authorization from the U.S. Fish and Wildlife Service. Determine the specific identity of any bats in a new-found colony before collecting any. Some nursery colonies are extremely sensitive to disturbance (e.g., movement, light, noise), and may abandon the location with adverse results to the population.

Several specialized techniques have been devised for capturing bats in conjunction with population, movement, and other life history studies. Many of these techniques were discussed and illustrated by Tuttle (1974), Greenhall and Paradiso (1968), Barbour and Davis (1969), Constantine (1958, 1969) and various volumes of Bat Research News (1960 to present).

Salvage

Many good records result from collecting mammals that are found dead on the road (abbreviated DOR). The specimen may be damaged, often the skull is crushed, but most tears can be sewn up to make a suitable preparation. Often the specimen will be bloated and have a foul odor, but a good preparation can still be made if the hair has not started to **slip** (that is if large patches cannot be easily pulled free from the underlying dermis).

Raptor pellets (both hawk and owl) frequently yield valuable mammal specimens. These are a particularly good source of shrews which are not often caught in traps. Only skeletal material and matted hair will remain, but specific identity of the prey species is usually possible.

Occasionally weathered skulls, leg bones, vertebrae, and other skeletal elements are found in the field. Such remains of animals that died of natural causes or predation can be identified and these often provide useful locality records.

Frequently it is possible to obtain hides, skulls, and/or other material from animals killed by hunters and trappers.

Whenever possible, salvage should be used to increase the size of series, document distribution, record variation, provide internal organs for analysis, etc. Salvage is the one collecting source that does not require that the scientist go out and further disturb a population.

Permits must be obtained and/or declarations made to salvage specimens of endangered species, certain marine mammals, or specimens found in national monuments and parks in the United States. Otherwise, you may be subject to arrest and fines for possession of these animals (See McGaugh and Genoways 1976:80-81; Berger and Phillips 1977).

Marking Mammals

It is frequently necessary to mark individual mammals for studies of population density, home range, homing, migration, or behavior. Generally, the techniques available for individual marking fall into four categories: multilation, tagging, coloring, and radioisotope implantation.

Figure 34–16.
Two systems for numbering toes to provide an identification number through amputation of selected digits (toe-clipping). A, four-digit system developed by French (1964) requires amputation of no more than one toe per limb, is easily read, and will provide up to 899 combinations using four toes on each front foot and five toes on each hind foot. B, system developed by Baumgartner (1940) gives consecutive numbers up to 158 with only one toe removed per foot. Additional consecutive numbers can be added by clipping more than one toe per foot.
(L.P. Martin)

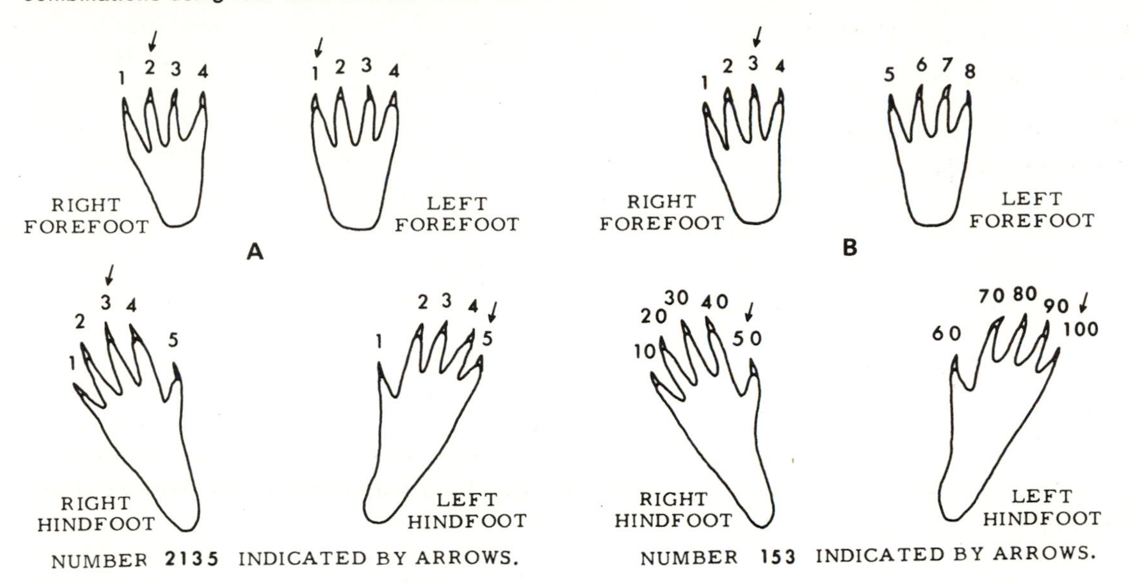

Mutilation

Mutilation for marking purposes usually involves toe-clipping, tail-docking, ear-cropping, ear-punching, fur-clipping, or branding. Toe-clipping, ear-cropping, and ear-punching are the most frequently utilized forms of mutilation marking. The animals recover quickly with little overall detriment. The marks are generally permanent but may be confused with natural injuries. Two schemes for toe-clipping are shown in Figure 34-16.

Tagging

Tagging involves the attachment of a metal or plastic tag to an animal. The tags are frequently numbered for individual recognition and may be attached to the ear, under loose skin, in blubber (whales), or around the neck or leg. Small rodents are frequently marked with fingerling tags attached to the ear. Bats are tagged (banded) with bird-type bands secured loosely around the distal portion of the forearm (Fig. 34-19). Greenhall and Paradiso (1968) and Bonaccorso et al. (1976) described banding techniques for use with bats and improvements in procedures. Twigg (1975c) provided a useful summary of procedures used for marking mammals.

Collar tags are frequently marked with reflective tape or plastic streamers so that individuals of large species (usually carnivores or ungulates) can be identified at a distance. Taber and Cowan (1969) and Knowlton et al. (1964) discussed several of these methods in detail. Buchler (1976) designed a chemiluminescent tag for tracking bats and other small nocturnal mammals; Heidt et al. (1967) utilized an implanted magnet to tract the movements of voles (*Microtus*) in an enclosure.

Coloring and Freeze Marking

Dyes are useful in marking mammals for behavioral investigations. The black dye, Nyazol D, can be applied to any species that has pale pelage. Several individuals can be marked with the same color by applying the dye to different regions of the body. Refer to New (1958), Haresign (1960), Taber and Cowan (1969), and Taylor and Quy (1973) for additional dyes and methods.

A pressurized refrigerant (dichlorodifluoromethane, CCl_2F_2) was used by Lazarus and Rowe (1975) to apply permanent markings (by changing the natural pigmentation pattern of the hair) on individuals of several species of rodents. In contrast, marks made by pigment dyes such as Nyazol D, last only until the dyed hair is shed in a molt.

Radioisotope Implantation

Radioisotopes can be a valuable tool for studying movements and activity of small mammals. However, there are several factors that limit their usefulness. To prevent harm to the animal and investigator, the isotopes selected must have a low radioactivity output. Generally, only one individual at a time can be tagged with each type of isotope if accurate identification of the individual is desired. The use of radioactive isotopes in the United States is under the regulation of the Nuclear Regulatory Commission. Consequently, field studies generally must be conducted in enclosures where the test animals cannot escape into surrounding populations. A technique using isotopes and recording devices to continuously monitor *Microtus* in field enclosures was described by Graham and Ambrose (1967). Consult Peterle (1969) for additional information.

Radio-Location Telemetry

Collars with radio transmitters and battery packs (or solar cells) can be attached to the necks or backs of animals to provide information on movements, activity, and physiological states (e.g., resting, moving, etc.). Individuals can be distinguished by differences in frequencies or pulse rates (clicks). While the initial cost of the equipment (transmitters, receivers, antennas) is costly, the system offers the advantage of providing many location coordinates at low cost per location compared with mark-recapture methods (see Chapter 37).

Radio-location telemetry techniques were used by Banks et al. (1975) to study the activity and home range of the brown lemming (*Lemmus trimucronatus*) and by Mineau and Madison (1977) to study the movements of white-footed mice (*Peromyscus leucopus*). Brander and Cochran (1969:95-108) provided additional references on the use of radio-location telemetery in a variety of mammals and a summary of procedures and equipment.

Care of Collected Mammals

Care of Live Mammals

Live traps should be checked frequently and any captured mammals removed to a suitable holding cage. Mammals held in captivity impose an obligation upon the captor to provide adequate housing, water, and food. Check the animals frequently, especially during the first days after capture, when they are adjusting to captive conditions.

Most captive mammals require a constantly available source of water or succulent vegetation. Some mammals need a complex diet for good health but others will thrive on one that is relatively simple. Most rodents live well on a diet of commercial rat or mouse chow although a seed diet is required for some species. Green food such as lettuce should be provided occasionally. Most carnivores can be fed canned and/or dry dogfood. Rabbits will thrive on commerical food pellets, supplemented with fresh lettuce or carrots. Consult Crandall (1964) or Lane-Petter, et al., (1967) for more information on care of live mammals. Greenhall (1976) provided a very comprehensive treatment on the care of bats in captivity.

Killing Mammals Humanely (Euthanasia)

It is sometimes necessary to kill a mammal that has been in captivity, is severely wounded, or caught in a live trap. This can be done humanely by a variety of methods. Small mammals can be asphyxiated by tightly compressing the thoracic cavity until the heartbeat ceases.

Small or large mammals can be killed by placing them in a closed container and introducing a cotton wad saturated with chloroform or ether (**Caution:** these chemicals are highly volatile). Carbon monoxide fumes, as from automobile exhaust, can also be piped into the container (**Caution:** Carbon monoxide is an extremely toxic, colorless, and odorless gas that must be used only where there is adequate ventilation).

Care of Dead Mammals

Mammals that are to be prepared as museum specimens should be processed quickly to prevent spoilage of meat or hide. Large mammals (e.g., deer) should be eviscerated if skinning will be delayed for several hours. Small mammals must be skinned quickly, preserved in chemical solutions, or frozen for subsequent processing. If frozen or refrigerated, always enclose a full set of data within the bag that contains the specimen. The freezer bag should be air tight, fastened securely and have little air in it to minimize specimen dehydration and conserve space. Refer to chapter 35 for detailed specimen preparation techniques.

Supplementary Readings

Berger, T. J. and J. D. Phillips. (Compilers). 1977. *Index of U.S. Federal wildlife regulations.* Assoc. Syst. Collections, Lawrence, Kansas. [looseleaf]

Crandall, L. S. 1964. *The management of wild mammals in captivity.* Univ. Chicago Press, Chicago. 761 pp.

Genoways, H. H. and J. R. Choate. 1976. Federal regulations pertaining to collection, import, export, and transport of scientific specimens of mammals. *J. Mammal.* 57(2, supplement):1-9.

Giles, R. H., Jr. (Ed). 1969. *Wildlife management techniques,* 3rd ed. (revised). The Wildlife Society, Washington. 633 pp.

Greenhall, A. M. and J. L. Paradiso. 1968. *Bats and bat banding.* Resource Publ. No. 72, USDI Bur. Sport Fisheries and Wildlife. 47 pp.

Harthoorn, A. M. 1965. Application of pharmacological and physiological principles in restraint of wild animals. *Wildlife Monog.* 14:1-78.

Harthoorn, A. M. 1975. *The chemical capture of animals.* Bailliere-Tindall, London. 416 pp.

Lane-Petter, W. et al. (Eds.). 1967. *The UFAW handbook on the care and management of laboratory animals.* 3rd ed. Williams & Wilkins, Baltimore. 1015 pp.

McCracken, H. and H. Van Cleve. 1947. *Trapping: the craft and science of catching fur-bearing animals.* Barnes Co., New York. 196 pp.

McGaugh, M. H. and H. H. Genoways. 1976. State laws as they pertain to scientific collecting permits. *Museology,* Texas Tech University 2:1-81.

Stonehouse, B. (Ed.). 1978. *Animal marking: recognition marking of animals in research.* Univ. Park Press, Baltimore. 224 pp.

Twigg, G. I. 1975. Catching mammals. *Mammal Review* 5:83-100.

Twigg, G. I. 1975. Marking mammals. *Mammal Review* 5:101-116.

Twigg, G. I. 1977. Techniques with captive mammals. *Mammal Review* 7:42-62.

35 Specimen Preparation and Preservation

A museum specimen of a mammal usually consists of a **study skin** (filled or tanned) and a cleaned skull. Frequently the postcranial skeleton is also cleaned and saved. Sometimes the specimen consists of a complete cleaned skeleton or the entire animal or parts of it preserved in fluid. The following instructions should enable you, with practice, to prepare specimens of high quality. A demonstration of the preparation of a study skin by an experienced preparator will be a valuable addition to these instructions.

Supplies and Equipment

Supplies needed for preparation include the following:

Cotton. Long staple, unabsorbent. Cotton batts used in making quilts usually serve well.

Wire. For reinforcing the legs and tail. Use monel or other wire which does not corrode.

Labels. For skin, skull, and each additional portion preserved (see chapter 33).

Permanent ink. See chapter 33 for discussion of suitable brands.

White thread. Sizes 40 and 8 (button and carpet).

Borax. For drying and preserving specimens; salt is frequently used for very large specimens.

Cornmeal or hardwood sawdust. For absorbing fat, blood and other body fluids.

A list of the equipment that may be used follows. The tools preceded by an asterisk (*) are essential. The others are helpful to have. See Figure 35-1.

Figure 35–1.
Equipment used in the preparation of a mammal study skin. (R.E. Martin)

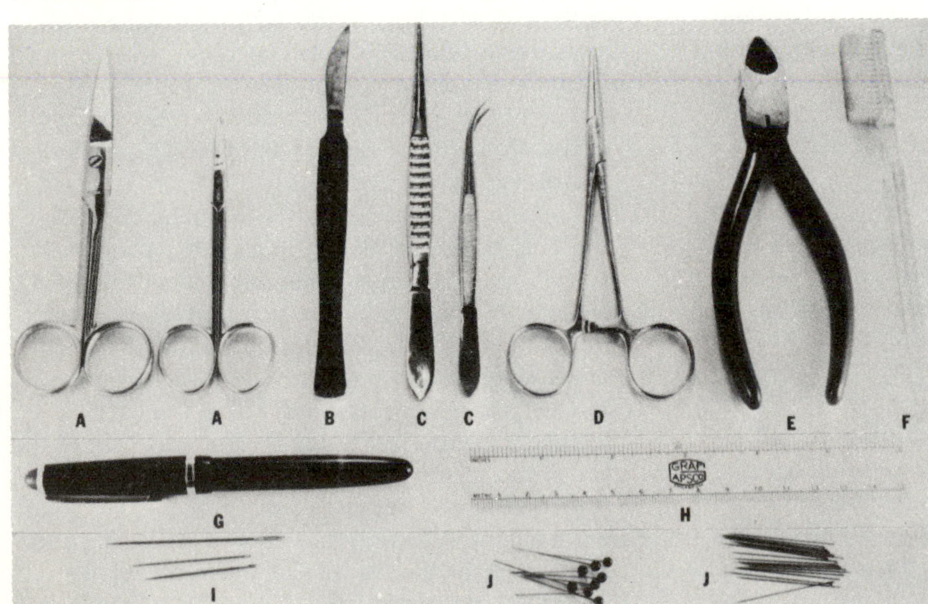

*Scissors (Fig. 35-1A). Good quality, surgical or dissecting, advisable to have a small pair with fine, sharp points and a larger pair with one sharp and one blunt point.

Scalpel (Fig. 35-1B).

***Forceps** (Fig. 35-1C). Straight or bent, fine or medium points.

Hemostat (Fig. 35-1D). Used to hold cotton body while inserting into skin.

***Pliers with wire cutter.** Or separate pliers and wire cutter (Fig. 35-1E).

***Toothbrush** (Fig. 35-1F). For brushing the fur.

***Pen.** For use with permanent inks. (See chapter 33.)

***Millimeter rule** (Fig. 35-1H). 6" and/or 12". Transparent plastic is helpful but not essential. For larger animals a steel tape marked in millimeters is handy.

***Needles** (Fig. 35-1I). Straight assorted sizes. Large eyes are advisable.

***Pins.** Glass headed (Fig. 35-1J), insect (Fig. 35-1K), or common straight pins. A long slender pin with a large head is best.

Prior to Preparation

If the specimen is frozen, thaw in a warm place. The time required for thawing will vary. Do not place specimens directly under a hot light bulb. This frequently causes differential thawing, with some areas starting to "slip" (the hair and epidermis separate easily from the dermis, resulting in bald spots on the finished skin), while others are still frozen.

Ectoparasites

Remove the fresh or freshly thawed specimen from its bag and examine for fleas, lice, ticks, mites, and other ectoparasites. See chapter 36 for techniques for collecting and preserving ectoparasites.

Catalog Entry

The preparator's catalog or field catalog discussed in chapter 33 is a numbered list of all specimens preserved. Enter location, date, and other pertinent data in the catalog and assign the specimen a number.

Sex and Reproductive Condition

The sex of most mammals is easily determined by examination of the external genitalia. In some species, particularly shrews, an internal examination for testes or ovaries may be necessary. Record sex in the catalog using the symbol ♂ for a male or ♀ for a female. Use a question mark (sex ?) if for any reason the sex cannot be determined.

Reproductive conditions such as lactation or descended testes should be noted in the catalog. Refer to Taber (1969) and Brown and Stoddart (1977) for further information on ascertaining the reproductive condition of a specimen.

Measurements

In the process of removing the skin from a carcass, it is inevitable that the skin will be stretched; and while drying, it will shrink. Thus a finished specimen will never be exactly the same size as the original animal. Since size and proportion of a mammal are important in identification and in many other ways, a set of standard measurements are taken of the specimen prior to skinning. Four measurements, total length, tail length, length of hind foot, and height of ear, are recorded for all species, where possible. Two additional measurements, length of tragus and length of forearm, are recorded for bats. These measurements are always recorded in millimeters and always in the order given above. Instructions for taking these measurements follow.

Total length. From the tip of the nose to the distal end of the tail vertebrae. Place the animal on its back so that the backbone is straightened but not stretched. Position the head with the rostrum extending straight forward in the same plane as the backbone. Measure the distance between the tip of the nose and the end of the last tail vertebra; exclude hairs extending beyond the tip of the tail (Fig. 35-2). If your millimeter rule is not long enough, place pins at the tip of the nose and tip of the tail vertebrae, remove the specimen and measure the distance between pins.

Tail length. Bend the tail up at a right angle to the body and measure the distance from the angle to the distal end of the last tail vertebra; exclude protruding hairs (Fig. 35-3).

Hind foot length. Measure from the back edge of the heel to the tip of the longest toe plus claw. Include the claw in the measurements (Fig. 35-4).

Ear length. Measure from the notch at the base of the ear to the furthermost point on the edge of the pinna (Fig. 35-5).

Tragus length. (Chiroptera only) The tragus is a leaflike structure projecting up from the base of the ear in most bats. Measure from the base to tip (Fig. 35-6).

Forearm length. (Chiroptera only) Fold the wing and measure from the outside of the wrist to the outside of the elbow.

Figure 35–2.
Measuring total length (TL).
(Setzer 1963)

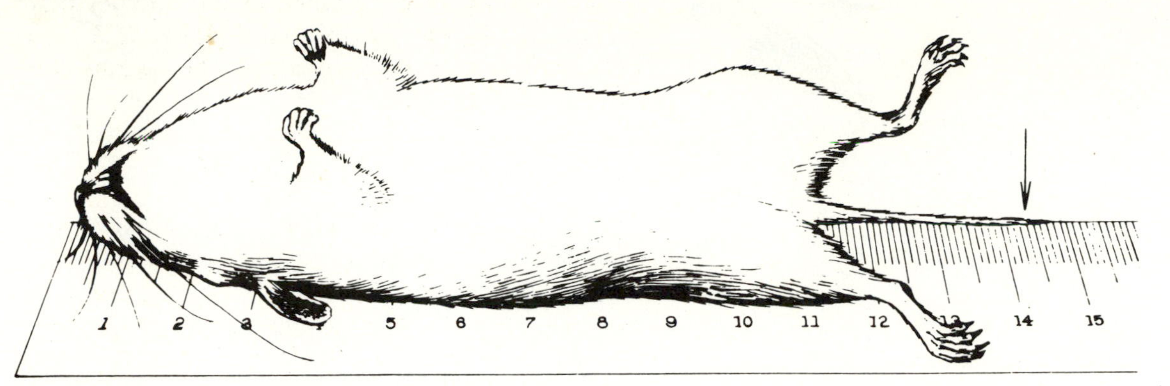

Figure 35–3.
Measuring tail length (T).
(Setzer 1963)

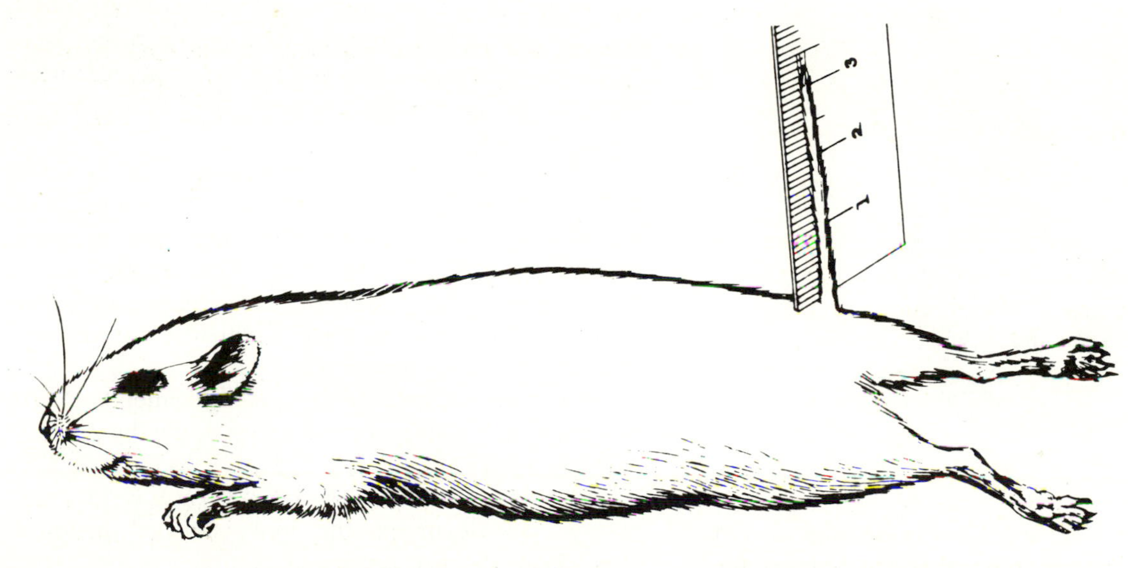

Figure 35–4.
Measuring the hindfoot (HF) of a rodent (A) and a hoofed mammal (B).
(A, R.E. Martin; B, after Hershkovitz 1954)

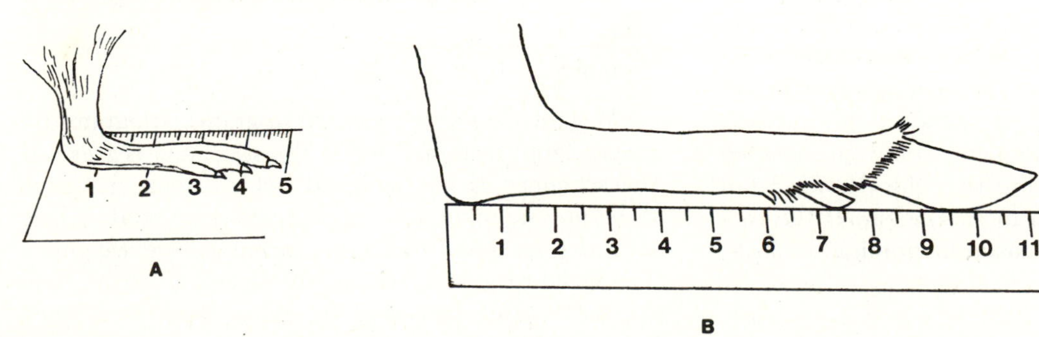

Figure 35–5.
Measuring the ear (E).
(Setzer 1963)

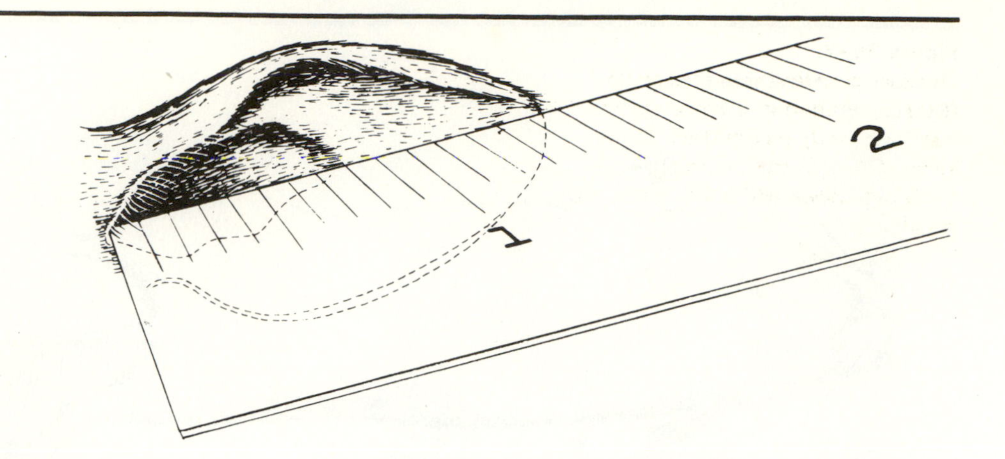

Figure 35–6.
Measuring the tragus (Tr).
(Setzer 1963)

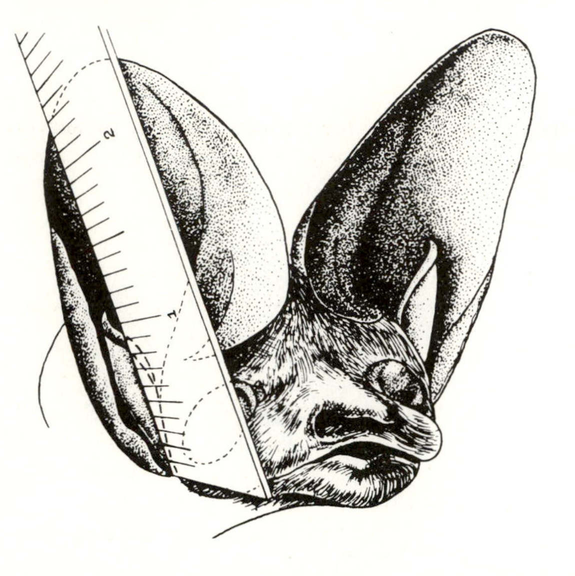

Record these measurements in the catalog. Be sure that they are in the proper order. The abbreviations TL, T, HF, E, Tr, and FA may be used, but since the order is constant, no abbreviations are generally necessary and the measurements may be given in sequence separated by dashes. Thus a rodent may be 310–150–40–12 and a bat may be 110–45–7–15–6–50.

If a portion of a mammal is missing or if for any other reason the exactness of a measurement is in doubt, we recommend that the measurement be set off in brackets or circled. Thus, if the tip of a mouse's tail was missing, the measurement would be recorded as [270]–[110]–40–12. Note that the damaged tail also affects the total length measurement. If the ears are torn or mutilated, the measurement might be 270–124–40–[10].

The measurements presented above are in standard use in North America. Collectors in other countries may take slightly different measurements. They record "head and body length" rather than total length. Thus the head and body length must be added to the tail length to get the total length; or the tail length must be subtracted from the total length to get head and body length. Regardless of which system is used, all three of these measurements are obtainable by simple arithmetic.

European and North American collectors also differ in the method used for recording hind foot length. North Americans include the claw in this measurement, while Europeans omit the claw. The initials c.u. for the Latin meaning "with claw" included or s.u. for the Latin meaning "without claw" frequently are included after the hind foot measurement.

Thus measurements of the same specimen by North American and European collectors might be

North American: TL 140, T 38, HF 14 (c.u.), E 12
European: HB 102, T 38, HF 12 (s.u.), E 12

Figure 35-7 illustrates the four standard measurements taken by Europeans.

Measurements of cetaceans and sirenians differ from those discussed above, since there are no hind feet or external ears to measure. Figure 35-8 illustrates some of the measurements most commonly taken on the smaller fully aquatic mammals.

Weight

Weight should be recorded in grams (kilograms for a very large mammal) while the specimen is fresh. If the specimen is not weighed before freezing, weigh it before skinning and indicate in the catalog how long the specimen had been frozen before weighing. The weight figure is frequently placed after the measurements and preceded by three horizontal lines, e.g., ≡ 310 g.

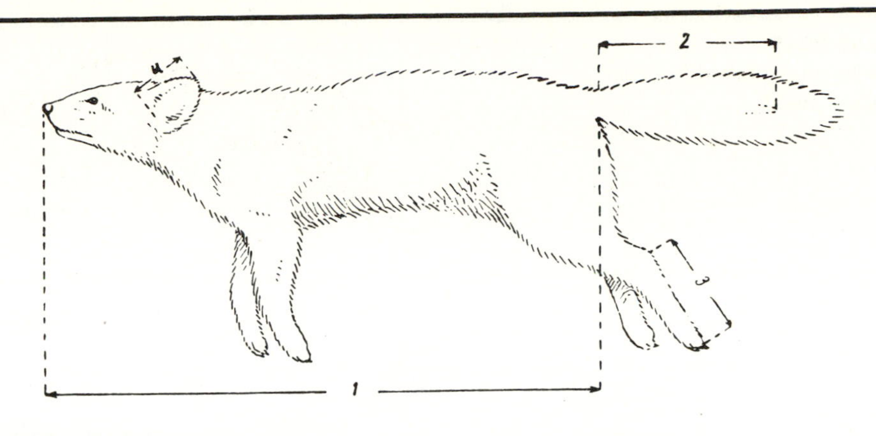

Figure 35–7.
The four standard measurements taken by European mammalogists. 1, head and body length (HB); 2, tail length (T); 3, hindfoot length excluding claws (HF s.u.); and 4, ear length (E).
(Novikov 1956)

Figure 35–8.
External measurements to be recorded for smaller cetaceans, as recommended by the American Society of Mammalogists' Committee on Marine Mammals. LENGTH: 1, total; 2, tip of upper jaw to center of eye; 3, tip of upper jaw to apex of melon boss; 4, gape; 5, tip of upper jaw to external auditory meatus; 6, center of eye to external auditory meatus; 9, tip of upper jaw to blowhole along midline or to midlength of two blowholes; 10, tip of upper jaw to anterior insertion of flipper; 11, tip of upper jaw to tip of dorsal fin; 12, tip of upper jaw to midpoint of umbilicus; 13, tip of upper jaw to midpoint of genital aperture; 14, tip of upper jaw to center of anus; 29, anterior insertion of flipper to tip; 30, axilla to tip of flipper; 33, dorsal fin base; 35, distance from nearest point on anterior border of flukes to notch. WIDTH: 31, flipper (maximum); 34, flukes (tip to tip). HEIGHT: 32, dorsal fin (fin tip to base). GIRTH: 21, on a transverse plane intersecting axilla; 22, maximum; 23, on a transverse plane intersecting the anus. Refer to Norris (1961) for further details.
(Norris 1961: 475)

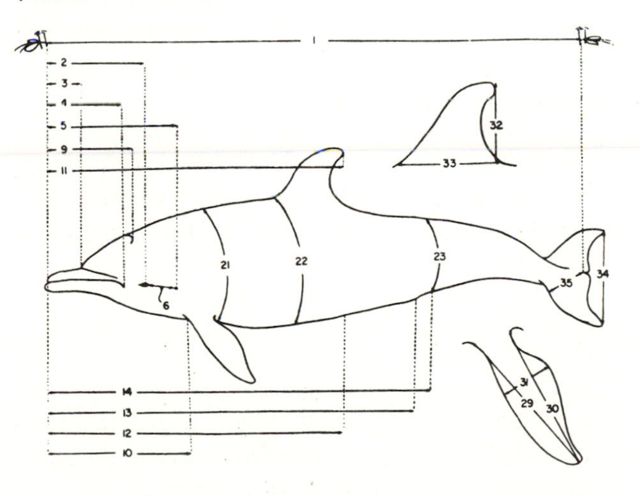

The Standard Study Skin

Most mammals the size of the striped skunk, *Mephitis mephitis,* or smaller are preserved as study skins. Larger animals are usually preserved as tanned skins; directions for preparing skins for tanning will be presented later in this chapter. Armadillos, rabbits, hares, porcupines, and certain other mammals that may be larger than a skunk are, for various reasons, preserved as study skins. Special instructions for each of these will be presented below.

1. Once ectoparasites have been removed and measurements and other data recorded in the catalog, place the specimen on its back in a shallow pan or on a newspaper-covered surface.
2. To make the first incision, pinch the skin of the lower abdomen between the fingers of one hand and lift it so that it is separated from the muscular body wall underneath. Use scissors to make a longitudinal cut through the skin (Fig. 35-9). Be careful to cut only through the skin and not through the muscles of the abdominal wall. Use the scissors to extend the cut backward to a point just in front of the anus, and forward to the posterior edge of the breast bone. If the abdominal wall is cut, use liberal amounts of corn meal or hardwood sawdust to absorb the body fluids that will leak out.
3. In males evert the penis and sever this organ between the skin and body as close to the body wall as possible, taking care not to cut through the baculum (if present). In small mammals leave the penis containing the baculum attached to the skin so that this taxonomically important organ and its contained bone will be available for study. In larger carnivores and other large mammals possessing a baculum, the penis must be cut from the skin and the baculum removed. It should be treated in much the same way as the skull (see Step 26).

Figure 35-9.
Making the midventral incision.
(Setzer 1963)

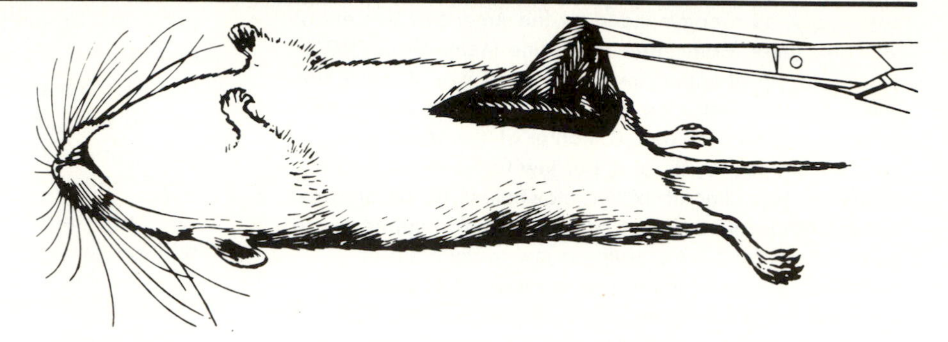

Figure 35-10.
Severing the hind leg at the knee joint.
(Setzer 1963)

4. Using fingers or forceps, loosen the skin along the sides of the incision until the knee joints are exposed. Use cornmeal or hardwood sawdust liberally to absorb blood or other body fluids.
5. Push the knee toward the incision and use fingers or forceps to free the skin from upper and lower legs. If the postcranial skeleton is not to be saved, cut through the knee joint (Fig. 35-10) and strip the muscle from the lower leg bones. Repeat for the other hind leg. If the postcranial skeleton is to be saved, separate the lower leg bones (tibia, fibula) from the foot bones (tarsals) at the ankle joint (*do not cut* the leg bones).
6. When both legs are free, cut through the rectum and urogenital duct(s) to free the skin from the abdomen. Use cornmeal to absorb fluids from the severed tracts. Then loosen the skin around the base of the tail.
7. With the fingers and fingernails of one hand hold the tail skin in position as you pull the tail out of the skin with the other hand. The tail skin should not be allowed to turn inside-out, but should fold up accordian-fashion behind the restraining fingers as the tail is pulled from it. If the tail is difficult to pull, it sometimes helps to roll it between the table top and the handle of a scalpel. (*Note*: The tails of armadillos, scaly-tailed mammals such as beavers and opossums, and very bushy-tailed mammals such as foxes and skunks require special techniques. Refer to the Special Techniques section.)
8. When the tail is free work anteriorly, using fingers to separate the skin from the body. To minimize stretching the skin, always pull the body from the skin rather than the skin from the body.
9. When the forelegs are reached, treat them in the same manner as the hind legs. (*Note*: With bats, to retain the forearm measurement, in the specimen, sever the humerus above the elbow joint *or* to prevent cutting of the humerus, disarticulate it from the shoulder joint).
10. Work forward until the bases of the ears are exposed. Fat and glandular tissue may have to be picked away so you can see where the ear cartilage enters the skull. In small mammals the base of the ear can be grasped between thumb and forefinger as close to the skull as possible and pulled free from the skull. With larger specimens it is necessary to cut the auditory canal (Fig. 35-11). The canal should be cut as close to its entrance into the skull as possible.

11. When both ears are free, work forward cautiously until the posterior edges of the eyes can just be seen through a layer of transparent tissue. Hold the skin slightly away from the head and cut through this membrane just over the eye (Fig. 35-12). Be careful not to cut into the eyeball or through the eyelids. When the membrane has been cut, the skin will be attached only at the front corner of the eye. Carefully sever this anterior attachment with a scalpel or fine-pointed scissors.
12. Using the fingernails or scalpel (depending on size of skin) work the skin forward until the lips are reached. Using scalpel or scissors, free the lips from their attachments to bone and connective tissue being careful not to cut through the bases of the vibrissae. Work the skin forward until the skin is attached to the body only at the nose. Use scalpel or scissors to sever the nasal cartilage, being careful not to damage the nasal bones (Fig. 35-13).
13. Lay the carcass to one side. Remove all fat and any large pieces of other tissues from the skin and turn it right side out. Check the fur for dirt or blood. Apply corn meal to dirty spots and brush with an old toothbrush. If the grease or stains remain, treat them in the following manner. To remove blood stains, sponge the area with cold water. It may be impossible to remove all traces of blood stains from white hair. To remove grease from a small area of the skin, sponge the area with soapy water or with a solvent such as white gasoline or benzene (**Caution:** These two solvents are highly flammable and poisonous). If the skin is excessively greasy, wash it in soapy water or one of the above solvents. Take particular care not to stretch the skin during washing. If the skin is washed, note this fact in the catalog and name the solvent used.

    To dry a small area that has been washed, work dry corn meal into the fur and brush it out. Repeat until the area is dry. If a major portion of the skin needs drying, place it in a large container with a quantity of cornmeal and shake. Remove the skin and brush out the wet cornmeal. Repeat this procedure until the hair is dry. Compressed air may also be used for drying instead of or in addition to cornmeal.
14. Turn the skin wrong side out and sew the mouth shut using a three-cornered stitch (Fig. 35-14). If conditions are humid or the skin is excessively greasy, dust the leg bones and inside of the skin with borax. (Note: Borax is not recommended for species with red pelage.) The borax will speed the drying process and reduce the chances of insect damage.

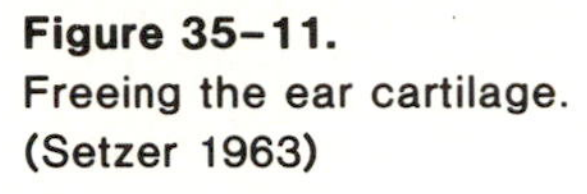

Figure 35–11.
Freeing the ear cartilage.
(Setzer 1963)

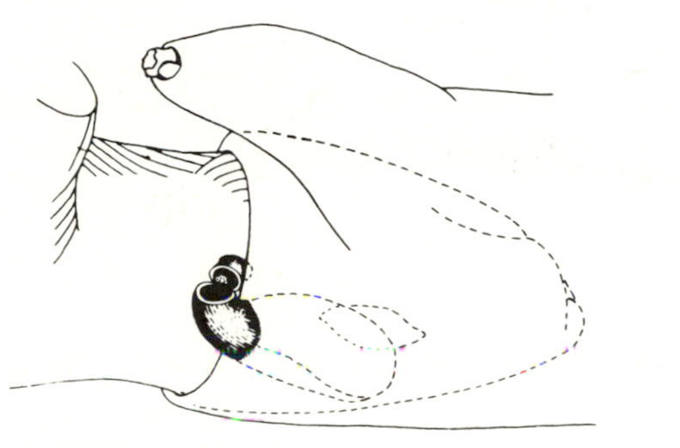

Figure 35–12.
Freeing the eyelids.
(Setzer 1963)

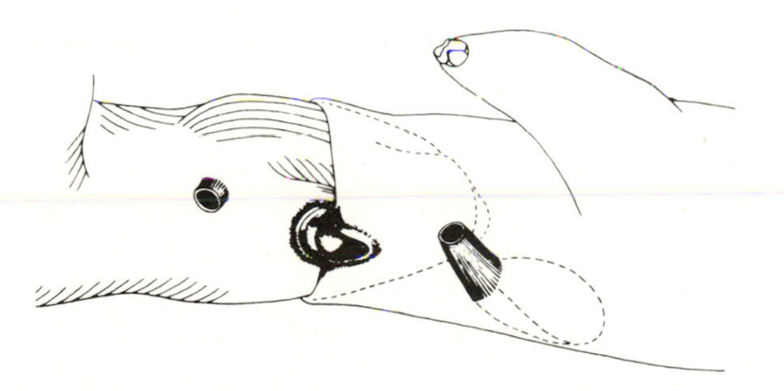

Figure 35–13.
Severing the nasal cartilage.
(Setzer 1963)

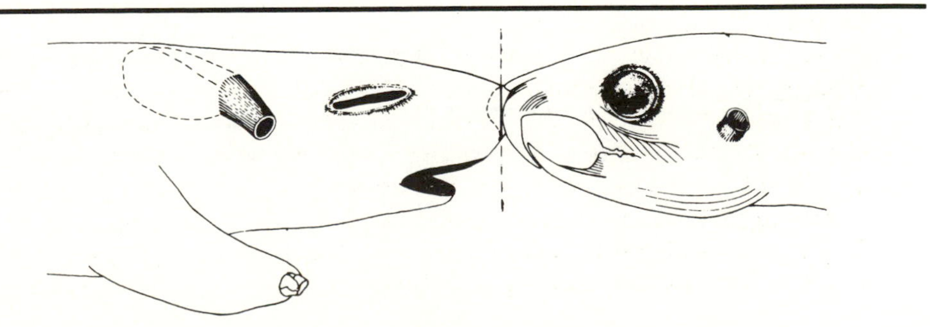

15. Select an appropriate diameter of monel wire for use in the specimen (e.g., 020 for pocket gophers and wood rats; 022 for large mice; 024 for smaller mice and bats; 026 for the smallest species). Generally it is more convenient to have a supply of several diameters available in your tool kit. To straighten monel wire, secure the end of a length of wire (a few meters) and pull on it until it no longer stretches. Then cut the wire to convenient lengths (12-18 inches) and store until needed.
16. Cut straight wires long enough to extend from the sole of the foot well into the body (Fig. 35-15). Insert a wire along the bone into the tip of the longest toe. Except for mouse-sized animals, wrap cotton around the wire and leg bones (if these were left in place). Turn the legs right side out. (Note: If the specimen is a rabbit or hare, refer to the Special Techniques section of this chapter before proceeding.) For small mammals (mouse-size or smaller), it is generally unnecessary to wrap the leg bones or wire with cotton.
17. For small mammals prepare the cotton for the body by loosely folding a thin sheet into a flattened cylinder that is longer and slightly larger in diameter than the skinned carcass. Use forceps or a hemostat to fold one end of the cylinder into a pointed cone.
18. Retaining a grip on the cotton point, place the point at the nose of the everted skin (Fig. 35-16) and slowly turn the skin right side out over the cotton body. Tear excess cotton from the posterior end of the roll and tuck the end into the rump area of the skin. Check to make sure that leg wires are correctly aligned (Fig. 35-15).
19. Cut a length of monel wire sufficient to reach from the tip of the tail to just beyond the anterior end of the ventral incision (Fig. 35-15). Use small wisps of cotton to tightly wrap the wire to a taper approximating that of the skinned tail, but of a slightly smaller diameter (Fig. 35-17). (Do not wrap tail wire for bats.) Insert the wrapped wire into the tail skin so that it will be on the ventral surface of the cotton body. (Note position of tail in Figure 35-19.) Moistening the cotton with water may ease entry.
20. Close the incision using a baseball stitch (Fig. 35-18). Be careful not to catch tufts of hair under the stitches.

Figure 35-14.
Sewing the mouth. After stitches are made draw the thread tight and knot.
(Setzer 1963)

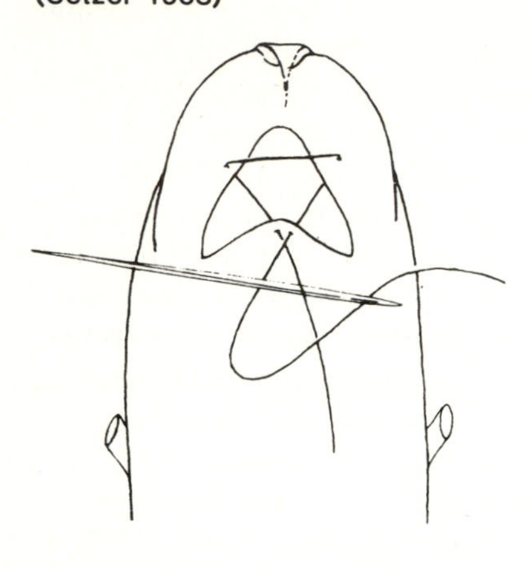

Figure 35-15.
Position of the leg and tail wires in the completed study skin prior to pinning. Refer to figures 35-19 and 35-20 for correct placement of feet.
(Setzer 1963)

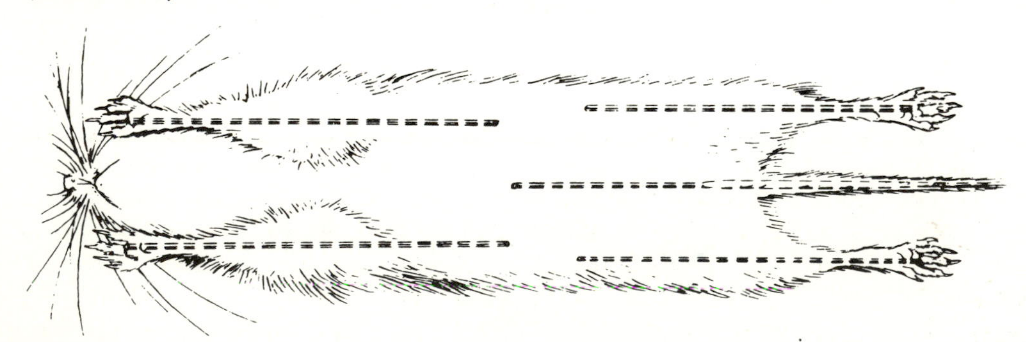

Figure 35–16.
Beginning the insertion of the cotton into the skin.
(Setzer 1963)

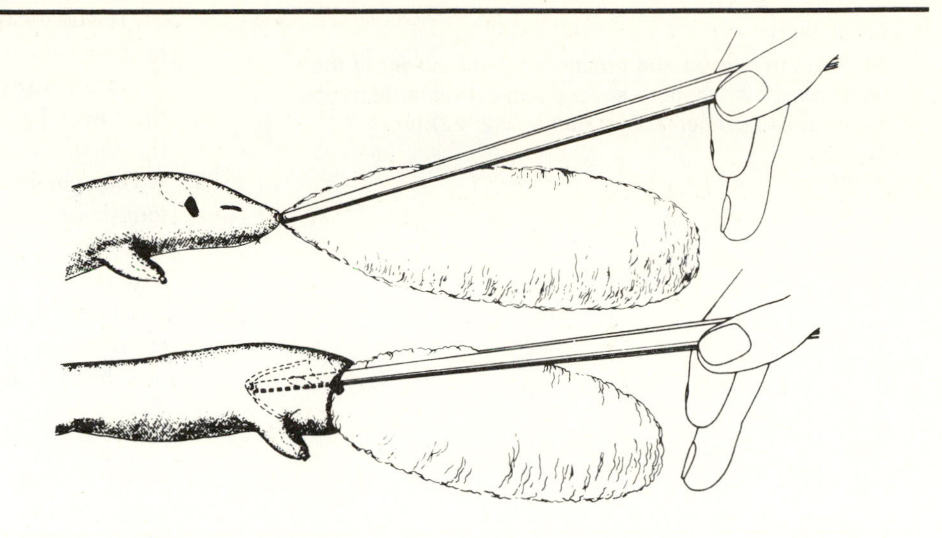

Figure 35–17.
Wrapping a tail wire.
(Setzer 1963)

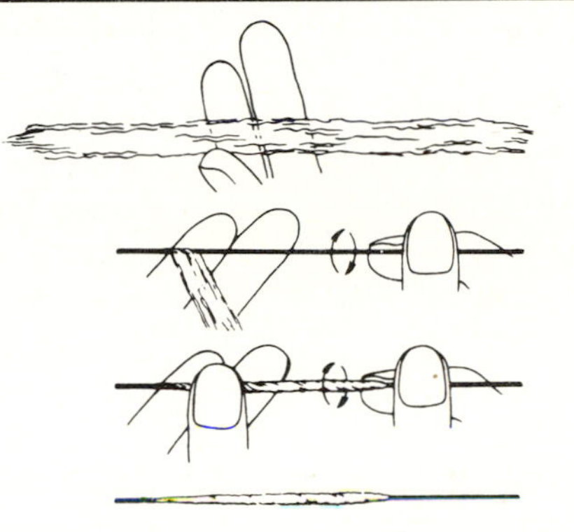

Figure 35–18.
Sewing up the incision.
(Setzer 1963)

Figure 35–19.
A mouse correctly pinned and positioned.
(Modified from Setzer 1963)

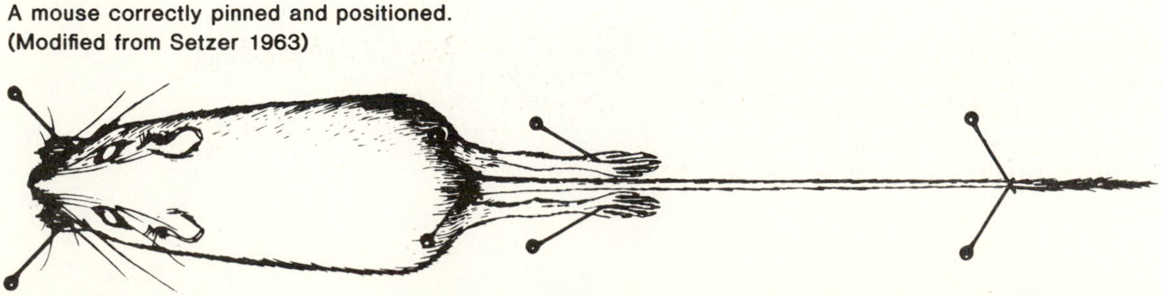

Figure 35–20.
A bat correctly pinned and positioned. Attachment of the string at point "A" is more secure and prevents distortion of the calcar (if present). See Hall (1962: 22) for elaboration.
(Hall 1962)

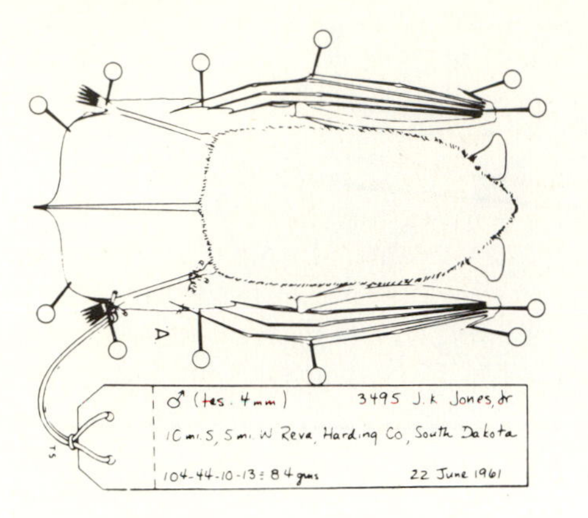

21. Fill out a specimen label and use a square knot to attach it *above the ankle* of the right hind foot. With bats the specimen label may be attached to the tibia (Fig. 35-20) to prevent distortion of the calcar.
22. Place the specimen belly-side down on a piece of cardboard, celotex, or other substrate that will hold pins and not hamper drying. Use your fingers to manipulate the specimen into a symmetrical shape with appropriate tapering of the head. Place the front feet close to the side of the head and pin them in place. Be sure that the two feet are placed evenly and that the pins do not crease the side of the face. Draw the hind feet posteriorly and pin them sole down, parallel and adjacent to the tail. Cross a pair of pins over the base of the tail and another pair near its tip. Be sure that the tail is in a straight line with the body and that the tail skin is right side up (Fig. 35-19).

    With bats the pinning technique is different. The wings are folded alongside the body and the wrist is pinned level with the nose. The phalanges and metacarpals are spread just enough so that each bone is accessible for measuring, and the wing is pinned in position. The thumb is pinned alongside the second metacarpel. The feet are pulled back and out, but not spread farther than the pinned position of the wings. A pin is used behind each of the calcars (spurs supporting the tail membrane) to draw the membrane tight. A properly pinned bat is illustrated in Figure 35-20.

    If the head of a pinned specimen has a tendency to raise up off the pinning surface, pin the head down with a pin at an angle through the lower lip. Be careful not to crease the side of the face.
23. If the specimen is a male, measure the greatest length and width of each testis, exclusive of the epididymis, and record these in the catalog and on the specimen tag. If the specimen is a female, check the uterus for embryos. If embryos are present, they should be counted and head-to-rump measurements taken of them. These data are recorded in the catalog and on the specimen tag. Embryos may be preserved by placing them in 10% buffered formalin (see Table 35-1). A tag with complete data should be inserted into the container with the embryos.
24. If stomach contents (see Chapter 31) or any portion of the internal anatomy is to be checked or saved, it should be done at this time.
25. If the postcranial skeleton is not to be saved, detach the skull by severing the neck. Take care not to damage the occipital condyles. Remove the eyeballs and cut away major muscle masses. Use a square knot to attach a labeled skull tag to one mandible (Fig. 35-21) and see step 27.
26. If the postcranial skeleton is to be saved, do not detach the skull of small to medium-sized animals. Remove viscera, fat, and major muscle masses. Be careful not to damage the baculum (if present). Also be careful not to remove the muscles that in some species (e.g., felids) contain the clavicles. Attach tags to the mandible and pelvis. Fold legs

Figure 35–21.
Labeled skull tag and proper method for attachment of tag to the skull. Note that the tag is secured to the mandible but some slack is left between the knot and the bone.
(R.E. Martin)

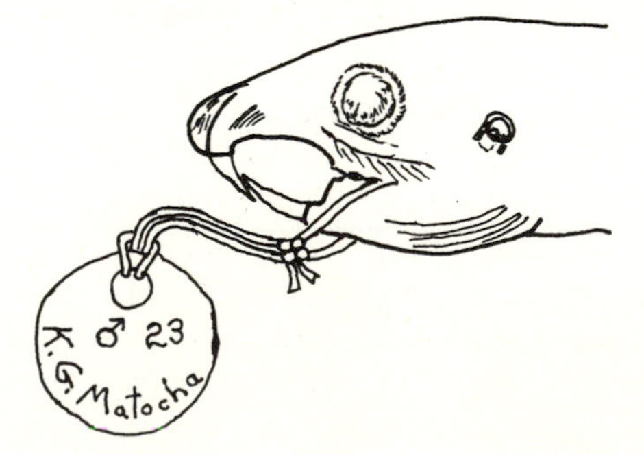

Figure 35–22.
Roughed out and labeled skeleton properly arranged for drying or shipment.
(Modified from Hershkovitz 1954)

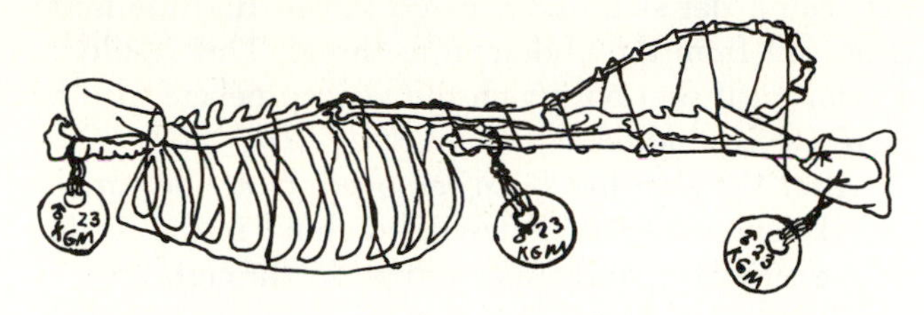

and tail along the body and if the neck is long, fold the head back as well. Use string to wrap the skeleton into a compact bundle with no protruding pieces that might be broken off (Fig. 35-22). Be sure that a tag is visible outside the wrapping.

27. For larger mammals, place the detached skull in a container of cold water and allow it to soak for 12 hours (in hot weather, change the water during this period to prevent excessive decomposition of the flesh and strong odors). Then, using an atomizer bulb and pipette or a hypodermic syringe fitted with a blunt needle, remove the brain by flushing the cranial cavity with water. In larger animals, the brain may also be removed with a brain spoon or hook that can be fashioned from a piece of heavy wire one end of which has been flattened and curved to form a hook. Anderson (1965) suggests a wooden spoon whittled out of a stick. Then flush the brain cavity with water.

    For smaller mammals, expose the foramen magnum by making a cut (without damaging the bones) between the skull and the first cervical vertebra so that cleaning by dermestids will be facilitated.

28. Rinse all borax and corn meal from the skull, skeleton, and baculum and place them in a safe place to dry. Uncleaned skull or skeletal materials should *never* be enclosed in an airtight container, since such treatment may cause the flesh to mold or macerate and make subsequent cleaning by dermestid beetles difficult. Hang in a tree in the shade to dry, and do not allow skulls to become infested with fly larvae. For techniques to use in cleaning skulls and skeletons, refer to the section on "Cleaning Skeletal Material."

Special Techniques

Armored Mammals

Armadillos, pangolins, porcupines and other armored mammals are generally prepared as standard study skins. Since the skins of these animals cannot be turned inside out, it is necessary to extend the primary incision from the neck to the tip of the tail and work the skin free progressing from the mid-ventral to the mid-dorsal line. In some armadillos the tail is encased in complete bony rings that must be cut with tin shears or a hammer and chisel to remove the tail. If it is impossible to remove the distal portion of an armadillo's tail (and with some species it always will be) inject the the area with formalin and carefully pin it so it will dry in a straight position.

Splitting Tails

In addition to the armored mammals mentioned above, many medium- to large-sized species having scaly tails (*e.g.*, beaver, muskrat, opossum) or very bushy tails (e.g., skunks, foxes) must have the tail skin cut from the tail vertebrae. Slit the skin of round or laterally flattened tails along the mid-ventral line, being careful not to cut the hair. Then use a knife or scalpel to free the skin from the tail. A beaver tail is split along a lateral edge. If the specimen is to be prepared as a standard study skin, the tail wire is formed, wrapped, and inserted and the tail sewn shut. If the skin is to be tanned, it should be treated as described in the section on Preparation of Skins to Be Tanned.

Rabbits

Rabbits and hares require special treatment since the hind feet are large and heavy and the skin is thin and tears easily. Proceed as described above through step 16, then spread the skin out on a table belly side down. Cut a piece of cardboard to fit the skin. The head will be narrower than the body and should be roundly tapered in front and slightly tapered behind. Cut a piece of 1/4″ dowel (or a similarly sized straight twig) to a length sufficient to extend from the tips of the hind toes to the midpoint of the body. Wire this stick in position along the midline of the cardboard body (Fig. 35-23). Then wrap the cardboard with a thin sheet of cotton. Cut and wrap a tail wire (see step 19 above) and insert into the tail. Insert the cardboard body into the skin, leaving the dowel extending out between the hind legs. Sew up the incision carefully (lagomorph skins are thin and tear easily). Tie the hind legs to the dowel at two points.

Figure 35–23.
Cardboard form and wooden leg support for rabbit or hare skin.
(R.E. Martin)

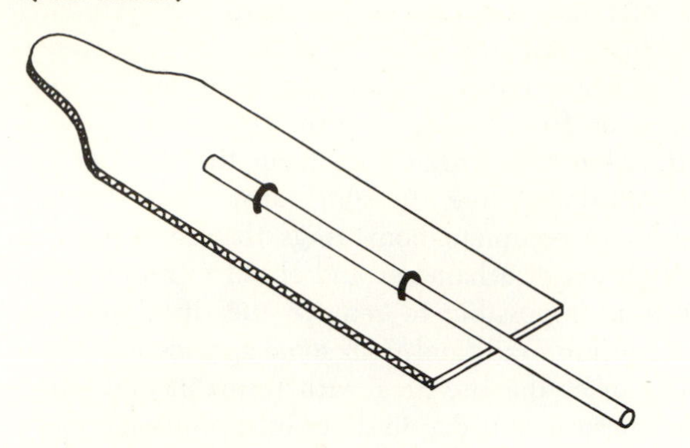

Orient the front feet anteriorly and with a stitch or two attach them to the underside of the neck. Lay the ears back and use a single stitch to loosely attach them to the skin of the back. Attach a specimen tag securely above the ankle of the right foot. Return to step 23 above and continue through step 28. An alternative technique for preparing rabbits and hares was presented by Anderson (1961).

Standard Study Skin with Cardboard Body

A study skin can be prepared using a cardboard body rather than a cotton body. This technique is widely utilized in Europe and, to some extent, in the United States. A slight amount of time may be saved in preparing skins using this method although space is not often saved in storage cabinets since the specimens have wide lateral dimensions and tend to occupy more tray space in comparison with round skins. But some institutions (Univ. Kansas, British Museum of Natural History) store these skins in file cabinets. Refer to Brown and Stoddart (1977) for a description of this technique.

Cased Skins

The smaller mammals for which tanned skins are desired (foxes, raccoons, etc.) may be prepared as modified **cased skins.** Cased skins for the fur trade are stretched and usually have the feet removed. Obviously this is not satisfactory for specimens for scientific study. To prepare an acceptable cased skin for tanning make the initial incision as in step 2 of "The Standard Study Skin." Extend this cut posteriorly to the tip of the tail but do not extend it anteriorly. From this incision make cuts along the insides of the hind legs to the feet. Remove the skin from the carcass turning it inside out over the body. When the forefeet are reached, make an incision from the underside of the upper leg along the inside to the foot and remove the skin. Proceed as above. The only difference between a flat skin and a cased skin is that the latter is not slit from the abdomen to throat. This results in a cylindrical skin on which the ventral pelage may be more easily studied.

After the skin has been removed, treat the carcass and skeletal material as described above in the section on the standard study skin (steps 25 through 28).

Preparation of Skins to Be Tanned
by Keith A. Carson*

The skins of mammals larger than raccoons or foxes are usually preserved by tanning. This process changes the skin into leather.

Skinning Large Mammals

29. Make a ventral incision from the tip of the tail to and through the middle of the lower lip. Extend the cuts up each of the legs to a point at the base of the toes (on the palm side) to the ventral incision (Fig. 35-24). To insure symmetry, it is important to make all cuts before beginning to free the skin from the carcass.
30. With a knife or scalpel, free the skin from the Achilles tendon and the adjacent portion of the hind leg. Then pull the skin away from the leg. A slight flexing of the leg will produce enough slack in the skin to permit the inversion of the skin over the foot. Insert a scalpel (bladeless) handle between and around each of the toes and pry them out of the skin up to the last joint. This joint is then cut through, leaving the terminal phalanx in the skin. See step 41 for special instructions for skinning out the feet.
31. Skin carefully around the tail vertebrae and work the skin over the rump. Pull the skin tight and cut through the rectum.
32. Skin the forelegs in the same manner as the hind legs and then work the skin free along each side of the ventral incision. As you skin, leave fat and dermal muscle on the carcass.
33. If the mammal has no horns or antlers, proceed to step no. 34. If it has horns or antlers, make a cut beginning about midway on the back of the

*This section written, in part, by Keith A. Carson, former tanner, Field Museum of Natural History.

Figure 35-24.
Incisions used when preparing a skin for tanning.
(Anderson, R.M. 1965. Methods of collecting and preserving vertebrate animals, 4th ed., revised. Bull. Nat. Mus. Canada, 69:1-199, fig. 23. With permission of National Museums of Canada.)

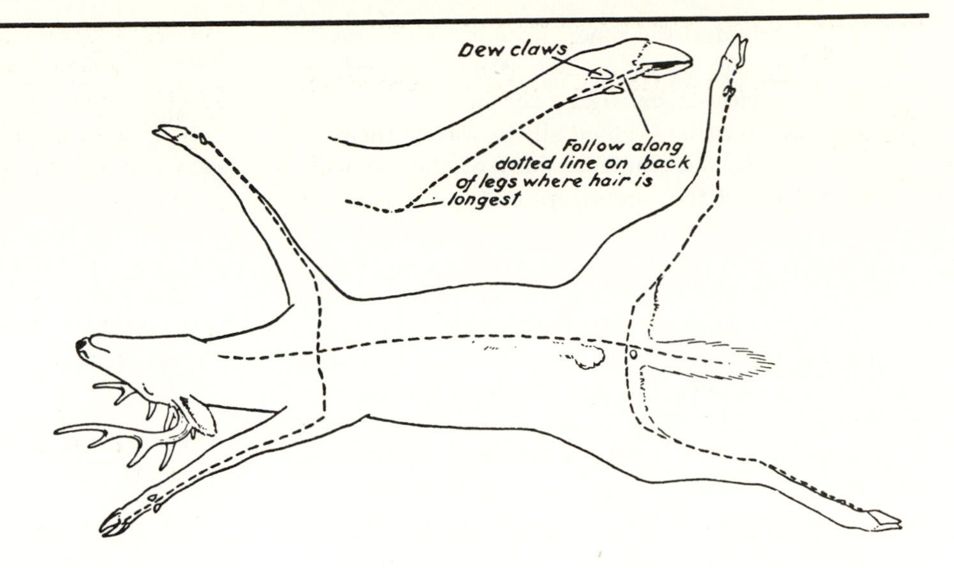

Figure 35-25.
Y-shaped incision through which horns or antlers are removed.
(Anderson, R.M. 1965. Methods of collecting and preserving vertebrate animals, 4th ed., revised. Bull. Nat. Mus. Canada, 69:1-199, fig. 24. With permission of National Museums of Canada.)

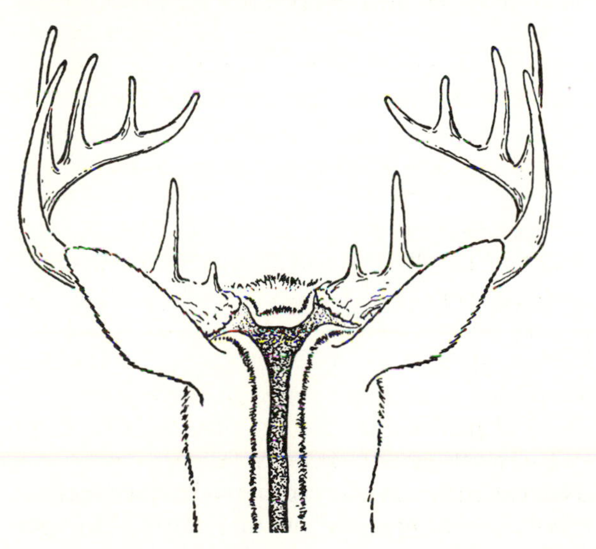

neck and extending straight to a point directly above the occipital condyles. Then extend the incision to the base of each horn or antler (Fig. 35-25). From the flesh side, cut the ears loose from the skull as close to the bone as possible and work the skin forward. To attempt to skin the remainder of the head at this point would be very difficult since the skin is still attached to the body.

34. With most specimens, especially large ones, it is helpful to suspend the carcass. Insert a hook between the tough Achilles tendon and the rest of the leg and hang the carcass from a hoist or convenient tree. Grasp the skin at the level of the tail (use an old towel or other cloth to provide a good grip) and pull the skin away (or cut free with knife) from the body all the way to the base of the skull.
35. Proceed with the skinning of the head. On mammals with horns or antlers, pry or cut the skin free at the base of these structures. Insert the index finger into the eye socket from the hair side of the skin and pull the skin gently away from the skull, stretching the tissues that hold the eyelids to the skull. These tissues can then be cut close to the bone with no danger of damaging the lids.
36. On some artiodactyls, there is a deep depression (the antorbital pit) located immediately in front of each eye. Keep tension on the skin and cut close to the bone to free the skin.
37. Cut through the lower lip at the midline all the way to the bone. Then cut posteriorly along each dentary (retaining as much of the inner surface of the lip as possible on the skin) until the skin is completely freed from the lower jaw. Continue skinning the head. Keep an even tension on the skin and work toward the rostrum. If care is taken, the skin can even be separated from much of the nasal cartilage. At this point cut straight down through the remaining nasal cartilage leaving about 1/4 inch on the skin. Continue to free the upper lip until the entire skin is cut loose. The lips should be "pocketed" to prevent slippage (loss of hair) due to decomposition. This is done by splitting the lips all the way to the bottom of the fold that is formed

by the skin and the inner mucous membrane. If the lips are thick remove as much muscle as possible without damaging the lips.

38. The ears must be skinned all the way to their tips (on the back or hair side) or the ear will shrivel up in the tanning process. Much of this work is done with the fingers (although a round-tipped butter knife or similar tool works well) inserting them between skin and cartilage and separating the connective tissue. Invert the ear as the work proceeds until the ear is inside out (Fig. 35-26). It is usually not necessary to remove the cartilage from the front of the ear unless the skin of the ear is fatty or densely haired.
39. The thick tissue on the inside of eyelids must be "slit" or opened to prevent the loss of the eyelashes due to decomposition. Make a number of small cuts (extend parallel to the edge of the eye lid) through the tissue back of the eyelid but not through the skin itself. Be careful not to cut into the papillae that contain the roots of the eyelashes. These appear as a line of small yellow bumps near the rim of the eyelid.
40. Split the quarter inch of nasal cartilage left on the inside of the skin down the middle almost to the skin of the nose. This cut will help to maintain a "natural" shape to the nose. On many mammals, there are large pads of tissue adhering to the inside of the nasal skin and covering the base of the whiskers. Remove this tissue but be careful not to damage the roots of the whiskers.
41. On nonungulates, make an incision on the ventral side of each foot (Fig. 35-27A). Then skin out the foot to the base of each distal phalanx. (Fig. 35-27B). The distal phalanges are left attached to the skin. In ungulates, skin the digits out completely and remove the distal phalanges from the skin. (Occasionally, the distal phalanges adhere so tightly to the hoof that their removal is not necessary). Leave the skin of the feet inverted for degreasing, defleshing, and drying. The skin is now ready for tanning.
42. After the skin has been removed, treat the carcass and skeletal materials as described above in the section on the standard study skin (steps 23-28).

Tanning Skins

Most institutions ship their salted and/or dried hides out to commercial tanneries specializing in hair-on tanning (See Dowler and Genoways, 1976, for list of commercial firms).

Figure 35-26.
Inverted ear of larger mammal showing adherent cartilage (A, stipple = cartilage) and the process of the cartilage being separated from the skin of the pinna (B).
(Redrawn from Hershkovitz 1954)

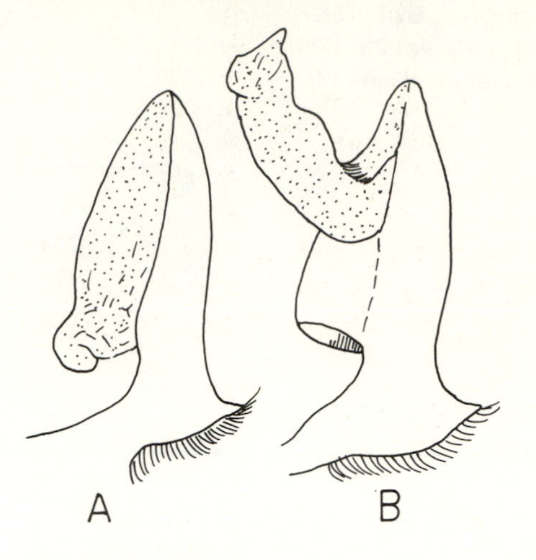

Figure 35–27.
Foot of carnivore showing incision on ventral side (A) and skinned digits (B).
(Redrawn from Hershkovitz 1954)

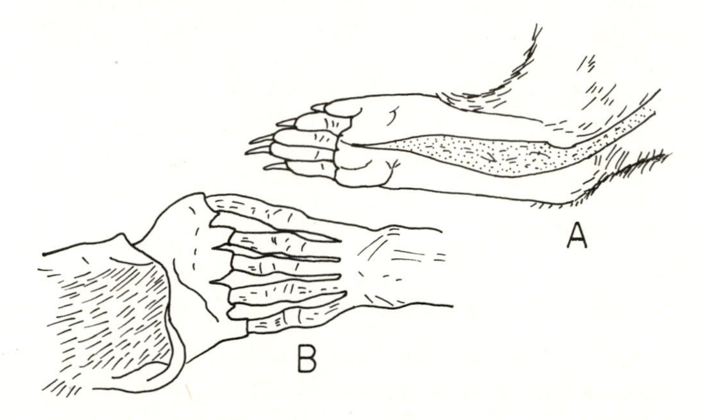

If tanning is not to proceed immediately (such as, under field conditions), allow the skin to dry partially and then fold it into a loose bundle (feet inside) with the flesh side on the outside. (In humid weather, the flesh side of the skin may be rubbed with borax or salt to hasten drying. According to some tanners, borax-treated hides result in slightly less flexible skins following tanning).

Each skin should have the catalog number punched into the skin from the flesh side to avoid confusion (Fig. 35-28).

Figure 35–28.
Punch code for marking skins.
(R.E. Martin)

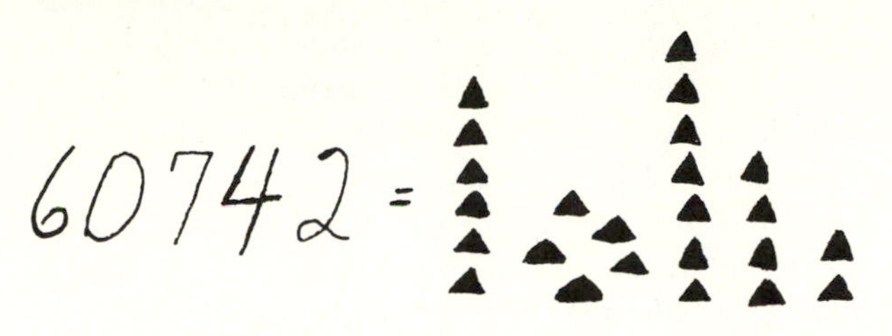

A three-cornered file that has been ground to a point is ideal for the purpose since the triangular hole it makes will not close up during the tanning process.

Cleaning Skeletal Material
by Laurie Wilkins*

Methods for cleaning skeletal material include boiling, maceration, the use of chemicals, and the use of various arthropods. The procedure for cleaning a mammal skeleton depends upon the size, age, and condition of the specimen; the number of specimens to be cleaned; the ultimate use of the specimen; and the facilities available for processing. There are variations in the procedures used, and some experimentation and judgement will be required to develop the most suitable methods for your needs. Presented here is a summary of various methods that have been described in the literature (see references at the end of the chapter), with emphasis on those that have produced the most satisfactory results at the Field Museum.

Cleaning with Dermestid Beetles

Although mealworms, ants, and crustaceans have been used for skeletal preparation, the use of beetles of the genus *Dermestes* (family Dermestidae) has been a very successful method. The size of the individual beetle colony maintained depends upon the type and amount of skeletal material to be processed. The use of beetles for cleaning is often restricted to small and medium-sized animals, but it is possible to develop a colony that will clean large specimens. A well-maintained colony will result in meticulously cleaned and articulated skeletons. There is generally no loss of teeth, minimal or no damage to the skeleton and very little effort required on the part of the preparator. The only disadvantage to this method is the length of time (one to two months) required to build a colony.

*Prepared by Laurie Wilkins, former Technical Assistant, Division of Mammals, Field Museum of Natural History.

Figure 35–29.
Adult (A) and larva (B) of dermestid, *Dermestes maculatus*, a species frequently used in museums for cleaning skeletal material. The adults of this and other species are generally dark with some sort of pattern and are covered with scales or "hairs." The antennae are generally short and club-shaped. Larvae are covered with many long "hairs."
(Hinton, H.E. 1945. A monograph of the beetles associated with stored products. British Museum (Natural History), figs. 341 & 391. With permission of the Trustees of the British Museum (Natural History).

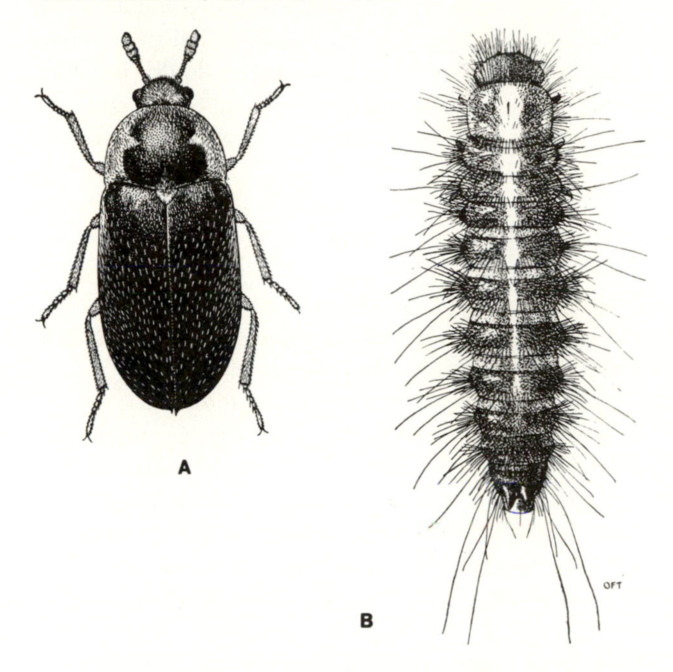

Establishing and Maintaining a Beetle Colony

A starter population of dermestid beetles can be collected from almost any dried carcass during warmer months (or obtained from an existing colony). *Dermestes maculatus* (=*D. vulpinus*) is frequently used in established colonies, although species of *Anthrenus* have also been reported (Voorhies 1948). In general, the adults of *Dermestes* are 5 to 12 mm in length, oval-shaped, with shiny reddish-brown to black cuticles and predominantly white ventral surfaces (Figure 35-29A). Larvae (Figure 35-29B) are segmented, elongated, bristly and vary in size from 2 to 12 mm depending upon the number of molts each has undergone. To establish the colony collect a minimum of 25 or 30 adults and larvae. Adults are needed for reproductive purposes but it is the larvae that do most of the cleaning. For further information on the biology and identification of Dermestidae, see Hinton (1945) and Russell (1947).

The container for the colony should be of glass, metal or plexiglass with tightly sealed seams and smooth vertical walls (Voorhies 1948). Wood, styrofoam and cardboard are not suitable because the larvae will excavate sites in these materials to pupate. For a small colony, a coffee can or wide-mouth gallon jar will suffice. Replace portions of the lid or sides of the container with fine wire screening for ventilation and to prevent the escape of the insects. Aquaria of various sizes with screened lids are excellent containers. A large metal storage container 20 x 60 inches has been used to successfully establish a colony for cleaning animals the size of a white-tailed deer (*Odocoileus virginianus*). The container should be well-ventilated and escape-proof. Several layers of sheet cotton placed at the bottom of the container will provide pupation sites for the larvae. A metal screen with 1/4 inch openings placed inside the container and supported slightly above the cotton will prevent the "bug" frass (feces and shed larval skins) from accumulating immediately around specimens (see Sommer and Anderson 1974 for details). It will also permit air flow around specimens. The screen can be raised from time to time as the debris accumulates.

Place containers in a darkened area or separate room far away from other biological specimens, especially skins. Every precaution must be taken to prevent the beetles from escaping because they can cause a serious problem in a collection of specimens. The room should be well-ventilated with frequent changes of air. This will remove odors and keep the air relatively free of "bug dust" that can be a strong irritant to some persons. High humidity ($> 70\%$) is not desirable because it creates conditions that might cause the formation of bug-repellant molds or mite infestations that could destroy a beetle colony. A room temperature of 70° to 80° F. is satisfactory. Fluctuations of humidity, temperature and light intensity, within reasonable limits, do not appear to hamper the colonies to any significant degree.

Initially, any fresh material can be used to build the colony, although the head of a sheep or pig that may be obtained from a butcher works very well. Smaller specimens (e.g., raccoon or opossum skulls) can be used for small colonies. Introduce the starter colony directly onto the specimen after it has been properly prepared (see next section on Preparing Specimens for Cleaning by Dermestids). It may take a month or two, depending upon the desired size of the colony, before the population has increased to the point where the beetles will be ready to clean specimens at a regular rate. Thereafter, it is necessary to introduce fresh material at 2 to 4 week intervals (more often in a very active colony) because most specimens, especially dried material from the field, do not have the fat or moisture content necessary to maintain a colony.

Place each specimen in a separate container (cardboard box lids, metal trays or mesh wire baskets) to insure against loss of elements or mixing of specimens. Mesh baskets seem to work best (cardboard boxes are quickly destroyed by the larvae as they excavate sites for pupation) but be sure to select a mesh that is fine enough to retain any elements that may become disarticulated (phalanges, claws, bacula, etc.).

The time required to clean a skull or skeleton depends on the number of beetles. In an active colony, a small skull or skeleton will be cleaned in a day. Hooper (1950) estimated approximately 5.45 minutes each for cleaning 300 small skulls. Larger skeletons the size of a raccoon may take 2 or 3 days. It is important to check the condition of the skeletal material daily. This may make the difference between a well-articulated skeleton and a heap of bones.

Preparing Specimens for Cleaning by Dermestids

Each specimen must be properly prepared before it is placed in the colony. The procedure differs slightly depending upon the condition of the specimen; that is, whether it is fresh, dried from the field, or previously preserved in fluids.

The first steps in preparing fresh specimens are similar to those outlined in Steps 25-28 under "The Standard Study Skin." When these are completed soak specimens in several changes of cold water for 6 to 24 hours, depending on size, to remove excess blood and produce a whiter skeleton. Never use warm water since this hastens the decomposition process and will set blood in the bone.

Allow specimens to dry but do not allow them to become completely dehydrated. Proper drying of specimens improves the quality of the preparation and hastens the cleaning. Proper amounts of moisture in the specimens will attract the beetles without creating conditions favoring mold or decomposition. Quickly dry specimens in indirect sunlight, with a moderate amount of artificial heat, or moving air from a fan, and protect them from house pets or wild animals (e.g., raccoons). In the field, hang the carcass from a tree (in the shade) out of reach of climbing animals. As an added precaution, specimens may be placed in ventilated containers. In the summer months, keep specimens free of flies or you may end up with a maggot infestation. Should this situation arise, a quick but thorough dousing with household bleach will kill the maggots on contact. Wash the specimens immediately with water and dry.

Dehydrated and mummified specimens that have been stored for long periods of time must be softened to make them more acceptable to the dermestids. Soak them in several changes of cold water for 6 to 24 hours. Soak fragile skeletons (e.g., bats or immature animals) only a short time and dry them well. Larger specimens require much longer soaking periods and may be still damp when given to the beetles. In this case, tissue that is barely sticky to the touch is perfect. In extreme cases, simmer mummified specimens in water for several hours and dry before introducing them to the colony. Case (1959) suggested that pre-soaking specimens in concentrated ammonium hydroxide (**Caution:** fumes are toxic) prior to the water treatment, gives good results.

Occasionally, it is desirable to clean material that has been preserved in alcohol or formalin. Remove the specimen from the preservative, immerse it in running water for about two days, and then dry (de la Torre 1951). If the dermestids fail to clean the skeleton adequately (especially with formalin-preserved material), several replications of the above treatment may be necessary.

An added incentive for the beetles in cleaning mummified or fluid-preserved specimens is the application of a thin layer of cod liver oil, vegetable oil or bacon drippings after soaking and drying. Tests by Hooper (1956) indicated that dermestids preferred specimens treated with cod liver oil.

Special Considerations

Immature animals create special problems. They will be quickly and completely disarticulated (including the epiphyses of long bones and all the elements of the skull in which the sutures have not fused) by a densely populated, active colony. It may take longer, but you will get better results if the specimens are put into a less active colony consisting mostly of smaller larvae.

It is sometimes necessary or desirable to limit or prevent cleaning by beetles in certain areas of the skeleton where disarticulation of delicate bones seems likely. A strong formalin solution (one part water to one part stock formalin) applied to the specific area by dipping or painting with a brush will discourage the dermestids from cleaning that portion of the skeleton (Sommer & Anderson 1974).

Never put a horned animal into a bug colony unless the horn sheaths have been removed. In addition to the damage the beetles will do to the horn sheaths, skulls in this state are difficult to fumigate and you will increase the risk of introducing beetles into a collection.

Fumigation

When material is cleaned to your satisfaction, transfer it to a bug-free tray or muslin bag, after checking to be certain all elements are present. Quickly move specimens to a fumigation container to kill any remaining live insects. The container is best housed in the bug room or in an adjoining room sealed off from your lab or collection area. Fumigate specimens in a well-sealed glass jar or an air-tight metal box or specimen case. The fumigant presently being used to protect your collection may be used. Check with your professor or collection manager for advice about the most effective, approved fumigant available. Fumigate small specimens for three days and large specimens for a week. Carefully inspect specimens removed from the fumigation chamber for signs of live beetles—an indication that the fumigant must be replenished and that you must refumigate.

Subsequent Cleaning

The final stage of cleaning involves a treatment with ammonium hydroxide and water. Remove any corrosive pins or wire that might have been used in the field to attach labels and replace these with string. If you are uncertain about the durability of the label or ink, augment the original tag with a permanent label before proceding. (*Important*: Keep the original tag associated with the specimen—to prevent the original tag from being damaged by the solution, attach it to the outside of the container during the treatment.) Rinse skeletal material and flush out the foramen magnum with water to remove any "bug" debris. All rinsing should be done over wire screening. Then place the skeleton in a solution of half-strength ammonium hydroxide (using 28-30% stock solution). Use vials, jars or stainless steel trays (one skull or skeleton per container). Cover the container and soak for 8 hours (4 hours for small skulls or skeletons). This treatment softens the remaining tissue for subsequent cleaning. The ammonium hyroxide is then drained off (the same solution may be used several times), the container refilled with water, and the specimen allowed to soak for an additional 4 hours. Large specimens may require a longer soaking period with several changes of water.

CAUTION: Work in a well ventilated area or under a fume hood. Ammonium hydroxide fumes are toxic. Quickly and thoroughly wash areas of the skin that come in direct contact with the solution. Use rubber gloves and eye protection.

Remove any tissue remaining on the skeleton by carefully scraping with a scalpel or a stiff-bristled brush. A fine pair of scissors and forceps are useful tools for removing resistant tissue. A fine stream of water with a cross stream of compressed air may be preferable to the use of tools, especially for small or fragile skulls (Hall and Russell 1933). Once cleaned, each specimen is placed in an individual tray to prevent mixing of elements. The specimen is then allowed to dry slowly at room temperature.

Degreasing

The treatment in ammonium hydroxide may adequately degrease small skeletons, but medium-sized to large material often requires additional attention.

Agents that have been used for degreasing are acetone, chloroform, benzene, carbon tetrachloride, ethylene trichloride, methylene chloride and trichloroethylene (See Sommer and Anderson 1974, for degreasing procedure). Hildebrand (1968) has given the subject of degreasing a thorough treatment and his account is recommended reading before undertaking this phase of the operation should it be necessary. But use caution, since EPA regulations may restrict or prohibit the use of such agents described in Hildebrand (1968).

CAUTION: Degreasing agents are hazardous and extreme care should be exercised. They should only be used in well ventilated areas or under fume hoods with appropriate respirators and filters.

Bleaching

Bleaching is not recommended for study materials because it loosens teeth and tends to disarticulate skeletons although it may be desirable to bleach skeletons that will be used for exhibit purposes.

Cleaning by Boiling, Maceration and Enzyme Digestion

If conditions do not favor the use of beetles, or if there is only an occasional specimen to be cleaned, other methods can be used for cleaning skeletons. These include boiling and maceration. For all three methods, prepare the specimen in the same way as for cleaning by dermestid beetles.

Boiling

Boiling is a good method for cleaning fresh, dried, or fluid-preserved specimens of medium to large size. This method usually results in a completely disarticulated skeleton. While this may be an asset for convenient storage or large skeletons, a small completely disarticulated skeleton is often not desirable. There are other disadvantages to the process. Cartilaginous elements are destroyed, teeth may fall out or crack, and the thin bones of immature animals may warp. In spite of these drawbacks, boiling can be used successfully if the preparator is careful.

Presoak prepared specimens in cold water for 6 to 24 hours. This step is not essential, but it will remove blood from the fresh carcass and soften the tissue of dried or fluid-preserved specimens. Dismember large animals into conveniently-sized pieces for handling. Choose a container that will allow the specimen to be completely immersed. This will keep grease (that rises to the top) from accumulating on the bones. Wrapping the feet tightly in cheesecloth will keep the elements of the feet together if they become disarticulated. For a skull with horns, position the head so that the skull (but not the horns) is immersed. Simmer in water, to which a little detergent or non-residue forming washing powder has been added, for up to two hours. Large, dried, and fluid-preserved specimens take longer than fresh specimens. When the tissue takes on a gelatinous appearance, remove the container from the fire, cool slowly to prevent the teeth from cracking, and wash under running tap water. The loosened flesh should fall away from the bone easily, but it may be necessary to use a brush, scraper, scissors or forceps to remove tougher ligaments, fascia, or tendons. Be careful not to cut into the bone. Always work over a screened drain, and be sure to retrieve any separate bone elements and teeth. It may be necessary to repeat the treatment. Allow the bones to air dry at room temperature. Long bones dried too quickly may crack. Hoffmeister and Lee (1963) describe a method of cleaning difficult specimens by boiling in ammonium hydroxide. This method may damage small fragile skulls and should be used with discretion.

Maceration

There are several maceration procedures for cleaning skeletons, but only bacterial maceration will be discussed here. Chemical maceration is faster but requires the addition of chemicals—a procedure that makes it impractical for many students. If allowed to proceed too far, chemical maceration may damage the skeleton and it is therefore *not recommended.* Nevertheless, one such technique using antiformin solution has been successfully used in the past for the preparation of skeletons from preserved materials (Green 1934; Harris 1959). For more information on chemical maceration methods, refer to Hildebrand (1968) and Mahoney (1966).

For bacterial maceration, soak the carefully roughed-out skeleton in water. Use a glass, enamel or earthenware container with a tight-fitting cover. Allow to stand at room temperature for 1 to 4 weeks, agitating frequently and change the water occasionally. The length of time required for soaking depends on the type of preparation desired.

It is possible to obtain an articulated skeleton using this process if the maceration is controlled by removing the skeleton before the ligaments are destroyed. This may take up to a week, but check the specimen daily. To obtain a disarticulated skeleton, maceration can be complete and may take from two to four weeks. Wrap the feet in cheesecloth to keep the elements of each foot together. Always use a screen or strainer when pouring off the water. Rinsing the skeleton in a dilute solution of household bleach and water will eliminate odors. Thoroughly rinse in water to stop the action of the bleach and dry slowly.

Enzyme Digestion

The use of enzymes for cleaning is much faster than bacterial maceration and if done properly is less damaging than chemical maceration. It is recommended only for small skeletons. Disarticulation is so complete with trypsin and papaine that these are not recommended for immature animals. Pancreatin produces articulated skeletons and has been successfully used in the preparation of small and fragile skeletons. The cost or difficulty in obtaining the materials for this process may be prohibitive, but if this method is desirable, refer to Hildebrand (1968), Harris (1959) and Mahoney (1966) before proceeding.

Fluid Preservation

Frequently it is advisable to preserve all or a portion of a mammal in a liquid preservative. Whole mammal specimens should first be hardened or fixed in ten percent formalin (prepared by adding nine parts of water to one part of thirty-seven to forty percent stock solution formaldehyde). The fixative may be injected into the specimen or allowed to enter through an incision in the abdominal wall. Body fluids will dilute the formalin so the volume of fixative should exceed the mass of the animal by at least ten times. If many specimens will occupy a container, the strength of the formalin must be increased to prevent deterioration of tissue.

Unbuffered formalin is acidic and will degrade bone if specimens are left in it for long periods. Thus specimens are soaked in running water for 24 hours then transferred to ten percent neutral buffered formalin (Table 35-1), to seventy percent ethanol, or fifty percent isopropyl alcohol for permanent storage.

Table 35-1

Some commonly used fixatives and preservatives for mammalian tissue. Based on information in Guyer (1953), Humason (1967), and Mosby and Cowan (1969).

Material	*Ingredients*	*Quantity*	*Nature and Suggested Use*
Alcohol, ethyl, 70%	95% ethanol Water	70 ml 25 ml	Preservative; fix and harden organs; storage
Alcohol-acetic acid-formaldehyde (AFA)	70% ethanol Acetic acid, glacial Formalin (40% formaldehyde)	90 ml 5 ml 5 ml	Preservative and fixative, especially organs; storage time not critical
Bouin's fluid	Picric acid, sat. aqueous sol. Formalin (40% formaldehyde) Acetic acid, glacial	750 ml 250 ml 50 ml	Fixative; organs, tissues; several weeks storage
Embalming fluid	Formalin (40% formaldehyde) Glycerin Phenol Water	5 parts 5 parts 5 parts 85 parts	Preservative; especially whole mammals used for anatomical dissection
Formalin, 10%, neutral buffered	Formalin (40% formaldehyde) Distilled water Sodium acid phosphate ($NaH_2PO_4 \cdot H_2O$) Anhydrous disodium phosphate (Na_2HPO_4)	100 ml 900 ml 4.0 g. 6.5 g.	Preservative and fixative; organs, tissues, whole mammals; storage time indefinite

The containers should be as airtight as possible and tags of high rag paper or parchment should be securely attached around the ankle of the right hind leg of each specimen. A label should never be attached to the outside of a container but always placed inside with the specimen.

Histological Preparations

Although formalin is an excellent fixative, it may be desirable to use other fluids for special histological procedures. Standard fixatives, their suggested uses, and preparation and storage times are given in Table 35-1 (Humason 1967; Guyer 1953).

A labeled 100% rag stock tag should be attached to the specimen which is then placed in a container along with the fixative. In addition to the standard specimen data (see chapter 33), record the type of fixative and the date it was added.

The Committee on Anatomy and Physiology, American Society of Mammalogists reviews guidelines for the proper preservation of mammalian specimens and tissues. Further information can be obtained from the current chairman of this committee (see a recent issue of the Journal of Mammalogy).

Supplementary Readings

Anderson, R. M. 1965. Methods of collecting and preserving vertebrate animals, 4th ed. *Bull. Nat. Mus. Canada* 69 (Biol. Ser. 18):1-199.

British Museum (Nat. Hist). 1968. *Instructions for collectors No. 1, Mammals (Non-marine)*, 6th ed. Publ. No. 665, British Museum (Nat. Hist.), London. 55 pp.

Brown, J. C., and D. M. Stoddart. 1977. Killing mammals and general post-mortem methods. *Mammal Review* 7:64-94.

Dowler, R. C., and H. H. Genoways. 1976. Supplies and suppliers for vertebrate collections. *Museology,* Texas Tech Univ. 4:1-83.

Hall, E. R. 1962. Collecting and preparing study specimens of vertebrates. *Univ. Kansas Mus. Natur. Hist., Misc. Publ.* 30:1-46.

Hildebrand, M. 1968. *Anatomical preparations.* Univ. California Press, Berkeley. 100 pp.

Knudsen, J. W. 1966. *Biological techniques.* Harper and Row, Publishers, New York. 511 pp.

Lucas, F. A. 1950. The preparation of rough skeletons. *American Mus. Nat. Hist., Sci. Guide* 59:1-19.

Mahoney, R. 1966. *Laboratory techniques in zoology.* Butterworth, Inc., Washington. 404 pp.

Setzer, H. W. 1963. Directions for preserving mammals for museum study. *U.S. Nat. Mus. Info. Leaflet* 380: 1-19.

Williams, S. L., R. Labach, and H. H. Genoways. 1977. A guide to the management of Recent mammal collections. *Carnegie Mus. Natur. Hist. Spec. Publ.* No. 4. 105 pp.

36 Collecting Ectoparasites of Mammals

When a mouse is trapped in the field, the collector frequently thinks that he has secured one specimen. But in reality, by trapping a single mouse, numerous zoological specimens have been captured, for a diverse population of insect and arachnid external parasites may be found on the mouse's body. For the purpose of this chapter, we are using the term ectoparasite loosely to mean all arthropods that are either found externally on the host or that are at least partially visible upon examination of the external surface of the host (e.g. bot fly larvae).

The ectoparasites ("ectos" for short) are valuable in many ways. They may provide data that improves our understanding of the ecology of the host. Many parasites are important vectors for diseases affecting man and his domestic animals. Because of this economic importance some of the organizations that have funded the study of vertebrates are more interested in the parasites than in the hosts themselves. Ectoparasites are thus well worth the little extra time it takes to properly collect and preserve them.

Segregation of Hosts

In chapter 34 we recommended placing each mammal specimen collected in a separate container. This practice is essential if maximum data are to be gathered on an ectoparasite fauna. Specimens should never be placed together in the same container until they have been thoroughly examined for ectoparasites. The parasites collected from one host must be kept in a separate vial and never mixed with those from other hosts.

Immediately place a small mammal killed by a trap or other method in a clean bag. Paper bags are best since these are inexpensive and may be used once and then discarded. While this may appear wasteful, it insures that a parasite will not be transferred via the bag from one host to another and thus contaminate a later sample. Cloth bags may be used but they must be very carefully searched, thoroughly washed, and dried between each use.

Live mammals may be placed in paper or cloth bags. Obviously paper bags cannot be used for specimens that will quickly gnaw their way out. To kill such specimens place the bags, along with a piece of cotton saturated with ether or chloroform, into a container with a tight-fitting lid.

Large mammals that cannot be placed in a bag must be searched for ectoparasites in the field immediately upon being collected.

Kinds of Ectoparasites
by Dr. Eric H. Smith*

Numerous kinds of insect and arachnid ectoparasites occur on mammals. Some groups are quite restricted, being found only on certain groups of mammals, while other parasite groups are widespread. Some parasites occupy very restricted parts of the host's body, while others may occur anywhere. Below is a list of the major groups of ectoparasites found on mammals.

In addition, you may wish to identify the kinds of ectoparasites that you have collected. A key to the major groups of ectoparasites found on mammals has been prepared for this purpose and follows the list below. All illustrations are of mammal ectoparasites with the following exceptions: 36-1A, B, 36-2B, 36-4A, B, D, 36-5, and 36-11A. For these exceptions, the species illustrated resembles as closely as possible those of these groups which are ectoparasites.

*Prepared by Dr. Eric H. Smith, Division of Insects, Field Museum of Natural History, Chicago.

Insecta

Fleas (Siphonaptera; Fig. 36-6) are laterally flattened, wingless insects found on almost all mammals. They may be present anywhere in the fur but some kinds are found embedded in certain areas of the skin. Fleas are among the first parasites to leave a dead mammal, so the host specimens must be captured alive or taken from traps soon after death if fleas are to be collected. The nests of mammals are also good sources of fleas because this is where the slender, whitish, and legless larvae usually breed and adults can spend extended periods of time off the host.

Chewing/biting lice (Mallophaga Fig. 36-9) and **sucking lice** (Anoplura Fig. 36-10) are dorsoventrally flattened wingless insects. One order or the other is found on most mammals (except bats) but no mammalian species is known to have both orders of lice. Their presence is often indicated by eggs or egg cases glued to hairs. The entire life cycle is spent on the host and they do not readily leave the host when it dies. Sometimes dead lice can be found on old skins in museum collections.

Bat bugs and **bed bugs** (Hemiptera: Cimicidae, Fig. 36-8) are relatively large, wingless, dorsoventrally flattened insects. They can be found in the nests of certain rats, mice, bats, and humans and occasionally on the mammals themselves. Another hemipterian family, Polyctenidae, is rare and is found only on bats and in bat guano.

Louse flies (Diptera: Hippoboscidae, Figs. 36-3C, 36-7A) are winged or wingless insects; some species are winged when they first emerge but shed their wings when a suitable host is found. They are fairly rare on most wild mammals in most areas and difficult to capture because they readily leave the host when it dies or is disturbed.

Bat flies (Diptera: Streblidae, Figs. 36-3D, 37-7C, D, and Nycteribiidae, Fig. 36-7B) are insects found only on bats and in bat roosts. Most emerging streblids are winged and, like hippoboscids, are very agile. Since bats frequently can be plucked from a roost and placed in a bag with a minimum of disturbance (see precautions, Chapter 34), these flies are more easily collected than are louse flies. Upon finding a suitable host, winged streblids shed their wings and embed themselves behind the ears, on wing membranes, and near genital openings. Nycteribiid flies are all wingless and are spider like in appearance. They are found in the fur and on the patagia of the host.

Bot flies (Diptera, Figs. 36-3A,B, 36-4C: Cuterebridae, Gasterophilidae, Hypodermatidae, and Oestridae) have larvae that are parasitic on rodents, artiodactyls, and certain other mammals. Warbles, one type of bot fly larvae, are located subcutaneously and breathe through a hole in the host's skin. Their presence can usually be detected by a lump in the skin or by a denuded patch around the breathing hole.

Earwigs (Dermaptera, Fig. 36-1A) which are ectoparasitic are rarely encountered. The suborder Arixenia contains Malayan species that are ectoparasites of bats and the suborder Diploglossata contains South African species that are ectoparasites of rodents.

Fur moths (Lepidoptera: Pyralidae, Fig. 36-4B) include several species that are found in the fur of sloths. The immature stages (larvae or caterpillars) feed on the algae that grows on the hair of these mammals.

Beetles (Coleoptera) include 5 families that contain at least some species which are found on mammals, as follows:

—**Silken fungus beetles** (Cryptophagidae) are found in the fur of mice; questionably parasitic.

—**Mammal-nest beetles** (Leptinidae, Figs. 36-2A, 36-4E) include a number of beetles that occur in the fur of beavers, mice, shrews, moles, and other insectivores. These beetles are often found in the mammal nests. Those on beavers are parasitic, and other species are questionably parasitic.

—**Small carrion beetles** (Leptodiridae) are found on rabbits, questionably parasitic.

—**Dung beetles** (Scarabaeidae, Fig. 36-2B) occur on wallabies, rat-kangaroos, and sloths. They are found in the fur, especially near the anus, and drop off when the mammal defecates to deposit eggs in the excrement or droppings. These beetles are not parasitic on the mammal.

—**Rove beetles** (Staphylinidae, Figs. 36-1B, 36-4D) are found in the fur of rodents and marsupials. Their most common place of attachment is behind the ears, but they will detach and scurry through the fur if the animal is distressed. They leave the dead host when it begins to cool but are then sluggish and do not wander far from the host. These beetles are parasitic.

Arachnida

Hard ticks (Acarina: Ixodidae; Fig. 36-15A, B) are usually firmly attached to the host with the small head embedded into the skin and the large abdomen protruding above the surface. They are common on mammals and may be found anywhere on the body, especially in the region of the ears. They do not leave the host until their feeding is completed and are sometimes difficult to detach without damage to either tick or host.

Soft ticks (Acarina: Argasidae; Fig. 36-15C, D) are sometimes found on bats and on certain burrow-nesting rodents in arid regions. They more commonly occur in crevices in caves, hollow trees, and burrows occupied by their hosts.

Chiggers and **follicle mites** (Acarina: Prostigmata; Fig. 36-18). Chiggers are parasites as larvae and are common on many mammals, forming clumps in, on, or around the ears, inside the nasal passages, around the lips, eyes, and urogenital area, or in other constricted regions. They do not readily leave the host. Follicle mites occur in the apocrine sweat glands of humans or the Meibomian glands of certain bats and rodents.

Body mites (Acarina: Mesostigmata; Fig. 36-14) are commonly found on many marsupials, bats, and rodents. They occur in the fur and on naked areas of skin such as the bat's patagium. Many species occur only in very specific body locations such as the ears, nasal cavities, and anal opening. They leave the host fairly soon after death.

Mange and **fur mites** (Acarina: Astigmata; Fig. 36-17) are extremely minute and frequently overlooked because of their small size. They burrow under the skin or attach to hairs. They are slow to leave the host after death.

Key to Arthropod Ectoparasites of Mammals
by Dr. Eric H. Smith*

1 With 1 pair of antennae, may be short and/or concealed in grooves on head (Fig. 36-6, 36-7); at most and usually with 3 pairs of jointed legs (rarely with additional tubular leg-like appendages on venter of abdominal segments) .... **(INSECTA,** in part) 3

1′ Without antennae; legs variable, usually not 3 pairs 2

2 (1′) Legless, without jointed legs or prolegs tube-like legs) .... **(INSECTA,** in part) 13

2′ Usually 4 pairs of jointed legs (some immatures with 3 pairs; rarely with only 2 pairs) .... **(ACARINA; Arachnida)** 15

3 (1) One or two pairs of well developed wings, may be membranous or horny/leathery; if front wings horny/leathery, then hind wings may be absent 4

3′ Wingless or if with vestigial or rudimentary wings, then front wings not horny/leathery 8

4 (3) Wings membranous (like cellophane), but may also be covered with scales 7

4′ Front wings horny/leathery 5

5 (4) Front wings short, leaving most of abdomen exposed (Fig. 36-1) 6

5′ Front wings long, covering most of abdomen (Fig. 36-2) Cryptophagidae, Leptinidae in part, Leptodiridae, Scarabaeidae .... **Coleoptera** beetles

*Prepared by Dr. Eric H. Smith, Division of Insects, Field Museum of Natural History, Chicago.

Figure 36–1.
A, earwig (Dermaptera): *Chelisoches morio*, forceps-like cerci variable in form; B, rove beetle (Coleoptera: Staphylinidae): *Myllaena dubia*.
(A, Helfer 1953: 14; B, Jaques 1951: 104)

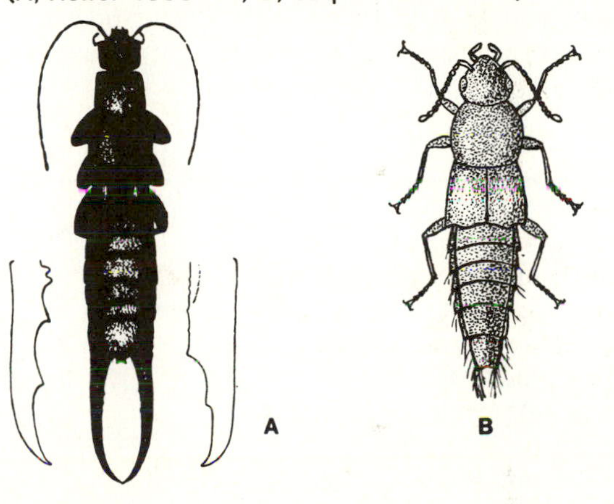

Figure 36–2.
A, mammal-nest beetle (Coleoptera: Leptinidae): *Leptinus testaceus*; B, dung beetle (Coleoptera: Scarabaeidae): *Aphodius distinctus*.
(A & B, Jaques 1951: 68 & 237)

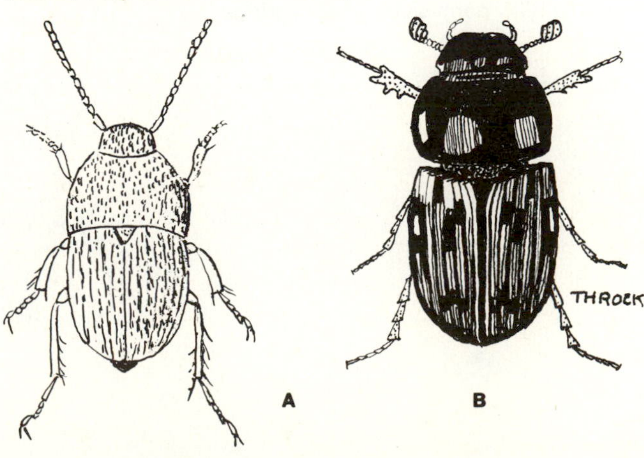

6 (5) Abdomen with relatively long, heavily sclerotized cerci (Fig. 36-1A), cerci either straight (on rodents) or forceps-like (on bats); hind wings absent .... **Dermaptera**
earwigs

6′ Abdomen without such cerci (Figs. 36-1B, and 36-2A); hind wings membranous and folded beneath front wings (Staphylinidae, Fig. 36-1B) or absent (Leptinidae in part; on beavers, Fig. 36-2A) .... **Coleoptera**
beetles

7 (4) With only 1 pair of wings, these not covered with scales (Fig. 36-3) .... **Diptera**
flies

7′ With 2 pairs of wings, largely or entirely covered with scales (on sloths) .... **Lepidoptera**
Pyralidae

8 (3′) Body appearing insect-like, with distinctive head, thorax, and abdomen, and with jointed legs suitable for locomotion 9

8′ Body larviform, appearing caterpillar-like (Fig. 36-4A, B) or maggot-like (Fig. 36-4C, with no apparent head and legs) or like an elongate grub (Fig. 36-4D, E) 13

9 (8) Tarsi 5-segmented (Fig. 36-5); antennae short and usually concealed in grooves on head; mouth parts piercing-sucking, forming a beak 10

9′ Tarsi with fewer than 5 segments; antennae and mouth parts variable 11

10 (9) Body (at least thorax) flattened laterally; usually jumping insects with large coxae and relatively long legs (Fig. 36-6) .... **Siphonaptera**
fleas

10′ Body flattened dorsoventrally; not jumping insects (Fig. 36-7) .... **Diptera**
bat flies and louse flies

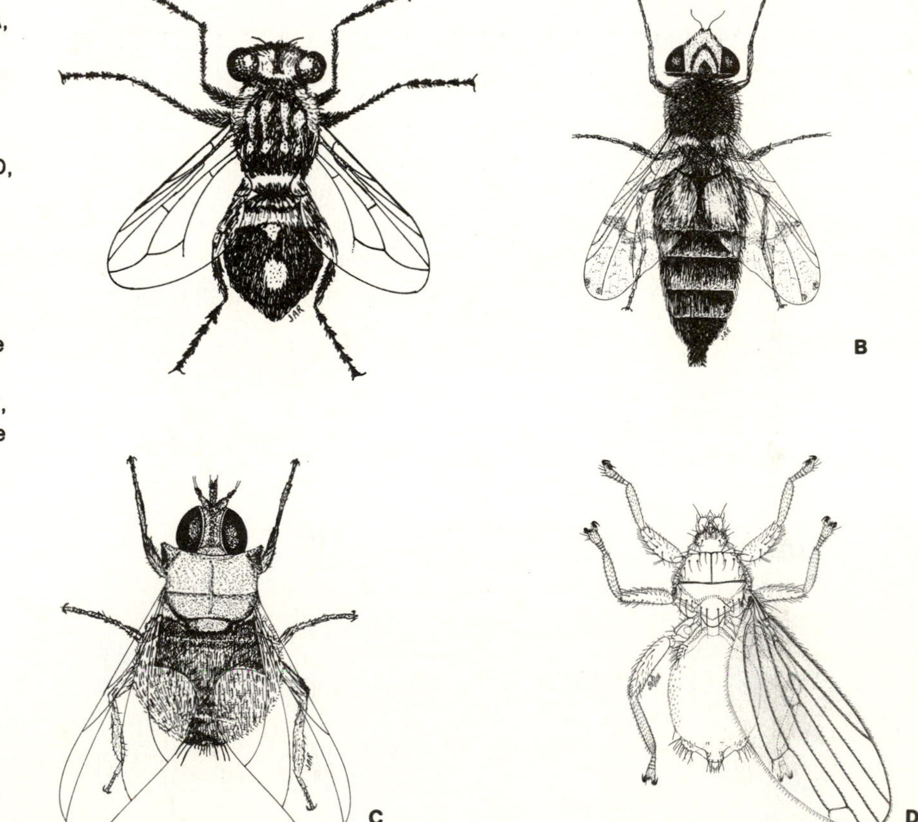

Figure 36–3.
Representatives of winged Diptera: A, common cattle grub (Oestridae); *Hypoderma lineatum*; B, horse bot fly (Gasterophilidae): *Gasterophilus intestinalis*; C, louse fly (Hippoboscidae): *Icosta americana;* D, bot fly (Streblidae): *Trichobius sphaeronotus,* left wing removed. (A,B&C, Bland & Jaques 1978: 350, 354; D, Jobling, B. 1939: 495. On some American genera of the Streblidae and their species, with the description of a new species of *Trichobius*. Parasitology, 31:486–497, fig. 4A. With permission of Cambridge University Press.)

Figure 36–4.
Representatives of larvae: A, caterpillar (Lepidoptera) showing prolegs on abdominal segments 3–6; B, garden webworm (Lepidoptera: Pyralidae): *Loxostege similaris*; C, bot fly larva (Diptera: Oestridae); D, rove beetle larva (Coleoptera: Staphylinidae): *Oligota oviformis*; E, mammal-nest beetle larva (Coleoptera: Leptinidae): *Platypsyllus castoris*.
(A,B&D, Chu 1949: 164,154 & 83; C, Bland & Jaques 1978: 354; E, Wood 1965: 53)

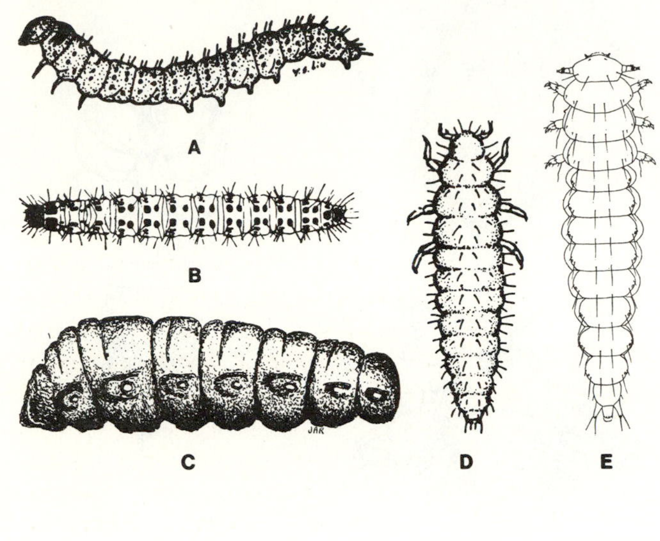

Figure 36–5.
Generalized insect leg: cx, **coxa**; fm, **femur**; pt, **pretarsus** (claws); tb, **tibia**; tr, **trochanter**; ts, **tarsus**.
(Bland & Jaques 1978: 32)

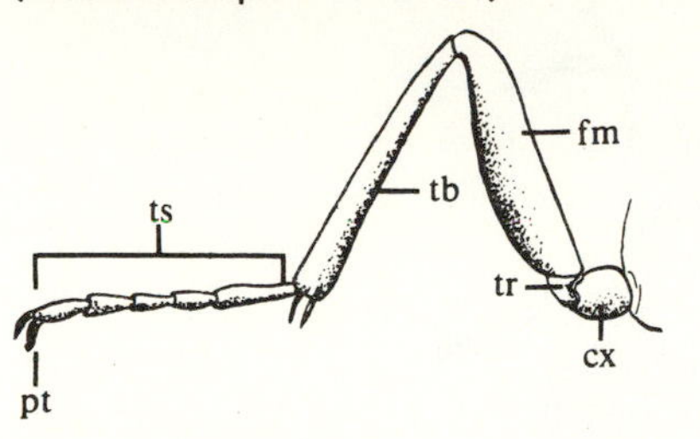

Figure 36–6.
Cat flea (Siphonaptera): *Ctenocephalides felis*.
(Bland & Jaques 1978: 357)

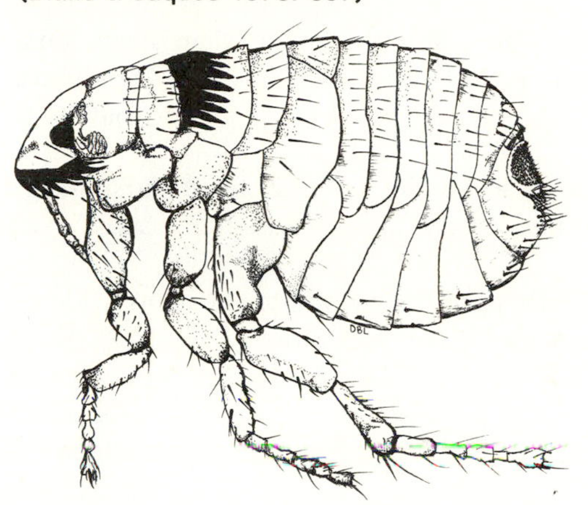

Figure 36–7.
Representatives of Diptera with reduced or no wings: A, sheep ked (Hippoboscidae): *Melophagus ovinus*; B, nycteribiid bat fly (Nycteribiidae): C and D, streblid bat flies (Streblidae): *Mastoptera guimaraesi* (enlargement of wing on right) and *Neotrichobius stenopterus*.
(A, Bland & Jaques 1978: 350; B, From Manual of Medical Entomology, Third Edition, by Deane P. Furman and Elmer P. Catts by permission of Mayfield Publishing Company. Copyright © 1961 and 1970, Deane P. Furman; C & D, Wenzel & Tipton 1966: 513 & 537)

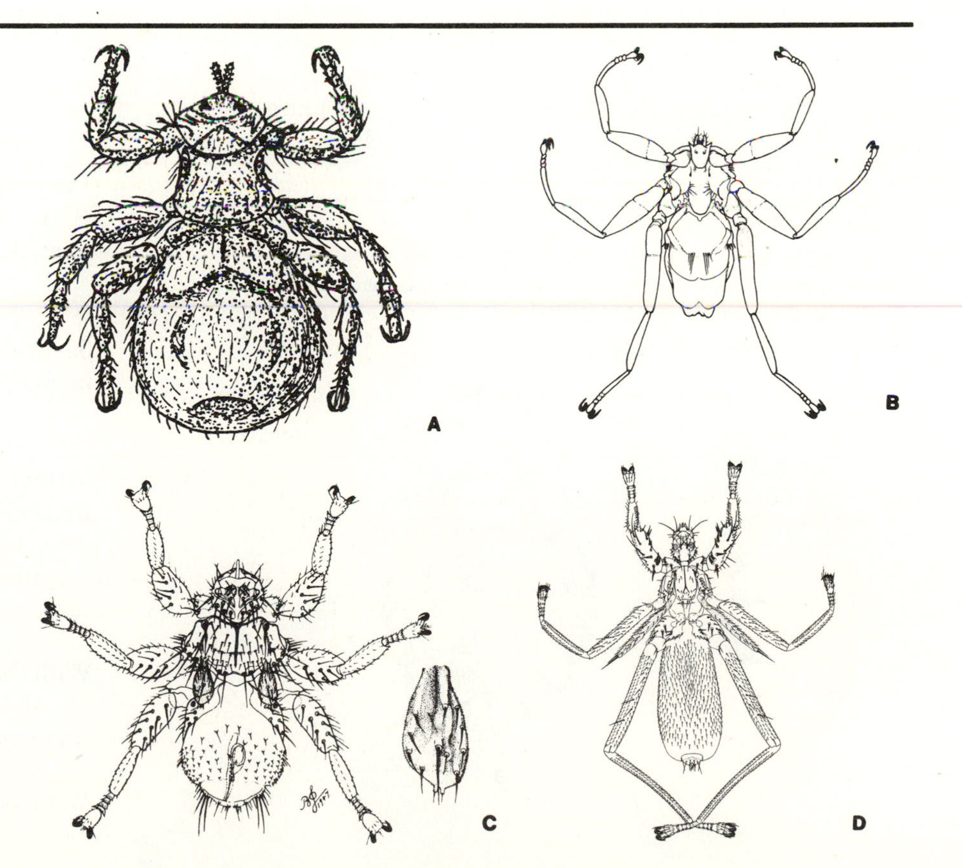

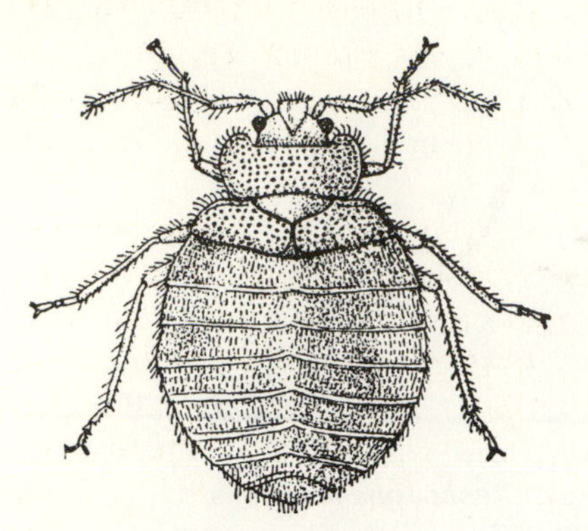

Figure 36–8.
Bed bug (Hemiptera): *Cimex lectularius.*
(Bland & Jaques 1978: 42)

11 (9′) Antennae distinctly longer than head; tarsi 3-segmented (Fig. 36-8) .... **Hemiptera**
bed bugs and bat bugs

11′ Antennae not longer than head; tarsi 1-segmented 12

12 (11′) Mouth parts mandibulate; head as wide as or wider than prothorax (Fig. 36-9); or if head narrower then mandibles borne at end of long proboscis (on elephants and warthogs) .... **Mallophaga**
chewing lice

12′ Mouth parts piercing-sucking; head usually narrower than prothorax (Fig. 36-10) .... **Anoplura**
sucking lice

Figure 36–9.
Representatives of chewing lice (Mallophaga): A, cattle chewing louse, *Bovicola bovis*; B, dog chewing louse, *Trichodectes canis*.
(Bland & Jaques 1978: 125)

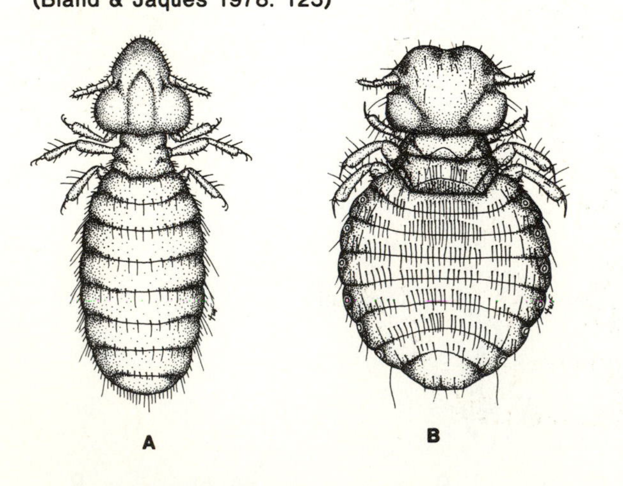

Figure 36–10.
Representatives of sucking lice (Anoplura): A, body louse, *Pediculus humanus humanus*; B, crab louse, *Pthirus pubis*; C, longnosed cattle louse, *Linognathus vituli*.
(Bland & Jaques 1978: 127,127 & 128)

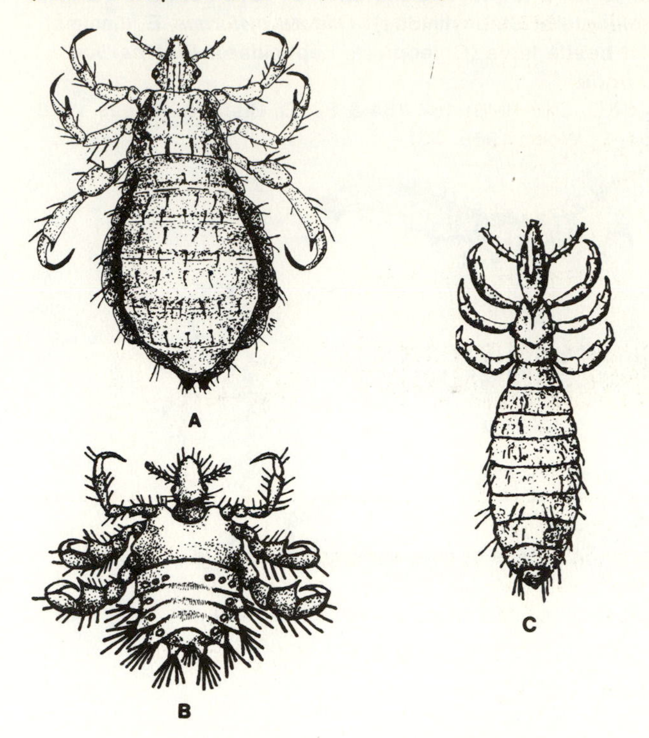

13 (2/8′) Body without head capsule, head represented by 1 or 2 median sclerotized hooks that move vertically (Fig. 36-11A); legless (Fig. 36-11B .... **Diptera**
bot flies

13′ Body with distinct head capsule (Fig. 36-4A, B, D, E), mouth parts mandibulate; with jointed legs 14

14 (13′) With prolegs (tube-like unjointed legs, Fig. 36-4A) on venter of some abdominal segments (on sloths) .... **Lepidoptera**
Pyralidae

14′ Without prolegs but may have apical appendages on abdomen (Leptinidae, on beavers, Fig. 36-4E; Staphylinidae, on rodents and marsupials, Fig. 36-4D) ...... **Coleoptera**
beetles

15 (2′) With body stigmata located behind coxae IV or laterad between coxae III-IV, each surrounded by a stigmal plate 16

15′ Without such stigmata 17

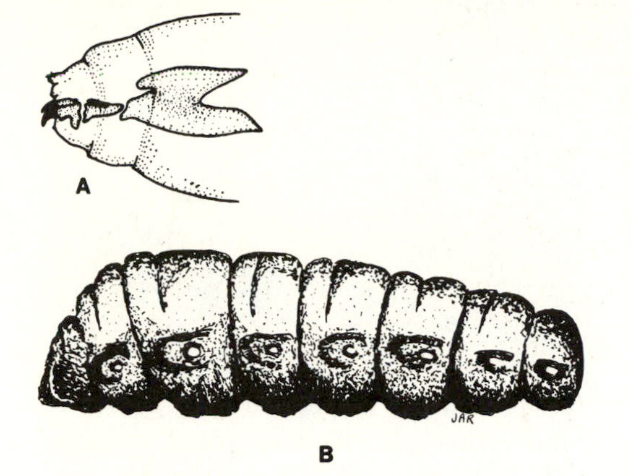

Figure 36–11.
Fly larvae (Diptera) which are without head capsule and legs: A, enlargement of anterior end where head is represented by 2 sclerotized hooks; B, bot fly larva (Oestridae).
(A, Chu 1949:29; B, Bland & Jaques 1978: 354)

16 (15) With body stigmata located laterad between coxae III-IV; with a terminal, subterminal or basal apotele on palpal tarsus (Fig. 36-12A); hypostome without recurved teeth (Fig. 36-13A); without Haller's organ; Fig. 36-14 .... **Mesostigmata** body mites

16′ With body stigmata located behind coxa IV; palpal tarsus without apotele; hypostome well developed, with recurved teeth ventrally (Fig. 36-13B); with Haller's organ on tarsus of 1st pair of legs (Fig. 36-12B); Fig. 36-15 .... **Metastigmata** ticks

Figure 36–12.
A, papal tarsus of mite showing an apotele, *Haemogamasus* sp.; B, tarsus of first pair of legs of tick showing Haller's organ, *Dermacentor* sp.

(From Manual of Medical Entomology, Third Edition, by Deane P. Furman and Elmer P. Catts by permission of Mayfield Publishing Company. Copyright © 1961 and 1970, Deane P. Furman.

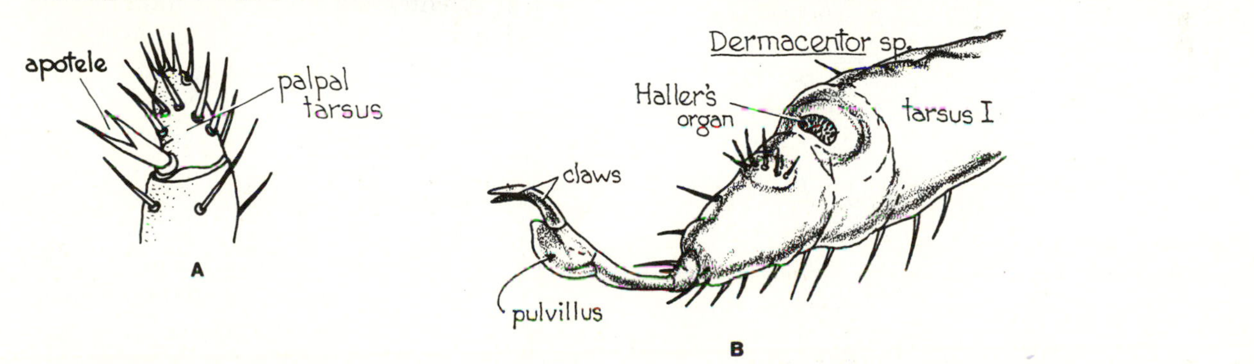

Figure 36–13.
A, Dorsum of tick head showing position of hypostome, *Dermacentor andersoni*; B, enlargement of tick hypostome with recurved teeth, ventral view. Numerals I, II, and III collectively are termed the **pulp**.
(A, from Manual of Medical Entomology, Third Edition, by Deane P. Furman and Elmer P. Catts by permission of Mayfield Publishing Company. Copyright © 1961 and 1970, Deane P. Furman. B, Baker et al. 1958: 13)

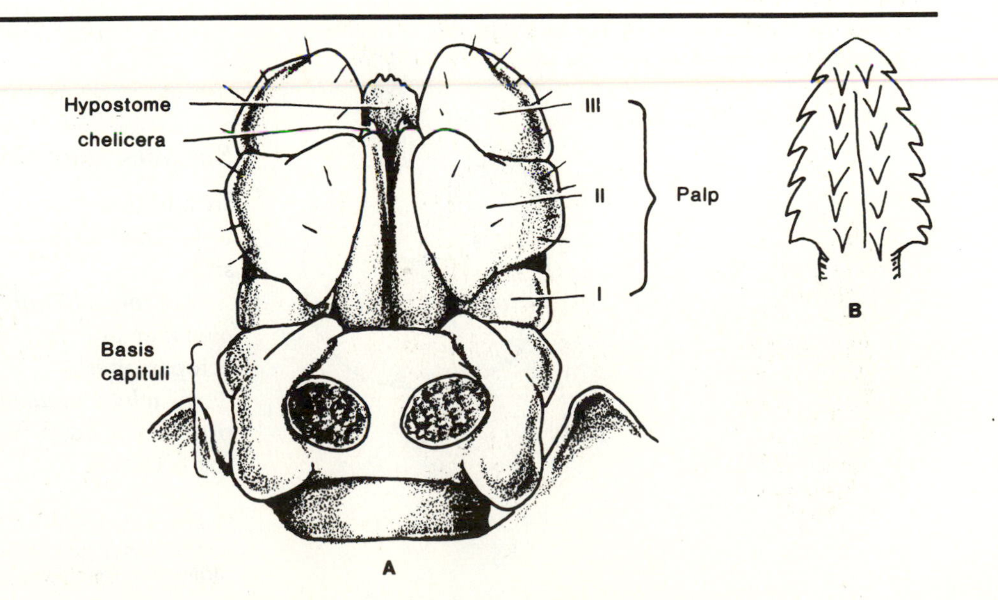

Figure 36–14.
Body mite (Mesostigmata: Laelapidae): female, A, dorsal view, and B, ventral view.
(Baker *et al.* 1958: 47)

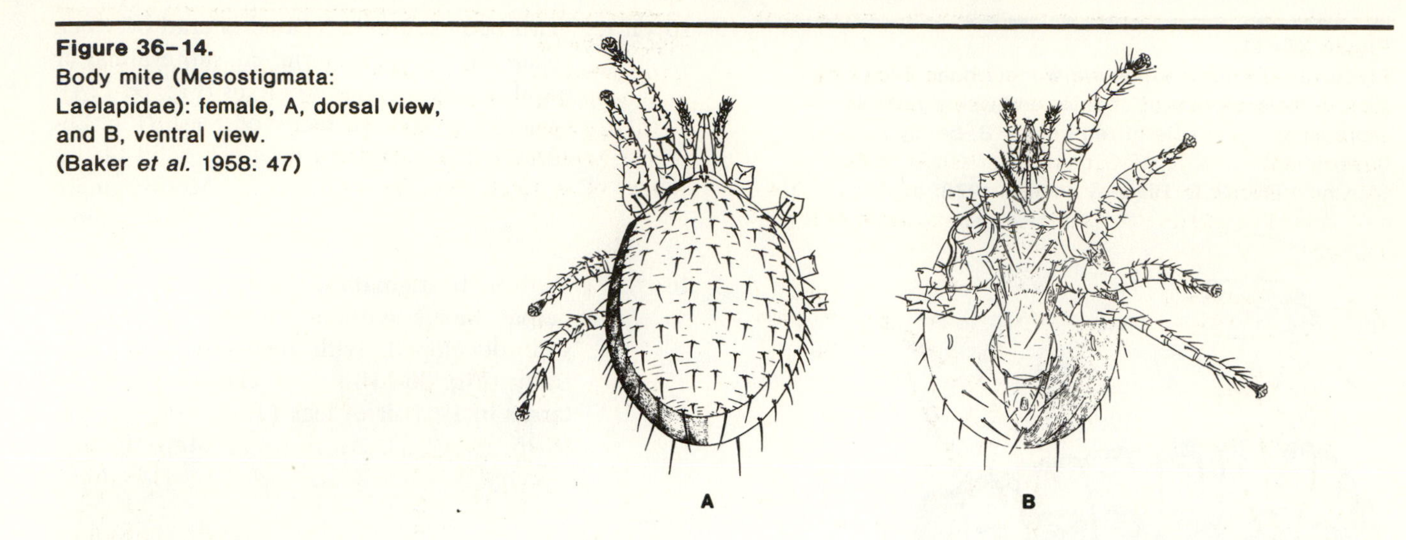

Figure 36–15.
Ticks (Metastigmata): hard ticks (Ixodidae): A, female, and B, male; soft ticks (Argasidae): C, dorsum, and D, venter, of female.
(Baker *et al.* 1958: 81)

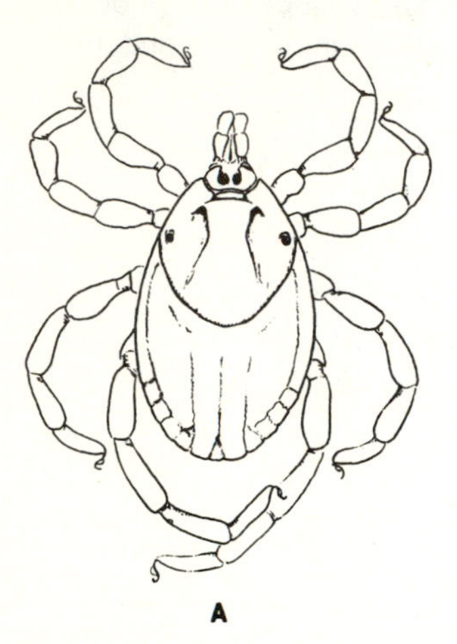

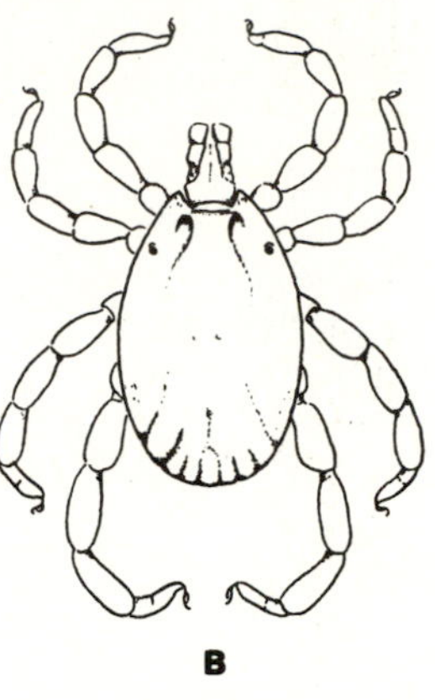

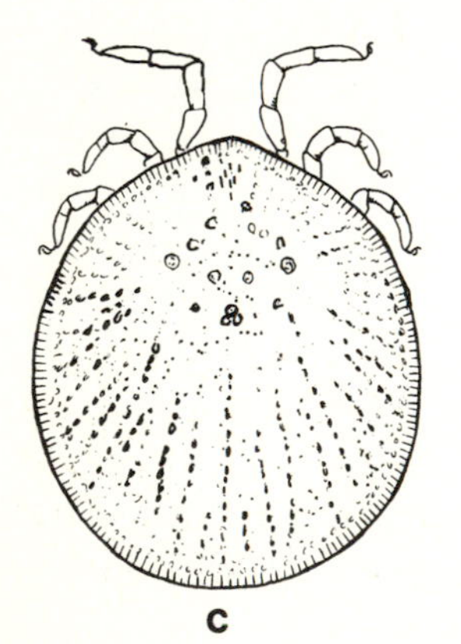

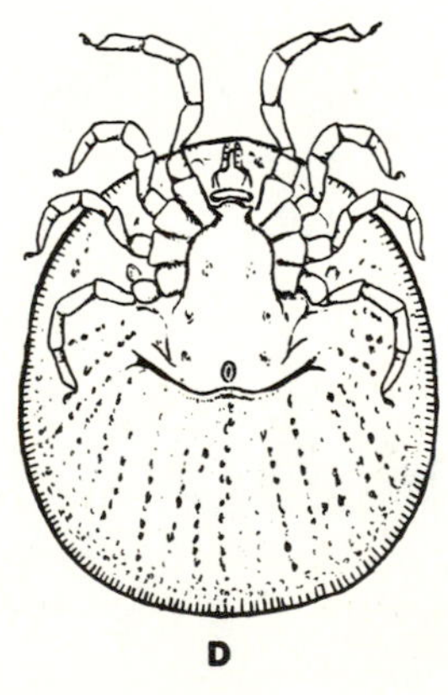

17 (15′) Pretarsus without paired lateral claws, empodium either claw-like or usually sucker-like (Fig. 36-16A); chelicerae typically chelate-dentate; genital papillae usually present; genital field/pore located ventrally, anterior to coxae IV; palpi small, 2-segmented; dorsum of body never covered by overlapping sclerites and never vermiform (Fig. 36-17) .... **Astigmata***
mange and fur mites

17′ Pretarsus with paired lateral claws (Fig. 36-16B), empodium rarely claw-like or sucker-like; chelicerae typically stylettiform or hook-like; genital papillae usually absent; genital field/pore terminal; palpi usually 3 to 5-segmented and conspicuous; if palpi small and with fewer segments, then body dorsum either vermiform or with overlapping sclerites; chiggers (Fig. 36-18A) and follicle mites (Fig. 36-18B-D) ...... **Prostigmata**
follicle mites and chiggers

Supplies and Equipment Needed

In addition to the supply of bags discussed above, a few other items are needed for collecting ectoparasites.

Cotton, chloroform or *ether,* and a *tightly-closed chamber* are needed to anesthetize live hosts and the ectoparasites.

A *white enameled tray* approximately 12″ x 18″ provides a good background against which to search for ectos.

*Some with a specialized subcutaneous parasitic deutonymph, called a hypopus, that looks completely different from all other developmental stages; see Krantz (1970).

Figure 36–16.
Tarsi of mites showing empodium: A, tarsus II showing sucker-like empodium; B, tarsus II showing pad-like empodium.
(Krantz, G.W. A manual of acarology. Oregon State University Book Stores, Inc., Corvallis. 335 pp., figs 5-3 & 5-7. Copyright 1970 G.W. Krantz, with permission of the author.)

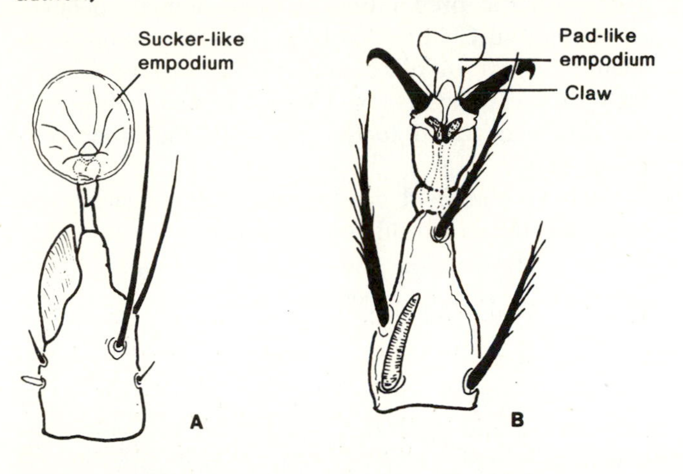

Figure 36–17.
Fur mite (Astigmata: Listrophoridae): A, male; B, female.
(Baker *et al.* 1958: 185)

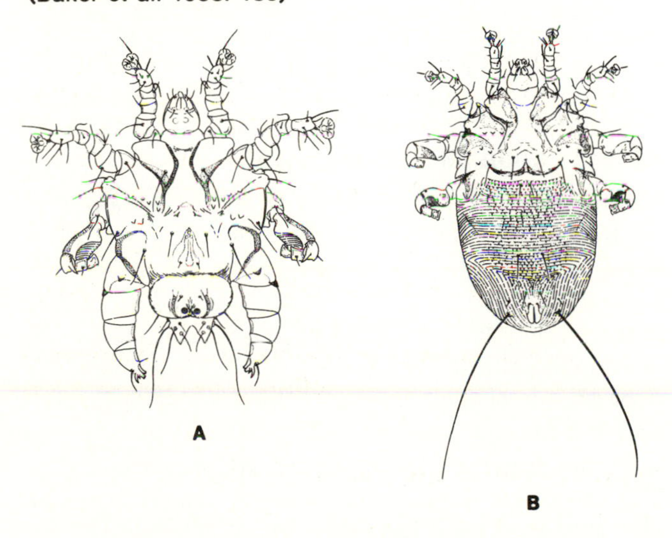

Figure 36–18.
Representatives of Prostigmata: A, chigger (Trombiculidae; dorsum on left and venter on right of larva); B-D, follicle mites (Demodicidae: B, male dorsum, and C, female venter; Myobiidae: D, female dorsum).
(Baker *et al.* 1958: 153, 144, 141)

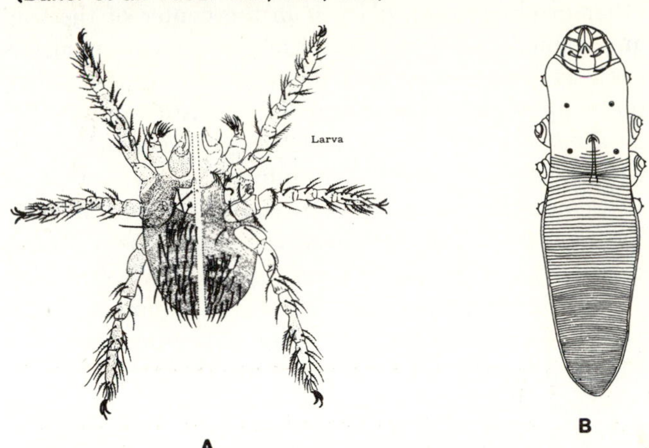

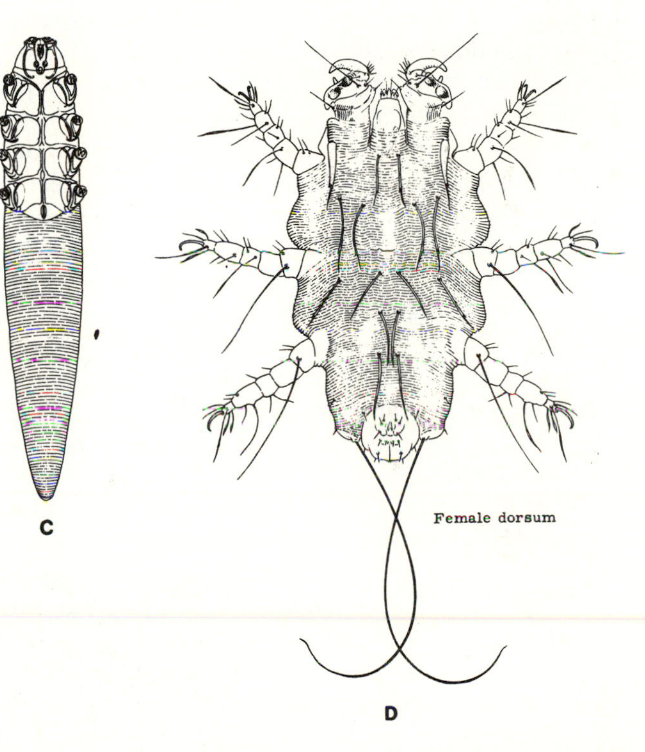

An old *toothbrush* or a similar implement is useful for brushing the fur to remove concealed parasites.

Very fine-pointed *jeweler's forceps,* a *camel's hair brush,* a *dissecting needle,* an *aspirator,* or a combination of these is needed to pick up the ectos and transfer them to alcohol vials.

Numerous *small vials* of *seventy percent ethanol* are needed for storage of ectoparasites. The vials should be equipped with tight-fitting caps or stoppers that will not allow the alcohol to leak or evaporate. One vial is needed for each host specimen examined.

Small *tags* of good quality paper (see Chapter 33) are needed for inserting data into the vials with the parasites.

A solution of *detergent and water* may be needed to thoroughly wash dead hosts and remove adherent ectoparasites.

Searching for and Removing Ectoparasites from the Host

Carefully remove the dead host from the bag, holding it over a clean white tray. Reseal the bag and set it to one side. If the specimen has not been placed in ether or chloroform, place it in the center of the tray and immediately begin to search for moving parasites such as fleas, mites, and the various dipterans. Pick these up with jeweler's forceps, a dissecting needle, or a camel's hair brush which have been moistened with alcohol, or with an aspirator.

Some collectors do not etherize or chloroform their specimens, feeling that they can more easily locate moving parasites than dead ones. However, since winged insects, fleas, and other active parasites may escape during the search, other collectors prefer to kill all parasites before looking for them.

Once all visible ectos have been gathered, search the areas of the ears, eyes, lips, axillae, and urogenital openings for ticks and other embedded parasites. These embedded creatures cannot be removed without damaging either the parasite or the mammal skin. Since the damage to the mammal is usually only a small hole in the skin, while the damage to the parasite could render it impossible to identify, it is best to remove embedded ectos by cutting free the small piece of skin to which they are attached.

Rub through the mammal's fur with your finger, forceps, or dissecting needle to find concealed organisms. Then, hold the host above the tray and brush the hair vigorously (including all directions) with a toothbrush or similar utensil. Carefully pick through the debris in the tray and don't neglect to examine the brush itself.

When you are certain that you have removed all ectoparasites from the host, examine the bag that it was in. Pay particular attention to seams. If it is paper tear it open for close examination. If it is cloth, turn it inside out and shake or brush it over the tray.

Ectoparasites may also be removed by placing a dead host in a closed container (e.g., glass jar) partially filled with a solution of liquid detergent (1 or 2 drops) and water. By vigorously shaking the container, the parasites may be dislodged from the host. The host is then removed from the container and the solution allowed to stand for a few minutes. (It may be necessary to add a few drops of 95% alcohol to disperse bubbles in the solution.) The supernatant fluid is then decanted several times, and the remaining fluid examined for ectoparasites with the aid of a binocular dissecting microscope. Hosts that must be kept alive can be placed on a wire screen above a detergent solution for 24 to 48 hours. Detaching or mobile ectoparasites will often fall into the solution and can then be recovered.

When you have located all ectoparasites and transferred them to a vial of seventy percent ethanol, proceed with the preparation of the mammal specimen as outlined in Chapter 35. If warbles are found, these should be left in place until the animal is skinned, then they can be removed easily from the inside of the skin and added to the vial with the other ectoparasites.

Note in the field catalog that ectoparasites were collected. If you can identify these to order or family, do so in the catalog and give at least approximate numbers. If certain types of ectos were found only in certain regions of the body (e.g., mites in the anus), note this fact in the catalog. If pieces of skin bearing embedded parasites have been cut away, note this in the catalog and give the location on the host from which they were removed. On a small durable tag write the catalog number of the host (including collector's name). If space permits, add date and place collected. Insert this tag *into* the vial with the parasites and seal the vial.

Before proceeding to the next specimen, be sure that no contamination remains on the tray, on the brushes, or on any of the other tools.

Identification of Ectoparasites

Identification of ectoparasites to order or family is relatively easy, but beyond the family level it is difficult or impossible for the novice to do. There are, however, experts in the United States and elsewhere in the world who are willing to receive collections and provide identifications (Arnett, 1978). The entomology departments of most large universities and major natural history museums usually are willing to help you locate an expert or group of experts on various taxa. The specimens themselves, however, should never be sent until the expert has expressed an interest in them and a willingness to work with them.

Supplementary Readings

Baker, E. W., and G. W. Wharton. 1952. *An introduction to acarology*. The Macmillan Co., New York. 465 pp.

Borror, D. J., D. M. DeLong, and C. A. Triplehorn. 1976. *An introduction to the study of insects*. 4th ed. Holt, Rinehart and Winston, New York. 852 pp.

Cheng, T. C. 1964. *The biology of animal parasites*. W. B. Saunders Co., Philadelphia. 727 pp.

Davis, J. W. and R. C. Anderson (Eds.). 1971. *Parasitic diseases of wild mammals*. Iowa State Univ. Press, Ames. 364 pp.

Ewing, H. E. and I. Fox. 1943. The fleas of North America. *USDA Misc. Publ.* 500:1-128.

Ferris, G. F. 1951. The sucking lice. *Memoir Pacific Coast Entomological Society*. 1:1-320.

Holland, G. P. 1964. Evolution, classification, and host relationships of Siphonaptera. *Ann. Rev. Entomol.* 9: 123-146.

Horsfall, W. R. 1962. *Medical entomology*. Ronald Press, New York. 467 pp.

Meyer, M. C. and O. W. Olsen. 1971. *Essentials of parasitology*. Wm. C. Brown Co., Dubuque. 305 pp.

Nelson, W. A., J. E. Keirans, J. F. Bell., and C. M. Clifford. 1975. Host-ectoparasite relationships. *J. Med. Entomol.* 12:143-166.

Richmond, N. D. 1951. Field methods for collecting mammal ectoparasites. *J. Mammal.* 32:123-125.

Strickland, R. K., R. R. Gerrish, J. L. Hourrigan, and G. O. Schubert (revisers). 1976. Ticks of veterinary importance. *USDA Agric. Handbook* 485:1-122.

37 Analysis of Spatial Distribution

An understanding of the spatial distribution of mammals and the movements that they make within an area is important for interpreting many ecological and evolutionary processes. It also aids in formulating management plans for various species. In this chapter, we will examine the types of spatial organization and movements in populations of mammals and how these phenomena can be studied and measured. General reviews of this subject can be found in Brown (1966), Jewell (1966), Sanderson (1966), Fisler (1969), Brown and Orians (1970), and Flowerdew (1976).

Spatial Organization

The **dispersion** or distribution of animals in an area can be categorized as random, uniform, or clumped (Brown and Orians (1970). In a **uniform** distribution (Figure 37-1A), the points in space occupied by an individual are approximately equidistant from one another. In a **clumped** or patchy distribution (Figure 37-1B), individuals are concentrated in some areas and absent from others. In **random** dispersion (Figure 37-1C), there is equal probability that an individual will occupy any given point in space and the presence of another individual nearby will not affect this probability. One can test for the type of dispersion pattern present by using various statistical tests. Morisita (1962) provides an **index of dispersion** (I_δ) that can be used to test these patterns:

$$I_\delta = N \frac{\Sigma n_i\ (n_i - 1)}{\Sigma x\ (\Sigma x - 1)},$$ where N equals the total number of observations, n_i equals the number of animals observed in the i_{th} observation, and Σx equals the total number of animals found in all observations. Randomness in dispersion is indicated by a value of one while values less than or greater than one indicate, respectively, that the distribution is uniform or clumped. The value obtained can be checked for significance using the following F-statistic:

$$F = \frac{I(\Sigma x - 1) + N - \Sigma x}{N - 1}$$

The F_{cal}-value is then compared with an F_{tab}-value (Rohlf and Sokol 1969:168-195) using the following degrees of freedom (df):

numerator $df = N - 1$
denominator $df = \infty$ (infinity)

Refer to Morisita (1962) and Flowerdew (1976) for more details on this procedure.

Figure 37–1.
Uniform (A), clumped (B), and random (C) distribution patterns.
(Modified from Brower and Zar 1974: 118)

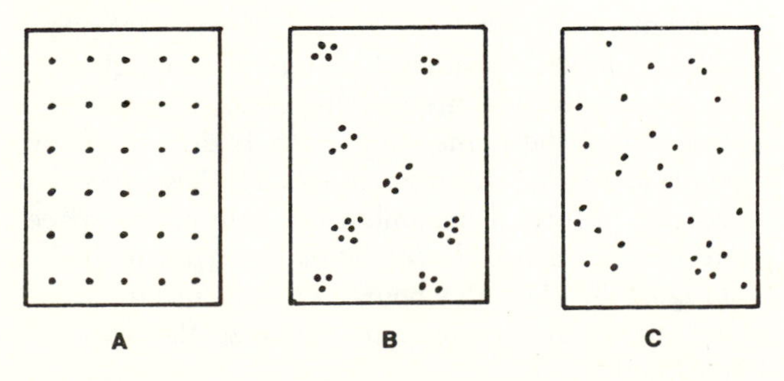

Home Range

The **home range** of a mammal is the area that it occupies during the course of its life, exclusive of migration, emigration, or unusual erratic wanderings (Brown and Orians 1970). Burt (1943), in differentiating the concept of home range from the related concept of territory, defined home range as the area traversed by an individual in its normal activities of food gathering, mating, and caring for young, and specifically excluded . . . "Occasional sallies outside the area." Brown (1966), however, suggested that one should use caution in not labeling longer movements of mammals in their home range as "occasional sallies" and thus excluding these movements from the area that the animals actually occupy.

The sizes of mammalian home ranges (Figure 37-2) are related to the energy demands of the species, with 'croppers' having smaller home ranges than 'hunters' (McNab 1963; Harestad and Bunnell 1979). In addition, home ranges may also involve a third dimension—e.g., such as vertically into shrubs or trees or down into burrows (Meserve 1977; Koeppl et al. 1977a) indicating that measuring area alone may be insufficient for determining ecological relationships of certain species.

The home ranges of individuals may overlap partially (Figure 37-3) or may be entirely separate from one another. The nonoverlap of home ranges may be due to territoriality, mutual avoidance, preferences for particular food sources or habitats, or physical barriers (Brown and Orians 1970). Jorgensen (1968) provides formulae for estimating the probability that occupants of two home ranges will meet by chance and Adams and Davis (1967) give a method for estimating the amount of overlap of home ranges.

The area of a home range can be estimated using a variety of techniques (see section on Methods for Studying Movements, this chapter). Stickel (1954), Sanderson (1966), and Jennrich and Turner (1969) reviewed the procedures for estimating home ranges using data from grid trapping. Methods for estimating the area (volume) of a home range include the following:

Figure 37-2.
Relationship of home range size to body weight in "hunters" (solid symbols) and "croppers" (open symbols). The thin line represents the curve for the pooled data. (McNab 1963: 135)

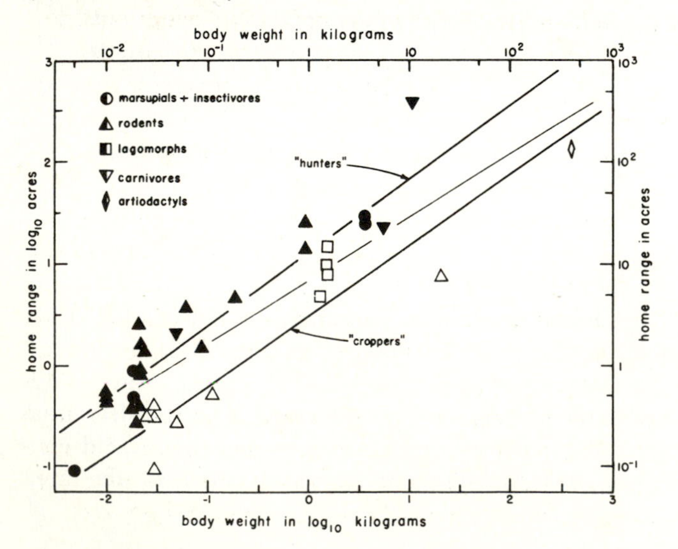

Figure 37-3.
Individual home ranges of four adult male asiatic elephants (*Elephas m. maximus*) superimposed upon a portion (solid line) of the group home range. (McKay 1973: 82)

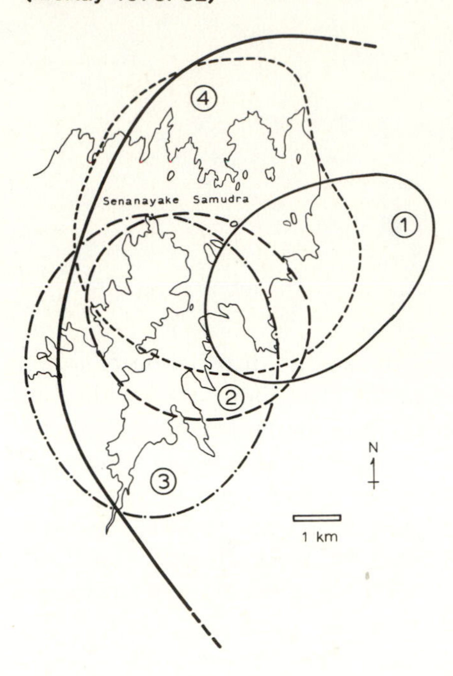

Polygon Methods

1. **Minimum area** (Figure 37-4A). Capture points are connected to enclose a polygon. Jennrich and Turner (1969) suggested that points be connected in counterclockwise fashion so that a unique polygon is obtained.
2. **Convex polygon** (Figure 37-4B). Capture points are connected to form the smallest convex polygon (Southwood 1966).
3. **Boundary strip.** A boundary strip equal in width to half the distance between traps is drawn around the minimum area. In the *inclusive boundary strip* method (Figure 37-4C), the peripheral points of capture are considered centers of rectangles (each side of which equal the distance between traps) and the home range area is delineated by connecting the exterior corners of these rectangles to form a maximum estimate of the space utilized. In the *exclusive boundary strip* method (Figure 37-4D), the boundary strip rectangles are drawn in a manner to minimize the area enclosed.

Figure 37–4.
Polygon (A-D) and range length (E-F) methods for estimating the two-dimensional home ranges and movements of animals. A, minimum area method (dotted lines); and convex polygon method (solid lines); B, inclusive boundary strip method; C, exclusive boundary strip method; D, range length; E, adjusted range length. The cross within a circle denotes the center of activity. (Mary Ann Cramer)

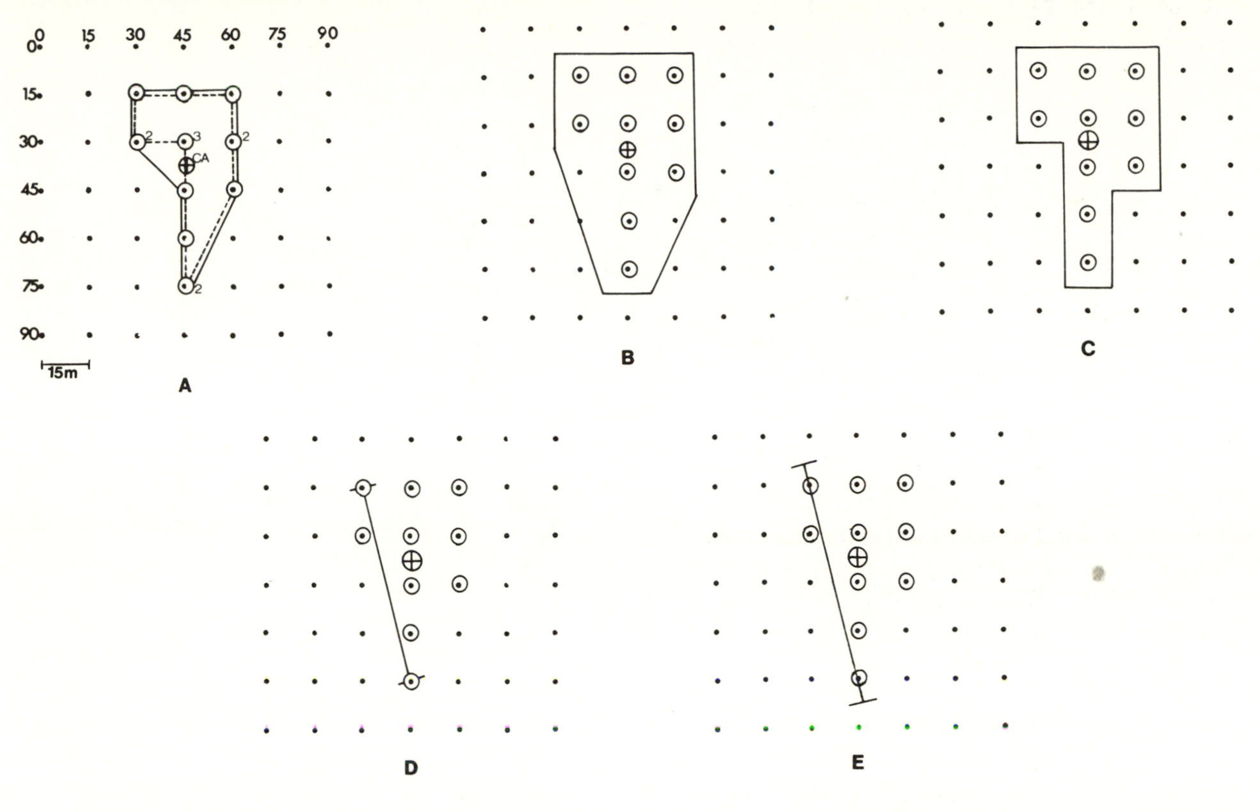

4. **Range length** (Figure 37-4E). This measure is the distance between the most widely separated capture points.
5. **Adjusted range length** (Figure 37-4F). The range length *plus* one-half the distance to the next trap is added onto each end.

Stickel (1954), in empirical experiments, found that the exclusive boundary strip method was the most accurate measure of the actual area of the home range while the adjusted range length most closely approximated the true range length. Jennrich and Turner (1969) pointed out that methods based on connecting points are biased by the size of the sample (i.e., the observed area and range length increase with an increase in N) Hayne (1950) found that the distance between traps also affected the estimated size of the home range.

Methods Based on Recapture Radii

6. Since the number of captures (N) affects the estimated size of the home range, various workers (Hayne 1949a; Calhoun and Casby 1958; Jennrich and Turner 1969) have sought to minimize this bias. The **center of activity** (Figure 37-5) is not necessarily the home site of an animal but rather the mean of a set of capture coordinates (Hayne 1949a). It is computed by summing each coordinate (X, Y) of the capture points to arrive at an average or geometric center. Smith et al. (1973) found that the burrows of old-field mice (*Peromyscus polionotus*) were outside the estimated home range of the animals and peripheral to the above ground center of activity as revealed by trapping.
7. The **standard deviation of capture radii.** This measure was used by Calhoun and Casby (1958) and other workers to estimate the dimensions of the home range of an animal. The home range shape must be circular in order to utilize this technique. Calhoun and Casby (1958) and Maza et al. (1973) provide tests and examples of the use of this method.
8. The **determinant of the covariance matrix.** This estimate was used by Jennrich and Turner (1969) to eliminate biases due to lack of circular symmetry in home range shape and

Figure 37–5.
Location data (A) and three-dimensional representation of home range (B), including 95% confidence ellipses for each dimension, of an individual gray squirrel, *Sciurus carolinensis*.
(Koeppl et al. 1977a: 214, 215)

A

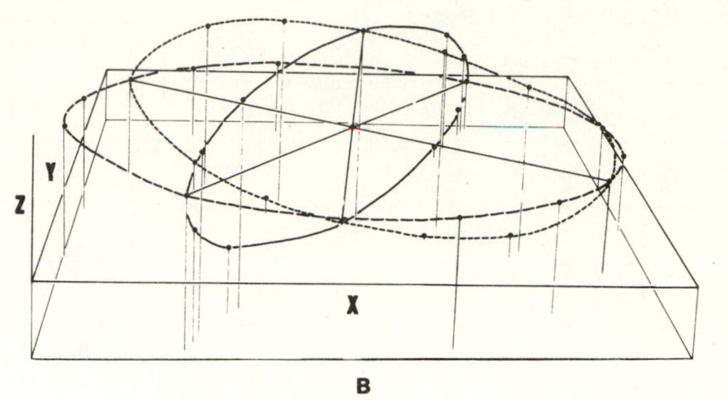

variations due to differing sample sizes. Refer to Jennrich and Turner (1969) for details and computational procedures.

9. **Bivariate home range.** Developed by Koeppl et al. (1975), extending the concepts of Calhoun and Casby (1958) and Jennrich and Turner (1969) into a general model. The method of Koeppl et al. (1975) enables calculation of standardized distances from activity centers and calculates probability ellipses around these centers.
10. **Three-dimensional home range.** Since many species utilize a vertical dimension in their movements, Koeppl et al. (1977a) extended the methodology of the bivariate home range model (Koeppl et al. 1975) to include this extra dimension. Location data obtained by trapping or observation (Figure 37-5A) can then be projected into three-dimensional space and confidence ellipses plotted (Figure 37-5B) for each dimension.
11. **Distance between observations indices.** Koeppl et al. (1977b) developed this method to measure the average size of the home range when the assumptions of the bivariate and three-dimensional models could not be met (e.g., lack of adequate sample sizes from a single individual).
12. **Index of home range size.** Metzgar and Sheldon (1974) developed another index that is free from the assumptions of large sample sizes. Strictly applied, the method does not provide an area estimate of home range size.

37-A. Examine several papers (including McNab 1963; French et al. 1975) that report sizes of home ranges of mammals. What trends are apparent from these data?

37-B. Use McNab (1963) to obtain values of home range size and then prepare a circular and oval transparent template of the home ranges of a "hunter" and a "cropper.' Using the same scale, prepare a map of a trapping (observation) grid that is several times larger than the area of each home range. The grid should be drawn to permit the testing of various trap spacings. *Alternatively,* the instructor may wish to prepare templates and the grid in advance using the suggestions in Stickel (1954). For each toss of the template onto the grid, record (Table 37-1) the observed range length and adjusted range length (range length plus one-half the distance of the spacing between stations added to each end of the range length). Continue for 25 trials.
Which method for computing range length came closest to the actual value (population parameter) of range length? Plot observed range length (Y-axis) as a function of trial number (X-axis). What effect does the number of trials ("captures") have on the shape of this curve? How could this result be used to plan a sampling program?

Table 37-1

Form for results of Exercise 37-B.

		Circular Template			*Oval Template*	
		Observed			*Observed*	
Trial No.	*Actual Range Length*	*Range Length*	*Adj. Range Length*	*Actual Range Length*	*Range Length*	*Adj. Range Length*
1						
2						
3						
4						
5						
6						
7						
8						
9						
10						
11						
12						
13						
14						
15						
16						
17						
18						
19						
20						
21						
22						
23						
24						
25						
	Σ			Σ		
	$\overline{X}$			X		

37-C. Verify for the following set of capture coordinates that the center of activity is 9.7, 7.2 (X, Y).

Capture No.	X, Y	Capture No.	X, Y
1	10,8	11	10,6
2	10,7	12	9,8
3	11,5	13	9,9
4	9,8	14	9,7
5	10,5	15	10,8
6	9,9	16	10,7
7	9,7	17	10,7
8	9,7	18	10,8
9	11,6	19	10,7
10	10,7	20	9,8

Compute the standard deviation (s) for these data using the following formula (Calhoun and Casby 1958):

$$s = \sqrt{\frac{\Sigma(\bar{x} - x_i)^2 + \Sigma(\bar{y}-y_i)^2}{2\ (N-n)}}$$

37-D. Using the capture coordinate data in 37-C or another set of data, plot two-dimensional home ranges using the minimum area, convex polygon, and inclusive and exclusive boundary strip methods. Which method produces the smallest estimate of home range area? The largest? What is the basis for adding a boundary strip? Is this valid?

Not all areas of a home range are utilized as intensively as other areas. Thus, the methods for defining home range dimensions based on the probability of occurrence in various portions of the range have a strong biological basis.

Most of the discussion on home range has assumed that it is characteristic of an *individual* mammal. However, chimpanzees, gorillas, giraffes, female red deer (*Cervus elaphas*) and other species often form **group home ranges** that are not defended against other groups (Fisler 1969). **Core areas** of intensive use in the group home range are found in some baboons, wild sheep, and coatimundis (Kaufmann 1962; Fisler 1969).

Territory

By the most generally accepted definition, a **territory** is an area defended by an individual or group. According to Brown and Orians (1970), a territory should possess the following characteristics: (1) a fixed area that may change slightly with time (see Fisler 1969); (2) the possessor of the territory exhibits acts of territorial defense that are overt (attacks, vocalizations, or displays) or indirect (scent marking—very characteristic of mammals); (3) these acts are effective in keeping out rivals. Pitelka (1959) argued that a territory should be defined solely as an exclusive area without necessarily involving overt defense. Although this interpretation differs from the classic definition it is probably functionally correct (see further discussion in Fisler 1969).

How does the concept of home range differ from that of a territory? A species may have movements that are part of its home range yet exhibit **territoriality** only during a portion of the year or only defend an area near a particular resource (females, food, nest site). The papers in Stokes (1974) presented a balanced view of the concept of territory and together with Fisler (1969) and Brown and Orians (1970) will give much insight into this subject.

Types of Territories

Fisler (1969) discussed the dynamic nature of territorial systems in mammals. A species may possess two or more types of territories during the course of a year. Territories may be **individual** and involve a fixed or changing area (**spatial territory,** Figure 37-6). A **nidic territory** involves only the immediate area

Figure 37–6.
Individual spatial territories of three adult males and yearling males of the black-tailed prairie dog, *Cynomys ludovicianus*.
(King 1955)

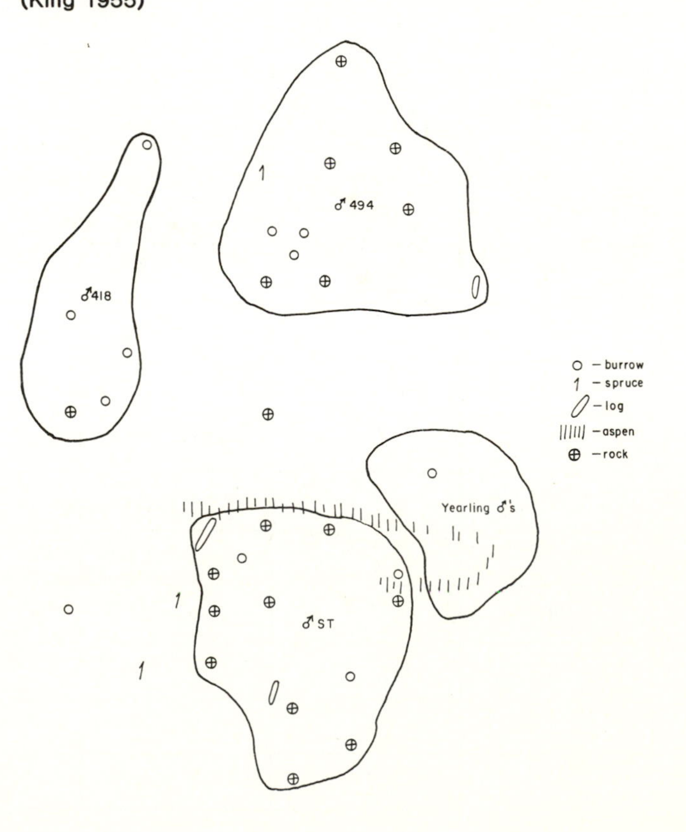

around a home site and is often only possessed by females. **Arena territories** are found in species with polygynous mating systems where the males exhibit overt territorial aggressive behavior at traditional breeding grounds. A **lek** is a special type of arena territory where males aggregate for the purpose of attracting females. Leks were described by Buechner (1961) and Leuthold (1966) in classic papers on the Uganda kob (*Adenota kob thomasi*). Leks are also known in wildebeest (*Connochaetes taurinus*), two small species of gazelles (*Gazella thomsonii* and *G. granti*), and the gray seal (*Halichoerus grypus*). Most pinniped species with polygynous mating systems have a *harem type* of *arena territory* (Figure 37-7). Males vigorously defend their territories from intrusions by other males and try to maintain an assemblage of females inside the territory.

Group spatial territories (Figure 37-8) are found in many highly social mammals such as gibbons (*Hylobates lar*), prairie dogs (*Cynomys ludovicianus*) and several other species (see review in Fisler 1969). In prairie dogs, the **coterie** (a "family" group consisting of a male, several females, and their young), typically occupy a group territory (Figure 37-8) that is defended by the group against other coteries.

Figure 37–7.
Distribution of arena territories (harem type) of the Alaska fur seal, *Callorhinus ursinus*.
(Bartholomew and Hoel 1953: 430)

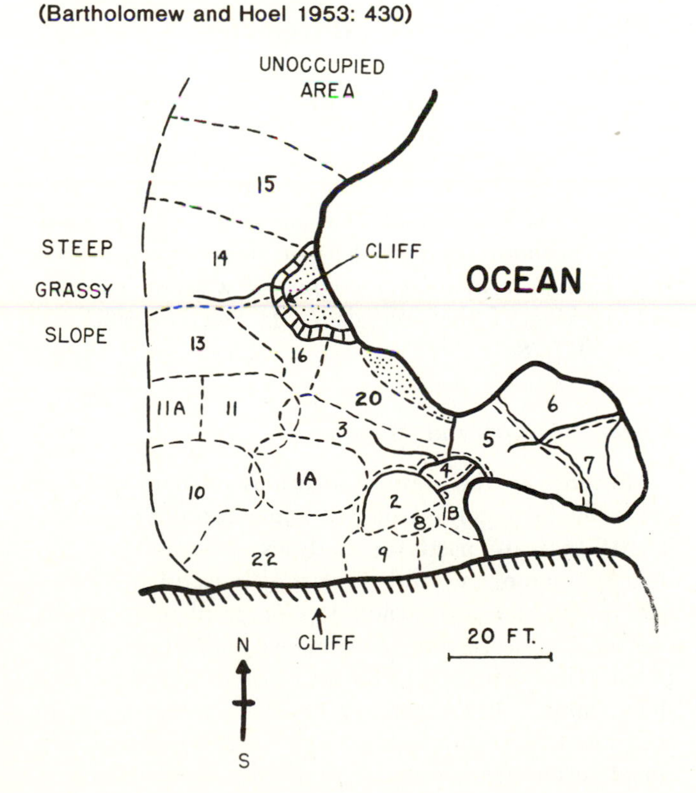

Figure 37–8.
Group territories of four coteries of black-tailed prairie dogs, *Cynomys ludovicianus*.
(King 1955: 55)

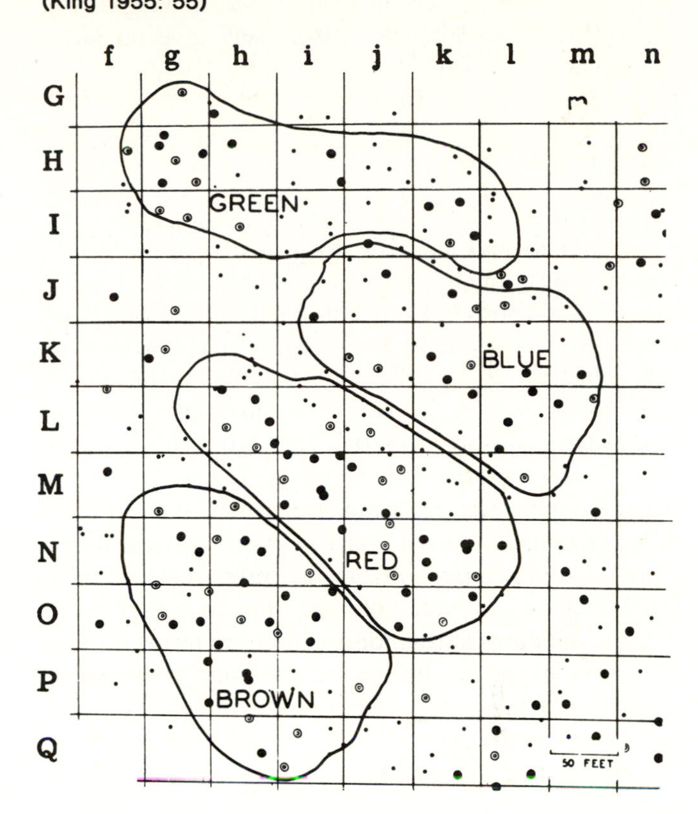

Colonies

Colonial species occur in aggregations whose density is greater than that expected for the nature of the resource (Brown and Orians 1970). Knowledge that a species lives in colonies tells us little about the actual social organization of the species since individual or group territories may be present or the species may be essentially solitary (Fisler 1969). Thus, this description is a general one that should be used with caution. Some fossorial species are often categorized as being colonial since evidence of their spatial distribution is readily discernable.

Habitat Utilization and Preference

According to M'Closkey and Lajoie (1975), studies of the distribution of animals may allow us to make definitive statements on the evolution of habitat selection, predict colonization or range extensions, and permit examination of species-packing and diversity.

The utilization of particular areas and habitats within the range of a species may depend on the distribution of available resources, climatic conditions, and the presence of other species (Krebs 1972).

Due to space limitations, this subject will not be covered further in this manual. However, the papers by Harris (1952), Rosenzweig and Winakur (1969), Brown et al. (1972), Grant (1972), Rosenzweig (1973), and Colwell and Fuentes (1975) will give an introduction to this field.

Movements

Migration

Mammals may migrate seasonally to reach feeding or breeding grounds or more favorable climatic conditions (Sanderson 1966). **Latitudinal migration** occurs in several species including the gray whale (*Eschrichtius robustus*) that travels some 9000 km from summer feeding grounds in arctic waters to winter breeding and calving grounds off the coast of Baja California (Rice and Wolman 1971). Alaska fur seals (*Callorhinus ursinus*) migrate some 5000 km from the Pribilof Islands to Pacific coastal areas. Many species of bats also make extensive latitudinal migrations (Griffin 1970).

Altitudinal migration occurs in some populations of elk (*Cervus canadensis*) that move from winter ranges at lower elevations to summer feeding ranges at higher elevations.

Seasonal migration also occurs in a variety of other terrestrial mammals, particularly ungulates. For example, caribou (*Rangifer tarandus*) move in large herds from winter ranges in central Manitoba and northwestern Ontario to summer ranges some 500-600 miles northward (Harper 1955). In the Serengeti plains of Africa, blue wildebeest (*Connochaetes taurinus*) move in large aggregations (tens of thousands) over a wide area while in the Ngorongoro area they are sedentary (Leuthold 1977).

Immigration and Emigration

The composition of a population (see Chapter 38) changes due to the influx or **immigration** of individuals from outside the population and the **emigration** of adults or young from a population. When the individuals leave the population this phenomenon is sometimes termed **dispersal.**

Special Movements

Mammals experimentally removed to areas well away from home territories or home ranges may return to their areas within a short time (Figure 37-9). This phenomenon is known as **homing** (Henshaw and

Figure 37–9.
Relationship of distance and per cent homing success in *Peromyscus maniculatus gambelii.*
(Furrer 1973: 471)

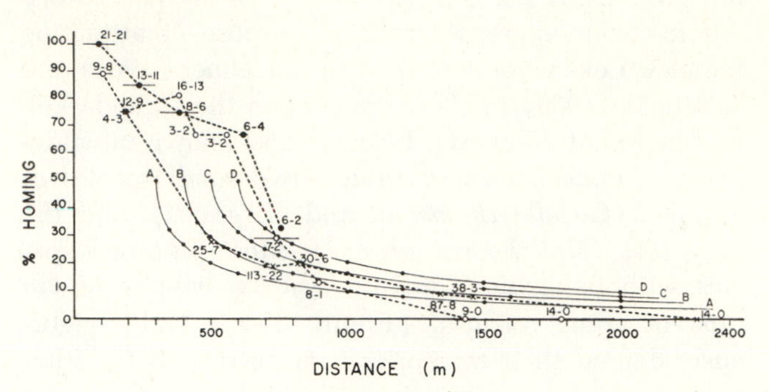

Stephenson 1974; Anderson et al. 1977, for examples) and indicates that individuals recognize, and perhaps prefer, these familiar areas (Brown 1966).

Nomadism has been defined as the tendency for individuals to make erratic or wandering types of movements outside of their "normal" territories or home ranges. Brown (1966) suggested that understanding these movements may aid in interpreting the population dynamics of a species. **Exploratory movements** are made by most mammals when placed in a novel environment or when moving about a familiar territory or home range.

Methods for Studying Movements

Tagging

Accurate estimation of individual or group movements and space utilization often requires that individuals be marked in a distinctive fashion (see Chapter 34). The individual characteristics of fins and bodies (porpoises and whales) or other body features may enable individual recognition without capture and marking of the animals (e.g., Schaller 1963:24; Hrdy 1977:78-81).

Record Keeping and Mapping

Records on the capture locations of mammals can generally be recorded on the same forms (e.g., Fig. 33-10) that are used for analyses of populations. A detailed map of the study area is also useful. This map can be made from an aerial photograph onto which coordinates or special features are indicated and labeled. Then, copies of the map can be used to plot daily, hourly, individual, or group movements of an individual or population. Mosby (1969b) gives additional recommendations on preparing these maps.

Trapping

Many estimates of home range size are based on data obtained from grid trapping. This method is time consuming but can produce reasonable quantities of location data. One disadvantage, if traps are checked only once or twice per day, is that many days must elapse before sufficient data are available for accurate estimates.

Radioisotopes

Radioisotopes can be implanted in the skin of mammals to follow the movements of an individual at frequent intervals (See Peterle 1969 and Chapter 34). The main disadvantages of this technique are the limited range for detection of the radioactivity and the inability to mark many individuals (due to possible confusion of the signal) in a given area (Sanderson 1966).

Radio-location Telemetry

This technique offers the capability of obtaining large quantities of location data on a sizeable number of individuals (e.g., one individual for each channel of a receiver). Some of these transmitters weigh little (1½-2 grams) and will work on small rodents (as small as *Microtus*, 30-40 grams). They can be powered by batteries or solar cells (in diurnal species like a tree squirrel or ground squirrel). The major disadvantage of this technique is the cost of the transmitters and receivers and the relatively short life (1-2 weeks) of the power packs for some transmitters used on small mammals. This system is important for accumulating large quantities of location data for a relatively small number of individuals. Brander and Cochran (1969) provide additional information on this technique.

Individual Observation

For diurnal mammals living in aggregations and groups, direct observation is a valuable means for determining space utilization. However, this method is time-consuming and may be misleading if proper precautions are not made to reduce bias in the observation procedure. Altmann (1974) provides valuable suggestions for making observations in a manner that minimizes bias. Armitage (1974) used this technique to study the territorial behavior of marmots (*Marmota flaviventris*) and it is widely used to study the movements of many primates (e.g., Schaller 1963; Hrdy 1977).

Special Techniques

Sanderson (1966) provides examples of the use of dyes in urine and feces to study movements. Justice (1961) and Metzgar (1973) used smoked paper to study the movements of rodents and found that trap and track-revealed home range estimates differed. Photographic techniques have also been used to study the movements of small mammals (Pearson 1960; Wiley 1971).

Supplementary Readings

Altmann, J. 1974. Observational study of behavior: sampling methods. *Behaviour* 49:227-267.

Armitage, K. B. 1974. Male behaviour and territoriality in the yellow-bellied marmot. *J. Zool., London* 172:233-265.

Brown, J. L., and G. H. Orians. 1970. Spacing patterns in mobile animals. *Ann. Rev. Ecol. Syst.* 1:239-262.

Brown, L. E. 1966. Home range and movement of small mammals. *Symp. Zool. Soc. London* 18:111-142.

Colwell, R. K., and E. R. Fuentes. 1975. Experimental studies of the niche. *Ann. Rev. Ecol. Syst.* 6:281-310.

Fisler, G. F. 1969. Mammalian organizational systems. *Los Angeles Co. Mus. Cont. Sci.* 167:1-31.

Jewell, P. A. 1966. The concept of home range in mammals. *Symp. Zool. Soc. London* 18:85-109.

Lidicker, W. Z., Jr. 1975. The role of dispersal in the demography of small mammals, pp. 103-128. *In* F. B. Golley, *et al.* (Eds.). *Small mammals: their productivity and population dynamics.* Cambridge Univ. Press, Cambridge.

McNab, B. K. 1963. Bioenergetics and the determination of home range size. *Amer. Natur.* 97:133-140.

Rosenzweig, M. L. 1973. Habitat selection experiments with a pair of co-existing heteromyid rodent species. *Ecology* 54:111-117.

Sanderson, G. C. 1966. The study of mammal movements: a review. *J. Wildl. Manag.* 30:215-235.

Stickel, L. F. 1954. A comparison of certain methods of measuring ranges of small mammals. *J. Mammal.* 35:1-15.

Stokes, A. W. (Ed.). *Territory.* Dowden Hutchinson and Ross, Stroudsburg, Penn. 398 pp.

38 Estimation of Relative Abundance and Density

Accurate estimates of population densities are important for answering many empirical and management questions in ecology and population biology. Presented is an introduction to methods for estimating densities of mammal populations. Excellent comprehensive reviews of these methods include Cormack (1968), Southwood (1966), Seber (1973), Delany (1974), Smith *et al.* (1975), Flowerdew (1976), and Caughley (1977).

Estimation of Population Density: General Comments

Prior to an examination of specific methods, it is important to consider aspects that are basic to any analysis of population densities.

Types of Population Estimates

Population density is a measure of the number of individuals that occupy a given area. For mammals, this statistic is generally reported as the number of individuals per hectare. Generally, **density** is an **absolute** measurement since it has a standardized base (area) as a point of reference. A **relative estimate** of population size (e.g., number of animals in trapline; animals caught per trap night) is not a true estimate of density.

Sampling Configurations for Estimating Sizes of Populations

In many studies, particularly those involving small mammals, a grid (Figure 38-1A) of traps is established in an area of uniform habitat. Generally, it is best to establish a minimum of two grids (more replicates are better) so that information on the variability of the estimate can be obtained (Hayne 1978). Practically, this suggestion is sometimes difficult to meet if the grids are large and manpower limited. The **standard-minimum grid** (16 rows by 16 columns) with 256 trap stations (each with 2 traps per station = 512 traps total) or a 12 x 12 grid are widely used in studies of small mammal population biology. The trap spacing in the standard-minimum grid is 15 meters and this seems to be a good compromise for most studies of small mammals. Distance between traps does affect estimates of home range size (Hayne 1950). Smith *et al.* (1975) indicated that more work needs to be done to determine the optimal grid spacings for estimating population densities of mammals.

A **transect** (Figure 38-*B*) is a line of traps spaced at regular intervals through a habitat. If an **assessment line** (Figure 38-*B*) of traps is placed at an angle to the transect then some measure of area and a resultant measure of density can be obtained (Smith *et al.* 1975). A **trap line** may be equivalent to a transect if the direction of travel is a straight line and the spacing of the traps is uniform. In more general terminology, a trap line is a line of traps placed at regular or irregular intervals to secure specimens for identification, study skins, or autopsy purposes.

Many mammalian populations cannot adequately be censused using grids or transects. Thus, population numbers of bats and many colonial species are often estimated by photographic means (Humphrey 1971) or direct enumeration (Peterson and Bartholomew 1967; Mills *et al.* 1975). Large terrestrial, semiaquatic, and aquatic species are generally counted by aerial census techniques (Martinka 1976; Caughley 1977). Additional information on specialized census configurations can be found in Overton (1969) or Caughley (1977).

Figure 38–1.
A, example of a 7 × 7 grid with 8 assessment lines (solid); B, transect (solid line) with assessment line (dotted) placed at an angle. The arrow marks the location of trap station 27 (i.e., 2nd row, 7th column).
(Mary Ann Cramer)

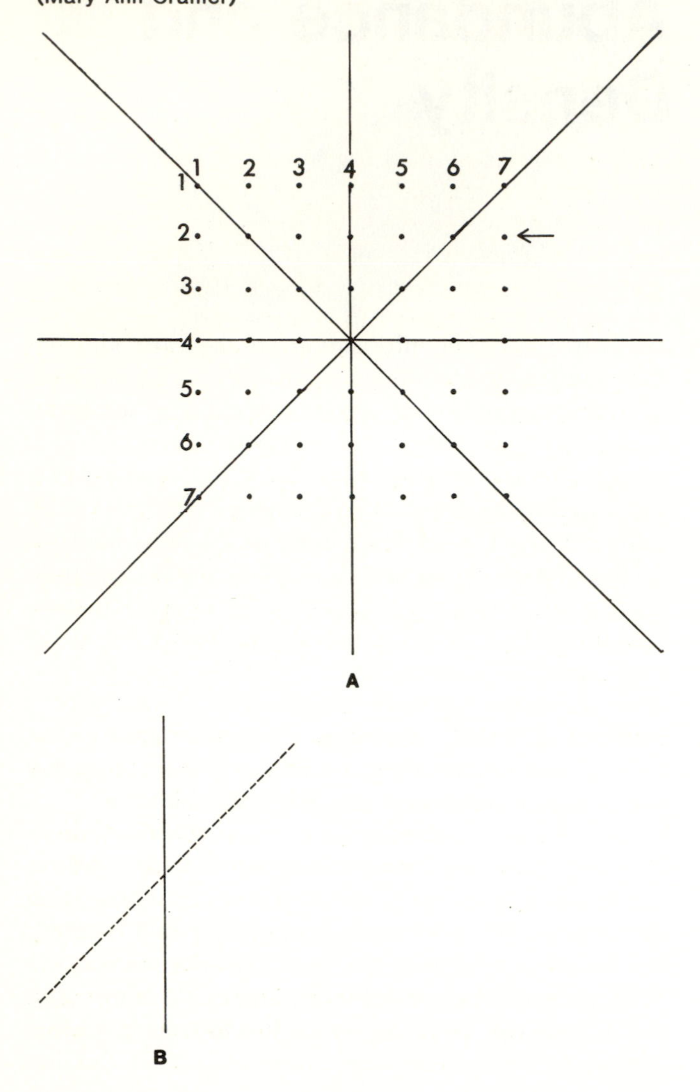

Figure 38–2.
Probability of capture (number of animals caught during the study divided by the number of traps) as a function of the distance of the traps from the outer edge of the grid. The width of the boundary strip is indicated by the arrow; *r* is the regression coefficient.
(After Smith et al. 1969: 28)

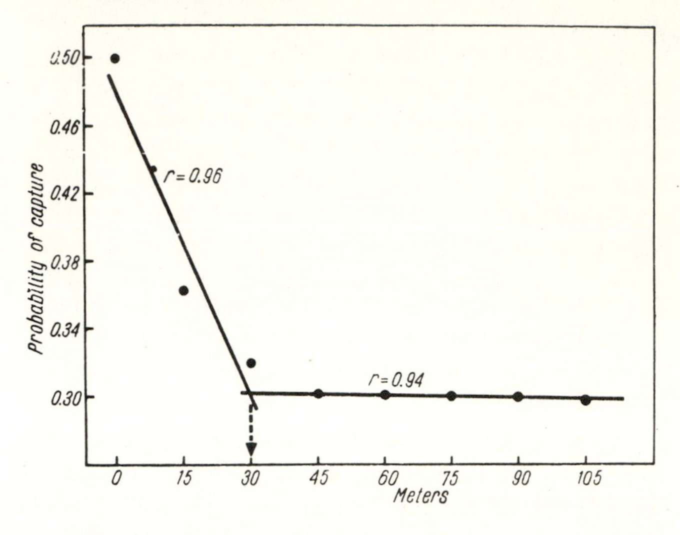

Estimating Size of Sampling Area

In grid studies, the simplest method of determining the size of the trapping area is to use the perimeter of the grid as the limit for determining the sampling area. Generally, this technique overestimates population densities since mammals are not sedentary and may have movements that are only partly within the grid area (although they may be trapped on the grid). The discrepancy between the population estimates of animals living on the grid and those living partly on the edges of the grid has been termed the **edge effect.**

Since the area of the grid is not the area sampled (see reviews by Stenseth *et al.* 1974; Smith *et al.* 1975), other approaches have been used to estimate the sampling area. One method is to compare the number of captures at outer grid stations with those of a "smaller grid" inside the larger grid (Hansson 1969; Smith *et al.* 1969; Pelikan 1970). In this system, the **probability of capture** (number of animals caught during the study divided by the number of traps) is plotted as a function of the distance of the traps from the outer edge of the grid (Figure 38-2). Then, the point in meters where the probability of capture changes abruptly to the horizontal represents the **boundary strip** that should be added to the size of the overall grid. This method assumes random distribution of individuals, no migration, circular home ranges, circular area of influence for each trap, and traps spaced such that each home range includes more than one trap (Smith *et al.* 1969). In practice , few of these assumptions can be met and thus the solution is not ideal.

Marking resident animals on a grid with the use of colored bait prior to the time of trapping (i.e., **prebaiting**) is another technique used to determine the size of the trapping or sampling area (see Smith *et al.* 1975 and Flowerdew 1976 for review).

The use of one or more *assessment lines* (Figure 38-1A) has been recommended by a number of investigators (Stenseth *et al.* 1974; Smith *et al.* 1975) to detect movement of animals onto (immigration) or away from (emigration) the trapping grid. These assessment lines are trapped immediately after the regular census. Stenseth *et al.* (1974) provide a FORTRAN program that can be used to calculate the size of the trapping area.

Accuracy of Population Estimates

The intensity of trapping (days, number of traps), the number of individuals marked, and many other factors affect the accuracy of population estimates. Roff (1973a) recommends that reliable estimates of population size should have a coefficient of variation (see Chapter 29) less than 0.05 (5%) and confidence limits of N ± 0.1N. The sampling intensities required to obtain these ranges (see Overton 1969; Roff 1973a; Smith *et al.* 1975) limits the usefulness of mark and recapture approaches for estimating population densities.

Recording Data

General information on recording data of varied types can be found in Chapter 33. For all types of population estimation techniques, accurate records must be kept on the number of animals observed or handled. Special forms (e.g., Fig. 33-9) are often used to indicate the species, sex, identifying number (if the estimation technique requires a unique number for each individual), location of capture, and condition of the animal captured. For mark and recapture studies, it is generally necessary to indicate whether an animal is a **new capture** (i.e., first time ever captured on grid) or a **recapture** (i.e., marked or tagged individual). Techniques for marking mammals with individual numbers are described in Chapter 34.

The location of a capture is frequently recorded using a grid coordinate system (Figure 38-1A) or matrix notation. For example, the location of an animal captured at the third station (i.e., row 3) on the second line (i.e., column 2) would be recorded as location "32." Similarly, an animal captured on row two and column three would be recorded as location "23." This procedure will facilitate data processing by computers. Alternatively, the cumulative distance between the traps may be used to indicate the grid coordinate (Jennrich and Turner 1969; Fig. 37-4). With this system, grid coordinate "23" might be represented as 3045 (i.e., 15 meters × 2 = 30; 15 m × 3 = 45).

Estimation of Population Densities By Direct Counts

It is sometimes feasible to count the total number of animals inhabiting an area. This procedure has been utilized for African ungulates (Talbot and Stewart 1964) and for pinnipeds (Peterson and Bartholomew 1967). This method of making **total counts** is often not practical for noncolonial species because (1) it requires too much effort, (2) some animals may be counted twice, and (3) it disturbs the entire population (Caughley 1977). In addition, a total count in almost all cases is still an estimate and thus confidence limits should be calculated on the value obtained by replicating the census several times (see Caughley 1977).

Sampled counts eliminate most of the practical problems associated with total counts. Sample areas are located at random (or, if not truly random, in a manner that minimizes bias) and then counts made on the sample areas. Caughley (1977) recommends that sampling intensity in high density areas should be greater than that in low density areas. Thus, an error of estimation in the low density area will not affect the error rate greatly. But undersampling in the high density area would cause the error rate to increase significantly.

Aerial censusing of larger species of mammals is an important technique for managing wildlife. According to Caughley (1977), random sampling of areas is more difficult for the navigator and takes more time per unit of area covered. As a result, systematic sampling of areas is more commonly utilized for aerial censusing.

38-A. Establish a theoretical population of mammals using models (or marbles) on a tray. Alternatively, the instructor may wish to establish the "population" in advance. Include a minimum of two and a maximum of four species in the population with values of population size (N) ranging from 150 to 300. Count the total number of individuals of each "taxon" 6 times and record your results in the following table:

Taxon	*1*	*2*	*3*	*4*	*5*	*6*	*Average N*

How precise were your counts of the individuals of each species (i.e., did you miss some individuals and count some individuals twice)? Repeat the count imposing a time limit (to simulate aerial census) on each counting session. Did this affect the precision?

38-B. Prepare two transparent overlays (one rectangular and one square in shape) whose areas each equal 1/50 or 1/100 of the total area occupied by the population in Exercise 38-A. Prepare two sampling schemes: (1) random, where 20 sampling areas are located by means of a table of random numbers (see Table 29-1) and (2) stratified, where the 20 sampling areas are placed somewhat uniformly in the area of the population. Select a "species" (e.g., one for each student in group) and count the number of individuals of the species that are sampled by the overlays. Record your results in the table below:

Taxon	*Random*		*Stratified*	
	Rectangular	*Square*	*Rectangular*	*Square*
Replicate				
1				
2				
3				
4				
5				
6				
7				
8				
9				
10				
11				
12				
13				
14				
15				
16				
17				
18				
19				
20				
$\Sigma =$				
$\hat{N} = \Sigma \times$ area $\hat{N} =$				

How similar are these estimates to one another? You might want to compare these estimates using nonparametric statistics (Chapter 29).

Relative Estimates of Populations

Trap Lines and Trap Nights

Data obtained from trap lines or transects may yield relative estimates of population numbers. Generally, it is not possible to obtain density estimates since the size of the trapping area is not known. An assessment line placed at an acute angle to the transect (Figure 38-1B) can be used to obtain an estimate of the area sampled (Calhoun and Casby 1958 and review in Smith *et al.* 1975). A removal procedure seems best for estimating population size using this method (see Removal Trapping). A very crude estimate of population size is the number of animals captured per **trap night** (trap night = one trap set for one night). A determination of relative numbers is less useful than a determination of absolute density since no estimate is available of the area sampled, and differences in the behavior of the animals, the habitat, and the weather may make inter-area comparisons meaningless.

Other Relative Estimates

Population size may also be estimated by counting the number of animals that pass a single point (**point-area count**), those that are flushed along a transect (**flush census**), counted by the roadside (**roadside counts**), or by counting the maximum number of individuals observed along a transect at a point in time (**bounded counts**). The signs of mammals may also give an indication of relative population numbers. For example, counts of feces, runways, feeding residues, mounds, and other sign may serve as indices to relative population numbers (Neff 1968; Lord, *et al.* 1970). Many of these techniques of great importance for wildlife managers are discussed in detail by Overton (1969) and Caughley (1977). Sarrazin and Bider (1973) developed an activity index as a measure of relative density.

Population Estimates Based on Mark and Recapture

Estimates of population size can also be made by capturing animals and, after marking them with unique numbers, releasing them back into the population where a proportion will be subsequently recaptured. Although this technique is widely used for estimating population sizes of small mammals, the estimates obtained may not be very accurate (Roff 1973a; Caughley 1977). General reviews of these techniques can be found in Cormack (1968), Southwood (1968), Eberhardt (1969), Overton (1969), Roff (1973a, 1973b), Seber (1973), Delany (1974), Smith *et al.* (1975), Flowerdew (1976), and Caughley (1977).

Mark and recapture techniques not only provide estimates of population densities but information on other biological parameters. One can gather data on movement patterns, growth rates, age-specific fecundity and morality, combined rates of birth and immigration, combined rates of death and emigration, and rate of increase (Caughley 1977). Generally, these methods require a considerable amount of effort and expense. Thus, one should determine if these techniques are required to answer the biological questions posed or whether other less time-consuming and costly methods could be applied.

Assumptions for Use of Mark and Recapture Methods

All mark and recapture methods require that the following three assumptions be met (Delany 1974; Smith *et al.* 1975; Flowerdew 1976):

1. That the animals do not lose their marks.
2. That the animals are correctly recorded as marked or unmarked individuals.
3. That marking does not affect the probability of survival of the marked individuals as compared with that of unmarked individuals.

In addition, most methods may require that some of the following assumptions be met:

4a. That the population is *closed* and no gain or loss of members occurs due to natality, mortality, emigration, or immigration, *or*

4b. That the population is *open* with natality and immigration occurring but with mortality and emigration affecting marked and unmarked animals equally.

5a. That marked animals disperse randomly into the population and that every animal, both marked and unmarked, has the same probability of capture, *or,*

5b. That if different probabilities of capture exist, they are proportionally distributed among all marked and unmarked animals in the population.

The assumptions of 4a or 4b can be checked by experimental techniques (Smith *et al.* 1975) or an appropriate model chosen. Generally, the assumptions of 5a and 5b are most frequently violated (see review in Smith *et al.* 1969, 1975), and often cannot be statistically tested (Roff 1973b; Caughley 1977). Seber (1973) describes the available statistics and computational procedures for testing these assumptions.

Petersen (Lincoln) Index (Single Marking)

The **Petersen** or **Lincoln Index** is one of the simplest methods for estimating population size with mark and recapture data. Unfortunately, it also requires that a stringent set of assumptions (1,2,3, 4a, 5a) be met. In the single mark Petersen Index method, population size (N) at the time of marking is estimated by capturing and marking a sample of animals (M) on one occasion and then capturing a second sample (n) on a subsequent occasion and checking for the number of animals marked (m, i.e., recaptured). Then the population size is estimated using the following formula:

$$\hat{N} = \frac{Mn}{m}$$

Because $\hat{N}$ overestimates N by $1/m$, Bailey (1952) and Roff (1973a) recommended the use of the following formula:

$$\hat{N} = \frac{M\ (n + 1)}{m + 1}$$

For example, suppose that 50 animals were captured, marked, and released back into a population. Then, a short time later (to reduce bias due to immigration), the population is resampled and 100 unmarked and 25 marked individuals captured. Then, using the formula $\hat{N} = M\ (n + 1)/m + 1$, the population is estimated:

$$\hat{N} = \frac{50(125 + 1)}{25 + 1} = 242.$$

The standard error of this estimate can be approximated using the formula

$$s_{\bar{x}} = \sqrt{\frac{M^2\ (n + 1)\ (n - m)}{(m + 1)^2\ (m + 2)}}$$

Thus, for this example,

$$s_{\bar{x}} = \sqrt{\frac{(2500)\ (126)\ (100)}{(625)\ (27)}} = \sqrt{1866.67} = 43.2.$$

and $\hat{N} = 242 \pm 43$.

If the number of recaptures is determined in advance (i.e., inverse sampling), then see the formulae in Bailey (1951, 1952) and Caughley (1977) for computing $\hat{N}$ and $s_{\bar{x}}$.

How many animals should be marked and recaptured to provide an estimate of the population size with a given standard error? This varies and formulae for estimating adequate sample sizes may be found in Overton (1969), Smith *et al.* (1975), and other sources. Caughley (1977:144) provides a graph that can be used to determine the number of recaptures required to have a 10% standard error of the Petersen estimate of the population.

38-C. Obtain a bag containing at least 200 beans. Then, using a table of random numbers (Table 29-1), select a two-digit number. Draw this number of beans out of the bag, mark them with a distinctive dot (with pen, crayon, paint, etc.). Return them to the bag. Select another two-digit number from the table of random numbers, mix the marked and unmarked beans well, and then draw out (without looking!) a number of beans equal to the second two-digit number. Then, compute an estimate of the population size ($\hat{N}$) using the modified formula for the Petersen index.

$\hat{N} = M\ (n + 1)/m + 1$

$\hat{N} =$ ______.

What value did you obtain for $\hat{N}$? How close was this estimated value to the population parameter (N)? Is this an open or a closed population? Is this a realistic simulation of a natural population? Why or why not? Remove the marked beans, replace them with unmarked ones and repeat the exercise *without randomly mixing* the marked beans in with the unmarked (just drop them in). Did this affect the result (replicate several times for each method)?

Estimates Based on Multiple Marking Occasions (Deterministic)

The Petersen Index method requires that animals be marked on a single occasion. Often, this requirement results in too few marked animals for accurate estimates. To overcome this difficulty, Schnabel (1938) devised a method that is essentially a series of Petersen estimates. For ease in computation, the formula of Schumacher and Eschmeyer (1943) provides an explicit solution:

$$\hat{N} = \frac{\Sigma M_i^2\ n_i}{\Sigma M_i\ m_i}$$

This procedure, sometimes termed the **Hayne Method** was developed independently by Hayne (1949b) and requires the same set of assumptions as the Petersen Index Method. Refer to Overton (1969) and Caughley (1977) for additional information and formulae for calculating standard errors for these estimates.

Changes in the size of the population due to immigration and natality during the sampling period violate the assumptions of the Petersen-type estimates. **Bailey's triple-catch method** detects and counteracts the effect of immigration and provides estimates of birth rates and death rates in addition to population size (Caughley 1977). Roff (1973a) recommended against its use since a better estimator, the Jolly-Seber method, is available.

The Jolly-Seber Stochastic Method

All of the models for estimating population size that have been discussed thus far can be classified as *deterministic* since they assume constant rates of death, birth, immigration, and emigration. In contrast, *stochastic* models are more realistic since the rates of death (with emigration) and birth (with immigration) are estimated independently on each sampling occasion.

The **Jolly-Seber Method,** developed independently by Jolly (1965) and Seber (1965), is the most well known of the stochastic population models. This method requires that assumptions 1, 2, 3, 4b, and 5a be met. Again, the requirement for equal probability of capture is the most serious limitation for the use of this and all other mark and recapture methods. The arithmetic computations required for this method are somewhat tedious. White (1971) provides a FORTRAN program listing for making these calculations. Programs are also available at a number of research institutions and universities. Good reviews of the Jolly-Seber technique and computation procedures can be found in Southwood (1966), Overton (1969), Seber (1973), and Caughley (1977).

Removal Trapping and Catch Per Effort Methods

Assuming no births, deaths, immigration, or emigration and an equal probability of capture of individuals, the proportion of animals trapped in a population (assuming constant trapping effort) on succeeding occasions should decrease (Hayne 1949b; Zippin 1956; Grodzinski *et al.* 1966). These procedures can be used with studies involving actual removal of the individuals from the population (e.g., by kill-traps) or by mark and recapture methods. In the latter case, the marked individuals represent the animals "removed" from the population.

The **Zippin** (1956) **Method,** based on two trapping occasions, is one of the simplest removal trapping (catch per effort) methods. In this procedure, population size (N) is estimated using the following formula:

$$\hat{N} = \frac{n_1^2}{n_1 - n_2},$$

where n_1 = number of animals caught on the first day and n_2 = number of animals caught on the second day. For example, 200 traps were set for two nights. On the first day, the catch was 50 (n_1) and on the second day, 20 (n_2). Substitution into the above equation yielded the following estimate of population size (N):

$$\hat{N} = \frac{(50)^2}{50\text{-}20} \cong 83.$$

Seber (1973) provided additional discussion of this procedure.

The **Hayne** (1949b) **Method** uses data obtained over a number of trapping occasions. Smith *et al.* (1975) cautioned that for the use of this technique, or any other regression technique, a significant r-value must be obtained. Hansson (1969) discussed another technique based on removal catches that also provides a method for calculating the effective trapping area. The method of Kaufman *et al.* (1971), using assessment lines, compensates for unequal probabilities of capture.

Enumeration Method

Equal probability of capture is a severe practical limitation for the use of most mark and recapture methods of estimating population size. Consequently, several investigators, including Petrusewicz and Andrzejewski (1962), Krebs (1966), and Krebs *et al.* (1969) sought instead to conduct a complete census of resident animals on their trapping grids. Roff (1973b) sharply criticized such an approach since "The problems involved in a complete enumeration are vastly greater than even in the mark and recapture method and the complete inability to measure the accuracy of the estimates makes it a technique that cannot be accepted." This method is generally referred to as the **Calendar of Captures Method** or the **Minimum Number Known Alive.** Flowerdew (1976) provides a clear description of the technique.

Change in Ratio Methods

Changes in the relative abundance of two categories of animals (e.g., adults or subadults; males and females) can be used to estimate population numbers (Seber 1973). Hanson (1963) and Rupp (1966) presented a review of these **Change-in-Ratio Methods** but pointed out the general lack of procedures for determining confidence limits for these estimates. Paulik and Robson (1969) remedied this problem by providing formulae to estimate standard errors. These methods are widely utilized to study populations of game animals. Consult the above papers, Overton (1969) or Seber (1973) for more details.

Supplementary Readings

Caughly, G. 1977. *Analysis of vertebrate populations.* John Wiley and Sons, New York. 234 pp.

Cormack, R. M. 1968. The statistics of capture-recapture methods. *Oceanography and Marine Biology* 6:455-506.

Delany, M. J. 1974. The ecology of small mammals. *Inst. Biol. Studies Biology* 51:1-60. (Edward Arnold, Publishers, London)

Eberhardt, L. L. 1969. Population analysis, pp. 457-495. *In* R. H. Giles, Jr. (Ed.). *Wildlife management techniques,* 3rd ed. The Wildlife Society, Washington.

Flowerdew, J. R. 1976. Ecological methods. *Mammal Review* 6:123-159.

Overton, W. S. 1969. Estimating the numbers of animals in wildlife populations, pp. 403-455. *In* R. H. Giles, Jr. (Ed.). *Wildlife management techniques,* 3rd ed. The Wildlife Society, Washington.

Paulik, G. J., and D. S. Robson. 1969. Statistical calculations for change-in-ratio estimators of population parameters. *J. Wildl. Manag.* 33:1-27.

Roff, D. A. 1973a. On the accuracy of some mark-recapture estimators. *Oecologia* 12:15-34.

Roff, D. A. 1973b. An examination of some statistical tests used in the analysis of mark-recapture data. *Oecologia* 12:35-54.

Seber, G. A. F. 1973. *The estimation of animal abundance and related parameters.* Griffin, London. 506 pp.

Smith, M. H., R. H. Gardner, J. B. Gentry, D. W. Kaufman, and M. H. O'Farrell. 1975. Density estimations of small mammal populations, pp. 25-53. *In* F. B. Golley, *et al.* (Eds.). *Small mammals: their productivity and population dynamics.* Cambridge Univ. Press, London.

Southwood, T. R. E. 1968. *Ecological methods.* Methuen and Co., London. 391 pp.

Stenseth, N. C., A. Hagen, E. Ostbye, and H. J. Skar. 1974. A method for calculating the size of the trapping area in capture-recapture studies on small rodents. *Norwegian J. Zool.* 22:253-271.

39 Performing a Literature Search

The literature pertaining to mammals is voluminous. In 1976 Jones, Anderson and Hoffman reported that it included 115,000 separate titles and was increasing at a rate of five to six thousand papers per year. Obviously it is impossible for a mammalogist to be familiar with all that has been published; but when a scientist is engaged in a particular study, it is necessary for him to know all that others have written in that field. Thus the mammalogist frequently finds it necessary to conduct a literature search.

The Literature of Mammalogy

Many full-length books have been published about mammals but most of the literature of mammalogy is in technical journals. Jones and Anderson (1970) reported that fifty percent of the existing literature on mammals is contained in the volumes of about forty journals. The remaining fifty percent is widely scattered with at least 150 different journals needed to encompass only seventy percent of the literature.

Several journals are devoted exclusively to publications in the field of mammalogy. Some of these are highly specialized and encompass only one aspect of mammalogy or one taxonomic group of mammals. Others publish articles on all aspects of mammalogy. Some of these mammalogical journals are listed below:

Acta Theriologica. 1955-Present. Bialowieza, Poland. General.

Australian Bat Research News. 1964-Present. Lyneham, Australia. Bats.

Australian Mammalogy. 1961-Present. Lyneham, Australia. General.

Bat Research News. 1960-Present. Potsdam, New York. Bats.

Bibliotheca Primatologica. 1962-Present. Basel, Switzerland. Primates.

Cetology. 1971-Present. St. Augustine, Florida. Whales.

Folia Primatologica. 1963-Present. Basel, Switzerland. Primates.

Journal of Mammalogy. 1919-Present. Lawrence, Kansas. General.

Journal of the Mammalogical Society of Japan. 1952-Present. Tokyo, Japan. General.

Laboratory Primate Newsletter. 1962-Present. Providence, Rhode Island. Primates.

Lutra. 1959-Present. Gouda, Netherlands. General.

Lynx. 1960-Present. Praha, Czechoslovakia. General.

Mammalia. 1937-Present. Paris, France. General.

Mammal Review. 1970-Present. London, England. Review papers.

Myotis. 1953-Present. Bonn, Germany. Bats.

Nyctalus. 1969-Present. Halle, Germany. Bats.

Primate News. 1963-Present. Beaverton, Oregon. Primates.

Primates. 1958-Present. Aichi, Japan. Primates.

Säugetierkündliche Mitteilungen. 1953-Present. Munich, Germany. General.

The Scientific Report of the Whales Research Institute. 1948-Present. Tokyo, Japan. Whales.

Teriologia. 1972-Present. Novosibirsk, *USSR*. General.

Zeitschrift für Säugetierkünde. 1926-Present. Hannover, Germany. General.

Kosin (1972) reported that of the animal science papers covered by the 1969 *Biological Abstracts* 63.1% were originally published in English, 14.8% in Russian, 6.5% in German, 4.1% in French, 2.1% in Japanese, 2% in Spanish and the remaining 7.4% in other languages. Unfortunately similar data for the literature in mammalogy are not available but English is the dominant language, with Russian, German and French all being very important. Japanese and Spanish constitute significant percentages with the remaining languages of the world only very poorly represented. Many of the journals in the list above publish articles in English, German, or French. Some of the journals published in non-English speaking nations include a high percentage of papers in English (e.g., *Acta Theriologica*), but usually the native language of the country predominates. English summaries are frequently included with articles published in other languages.

Many diverse journals published in all parts of the world include papers on mammals as well as on other groups. In the United States alone the number of such journals is large. Most states and many cities have academies of science in whose publications mammal papers may appear (e.g., *Transactions, Kansas Academy of Science; Proceedings, California Academy of Science; Proceedings, Biological Society of Washington,* [D.C.]). Similarly, many major universities and museums have one or more publication series that may include mammal papers, e.g., *Occasional Papers, Museum of Zoology, University of Michigan; University of Kansas Publications, Museum of Natural History* (now replaced by *Occasional Papers* and *Miscellaneous Publications, University of Kansas Museum of Natural History*); *Bulletin* and *Novitates, American Museum of Natural History; Fieldiana: Zoology,* Field Museum of Natural History; *Proceedings* and *Bulletin, U.S. National Museum* (now replaced in part by *Smithsonian Contributions to Zoology*); *American Midland Naturalist,* University of Notre Dame. *Ecological Monographs, Ecology, Evolution, The Journal of Wildlife Management, Murrelet, Southwestern Naturalist,* and *Systematic Zoology* are a sample of U.S. journals that are published by various scientific and professional societies and that frequently include papers of direct interest to mammalogists.

39-A. Which of the journals devoted exclusively to mammals are present in your library? Examine the two most recent volumes of each. What languages predominate in each? Which include English summaries with articles published in languages other than English? Do any of the "general mammalogy" journals seem to have a predominance of articles on one particular aspect of mammalogy (e.g., systematics, ecology)?

39-B. Examine the most recent complete volume of each of several other biological journals in your library. List those that include articles that are within the scope of mammalogy.

39-C. List the ten journals which you think are most important to the aspect of mammalogy in which you are primarily interested.

Bibliographies and Abstracts

Several types of bibliographies are published that aid the mammalogist in keeping up with the literature of his field. Some of these are also useful in performing systematic literature searches.

Many of the mammalogical journals mentioned above include lists of articles on mammals published in other journals. Probably the most complete of these is the "Recent Literature" section of the *Journal of Mammalogy.* This is an alphabetical listing, by author, of articles that have been brought to the attention of the editors of this section of the journal. The coverage is quite extensive since many volunteers check specific journals and report on pertinent mammalogical literature. The "Recent Literature" list serves primarily to keep the professional mammalogist appraised of recent publications. And it functions well in this respect if the mammalogist thoroughly scans each issue as it appears to locate papers of interest to him. However, even though this list has been subdivided by general topics (e.g., behavior, disease and parasitism, distribution and faunas) since 1972, the absence of cross-indexing makes the list essentially useless for a complete search of the past literature on a particular topic.

Biological Abstracts is one of the most complete and up-to-date bibliographies in the biological sciences. As the title implies, **abstracts** of papers are included along with the bibliographic references. While *Biological Abstracts* (*BioAbstracts,* for short) is extremely useful for some fields, as Jones and Anderson (1970) point out, it is "quite incomplete for some branches of mammalogy." In addition a literature search using only *BioAbstracts* can be cumbersome and time consuming.

Current Contents® is a weekly publication of the Institute for Scientific Information (Philadelphia). It's speciality is reproduction of the tables of contents of a large number of scientific journals within a short period of time following their publication. The version *CC® Agriculture, Biology & Environmental Sciences* provides a good (though not complete) selection of journals that are of interest to mammalogists.

Wildlife Review is a publication of the U.S. Department of Interior that includes abstracts along with most of the bibliographic entries. Entries are arranged alphabetically by author under several categories and subcategories. *Wildlife Review* covers many American and foreign journals, and while it emphasizes game species, nongame species are also included. This source is particularly good for Soviet and East European journals. *Wildlife Abstracts* is issued at about ten-year intervals by the publishers of *Wildlife Review*. The title is a misnomer since this publication contains no abstracts, but is a subject cross-index to *Wildlife Review*.

Probably the most complete and useful bibliography of publications on mammals is *Zoological Record, Section 19, Mammalia*. Each section of *Zoological Record* is divided into three parts. The first is a list of complete bibliographic entries arranged alphabetically by author. The second part is a subject index to the papers listed in the first part and the third is a systematic (taxonomic) index to these same papers. The main drawback to *Zoological Record* has been the time involved in its production. Volume 110, for instance, covered the publications of 1973 but was not available until 1978. This delay has allowed the editors time to compile a very thorough list (*Section 19* of Volume 110, for example, covers about 4,557 papers).

Several other bibliographic sources are useful to the student of mammalogy. These include such diverse publications as printed card catalogs of great libraries (e.g., *The General Catalogue of Printed Books* of the Library of the British Museum, London), old bibliographic series now defunct (e.g., *Bibliotheca Historico-Naturalis* covering 1700-1846, *Bibliotheca Zoologica* covering 1846-1860, and *Bibliotheca Zoologica II* covering 1861-1880), and still active bibliographies of peripheral interest to mammalogists (e.g., *Index Medicus*). For a detailed discussion of these and other bibliographic sources see Smith and Reid (1972) *Guide to the Literature of the Life Sciences*.

39-D. Examine the "Recent Literature" section of the *Journal of Mammalogy*. Skim one complete volume for articles in your field of interest. How long did it take? How many articles did you find?

39-E. Examine *Biological Abstracts*. If you have never used this, have your instructor or librarian show you how to use the various indexes, including the author, taxonomic, and computer-rearranged subject (B.A.S.I.C.) indexes. While *BioAbstracts* can be cumbersome, it is an important resource. Every biologist should be able to use it.

39-F. Examine *Wildlife Review* and *Wildlife Abstracts*. How does this listing compare with the "Recent Literature" section of the *Journal of Mammalogy?* Compare the number of entries for game and nongame species of mammals.

39-G. Examine *Zoological Record, Section 19, Mammalia*. In what year did publication of this bibliography begin? What is the most recent year available? Compare the method of indexing in this resource with those of *Biological Abstracts* and *Wildlife Abstracts*.

39-H. Examine other similar resources that may be available in your library (e.g., *Current Contents, Citations Index, Index Medicus)*. Become familiar with the use of each of these.

Performing a Search

One of the most commonly used methods of performing a literature search is to find a recent, fairly thorough article on the subject in which you are interested and examine the bibliography at the end of the article. Then look up all of the articles listed in the first paper and examine their bibliographies. If this procedure is continued long enough, it is possible to locate most of the pertinent publications on a given topic. However, there may well be important papers that will be missed, and this type of seach can be extremely time consuming and repetitious.

To perform a thorough literature search, we recommend beginning with the most recent issue of *Zoological Record*. Check the appropriate index or indexes and prepare a 3 x 5 card on each pertinent reference. It is good practice to begin immediately to use a standardized system of bibliographic entry such as those outlined in the *CBE* (Council of Biology Editors) *Style Manual* (1972). For examples of bibliographic entries for a variety of kinds of publications, see the Bibliography of this manual.

After checking the most recent issue of *Zoological Record*, continue backward in time using this same source. *Zoological Record* has been published for over 100 years. A complete literature search would thus require examination of over 100 volumes, but this is usually neither practical nor necessary. The nature of the topic being researched will dictate the depth of time that must be explored. We recommend examining at least the latest ten volumes of *Zoological Record* and often it may be necessary to examine up to twenty-five volumes. Important papers older than this will usually, but not always, be mentioned in the bibliographies included in more recent papers.

Once you have completed with *Zoological Record,* it will be necessary to use another source to fill in the gap between the most recent issue of *Zoological Record* and the present data. Again depending upon the nature of the topic being researched, a variety of possibilities are available. *Biological Abstracts* may prove to be the most useful, but *Wildlife Review* or sources such as *Citations Index* may be helpful.

When you have completed the search of the bibliographic sources mentioned above, you will probably have a sizable stack of reference cards. Sort the cards by journal and locate each of the articles. You will probably be able to discard several references that are of only marginal interest. As you read each pertinent article, examine the bibliography included in it to locate important references that you may have missed in your search.

The procedure for performing a literature search described above is only one way of going about this activity. A truly complete literature search is essentially impossible and always impractical. Whitehead (1971:216) said that an average but conscientious systematic zoologist with access to a large library will retrieve only about 80% of the existing information.

As you will see after performing the exercises that follow, the coverages of *Zoological Record* and *Biological Abstracts* are not identical. Many papers will be included in one but missed by the other. Some papers will be missed by both but will appear in other bibliographic sources. The greater the number of bibliographic sources searched and the greater the depth of time examined, the more complete a search will become. However the time required to perform a full search of many sources is prohibitive. Because of the relatively better coverage by *Zoological Record* of papers of a systematic, anatomical, or ecological emphasis and because of the relative ease of using this source, we recommend the above literature search for basic use. However, if the primary research interest centers on other topics, such as genetics or physiology of mammals, *Biological Abstracts* may be a better tool for the in-depth search than *Zoological Record.* Availability may be another factor dictating which bibliographic sources will be used. *Biological Abstracts* will be present in the libraries of most U.S. institutions where biological research is being conducted, but *Zoological Record* may not be available. Thus, *Biological Abstracts* will be used as the main bibliographic source.

The main thing you must do is become thoroughly familiar with the literature and the bibliographic tools in your field. The scientific literature exists for the use of other scientists. If your publications are to be of merit, you must examine and utilize all previous pertinent literature.

39-I. Examine the most recent ten volumes of *Zoological Record, Section 19, Mammalia* and prepare a set of bibliography cards on one of the topics listed below or another topic approved by your instructor.

a. Gestation and care of the young in the polar bear.
b. The use of vision in microchiropteran bats.
c. Use of the eye lens as a tool for age determination in mammals.
d. Past and present distribution of the cheetah in Asia.
e. Methods of marking large ungulates for recognition at a distance.
f. Echolocation in mammals other than bats and cetaceans.
g. Fighting behavior of male wild goats.
h. The structure of the tympanic regions of the skulls of desert-dwelling rodents.
i. Seasonal variation in the diets of foxes.

39-J. Repeat the literature search in 38-I above for the same ten-year period using *Biological Abstracts.* How do the references obtained from the two different sources compare? How much time was required for each of the two searches? What are the advantages and disadvantages of each source?

39-K. Repeat the search conducted in 38-I and 38-J using *Wildlife Review.* How did time spent and results obtained compare with those of the previous searches?

39-L. Use all of the sources available to perform a relatively complete literature search on one of the topics listed in 38-I above or another topic approved by your instructor.

Computerized Searches

In recent years many organizations have begun to make computerized literature searches available to scientists in diverse fields. The data banks upon which most of these systems can draw are relatively small, but as time passes, they will increase in size and their potential value will increase correspondingly. The publishers of *Biological Abstracts* and many similar organizations offer individualized computer search services on various subjects.

Supplementary Readings

Blackwelder, R. E. 1972. *Guide to the taxonomic literature of vertebrates.* Iowa St. Univ. Press, Ames. 259 pp.

Council of Biology Editors, Committee on Form and Style. 1972. *C B E style manual,* 3rd ed. *Amer. Inst.* Biol. Sci., Washington. 297 pp.

Smith, R. C., and W. M. Reid. 1972. *Guide to the literature of the life sciences,* 8th ed. Burgess Publ. Co., Minneapolis. 166 pp.

Whitehead, P. J. P. 1971. Storage and retrieval of information in systematic zoology. *Biol. J. Linn. Soc.* 3:211-220.

Glossary

abdomen. In insects, the posterior or third division of the insect body, consisting of usually at least four segments; with no functional legs in the adult stage. See **head, thorax.**
Achilles tendon. Large tendon connecting the muscles of the calf to the calcaneum.
abdominal cavity. The largest body cavity. It contains the digestive and urogenital organs. In mammals, it is separated from the thoracic cavity by a muscular diaphragm.
absolute age. Age of individual in terms of real time. Compare **relative age.**
abstract. A brief, concise statement or set of statements summarizing the significant content and conclusions of a scientific publication.
accuracy. The nearness of a measurement to the true value.
acetabulum. Socket in pelvic girdle at point where ilium, ischium, and pubis meet, and into which the head of the femur articulates.
adjusted range length. The range length plus one-half the distance to the next trap added onto each end. See Fig. 37-4E.
adult. Generally, the sexually mature and breeding individuals in a population.
adult pelage. The type of hair covering characteristic of adults of the species.
aerial. Pertaining to mammals that have the capacity for sustained flight. Bats are the only aerial mammals. See Figs. 8-18, 8-19.
age-specific life table. A life table prepared by following a cohort of individuals through time. Compare with **time-specific life table.**
agglutinated. Stuck together, clumped, e.g., the agglutinated hairs of rhino horn.
agouti hair. A hair with alternate light and dark color banding.
alarm pheromone. A pheromone that signals an alarm or danger. See **pheromone.**
albinism. The absence of all external pigmentation, typically inherited as a Mendelian recessive.
albino. An animal lacking all external pigmentation. See **albinism.**
alisphenoid bone. One of the paired bones that form part of the lateral walls of the braincase, located anterior to the squamosal in the temporal fossa. See Fig. 2-1B, C.
alisphenoid canal. A tubular passageway beneath an arch of bone near the base of the alisphenoid bone, through which a blood vessel passes. Found only in some groups of mammals. See Figs. 2-1B, C; 23-8.
allantois. An extraembryonic membrane of reptiles, birds, and mammals. A saclike outgrowth of tissue that extends outward from the gut of a developing embryo. In mammalian embryos it is situated between the **chorion** and **amnion** and functions in respiration, excretion, and nutrition. Fig. 9-10.
allopatric. Pertaining to disjunct (Fig. 11-2C) geographic ranges of two or more taxa.
allotype. A type specimen of the sex opposite to that of the holotype.
alternative hypothesis (abbreviated $\mathbf{H_A}$). The operational statement or hypothesis that the researcher is testing. See also **null hypothesis.**
altitudinal migration. Migration by mammals to different elevations in response to seasonal availability of food or climatic conditions.
altricial. Pertaining to newborn mammals that require prolonged parental care for survival. Usually born blind and naked.
alveolar sheath. See **incisive alveolar sheath.**
alveolus. A socket in the jaw bone for the root or roots of a tooth.
ambulatory. Pertaining to walking, e.g., the ambulatory locomotion of a bear.
ammonium hydroxide (NH_4OH). Caustic chemical that in diluted form is used to clean and partially degrease skeletal material.
ampullary glands. Small accessory reproductive glands in some male mammals, generally paired, with ducts that enter the vas deferens near the vestibule of the ampulla. See Fig. 9-5.
analysis of variance (abbrev. **AOV, ANOVA**). A statistical technique for segregating the sources of variability affecting a set of observations.
angora. Pertaining to hair that has continuous growth.

angular process. The posterior ventral projection of the dentary. See Fig. 2-2.

ankle. The joint in the hind limb between the tibia and fibula of the lower leg and the tarsal bones of the foot.

annual molt. A once yearly shedding and replacing of hair in mammals.

antebrachium. Forearm, between wrist and elbow joints.

antennae. In Arthopoda, the paired segmented sensory organs located on the head. In insects, one is attached on each side of the head; commonly called feelers.

anterior. Of, pertaining to, or toward the front end.

anterior nares. Anterior openings of the nasal passages in the skull.

anterior prostate glands. See **coagulating glands.**

antitragus. Small fleshy projection on the postero-ventral margin of the pinna of some bats. See Fig. 18-3D.

antler. A branched bony head ornament found on deer. Covered with skin (velvet) during growth. Shed annually. See Figs. 5-5 through 5-9.

antrum. The fluid-filled space in the Graafian follicle in which the egg develops. See Fig. 9-8.

awns. Guardhairs of relatively uniform length with slender bases and expanded tips.

baculum. A bone in the penises of certain mammals. See Fig. 9-7.

Bailey's triple-catch method. A deterministic mark and recapture method of population estimation that corrects for immigration, mortality, and natality during the period of sampling.

baleen. Whalebone. The cornified epithelial plates suspended from the upper jaws of mysticete whales. Used to strain food from the water. See Figs. 22-7, 22-8.

barbs. Small projections on a spine that prevent the spine, once embedded, from being easily removed. See Fig. 4-4.

bar graph. Vertical bar graphical representation of discrete, ranked, or attribute frequency data where each bar is separated from adjacent bars.

basioccipital. See **occipital bone.**

basisphenoid. The bone forming the floor of the braincase. Anterior to the basioccipital and posterior to the presphenoid and pterygoids. See Fig. 2-1B.

beam. The main trunk of an antler. See Fig. 5-8.

beast of burden. An animal used to carry cargo or other loads.

bez tine. The first tine above the brow tine of an antler. See Fig. 5-8.

bibliography. A collection of references to publications, usually arranged in alphabetical order by the last name of the primary or only author and usually covering a stated topic, time period, or both.

bicornuate uterus. Type of uterus in some eutherian mammals that has a single cervix and the two uterine horns fused for a part of their length. See Fig. 9-4C.

bicuspid. 1. A tooth having two major cusps. **2.** In human anatomy, a premolar.

binomen. In nomenclature, a set of two Latin or Latinized words that form the technical name of a species. The generic name and specific name together.

binomial distribution. The theoretical distribution or probability distribution of events that can occur in two classes.

binomial nomenclature. A system of providing a unique, two-word name for a species, consisting of a generic and a specific name.

biological species. See **species.**

bipartite uterus. Type of uterus in eutherian mammals which is almost completely divided along the median line, with a single cervical opening into the vagina. See Fig. 9-4B.

bipedal. Pertaining to locomotion on only two legs.

birth-flow. Model describing birth frequencies in mammals where breeding is constant throughout the year. See Caughley (1977).

birth-pulse. Model describing birth frequencies in mammals where the period of births is seasonal. See Caughley (1977).

bivariate home range. Home range estimate in two dimensions, formulated into general model by Koeppl et al. (1975).

blastocyst. An early embryo consisting of 8 to 16 cells.

borax. A crystalline, slightly alkaline borate of sodium, $Na_2B_4O_7$, used for drying and preserving skins. See Chapter 35.

boundary strip. An area that should be added to the basic size of a trapping grid so that an effective trapping area can be determined. The width of the boundary strip can be estimated using assessment lines or by other means.

boundary strip method. In home range estimates, an area equal in width to half the distance between traps is added around the minimum area. In the *inclusive boundary strip* method (Fig. 37-4B), the peripheral points of capture are considered centers of rectangles (each side of which equal the distance between traps) and the home range area is delineated by connecting the exterior corners of these rectangles to form a maximum estimate of the space utilized. In the *exclusive boundary strip* method (Fig. 37-4C), the boundary strip rectangles are drawn in a manner to minimize the area enclosed.

bounded counts. Census made by counting the maximum number of individuals observed along a transect at a point in time.

brachiation. A method of locomotion involving movement by swinging from one handhold to another, usually through trees. See Fig. 8-14.

brachium. Upper arm between shoulder and elbow.

brachyodont. Pertaining to teeth that have low crowns.

bracket key. A key in which the parts of each couplet are in immediate succession. See Fig. 12-1.

braincase. 1. That portion of the skull that encloses and protects the brain. **2.** The skull posterior to a plane drawn vertically through the anterior margins of the orbits. See **rostrum.**

bristles. Overhairs with angora growth; frequently stiff and wiry.

browsers. Mammals that feed on tender shoots or twigs of shrubs and trees.

brow tine. The first tine above the base of an antler. See Fig. 5-8.

buccal. 1. Pertaining to the cheeks, e.g., the buccal side of a tooth is that side closest to the cheek. **2.** In some usages, pertaining to the mouth as a whole.

bulbo-urethral glands. Cowper's glands. In male mammals, the small paired glands that secrete mucus into the urethra at the time of sperm discharge (Fig. 9-5).

bulla. See auditory bulla.

bunodont teeth. Low-crowned teeth with roughly hemispherical cusps. See Fig. 3-13.

burr. Enlarged, rugose area at the base of an antler where it joins the pedicel. See Fig. 5-8.

burrow. A tunnel excavated and inhabited by an animal.

caecotrophic pellets. Soft fecal pellets, rich in vitamin B_1, produced in the caeca of most species of lagomorphs. After the pellets are voided, they are reingested by the animal so that the vitamin can be absorbed. See **reingestation.**

calcaneum. Calcaneus. The heel bone, the largest and posteriormost tarsal bone. See Fig. 7-4.

calcar. A spur (of cartilage or bone) that projects medially from the ankle in many species of bats and from the wrist of many gliding mammals and that helps support a patagium. See Fig. 18-1.

calendar of captures method. See **minimum number known alive.**

calipers. Instrument for taking precise measurements. Available in two basic types–those in which the reading is taken from a dial and those in which it is read from a vernier scale.

callus. Any hard, thickened area of the epidermis.

camouflage coloration. A color pattern that allows a mammal to blend with the background and renders it less visible to predators or prey.

canal. A tubular passage or channel in a bone or other tissue.

canine. One of the four basic kinds of teeth found in mammals. The anteriormost tooth in the maxilla (and its lower counterpart in the dentary). Frequently elongated, unicuspid, and single rooted. Never more than one per quadrant.

caniniform. Pertaining to an incisor or premolar that has the shape and appearance of a canine.

cannon bone. Fused metatarsals or metacarpals. See Fig. 28-2B, C.

capacitation. Physiochemical changes that take place in spermatozoa to enable them to penetrate the protective covering of cells surrounding an oocyte (see Austin 1972).

carcass. The portion of the body of an animal that remains after the removal of the skin.

carnassial dentition. See **secondont dentition.**

carnassial pair. The last upper premolar and first lower molar of living mammals with secondont (=carnassial) dentition. The largest pair of bladelike teeth that occlude with a scissorslike action. Possesed by most members of the order Carnivora. See Fig. 23-3.

carnassial teeth. See **carnassial pair.**

carnivore. A mammal that consumes meat as the primary component of its diet.

carpal. Any one of the group of bones forming the wrist joint in the skeleton of the forelimb. Distal to the radius and ulna and proximal to the metacarpals. See Figs. 7-1, 7-2.

cased skin. A mammal skin prepared for tanning by making an incision from the heel of one hind foot across the urogenital region and down the heel of the leg on the opposite side.

catalog. The portion of the field notes in which full data are entered for each specimen preserved. All entries are consecutively numbered. See Fig. 33-2.

category. See **taxonomic category.**

caviomorphs. Rodents of the New World (especially of South America) that have a greatly enlarged infraorbital foramen.

cementum. The layer of bonelike material covering the root of a tooth. Sometimes termed cement. See Fig. 3-1.

center of activity. The mean of a set of capture coordinates or observation coordinates.

central tendency. In statistics, the localization of values near a central point.

cercus (pl. cerci). In arthopods, an appendage, usually paired, of the terminal abdominal segment, usually slender, filamentous and segmented, sometimes heavily sclerotized. See Fig. 36-1A.

cervix. The tip of the uterus that sometimes projects into the vagina. See Figs. 9-2, 9-4.

cf. confer. Compare

cheek teeth. Collectively, the premolars and molars or any teeth posterior to the position of the canines.

change-in-ratio method. A procedure for determining population densities by observing changes in the relative abundance of two categories of animals such as adults and subadults or males and females.

character. See **variable.**

chelate-dentate. Mouth parts of Acarina consisting of a movable digit or chelicera and an immovable digit, one or both digits bearing "teeth."

chelicerae. The pincer-like first pair of appendages of adult Acarina.

Chordata. The animal phylum distinguished by a hollow dorsal nerve cord, notochord, and pharyngeal gill slits or pouches at some stage in the life history includes some primitive marine forms, fishes, amphibians, reptiles, birds, and mammals.

chorioallantoic placenta. Type of placenta composed of two extraembryonic membranes—an outer chorion and an inner vascularized allantois. Found in all eutherian mammals and bandicoots of the order Peramelina.

chorion. The outermost extraembryonic membrane of reptiles, birds, and mammals. In eutherian mammals the chorion contributes to the formation of the placenta. See Fig. 9-10.

choriovitelline placenta. Type of placenta involving the chorion and yolk sac and found in all marsupials except bandicoots of the order Peramelina.

cingulid. See **cingulum.**

cingulum. An enamel shelf, frequently with cusps, that borders one or all margins of an upper tooth. A *cingulid* is its counterpart in the lower tooth. See Fig. 3-10.

circular overlap. In polytypic species, when a chain of contiguous and intergrading populations curve back until the terminal populations overlap one another and, in the area of sympatry, exhibit the characteristics of distinct species.

cladism. See **cladistics.**

cladist. A person engaged in the practice of cladistics.

cladistics. A method of developing phylogenies based upon the branching sequences of evolution.

cladogram. A two-dimensional, pictorial representation of the inferred branching sequences of a phylogeny, based on the methods of cladistics. See Fig. 11-7.

class. A major subdivision of a phylum, consisting of one or more orders. See Fig. 11-1.

classification. The assignment of groups to taxa.

clavicle. A ventral bone of the pectoral girdle. Reduced or absent in many mammals. The collarbone in man.

claw. The most common form of digital keratinization found in mammals. Usually long, curved, and sharply pointed. See Fig. 6-1.

cline. A character gradient; a gradual change of a character through a series of interconnecting populations.

clitoris. Female homolog of the glans penis. In some mammals, supported by a small bone, the *os clitoris.* See Fig. 9-1.

cloaca. A chamber into which the digestive, reproductive, and urinary systems empty and from which the products of these systems leave the body. See Fig. 9-2.

clumped dispersion. Individuals are concentrated in some areas and absent from other areas. See Fig. 37-1B.

coagulating glands. Anterior prostate glands of some mammals; when present, the most anterior of the prostate glands, with ducts that enter the anterior portion of the prostatic urethra; upon ejaculation, the secretion of these glands, when mixed with the secretions of the vesicular glands, sometimes form a viscous substance that constitutes a "copulation plug" left in the vagina. See Fig. 9-5.

coefficient of variation. The standard deviation expressed as a percentage of the mean $CV = (s)(100/\bar{X})$. Used to minimize effects of size when comparing the variability of organisms that differ greatly in size (e.g., the variability of dimensions of a mouse can be compared with that of a horse).

coil spring trap. A type of steel leg trap which utilizes coil springs to maintain pressure on the jaws.

coitus. See **copulation.**

colostrum. Fluid produced by the mammary glands during the first few days before and after parturition. Rich in proteins and antibodies.

columella. Bone that transmits vibrations from the tympanum to the inner ear in reptiles, birds, and anurans, homologous with the hyomandibular of fishes and the stapes of mammals.

commissure. Any sharp, crescent-shaped crista found on cheek teeth.

comparison. See **differential diagnosis.**

composite life table. Life table prepared by using data obtained from time-specific and age-specific procedures. See **life table.**

concealing coloration. Protective coloration facilitating concealment.

condylarth. One of the primitive ungulate mammals of the order Condylartha, from the Paleocene and Eocene epochs, having a slender body, low-crowned teeth, and five-toed feet, each toe ending in a small hoof.

Conibear trap. A steel kill trap used primarily to collect squirrels and the smaller carnivores. See Fig. 34-4.

conspecific. A condition where populations or specimens are not sufficiently distinct to be considered separate species.

continental drift. The movement of the continental plates over the earth's mantle. See **plate tectonics.**

continuous estrus. Situation where females are receptive to males throughout the year and mating may induce reproductive and hormonal changes.

continuous variable. See **variable.**

Convention on International Trade in Endangered Species of Wild Fauna and Flora. International convention regulating the importation of threatened and endangered species of wildlife. See Chapter 34.

convergence. The occurrence of two derived characters arising independently in two only distantly related groups. See Fig. 11-6D

convex polygon. Home range estimate where the capture points are connected to form the smallest convex polygon. See Fig. 37-4A.

copulation. Coitus. The union of male and female reproductive organs to facilitate reception of sperm by the female.

copulation plug. A plug of coagulated semen formed in the vagina after copulation. Found only in certain mammalian species. See **coagulating glands.**

coracoid. One of the elements of the pectoral girdle in lower vertebrates. Rudimentary and fused to the scapula in most marsupial and placental mammals.

core area. Area of intensive use by a group within the overall group home range.

cornification. See **keratinization.**

cornu. A horn or hornlike projection, e.g., horn of the uterus.

coronoid process. The projection of the posterior portion of the dentary that is dorsal to the mandibular condyle. See Fig. 2-2.

corpora quadrigemina. Four oval masses that serve as centers of optic and auditory reflexes and form the dorsal part of the mesencephalon in the brain of mammals.

corpus albicans. The degenerated corpus luteum formed after birth of the fetus or after the egg fails to implant in the uterus.

corpus luteum. A mass of yellowish, glandular tissue formed from the Graafian follicle after ovulation. See Fig. 9-8.

correlation analysis. An investigation of the degree of association between pairs of variables, but implying no cause and effect relationship between these variables.

cortex. Any outer layer or rind, such as the outer layer of the mammalian ovary or the cortex layer of the hair.

coterie. A "family" group consisting of a male, several females, and their young.

cotyledonary placenta. Type of chorioallantoic placenta that is distributed in discrete patches around the embryo. See Fig. 9-10B.

cotyledons. Distinct patches of attachment between the maternal and embryonic portions of the placenta.

cotype. See **syntype.**

countershading. Coloration of an animal with parts normally in shadow being light or parts normally illuminated being dark.

couplet. Each pair of mutually exclusive alternatives that collectively compose a biological key.

Cowper's gland. See **bulbo-urethral gland.**

cranium. See **skull.**

crepuscular. Pertaining to the twilight periods of dusk and dawn.

crista. A crest or ridge on a tooth, frequently identified by a distinctive name, such as entocrista. A *cristid* is the corresponding structure on a lower tooth.

cristid. See **crista.**

cropper. According to McNab (1963: 136), mammals that are principally grazers or browsers.

crown. The portion of a tooth extending above the gumline. See Fig. 3-1.

cryptic coloration. See **concealing coloration.**

cumulative frequency polygon. Continuous frequency data represented by line connecting points and showing the contribution of particular values to overall totals. See Fig. 29-4.

cursorial. Pertaining to running. Cursorial locomotion is running locomotion.

cusp. A point, projection, or bump on the crown of a tooth, usually distinguished by a particular name, such as hypocone. See Fig. 3-10.

cuticle. The thin outer layer of a hair. See Fig. 4-1.

dactylopatagia. In a bat, thin webs of skin that fill the spaces between the digits of the forelimb as follows: *dactylopatagium minus* (between digits 2 and 3), *dactylopatagium longus* (between digits 3 and 4), and *dactylopatagium latus* (between digits 4 and 5). See Fig. 18-1.

data (singular datum). Observations or measurements taken on a sampling unit.

deciduate placenta. The type of placenta in which a portion of the uterine wall is torn away at parturition.

deciduous dentition. The juvenal or milk dentition of mammals. Consists of incisors, canines, and most premolars. Replaced by permanent or adult dentition.

deciduous teeth. Milk teeth; teeth that appear first in the lifetime of a mammal and that generally are replaced by permanent teeth. See **deciduous dentition.**

definitive hair. Hair which grows to a certain length and is then shed and replaced.

degreasing. Process whereby fats and oils are removed from bones with the use of various chemical agents. See Chapter 35.

degrees of freedom. A quantity, generally *N*-1, that corrects for bias when certain sample statistics are computed. Abbreviated *df.*

delayed development. Found in some Neotropical bats and characterized by a slowing down of the growth rate of the embryo following implantation.

delayed fertilization. An adaptation of certain species of hibernating bats where mating occurs in late summer but fertilization does not occur until the following spring.

delayed implantation. Phenomenon found in most temperate species of the Mustelidae and several other kinds of mammals where implantation of the embryo, and consequent cessation of further growth, is delayed for several months.

den. A cave, hollow log, burrow, or other cavity used by a mammal for shelter.

density. See **population density.**

density dependent. The effect on the size of a population that varies with the number of individuals per unit area in the population.

density independent. The effect on the size of a population that does not vary with the number of individuals per unit area in the population.

dental formula. A convenient way of designating the number and arrangement of mammalian teeth; e.g., I 3/3, C 1/1, P 4/4, M 3/3. The letters indicate incisors, canines, premolars, and molars, respectively. The numbers above the line indicate the number of teeth on one side of the upper jaw; those below the line indicate the number of teeth on one side of the lower jaw.

dentary. One of the pair of bones which comprise the lower jaw (mandible) of mammals. See Fig. 2-2.

dentine. A hard, generally acellular material between the pulp and enamel portions of a tooth. See Fig. 3-1.

dermal bone. Bone that forms without a cartilage precursor, e.g., some skull bones.

dermal papilla. An extension of the dermis up into one of the hollow, hairlike fibers that compose a rhinoceros horn. See Fig. 5-11.

dermal scale. A bony scale that originates in the dermal layer of the skin. Found primarily in fishes.

dermestid. A beetle of the family Dermestidae that feeds on dried flesh and tissues. The beetles are utilized in many institutions since the larvae are very efficient in cleaning skeletal material. See Fig. 35-29.

dermis. The deeper of the two layers of the integument. Composed primarily of connective tissue. Also contains vascular, adipose, nervous, and other types of tissue. See Fig. 4-1.

description. A statement of characters and supplementary information that accompanies the naming of a new taxon. See also diagnosis and differential diagnosis.

descriptive statistics. Statistics that provide a useful numerical summary on the properties of an observed frequency distribution. Included are such measures as the **sample mean, sample size, degrees of freedom, range, variance, standard deviation, standard error of the mean,** and **coefficient of variation.**

determinant of covariance matrix. Estimate used by Jennrich and Turner (1969) to measure noncircular home ranges of animals.

diagnosis. A statement of the characters that serve to distinguish a taxon from similar or closely related taxa.

diaphragm. The septum dividing the abdominal and thoracic cavities, muscular in mammals.

diastema. A space between adjacent teeth. For example, the space between incisors and premolars in species lacking canines. See Fig. 2-6 between points A and B.

Dice-gram. A two-dimensional, graphic representation of descriptive statistics, generally showing mean, range, standard deviation, and standard error of the mean. See Fig. 29-5.

Dice-Leraas diagram. See **Dice-gram.**

dichotomous. Pertaining to a division into two equivalent or similar branches.

didactylous. Pertaining to manus or pes that contains only two digits, such as those occurring in the two-toed sloth, *Choloepus hoffmanni.*

diestrous. Pertaining to those species that have two estrous cycles each year.

diestrus. Final stage of estrous cycle where cornified cells are rare and some mucus may be present in a vaginal smear. In this stage, progesterone levels increase, reach a peak, and then decline.

dietary overlap. The extent to which the diets of two organisms, populations, or species are similar.

differential diagnosis. A comparison of characters of a taxon being named with the characters of other specifically mentioned taxa of equivalent rank.

diffuse placenta. Type of chorioallantoic placenta that completely surrounds the embryo. See Fig. 9-10A.

digit. Any finger or toe.

digitigrade. Pertaining to walking on the digits, with the wrist and heel bone held off the ground. See Fig. 8-1B, 8-2.

dilambdodont tooth. Type of tooth characterized by a W-shaped ectoloph on the occlusal surface. See Fig 3-12.

diphyodont. Having two sets of teeth: a milk or deciduous set and then a permanent set.

diprotodont. The condition that exists in the Paucituberculata and Diprotodonta. The lower jaw is shortened and the first lower incisors are greatly elongated to meet the upper incisors. Fig. 15-2. Compare **polyprotodont.**

discoid placenta. Type of chorioallantoic placenta that resembles a disc, plate (Fig. 9-10D), or cup (Fig. 9-10E).

discontinuous variable. See **variable.**

dispersal. The permanent emigration of individuals from a population.

dispersion. In statistics, the scatter of values from a central point. In spatial terms, the distribution of animals in an area. See **random dispersion, uniform dispersion, clumped dispersion.**

disruptive coloration. Coloration pattern that causes visual disruption because the pattern does not coincide with the shape and outline of the animal's body.

distal. Distad. Situated away from the base or point of attachment.

distance between observations index. Home range index developed by Koeppl et al. (1977b) for use when data are not normally distributed.

diurnal. Pertaining to the daylight hours. Opposite of nocturnal.

DOR. Abbreviation for "Dead on Road." Used in field records to indicate specimens found in this condition.

dorsal. Pertaining to the back or upper surface.

dorsal fin. In mammals, a mid-dorsal projection of fibrous connective tissue found only in certain cetaceans. See Fig. 22-33.

duplex uterus. Type of uterus in which the right and left parts are completely unfused and each has a distinct cervix. See Fig. 9-4A.

dynamic life table. See **age-specific life table.**

eccrine sweat glands. See **sweat glands.**

echolocation. Sonar; the process of locating objects by emitting sound pulses and receiving and identifying the echos of those sounds reflected by the objects. Used by most bats and cetaceans.

ecto. Abbreviation for ectoparasite.

ectoloph. A crista, generally connecting parastyle, paracone, metacone, and metastyle. See Figs. 3-11, 3-12, 3-14.

ectoparasite. A parasite on, or communicating in direct contact with, the external surface of an animal, e.g., fleas, lice, ticks.

edentulate. Lacking teeth.

edge effect. The discrepency between the population estimates of animals living on the grid and those living partly on the edges of the grid.

edge sort card. A stiff card that has holes punched at regular intervals along its edges. By notching particular locations according to a definite scheme, small quantities of data can be coded.

elbow. The joint in the forelimb between the humerus and the radius and ulna.

embrasure. The space between two teeth in one half of one jaw into which a tooth in the opposing jaw fits when the teeth are occluded. See Fig. 3-8B.

embrasure shearing. A cutting or scissorslike shearing action produced by two teeth moving past each other in occlusion. Characteristic of the teeth of several early groups of mammals (see Fig. 3-5) and the carnassial teeth of carnivores.

embryonic diapause. See **aseasonal delayed implantation.**

emigration. The process whereby individuals move away from a population.

empodium. The median structure between the claws of arthropods arising from the pretarsus, pad-like, sucker-like, etc. See Fig. 36-16.

enamel. Extremely hard outer layer on the crown of a tooth, consisting of calcareous compounds and a small amount of organic matrix. Usually white but sometimes brown, red, or yellow in rodents and some other mammals. See Fig. 3-1.

enamel island. A ring of enamel surrounded by dentine. Typically found in the teeth of some rodents, particularly the hystricomorphs. See Fig. 25-20A.

Endangered Species Act. U.S. Law, passed in 1973 and subsequently amended, that regulates the capture, possession, and sale of threatened and endangered species of wildlife. See Chapter 34.

entoconid. In lower cheek teeth, a cusp on the anterior, lingual side of the talonid. See Fig. 3-10.

entotympanic. Bone surrounding the middle ear cavity. Sometimes fused with the tympanic to form a compound auditory bulla.

enucleate. Lacking a nucleus. Mammalian red blood cells (erythrocytes) are enucleate.

enumeration data. See **variable.**

enzyme digestion. Use of enzymes, such as trypsin or papaine, to clean bones. See Chapter 35.

epidermal scales. Keratinized scales formed from the epidermis.

epidermis. The surface layer of the integument exterior to the dermis. Composed entirely of epithelial tissue. See Fig. 4-1.

epididymis. Coiled duct that receives sperm from the seminiferous tubules of the testis and transmits them to the vas deferens. See Fig. 9-5.

epipubic bones. Paired bones that project anteriorly from the pelvic girdle into the abdominal body wall of most marsupials and monotremes. See Fig. 15-5.

epithelial tissue. The general tissue type that covers a body or structure or lines a cavity.

equal probability of capture. When the probability of capturing any individual is equal to the probability of capturing any other indivdual.

erythrocytes. Red blood cells.

estrous cycle. A sequence of reproductive events, including hormonal, physiological, and behavioral, that typically occur at regular intervals in a female mammal. Generally divided into four stages: **proestrus, estrus, metestrus, diestrus.**

estrus. A period of time when female mammals may accept males and mating occurs. In specific terms, when ovulation occurs. At this time, the pituitary output is predominantly LH and the newly formed corpus lutêum is producing large quantities of progesterone. In vaginal smears, the cells are cornified.

Ethiopian Region. The faunal region which includes all of Africa south of the Sahara Desert, the southern portion of the Arabian Peninsula, and the island of Madagascar. See Fig. 1-3.

eumelanin. The pigment in the mammalian hair and integument that produces various browns and black.

euthemorphic tooth. A modified tribosphenic tooth. Frequently square in outline, but modified in other ways in different groups of mammals. See **dilambdodont** and **zalambdodont teeth.**

Eutheria. The one of the three infraclasses of mammals in the subclass Theria that includes the true placental mammals.

evolutionary or **traditional systematics.** Philosohpy of systematics where the goal is to discover the natural groups that are the current products of evolutionary lines.

exoccipital. See **occipital bone.**

exploratory movements. Movements by individuals when they are placed in a novel or strange environment.

exponential growth. Growth, especially in the number of organisms, that is a simple function of the size of the growing entity; the larger the entity, the faster it grows (Wilson 1975).

expressed. Pertaining to the mechanism by which milk is released from the nipple (or teat) through action of the tongue of the young pressing against the hard palate.

external auditory meatus. Canal leading from the surface of the head to the tympanic membrane. See Fig. 2-1C.

fallopian tube. See **oviduct.**

family. A category within an order, consisting of one or more genera. See Fig. 11-1.

family-group. Collectively, the categories tribe, subfamily, family, and superfamily.

Faunal Regions. Divisions of the earth's land surface based upon broad faunal differences and similarities. See Fig. 1-3.

fecundity, m_x. Rate at which an individual produces offspring, usually expressed only for females.

fecundity table. A table, frequently part of a life table, that lists fecundity (m_x) schedules for different ages.

femur. The single bone of the upper (proximal) portion of each pelvic limb. See Fig. 7-3.

fibula. The lateralmost of the two bones in the lower (distal) portion of each pelvic limb. See Fig. 7-3.

field journal. See **journal.**

field notes. Collectively, the journal, catalog, and species accounts.

finite rate of increase, λ. The ratio of population densities in two successive years. Sometimes termed the growth multiplier.

first reviser. The first taxonomist who selects and publishes one of two names for a taxon that were previously equally available under the International Code.

Fitch trap. Live trap constructed of hardware cloth, galvanized metal, and a metal can. Easily made, using economical materials and generally useful for capturing most kinds of small rodents. See Fitch (1950) for details.

flippers. Feet fully adapted for an aquatic life. Digits elongate and fully webbed. See Figs. 8-12, 8-13, and 26-9, for examples.

fluid-preserved. Pertaining to specimens fixed and/or preserved in a fluid such as formalin or alcohol. See Table 35-1.

flukes. Lateral projections of a whale's tail. Supported entirely by fibrous connective tissue (no skeletal support). See Fig. 22-1.

flush census. Census of individuals that are flushed along a transect.

follicle. A small cavity or pit, e.g., Graafian follicle, hair follicle.

follicle stimulating hormone, FSH. A pituitary hormone that causes follicles to increase in size and, together with LH, to produce estrogen and bring about ovulation.

foramen (pl. foramina). Any opening, orifice, or perforation, especially through bone.

foramen magnum. The large opening at the rear of the skull through which the spinal cord enters the braincase. See Fig. 2-1B.

form. A depression in vegetation or soil used as a resting place by certain kinds of mammals.

fossa. A pit or depression in a bone. Frequently a site of bone articulation or muscle attachment.

fossorial. Pertaining to life under the surface of the ground. See Figs. 8-7, 8-8.

frass. Feces and shed larval skins that accumulate in a colony of dermestid beetles.

frequency distribution. A systematic arrangement of statistical data that exhibits the division of the values of the variables into classes and that indicate the frequencies or relative frequencies that corresponds to each of the classes.

frequency polygon. Continuous frequency data represented by line connecting points. See Fig. 29-3.

frequency table. A tabular representation of frequency data arranged by classes and frequency of occurrence. See **frequency distribution.**

friction ridges. Ridges in the epidermis that serve to increase friction (and thus traction) on naked areas, such as the soles of the feet.

frog. The pad in the central area of a hoof. See Fig. 6-3.

frontal bones. The anteriormost pair of bones in the roof of the braincase. Situated between the orbits anterior to the parietals, and posterior to the nasals and maxillae. See Fig. 2-1A, C.

frugivorous. Feeding on fruit.

FSH. See **follicle stimulating hormone.**

fundamental niche. The theoretical niche of a species in the absence of all other species. See Hutchinson (1957).

fur. Dense underhair with definitive growth. Serves primarily for insulation.

fusiform. Compact, tapered. Pertaining to body with shortened projections and no abrupt constrictions. See Figs. 8-7 and 8-13, for example.

gamete. General term for a sex cell (egg or sperm).

generalist. A species with broad preferences of food, habitat, or other factor(s). Compare with **specialist.**

generic name. First word of a binomen or trinomen. The name of a genus. See Fig. 11-1.

genital field/pore. The region immediately surrounding and encompassing the genitalia of Acarina.

genital papillae. The small finger-like/sucker-like protuberances within the region of the genital field of Acarina.

Gen trap. Live trap consisting of a plastic cylinder and a hinged metal door. See Fig. 34-11 and Shemanchuk and Berger (1968) for details.

genus. A category within a family, consisting of one or more, generally similar species. See Fig. 11-1.

germinative epithelium. Outer layer of cells of the ovary in which meiosis occurs.

gestation. The length of time from fertilization until birth of a fetus.

glans penis. Sensitive distal portion of the penis.

glenoid fossa. Depression in the pectoral girdle, into which the head of the humerus articulates.

Gloger's Rule. An ecological "rule" which states that races of mammals in arid regions are lighter in color than related races in humid regions.

gnathosoma. The anterior body region of Acarina bearing the food-gathering apparatus.

Gondwanaland. The southern land mass that began to separate near the end of the Paleozoic Era and which ultimately split into South America, Africa, Madagascar, India, Antarctica, and Australia. See **continental drift** and Fig. 1-4.

Graafian follicle. A structure in the ovary that contains the developing egg and surrounded by a large, fluid-filled cavity. See Fig. 9-8.

granivorous. Feeding on grains or seeds.

graviportal. Pertaining to a limb structure adapted for supporting great weight. Found in very heavy mammals, e.g., elephants. See Fig. 8-6.

grazers. Mammals that feed on grass and other herbage by cropping and nibbling.

group home range. The home range of a group of individuals such as that of a herd or family group.

group spatial territory. Territory similar to spatial territory but possessed by members of a group. See Fig. 37-8.

growth lines. Absolute, incremental growth lines present in teeth, bones, and other tissues of mammals and useful for aging purposes.

growth multiplier. See **finite rate of increase.**

guard hairs. Outer coat of coarse, protective hairs found on most mammals. See **spines, bristles,** and **awns.**

hair. Cylindrical, filamentous outgrowths of the epidermis that consist of numerous cornified epidermal cells. Found only in mammals. See Fig. 4-1.

Haller's organ. A structure located on tarsus 1 of Metastigmata (Acarina). See Fig. 36-12B.

hallux. The first (medial) digit of the pes. Is frequently opposable in arboreal mammals. See Fig. 7-4.

hard palte. The bony septum between the nasal passages and the oral cavity. Formed by the palatine bones and portions of the maxillae and premaxillae. See Fig. 2-1B.

haustellate. In Arthropoda, pertains to forming a beak suitable for piercing.

Hayne Method. Deterministic method of population estimation for use when individuals are marked on several occasions. The Schnabel (1938) and Schumacher and Eschmeyer (1943) methods are similar.

head. In insects, the anterior body region on which are found the eyes, mouth parts, and antennae.

herbivore. An animal that consumes plant material as the primary component of its diet.

herbivorous. Feeding primarily or principally on vegetation.

Hertwig's solution. Reagent for clearing plant tissues during preparation of microscope slides for dietary analysis. See Chapter 31.

heterodont. Pertaining to a dentition with teeth differentiated into various types such as incisors, canines, premolars, and molars. See Fig. 2-1C, for example. Opposite of homodont.

histogram. A two-dimensional, graphic representation of frequency data, with the height of a particular bar representing the frequency. See Figs. 11-8, 29-2.

Holarctic Region. Collectively, the Nearctic and Palearctic Faunal Region. See Fig. 1-3.

holotype. A single specimen that is the type-specimen of a particular named species or subspecies.

home range. The area that a mammal occupies during the course of its life, exclusive of migration, emigration, or unusual erratic wanderings (Brown and Orians 1970).

homing. The tendency of animals to return to their home area when experimentally displaced to another area. See Fig. 37-9.

homodont. Pertaining to a dentition with all teeth very similar in form and function. See Figs. 20-11, 22-32, for examples. Opposite of heterodont.

homogeneity of variances. Statistical test to evaluate whether the variances of two sample groups are approximately (at the level of significance chosen) equal.

homonym. In general, one of two or more identical names proposed for the same taxon or different taxa. Specifically, a *primary homonym*, as applied to (1) genus-group and family-group names, is one of two or more identical names used for the same or different taxa, or (2) identical species-group names applied to the same taxon or different taxa within the same superior taxon. A *secondary homonym* results from the transfer of a species-group name from the immediately superior taxon in which it was originally proposed to another immediately superior taxon that already contains an identical name. A *senior homonym* is a homonym with the earliest date of appearance in the literature; a *junior homonym* is a homonym proposed after the date of publication of the senior homonym.

hoof (pl. hoofs or hooves). The digital keratinization in unguligrade mammals, a horny sheath completely encasing the tip of a phalanx and usually providing the animal's only point of contact with the substrate. See Fig. 6-3.

horizontal life table. See **age-specific life table.**

horn. 1. Structure projecting from the head of a mammal and generally used for offense, defense, or social interaction. Cattle, sheep, Old World antelopes, etc. (family Bovidae) have horns formed by permanent, hollow, keratin sheaths growing over bone cores. See Fig. 5-2. **2.** The keratinized material that forms the sheaths of these horns. See Fig. 5-1. **3.** One of a pair of projections, e.g., horn of the uterus.

host. The animal parasitized by a parasite.

Hoyer's medium. Aqueous mounting medium used in microscope slide preparations. See Chapter 31.

humerus. The single bone in the upper (proximal) portion of each pectoral limb. See Fig. 7-1.

hunter. According to McNab (1963: 136), mammals that are principally granivorous, frugivorous, insectivorous, or carnivorous.

hybrid. The offspring resulting from a cross between individuals of two species.

hypocone. In upper cheek teeth, a cusp on the posterior lingual side of the crown, sometimes situated on a distinct talon. See Fig. 3-13.

hypoconid. In lower cheek teeth, a cusp on the posterior, labial side of the talonid area of the crown. See Fig. 3-10.

hypoconulid. In lower cheek teeth, the posteriormost cusp in the talonid area of the crown. See Fig. 3-10.

hypostome. The anterior region of the ventral surface of the gnathosoma of Acarina. See Fig. 36-13.

hypsodont. Pertaining to a high-crowned tooth. See Figs. 27-3, 28-10, for examples. Opposite of brachydont.

hysternosomal. The entire region of the body of Acarina posterior to the insertion of legs III.

hystricognath mandible. Type of mandible found in certain rodents (Bathyergidae, Old and New World hystricomorphs) in which the angular process arises laterad to the outer border of the incisive alveolar sheath. See Fig. 25-9B.

hystricomorphs. Rodents from the Old and New World in which the infraorbital foramen is greatly enlarged. It includes the caviomorphs (New World) and Old World forms such as Hystricidae, Thryonomyidae, and Petromyidae.

-idae. A suffix added to the stem of the name of the type-genus to form the name of a family. See Fig. 11-1.

identification. In taxonomy, the assignment of particular specimens or individuals to already established or named taxa.

idisoma. The body region of Acarina posterior of the gnathosoma including the leg bearing region and the region posterior to the legs.

ilium (pl. ilia). Most dorsal of the three bones in each half of the pelvic girdle. The pelvic bone which articulates with the sacral vertebrae. See Fig. 7-3.

imbricate. Overlapping, as the shingles of a roof.

immigration. The process whereby individuals move into a population.

implantation. The attachment of the embryo to the uterine wall of the female mammal. See also **delayed implantation.**

-inae. A suffix added to the stem of the name of the type genus to form the name of a subfamily. See Fig. 11-1.

incisiform. Pertaining to a canine or other tooth that has the shape and appearance of an incisor.

incisive alveolar sheath. The bony socket or alveolus that receives the root of an incisor.

incisive foramina. See **palatal foramina.**
incisor. The anteriormost of the four basic kinds of teeth found in mammals. Usually chisel-shaped. Upper incisors are always rooted in the premaxillae.
incremental lines. See **growth lines.**
Inc. Sed. (Incertae Sedis). In an uncertain position. In classification, relationship of taxon to other taxa not known.
incus. The middle ear ossicle of mammals, situated between the malleus and stapes. Derived from the quadrate bone of more primitive vertebrates. See Fig. 1-1.
indented key. A format of a biological key in which the two alternatives of each couplet may be widely separated by intervening couplets. See Fig. 12-2.
index of dispersion. Statistical measures of the nature of spatial dispersion, such as Morisita's (1962) Index ($I\delta$).
index of home range size. Index of home range size developed by Metzgar and Sheldon (1974).
individual home range. The home range of an individual animal.
individual territory. Territory defended by a single individual.
inflected. Deflected or bent inward.
infraorbital aperture. See **infraorbital foramen.**
infraorbital canal. A canal through the zygomatic process of the maxilla from the anterior wall of the orbit to the side of the rostrum. See Fig. 2-1A, C.
infraorbital foramen. A foramen through the zygomatic process of the maxilla.
inguinal. Pertaining to the region of the groin.
inguinal canal. In male mammals, a small opening in the musculature of the abdominal wall on either side at the base of the scrotum through which the testis moves out of the abdominal cavity into the scrotum.
-ini. A suffix added to the stem of the name of a type-genus to form the name of a tribe. See Fig. 11-1.
inner fold. An elongated enamel island present in a tooth.
insectivorous. Feeding on insects.
integument. The skin, consisting of two layers, the dermis and epidermis, and the derivatives of these layers such as scales and hair. See Fig. 4-1.
interactive terminal. A video display terminal used for data entry and manipulation.
interclavicle. Median bone lying between the anterior end of the clavicles and the ventral surface of the sternum. In mammals found only in monotremes.
internal nares. The posterior openings of the nasal passages of the skull.
International Code. See **zoological nomenclature.**
International Commission. See **zoological nomenclature.**
interparietal. An unpaired bone on the dorsal part of the braincase between the parietals and just anterior to the supraoccipital. Absent in some mammals.
interpopulation variation. Degree of variability between populations.
interstitial cell. Any one of the cells filling the spaces between seminiferous tubules in the testes. Functions as endocrine gland in the secretion of hormones. See Fig. 9-9.
intrapopulation variation. Degree of variability within a population.
intrinsic rate of increase, *r*. The instantaneous rate of increase in a population, sometimes termed r_m, where the age distribution is stable and no resource is limiting (Caughley 1977).
ischial callosity. A large, keratinized thickening (callus) on each buttock of some Old World monkeys and gibbons. Frequently brightly colored in males.
ischium (pl. ischia). Most posterior and ventral of the three bones in each half of the pelvic girdle. See Fig. 7-3.
iteroparous (noun is iteroparity). Pertaining to the production of offspring by an organism in successive batches or litters. Most mammals are iteroparous.
Jolly-Seber Method. A stochastic method of population estimation that corrects for mortality and natality effects on each sampling occasion.
journal. 1. A serial publication. Usually of a professional or scientific society, e.g., Journal of Mammalogy. **2.** A portion of the field notes, a chronologically arranged account of all activities. See Fig. 33-1.
jugal. The bone which forms the midsection of the zygomatic arch between the zygomatic processes of the maxilla and squamosal. See Fig. 2-1.
jump trap. A type of steel leg-trap which is equipped with a steel strap spring located beneath the bait-pan.
junior homonym. See **homonym.**
junior synonym. See **synonym.**
juvenal. See **juvenile.**
juvenile. An individual that is physiologically immature or undeveloped. In mammals, often have distinctive pelage coloration and texture.
juvenile pelage. The type of pelage characteristic of a juvenile of a species.
karyotype. 1. Morphological characteristics of the chromosomes of a cell. **2.** An arrangement of chromosomes of a cell according to shape, centromere position, and number.
keeled sternum. A breastbone (sternum) with a medial ventral ridge that provides an expanded surface area for muscle attachment. Present in most bats.
keratinized. Impregnated with keratin —a tough fibrous protein especially abundant in the epidermis and epidermal derivatives.
key. Tabular, usually dichotomous arrangement of diagnostic characters used to identify organisms to various taxonomic levels. See Figs. 12-1, 12-2, 12-3.
key characters. Characters used in the construction of a biological key to distinguish different organisms.
keypunch machine. A terminal, with a keyboard similar to that of a typewriter, that is used to punch data onto tabulating cards.
knee. The joint in the pelvic limb between the femur and the tibia and fibula.

known age. Age established for individual based on observation of birth or other equally reliable criteria.

***K*-selected.** Selection favoring superiority in stable, predictable environments in which rapid population growth is unimportant (Wilson 1975).

***K*-strategist** (equilibrium species). A species living in a more predictable environment where individual longevity, care of young, and other factors are selected for. Compare with **r-strategist.**

labia. Fleshy folds of skin found in vuvular region of female primates and most ungulates. The largest of these folds, the *labia majora,* secrete lubricating fluids at the time of mating and are the female homolog of the male scrotum.

labial. Pertaining to the lips, e.g., the labial side of a tooth is the side closest to the lips.

labyrinthine. Pertaining to elaborate, complex projections of the placenta. See also **villus.**

lacrimal bone. A small bone in the anterior wall of each orbit. See Fig. 2-1C.

lacrimal duct. A small duct, usually perforating the lacrimal bone, extended from the inner corner of each eye to the nasal cavity; serves as drain for tears, the secretions of the lacrimal gland.

lactation. Production of milk by the mammary glands.

latitudinal migration. Migration by mammals from one latitude to another latitude, as for example, the migration of gray whales.

lactophenol. Reagent for clearing sclerotized tissues during preparation of microscope slides. See Chapter 31.

lambdoidal crest. See **occipital crest.**

lamina (pl. laminae). On some teeth plate-like structures that may (or may not) bear cusps.

lanugo. Fine, soft hair on the fetus. A type of vellus.

lateral. Located away from the midline; at or near the side(s).

Laurasia. A single landmass or compact group of landmasses that existed in the northern hemisphere (Fig. 1-4B).

Law of Homonymy. In nomenclature, any name that is a junior homonym of an available name must be rejected and replaced.

Law of Priority. In general, the valid name of a taxon is the oldest available name applied to it, subject to certain provisions of the International Code and opinions of the International Commission.

lectotype. One of several syntypes, which is subsequently designated as the type-specimen of a particular taxon.

left aortic arch. The structure through which the blood leaves the left ventricle of the heart and passes to the body proper in mammals. Both left and right arches are present in reptiles and more primitive vertebrates, only the right is present in birds. See Fig. 1-2.

lek. See **arena territory.**

LH. See **lutenizing hormone.**

life table. Tabulation presenting complete time-specific data on survival, mortality, fecundity, and other parameters of a particular cohort of individuals. See **age-specific life table** and **time-specific life table.** See also **composite life table.**

Lincoln Index. See **Petersen Index.**

lingual. Pertaining to the tongue, e.g., the lingual side of a tooth is the side nearest the tongue.

litter. The set of young born of a female mammal following a pregnancy.

litter size. The number of young delivered by a female from one pregnancy.

logistic growth. Growth, especially in the number of organisms constituting a population, that slows steadily as the entity approaches its maximum size (Wilson 1975).

long-spring trap. A type of steel leg-trap in which the spring is a long metal strap projecting away from the jaws. See Fig. 34-2.

Longworth trap. A brand of live trap that has a detachable nest box. See Fig. 34-10.

loph. A ridge on the occlusal surface of a tooth formed by the elongation and fusion of cusps. See Fig. 3-14.

lophodont. Pertaining to a tooth which has an occlusal surface pattern consisting of lophs. See Fig. 3-14.

lordosis. 1. In locomotion, suppleness of the spinal column that allows the hind feet to be placed well anterior to the forefeet when the animal is running. **2.** In behavior, a posture assumed by many female mammals, particularly rodents, during copulation.

lutenizing hormone, LH. Pituitary hormone responsible for the formation of the corpus lutem; together with FSH, this hormone stimulates the follicle to secrete estrogen.

maceration. Use of bacteria to clean bones. See Chapter 35.

malar (=jugal). In human anatomy, the cheek bone.

malleus. The outermost of the three middle ear ossicles. Derived from the articular bone. See Fig. 1-1.

mamma (pl. mammae). See **mammary gland.**

Mammalia. Vertebrates with the mandible composed of only one pair of bones, the dentaries, that articulate directly with the squamosal of the cranium.

mammary gland. Milk-producing gland unique to mammals. Growth and activity governed by hormones of the ovary, uterus and pituitary.

mandible. The lower jaw. In mammals the mandible is composed of a single pair of bones, the dentaries. See Fig. 2-2.

mandibular condyle. The knob by which each mandible articulates with the cranium. See Fig. 2-2.

mandibular fossa. One of the two concavities on the cranium in which the mandibular condyles articulate. In mammals, usually confined to the squamosal bone. See Fig. 2-1B.

mandibular symphysis. The suture between the paired dentaries.

mandibulate. In Arthropoda, with mandibles, fitted for chewing.

manual. Pertaining to the manus or forefoot.

manus. The forefoot or hand. Collectively, the carpus, metacarpus, and digits. See Fig. 7-2.

Marine Mammal Protection Acts. U.S. laws regulating the salvage and capture of marine mammals. See Chapter 34.

marking. The behavioral process by which a mammal intentionally leaves an indication of its presence in its environment. Often accomplished by deposition of scent gland secretions.

marsupial. A mammal in the infraclass Metatheria. Includes the Orders Marsupicarnivora, Paucituberculata, Peramelina, and Diprotodonta.

marsupium. External pouch formed by folds of skin in the abdominal wall. Found in many marsupials and in some monotremes (echidnas). Encloses mammary glands and serves as incubation chamber.

masseteric canal. A perforation through the mandible at the base of the masseteric fossa.

masseteric fossa. A concavity in the outer side of the mandible at the posterior end of the ramus. See Fig. 2-2.

masseter muscle. Adductor muscle of the mandible of mammals. Generally divided into external, middle, and deep portions. See Fig. 25-3A to D for examples.

mastoid process. The exposed portion of the petromastoid bone, most of which is concealed within the auditory bulla, that forms the otic capsule enclosing the inner ear. The mastoid process, if present, is located just posterior to the auditory bulla. See Fig. 2-1C.

maxilla (pl. maxillae). Either of the pair of relatively large bones that forms a major portion of the side of the rostrum, contributes to the hard palate, forms the anterior root of the zygomatic arch, and bears all upper teeth except the incisors. See Fig. 2-1A, B, C.

maxillary. Pertaining to the maxilla.

maxillary process. A projection of bone from the maxilla that forms the anterior root of the zygomatic arch. Process frequently divided into an upper portion (above infraorbital foramen) and a lower portion (below infraorbital foramen).

mean. For a sample, the arithmetic average of a set of observations. $\bar{X} = \Sigma X/N$. The sample mean ($\bar{X}$) is the best estimator of the population mean (μ).

mean expectation of life, e_x. The average duration of life expected of individuals at the start of a particular age, x.

medial. Lying in or near the plane dividing a mammal into two mirror-image halves.

medulla. The central portion of a structure composed of distinct concentric layers or regions, e.g., the medulla of a hair, ovary, kidney, or adrenal gland.

melanism. Unusual darkening due to the deposition of large amounts of melanins in the integument.

menstrual cycle. A special type of estrous cycle found in certain female primates where the endometrium of the uterus regresses slowly after ovulation and then, in the process termed *menstruation,* rapidly breaks down with the passage of blood and cellular debris through the vaginal opening.

menstruation. See **menstrual cycle.**

mental foramen. Small opening (sometimes two openings) near the anterior end of each mandibular ramus; carries a nerve and blood vessel. See Fig. 2-2.

mesaxonic foot. Type of foot structure where the main axis of weight is supported by a single digit. See Figs. 8-2, 27-6.

metacarpal. In the metacarpus, any of the bones in the manus between the carpals and phalanges. One metacarpal per digit. See Figs. 7-1, 7-2.

metacone. In upper cheek teeth, a cusp on the posterior, labial side of the trigon area of the crown. See Fig. 3-10.

metaconid. In lower cheek teeth, a cusp on the posterior, lingual side of the trigonid area of the crown. See Fig. 3-10.

metaconule. In upper cheek teeth, a cusp on the posterior portion of the crown between the metacone and hypocone. See Fig. 3-10.

metapodial. General term used for either a metacarpal or metatarsal.

metatarsal. Any one of the bones in the pes between the tarsals and the phalanges, one metatarsal per digit. See Figs. 7-3, 7-4.

Metatheria. One of the infraclasses of the subclass Theria. Contains only the marsupials.

metestrus. Third stage of estrous cycle following estrus, where leucocytes appear among the cornified epithelial cells in a vaginal smear and when the corpora lutea are fully formed and progesterone levels are high.

middle ear. The area of the ear between the tympanic membrane and the otic capsule. Contains the malleus, incus, and shapes. Usually enclosed in an auditory bulla. See Fig. 1-1.

milk. The substance secreted by mammary glands. Provides nourishment to the young of mammals.

milk tooth. Any tooth in the deciduous set of teeth of mammals with diphyodont dentition. Replaced by the permanent teeth.

minimum area. Home range estimate where capture points are connected in counterclockwise fashion. See Fig. 37-4A.

minimum number known alive. An enumeration of individuals captured on a trapping grid and including those marked individuals not trapped on one occasion but appearing at a later occasion. Also known as the **calendar of captures method.**

mistnet. A net of fine mesh used to capture birds and bats. Usually three meters high and ranging from 6 to 30 meters long. See Fig. 34-14.

molar. Any cheek tooth situated posterior to the premolars and having no deciduous precursor. One of the four kinds of teeth in mammals.

molariform teeth. Teeth that have the shape and appearance of molars regardless of whether they are true molar teeth.

molt. The process by which hair is shed and replaced. Se Fig. 4-5.

molt pattern. The pattern by which the hair is replaced over the body during a molt. The pattern is frequently species specific. See Fig. 4-5.

monestrous. Pertaining to species that have only one period of estrus or heat per year.

monophyletic. Pertaining to a taxon the members of which are all part of a single immediate line of descent.

monophyodont. Having a single set of teeth, with none being replaced. In contrast, see **diphyodont.**

monotypic. A taxon containing only a single taxon at the next lower major category.

mortality. The decrement of individuals to a population through deaths. In life tables, time-specific mortality is indicated by the symbol d_x. See also **mortality rate,** q_x.

mortality rate, q_x. The proportion of individuals alive at age x that die before $x + 1$. This statistic is calculated using the formula $q_x = d_x/l_x$.

mousetrap. A commercial snap trap designed to kill the house mouse, *Mus musculus.* Due to its small size it usually kills by crushing the skull and thus is not useful as a collecting tool by mammalogists. See Fig. 34-1.

multiparous. State of female mammals that show evidence of placental scars of different ages and representing several litters.

multivariate statistics. Simultaneous analysis of two or more variables in a sample or samples.

mummified. Pertaining to a specimen that has been preserved by natural dehydration.

muscular diaphragm. The muscular septum between the thoracic and abdominal cavities of mammals.

Museum Special. A snap trap specifically designed for collecting mammals for museums. Intermediate in size between the rat- and mousetraps. See Fig. 34-1.

musk. Scent secretion from any one of a variety of special scent glands in many kinds of animals.

myomorphs. Members of one of the three suborders in the order Rodentia; generally, those rodents that have an oval, round, or V-shaped infraorbital foramen. Members of the superfamilies Muroidea, Gliroidea, and Dipodoidea.

nail. Flat, keratinized, epidermal, translucent growth protecting the upper portion of the tip of a digit; a nail is a modified claw. See Fig. 6-2.

nasal bone. Either of the paired bones on the rostrum that form a roof over the nasal passages. Usually situated between the dorsal margins of the premaxillae and maxillae. See Fig. 2-1A, C.

nasal branches. See **premaxillae.**

nasal passages. The passages, located dorsad to the secondary (hard) palate, that connect the external and internal nares.

natality. The increment of young into a population through births. See **fecundity.**

Nearctic Region. The faunal region that includes North America south to central Mexico. See Fig. 1-3.

nectivorous. Feeding on nectar.

neopallium. Nonolfactory portion of the cerebral cortex.

Neotropical Region. The faunal region that includes South America, the West Indies, and Central America north to central Mexico. See Fig. 1-3.

neotype. A single specimen designated as the type-specimen for a named species or subspecies when the original holotype, the lectotype, or syntypes have been lost or destroyed. The neotype can be any specimen collected at the type locality.

nest. A structure (of grass, leaves, or some other material) built by a mammal for shelter or insulation.

nestling. An individual, generally recently born, that is still confined to a nest.

net reproductive rate, R_0. The average number of female offspring produced by each female during her entire lifetime (Wilson 1975).

new capture. In population studies, the first capture of an individual.

niche. All the components of the environment with which an organism or population interacts. See **realized niche** and **fundamental niche.**

niche breadth. A measure of the magnitude of the niche dimensions possessed by a species.

niche overlap. Measurable portion of the niche of an organism or population that overlaps with that of another individual or population.

nictitating membrane. Thin membrane at the inner angle of the eye in some species, which can be drawn over the surface of the eyeball. The "third eyelid."

nidic territory. A territory that involves only the immediate area around a home site and often only possessed by a female.

nipple. Protuberance of a mammary gland that has numerous ducts leading directly to the surface to serve as outlets for the milk. See Fig. 9-13B.

NIRM standards. Data standards adopted for use in possible National Network for Information Retrieval in Mammalogy.

nocturnal. Pertaining to nighttime, the hours without daylight. In particular, pertaining to animals that are active at night. Opposite of diurnal.

nomadism. The tendency for individuals to make erratic or wandering types of movements.

nomen nudum. One kind of name that does not meet the provision of availability under the International Code (see Articles 12 to 16). A name without a description or indication.

nomen oblitum. A forgotten name. A senior synonym that is not available because it has not been in use in the primary zoological literature during the 50-year period preceding its rediscovery (see chapter 11 and Article 23b of International Code).

nomenclature. The system for giving distinctive names to different taxa.

nondeciduate. Term applied to placentas that separate easily between embryonic and maternal tissue at parturition, resulting in little or no damage to the uterine wall.

nonparametric. Pertaining to data that do not conform to a normal distribution and thus must be tested using *nonparametric statistics.*

normal deviates. See **V-values.**

normal distribution. A theoretical frequency distribution that is bell-shaped, symmetrical, and of infinite extent.

nose leaf. A structure on the noses of some bats. Ranges from a small, simple flap to a highly complex structure of numerous projections and chambers. Believed to aid in echolocation. See Fig. 18-4, for examples.

notochord. In the phylum Chordata, a long, slender skeletal rod composed of large, vacuolated cells and lying between the dorsal nerve cord and digestive system. In mammals it is present only in the embryonic stages and later is surrounded and supplanted by the vertebral column.

nuliparous. State of female mammal that shows no evidence of placental scars or pregnancy.

null hypothesis (abbrev. $\mathbf{H_0}$). Statement that there is no difference (e.g., between sample groups) and formulated for the purpose of being rejected.

numerical phenetics. See **phenetics.**

observed rate of increase, $\bar{r}$. The instantaneous rate of increase in a population when the assumptions of a stable age distribution, a constant rate of increase, or unlimited resources cannot be met (Caughley 1977).

occipital bone. The bone surrounding the foramen magnum and bearing the occipital condyles. Formed from four embryonic elements: a ventral *basioccipital,* a dorsal *supraoccipital,* and two lateral *exoccipitals.* See Fig. 2-1B, C.

occipital condyle. In mammals, either of the two knobs on the occipital bone flanking the foramen magnum and articulating with first cervical vertebra. See Fig. 2-1B, C.

occipital crest. Confluent ridges extending laterally from the sagittal crest near the dorsal edge of the occipital bone. See Fig. 2-1A, C.

occiput. General term for the posterior portion of the skull. See Fig. 2-1.

occlusal. Pertaining to the surfaces of contact in the upper and lower teeth. The contact of these tooth surfaces is termed *occlusion.*

occlusion. See **occlusal.**

Oceanic. Region consisting of oceanic islands not included in one of the defined faunal regions, e.g., New Zealand and Hawaii. See Fig. 1-3.

-oidea. Suffix added to the stem of the name of a type-genus to form the name of a superfamily. See Fig. 11-1.

olecranon process. Process of the ulna which projects proximally beyond the articulation point with the humerus. See Fig. 7-1.

omnivorous. Pertaining to those animals that eat quantities of both animal and vegetable food.

optic foramen. Opening (in the medial wall of the orbit) through which the optic nerve and ophthalmic artery pass.

orbit. The socket in the skull in which the eyeball is situated. See Fig. 2-1A, B.

orbitosphenoid. Portion of the presphenoid that is visible in the wall of the orbit. See Fig. 2-1C.

order. A category within a class, consisting of one or more families. See Fig. 11-1.

Oriental Region. The faunal region that includes the tropical portion of Asia. Extends from Pakistan south of the Himalayas through Indo-China and Indonesia to Wallace's Line. See Fig. 1-3.

os clitoris. A small bone present in the clitoris in some mammal species. Homologous to the baculum in males.

os penis. See **baculum.**

ossicle. Any small bone, e.g., the three middle ear ossicles of mammals.

ossify. To become bony.

otic capsule. The bony capsule enclosing the organs of hearing and balance of the inner ear.

OTU. Operational Taxonomic Unit. The basic taxonomic unit of phenetic classification studies. May or may not be equivalent to some taxon in another taxonomic scheme.

ova. Plural of ovum. See **ovum.**

ovarian scar. See **corpus albicans.**

ovary. (pl. ovaries). The female gonad. The site of egg production and maturation. See Figs. 9-1, 9-8.

oviduct. The duct that carries the eggs from the ovary to the uterus. In human anatomy, the term Fallopian tube is used. See Figs. 9-1 through 9-4.

ovulation. The releasing of an egg (by the ovary) into the oviduct.

ovum (pl. ova). Mature but unfertilized egg. See Fig. 9-8.

oxytocin. A pituitary hormone that causes rhythmical contractions of the uterus during copulation and at parturition and enhances milk "letdown."

pad. Usually naked structure on the ventral surface of the manus or pes, that comes in contact with the ground. May be several per manus or pes. See Figs. 6-1, 6-2.

palatal branches. Portions or projections of the premaxillae and maxillae that contribute to the formation of the secondary palate. See Fig. 2-1B.

palatal foramina. Incisive foramina. Paired perforations of the anterior end of the hard palate at the point where the premaxillae and maxillae meet. See Fig. 2-1B.

palate. The bony plate formed by the palatine bones and palatal branches of the maxillae and premaxillae. Separates the nasal passages from the oral cavity. See Fig. 2-1B.

palatine. Either of the pair of bones that forms the posterior portion of the palate. See Fig. 2-1B.

Palearctic Region. The faunal region that includes Europe, Africa north of the Sahara, and Asia (except for the southern portion of the Arabian Peninsula and the tropical regions included in the Oriental Region). See Fig. 1-3.

palmate. Pertaining to the presence of webbing between the digits.

palpal apotele. The modified claw of the palp pretarsus of Mesostigmata (Acarina). See Fig. 36-12A.

Pangaea. A supercontinent or aggregation of continents that existed some 180 million years ago (Windley 1977: 203). See Fig. 1-4A.

papilla. Any blunt, rounded, or nipple-shaped projection.

paraconid. In lower cheek teeth, a cusp on the anterior, lingual side of the trigonid portion of the crown. See Fig. 3-10.

paraconule. A cusp in some upper cheek teeth that is situated between the protocone and paracone. Sometimes termed the protoconule.

parameter. Population statistic. Any one of the true values for various statistical properties, such as mean of the population, variance of the population. Generally these values are never known but are estimated by sample statistics.

parametric. Pertaining to data that conform reasonably well to a normal distribution and that can be tested using *parametric statistics.*

parapatric. Pertaining to the ranges of species that are contiguous but not overlapping. See Fig. 11-2B.

parastyle. The anteriormost cusp on the stylar shelf. See Fig. 3-10.

paratype. A specimen other than the holotype that a taxonomist assigns to a supposed new species or subspecies in its description.

paraxonic foot. Type of foot structure where the main axis of weight passes between a pair of similarly-sized digits.

parietal. Either of the pair of bones contributing to the roof of the cranium posterior to the frontals and anterior to the occipital. See Fig. 2-1A, C.

paroccipital process. A process on the exoccipital portion of the occipital bone. It extends ventrally from just posterior to the tympanic bulla and mastoid process. See Fig. 2-1C.

parous. Pertaining to female mammal that is pregnant or shows evidence of previous pregnancies (e.g., with placental scars).

parturition. Birth.

patagium (pl. patagia). A web of skin such as the parachutelike skin extensions of a colugo, the gliding membranes of a flying squirrel, or the membranes of the wing of a bat. See Figs. 8-16, 8-17, and 18-1, for examples.

patchy distribution. See **clumped distribution.**

patronym. A taxonomic name based on the name of a person or persons.

pectinate. Comb-shaped. With or consisting of a series of projections like the teeth of a comb.

pectoral girdle. The shoulder girdle. In most mammals, formed by the scapula and clavicle or scapula alone. See Fig. 7-1.

pectoral limb. The forelimb. Includes the humerus, radius and ulna, carpals, metacarpals, and phalanges. See Fig. 7-1.

pedal. Pertaining to the pes or hind foot.

pedicel. pedicle. Any stalk or stem supporting an organ or other structure e.g., the pedicel of an antler. See Fig. 5-7.

pelage. Collectively, all the hairs on a mammal.

pelvic girdle. The hip girdle. Composed of the paired ischia, ilia, and pubic bones. See Fig. 7-3.

pelvic limb. The hind limb. Composed of the femur, patella, fibula and tibia, tarsals, metatarsals, and phalanges. See Fig. 7-3.

penis. The intromittent organ that the male mammal inserts into the vagina of the female for the purpose of insemination.

pentadactyl. Five-digited. See Figs. 7-2 and 7-4, for examples.

periotic bone. One of the bones forming the otic capsule.

permanent teeth. Teeth in the second of the two sets of dentition in diphyodont mammals. Succeed the milk or deciduous teeth.

pes. The hind foot. Collectively, the tarsals, metatarsals, and phalanges of the pelvic limb. See Fig. 7-4.

Petersen Index. A method for estimating population size using the formula $N = Mn/m$, where only one marking occasion is employed.

phalanx (pl. phalanges). Any one of the distal two or three bones in each manual and pedal digit. See Figs. 7-1 through 7-4.

phenetics. Numerical phenetics. A philosophy of classification based upon the degree of overall phenotypic similarity between taxa or between operational taxonomic units (OTU's). See chapter 11.

phenogram. A two-dimensional, pictorial representation of a classification scheme, based on the methods of phenetics or numerical phenetics. See Fig. 11-4.

pheomelanin. The pigment in the mammalian integument that produces shades and tints of red and yellow.

pheromone. A substance that is secreted by an animal and that influences the behavior of other individuals of the same species.

phyletic classification. See **phylogenetic classification.**

phyletic systematics. See **phylogenetic systematics.**

phylogeneticist. An individual that studies extant and/or fossil material and attempts to determine their phylogenetic history.

phylogenetic systematics. The type of systematics practiced by an evolutionary or traditional systematist *or* (depending on the author) the type of systematics practiced by a cladist.

phylogeny. The evolutionary history of an organism or groups of related organisms.

phylogram. Graphical representation of hypothesized phylogeny based on data evaluated by an evolutionary systematist. See Fig. 11-3.

piercing-sucking. In arthropods, mouth parts suitable for piercing tissue and for sucking fluid.

pigment. The minute granules that impart color to an organism and that are especially abundant in the integument. Such granules are usually metabolic wastes and may be black, brown, yellow, or red.

pinna (pl. pinnae). External ear. The flap located around the external auditory meatus. It gathers sound vibrations and channels them toward the tympanum. Absent in many aquatic and fossorial mammals.

piscivorous. Feeding on fish.

placenta. An apposition or fusion of the fetal membranes to the uterine mucosa for physiological exchange (Mossman 1937).

placental scar. The scar that remains on the uterine wall after a deciduate placenta detaches at parturition.

plagiaulacoid. Pertaining to an enlarged shearing cheek tooth found in the rat kangaroos, the pygmy phalanger, and certain extinct caenolestids and multituberculates. See Figs. 15-20, 15-26.

plagiopatagium. The patagium forming the wing of a bat. See Fig. 18-1.

plantigrade. Pertaining to feet in which the parts enclosing phalanges and metatarsals or metacarpals all touch the ground. The basic structure for ambulatory (walking) locomotion.

plate tectonics. The movement of a series (six major and various minor) rigid plates whose margins are located along volcanic ridges and deep trenches. The motion of these plates create ocean basins and result in new continental configurations.

plesiomorphic. In cladistics, a character that is said to be primitive.

point-area count. Census of individuals that pass a single point.

pollex. The first (most medial) digit on the manus. In humans, the thumb. See Fig. 7-2.

polyestrous. Pertaining to species that have three or more estrous cycles each year.

polyprotodont. Condition found in marsupicarnivore and peramelid marsupials in which the lower jaw is not shortened and the anterior lower incisors are not greatly elongated. Fig. 15-2. Compare **diprotodont.**

polytypic. Pertaining to a taxon that contains two or more taxa at the immediately lower level, e.g., a species with two or more and recognized named subspecies.

population. **1.** In statistics, all possible values (observations) of a particular variable (e.g., character) in all individuals of a particular group within a specified space or time interval. **2.** In biology, all the individuals that form a single interbreeding group.

population density. A measure of the number of individuals that occupy a given area.

population mean. See **mean.**

population statistic. See **parameter.**

postcanine teeth. See **cheek teeth.**

posterior. Of or pertaining to the rear portion.

postorbital bar. A bony rod separating the orbit and temporal fossa. Formed by a union of the postorbital processes of the frontal bone and zygomatic arch.

postorbital process. A projection of the frontal bone which marks the posterior margin of the orbit. See Fig. 2-1C.

prebaiting. Placing bait at trapping stations prior to the initiation of trapping.

precision. The nearness of values of successive measurements of the same specimen.

precocial. Pertaining to young that upon birth are capable of moving about and feeding with little parental assistance.

precoracoid. A bone in the pectoral girdle of reptiles. Found in mammals only in Monotremata.

preference index. A measure of the food preference of a population. Determined by comparing the relative occurrence of dietary items in the organism with the relative occurrence of the resource upon which they feed. The *electivity index* is another statistical measure of dietary preference. See chapter 31.

prehensile. Pertaining to structures adapted for grasping or seizing by curling or wrapping around, such as the tail of some American monkeys and opossums.

premaxilla (pl. premaxillae). One of the paired bones at the anterior end of the rostrum that frequently bear teeth. *Palatal branches* of the premaxillae form the anterior part of the secondary palate and the *nasal branches* contribute to the sides of the rostrum. See Fig. 2-1A, B, C.

premolar. One of the four kinds of teeth in mammals. Located anterior to the molars and posterior to the canine. The only cheek teeth normally present in both the permanent and milk dentitions. Usually do not exceed four per quadrant.

prepuce. Fold of skin covering the glans penis.

preputial glands. Modified sebaceous glands that in males of some species contribute to the formation of the semen and in others secrete a scent used in "marking."

presphenoid. A complex bone that contributes to the ventral and lateral walls of the braincase. Ventrally it is visible between the pterygoids from where it then passes beneath other bones to reappear in the wall of the orbit where it is termed the orbitosphenoid. See Fig. 2-1B, C.

primary follicles. Structures in the ovary that contain eggs awaiting development.

primary homonym. See **homonym.**

primary oocytes. Egg cells contained in primary follicles in the ovary.

primary spermatocyte. A cell at one of the stages in sperm production. It is formed by division of a spermatogonium, and divides to form secondary spermatocytes.

probability. The chance for the occurrence of a particular event given the total number of possible outcomes of all events. Probability values always range from 0 to 1.

probability distribution. A function of a discrete or continuous variable that gives the probability that a specified value or interval will occur.

probability of capture. The number of animals caught during a study divided by the number of traps.

proboscis (pl. probosces or proboscides). A long, more or less flexible snout, as in tapirs and elephants.

procumbent. Pertaining to teeth that slant forward, such as the incisor teeth of a horse.

proestrus. Beginning stage of the estrous cycle when nucleated cells are present in a vaginal smear and when estrogen, progesterone, and lutenizing hormone levels reach their peak.

progestogens. A family of hormones, including *progesterone,* produced in small quantities by the follicle and in larger quantities by the corpus luteum. These hormones promote changes in the body, particularly in the uterus, to prepare the female for pregnancy.

prolactin. A pituitary hormone associated with lactation and maternal behavior.

prolegs. The tubular, fleshy leg-like appendages located on the venter of some abdominal segments of some immature insects. See Fig. 36-4A.

pronghorn. 1. A type of horn (in both sexes of Antilocapridae) which grow over permanent bony cores and are shed annually. Each horn slightly curved, with one antero-lateral prong. See Fig. 5-4. **2.** *Antilocapra americana*, the North American pronghorn "antelope." See Fig. 28-1.

propatagium. In bats, thin web of skin that extends from the shoulder to the wrist anterior to the upper arm and forearm. See Fig. 18-1.

prostate glands. Groups of muscular and glandular tissue surrounding the base of the urethra in male mammals. At the moment of ejaculation, they secrete an alkaline fluid that has a stimulating effect on the action of sperm. The anterior prostate gland is also termed the coagulating gland. See Fig. 9-5.

prothorax. The anterior segment of the insect thorax, bearing the anterior pair of legs but no wings.

protocone. In upper cheek teeth, a cusp on the lingual side of the crown at the apex of the trigon area of the tooth. See Fig. 3-10.

protoconule. See **paraconule.**

protoconid. In lower teeth, a cusp on the posterior lingual side of the trigonid area of the crown. See Fig. 3-10.

Prototheria. A primitive subclass of mammals. Contains only the monotremes or egg-laying mammals.

protoungulate. A representative of a very primitive pre-ungulate mammal. According to Wendt (1975b:466) and Melton (1976:81), the aardvark is the only living representative of this group.

proximal. Situated nearest body, e.g., the proximal end of a limb is the end closest to the body. Contrast with **distal.**

pseudopregnancy. A false pregnancy induced by cervical stimulation or mating without fertilization where the corpora lutea remain active for a considerable period of time.

pterygoid. Either of the paired bones in the ventral wall of the braincase that are posterior to the palatines. See Fig. 2-1B, C.

pubic bone. See **pubis.**

pubic symphysis. Midventral plane of contact between the two halves of the pelvic girdle.

pubis. Either of the pair of bones forming the anterior ventral portion of the pelvic girdle. See Fig. 7-3.

pulp. Collectively, the nerves, blood vessels, and connective tissue occupying the pulp chamber and root canals of a tooth. See Fig. 3-1.

punch code. Identification number that is punched into the dried skin of a mammal. Used to mark catalog numbers on skins sent to tannery. See Fig. 35-28.

quadrant. In mammalian dentition, pertaining to the four tooth-bearing sections of the skull: the right and left premaxillae/maxillae and the dentaries.

quadrat. A plot used for ecological sampling; strictly speaking always square in outline, practically may be of other shapes.

quadrate. A bone of the mandibular fossa of lower vertebrates which, in mammals, becomes the incus, one of the middle ear ossicles.

quadrate tooth. A euthemorphic cheek tooth that has a square outline. See Fig. 3-13.

quadritubercular tooth. Any upper cheek tooth with four major cusps, the paracone, metacone, protocone, and hypocone.

quadrupedal. Pertaining to an animal that uses all four limbs for terrestrial locomotion.

rack. The pair of antlers on a cervid. See Fig. 5-9.

radio-location telemetry. Technique for obtaining location data on an animal by the use of a transmitter (on the animal) and a receiver (observer).

radioisotope. A radioactive isotope.

radius (pl. radii). One of the two bones in the lower foreleg. Generally the more medial of the two. See Fig. 7-1.

ramus (pl. rami). The horizontal portion of each lower jaw, the portion in which the teeth are rooted. See Fig. 2-2. (This defines *ramus* as used in this manual. Other authors sometimes use the term to refer to the entire mammalian dentary whereas some use it to refer only to the vertical portion of the dentary.)

random. Having the same probability of occurrence as every other member of a set.

random dispersion. Distribution of individuals in a given area such that there is an equal probability that an individual will occupy any given point in space and the presence of another individual nearby will not affect this probability. See Fig. 37-1C.

range. 1. The difference between the smallest and largest values in a statistical distribution or set of data. **2.** The geographic area inhabited by a particular taxon.

range length. The distance between the most widely separated capture points. See Fig. 37-4D.

ranked variable. See **variable.**

relative age. The age of an individual compared with other individuals in a sample.

rattrap. A commercial snap trap that can kill mammals the size of the Norway rat, *Rattus norvegicus.* See Fig. 34-1.

realized niche. The existing niche of a species. See Hutchinson (1957).

recapture. In population studies, the capture of a previously marked or tagged individual.

re-entrant angle. Inward-pointing angle along the margin of a cheek tooth. See Fig. 25-14A.

re-entrant fold. Invagination along the margin of a cheek tooth. See Fig. 25-41.

regression analysis. An investigation of the dependence of one variable on another (the independent) variable.

reingestation. The ingestation by an individual of caecotrophic fecal pellets that the animal voided previously. Occurs in many lagomorphs. See **caecotropic pellet.**

relative age. Age of an individual in terms relative to others in the population. Compare **absolute age.**

relative estimate. In contrast to density estimates, a measure of numbers of individuals that has not been standardized to number of individuals per unit area.

removal trapping. Trapping scheme where individuals are permanently (thru kill trapping) or temporarily (thru live trapping) removed from an area.

reproductive effort. The effort required to reproduce, measured in terms of the decrease in the ability of the organism to reproduce at later times (Wilson 1975).

reproductive value. Symbolized by v_x, the relative number of female offspring remaining to be born to each female of age x (Wilson 1975).

resorption. Process where certain embryos do not complete development and may leave a scar on the uterine wall.

resource matrix. Tabular method for assembling quantitative data on resource categories (column headings) for sets of species or populations (row headings). See Table 31-1.

ricochetal. Pertaining to bipedal movement involving propulsion by both pelvic limbs simultaneously and landing on both pelvic limbs without involvement of pectoral limbs, e.g., locomotion in a kangaroo. See Fig. 8-4.

roadside count. Census of individuals seen along a roadway.

root. The portion of a tooth that lies below the gum line and fills the alveolus. See Fig. 3-1.

rooted tooth. A tooth that has definitive growth (not evergrowing). See Fig. 3-1.

rootless tooth. A tooth that is evergrowing, having a continuously open root canal.

rostrum. The facial region of the skull anterior to a plane drawn through the anterior margins of the orbits. See Fig. 2-1A, C.

rough out. Removal of large muscles and viscera when preparing a skeleton for cleaning by dermestids or other means.

r-selected. Selection favoring rapid rates of population increase, especially prominent in species that specialize in colonizing short-lived environments or undergo large fluctuations in population size (Wilson 1975).

r-strategist (opportunistic species). A species that takes advantage of temporary or local conditions. Compare with **K-strategist.**

rudimentary. Undeveloped.

thorax. In insects, the median body region bearing the jointed legs and if present, the wings.

runway. A worn or otherwise detectable pathway caused by the repeated usage of mammals.

rutting season. Season when mating occurs. Particularly applied to deer and other artiodactyls.

sagittal crest. The medial dorsal ridge on the braincase formed by a coalescense of the temporal ridges. See Fig. 2-1C.

Saharo-Sindian Region. The arid and semiarid areas extending across northern Africa and southwest Asia to the Sind area of northwestern India. Sometimes recognized as a distinct faunal region.

saltation. Form of locomotion characterized by leaping. Includes **ricochetal** and **spring** locomotor movements.

saltatorial. Adapted for saltation (leaping).

sample. In statistics, a collection of individuals, observations, or values selected from a population. See **population** and **sample size.**

sample mean. See **mean.**

sample size, N. The number of individuals, observations, or values that comprise a sample. See **sample** and **population.**

sample statistics. Computed quantities or statistics that are estimates of population statistics or parameters.

sampled count. Direct count of individuals occupying several sample areas within a larger area.

sampling unit. An object, specimen, or other entity upon which measurements are taken.

sanguinivorous. Feeding on blood.

scale. 1. In mammals, one of many flattened, epidermal plates that may cover the tail, feet, and (in pangolins and armadillos only) most of the body. **2.** In insects, a flat unicellular outgrowth of the body-wall.

scansorial. Pertaining to arboreal animals that climb by means of sharp, curving claws, e.g., tree squirrels.

scapula. The shoulder blade. The dorsalmost bone in the pectoral girdle of mammals. See Fig. 7-1.

scat. Mammal feces or droppings. See Fig. 32-3.

scatter diagram. A bivariate or multivariate graphical representation of measurements on sets of characters. See Fig. 11-9.

scavenger. An animal that feeds on dead animal matter that it has not killed.

scent glands. Sweat, sebaceous, or a combination of these two gland types modified for the production of odoriferous secretions.

sciurognath mandible. Type of mandible found in sciuromorph rodents in which the angular process arises medial to the outer border of the incisive alveolar sheath. See Fig. 25-8.

sciuromorph. One of the three suborders in the order Rodentia. Generally, those rodents that have a small to moderately enlarged infraorbital foramen and a sciurognath mandible. Typically, members of the families Aplodontidae, Sciuridae, Anomaluridae, Pedetidae, Geomyidae, Heteromyidae, and Castoridae.

sclerite. Any piece of the insect body wall bounded by sutures which are seams or impressed lines indicating the division of the distinct parts of the body wall.

scrotum. The pouch of skin in which the testes may be situated outside of the abdominal cavity. Permanently present in some mammalian species, seasonally present in others, and never present in still others.

seasonal migration. Movements of individuals, groups, or populations of animals in response to seasonal patterns of resource availability or other factors.

seasonal molts. Molts in species that replace all hair more than once a year. For example, the winter and summer molts of some long-tailed weasels or the Arctic hare.

seasonally polyestrous. Situation where all estrous cycles of a species are restricted to a certain season or period of the year.

sebaceous gland. One of a number of epidermal glands that secrete a fatty substance and usually open into a hair follicle. See Fig. 4-1.

secondont dentition. Cheek teeth that have a cutting or shearing action adapted for a carnivorous diet. The carnassial teeth, found in the order Carnivora, are examples of this type of dentition. See Fig. 23-3, for example.

secondary homonym. See **homonym.**

secondary sex characteristics. External characteristics that distinguish the two sexes but that have no direct role in reproduction.

secondary spermatocyte. A cell (in the wall of a seminiferous tubule of a testis) that is an early stage in the production of a sperm cell. Formed by division of a primary spermatocyte and later divides to form spermatids.

selenodont. Type of dentition characterized by molariform teeth with a crown pattern of longitudinally oriented, crescent-shaped ridges formed by elongated cusps. See Fig. 3-15.

semelparous (noun is **semelparity**). The production of a single offspring or litter during the lifetime of an individual.

semen (adj., seminal). Collectively, sperm and the secretions of various glands associated with the male reproductive tract. The ejaculate.

semiaquatic. Pertaining to mammals partially, but not fully, adapted to life in water—for example, otters, beavers, water shrews.

semifossorial. Pertaining to mammals that are partially, but not fully, adapted for life underground—for example, ground squirrels, badgers, European rabbits.

seminal vesicles. Vesicular glands; swollen portions of the male reproductive tract that secrete a thick, liquid component of the semen. See Fig. 9-5.

seminiferous tubule. Any one of the numerous tubules (in the testes) that are the sites of sperm formation. See Fig. 9-9.

senior homonym. See **homonym.**

senior synonym. See **synonym.**

Sertoli cell. The supporting cells in the seminiferous tubules of the testis. Distinguished from the germinal cells (e.g., spermatogonium) by having a vesicular nucleus and a compound nucleolus. See Fig. 9-9.

sesamoid bone. Any ossification in a tendon. For example, the patella.

sexual dimorphism. The condition that exists when there is an externally apparent difference other than external genitalia between the males and females of a particular species.

Sherman live trap. A commercial brand of live trap, constructed of galvanized iron or aluminum. See Fig. 34-9.

sibling species. Species that are reproductively isolated but morphologically indistinguishable, or virtually so.

sign. Any indication of an animal's presence, e.g., footprints, scats, burrows, runways.

significance level. The probability (α) of making a Type-I error, expressed as a percentage. For example, if $\alpha = 0.01$, then the significance level is 1% for a given test.

simplex uterus. A uterus that consists of a single medial structure. See Fig. 9-4D.

sister groups. In cladistics, the two groups resulting from each bifurcation of the phylogeny.

skull. The skeleton of the head. Includes the cranium and mandible. See Figs. 2-1, 2-2.

slip. A skin is said to "slip" when it has decomposed to the stage where large patches of hair and epidermis can be easily pulled away from the dermis.

sloth movement. A quadrupedal walk by an animal that is suspended from all four limbs. See Fig. 8-15.

snap trap. Kill traps that usually consist of a wooden base with a wire bail, a spring, and a trigger mechanism. Designed primarily to catch small rodents and insectivores. See **rattrap, mousetrap** and **Museum Special.** See Fig. 34-1.

spatial territory. Basic type of territory involving a fixed or changing area.

specialist. A species with narrow preferences of food, habitat, or other factors. Compare with **generalist.**

speciation. The evolutionary process that produces new species. See chapter 11.

species. A group of interbreeding natural populations that are reproductively isolated from other such groups (Mayr 1969).

species accounts. A chronological listing of all observations pertaining to a species; a portion of the field notes. See Fig. 33-5.

species-group name. A name that has been proposed for a species or subspecies.

specific name. The second word of a binomen or trinomen which, in conjunction with the generic name, forms the name of a species. See Fig. 11-1.

sperm (pl. sperm or sperms). See **spermatozoon.**

spermatid. The cell that elongates and grows a tail to form a spermatozoon (sperm cell). See Fig. 9-9.

spermatogonium (pl. spermatogonia). Any one of the large primary germinal cells (in the seminiferous tubules) that begin dividing and ultimately form spermatozoa. See Fig. 9-9.

spermatozoon (pl. spermatozoa). Sperm. The kind of gamete produced by the male.

sphincter. Any muscle having its fibers in a circular arrangement around an opening or passage so that upon contraction that opening or passage is closed.

spine. A thick, sharply pointed guard hair that serves primarily for protection.

spring. A quadrupedal leap in which a mammal propels itself forward with both pelvic limbs and lands first on both pectoral limbs, for example, the leap of a rabbit.

spur. In mammals, a pointed projection of keratinized material from the ankle of male monotremes.

squamosal. The bone that forms the major portion of the lateral wall of the braincase and the posterior root of the zygomatic arch. See Fig. 2-1B, C.

standard deviation. The square root of the variance, expressed in the same units as the original observations.

$$(s = \sqrt{s^2})$$

standard deviation of capture radii. Used by Calhoun and Casby (1958) and others to estimate the dimensions of the home range of an animal. See **standard deviation.**

standard error of the mean. The square root of the quotient obtained by dividing the variance by the sample size.

$$(s_{\bar{x}} = \sqrt{s^2/N})$$

standardized normal distribution. A normal distribution in which $\mu = 0$ and $\sigma = 1$.

standard-minimum grid. A trapping grid consisting of 16 rows and 16 columns, yielding 256 trap stations each of which has two traps.

standard scores. See **Z-values.**

stapes. The innermost of the three middle ear ossicles. A small, stirrup-shaped bone derived from the columella of reptiles.

static life table. See time-specific life table.

statistics. **1.** The scientific study of numerical data concerning natural phenomena (Sokal and Rohlf 1969). **2.** Any quantities estimated or computed by using the methods of statistics.

stratum corneum. The relatively well-cornified surface layer of the epidermis. See Fig. 4-1.

stratum germination. The basal epidermal layer where mitosis is actively producing the outer layers of cells. See Fig. 4-1.

stigmata. In arthropods, a spiracle or breathing pore.

study skin. A mammal (or bird) skin prepared for placement in a research collection. See Figs. 35-19, 35-20.

stylar shelf. Expanded horizontal portion of the labial cingulum. Frequently bears stylar cusps. See Fig. 3-10.

stylettiform. In arthropods, blade-like mouth parts, long thin piercing structures.

subadult. An individual, generally smaller than an adult, that may be a young-of-the year and may or may not be in breeding condition.

subadult pelage. A pelage that is characteristic of subadult mammals. This term is used only when a distinctive pelage exists between the juvenile and adult pelages.

subcutaneous. Pertaining to a location immediately beneath the skin.

subgeneric name. The name of a subgenus. Subgenus is an optional category between the genus and the species. The name of a subgenus is enclosed in parentheses after the name of a genus, and is not considered a part of the trinomen or binomen. See Fig. 11-1.

subspecies. A relatively uniform and genetically distinct portion of a species representing a separately or recently evolved lineage with its own evolutionary tendencies, definite geographic range, and a narrow zone of intergradation (Lidicker 1962).

subspecific name. The third word of a trinomen. Used with the generic and specific names as the name of a subspecies. See Fig. 11-1.

subunguis. The ventral, cornified portion of a claw, nail, or hoof. It is softer than the unguis. See Figs. 6-1 through 6-3.

subungulates. Proboscideans, sirenians, and hyracoidians, collectively. Tubulidentates are also sometimes included in this group.

sudoriferous (=apocrine sweat) **gland.** See **sweat glands.**

sum of squares (abbreviated SS). A measure of the dispersion of a set of values around a central point prior to division by the degrees of freedom (generally N-1).

supraoccipital. See **occipital bone.**

survival-fecundity rate of increase, r_s. The instantaneous rate of increase in a population where resources may be limiting (Caughley 1977).

survival rate, p_x. The proportion of individuals alive at age x that survive to age $x + 1$. This statistic is calculated using the formula $p_x = 1\text{-}q_x$.

survivorship. The probability (l_x) **that** an individual in a population will survive to any particular age (x).

suture. An immovable contact zone between two bones, particularly in the skull. See Fig. 2-1.

sweat. Perspiration. A very dilute aqueous solution that contains small amounts of inorganic salts and certain nitrogenous excretory products.

sweat glands. Long, tubular epidermal glands that extend into the dermis and that secrete perspiration and/or various scents. *Apocrine sweat glands* or *sudoriferous glands* generally empty their secretions into a hair follicle and produce the odor component of perspiration. *Eccrine sweat glands* (Fig. 4-1), open directly onto the skin surface, and produce most of the fluid component of perspiration and function primarily to regulate body temperature by evaporation of this liquid.

sympatric. Pertaining to two or more populations which occupy overlapping geographical areas. See Fig. 11-2A.

symphysis. A relatively immovable articulation between two bones, e.g., the pubic symphysis of the pelvic girdle.

synaptomorphy. In cladistics, the presence of shared derived characters. See Fig. 11-6C.

syndactylous. Pertaining to two or more digits that are fused together. See Figs. 15-3, 15-4.

synonym. Each of two or more names applied to the same taxon. The *senior synonym* is the name with the earliest date of publication. A *junior synonym* has a later date of publication.

synonymy. A list of the scientific names that have been applied to a particular taxon along with pertinent bibliographic details and comments.

synopsis. Summarized description of a taxon.

synplesiomorphy. In cladistics, the presence of shared primitive characters. See Figs. 11-5A, B.

syntype. Any one of two or more specimens that serve as "types" for a nominal species or subspecies because no holotype was designated. Designation of syntypes is discouraged. See Chapter 11.

systematics. The field of science concerned with taxonomy and phylogeny.

systematist. A scientist who does research in systematics.

t-test. Statistical test, named after Student, to compare two sample means and utilizing the two-tailed *t*-distribution for the evaluation. See Table 29-4.

table of random numbers. Table of random digits used to select an unbiased set of numbers for determining sample locations, order of experimental procedures, or other purpose where randomness is a requirement of the experimental design. See Table 29-1.

tabulating card. A stiff rectangular card, generally measuring 7 3/8 x 3 1/4 inches, that has spaces for 80 numeric or alphabetic characters. See **keypunch machine.**

tagging. The process of uniquely marking animals so that they can be identified when recaptured or observed.

talon. An expansion of the posterior cingulum of an upper cheek tooth. Frequently identifiable only as a cusp, the hypocone.

talonid. An extension of the posterior cingulum of a lower cheek tooth. It squares the outline of the tooth. See Fig. 3-10.

talonid basin. A region of the talonid, frequently forming a basin, surrounded by the hypoconulid, hypoconid, and entoconid.

tarsal bones. Series of bones in the ankle. They are distal to the fibula and tibia and proximal to the metatarsals. See Figs. 7-3, 7-4.

tarsus (pl. tarsi). In arthropods, the jointed appendage attached to the apex of the tibia, consisting of 1 to 5 segments and bearing the pretarsus. See Fig. 36-5.

taxon (pl. taxa). Any group that is distinguished from other groups and believed to represent a distinct species, phylum, etc.

taxonomic category. A rank or level in the classification hierarchy. See Fig. 11-1 for examples.

taxonomist. A scientist who classifies organisms.

taxonomy. The discipline devoted to classifying organisms.

teat. A protuberance of the mammary gland in which numerous small ducts empty into a common collecting structure that in turn opens to the exterior through one or a few pores (through which milk emerges during lactation).

temporal. A paired bone in the braincase of certain mammals. It results from the fusion of the squamosal and tympanic bones.

temporal fossa. The portion of the space bounded laterally by the zygomatic arch and which is posterior to the orbit. See Fig. 2-1A, B.

temporal ridges. A pair of ridges on the top of the braincase of many mammals. Usually originate on the frontal bones near the postorbital processes and converge posteriorly to form the mid-dorsal sagittal crest. See Fig. 2-1A. They mark the dorsal edges of the origins of the external portions of the masseter muscles.

tendon. Cord or band of dense connective tissue attaching a muscle to a skeletal element or to another muscle.

terete. Cylindrical and tapering.

territoriality. The pattern of behavior associated with the defense of a territory. The persistent attachment to a specific territory.

territory. An area defended by an individual or group. By some ecologists (e.g., Pitelka 1959), an exclusive area possessed by an individual or group.

test of significance. A statistical test designed to evaluate the probability of rejecting the null hypothesis (H_0) when it is true.

testis (pl. testes). Gonad of the male. The organ of sperm formation. See Fig. 9-5.

tetrapod. A mammal, bird, reptile, or amphibian; vertebrates other than fishes.

therapsid reptiles. See **†Therapsida.**

†Therapsida. The order of Permian and Triassic reptiles (of the subclass †Synapsida) considered to be ancestral to mammals.

Theria. The subclass that includes marsupial (metatherian) and placental (eutherian) forms of mammals.

three-dimensional home range. Home range estimate in three dimensions, formulated into a general model by Koeppl et al. (1977a). See Fig. 37-5A, B.

tibia. More medial of the two bones between the knee and ankle in the lower hind limb. The shin bone. See Fig. 7-3.

time-specific life table. Life table prepared by aging a sample (imaginary cohort) of individuals that died in a brief period or by knowing the ages of death (e.g., in human population). Compare with **age-specific life table.**

tine. A spike or prong on an antler. See Fig. 5-8.

topotype. A specimen of a particular taxon taken at the type locality of that taxon.

total count. A complete enumeration of all individuals occupying a given area.

tragus. The projection from the lower medial margin of the pinna in most microchiropteran bats. See Fig. 18-3A-C.

trail. A pathway created by repeated use by animals.

transect. In population studies, a line of traps spaced at regular intervals through a habitat. See Fig. 38-1B.

trap line. A line of traps placed at regular (=**transect**) or irregular intervals to secure specimens for identification, study skins, or autopsy purposes.

trap night. A trap set for one night.

trap nights. The number of traps set multiplied by the number of nights on which they were set.

tribosphenic tooth. An upper molar with three main cusps (the trigon) or a lower molar with a trigonid and talonid. Modifications of this type of tooth form the great diversity of many cheek teeth of modern mammals. See Fig. 3-10.

triconodont tooth. Cheek tooth characterized by three to five main cusps, particularly in members of the extinct order †Triconodonta. See Fig. 3-5.

tricuspid. 1. A tooth with three major cusps. 2. Having three points or cusps.

trigon. A triangle formed by three main cusps of an upper molar with the protocone side of the triangle oriented along the lingual edge of the tooth. See Figs. 3-8A, 3-9, 3-10A.

trigonid. A triangle formed by three main cusps of a lower molar with the protoconid side of the triangle oriented along the labial edge of the crown. See Figs. 3-9, 3-10B.

trinomen. A series of three words (the generic, specific, and subspecific names) that constitute the name of a subspecies.

tritubercular tooth. An upper tribosphenic molar or premolar.

tuberculosectorial tooth. A lower tribosphenic molar or premolar.

tunica albuginea (pl. tunicae albugineae). A tough, connective tissue capsule encasing a testis or ovary.

turbinals. Convoluted or scroll-shaped bones in the nasal passages of the skull.

tympanic bone. The bone which forms the ring holding the ear drum or tympanic membrane and which usually forms the major portion of the auditory bulla. See Fig. 2-1B, C.

tympanic bulla. See **auditory bulla.**

tympanic membrane. Eardrum. The thin membranous structure which receives external vibrations from the air and transmits them to the middle ear ossicles.

type. The **holotype.**

Type-I error (symbolized by α, alpha). The rejection of the null hypothesis (H_0) when it is true or accepting a difference when there is none.

Type-II error (symbolized by β, beta). Acceptance of the null hypothesis (H_0) when it is false or failing to find a difference when there is a difference.

type-genus. A genus that is designated as the type of a family or subfamily. The stem of the name of the type genus in combination with various suffixes, forms the names of certain higher categories.

type-species. A species that is designated as the type of a genus or subgenus.

ulna. One of the two bones in the lower pectoral limb. Generally the outermost of the two. See Fig. 7-1.

underhairs. Kinds of hair that serve primarily for insulation: **fur, wool,** and **velli.**

unguligrade. Hoofed. Having a foot structure in which only the unguis (or hoof) is in contact with the ground. See Fig. 8-2, for example.

unguis. The hard, dorsal, keratinized portion of a claw, nail, or hoof. See Figs. 6-1, 6-2, 6-3.

unicuspid. Having a single cusp.

uniform dispersion. The points in space occupied by an individual are approximately equidistant from one another. See Fig. 37-1A.

uninomial. In nomenclature, a name consisting of a single word, e.g., the name of a genus, family, or higher taxonomic category.

univariate statistics. Independent analysis of one variable at a time in a sample or samples.

universe. See **population.**

urethra. In mammals, the tube that carries urine from the urinary bladder to the exterior.

urogenital sinus. A common chamber for the reception of products from the reproductive and urinary systems. In mammals, found in monotremes and marsupials.

uropatagium. A web of skin extending between the hind legs and frequently enclosing the tail—especially in bats. See Fig. 8-19.

uterine migration. Phenomenon occurring in certain mammals with bipartite and bicornuate uteri whereby eggs produced by one ovary may migrate to the uterine horn of the opposite side.

uterus (pl. uteri). In female mammals a muscular expansion of the reproductive tract in which the embryo and fetus develop; opens externally by way of the vagina. Usually paired, but may be partially or completely fused. See **duplex, bipartite, bicornuate,** and **simplex uteri.** See Figs. 9-1 through 9-4.

vagina. That portion of the female reproductive tract that receives the male's penis during copulation, and through which the foetus passes at parturition. See Figs. 9-1 through 9-4.

vaginal smear technique. Procedure for monitoring different stages of the estrous cycle by observing changes in the types of cells lining the vaginal canal. See Chapter 9.

valvular. Capable of being closed, e.g., a valvular nostril.

variable. A quantity or characteristic that may assume any given value or set of values. A *continuous variable* is one for which an infinite number of values for a particular range are possible (e.g., between 1 and 2 there may exist 1.10, 1.117, 1.1129.) A *discontinuous variable* has only a fixed numerical value with no intermediate value possible (e.g., 1, 2, or 3 toes, but not 1.4 toes). An *attribute* is a qualitative measure or property such as black, brown, or blue eyes. *Enumeration* data are attribute data arranged by frequencies. *Ranked variable* data reflect relative (e.g., Class "1," Class "2") rather than absolute differences.

variance. In statistics, the sample variance is a measure of the dispersion of a set of data about the mean, always expressed in squared units. $s^2 = (\Sigma X^2 - (\Sigma X)^2/N)/N-1$. The sample variance ($s^2$) is the best estimate of the population variance (σ^2).

vas deferens (pl. vasa deferentia). Ductus deferens. The tube that carries sperm from the epididymis to the cloaca or urethra in male mammals. See Fig. 9-5.

velli. Very fine, short, dense underhairs.

velvet. The skin covering a growing antler. See Fig. 5-5.

ventral. Pertaining to the under or lower surface.

vermiform. Wormlike, worm-shaped. e.g., the vermiform tongue of many ant eating mammals. See Figs. 20-1B, 21-1, 36-18B, C.

vernier calipers. See **calipers.**

Vertebrata. The subphylum of the phylum Chordata that includes all fishes, amphibians, reptiles, birds, and mammals. The brain is enclosed in a cranium or braincase, and a segmented vertebral column supports the body.

vertical life table. See **time-specific life table.**

vesicular glands. See **seminal vesicles.**

vestigial. Small or degenerate.

vibrissae. Long, stiff hairs that serve primarily as tactile receptors. See Fig. 4-3.

villus (pl. villi). Fingerlike projections of the embryonic portion of the placenta that penetrate into the maternal portion.

volant. Pertaining to the ability to fly. Formerly used (incorrectly) to describe the locomotion of mammals that glide but are not capable of sustained flight.

vomer. Unpaired bone that forms the septum between the nasal passages of the skull.

Wallace's Line. Imaginary line through Indonesia. It separates the Australian from the Oriental Faunal Regions and runs between Bali and Lombok, between Borneo and Celebes, and then continues east of the Philippines. Conceived of by Alfred Russell Wallace (1823-1913) who independently developed the same theory of evolution as Charles Darwin. See Fig. 1-3.

webbed. Having a patagium or patagia; e.g., webbed toes have patagia extending between the digits.

Wiley mill. Machine for grinding food samples into uniform particle sizes for microscopic dietary analysis.

wing. A forelimb modified for sustained flight. Among mammals found only in bats. See Figs. 8-18, 8-19.

wool. Hair or underhair with angora growth. Serves primarily for insulation.

wrist. The joint between the manus and the rest of the forelimb.

xenarthrales. Extra articular surfaces found on the posterior trunk vertebrae of edentates. See Fig. 20-8.

yolk-sac placenta. See choriovitelline placenta.

young of the year. General age description for an animal that was born in the most recent breeding season and is less than a year old. Frequently used when it is difficult to assign individuals to more precise age categories or when a collective term is needed for this group of young animals.

Z-value. A statistic that indicates the number of standard deviations from the mean that an X_i value is located. Sometimes termed **standard scores** or **normal deviates.**

zalambdodont tooth. A tooth characterized by a V-shaped ectoloph on the occlusal surface. See Fig. 3-11.

Zippin method. A census method based on removal trapping on two occasions. See Chapter 38.

zoological nomenclature. A system for providing distinctive names for certain groups of animals. Governed by the *International Code of Zoological Nomenclature* and the opinions and declarations of the *International Commission on Zoological Nomenclature* (*I.C.Z.N.*)

zonary. A placenta with villi and with contyledons arranged in distinct bands or zones.

zonary placenta. Type of chorioallantoic placenta that surrounds the embryo in a band (See Fig. 9-10C).

zygapophysis (pl. zygapophyses). Any one of the articulation points between the vertebrae.

zygomatic arch. An arch of bone that encloses the orbit and temporal fossa laterally and is formed by the jugal bone and processes of the maxilla and squamosal. See Fig. 2-1.

zygomatic plate. In rodents, the expanded and flattened lower maxillary process. See Fig. 25-2C(dd).

zygomatic process of maxilla. Maxillary process; the projection of the maxilla that forms the anterior root of the zygomatic arch. See Fig. 2-1.

zygomatic process of squamosal. Zygomatic process; the projection of the squamosal bone that forms the posterior root of the zygomatic arch. See Fig. 2-1.

zygote. The fertilized egg.

Literature Cited

Adamczewska-Andrzejewska, K. 1973. Growth, variations and age criteria in *Apodemus agrarius* (Pallas, 1771). *Acta Theriologica* 18:353-394.

Adams, C. E. 1972. Ageing and reproduction, Chap. 5, pp. 128-156. *In* C. R. Austin, and R. V. Short. *Reproduction in mammals, book 4: Reproductive patterns.* Cambridge Univ. Press, London.

Adams, L. 1957. A way to analyze herbivore food habits by fecal examination. *Trans. N. American Wildlife Conf.* 22:152-159.

———, and S. D. Davis 1967. The internal anatomy of home range. *J. Mammal.* 48:529-536.

Afanasev, A. V., V. S. Barzhanov, et al. 1953. [*The animals of Kazakhstan.*] Alma-Alta. Akademia Nauk Kazakhskoi SSR. 536/pp. [In Russian]

Alcoze, T. M., and E. G. Zimmerman 1973. Food habits and dietary overlap of two heteromyid rodents from the mesquite plains of Texas. *J. Mammal.* 54:900-908.

Allen, G. M. 1939. A checklist of African mammals. *Bull. Mus. Comp. Zool., Harvard* 83:1-763. (Reprinted 1954).

Allen, J. A. 1924. Carnivora collected by the American Museum Congo expedition. *Bull. Amer. Mus. Nat. Hist.* 47:1-283.

Altevogt, R. 1975. Elephants, pp. 478-480. *In* B. Grzimek (Ed.). *Animal life encyclopedia, Vol.* 12, *Mammals III.* Van Nostrand Reinhold, New York.

———, and F. Kurt 1975. Asiatic elephants, pp. 484-500. *In* B. Grzimek, (Ed.). *Animal life encyclopedia,* Vol. 12, *Mammals III.* Van Nostrand Reinhold, New York.

Altmann, J. 1974. Observational study of behavior: sampling methods. *Behaviour* 49:227-267.

Andersen, K. 1912. *Catalog of the Chiroptera in the collection of the British Museum. Vol. 1. Megachiroptera.* British Museum (Nat. Hist.), London. 854 pp.

Anderson, P. K., George E. Heinsohn, P. H. Whitney, and Jean-Pierre Huang. 1977. *Mus musculus* and *Peromyscus maniculatus*: homing ability in relation to habitat utilization. *Can. J. Zool.* 55:169-182.

Anderson, R. M. 1965. Methods of collecting and preserving vertebrate animals. 4th ed., revised. *Bull. Nat. Mus. Canada* 69:1-199.

Anderson, S. 1961. A new method for preparing lagomorph skins. *J. Mammal.* 42:409-410.

———. 1972. Two semiautomatic systems for linear measurements. *Curator* 15:220-228.

———. 1969. Taxonomic status of the woodrat, *Neotoma albigula,* in southern Chihuahua, Mexico. *Misc. Publ. Mus. Natur. Hist., Univ. Kansas.* 51:25-50.

———. 1967. Introduction to the rodents, pp. 206-209. *In* S. Anderson, and J. K. Jones, Jr. (Eds.). *Recent mammals of the world: a synopsis of families.* Ronald Press Co., New York.

———., and J. K. Jones, Jr. (Eds.). 1967. *Recent mammals of the world: a synopsis of families.* Ronald Press Co., New York. 453 pp.

Anderson, T. W. 1958. *An introduction to multivariate statistical analysis.* John Wiley and Sons, New York. 374 pp.

Armitage, K. B. 1974. Male behavior and territoriality in the yellow-bellied marmot. *J. Zool., London* 172:233-265.

Arnett, R. H. (Ed.). 1978. *The naturalists' directory and almanac (International),* 43rd ed. World Natural History Publications, Baltimore. 310 pp.

Asdell, S. A. 1964. *Patterns of mammalian reproduction,* 2nd ed. Cornell Univ. Press, Ithaca. 670 pp.

———. 1965. Reproduction and development. pp. 1-41. *In* W. V. Mayer, and R. G. Van Gelder (Eds.) *Physiological mammalogy,* Vol. II. Academic Press, New York.

Ashlock, P. D. 1971. Monophyly and associated terms. *Syst. Zool.* 20:63-69.

Atchley, W. R., and E. H. Bryant. 1975. Multivariate statistical methods: among groups covariation. *Benchmark Papers Systematic Evolutionary Biology* 1:1-464. Dowden, Hutchinson and Ross, Inc., Stroudsburg, PA.

Austin, C. R. 1972. Fertilization, pp. 103-133. *In* Austin, C. R., and R. V. Short (Eds.). *Reproduction in mammals: Book 1, Germ cells and fertilization.* Cambridge Univ. Press, London.

Bailey, N. T. J. 1951. On estimating the size of mobile populations from recapture data. *Biometrika* 38:293-306.

———. 1952. Improvements in the interpretation of recapture data. *J. Anim. Ecol.* 21:120-127.

Baird, S. F. 1859. *Mammals of North America.* J. B. Lippincott and Co., Philadelphia. 764 pp. + 87 pls.

Baker, E. W., J. H. Camin, F. Cunliffe, T. A. Woolley, and C. E. Yunker. 1958. Guide to the families of mites. *Institute of Acarology Contribution* 3:1-242.

Baker, R. J., and S. L. Williams. 1972. A live trap for pocket gophers. *J. Wildlife Manag.* 36:1320-1322.

Baker, T. G. 1972. Oogenesis and ovulation, pp. 14-45. *In* C. R. Austin, and R. V. Short (Eds.). *Reproduction in mammals: Book 1, Germ cells and fertilization.* Cambridge Univ. Press, London.

Bander, R. B., and W. W. Cochran. 1969. Radio-location telemetry, pp. 95-103. *In* R. H. Giles, Jr. (Ed.). *Wildlife management techniques,* 3rd ed. The Wildlife Society, Washington.

Banks, E. M., R. J. Brooks, and J. Schell. 1975. A radiotracking study of home range and activity of the brown lemming (*Lemmus trimucronatus*). *J. Mammal.* 56:888-901.

Barbour, R. A. 1977. Anatomy of marsupials, pp. 237-272. *In* B. Stonehouse, and D. Gilmore (Eds.). *The biology of marsupials.* Univ. Park Press, Baltimore.

Barbour, R. W., and W. H. Davis. 1969. *Bats of America.* Univ. Press of Kentucky, Lexington. 286 pp., 24 pls.

Barrett-Hamilton, G. E. H. 1910. *A history of British mammals. Vol. 1. Bats.* Gurney and Jackson, London. 263 pp.

Bartholomew, G. A., and P. G. Hoel. 1953. Reproductive behavior of the Alaska fur seal, *Callorhinus ursinus. J. Mammal.* 34:417-436.

Baumgartner, L. L. 1940. Trapping, handling and marking fox squirrels. *J. Wildlife Manag.* 4:444-450.

———, and A. C. Martin. 1939. Plant histology as an aid in squirrel food-habit studies. *J. Wildl. Manag.* 3:266-268.

Beddard, F. E. 1902. Mammalia, Vol. 10. *In* S. F. Harmer, and A. E. Shipley (Eds.). *The Cambridge Natural History.* Macmillan and Co., Ltd., New York. 604 pp.

Benedict, F. A. 1957. Hair structure as a generic character in bats. *Univ. California Publ. Zool.* 59:285-548.

Berger, T. J. and Phillips, J. D. (compilers) 1977. *Index of U.S. Federal Wildlife Regulations.* Assoc. Syst. Collections, Lawrence, Kansas. looseleaf.

Biggers, J. D. 1966. Reproduction in male marsupials. Symp. Zool. Soc. London 15:251-280.

Birney, E. C. 1973. Systematics of three species of woodrats (genus *Neotoma*) in central North America. *Univ. Kansas Mus. Nat. Hist. Misc. Publ.* 58:1-173.

———, R. Jenness and D. D. Baird. 1975. Eye lens proteins as criteria of age in cotton rats. *J. Wild. Manag.* 39:718-728.

Bland, R. G., and H. E. Jaques. 1978. *How to know the insects,* 3rd ed. Wm. C. Brown Co., Dubuque. 409 pp.

Bobrinskii, N. A., B. A. Kuznekov and A. P. Kuzyakin. 1965. [*Key to the mammals of the USSR,* 2nd ed.]. Moscow, 382 pp. [In Russian]

Bock, W. J. 1963. Evolution and phylogeny in morphologically uniform groups. *Amer. Natur.* 97:265-285.

———. 1974. Philosophical foundations of classical evolutionary classification. *Syst. Zool.* 11:375-392.

———. 1977. Adaptation and the comparative method. pp. 57-82. *In* M. K. Hecht, *et al.* (Eds.). *Major patterns in vertebrate evolution.* Nato Advanced Study Institute Series A: Life Sciences Vol. 14.

Bonaccorso, F. J., N. Smythes, and S. R. Humphrey. 1976. Improved techniques for marking bats. *J. Mammal.,* 57:181-182.

Bonde, N. 1977. Cladistic classification as applied to vertebrates, pp. 741-804. *In* M. K. Hecht, *et al.* (Eds.) *Major patterns in vertebrate evolution.* Nato Advanced Study Institute Series A: Life Sciences. Vol. 14.

Braithwaite, R. W., and A. K. Lee. 1979. A mammalian example of semelparity. *Amer. Natur.* 113:151-155.

Bram, R. A. (compiler). 1978. Surveillance and collection of arthropods of veterinary importance. *USDA Agric. Handbook* 518:1-125.

Brander, R. B., and W. W. Cochran. 1969. Radio-location telemetry, pp. 95-108. *In* R. H. Giles, Jr. (Ed.). *Wildlife management techniques.* 3rd. ed. The Wildlife Society, Washington.

Brandt, J. F. 1855. Untersuchugen uber die kraniologischen Entwicklungsstufen und klassifikation der Nager der Jetztwelt. *Mem. Acad. St. Petersbourg.*

Brazenor, C. W. 1950. *The Mammals of Victoria.* Nat. Mus., Victoria, Australia 125 pp.

Broom, R. 1916. On the structure of the skull in Chrysochloria. *Proc. Zool. Soc. London* 1916:449-459 + 2 pls.

Brower, J. E. and J. H. Zar. 1974. *Field and laboratory methods for general ecology.* Wm. C. Brown Co., Dubuque. 194 pp.

Brown, J. C., and D. W. Yalden. 1973. The description of mammals—2. Limbs and locomotion of terrestrial mammals. *Mammal Review* 3:107-134.

Brown, J. C., and D. M. Stoddart. 1977. Killing mammals and general post-mortem methods. *Mammal Review* 7:63-94.

Brown, J. H., and A. K. Lee. 1969. Bergman's rule and climatic adaptation in woodrats (*Neotoma*). *Evolution* 23:329-338.

———, G. A. Lieberman, and W. F. Dengler. 1972. Woodrats and cholla: dependence of a small mammal community on the density of cacti. *Ecology* 53:310-313.

Brown, J. L., and G. H. Orians. 1970. Spacing patterns in mobile animals. *Ann. Rev. Ecol. Syst.* 1:239-262.

Brown, L. E. 1966. Home range and movement of small mammals. *Symp. Zool. Soc. London* 18:111-142.

Bryant, E. H., and W. R. Atchley. 1975. Multivariate statistical methods: within-groups covariation. *Benchmark Papers in Systematic and Evolutionary Biology* 2:1-436. Dowden Hutchinson & Ross, Inc. Stroudsburg, Pennsylvania.

Buchler, E. R. 1976. A chemiluminescent tag for tracking bats and other small nocturnal animals. *J. Mammal.* 57:173-176.

Buechner, H. K. 1961. Territorial behavior in Uganda kob. *Science* 133:698-699.

Burrell, H. 1927. *The platypus.* Angus and Robertson, Ltd., Sydney. 227 pp.

Burt, W. H. 1927. A simple live trap for small mammals. *J. Mammal.* 8:302-304.

———. 1943. Territoriality and home range concepts as applied to mammals. *J. Mammal.* 24:346-352.

———. 1960. Bacula of North American mammals. *Univ. Michigan Mus. Zool. Misc. Publ.* 113:1-76.

———, and R. P. Grossenheider. 1964. *A field guide to the mammals,* 2nd ed. Houghton Mifflin Company, Boston. 284 pp.

Butler, P. M. 1941. A theory of the evolution of mammalian molar teeth. *Amer. J. Sci.* 239:421-450.

———. 1956. The skull of *Ictops* and the classification of the Insectivora. *Proc. Zool. Soc. London* 126:453-481.

———. 1972. The problem of insectivore classification, pp. 253-265. *In* K. A. Joysey, and T. S. Kemp (Eds.). *Studies in vertebrate evolution.* Oliver and Boyd, Edinburgh.

———. 1978. A new interpretation of the mammalian teeth of tribosphenic pattern from the Albian of Texas. *Breviora* 446:1-27.

Butler, S. R., and E. A. Rowe. 1976. A data acquisition and retrieval system for studies of animal social behaviour. *Behaviour* 57:281-287.

Cabrera, A. 1914. *Fauna Iberica. Mamiferos.* Museo Nacional de Ciencias Naturales, Madrid. 441 pp. + 22 pls.

———. 1919. *Genera Mammalium. Monotremata Marsupialia.* Museo Nacional de Ciencias Naturales, Madrid. 177 pp. + 17 pls.

———. 1925. *Genera Mammalium: Insectivora Galeopithecia.* Museo Nacional de Ciencias Naturales, Madrid. 232 pp. + 18 pls.

———. 1932. Los mamiferos de Marruecos. *Trabajos del Museo Nacional de Ciencias Naturales (ser. Zool.)* 57:1-361 + 12 pls.

Cahalane, V. H. 1932. Age variation in the teeth and skull of the whitetail deer. *Cranbrook Inst. Sci., Scient. Publ.* 2:1-14.

Calaprice, J. R., and J. S. Ford. 1969. Digital calipers—an inexpensive electronic measuring and recording device. *Fisheries Res. Board Canada, Tech. Rept.* 141:1-6.

Calhoun, J. B., and J. U. Casby. 1958. Calculation of home range and density of small mammals. *USDHEW, Pub. Health Monog.* 55:1-24.

Cameron, G. N., and D. G. Rainey. 1972. Habitat utilization by *Neotoma lepida* in the Mohave Desert. *J. Mammal.* 53:251-266.

Camin, J. H., and R. R. Sokal. 1965. A method for deducing branching sequences in phylogeny. *Evolution.* 19:311-326.

Campbell, C. B. G. 1974. On the phyletic relationships of the tree shrews. *Mammal Review* 4:125-143.

Carter, D. C. 1970. Chiropteran reproduction, pp. 233-246. *In* B. H. Slaughter, and D. W. Walton (Eds.). *About bats.* Southern Methodist Univ. Press, Dallas.

Case, L. D. 1959. Preparing mummified specimens for cleaning by dermestid beetles. *J. Mammal.* 40:620.

Caughley, G. 1965. Horn rings and tooth eruption as criteria of age in the Himalayan thar, *Hemitragus jemlahicus. New Zealand J. Sci.* 8:333-351.

———. 1966. Mortality patterns in mammals. *Ecology* 47:906-918.

———. 1977. *Analysis of vertebrate populations.* John Wiley & Sons, New York. 234 pp.

———, and L. C. Birch. 1971 Rate of increase. *J. Wildl. Manag.* 35:658-663.

Charlesworth, B., and J. A. Leon. 1976. The relation of reproductive effort to age. *Amer. Natur.* 110:449-459.

Chiarelli, A. B. 1972. *Taxonomic atlas of living primates.* Academic Press, London. 362 pp.

Chiasson, R. B. 1969. *Laboratory anatomy of the white rat.* 3rd ed. Wm. C. Brown Co., Dubuque. 81 pp.

Chitty, D. 1960. Population processes in the vole and their relevance to general theory. *Canadian J. Zool.* 38:99-113.

Christensen, I. 1973. Age determination, age distribution and growth of bottlenose whales, *Hyperoodon ampullatus* (Forster), in the Labrador Sea. *Norwegian J. Zool.* 21:331-340.

Choate, J. R. 1970. Systematics and zoogeography of Middle American shrews of the genus *Cryptotis. Univ. Kansas Publ., Mus. Natur. Hist.* 19:195-317.

———. 1975. Review of: A. F. DeBlase and R. E. Martin. 1974. A manual of mammalogy: with keys to families of the world. *In J. Mammal.* 56:281-283.

Chu Ching. 1974. On the systematic position of the giant panda, *Ailuropoda melanoleuca* (David). *Acta Zoologica Sinica* 20:187-190. [In Chinese with English summary]

Chu, H. F. 1949. *How to know the immature insects.* Wm. C. Brown Co., Dubuque. 234 pp.

Clifford, H. T., and W. Stephenson. 1975. *An introduction to numerical classification.* Academic Press, New York. 229 pp.

Cody, M. L. 1974. Competition and the structure of bird communities. *Monog. Pop. Biol.* 7:1-318.

Coffey, D. J. 1977. *Dolphins, whales and porpoises: an encyclopedia of sea mammals.* Macmillan Publishing Co., Inc., New York. 223 pp.

Cole, L. C. 1954. The population consequences of life history phenomena. *Quart. Rev. Biol.* 29:103-137.

———. 1957. Sketches of general and comparative demography. *Cold Springs Harbor Symp. Quant. Biol.* 22:1-15.

Colwell, R. K., and E. R. Fuentes. 1975. Experimental studies of the niche. *Ann. Rev. Ecol. Syst.* 6:281-310.

———, and D. J. Futuyma. 1971. On the measurement of niche breadth and overlap. *Ecology* 52:567-576.

Constantine, D. G. 1958. An automatic bat-collecting device. *J. Wildlife Manag.* 22:17-22.

———. 1969. Trampa portatil para vampiros usada en programas de campana antirrabica. *Bol. Of. Sanit. Panamer.* 67:39-42.

———. 1970. Bats in relation to the health, welfare, and economy of man, pp. 65-96. *In* W. A. Wimsatt (Ed.) *Biology of bats, Vol. 1.* Academic Press, New York.

Cooley, W. W., and Lohnes, P. R. 1971. *Multivariate data analysis.* John Wiley & Sons, Inc., New York. 364 pp.

Corbet, G. B. 1978. *The mammals of the Palaearctic Region: a taxonomic review.* British Museum (Nat. Hist.) & Cornell Univ. Press, London and Ithaca. 314 pp.

Corliss, J. O. 1972. Priority and stability in zoological nomenclature: resolution of the problem of Article 23b at the Monaco Congress. *Science* 178:1120.

Cormack, R. M. 1968. The statistics of capture-recapture methods. *Oceanography and Marine Biology* 6:455-506.

Couch, L. K. 1942. Trapping and transplanting live beavers. *U.S.D.I., Fish and Wildlife Service, Cons. Bull.* 30:1-20.

Council of Biology Editors, Committee on Form and Style. 1972. *CBE style manual,* 3rd ed. Washington: American Inst. Biol. Sciences. 297 pp.

Cowie, A. T. 1972. Lactation and its hormonal control, pp. 106-143. *In* C. R. Austin, and R. V. Short (Eds.). *Reproduction in mammals: Book 3. Hormones in reproduction.* Cambridge Univ. Press, London.

Cox, G. W. 1976. *Laboratory manual of general ecology,* 3rd ed. Wm. C. Brown Co., Dubuque, Ia. 232 pp.

Cracraft, J. 1974a. Phylogenetic models and classification. *Syst. Zool.* 23:71-90.

———. 1974b. Continental drift and vertebrate distribution. *Ann. Rev. Ecol. Syst.* 5:215-261.

Crandall, L. S. 1964. *The management of wild mammals in captivity.* Univ. Chicago Press, Chicago. 761 pp.

Crompton, A. W., and F. A. Jenkins, Jr. 1967. American Jurassic symmetrodonts and Rhaetic "pantotheres." *Science* 155:1006-1009.

———. 1968. Molar occlusion in late Triassic mammals. *Biol. Rev.* 43:427-458.

Crouch, W. E. 1933. Pocket-gopher control. *U.S.D.A. Farmer's Bull.* 1709:1-20.

Dapson, R. W. 1973. Letter to the editor. *J. Mammal.* 54:804.
———., and J. M. Irland. 1972. An accurate method of determining age in small mammals. *J. Mammal.* 53:100-106.
Davis, D. D. 1964. The giant panda; a morphological study of evolutionary mechanisms. *Fieldiana: Zool. Mem.* 3:1-339.
Davis, W. B. 1974. *The mammals of Texas.* 2nd ed. Texas Parks and Wildlife Dept., Austin. 294 pp.
Davis, W. H. 1970. Hibernation: ecology and physiological ecology, pp. 265-300. *In* W. A. Wimsatt (Ed.). *Biology of bats, Vol. 1.* Academic Press, New York.
Dawson, M. R. 1958. Later Tertiary Leporidae of North America. *Univ. Kansas Paleont. Contrib. Vertebrata* 6:1-75 + 2 plates.
———. 1967. Fossil history of the families of Recent mammals, pp. 12-53. *In* S. Anderson and J. K. Jones, Jr. (Eds.). *Recent mammals of the world: a synopsis of families.* Ronald Press Co., New York.
Day, M. G. 1966. Identification of hair and feather remains in the gut and faeces of stoats and weasels. *J. Zoology, London.* 148:201-217.
Deevey, E. S. 1947. Life tables for natural populations of animals. *Quart. Rev. Biol.* 22:283-314.
DeKeyser, P. L. 1955. *Les mammiferes de l'Afrique Noire Francaise,* 2nd ed. Initiations Africaines, Institut Francais D'Afrique Noire. 426 pp.
Delany, M. J. 1974. The ecology of small mammals. *Inst. Biol. Studies Biology* 51:1-60. Edward Arnold (Publishers), London.
de la Torre, Luis. 1951. A method for cleaning skulls of specimens preserved in alcohol. *J. Mammal.* 32:231-232.
Diersing, V. E. 1980. Systematics and evolution of pygmy shrews (subgenus *Microsorex*) of North America. *J. Mammal.* 61:76-101.
Dietz, R. S., and J. C. Holden. 1970. Reconstruction of Pangea: breakup and dispersion of continents, Permian to present. *J. Geophy. Res.* 75:4939-4956.
Dobson, G. E. 1876. On the peculiar structures in the feet of certain species of mammals which enable them to walk on smooth perpendicular surfaces. *Proc. Zool. Soc. London* 1876:526-535.
———. 1878. *Catalogue of the Chiroptera in the collection of the British Museum.* British Museum (Nat. Hist.), London. 567 pp.
Dorst, Jean, and P. Dandelot. 1970. *A field guide to the larger mammals of Africa.* Houghton Mifflin Company, Boston. 287 pp.
Doty, R. L. (Ed.) 1976. *Mammalian olfaction, reproductive processes, and behavior.* Academic Press, New York. 344 pp.
Doutt, J. K. 1967. Polar bear dens on the Twin Islands, James Bay, Canada *J. Mammal.* 48:468-471.
Dowler, R. C., and H. H. Genoways. 1976. Supplies and suppliers for vertebrate collections. *Museology,* Texas Tech University 4:1-83.
Ducker, G. 1975. Viverrids and aardwolves, pp. 144-184. *In* B. Grzimek (Ed.).*Animal life encyclopedia. Vol. 12, Mammals III.* Van Nostrand Reinhold, New York.
Duncan, P. M. 1877-83. *Cassell's natural history.* Cassell & Co., London. 3 Vols.
Dusi, J. L. 1949. Methods for the determination of food habits by plant microtechnics and histology and their application to cottontail rabbit food habits. *J. Wildlife Manag.* 13:295-298.
Eberhardt, L. L. 1969. Population analysis, pp. 457-495. *In* R. H. Giles, Jr. (Ed.). *Wildlife management techniques,* 3rd ed. The Wildlife Society, Washington.
Egoscue, H. J. 1962. The bushy-tailed wood rat: a laboratory colony. *J. Mammal.* 43:328-337.
Ehrlich, P. R., and A. H. Ehrlich. 1970. *Population, resources, environment: issues in human ecology.* W. H. Freeman Co., San Francisco. 383 pp.
Eisenberg, J. F. 1966. The social organizations of mammals. *Handbuch der Zoologie* 10(7):1-92.
———., and E. Gould. 1970. The tenrecs: a study in mammalian behavior and evolution. *Smithsonian Cont. Zool.* 27:1-137.
———, and D. G. Kleiman. 1972. Olfactory communication in mammals. *Ann. Rev. Ecol. Syst.* 3:1-32.
Eisentraut, M. 1975. Bats, pp. 67-148. *In* B. Grzimek (Ed.). *Animal life encyclopedia. Vol. 11 Mammals II.* Van Nostrand Reinhold, New York.
———. 1976. Das Gaumenfaltenmuster der Saugetiere und seine Bedeutung fur stammesgeschichtliche und taxonomische Untersuchungen. *Bonn. Zool. Monog.* 8:1-214.
Elder, W. H. 1951. The baculum as an age criterion in mink. *J. Mammal.* 32:43-50.
Ellerman, J. R. 1940-1941. *The families and genera of living rodents.* British Museum (Nat. Hist.), London. 2 vols.
———, and T. C. S. Morrison-Scott. 1951. *Checklist of Palearctic and Indian mammals 1758 to 1946.* British Museum (Nat. Hist.), London. 810 pp.
Emry, R. J. 1970. A North American Oligocene pangolin and other additions to the Pholidota. *Bull. Amer. Mus. Nat. Hist.* 142:459-510.
Enders, A. C. 1963. *Delayed implantation.* Univ. Chicago Press, Chicago. 318 pp.
Erickson, J. A., and W. G. Seliger. 1969. Efficient sectioning of incisors for estimating ages of mule deer. *J. Wildlife Manag.* 33:384-388.
Ewer, R. F. 1968. *Ethology of mammals.* Plenum Press. New York. 418 pp.
———. 1973. *The carnivores.* Cornell Univ. Press, Ithaca. 494 pp.
Farris, J. S. 1976. Phylogenetic classification of fossils with Recent species. *Syst. Zool.* 25:271-282.
———. 1977. On the phenetic approach to vertebrate classification, pp. 823-850. *In* M. K. Hecht, *et al.* (Eds.) *Major patterns in vertebrate evolution.* NATO Advanced Study Institute Series, A: Life Sciences, Vol. 14.
Fiedler, W., et al. 1972. Guenons and their relatives, pp. 396-441. *In* B. Grzimek (Ed.) *Animal life encyclopedia. Vol. 10, Mammals I.* Van Nostrand Reinhold, New York.
Findley, J. S. 1967. Insectivores and dermopterans, pp. 87-108. *In* S. Anderson and J. K. Jones, Jr. (Eds.). *Recent mammals of the world: A synopsis of families.* Ronald Press Co., New York.
Finn, F. 1929. *Sterndale's mammals of India.* Thacker, Spink and Co., Bombay. 347 pp.
Fisler, G. F. 1969. Mammalian organizational systems. *Los Angeles County Museum Contrib. Sci.,* 167:1-31.
Fitch, H. S. 1950. A new style live-trap for small mammals. *J. Mammal.* 31:364-365.
Flake, L. D. 1973. Food habits of four species of rodents on a short-grass prairie in Colorado. *J. Mammal.* 54:636-647.
Fleming, T. H. 1971. *Artibeus jamaicensis*: delayed embryonic development in a Neotropical bat. *Science* 171:402-404.
Flower, W. H. 1885. *An introduction to the osteology of the Mammalia.* Macmillan and Co., London. 382 pp.
———, and R. Lydekker. 1891. *An introduction to the study of mammals living and extinct.* Adam and Charles Black, London. 763 pp.

Flowerdew, J. R. 1976. Ecological methods. *Mammal Review* 6:123-159.

Fooden, J. 1972. Breakup of Pangaea and isolation of relict mammals in Australia, South America, and Madagascar. *Science* 175:894-898.

Fracker, S. B., and H. A. Brischle. 1944. Measuring the local distribution of *Ribes. Ecology* 25:283-303.

Fradrich, H. 1972. Tapirs, pp. 17-33. *In* B. Grzimek (Ed.). *Animal life encyclopedia. Vol. 13, Mammals IV.* Van Nostrand Reinhold, New York.

Franson, J. C., P. A. Dahm, and L. D. Wing. 1975. A method for preparing and sectioning mink (*Mustela vison*) mandibles for age determination. *Amer. Midl. Natur.* 93:507-508.

Franz, C. E., O. J. Reichman, and K. M. Van DeGraff. 1973. Diets, food preferences and reproductive cycles of some desert rodents. *US/IBP/Desert Biome Research Memorandum RM* 73-24:1-128.

Fraser, F. C., and P. E. Purves. 1960. Hearing in cetaceans. Evolution of the accessory air sacs and the structure and function of the outer and middle ear in Recent cetaceans. *Bull. British Mus. (Nat. Hist.)* 7:1-140 + 53 pls.

Free, J. C., R. M. Hansen, and P. L. Sims. 1970. Estimating dryweights of foodplants in feces of herbivores. *J. Range Manag.* 23:300-302.

Freeman, P. W. 1977. A multivariate study of the Molossidae (Mammalia: Chiroptera): morphology, ecology, evolution, *Ph.D. Dissertation. Univ. New Mexico,* 269 pp.

French, N. R. 1964. *Description of a study of ecological effects on a desert area from chronic exposure to low level radiation.* AEC Report No. UCLA 12-532, Biology and Medicine, TID-4500.

———, Stoddart, D. M., and B. Bobek. 1975. Patterns of demography in small mammal populations, pp. 73-102. *In* F. B. Golley, *et al.* (Eds.). *Small mammals: their productivity and population dynamics.* I.B.P. Handbook No. 5, Cambridge Univ. Press, Cambridge.

Fretwell, S. D. 1972. Populations in a seasonal environment. *Princeton Monog. Pop. Biol.* 5:1-217.

Friend, M. 1968. The lens techniques. *Trans. 33rd. N. Amer. Wild. Nat. Res. Conf.* 33:279-298.

Friley, C. E., Jr. 1949. Use of the baculum in age determination of Michigan beavers. *J. Mammal.* 30:261-266.

Furman, D. P., and E. P. Catts. 1970. *Manual of medical entomology.* National Press Books, California. 163 pp.

Furrer, R. K. 1973. Homing of *Peromyscus maniculatus* in the channeled scablands of east-central Washington. *J. Mammal.* 54:466-482.

Gandal, C. P. 1954. Age determination in mammals. *New York Acad. Sci. Trans.* (Ser. 2) 16:312-314.

Gardner, A. L. 1977. Feeding habits, pp. 293-350. *In* R. J. Baker, *et al.* (Eds.). Biology of bats of the New World family Phyllostomatidae. Part II. *Texas Tech Univ. Spec. Publ. Mus.* 13:1-364.

Garlough, F. E., J. F. Welch, and H. J. Spencer. 1942. *Rabbits in relation to crops,* U.S.D.I. Fish and Wildlife Serv., Cons. Bull. 11, 20 pp.

Gebcynska, Z., and A. Myrcha. 1966. The method of quantitative determining of the food composition of rodents. *Acta Theriologica* 11:385-390.

Geist, V. 1966. Validity of horn segment counts in ageing bighorn sheep. *J. Wildlife Manag.* 30:634-635.

Genoways, H. H., and J. R. Choate. 1972. A multivariate analysis of systematic relationships among populations of the short-tailed shrew (genus *Blarina*) in Nebraska. *Syst. Zool.* 21:106-116.

———, and J. R. Choate. 1976. Federal regulations pertaining to collection, import, export, and transport of scientific specimens of mammals. *J. Mammal.* 57 (2, supplement):1-9.

———, and J. K. Jones, Jr. 1971. Systematics of southern banner-tailed kangaroo rats of the *Dipodomy phillipsii* group. *J. Mammal.* 52:265-287.

Gerber, J. D., and C. A. Leone. 1971. Immunologic comparisons of the sera of certain phyllostomatid bats. *Syst. Zool.* 20:160-166.

Gervias, M. P. 1855. *Histoire naturelle des mammiferes.* L. Curmer, Paris. 344 pp.

Getz, L. L., and G. O. Batzli. 1974. A device for preventing disturbance of small mammal live-traps. *J. Mammal.* 55:447-448.

Getz, L. L., and M. L. Prather. 1975. A method to prevent removal of trap bait by insects. *J. Mammal.* 56:955.

Gewalt, W. 1975. Phalangers: other phalangers, pp. 110-120. *In* B. Grzimek (Ed.). *Animal life encyclopedia Vol. 10, Mammals III.* Van Nostrand Reinhold, New York.

Gidley, J. W. 1912. The lagomorphs an independent order. *Science* 36:285-286.

Giebel, C. G. 1859. *Die Naturgeschichte des Thierreichs. Book 1: Die Saugethiere.* Verlag von Otto Wigand, Leipzig. 522 pp.

———, and W. Leche. 1874-1900. *Dr. H. G. Bronn's Klassen und Ordnungen des Thier-Reichs wissenschaftlich dargestelltin wort und Bild. Sechster Band. V. Abtheilung. Saugethiere: Mammalia.* C. F. Wintersche Verlagshandlung, Leipzig. 1169 pp. + 121 pls.

Gilmore, R. M. 1943. Mammalogy in an epidemiological study of jungle yellow fever in Brazil. *J. Mammal.* 24:144-162.

Glass, B. P. 1970. Feeding mechanisms of bats, pp. 84-92. *In* B. H. Slaughter, and D. W. Walton (Eds.). *About bats.* Southern Methodist Univ. Press, Dallas.

Golley, F. B., Petrusewicz, and L. Ryszkowski. (Eds.). 1975. *Small mammals: their productivity and population dynamics.* International Biol. Prog. Handbook No. 5, Cambridge Univ. Press, Cambridge. 451 pp.

Goodwin, G. G., and A. M. Greenhall. 1961. A review of the bats of Trinidad and Tobago: descriptions, rabies infection, and ecology. *Bull. Amer. Mus. Natur. Hist.* 122:187-302 + 40 pls.

Graham, W. J., and H. W. Ambrose, III. 1967. A technique for continuously locating small mammals in field enclosures. *J. Mammal.* 48:639-642.

Grant, P. R. 1970. A potential bias in the use of Longworth traps. *J. Mammal.* 51:831-835.

———. 1972. Interspecific competition among rodents. *Ann. Rev. Ecol. Syst.* 3:79-106.

Grasse, P. P., and P. L. Dekeyser. 1955. Ordre des rongeurs, pp. 1321-1525. *In* P. P. Grasse (Ed.). *Traite de zoologie: Anatomie, systematique, biologie, Vol. 17.* Masson et Cie, Paris.

Gray, J. E. 1865. Notices of some apparently undescribed species of sapajous (*Cebus*) in the collection of the British Museum. *Proc. Zool. Soc. London* 1865:824-828 + pl. 45.

———. 1869. *Catalogue of carnivorous, pachydermatous, and edentate Mammalia in the British Museum.* British Museum (Nat. Hist.), London. 398 pp.

———. 1873. *Hand-list of the edentate, thick-skinned and ruminant mammals in the British Museum.* British Museum (Nat. Hist.), London. 176 pp. + 41 pls.

Greenhall, A. M., and J. L. Paradiso. 1968. *Bats and bat banding.* U.S. Dept. Interior, Bureau Sport Fisheries and Wildlife, Resource Publ. 72:1-47.

Gregory, W. K. 1934. A half century of trituberculy: the Cope-Osborn theory of dental evolution. *Proc. American Philos. Soc.* 73:169-317.

Green, H. L. H. H. 1934. A rapid method of preparing clean bone specimens from fresh or fixed material. *Anat. Rec.* 61:1-3.

Greenhall, A. M. 1976. Care in captivity, pp. 89-131. *In* R. J. Baker, *et al.*, (Eds.). Biology of bats of the New World family Phyllostomatidae. Part 1. *Texas Tech Univ. Spec. Publ. Museum* 10:1-218.

Gregory, W. K. 1910. The orders of mammals. *Bull. Amer. Mus. Nat. Hist.* 27:1-524.

Griffin, D. R. 1970. Migrations and homing of bats, pp. 233-264. *In* W. A. Wimsatt (Ed.). *Biology of bats, Vol. 1.* Academic Press, New York.

Griffiths, M. 1968. *Echidnas.* Pergamon Press, Oxford. 282 pp.

———. 1978. *The biology of the monotremes.* Academic Press, New York. 368 pp.

Grodzinski, W., Z. Pucek and L. Ryskowski. 1966. Estimation of rodent number by means of prebaiting and intensive removal. *Acta Theriologica* 11:297-314.

Gromov, I. M., A. A. Guryev, G. A. Novikov, I. I. Sokolov, P. P. Strelkov, and K. K. Chapskiy. 1963. *Mammalian fauna of the USSR.* USSR Academy of Science, Moscow. 2 vols. [In Russian.]

Gromova, V. I. (Ed.). 1962. *Fundamentals of paleontology: A manual for paleontologists and geologists of the USSR.* Vol. XIII. Mammals. trans. from Russian. 1968, Israel Prog. Scient. transl.

Grzimek, B. (Ed.) 1972 and 1975. *Animal life encyclopedia. Vol. 10, Mammals I; Vol. 11, Mammals II; Vol. 12, Mammals III; Vol. 13, Mammals IV.* Van Nostrand Reinhold, New York.

———. 1975. African elephants, pp. 500-512. *In* B. Grzimek (Ed.). *Animal life encyclopedia, Vol. 12, Mammals III.* Van Nostrand Reinhold, New York.

———., and D. Heinemann. 1975. Kangaroos, pp. 147-173. *In* B. Grzimek (Ed.). *Animal life encyclopedia, Vol. 10, Mammals I.* Van Nostrand Reinhold, New York.

Guggisberg, C. A. W. 1960. *Simba.* Hallwag, Berlin. (English translation, 1963, Chilton Books).

Gunderson, H. L. 1976. *Mammalogy.* McGraw-Hill Book Co., New York. 483 pp.

Gunther, M. 1977. Rearing human infants: breast or bottle. *Symp. Zool. Soc. London* 41:277-284.

Guryev, A. A. 1964. [*Fauna USSR. Vol. 3 No. 10, Lagomorpha.*] USSR Academy of Science, Moscow. 275 pp. [In Russian]

Guyer, M. F. 1953. *Animal micrology.* 5th ed. Univ. Chicago Press, Chicago. 327 pp.

Hall, A. V. 1970. A computer-based system for forming identification keys. *Taxon* 19:12-18.

Hall, E. R. 1962. Collecting and preparing study specimens of vertebrates. *Univ. Kansas Mus. Natur. Hist., Misc. Publ.* 30:1-46.

———, and K. R. Kelson. 1959. *The mammals of North America.* Ronald Press, New York. 2 vols.

———, and W. C. Russell. 1933. Dermestid beetles as an aid in cleaning bones. *J. Mammal.* 14:372-374.

Hamilton, W. J. III. 1973. *Life's color code.* McGraw-Hill Book Co., New York. 238 pp.

Hansen, R. M., and J. T. Flinders. 1969. Food habits of North American hares. *Colorado State Univ. Science Series, Range Science Dept.,* 1:1-18.

Hansen, R. M., and D. N. Ueckert. 1970. Dietary similarity of some primary consumers. *Ecology* 51:640-648.

Hanson, W. R. 1963. Calculation of productivity, survival, and abundance of selected vertebrates from sex and age ratios. *Wildlife Monog.* 9:1-60.

———, and F. Graybill. 1956. Sample size in food-habits analysis. *J. Wildlife Manag.* 20:64-68.

Hansson, L. 1969. Home range, population structure and density estimates at removal catches with edge effect. *Acta Theriologica* 14:153-160.

———. 1970. Methods of morphological diet micro-analysis in rodents. *Oikos* 21:255-266.

Haresign, T. 1960. A technique for increasing the time of dye retention in small mammals. *J. Mammal.* 41:528.

Harestad, A. S. and F. L. Bunnel. 1979. Home range and body weight—a reevaluation. *Ecology* 60:389-402.

Harper, F. 1955. The barren ground caribou of Keewatin. *Univ. Kansas Mus. Natur. Hist. Misc. Publ.* 6:1-164.

Harris, R. H. 1959. Small vertebrate skeletons. *Museums Journal* 58:-223-224.

———. 1978. Age determination in the red fox (*Vulpes vulpes*)—an evaluation of technique efficiency as applied to a sample of suburban foxes. *J. Zool., London* 184:91-117.

Harris, V. T. 1952. An experimental study of habitat selection by prairie and forest races of the deermouse, *Peromyscus maniculatus. Cont. Lab. Vert. Biol., Univ. Mich.* 56:1-53.

Harrison, D. L. 1964. *The mammals of Arabia. Vol. 1. Insectivora, Chiroptera, Primates.* Ernest Benn, London. 192 pp. + 59 pls.

Harthoorn, A. M. 1965. Application of pharmacological and physiological principles in restraint of wild animals. *Wildlife Monog.* 14:1-78.

———. 1975. *The chemical capture of animals.* Bailliere-Tindall, London 416 pp.

Hatt, R. T. 1934. The pangolins and aardvarks collected by the American Museum Congo expedition. *Bull. Amer. Mus. Nat. Hist.* 66:643-672 + 8 pls.

———. 1946. Guide to the hall of biology of mammals. *American Museum of Natural History, Science Guide* 76:1-49.

Hayne, D. W. 1949a. Calculation of size of home range. *J. Mammal.* 30:1-18.

———. 1949b. Two methods for estimating populations from trapping records. *J. Mammal.* 30:399-411.

———. 1950. Apparent home range of *Microtus* in relation to distance between traps. *J. Mammal.* 31:26-39.

———. 1978. Experimental design and statistical analyses, pp. 3-13. *In* D. P. Snyder (Ed.). *Populations of small mammals under natural conditions.* Univ. Pittsburgh Pymatuning Lab. Ecol. Spec. Publ. Ser., Vol. 5.

Hediger, H. 1975. Superfamily: river dolphins, pp. 502-503. *In* B. Grzimek, (Ed.). *Animal life encyclopedia. Vol. 11, Mammals II.* Van Nostrand Reinhold, New York.

Heidt, G. A., R. H. Baker, and I. O. Ebert. 1967. Magnetic detection of small mammal activity. *J. Mammal.* 48:330-331.

Heine, H. 1973. Das Herz der Procyonidae, Ailuridae und Ursidae (Carnivora, Fissipedia). Ein Beitrag zur vergleichenden funkitionellen morphologie des Saugetierherzen *Zeitschrift Wiss. Zool.* 186:1-51.

Helfer, J. R. 1953. *How to know the grasshoppers, cockroaches and their allies.* Wm. C. Brown Co., Dubuque. 353 pp.

Helm, J. D., III 1975. Reproductive biology of *Ototylomys* (Cricetidae) *J. Mammal.* 56:575-590.

Henderson, F. R. 1960. Beaver in Kansas. *Univ. Kansas Mus. Nat. Hist., Misc. Publ.,* 26:1-85.

Hennig, W. 1966. *Phylogenetic systematics.* Univ. Illinois Press, Urbana. 263 pp.

———. 1975. "Cladistic analysis or cladistic classification?" A reply to Ernst Mayr. *Syst. Zool.* 24:244-256.

Henshaw, R. E., and R. O. Stephenson. 1974. Homing in the gray wolf (*Canis lupus*). *J. Mammal.* 55:234-237.

Henson, O. W., Jr. 1970. The central nervous system, pp. 58-152. *In* W. A. Wimsatt (Ed.). *Biology of bats, Vol. 2.* Academic Press, New York.

Hershkovitz, P. 1954. *Collecting and preserving mammals for study: a provisional account for museum personnel and field associates.* Chicago Natural History Museum. 48 pp. [mimeo.]

———. 1962. Evolution of Neotropical cricetine rodents (Muridae) with special reference to the phyllotine group. *Fieldiana: Zoology* 46:1-524.

———. 1968. Metachromism or the principle of evolutionary change in mammalian tegumentary colors. *Evolution* 22:556-575.

———. 1970. Metachromism like it is. *Evolution* 24:644-648.

———. 1971. Basic crown patterns and cusp homologies of mammalian teeth, pp. 95-150. *In* A. A. Dahlberg (Ed.). *Dental morphology and evolution.* Univ. Chicago Press, Chicago.

———. 1977. *Living New World monkeys (Platyrrhini) with an introduction to the Primates.* Vol. 1. Univ. Chicago Press, Chicago. 1117 pp.

Herter, K. 1975. The insectivores, pp. 176-257. *In* B. Grzimek (Ed.). *Animal life encyclopedia, Vol. 10, Mammals I.* Van Nostrand Reinhold, New York.

Hildebrand, M. 1952. The integument of Canidae. *J. Mammal.* 33:419-428.

———. 1968. *Anatomical preparations.* Univ. Calif. Press, Berkeley. 100 pp.

Hill, J. E. 1974. A new family, genus and species of bat (Mammalia: Chiroptera) from Thailand. *Bull. British Museum (Nat. Hist.)* 27:301-336.

———. 1977. A review of the Rhinopomatidae (Mammalia: Chiroptera). *Bull. British Mus. (Nat. Hist.).* 32:29-43.

Hinton, H. E. 1945. *A monograph of the beetles associated with stored products. Bull. British Mus. (Nat. Hist.).* 443 pp.

Hoffmeister, D. F., and M. R. Lee. 1963. Cleaning mammalian skulls with ammonium hydroxide. *J. Mammal.* 44:283-284.

———, and E. G. Zimmerman. 1967. Growth of the skull in the cottontail (*Sylvilagus floridanus*) and its application to age-determination. *Amer. Midl. Natur.* 78:198-206.

Hooper, E. T. 1950. Use dermestid beetles instead of cooking pots. *J. Mammal.* 31:100.

———. 1956. Selection of fats by dermestid beetles. *J. Mammal.* 37:125-126.

Hopson, J. A. 1970. The classification of nontherian mammals. *J. Mammal.* 51:1-9.

———, and A. W. Crompton. 1969. Origin of mammals. *Evolutionary Biology* 3:15-72

Horn, H. S. 1966. Measurement of "overlap" in comparative ecological studies. *Amer. Natur.* 100:419-424.

Howard, W. E. 1952. A live trap for pocket gophers. *J. Mammal.* 33:61-65.

———. 1953. A trigger mechanism for small mammal live traps. *J. Mammal.* 34:513-514.

———, and E. M. Brock. 1961. A drift-fence pit trap that preserves captured rodents. *J. Mammal.* 42:386-391.

Hrdy, S. B. 1977. *The langurs of Abu: female and male strategies of reproduction.* Harvard Univ. Press, Harvard. 361 pp.

Hsia Wu-ping, et al. 1964. [*Illustrated description of animals in China-Mammals*] Peking. 104 pp. [In Chinese]

Hull, D. L. 1966. Phylogenetic numericlature. *Syst. Zool.* 15:14-17.

Humason, G. L. 1967. *Animal tissue techniques.* 2nd ed. W. H. Freeman & Co., San Francisco. 569 pp.

Humphrey, S. R. 1971. Photographic estimation of population size in the Mexican free-tailed bat (*Tadarida brasiliensis*). *Amer. Midl. Natur.* 86:220-223.

———. 1974. Zoogeography of the nine banded armadillo (*Dasypus novemcinctus*) in the United States. *Bioscience* 24:457-462.

Hurlbert, S. H. 1978. The measurement of niche overlap and some relattives. *Ecology* 59:67-77.

Hutchinson, G. E. 1957. Concluding remarks. *Cold Springs Harbor Symposia on Quantitative Biology* 22:415-427.

———. 1978. *An introduction to population ecology.* Yale Univ. Press, New Haven 260 pp.

Hyrtl, J. 1855. *Chlamydophori truncati* cum Dasypode gymnuro comparatum examen anatomicum. *Denkschr. K. Akad. Wiss. Wien* 9:1-66.

Inglis, J. M., and C. J. Barstow. 1960. A device for measuring the volume of seeds. *J. Wildlife Manag.* 24:221-222.

Jackson, H. H. T. 1915. A review of the American moles. *N. American Fauna* 38:1-100 + 6 pls.

Jaeger, E. C. 1955. *A source book of biological names and terms.* 3rd ed. Chas. C. Thomas, Springfield, Ill. 323 pp.

Japan Whaling Association. 1978. *Living with whales.* Tokyo. 14 pp.

Jaques, H. E. 1951. *How to know the beetles.* Wm. C. Brown Co., Dubuque. 372 pp.

Jardine, N., and D. McKenzie. 1972. Continental drift and the dispersal and evolution of organisms. *Nature* 235:20-24.

Jardine, R., and R. Sibson. 1971. *Mathematical taxonomy.* John Wiley and Sons, London.

Jennrich, R. I., and F. B. Turner. 1969. Measurement of non-circular home range. *J. Theoret. Biol.* 22:227-237.

Jepsen, G. L. 1966. Early Eocene bat from Wyoming. *Science* 154:1333-1339.

———. 1970. Bat origins and evolution, pp. 1-64. *In* W. A. Wimsatt (Ed.). *Biology of bats, Vol. 1.* Academic Press, New York.

Jewell, P. A. 1966. The concept of home range in mammals. *Symp. Zool. Soc. London* 18:85-109.

Jobling, B. 1939. On some American genera of the Streblidae and their species, with the description of a new species of *Trichobius. Parasitology* 31:486-497.

Johnson, L. A. S. 1968. Rainbow's end: the quest for an optimal taxonomy. *Proc. Linnean Soc. New South Wales.* 93:1-45. [Reprinted 1970. *Syst. Zool.* 19:203-239, with addendum.]

Johnson, W. E., and R. K. Selander. 1971. Protein variations and systematics in kangaroo rats (genus *Dipodomys*). *Syst. Zool.* 20:377-405.

Jolicoeur, P. 1959. Multivariate geographical variation in the wolf *Canis lupus* L. *Evolution* 13:283-299.

Jolly, G. M. 1965. Explicit estimates from capture-recapture data with both death and immigration—stochastic model. *Biometrika* 52:225-247.

Jones, F. W. The mammals of South Australian. Part 1, 1923: The monotremes and carnivorous marsupials; Part 2, 1924: The bandicoots and the herbivorous marsupials; Part 3, 1925: The monodelphia. *Handbook Flora Fauna South Australia,* Adelaide. 458 pp. (reprint as one volume 1968)

Jones, J. K., Jr., and S. Anderson. 1970. Readings in mammalogy. *Monog. Mus. Natur. Hist., Univ. Kansas.* 2:1-586.

———, S. Anderson, and R. S. Hoffman. 1976. Selected readings in mammalogy, *Univ. Kansas Monog. Mus. Nat. Hist.* 5:1-640.

———, and D. C. Carter. 1976. Annotated checklist, with keys to subfamilies and genera, pp. 7-38. *In* R. J. Baker, *et al.* (Eds.). Biology of bats of the New World family Phyllostomatidae Part 1. *Texas Tech Univ. Spec. Publ. Mus.* 10:1-218.

———, and H. H. Genoways. 1967. A new subspecies of the fringe-tailed bat, *Myotis thysanodes,* from the Black Hills of South Dakota and Wyoming. *J. Mammal.* 48:231-235.

———, and R. R. Johnson. 1967. Sirenians, pp. 366-373. *In* S. Anderson, and J. K. Jones, Jr. (Eds.). *Recent mammals of the world: a synopsis of families.* Ronald Press Co., New York.

Jonsgard, A. 1969. Age determination of marine mammals, pp. 1-30. *In* H. T. Andersen (Ed.). *The biology of marine mammals.* Academic Press, London.

Jorgensen, C. D. 1968. Home range as a measure of probable interactions among populations of small mammals. *J. Mammal.* 49:104-112.

Justice, K. E. 1961. A new method for measuring home range of small mammals. *J. Mammal.* 55:309-318.

Kaufman, D. W., G. C. Smith, R. M. Jones, J. B. Gentry, and M. H. Smith. 1971. Use of assessment lines to estimate density of small mammals. *Acta Theriol.* 16:127-147.

Kaufmann, J. H. 1962. The ecology and social behavior of the coati, *Nasua narica,* on Barro Colorado Island Panama. *Univ. Calif. Publ. Zool.* 60:95-222.

Kawamichi, T. 1976. Hay territory and dominance rank of pikas (*Ochotona princeps*). *J. Mammal.* 57:133-148.

Kaye, S. V. 1961. Movements of harvest mice tagged with Gold-198. *J. Mammal.* 42:323-337.

Keith, J. O., R. M. Hansen, and A. L. Ward. 1959. Effect of 2,4D on abundance and foods of pocket gophers. *J. Wildlife Manag.* 23:137-145.

Kellogg, R. 1936. A review of the Archaeoceti. *Carnegie Inst. Publ.* 482:1-366 + 37 pls.

Kielan-Jaworowska, Z. 1975. Late Cretaceous mammals and dinosaurs from the Gobi Desert. *Amer. Sci.* 63:150-159.

King, J. A. 1955. Social behavior, social organization, and population dynamics in a black-tailed prairiedog town in the Black Hills of South Dakota. *Cont. Lab. Vert. Biol., Univ. Michigan* 67:1-123 + 4 pls.

Kingsley, J. S. (Ed.). 1884. *The Riverside natural history. Vol. V. Mammals.* Houghton Mifflin, Boston. 541 pp.

Kirsch, J. A. W. 1977. The classification of marsupials, pp. 1-50. *In* D. Hunsaker II (Ed.). *The biology of marsupials.* Academic Press, New York.

———, and J. H. Calaby. 1977. The species of living marsupials: an annotated list, pp. 9-26. *In* B. Stonehouse, and D. Gilmore (Eds.). *The biology of marsupials.* Univ. Park Press, Baltimore.

Kirtpatrick, C. M., and R. A. Hoffman. 1960. Ages and reproductive cycles in a male gray squirrel population. *J. Wildlife Manag.* 24:218-221.

Klevezal, G. A. and S. E. Kleinenberg. 1967. *Age determination of mammals from annual layers in teeth and bones.* Acad. Sci. USSR. Trans. from Russian, 1969. Israel prog. Scient. Transl.

Knowlton, F. F., E. D. Michael, and W. C. Glazener. 1964. A marking technique for field recognition of individual turkeys and deer. *J. Wildlife Manag.* 28:167-170.

Koeppl, J. W., N. A. Slade, and R. S. Hoffman. 1975. A bivariate home range model with possible application to ethological data analysis. *J. Mammal.* 56:81-90.

———, N. A. Slade, K. S. Harris, and R. S. Hoffman. 1977a. A three-dimensional home range model. *J. Mammal.* 58:213-220.

———, N. A. Slade, and R. S. Hoffman. 1977b. Distance between observations as an index of average home range size. *Amer. Midl. Natur.* 98:476-482.

Koopman, K. F. 1967. Artiodactyls, pp. 385-406. *In* S. Anderson, and J. K. Jones, Jr. (Eds.). *Recent mammals of the world: a synopsis of families.* Ronald Press Co., New York.

———, and J. K. Jones, Jr. 1970. Classification of bats, pp. 22-28. *In* B. Slaughter, D. W. Walton (Eds.). *About bats.* Southern Methodist Univ. Press, Dallas.

Korschgen, L. J. 1969. Procedures for food-habits analyses, pp. 233-250. *In* R. H. Giles, Jr. (Ed.). *Wildlife management techniques,* 3rd ed. The Wildlife Society, Washington.

Kosin, I. L. 1972. The growing importance of Russian as a language of science. *BioScience* 22:723-724.

Kowalski, K. 1976. *Mammals, an outline of theriology.* Warsaw. 617 pp.

Krantz, G. W. 1970. *A manual of acarology.* Oregon State Univ. Book Stores, Corvallis. 335 pp.

———. 1978. Collection, rearing, and preparation for study, pp. 77-98. *In* G. W. Krantz. *A manual of acarology,* 2nd ed. Oregon State Univ. Bookstores, Corvallis.

Krebs, C. J. 1966. Demographic changes in fluctuating populations of *Microtus californicus. Ecol. Monog.* 36:239-273.

———. 1972. *Ecology: the experimental analysis of distribution and abundance.* Harper and Row, Publishers, New York. 694 pp.

———, M. S. Gaines, B. L. Keller, J. H. Meyers, and R. H. Tamarin. 1973. Population cycles in small rodents. *Science* 179:35-41.

———, B. L. Keller, and R. H. Tamarin. 1969. *Microtus* population biology: demographic changes in fluctuating populations of *M. ochrogaster* and *M. pennsylvanicus* in southern Indiana. *Ecology* 50:577-607.

Kurt, F., and H. Wendt. 1975. Sirenians, pp. 523-533. *In* B. Grzimek (Ed.). *Animal life encyclopedia. Vol. 12, Mammals III.* Van Nostrand Reinhold, New York.

Kuzyakin, A. P. 1950. [Bats]. Moscow, 443 pp. [In Russian]

Lackey, J. A. 1976. Reproduction, growth, and development in the Yucatan deer mouse, *Peromyscus yucatanicus. J. Mammal.* 57:638-655.

Lande, R. 1977. On comparing coefficients of variation. *Syst. Zool.* 26:214-217.
Landry, S. O., Jr. 1957. The interrelationships of the New and Old World hystricomorph rodents. *Univ. California Publ. Zool.* 56:1-118.
———. 1977. Review of: I. W. Rowlands, and B. J. Weir (Eds.). 1974. The biology of hystricomorph rodents. *In J. Mammal.* 58:459-461.
Lane-Petter, W., et al., (Eds.). 1967. *The UFAW handbook on the care and management of laboratory animals.* 3rd ed. Williams & Wilkins, Baltimore, 1015 pp.
Lang, E. M. 1972. Introduction [in part] and Asiatic rhinoceros, pp. 36-48. *In* B. Grzimek (Ed.). *Animal life encyclopedia. Vol. 13, Mammals IV.* Van Nostrand Reinhold, New York.
Lange, H., and J. P. Chapin. 1917. Notes on the distribution and ecology of central African Chiroptera. *In* American Museum Congo Expedition Collection of Bats. *Bull. Amer. Mus. Natur. Hist.* 37:479-563.
Lavocat, R. 1962. Réflexions sur l'origine et la structure du groupe des rongeurs. *Coll. Internat. Centre Nat. Recher. Sci.* 104:287-299.
Lawler, T. E. 1969. The principle of metachromism: a critique. *Evolution* 23:509-512.
Lawrence, B., and W. H. Bossert. 1967. Multiple character analysis of *Canis lupus, latrans,* and *familiaris,* with a discussion of the relationships of *Canis niger. American Zoologist* 7:223-232.
Lawrence, M. J., and R. W. Brown. 1967. *Mammals of Britain—their tracks, trails and signs.* Blanford Press, London. 223 pp.
Laws, R. M. 1952. A new method of age determination for mammals. *Nature* 169:972-973.
Layne, J. N. 1967. Lagomorphs, pp. 192-205. *In* S. Anderson, and J. K. Jones, Jr. (Eds.). *Recent mammals of the world: a synopsis of families.* Ronald Press, New York.
Lazarus, A. B., and F. P. Rowe. 1975. Freeze-marking rodents with a pressurized refrigerant. *Mammal. Rev.* 5:31-34.
Leuthold, W. 1966. Territorial behavior of Uganda Kob. *Behaviour* 27:255-256.
———. 1977. *African ungulates—a comparative review of their ethology and behavioral ecology.* Zoophysiology and Ecology 8:1-307. Springer-Verlag, Berlin.
Levins, R. 1968. Evolution in changing environments. *Monog. Pop. Biol.* 2:1-120.
Lewontin, R. C. 1966. On the measurement of relative varability. *Syst. Zool.* 15:141-142.
Lidicker, W. Z., Jr. 1962. The nature of subspecies boundaries in a desert roden and its implications for subspecies taxonomy. *Syst. Zool.* 11:160-171.
———. 1973. Regulation of numbers in an island population of the California vole, a problem in community dynamics. *Ecol. Monog.* 43:271-302.
Linnaeus, C. 1758. *Systema naturae per regna tri naturae, secundum classes, ordines, genera, species cum characteribus, differentiis, synonymis, locis.* Edito decima, reformata. Tom. I. Laurentii Salvii, Holmiae. 824 pp.
Little, F. J. 1964. The need for a uniform system of biological numericlature. *Syst. Zool.* 13:191-194.
LoBue, J. P., and R. M. Darnell. 1958. An improved live trap for small mammals. *J. Mammal.* 39:286-290.
Long, A., R. M. Hanson, and P. L. Martin. 1974. Extinction of the Shasta ground sloth. *Geol. Soc. Amer. Bull.* 85:1843-1848.
Lord, R. D., Jr. 1959. The lens as an indicator of age in cottontail rabbits. *J. Wildl. Manag.* 23:358-360.
———, A. M. Vilches, J. I. Maiztegui, and C. A. Soldini. 1970. The tracking board: a relative census technique for studying rodents. *J. Mammal.* 51:828-829.
Lovtrup, S. 1973. Classification, convention and logics. *Zool. Ser.* 2:49-61.
———. 1975. On phylogenetic classification. *Acta Zool.,* Cracow 20:499-523.
———. 1977. Phylogenetics: some comments on cladistic theory and method, pp. 805-822. *In* M. K. Hecht, *et al.* (Eds.). *Major patterns in vertebrate evolution.* NATO Advanced Study Institute, Series A: Life Sciences, Vol. 14.
Lutton, L. M. 1975. Notes on territorial behavior and response to predators of the pika, *Ochotona princeps. J. Mammal.* 56:231-234.
Lydekker, R. 1909. *Guide to the whales, porpoises and dolphins (order Cetacea) exhibited in the Department of Zoology British Museum (Natural History).* British Museum (Nat. Hist.), London. 47 pp.
MacArthur, R. H., and E. O. Wilson. 1967. The theory of island biogeography. *Monog. Popul. Biol.* 1:1-203.
Madson, R. M. 1967. *Age determination of wildlife—a bibliography.* U.S. Dept. Int., Bibliog. 2.
Mahoney, R. 1966. Techniques for the preparation of vertebrate skeletons, pp. 327-351. In *Laboratory techniques in zoology.* Butterworth, Inc., Washington.
Marshall, L. G. 1972. Evolution of the permelid tarsus. *Proc. Roy. Soc. Victoria* 85:51-60.
———. 1978. *Lutreolina crassicaudata. Mammalian Species* 91:1-4.
Martin, A. C. 1946. The comparative internal morphology of seeds. *Amer. Midl. Natur.* 36:513-660.
———, and W. D. Barkley. 1961. *Seed identification manual.* Univ. California Press, Berkeley. 221 pp.
Martin, R. E. 1970. Cranial and bacular variation in populations of spiny rats of the genus *Proechimys* (Rodentia: Echimyidae) from South America. *Smithsonian Contrib. Zool.* 35:1-19.
Martinka, C. J. 1976. Population characteristics of grizzly bears in Glacier National Park, Montana. *J. Mammal.* 55:21-29.
Marvin, U. 1973. *Continental drift: the evolution of a concept.* Smithsonian Institution Press, Washington. 239 pp.
Mathiak, H. A. 1938. A rapid method of cross-sectioning mammalian hairs. *J. Wildlife Manag.* 2:162-164.
Matthew, W. D. 1910. On the osteology and relationships of *Paramys,* and the affinities of the Ischyromyidae. *Bull. Amer. Mus. Natur. Hist.* 28:43-72.
Matthews, L. H. (Ed.) 1968. *The whale.* Simon and Schuster, New York. 287 pp.
———. (Ed.). 1971. *The life of mammals, Vol. 2.* Universe Books, New York. 440 pp.
May, R. M. 1976. Estimating *r:* a pedagogical note. *American Naturalist* 110:496-499.
Mayer, W. V. 1952. The hair of California mammals with keys to the dorsal guard hairs of California mammals. *American Midland Natur.* 48:480-512.
Mayr, E. 1963. *Animal species and evolution.* Harvard Univ. Press, Cambridge. 797 pp.
———. 1965. Numerical phenetics and taxonomic theory. *Syst. Zool.* 14:73-97.
———. 1969. *Principles of systematic zoology.* McGraw-Hill Book Company, New York. 428 pp.
———. 1974. Cladistic analysis or cladistic classification? *Z. Zool. Syst. Evolutions.* 12:94-128.

Maza, A. E., B. G. French, and N. R. Aschwanden. 1973. Home range dynamics in a population of heteromyid rodents. *J. Mammal.* 54:405-424.

M'Closkey, R. T., and D. T. Lajoie. 1975. Determminants of local distribution and abundance in white-footed mice. *Ecology* 56:467-472.

McGaugh, M. H., and H. H. Genoways. 1976. State laws as they pertain to scientific collecting permits. *Museology*, Texas Tech Univ. 2:1-81.

McKay, G. M. 1973. Behavior and ecology of the Asiatic elephant in southeastern Ceylon. *Smithsonian Contrib. Zool.* 125:1-113.

McKenna, M. C. 1972. Possible biological consequences of plate tectonics. *BioScience* 22:519-525.

———. 1975. Toward a phylogenetic classification of the Mammalia, pp. 21-46. *In* W. P. Luckett, and F. S. Szalay (Eds.). *Phylogeny of the primates: a multidisciplinary approach.* Plenum Press, New York.

McNab, B. K. 1963. Bioenergetics and the determination off home range size. *Amer. Natur.* 97:133-140.

Meester, J., and H. W. Setzer (Eds.). 1971. *The mammals of Africa: an identification manual.* Smithsonian Institution Press, Washington, D. C.

Melton, D. A. 1976. The biology of aardvark (Tubulidentata–Orycteropodidae). *Mammal Review* 6:75-88.

Merriam, C. H. 1895. Monographic revision of the pocket gophers family Geomyidae. *North Amer. Fauna* 8:1-220 + 19 pls.

Meserve, P. L. 1976. Food relationships of a rodent fauna in a California coastal sage scrub community. *J. Mammal.* 57:300-319.

———. 1977. Three-dimensional home ranges of cricetid rodents. *J. Mammal.* 58:549-558.

Metcalf, Z. P. 1954. The construction of keys. *Syst. Zool.* 3:38-45.

Metzgar, L. H. 1973. A comparison of trap- and track-revealed home ranges in *Peromyscus. J. Mammal.* 54:513-515.

———, and A. L. Sheldon. 1974. An index of home range size. *J. Wildlife Manag.* 38:546-551.

Michener, C. D. 1964. The possible use of uninominal nomenclature to increase the stability of names in biology. *Syst. Zool.* 13:182-190.

Miles, W. B. 1965. Studies of the cuticular structure of the hairs of Kansas bats. *Search, Univ. Kansas Publ.* 5:48-50.

Millar, J. S. 1973. Evolution of litter size in the pika, *Ochotona princeps* (Richardson). *Evolution* 27:134-143.

———, and F. C. Zwickel. 1972. Determination of age, age structure, and mortality of the pika, *Ochotona princeps* (Richardson). *Canad. J. Zool.* 50:229-232.

Miller, E. H. 1975. Walrus ethology. I. The social role of tusks and applications of multidimensional scaling. *Canad. J. Zool.* 53:590-613.

Miller, F. L. 1974. Age determination of caribou by annulations in dental cementum. *J. Wildlife Manag.* 38:47-53.

Miller, G. S., Jr. 1907. The families and genera of bats. *Bull. U.S. Nat. Mus.* 57:1-282, 14 pls.

———, and J. W. Gidley. 1918. Synopsis of the supergeneric groups of rodents. *J. Wash. Acad. Sci.* 8:431-448.

———, and R. Kellogg. 1955. List of North American Recent mammals. *Bull. U.S. Natl. Mus.* 205:1-954.

Mills, R. S., G. W. Barrett, and M. P. Farrell. 1975. Population dynamics of the big brown bat (*Eptesicus fuscus*) in southwestern Ohio. *J. Mammal.* 56:591-604.

Mineau, P. and D. Madison. 1977. Radio-tracking of *Peromyscus leucopus. Canadian J. Zool.* 55:465-468.

Mitchell, E. D. 1975. Parallelism and convergence in the evolution of Otariidae and Phocidae. *Rupp. P-V. Reun. Cons. Int. Explor. Mer.* 169:12-26.

Mitchell, E., and R. H. Tedford. 1973. The Enaliarctinae a new group of extinct aquatic Carnivora and a consideration of the origin of the *Otariidae. Bull. Amer. Mus. Nat. Hist.* 151:201-284.

Moeller, W. 1975. Edentates, pp. 149-181. *In* B. Grzimek (Ed.). *Animal life encyclopedia, Vol. 11, Mammals II.* Van Nostrand Reinhold, New York.

Moore, A. W. 1940. A live mole trap. *J. Mammal.* 21:223-225.

Moriarty, D. J. 1977. On the use of variance in logarithms. *Syst. Zool.* 26:92-93.

Morisita, M. 1962. Iδ index, a measure of dispersion of individuals. *Res. Pop. Ecol.* 4:1-7.

Morris, D. 1965. *The mammals. A guide to the living species.* Hodden and Stoughton, London. 448 pp.

Morris, P. 1972. A review of mammalian age determination methods. *Mammal Review* 2:69-104.

Morris, R. and D. Morris. 1966. *Men and pandas.* McGraw-Hill Book Co., New York. 223 pp.

Morrison, D. F. 1967. *Multivariate statistical methods.* McGraw-Hill Book Co., New York. 338 pp.

Morse, L. G. 1971. Specimen identification and key construction with time-sharing computers. *Taxon* 20:269-282.

———. 1974. Computer programs for specimen identification, key construction and description printing using taxonomic data matrices. Mus. *Publ., Michigan State Univ.,* 5:1-128.

Morzer Bruyns, W. F. J. 1971. *Field guide of whales and dolphins.* C. A. Mees Uitgeuerij, v.h., Amsterdam, 258 pp.

Mosby, H. S. 1969a. Making observations and records, pp. 61-72. *In* R. H. Giles, Jr. (Ed.). *Wildlife management techniques,* 3rd ed. The Wildlife Society, Washington.

———. 1969b. Reconaissance mapping and map use, pp. 119-134. *In* R. H. Giles, Jr. (Ed.). *Wildlife management techniques,* 3rd ed. The Wildlife Society, Washington.

———, and I. McT. Cowan (revised by Lars Karstad). 1969. Collection and field preservation of biological materials, pp. 259-275. *In* R. H. Giles (Ed.). *Wildlife management techniques,* 3rd ed. The Wildlife Society, Washington.

Mossman, H. W. 1937. Comparative morphogenesis of the fetal membranes and accessory uterine structures. *Contrib. Embryology, Carnegie Inst. Washington.* 158:133-246 + 24 pls.

Murie, A. 1944. The wolves of Mt. McKinley. U.S. Park Service, *Fauna National Parks* 5:1-238.

Murie, O. J. 1954. *A field guide to animal tracks.* Houghton Mifflin Co., Boston. 374 pp.

Musil, A. F. 1963. Identification of crop and weed seeds. *USDA Handbook* 219:1-171 + 43 pls.

Myers, G. T., and T. A. Vaughn. 1964. Food habits of the plains pocket gopher in eastern Colorado. *J. Mammal.* 45:588-598.

Myers, J. H., and C. K. Krebs. 1971. Genetic, behavioral and reproductive attributes of dispersing field voles, *Microtus pennsylvanicus* and *Microtus ochrogaster. Ecol. Monog.* 41:53-78.

Myers, K., J. Carstairs, and N. Gilbert. 1977. Determination of age of indigenous rats in Australia. *J. Wildlife Manag.* 41:322-326.

Myhze, R., and S. Myrberget. 1975. Diet of wolverines (*Gulo gulo*) in Norway *J. Mammal.* 56:752-757.

Nalbandov, A. V. 1976. *Reproductive physiology of mammals and birds,* 3rd. Ed. W. H. Freeman and Co., San Francisco. 334 pp.

Napier, J. R. and P. H. Napier. 1967. *A handbook of living primates.* Academic Press, London. 456 pp.

Nason, E. D. 1948. Morphology of hair of eastern North American bats. *Amer. Midland Natur.* 39:345-361.

Neff, D. J. 1968. The pellet-group count technique for big game—trend, census, and distribution: a review. *J. Wild. Manag.* 32:597-614.

Nelson, G. J. 1971. "Cladism" as a philosophy of classification. *Syst. Zool.* 20:373-376.

———. 1972. Phylogenetic relationship and classification. *Syst. Zool.* 21:227-231.

———. 1974a. Classification as an expression of phylogenetic relationships. *Syst. Zool.* 22:344-359.

———. 1974b. Darwin-Hennig classification: a reply to Ernst Mayr. *Syst. Zool.* 23:452-458.

Neuner, A. M. 1976. *Selgem workbook.* Association of Systematics Collections, Lawrence, Kansas. 51 pp.

New, J. G. 1958. Dyes for studying the movements of small mammals. *J. Mammal.* 39:416-429.

Nichols, J. D., W. Conley, B. Batt, and A. R. Tipton. 1976. Temporally dynamic reproductive strategies and the concept of *r*- and *K*-selection. *Amer. Natur.* 110:995-1005.

Nishiwaki, M. 1963. Taxonomical consideration on genera of Didelphinidae. *Sci. Rept. Whales Res. Inst.,* Tokyo 17:93-104.

Norris, K. S. (Ed.). Committee on Marine Mammals, American Society of Mammalogists. 1961. Standardized methods for measuring and recording data on the smaller cetaceans. *J. Mammal.* 42:471-476.

Novick, A. 1977. Acoustic orientation pp. 74-289. *In* W. A. Wimsett (Ed.). *Biology of bats,* Vol. III. Academic Press, New York.

Novikov, G. A. 1956. [*Carnivorous animals of the USSR*]. USSR. Academy of Sciences, Moscow. 293 pp. [in Russian].

O'Gara, B. W., and G. Matson. 1975. Growth and casting of horns by pronghorns and exfoliation of horns by bovids. *J. Mammal.* 56:829-846.

Ognev, S. I. 1940. *Mammals of the U.S.S.R. and adjacent countries.* Vol. IV. Rodents. English translation, 1966, Israel Prog. Scient. Trans.

———. 1947. *Mammals of the U.S.S.R. and adjacent countries. Vol. V. Rodents.* English translation 1966, Israel Prog. Scient. Trans.

———. 1951. [Papers on the ecology of mammals]. Moscow. 252 pp. [In Russian]

Olson, E. C. 1959. The evolution of mammalian characters. *Evolution* 13:344-353.

Oosting, H. J. 1956. *The study of plant communities,* 2nd ed. W. H. Freeman and Co., San Francisco. 440 pp.

Osborn, H. F. (W. K. Gregory, Ed.). 1907. *Evolution of mammalian molar teeth to and from the triangular type.* The Macmillan Co., New York, 250 pp.

Osgood, W. H. 1921. A monographic study of the American marsupial, *Caenolestes. Field Mus. Nat. Hist.* (*Zool. Ser.*) 14:1-162 + 22 pls.

———. 1924. Review of living Caenolestids with description of a new genus from Chile. *Field Mus. Nat. Hist.* (*Zool. Ser.*) 14:163-172 + 1 pl.

———. 1943. The mammals of Chile. *Field Mus. Nat. Hist.* (*Zool. Ser.*) 30:1-268.

Overton, W. S. 1969. Estimating the numbers of animals in wildlife populations, pp. 403-455. *In* R. H. Giles, Jr. (Ed.). *Wildlife management techniques,* 3rd ed. The Wildlife Society, Washington.

Owen, R. 1866. *On the anatomy of vertebrates* Vol. *II.* Longmans, Green, and Co., London. 592 pp.

———. 1868. *On the anatomy of vertebrates* Vol. III. Longmans, Green, and Co., London. 915 pp.

Pankhurst, R. J. 1971. Botanical keys generated by computer. *Watsonia* 8:357-368.

Parker, P. 1977. An ecological comparison of marsupial and placental patterns of reproduction, pp. 273-286. *In* B. Stonehouse, and D. Gilmore. *The biology of marsupials.* University Park Press, Baltimore.

Patterson, B. 1956. Early Cretaceous mammals and the evolution of mammalian molar teeth. *Fieldiana: Geol.* 13:1-105.

Patton, D. R., and W. B. Casner. 1970. *Port-A-Punch recording and computer summarization of pellet count data.* USDA Forest Service Research Note RM-170:1-7.

Patton, J. L., S. Y. Yang, and P. Myers. 1975. Genetic and morphologic divergence among introduced rat populations (*Rattus rattus*) of the Galapagos Archipelago, Ecuador. *Syst. Zool.* 24:296-310.

Paulik, G. J., and D. S. Robson. 1969. Statistical calculations for change-in-ratio estimators of population parameters. *J. Wildlife Manag.* 33:1-27.

Peaker, M. (Ed.). 1977. Comparative aspects of lactation. *Symp. Zool. Soc. London* 41:1-374.

Pearson, O. P. 1958. A taxonomic revision of the rodent genus *Phyllotis. Univ. California Publ. Zool.* 56:391-496.

———. 1960. Habits of *Microtus californicus* revealed by automatic photographic recorders. *Ecol. Monog.* 30:231-249.

Pedersen, A. 1975. Bears, pp. 117-143. *In* B. Grzimek (Ed.). *Animal life encyclopedia, Vol 12, Mammals III.* Van Nostrand Reinhold, New York.

Pelikan, J. 1970. Testing and elimination of the edge effect in trapping small mammals, p. 57-61. *In* K. Petrusewicz, and L. Ruszkowski (Eds.). *Energy flow through small mammal populations.* Polish Scientific Publishers, Warsaw.

Perkins, H. H. 1954. *Animal tracks; the standard guide for identification and characteristics.* Stackpole, Harrisburg. 63 pp.

Perry, J. W. 1972. *The ovarian cycle of mammals.* Hafner Publishing Co., New York. 219 pp.

———, and I. W. Rowlands (Eds.). 1969. Biology of reproduction in mammals. *J. Reprod. Fertility Suppl.* 6:1-531.

———, and I. W. Rowlands (Eds.). 1973. The environment and reproduction in mammals and birds. *J. Reprod. Fertility Suppl.* 19:1-613.

Peterle, T. J. 1969. Radioisotopes and their use in wildlife research, pp. 109-118. *In* R. H. Giles, Jr. (Ed.). *Wildlife management techniques,* 3rd ed. The Wildlife Society, Washington.

Peterson, R. S., and G. A. Bartholomew. 1967. *The natural history and behavior of the California Sea Lion.* Spec. Publ. No. 1, Amer. Soc. Mammalogists. 79 pp.

Petrides, G. A. 1946. Snares and deadfalls. *J. Wildlife Manag.* 10:234-238.

Petrusewicz, K., and R. Andrzejewski. 1962. Natural history of a free-living population of house mice (*Mus musculus* L.) with particular reference to groupings within the population. *Ekologia Polska,* Ser. A. 10:85-122.

Petter, F. 1964. Affinités du genre *Cricetomys.* Une nouvelle sous-famille de rongeurs Cricetidae, les Cricetomyinae. *C r. hebd. Seanc. Acad. Sci., Paris* 258:6516-6518.

Pettingill, O. S., Jr. 1956. *A laboratory and field manual of ornithology,* 3rd ed. Burgess Publishing Co., Minneapolis. 379 pp.

Pianka, E. R. 1976. Natural selection of optimal reproductive tactics. *Amer. Zool.* 16:775-784.

———, and W. S. Parker. 1975. Age-specific reproductive tactics. *Amer. Natur.* 109:453-464.

Pitelka, F. A. 1959. Numbers, breeding schedule, and territoriality in pectoral sandpipers of northern Alaska. *Condor* 61:233-264.

Pocock, R. I. 1924a. Some external characters of *Orycteropus afer. Proc. Zool. Soc. London* 1924:697-706.

———. 1924b. The external characters of the pangolins (Manidae). *Proc. Zool. Soc. London* 1924:707-723.

———. 1924c. The external characters of the South American edentates. *Proc. Zool. Soc. London* 1924:983-1031.

———. 1926. The external characters of the flying lemur (*Galeopterus temminckii*). *Proc. Zool. Soc. London* 1926:429-444.

Poglayen-Neuwall, I. 1975. Raccoons and pandas, pp. 90-116. *In* B. Grzimek (Ed.). *Animal life encyclopedia, Vol. 12, Mammals III.* Van Nostrand Reinhold, New York.

Poole, R. W. 1971. The use of factor analysis in modeling natural communities of plants and animals. *Ill. Nat. Hist. Surv., Biol. Notes* 72:3-14.

———. 1974. Measuring the structural similarity of two communities composed of the same species. *Researches Population Ecology* 16:138-151.

Popov, B. M. 1956. [*Fauna of the Ukraine. Vol. I.*]. Ukranian Academy of Science, Kiev. 446 pp. [In Russian]

Pucek, Z., and V. P. W. Lowe. 1975. Age criteria in small mammals, pp. 55-72. *In* F. B. Golley, et al (Eds.). *Small mammals: their productivity and population dynamics.* Cambridge I. B. Handbook No. 5. Cambridge Univ. Press.

Quay, W. B. 1970. Integument and derivatives, pp. 1-56. *In* W. A. Wimsatt (Ed.). *Biology of bats,* Vol. II. Academic Press, New York.

Rahm, V. 1975a. Pangolins, pp. 182-188. *In* B. Grzimek (Ed.) *Animal life encyclopedia, Vol. 11, Mammals II.* Van Nostrand Reinhold, New York

———. 1975b. Aardvarks, pp. 473-477. *In* B. Grzimek (Ed.). *Animal life encyclopedia, Vol. 12, Mammals III.* Van Nostrand Reinhold, New York.

———. 1975c. Hyraxes. pp. 513-522. *In* B. Grzimek (Ed.). *Animal life encyclopedia, Vol. 12, Mammals III.* Van Nostrand Reinhold, New York.

Ralls, K. 1971. Mammalian scent marking. *Science* 171:443-449.

Ramsey, C. W. 1968. A drop-net deer trap. *J. Wildlife Manag.* 32:187-190.

Randal, J. M., and G. H. Scott. 1967. Linnean nomenclature: an aid in data processing. *Syst. Zool.* 16:278-281.

Randolph, P. A., J. C. Randolph, K. Mattingly and M. M. Foster. 1977. Energy costs of reproduction in the cotton rat, *Sigmodon hispidus. Ecology* 58:31-45.

Rasweiler, J. J., IV. 1977. The care and management of bats as laboratory animals. pp. 519-617. *In* W. A. Wimsatt (Ed.). *Biology of bats,* Vol. III. Academic Press, New York.

Ray, G. C. 1973. Underwater observation increases understanding of marine mammals. *Mar. Technol. Soc. J.* 7:16-20.

Reichman, O. J. 1975. Relation of desert rodent diets to available resources. *J. Mammal.* 56:731-751.

Reig, O. A. 1970. Ecological notes on the fossorial octodont rodent *Spalacopus cyanus* (Molina). *J. Mammal.* 51:592-601.

———. 1977. A proposed unified nomenclature for the enameled components of the molar teeth of the Cricetidae (Rodentia). *J. Zool.* London 181:227-241.

Repenning, C. A. 1975. Otarioid evolution. *Rapp. P.-V. Reun. Cons. Int. Explor. Mer* 169:27-33.

Rice, D. W. 1967. Cetaceans, pp. 291-324. *In* S. Anderson, and J. K. Jones, Jr. (Eds.) *Recent mammals of the world: a synopsis of families.* Ronald Press Co., New York.

———. 1977. *A list of the marine mammals of the world.* Nat. Ocean. Atmospheric Admin. Tech. Rept. (NMFS/SSRF) 711:1-15.

———, and A. A. Wolman. 1971. The life history and ecology of the gray whale (*Eschrichtius robustus*). *Spec. Publ., Amer. Soc. Mammalog.* 3:1-142.

Ride, W. D. L. 1964. A review of Australian fossil marsupials. *J. Royal Soc. Western Australia* 47:97-131.

———. 1970. *A guide to the native mammals of Australia.* Oxford Univ. Press, Melbourne. 249 pp.

Riggs, E. S. 1933. Preliminary description of a new marsupial sabertooth from the Pliocene of Argentina. *Field Mus. Publ.* (*Geol. Ser.*) 6:61-65.

Robinson, J. W., and R. S. Hoffmann. 1975. Geographical and interspecific cranial variation in big-eared ground squirrels (*Spermophilus*): a multivariate study. *Syst. Zool.* 24:79-88.

Roe, H. S. J. 1967. Seasonal formation of laminae in the ear plug of the fin whale. *Discovery Rept.* 35:1-30.

Roff, D. A. 1973a. On the accuracy of some mark-recapture estimators. *Oecologia* 12:15-34.

———. 1973b. An examination of some statistical tests used in the analysis of mark-recapture data. *Oecologia* 12:35-54.

Rohlf, F. J. 1974. Methods of comparing classifications. *Ann. Rev. Ecol. Syst.* 5:101-113.

———, and R. R. Sokal. 1969. *Statistical tables.* W. H. Freeman and Co., San Francisco 253 pp.

Rolan, R. G., and H. T. Gier. 1967. Correlation of embryo and placental scar counts of *Peromyscus maniculatus* and *Microtus ochrogaster. J. Mammal.* 48:317-319.

Romer, A. S. 1966. *Vertebrate paleontology.* 3rd ed. Univ. Chicago Press, Chicago. 468 pp.

Rose, R. K. 1973. A small mammal live trap. *Trans. Kansas Acad. Sci.* 76:14-17.

Rosenzweig, M. L. 1973. Habitat selection experiments with a pair of co-existing heteromyid rodent species. *Ecology* 54:111-117.

———, and J. Winakur. 1969. Population ecology of desert rodent communities: habitats and environmental complexity. *Ecology* 50:558-572.

Rosevear, D. R. 1969. *The rodents of West Africa.* British Museum (Nat. Hist.), London. 604 pp., 11 pls.

Rowlands, I. W. (Ed.) 1966. Comparative biology of reproduction in mammals. *Symp. Zool. Soc. London* 15:1-559.

Rowlands, I. W., and B. J. Weir. 1974. The biology of hystrocomorph rodents. *Symp. Zool. Soc.* London. 34:1-482.

Ruschi, A. 1953. Algumas observacoes realizadas sobre os Quiropteros do E. Espirito Santo. *Palestra realizada na Faculdade de Ciencias e Filosofia, Rio de Janeiro.*

Russell, W. C. 1947. Biology of the dermestid beetle with reference to skull cleaning. *J. Mammal.* 28:284-287.

Rupp, R. S. 1966. Generalized equation for the ratio method of estimating population abundance. *J. Wildlife Manag.* 30:523-526.

Ryder, M. L. 1962. Rhinoceros horn. *Turtox News* 40:274-277.

Sadleir, R. M. F. S. 1969. *The ecology of reproduction in wild and domestic mammals.* Methuen & Co. 321 pp.

———. 1972. Environmental effects, pp. 69-93. *In* C. R. Austin, and A. V. Short (Eds.). *Reproduction in mammals: Book 4, Reproductive patterns.* Cambridge Univ. Press, London.

———. 1973. *The reproduction of vertebrates.* Academic Press, New York. 227 pp.

Sale, P. F. 1974. Overlap in resource use, and interspecific competition. *Oecologia* 17:245-256.

Sanderson, G. C. 1966. The study of mammal movements—a review. *J. Wildl. Manag.* 30:215-235.

Sarich, V. 1973. The giant panda is a bear. *Nature* 245:218-220.

Sarrazin, J. P. R., and J. R. Bider. 1973. Activity, a neglected parameter in population estimates—the development of a new technique. *J. Mammal.* 54:369-382.

Schaffer, W. M., and C. A. Reed. 1972. The co-evolution of social behavior, and cranial morphology in sheep and goats (Bovidae, Caprinii) *Fieldiana: Zoology* 61:1-88.

Schaller, G. B. 1963. *The mountain gorilla: ecology and behavior.* Univ. Chicago Press, Chicago. 431 pp.

Schevill, W. E. 1974. *The whale problem, a status report.* Harvard Univ. Press, Cambridge, Mass. 419 pp.

Schnabel, Z. E. 1938. The estimation of total fish in a lake. *Amer. Math. Mon.* 45:348-352.

Schultze-Westrum, T. 1975. Colugos or flying lemurs, pp. 64-66. *In* B. Grzimek (Ed.). *Animal life encyclopedia. Vol. 11, Mammals II.* Van Nostrand Reinhold Co., New York.

Schumacher, F. X., and R. W. Eschmeyer. 1943. The estimation of fish populations in lakes and ponds. *J. Tenn. Acad. Sci.* 18:228-249.

Sclater, W. L. 1900. *The mammals of South Africa, Vol. I: Primates, Carnivora and Ungulata.* R. H. Porter, London. 324 pp.

———. 1901. *The mammals of South Africa, Vol. II: Rodentia, Chiroptera, Insectivora, Cetacea and Edentata.* R. H. Porter, London. 241 pp.

———, and P. L. Sclater. 1899. *The geography of mammals.* Kegan Paul, Trench, Trubner & Co., London. 335 pp.

Scott, T. G. 1943. Some food coactions of the northern plains red fox. *Ecol. Monog.* 13:427-479.

Seal, H. L. 1964. *Multivariate statistical analysis for biologists.* John Wiley & Sons, New York. 207 pp.

Searle, S. R. 1966. *Matrix algebra for the biological sciences.* John Wiley & Sons, New York. 296 pp.

Seber, G. A. F. 1965. A note on the multiple-recapture census. *Biometrika* 52:249-259.

———. 1973. *The estimation of animal abundance and related parameters.* Griffin, London. 506 pp.

Seton, E. T. 1958. *Animal tracks and hunter signs.* Doubleday, New York. 160 pp.

Setzer, H. W. 1963. Directions for preserving mammals for museum study. *U.S. Nat. Mus. Info. Leaf.* 380:1-19.

Shadle, A. R., and D. Po-Chedley. 1949. Rate of pentration of porcupine spine. *J. Mammal.* 30:172-173.

Sharman, G. B. 1959. Marsupial reproduction. *Monog. Biol.* 8:332-368.

———, J. H. Calaby, and W. E. Poole. 1966. Patterns of reproduction in female diprotodont marsupials. *Symp. Zool. Soc. London* 15:205-232.

Shemanchuk, J. A., and H. J. Bergen. 1968. The gen trap, a simple, humane trap for Richardson's ground squirrels, *Citellus richardsonii* (Sabine). *J. Mammal.* 49:553-555.

Short, R. V. 1972a. Role of hormones in sex cycles, pp. 42-72. *In* C. R. Austin, and R. V. Short (Eds.). *Reproduction in mammals: Book 3: Hormones in Reproduction.* Cambridge Univ. Press, London.

———. 1972b. Species difference, pp. 1-33. In C. R. Austin, and R. V. Short (Eds.). *Reproduction in mammals: Book 4: Reproductive patterns.* Cambridge Univ. Press, London.

Siegel, S. 1956. *Nonparametric statistics for the behavioral sciences.* McGraw-Hill Book Co., New York. 312 pp.

Silver, J., and A. W. Moore. 1941. Mole control. *U.S. Fish & Wildlife Service, Cons. Bull.* 16:1-17.

Simpson, G. G. 1930. Post-Mesozoic Marsupialia. In Fossilium catalogus. *Animalia* 1:1-87. W. Junk, Berlin.

———. 1930b. Rodent giants. *Natural Hist.* 30:305-313.

———. 1931. *Metacheiromys* and the relationships of the Edentata. *Bull. Amer. Mus. Nat. Hist.* 59:295-381.

———. 1937. The Fort Union of the Crazy Mountain Field, Montana, and its mammalian faunas. *Bull. U. S. Natl. Mus.* 169:1-287.

———. 1945. The principles of classification and the classification of mammals. *Bull. American Mus. Natur. Hist.* 85:1-350.

———. 1961a. Evolution of Mesozoic mammals, pp. 57-95. *In Int. Colloq. Evol. Lower and Non-specialized mammals.* Kon Vlaamse Acad. Wetensch. Lett. Sch. Kunsten Belgie.

———. 1961b. *Principles of animal taxonomy.* Columbia Univ. Press, New York. 434 pp.

———. 1961c. Historical zoogeography of Australian mammals. *Evolution* 15:431-446.

———, A. Roe, and R. C. Lewontin. 1960. *Quantitative zoology,* rev. ed. Harcourt, Brace and Co., New York. 440 pp.

Sinclair, W. J. 1906. Mammalia of the Santa-Cruz beds. Marsupialia. *Rept. Princeton Univ. Exped. Patagonia* 4:333-460.

Sladen, B. K. 1969. Comparative demography, pp. 47-69. *In* B. K. Sladen, and F. B. Bang (Eds.). *Biology of populations: the biological basis of public health.* Amer. Elsevier Publ. Co., New York.

Slijper, E. J. 1962. *Whales.* Hutchinson & Co., London. 475 pp. English translation by A. J. Pomerans. of *Walvissen.* first published in Danish in 1956.

Slijper, E. J. 1976. *Whales and dolphins.* Univ. Mich. Press, Ann Arbor. 170 pp.

———, and O. Heinemann. 1975. Whales, pp. 457-476; Baleen whales, pp. 477-492; Toothed whales, pp. 493-524. *In* B. Grzimek (Ed.). *Animal life encyclopedia, Vol. 11, Mammals II.* Van Nostrand Reinhold, New York.

Slobodkin, L. B. 1961. *Growth and regulation of animal populations.* Holt, Rinehart and Winston, New York. 184 pp.

Smith, J. D. 1972. Systematics of the chiropteran family Mormoopidae. *Misc. Publ., Univ. Kansas Mus. Natur. Hist.* 56:1-132.

———. 1976. Chiropteran evolution, pp. 49-69. *In* R. J. Baker, *et al.* (Eds.). Biology of bats of the New World family Phyllostomatidae. Part 1. *Texas Tech Univ. Mus. Spec. Publ.* 10:1-218.

Smith, M. H., B. J. Boize, and J. B. Gentry. 1973. Validity of the center of activity concept. *J. Mammal.* 54:747-749.

———, Gardner, R. H., J. B. Gentry, D. W. Kaufman, and M. H. O'Farrell. 1975. Density estimation of small mammal populations, pp. 25-53. *In* F. B. Golley, *et al.* (Eds.). *Small mammals: their productivity and population dynamics.* Cambridge Univ. Press, London.

———, J. B. Gentry, and F. B. Golley. 1969. A preliminary report on the examination of small mammal census methods, pp. 25-29. *In* K. Petrusewicz, and L. Ryszkowski (Eds.). *Energy flow through small mammal populations.* Warsaw, Poland.

Smith, R. C., and W. M. Reid. 1972. *Guide to the literature of the life sciences.* 8th ed. Burgess Publishing Co., Minneapolis. 166 pp.

Sneath, P. H. A. 1975. Cladistic representation of reticulate evolution. *Syst. Zool.* 24:360-368.

———, and R. R. Sokal. 1973. *Numerical taxonomy,* 2nd edition. W. H. Freeman and Co., San Francisco. 573 pp.

Snedecor, G. W., and W. G. Cochran. 1967. *Statistical methods,* 6th ed. Iowa State Univ. Press, Ames. 593 pp.

Snyder, D. P. (Ed.). 1978. *Populations of small mammals under natural conditions.* Pymatuning Lab. Ecol. Spec. Publ. Ser., Univ. Pittsburgh, No. 5. 237 pp.

Sokal, R. R. 1965. Statistical methods in systematics. *Biol. Rev.* 40:337-391.

———. 1975. Mayr on cladism—and his critics. *Syst. Zool.* 24:257-262.

———, and J. H. Camin. 1965. The two taxonomies: areas of agreement and conflict. *Syst. Zool.* 14:176-195.

———, and F. J. Rohlf. 1969. *Biometry: The principles and practice of statistics in biological research.* W. H. Freeman and Co., San Francisco. 776 pp.

———, and P. H. A. Sneath. 1963. *Principles of numerical taxonomy.* W. H. Freeman and Co., San Francisco. 359 pp.

Sokolov, I. I. 1959. [*Fauna of the USSR. Vol. 1(3): Perissodactyla and Artiodactyla.*] USSR Acad. Sci., Moscow. 639 pp. [In Russian]

Sommer, H. G., and S. Anderson. 1974. Cleaning skeletons with dermestid beetles—two refinements in the method. *Curator* 17:290-298.

Sonntag, C. F. 1925. A monograph of *Orycteropus afer.* I. Anatomy except the nervous system, skin and skeleton. *Proc. Zool. Soc. London* 1925:331-437 + 1 pl.

Soper, J. D. 1944. On the winter trapping of small mammals. *J. Mammal.* 25:344-353.

Southwick, C. H. 1969. Fluctuations of vertebrate populations, pp. 101-115. *In* B. K. Sladen, and F. B. Bang (Eds.). *Biology of populations: the biological basis of public health.* Amer. Publ. Co., New York.

Southwood, T. R. E. 1966. *Ecological methods with particular reference to the study of insect populations.* Meuthen and Co., London. 391 pp.

Sparks, D. R., and J. C. Malechek. 1967. Estimating percentage dry weights in diets. *J. Range Manag.* 21:203-208.

Spencer, A. W., and H. W. Steinhoff. 1968. An explanation of geographic variation in litter size. *J. Mammal.* 49:281-286.

Spinage, C. A. 1973. A review of age determination of mammals by means of teeth, with especial reference to Africa. *E. African Wildl. J.* 11:165-187.

———, and G. M. Jolly. 1974. Age estimation of warthog. *J. Wildlife Manag.* 38:229-233.

Stains, H. J. 1962. *Game biology and game management: a laboratory manual.* Burgess Publishing Co., Minneapolis. 143 pp.

———. 1967. Carnivores and pinnipeds, pp. 325-354. *In* S. Anderson and J. K. Jones, Jr. (Eds.). *Recent mammals of the world: a synopsis of families.* Ronald Press Co., New York.

Steel, R. G. D., and J. H. Torrie. 1960. *Principles and procedures of statistics.* McGraw-Hill Book Co., New York. 481 pp.

Stehlin, H. G., and S. Schaub. 1951. Die Trigonodontie der simplicidenten Nager. *Schweiz. Palaeont. Abhand.* 67:1-385.

Stenseth, N. C., A. Hagen, E. Ostbye, and H. J. Skar. 1974. A method for calculating the size of the trapping area in capture-recapture studies on small rodents. *Norwegian J. Zool.* 22:253-271.

Stephenson, G. R., D. P. B. Smith, and T. W. Roberts. 1975. The SSR system: an open format event recording system with computerized transcription. *Behavior Research Methods and Instrumentation* 7:497-515.

Stickel, L. F. 1954. A comparison of certain methods of measuring ranges of small mammals. *J. Mammal.* 35:1-15.

Stokes, A. W. (Ed.). 1974. *Territory.* Dowden Hutchinson & Ross, Stroudsburg, Pennsylvania. 398 pp.

Stoneberg, R. P. and C. J. Jonkel. 1966. Age determination of black bears by cementum layers. *J. Wildlife Manag.* 30:411-414.

Stonehouse, B. (Ed.) 1978. *Animal marking: Recognition marking of animals in research.* Univ. Park Press, Baltimore. 224 pp.

Storr, G. M. 1961. Microscopic analysis of faeces, a technique for ascertaining the diet of herbivorous mammals. *Australian J. Biol.* 14:157-164.

Stroganov, S. U. 1957 [*Insectivora of Siberia*] Acad. Sci., SSSR, Moscow. 265 pp. [In Russian]

Stroganov, S. 1962. [*Carnivora of Siberia*] Acad. Sci. SSSR, Moscow 458 pp. [In Russian]

Sullivan, E. G., and A. O. Haugen. 1956. Age determination of foxes by X-ray of forefeet. *J. Wildlife Manag.* 20:210-212.

Szalay, F. S. 1969. Mixodectidae, Microsyopidae, and the insectivore-primate transition. *Bull. American Mus. Natur. Hist.* 140:193-330, pls. 17-57.

———. 1978. Phylogenetic relationship and a classification of the eutherian mammalia, pp. 315-374. *In* M. K. Hecht, *et al* (Eds.) *Major patterns in vertebrate evolution.* Plenum Press, New York.

Taber, R. D. 1969. Criteria of sex and age, pp. 325-401. *In* R. H. Giles, Jr. (Ed.). *Wildlife management techniques.* 3rd. The Wildlife Society, Washington.

———, and I. McT. Cowan. 1969. Capturing and marking wild animals, pp. 277-317. *In* R. H. Giles, (Ed.). *Wildlife manaagement techniques,* 3rd ed. The Wildlife Society. Washington.

Talbot, L. M., and D. R. M. Stewart. 1964. First wildlife census of the entire Serengeti-Mara region, *East Africa J. Wildl. Manag.* 28:815-827.

Tamarin, R. H. (Ed.). 1978. *Population regulations.* Benchmark Papers in Ecology, Vol. 7. Academic Press, Inc., New York. 416 pp.

Tarling, D. H., and M. Tarling. 1971. *Continental drift: a study of the earth's moving surface.* Doubleday and Co., Garden City. 140 pp.

Taylor, K. D., and R. J. Quy. 1973. Marking systems for the study of rat movements. *Mammal Review* 3:30-34.

Tedford, R. H. 1976. Relationship of pinnipeds to other carnivores (Mammalia). *Syst. Zool.* 25:363-374.

Thenius, E. 1972. Phylogeny of Rhinoceros, pp. 34-36. *In* B. Grzimek (Ed.). *Animal life encyclopedia, Vol. 13, Mammals IV.* Von Nostrand Reinhold, New York.

----. 1975a. Phylogeny of Primates, pp. 262-269. *In* B. Grzimek (Ed.). *Animal life encyclopedia, Vol. 11, Mammals II.* Van Nostrand Reinhold, New York.

---. 1975b. Phylogeny of Whales, pp. 457-476. *In* B. Grzimek, (Ed.). *Animal life encyclopedia, Vol. 11, Mammals II.* Van Nostrand Reinhold, New York.

---. 1975c. Phylogeny of Elephants, pp. 480-484. *In* B. Grzimek, (Ed.). *Animal life encyclopedia, Vol. 12, Mammals III.* Van Nostrand Reinhold, New York

---. 1975d. Phylogeny of Hyraxes, pp. 513-514. *In* B. Grzimek (Ed.). *Animal life encyclopedia, Vol. 12, Mammals III.* Van Nostrand Reinhold, New York.

---. 1975c. Phylogeny of Sirenians. p. 525. *In* B. Grzimek (Ed.). *Animal life encyclopedia, Vol. 12, Mammals III.* Van Nostrand Reinhold, New York.

Thomas, O. 1904. On the osteology and systematic position of the rare Malagasy bat *Myzopoda aurita. Proc. Zool. Soc. London* 1904:2-6.

Throckmorton, L. H. 1968. Concordance and discordance of taxonomic characters in *Drosophila* classification. *Syst. Zool.* 17:355-387.

Tiermeir, O. W., and M. L. Plenert. 1964. A comparison of three methods for determining the age of black-tailed jackrabbits. *J. Mammal.* 45:409-416.

Tin Thein, U. 1977. The Burmese fresh-water dolphin. *Mammalia* 41:233-234.

Tinkle, D. W. 1977. The dynamics of populations of squamates, crocodilians and rhynchorephalians, pp. 157-264. *In* C. Gans (Ed.). *Biology of the Reptilia,* Vol. 7. Academic Press, London.

Tomes, C. S. 1894. *A manual of dental anatomy, human and comparative,* 4th Ed. London.

Tomlin, A. G. 1962. *Mammals of the U.S.S.R. and adjacent countries. Vol. 9, Cetacea.* English translation 1967, Israel Prog. Sci. Translation, 717 pp.

Tullberg, T. 1899. *Uber das System der Nagethiere: eine phylogenetische Studie.* Akademischen Buckdrukerei, Upsala. 514 pp.

Turnbull, W. D. 1971. The Trinity therians: Their bearing on evolution in marsupials and other therians, pp. 151-179. *In* A. A. Dahlberg (Ed.) *Dental morphology and evolution.* Univ. Chicago Press, Chicago.

Turner, D. C. 1975. *The vampire bats: a field study in behavior and ecology.* Johns Hopkins Univ. Press, Baltimore. 145 pp.

Tuttle, M. D. 1974. An improved trap for bats. *J. Mammal.* 55:475-477.

Twigg, G. I. 1965. Studies on *Holochilus berbicensis,* a cricetine rodent from the coastal regions of British Guiana. *Proc. Zool. Soc. London* 145:263-283.

---. 1975a. Finding mammals—their signs and remains. *Mammal Review* 5:71-82.

---. 1975b. Catching mammals. *Mammal Review* 5:83-100.

---. 1975c. Marking mammals. *Mammal Review* 5:101-116.

---. 1977. Techniques with captive mammals. *Mammal Review* 7:42-62.

Tyndal-Briscoe, H. 1973. *Life of marsupials.* Edward Arnold Ltd., London. 254 pp.

USDA. 1948. Woody-plant seed manual. *Forest Service, USDA Misc. Publ.* 654:1-416.

van den Brink, F. H. 1967. *A field guide to the mammals of Britain and Europe.* Collins, London. 221 pp.

Vanderbroeck, G. 1967. Origin of the cusps and crests of the tribosphenic molor. *J. Dental Res.* 46:796-804.

Vandermeer, J. H. 1972. Niche theory. *Ann. Rev. Ecol. Syst.* 3:107-132.

Van Gelder, R. G. 1977. Mammalian hybrids and generic limits. *Amer. Mus. Novitates* 2635:1-25.

---. 1978. A review of canid classification. *Amer. Mus. Novitates* 2646:1-10.

Van Valen, L. 1960. Therapsids as mammals. *Evolution* 14:304-313.

---. 1966. Deltatherida, a new order of mammals. *Bull. American Mus. Natur. Hist.* 132:1-126. 8 pls.

Vaughn, T. A. 1972. Mammalogy. W. B. Saunders, Philadelphia. 463 pp.

Vaughn, T. A. 1978. *Mammalogy,* 2nd ed. W. B. Saunders, Philadelphia. 522 pp.

Vinogradov, B. S. 1937. [*Fauna of the USSR. Vol. 3 (4): Dipodidae*]. USSR Acad. Sci., Moscow. 196 pp. [In Russian]

---, and A. I. Argiropulo. 1941. *Fauna of the USSR: Mammals: Key to rodents.* USSR Acad. Sci., Moscow. English translation, 1968, Israel Prog. Scient. Transl. 241 pp.

Volf, J. 1975. Horses, pp. 539-580. *In* B. Grzimek (Ed.). *Animal life encyclopedia Vol. 12, Mammals III.* Van Nostrand Reinhold, New York.

Voorhies, C. T. 1948. A chest for dermestid cleaning of skulls. *J. Mammal.* 29:188-189.

Voss, E. G. 1952. The history of keys and phylogenetic trees in systematic biology. *J. Sci. Labs., Denison Univ.* 43:1-25.

Walker, E. P., F. Warnick, S. E. Hamlet, K. I. Lange, M. A. Davis, H. E. Uible, and P. F. Wright. 1964. Mammals of the world. Johns Hopkins Press, Baltimore. 3 vols.; 2nd ed., 1968, revised by J. L. Paradiso, 2 vols.; 3rd ed., 1975, revised by J. L. Paradiso, 2 vols.

Walker, E. P., F. Warnick, K. I. Lance, H. E. Uible, S. E. Hamlet, M. A. Davis, and P. F. Wright (3rd ed. revised by J. L. Paradiso). 1975. *Mammals of the world.* 3rd ed. John Hopkins Press, Baltimore, 2 vols.

Wallace, J. T., and R. S. Bader. 1967. Factor analysis in morphometric traits of the house mouse. *Syst. Zool.* 16:144-148.

Weber, M. 1928. *Die Saugetiere. Einfuhrung in die Anatomie und Systematik der Recenten und Fossilen Mammalia. Band 2, Systematischer Teil.,* 2nd ed. Gustav Fischer, Jena. 898 pp.

Weir, B. J. 1974. Reproductive characteristics of hystricomorph rodents. *Symp. Zool. Soc. London* 34:265-301.

---, and Rowlands, I. W. 1973. Reproductive strategies of mammals. *Ann. Rev. Ecol. Syst.* 4:139-163.

Wendt, H. 1975a. Goeldi's monkeys, marmosets, and tamarins, pp. 357-395. *In* B. Grzimek (Ed.). *Animal life encyclopedia. Vol. 10, Mammals I.* Van Nostrand Reinhold, New York.

———. 1975b. Introduction to the ungulates, pp. 465-466. *In* B. Grzimek (Ed.). *Animal life encyclopedia. Vol. 12, Mammals III.* Van Nostrand Reinhold, New York.

Wenzel, R. L., and V. J. Tipton (Eds.). 1966. *Ectoparasites of Panama.* Publication No. 1010. Field Museum of Natural History, Chicago. 861 pp.

Westoby, M., G. R. Rost, and J. A. Weis. 1976. Problems with estimating herbivore diets by microscopically identifying plant fragments from stomachs. *J. Mammal.* 57:167-172.

Wetzel, R. M. 1977. The Chacoan peccary, *Catagonus wagneri* (Rusconi). *Bull. Carnegie Mus. Natur. Hist.* 3:1-36.

White, E. G. 1971. A versatile Fortran computer program for the capture-recapture stochastic model of G. M. Jolly. *J. Fish. Res. Board Canada* 28:443-445.

Whitehead, P. J. P. 1971. Storage and retrieval of information in systematic zoology. *Biol. J. Linn. Soc.* 3:211-220.

Wight, H. M., and C. H. Conaway. 1962. A comparison of methods for determining age of cottontails. *J. Wildlife Manag.* 26:160-163.

Wiley, R. W. 1971. Activity periods and movements of the eastern woodrat. *Southwest. Natur.* 16:43-54.

Willcox, W. R., S. P. Lapage, S. Bascomb, and M. A. Curtis. 1973. Identification of bacteria by computer: theory and programming. *J. Gen. Microbiol.* 77:317-330.

Williams, O. 1962. A technique for studying microtine food habits. *J. Mammal.* 43:365-368.

Williams, S. L., R. Laubach, and H. H. Genoways. 1977. A guide to the management of Recent mammal collections. *Carnegie Mus. Natur. Hist., Spec. Publ.,* 4:1-105.

———, M. J. Smolen, and A. A. Brigida. 1979. *Documentation standards for automatic data processing in mammalogy.* The Museum of Texas Tech Univ., Lubbock. 48 pp.

Williamson, V. H. H. 1951. Determination of hairs by impressions. *J. Mammal.* 32:80-85.

Wilson, E. O. 1975. *Sociobiology: the new synthesis.* Harvard Univ. Press, Cambridge. 416 pp.

———, and W. H. Bossert. 1971. *A primer of population biology.* Sinauer Associates, Sunderland, Mass. 192 pp.

———, and W. L. Brown, Jr. 1953. The subspecies concept and its taxonomic application. *Syst. Zool.* 2:97-111.

Windley, B. F. 1977. *The evolving continents.* John Wiley and Sons, London. 371 pp.

Wimsatt, W. A., and A. Guerriere. 1962. Observations on the feeding capacities and excretory functions of captive vampire bats. *J. Mammal.* 43:17-27.

Wood, A. E. 1954. Comments on the classification of rodents. *Breviora* 41:1-9.

———. 1955. A revised classification of the rodents. *J. Mammal.* 36:165-187.

Wood, D. M. 1965. Studies on the beetles *Leptinillus validus* (Horn) and *Platypsyllus castoris* Ritsema (Coleoptera: Leptinidae) from beaver. *Proc. Entomol. Soc. Ontario* 95:33-63.

Woodburne, M. O., and R. H. Tedford. 1975. The first Tertiary monotreme from Australia. *Amer. Mus. Novitates* 2588:1-11.

Zar, J. H. 1974. *Biostatistical analysis.* Prentice-Hall, Englewood Cliffs, N. J. 620 pp.

Zerov, S. A., and E. N. Pavlovskii. 1953. [*Atlas of economic and game birds and animals of the U.S.S.R. Vol. 2. Beasts*] Acad. Sci. SSSR, Moscow. 371 pp. + 76 pls. [In Russian]

Zeuner, F. E. 1963. *History of domesticated animals.* Hutchinson, London. 560 pp.

Zippin, C. 1956. An evaluation of the removal method of estimating animal populations. *Biometrics* 12:163-189.

Index

The coverage in this index is a supplement to entries that appear in the Glossary. **Boldface** type indicates pages that provide substantial coverage of a particular subject while *italic* type refers to an illustration pertaining to that subject.

Aardvark. See Tubulidentata
Aardwolf. See Hyaenidae
Abrocomidae, 223, *241*, 249
Abundance, See population
Age
  absolute, 289
  determination, 10, 35, 37, **289-295**
Ailuridae, 214
Ailuropodidae, 199, 202, *212*, 213-214
Albinism, 32
Allopatric, 82
†Allotheria. See †Multituberculata
Altricial young, 69, 215
Alveolar length and width, 12
Ambulatory locomotion, 47
Analysis
  covariate, 286-*287*
  multivariate, 287-*288*
  univariate, 279-281
Ancodonta, 272
Anomaluridae, 222, **237**, 248
  arboreal and gliding adaptations, 54-55
Anomaluroidea, 248
Anoplura, 354,*358*
Anteater. See Myrmecophagidae
Anteater, banded. See Myrmecobiidae
Anteater, scaly. See Pholidota
Anteater, spiny. See Tachyglossidae
Antelope. See Bovidae
†Anthracotheriidae, 265
†Anthracotheroidea, 272
Anthropoidea, 169
Antilocapridae, 10, *263*, 264-266, 269, 272
  pronghorns, *36*
  scat, 306
Antlers, *36-38*, 40, 71, 264, 294, 309
  removing skin from animals with antlers, 344-345
Ape. See Cercopitheceidae
Ape, great. See Pongidae
Aplodontidae, 221, 222, 224, *238*, 248
  burrows, 307
  jaw musculature and associated structures, *222*
Aplodontoidea, 248
†Archaeoceti, 182, *183*
Argasidae, 355 *360*
Armadillo. See Dasypodidae
Artiodactyla, 102, 107, **263-273**
  bicornuate uterus, 61
  horns and antlers, 35-40
  mesaxonic limbs, 48
  placenta, 67
  precocial young, 69
  testes, position, 62
  tooth eruption and age determination, 291
Assessment line. See trap placement
Astigmata, 355, *361*-362
†Astrapotheria, 102
Australian Faunal Region, 3
Aye-aye. See Daubentoniidae

Baboon. See Cercopithecidae
Baculum, 62, *63*
  absent
    Artiodactyla, 265
    Dermoptera, 143
    Edentata, 173
    Hyracoidea, 255
    Lagomorpha, 216
    Marsupialia, 114
    Monotremata, 110
    Perissodactyla, 260
    Pholidota, 178
    Proboscidea, 254
    Sirenia, 257
    Tubulidentata, 252
  in age determination, 290
  present
    Carnivora, 201
    Chiroptera, 147
    Insectivora, 134
    Primates, 160
    Rodentia, 221
  study skin preparation, 337
Badger. See Mustelidae
Bait, 297, 321, 326
Balaenidae, 181, *182*, 184-188
Balaenopteridae, 184-188
Baleen, 184-*185*, 294
Bandicoots. See Peramelina
Barrier, zoogeographic, 3
Basal length, 12
Basilar length, 12
Bat. See Chiroptera
Bat bugs. See Cimicidae
Bat flies. See Streblidae and Nycteribiidae
Bathyergidae, 221, 223-224, *239*, 247, 249
  burrows, 307
Bathyergoidea, 249
Bathyergomorpha, 249
Bathyergomorphi, 247
Bear. See Ursidae
Beaver. See Castoridae
Beaver, mountain. See Aplodontidae
Bed bugs. See Cimicidae
Beetles. See Coleoptera
Behavior. See also communication, feeding, hibernation, home range, migration, reproduction, territory, etc.
  feeding, 309
  marking, 309
  recording data on, 317
Binomen, 89
Birth rate, 73
Biting lice. See Mallophaga
Body mites. See Mesostigmata
†Borhyaenidae, 117-*118*, 122
Bot flies. See Cuterebridae, Gasterophilidae, Hypodermatidae, and Oestridae
Bovidae, 263-266, *270-271*
  feet modified for soft substrates, 48
  horn in age determination, 294
  leks, in Uganda kob and gazelles, 371
  marking, 309
  migrations, wildebeest, 372
  scats, 306
  true horns, *35-36*
Bovoidea, 272
Bradypodidae, *171-176*
  arboreal adaptations, *54*
  claws, 41
  color and hair, 33
  diet and teeth, 23
Brain, 2
  Insectivora, 133-134
  Marsupialia, 113
  removal from skull, 343
  whales, 184
Braincase, breadth of, 12
Buffalo. See Bovidae
Burramyidae, 116, 125, *128-129*, 131
  arboreal, 114
  Burramys, history, 115
Burrows, *307-308*
Caecotrophic feces, 215
Caecum
  Artiodactyla, 265
  Carnivora, 201
  Dermoptera, 143
  Diprotodonta, 125
  Marsupicarnivora, 117
  Peramelina, 122
  Perissodactyla, 260
  Proboscidea, 254
  Tubulidentata, 252
Caenolestidae. See Paucituberculata
Caenolestoidea, 116
Calipers, 11
Callimiconidae, 160-161, 169
Callitrichidae, 159-161, *165*, 169
Cambrian, 4
Camelidae, 263-265, *268*, 271-272
  estrus, 65
  foot structure, 48
Cameloidea, 272
Canidae, 199-202, *207-208*, 213-214
  arrival in Australia, 115
  estrus, 65
  eye lenses and age determination, *292*, 293
  scents to attract, 326
  study skin tail preparation, 338, 343
  teeth and age determination, 291, 293
  tooth number, *Otocyon*, 25
Canines, 17
Caniniformia, 202
Cannon bone, 48, 263-*264*
Capromyidae, 223, *245*, 249
Capybara. See Hydrochoeridae
Carnivora, 102, 107-108, *199-214*
  baculum present, 62
  baits to use for capture, 326
  bipartite uterus, 59-61
  caching food, 309
  carnassials, 23
  feeding in captivity, 330
  marking, 329-330
  placenta, 67
  similarities of marsupicarnivores, 120
  testes, position, 62
Carnivore. See also diets
  claws and prey capture, 41
  tooth types, 23

Castoridae, 219, 222, 224, *226-227*, 248
dens, 308-309
special study skin tail preparation, 338, 343
tracks, 306
trapping, 321-322, 325-326
tree cutting, 309
Castorimorpha, 248
Castoroidea, 248
Cat. See Felidae
Catarrhini, 159-*161*, 169
Cattle. See Bovidae
Caviidae, 219, 223, 239-240, 249
Caviomorpha, 249
Cavoidea, 249
Cebidae, *159*, 160-161, *165-166*, 168-169
Ceboidea, 169
Cenozoic, 4
Cercopithecidae, 159-161, 166-169
Cercopithecoidea, 169
Cervidae, 263-*264*, 265-266, *269*, *271*-272
age determination, teeth, 291, 293
antlers, *36-38*, 294, 309
canines, 71
delayed implantation, roe deer, 66
feet modified for soft substrates, 48
group home range, red deer, 370
migrations, 372
netting, 327
scats, *306*
scent marking, 33
Cervoidea, 272
Cetacea, 108, 181. See also Mysticeta, Odontoceta and †Archaeoceti
†Chalicotheriidae, 261
Cheek teeth. See premolars and molars
Chewing lice. See Mallophaga
Chiggers. See Prostigmata
Chimpanzee. See Pongidae
Chinchillidae, 219, 223, *240*, 249
Chinchilloidea, 249
Chiroptera, 102, 106, 144, **145-158**
aerial adaptations, **55-56**
in Australian region, 115
baculum in most, 62
banding and tracking, 329
bicornuate uterus, 61
collecting and conservation, 320, 327-328
copulation plug, 62
embryo placement in uterus, 66
homing, 372
migrations, 372
reduced teeth, 23
roosting sites, 309
study skin, special pinning information, *342*
Chi-square test, 285-286
Chordata, characteristics of, 1
Chrysochloridae, 134, 138-140
fossorial behavior, 50-*51*
similarity to Notoryctidae, 120
Chrysochloroidea, 140
Cimicidae, 354, *358*
Circular overlap, 83
Civet. See Viverridae
Cladistics, 84-*85*, 86
Class, 83
Classification, 81-93
Claws, *41*, 172-173
Chalicotheriidae, 261
Cline, 82-83
Clitoris, *59*, 61
Cloaca, 59-60
Marsupialia, 114
Monotremata, 109-110
Coatimundi. See Procyonidae
Coefficient of variation, 279-280
Coleoptera, 354
Collecting, **319-331**
for diet analysis, 297-298
drug immobilization, 327
ectoparasites, 353, 360-362
hand capture, 327-328
hunting, 327
laws and conservation, 319-320
method recorded, 315
for population estimation, 375-377
salvage, 329
taxonomic samples, 86
trapping, 320-337
Colobidae, 169
Colugo. See Dermoptera
Columella, 1
Communication, coloration, 31-32
†Condylarthra, 102, 253
Condylobasal length, 12
Condylocanine length, 12
Continental drift, **5-6**
Copulation, 65
Copulation plug, 62
Coyote. See Canidae
Cranium, **7-9**
Craseonycteridae, 147, *152*, 158
†Creodonta, 201
Cretaceous, 4
Cricetidae, 219, 222, 224, *227*, *231-233*, 247, 248
age determination, 292-293
arboreal mice, 54
dens and nests, 308, 327
feet, lemming, 48
food caches, 309
marking and tracking, 329-330
runways and burrows, 307
special study skin tail preparation, muskrat, 343
testes, descent, 62
trapping, muskrat, 321-322
Cryptophagidae, 354
Ctenodactylidae, 223-224, *235-236*, 248-249
Ctenodactyloidea, 248-249
Ctenomyidae, 223, *244*, 249
Cuniculidae, 249
Cuterebridae, 353-354, 362
Cynocephalidae. See Dermoptera

Dasypodidae, *171-176*
delayed implantation, 66
fossorial, 50
scales, shells, 27-28
similarity of giant armadillo skull to aardvark skull, 253
study skin preparation, 337-338, 343
tooth number, giant armadillo, 25
Dasyproctidae, 223, *241-242*, 249
Dasyuridae, 116-117, 120, *121*, 122
ancestors to other Australian forms, 115
Dasyuroidea, 116
Dasyuromorphia, 116
Daubentoniidae, 159-161, *162*, 168-169
similarity to rodents, 247
Daubentonoidea, 169
Data, processing of, 317-318
Data, recording, **311-318**, 320, 334, 337
data standards for museum specimens, 318
for distribution studies, 372
ectoparasites, 362
for population density studies, 377
taxonomic, 86
Death rate, 73
Deer. See Cervidae
Degrees of freedom, 279
Delayed development, 66
Delayed fertilization, 66
Delayed implantation, 66
Delphinidae, *182-183*, *188-189*, 190, *193*, *195-196*
Dens, *305*, *308-309*
Density of individuals. See population and dispersion
Dental formulas. See teeth
Dermaptera, 354-355
Dermestid beetles, cleaning skeletal material with, 347-350
Dermis, 27, *28*
Dermoptera, 102, 106, 108, 140, **143-144**
arboreal and gliding adaptations, **54-55**
Desmodontidae, 158
†Desmostylia, 102, 251
Devonian, 4
Diaphragm, muscular, 2
Diastema length, 12
Dice gram, 87, *281*
Dice-Leraas diagram. See Dice Gram
Didactyla, 115
Didelphidae, 116-117, *118-119*, 120, 122
arboreal adaptations, 54, 114
economic importance, 114
fossil history, 115
mammae, 70
marsupium, 67
molars, 21
study skin tail preparations, 338, 343
web feet, *Chironectes*, 114
Didelphimorphia, 116
Didelphoidea, 116
Diet. See also insectivorous, herbivorous, etc.
analysis of, **297-304**
Artiodactyla, 263
Carnivora, 199
Chiroptera, 145
Edentata, 172
feeding residues, 309
Insectivora, 133
Lagomorpha, 215
Marsupicarnivora, 117
Mysticeta, 184
Odontoceta, 188
Primates, 159
Rodentia, 219
†Dinocerata, 102
Dinomyidae, 223-224, *245*, 249
Dipodidae, 223-224, *235*, 248
hairy feet, 48
ricochetal locomotion, 49
Dipodoidea, 248
Diprotodont dentition, 113, 123, 125
Diprotodonta, 102, 106, 113, 115, 116, **125-131**
arboreal and gliding adaptations, 54-55
gliding membranes compared to colugos, 144
similarity to Paucituberculata, 124
syndactylous digits, 114
Diprotodontidae, 125, *126*
Dispersion patterns of mammals, *365*
Distribution (geographic). See also faunal regions.

Artiodactyla, 265
Carnivora, 199, 201-202
Chiroptera, 147
Dermoptera, 144
Diprotodonta, 125
Edentata, 173-174
Hyracoidea, 255
Insectivora, 134
Lagomorpha, 216
Marsupicarnivora, 117
Monotremata, 110
Mysticeta, 184, 186
Odontoceta, 189
of orders, 2-3
Paramelina, 122
Paucituberculata, 123
Perissodactyla, 261
Pholidota, 178
Primates, 160
Proboscidea, 254
Rodentia, 221-223
Sirenia, 256-257
Tubulidentata, 252
Distribution (spatial), **365-373**
Distribution (statistical), 281-*282*
†Docodonta, 18, 101, 107
Dog. See Canidae
Dog, prairie. See Sciuridae
Dolphin. See Delphinidae
Dolphin, freshwater. See Platanistidae
Domestic animals
Artiodactyla, 263
Carnivora, *200*
diet, analysis of, 297
Perissodactyla, 259-260
Proboscidea, *253*
Rodentia, 219
Dormouse. See Gliridae
†Dryolestoidea, 107
Duckbill. See Ornithorynchidae
Dugonidae, *256-258*
Dung beetles. See Scarabaeidae
Duplicidentata, 217

Ear
of bats, *146-147*
middle ear bones (=ossicles), **1**
of whales, 188, 294
Earwigs. See Dermaptera
Echidna. See Tachyglossidae
Echimyidae, 223, *243-244* 249
tooth wear and age determination, 291
Echolocation
bats, 55-56, 145-147
whales, 188
Economic importance
Artiodactyla, 263
Carnivora, 200
Chiroptera, 145
Dermoptera, 143
Edentata, 173
Insectivora, 133
Lagomorpha, 216
marsupials, 114-115
Monotremata, 109
Perissodactyla, 259-260
Primates, 160
Proboscidea, 253
Rodentia, 219
Sirenia, 256
Tubulidentata, 252
whales, 183-184, 188
Edentata, 102, 106, **171-176**, 178, 253
simplex uterus, 61
testes, position of, 62
†Ektopodontidae, 125
Elephantidae. See Proboscidea,
Emballonuridae, 147, 153, 156
†Embrithopoda, 102, 251
Embryology, 66
resorption, 69
Embryos
counting and measuring, 342
preserving, 342
Emmigration, 73, 372
†Enaliarctidae, 201
†Eotheria, 102, 107
†Eozostrodontidae, 17, *18*
Epidermis, 27, *28*
Epipubic bones
in marsupials, 114
in monotremes, 110
†Epoicotheriidae, 178
Equiidae, *259-262*
canines, 71
incisors, 16
insemination, 65
limbs, *48*
mammae, 70
marking, 309
Erethizontidae, 223, *245-246*, 249
spines, *30*
study skin preparation, 337, 343
Erethizontoidea, 249
Erinaceidae, 134, *138*, 139
spines of hedgehog, 30
Erinaceoidea, 140
Erinaceomorpha, 140
Erythrocytes, enucleate, 2
Eschrichtiidae, 184, *186-188*
migrations, 372
Estrus, 64-65
Ethiopian Faunal Region, 3
†Eurymylidae, 216
Eutheria, 4, 5, 102, 107
evolution, 102
primitive dental formula, 24
reproduction, 59
uterus types, *60*
Eye
in bats, 56
in fossorial mammals, 50
lens weight and age determination, 292-293
in primates, 159
in whales, 188

Family, 83
Faunal regions, **3**
Fecundity, 75-76
Feet
anteaters, *173*
armadillo, *172*
†chalicotheres, 261
digitigrade, *47*
Hyracoidea, 255
Lagomorpha, 215-216
modified for soft substrates, 48-*49*
plantigrade, *47*
sloth, 172
Felidae, 199-202, *208-209*, 213-214
caching food, 309
claws, 41
limb movement, 48
mane of lion, 71
marking, 309
scents to attract, 326
Feliformia, 201-202
Ferae, 213
Ferret. See Mustelidae
Fertilization, 66
Fin, dorsal, in whales, 181, 372
Fissiped. See all Carnivora except Otariidae and Phocidae
Fissipedia, 213
Fleas. See Siphonaptera
Follicle mites. See Prostigmata
Fossorial
adaptations, **50-51**
claws, 41
Fox. See Canidae
Fur mites. See Astigmata
Fur moths. See Pyralidae
Furipteridae, 147, *157*, 158

Galagidae, 159-161, *163-164*, 168
Galeopithecidae, 140, 144
Gasterophilidae, 353-354, *356*, 362
Genet. See Viverridae
†Geniohyidae, 256
Genus, 83, 88
name, 89
type-genus, 91
Geomyidae, 222, *224-225*, 248
burrows, 307-*308*
cheek pouches and diet study, 297
gopher traps, *322-324*
Geomyoidea, 248
Gerbil. See Cricetidae
Gestation, 69
Gibbon. See Hylobatidae
Giraffidae, 264-266, *268*, 271-272
group home range, 370
horns, *39*
Giraffoidea, 272
Gland. See mammary glands, poison glands, sebaceous glands, sweat glands
Glires, 247
Gliridae, 223-224, *238*, 248
Gliroidea, 248
Globicephalidae, 196
Gloger's Rule, 32
†Glyptodonts, *174*
Goat. See Bovidae
Gondwanaland, *5*
Gopher, pocket. See Geomyidae
Gorilla. See Pongidae
Grampidae, 196
Guard hairs, **30**
in diet analysis, 298

Hair, 2, 309
anatomy, **28-29**
classification, **29-30**
color, **31-33**, 172
replacement, **30-31**
Haplorhini, 160
Hard ticks. See Ixodidae
Hare. See Leporidae
Heart, 2
†Heptaxodontidae, 247, 249
Herbivore. See also diets
Dermoptera, 143
Diprotodonta, 125
Sirenia, 256
sloths, 172
tooth types, 21-22
Heteromyidae, 219, 222, *225*, 248
cheek pouches and diet studies, 297
collecting from dens, 327
food caches, 309
fossorial behavior, 50
ricochetal locomotion, 49
Hibernation in bats, 145
Hippoboscidae, 354, *356-357*
Hippopotamidae, 264-265, *267*, 271-272
Hipposideridae, 158
Histogram, preparation of, 87-*88*
Hog. See Suidae
Holarctic Faunal Region, *3*
Home Range, 309, **365-370**
Homing, 372
Hominidae, 10, 159-161, *168*, 169
arrival in Australia, 115
breastfeeding, 71
incisors, 16
menstrual cycle, 64

  paraxonic limbs, 47
  placenta, 67
  primordial germ cells in female, 63
  secondary sex characters, 71-72
  sweat glands, 33
Hominoidea, 169
Homonyms, 91
Hoofs, *42*
Hormones. See also pheromone
  female reproductive, 64, 71
  influencing antlers, 38, 71
  influencing pelage, 31
Horns. See also antlers
  function, 40
  giraffe horns, *39*, 271
  pronghorns, *36*, 272
  removing skin from animals with horns, 344-345
  rhino horns, *39*, 260
  true (bovid) horns, *35-36*, 264, 272, 294
Horse. See Equidae
Hyaenidae, 199, 201-202, *208-210*, 213-214
  aardwolf teeth, 19
  external genitalia, 61
  limb structure, 48
Hybrids, 83
Hydrochoeridae, 219, 223, *239*, 249
  limb structure, 48
Hydrodamalidae, 256-258
Hylobatidae
  brachiation, *54*, 160-161, *167*, 169
  group territories, 371
Hypodermatidae, 353-354, 362
Hypotheses, testing in statistics, 282-283
Hyracoidea, 102, 107, 251, **255-256**
  precocial young, 69
  similarity to rodents, 247
  testes, position, 62
Hystricidae, 223, *245*, 249
  spines, 30
  study skin preparation, 337, 343
Hystricognathi, 247
Hystricoidea, 249
Hystricomorpha, 223-224, 247-249
  jaw musculature and associated structures, 221
  precocial young, 69
Hystricomorphi, 247

†*Icaronycteris index*, 148
Ice Age, 4
Immigration, 73, 372
Implantation, 66
Incisive foramina length, 12
Incisors, **16-17**
  in colugos, 143-*144*
  in elephants, 253
  in hyraxes, 255
  in rabbits, 216
  in rodents, 219-221
Incus, *1*
Indriidae, 159-160, *161-162*, 168-169
Iniidae, 197
Insectivora, 102, 106, 108, **133-141**, 169
  baculum in most, 62
  baits to attract, 326
  bicornuate uteri, 61
  copulation plugs, 62
  dental formulas, 25
  molars, 20-21
  similarities to marsupials, 120
  testes, position, 62
  vaginal opening, 61
Insectivore. See also diets
  aardvarks, 251-252
  anteaters and armadillos, 172-173
  Paucituberculata, 123
  Peramelina, 122
  tooth types, 23
Interorbital breadth, least, 12
Ixodidae, 354, *359-360*, 362

Jerboa. See Dipodidae
Jurassic, 4

Kangaroo. See Macropodidae
†*Kennalestes*, 134
Keys, biological identification
  to arthropod ectoparasites of mammals, 335-360
  to families of
    Artiodactyla, 266-271
    Carnivora, 201-210
    Chiroptera, 148-157
    Diprotodonta, 125
    Edentata, 175
    Insectivora, 135-138
    Lagomorpha, 217
    Marsupicarnivora, 118-120
    Monotremata, 111
    Mysticeta, 186
    Odontoceta, 190-195
    Peramelina, 123
    Perissodactyla, 262
    Primates, 161-167
    Rodentia of North America, 246-247
    Rodentia of world, 224-245
    Sirenia, 258
  how to use and write, 95-98
  use of those in this manual, 10, 99
Koala. See Phascolarctidae
Kogiidae, 196
†Kuehneotheriidae, 17
Labels
  data standards for museum specimens, 318
  diet analysis samples, 298
  ectoparasites, 362
  embryos, 342
  labeling skins for tanning, 346-*347*
  skull and/or skeleton, *342*-343
  specimen, 6, 312, **315-317**, 342
Lagomorpha, 102, 106, **215-217**, 247
  duplex uterus, 59
  estrus, 65
  incisors, 15
  scrotum anterior to penis, 62
  similarity to rodents, 247
Laurasia, *5*
Lemming. See Cricetidae
Lemur. See Lemuridae and Indriidae
Lemur, flying. See Dermoptera
Lemuridae, 159-161, *163*, 168-169
  incisors as relates to colugos, 144
Lemuriformes, 169
Lemuroidea, 169
Leporidae, *215-217*
  age determination, 290, 292
  baits to attract, 326
  food in captivity, 330
  forms, 307
  fossorial behavior, 50
  locomotion on soft substrates, 48
  precocial young, hares, 69
  saltatorial locomotion, 49
  study skin preparation, 337, 343-*344*
  territorial marking, 34
Leptinidae, 354-*355*, *357*
Leptodiridae, 354
Life table, 75, 76
Limb structure
  aerial adaptations, 55-56
  ambulatory limbs, *43-45*
  aquatic adaptations, 51-53, 181-*182*, 257
  arboreal adaptations, 53-55, 172, 255
  cursorial adaptations, 47-48, 259-260, 263-264
  fossorial adaptations, 50-51
  graviportal adaptations, 50, 254
  mesaxonic limbs, 47, 259
  paraxonic limbs, 47, 263-264
  pinniped limbs, 52-*53*
  saltatorial adaptations, 49
  soft substrate adaptations, 48-49
Lipotyphla, 133-134, 140
†Litoptera, 102
Litter size, 69
Llama. See Camelidae
Locomotion
  aerial, 55-56
  aquatic, 51-54, 181-182, 256-257
  arboreal, 54-55, 255
  fossorial, 50-51
  terrestrial, 47-50
Lordosis, 48
Lorisidae, 159-161, *164*, 168
Lorisiformes, 169
Louse flies. See Hippoboscidae

Macropodidae, *113*, 116, 125, *127-128*, 131
  marsupium, 67
  ricochetal locomotion, *49*, 114
  stomach complex, 125
  tooth replacement, 15-16
  urogenital sinus, 59-*60*
Macroscelidea, 140
Macroscelididae, 134, *137*
Macroscelidoidea, 140
Malleus, 1
Mallophaga, 354, *358*
Mammalia
  definition of, 1-2
  evolution of, 101-102
  origins of, 4, 101
Mammal-nest beetles. See Leptinidae
Mammals. See also Mammalia
  Age of, 4
Mammary glands, 2, 34, 70-*71*
  Monotremata, 109-110
  Sirenia, 256
Man. See Hominidae
Manatee. See Trichechidae
Mandible, 7, 9-10
  rodent mandible types, 225
Mandible length, 13
Mandibular diastema length, 13
Mandibular tooth row length, 13
Mange mites. See Astigmata
Manidae. See Pholidota
Manus, *44*. See also feet
Marking
  by mammals, 33, 309
  of mammals, 328-330, 372-373, 379-382
Marmoset. See Callitrichidae
Marmot. See Sciuridae
Marsupialia. See also Metatheria. 15, 107-108, **113-131**
  bifurcate glans penis, 62
  copulation plugs, 62
  dental formula, 17
  development at birth, 69

embryonic diapause, 66
gestation, 67, 69
mammae, 70
marsupium, 67
placentation, 67
reduced teeth, 23
scrotum anterior to penis, **62**
testes, position, 62
Marsupiata, 107
Marsupicarnivora, 102, 106-107, 113, 116, **117-122**
Marsupium, 67
in Diprotodonta, 125
in marsupials, 113-114
in Marsupicarnivora, 117
in monotremes, 109
in Paucituberculata, 123
in Peramelina, 122
Mastoid breadth, 12
Maxillary tooth row, 12
Mean, 279, 281-282
Measurements
cranial, **11-13**
in diet analysis samples, 301
of population levels, 301-302
recording, 317
testes and embryos, 342
weight, 336
of whole mammals, 313, *334-337*
Megachiroptera, 147-*148,* 158
claw on second digit, 145
ear, *146*
echolation, 56
skeleton, *56*
Megadermatidae, 147, *152,* 158
Melanism, 32
Menotyphla, 134
Menstrual cycle, 64-65
Mesostigmata, 355, *359-360,* 362
Mesozoic, 4
†Metacheiromyidae, 178
Metastigmata, 354, *359-360,* 362
Metatheria, 4-6, 107
evolution, 102
primitive dental formula, 24-25
reproduction, 59
uterus, structure, 60
†Miacidae, 201
†Miacoidea, 201
Microbiotheridae, 116-*118,* 122
arboreal, 114
Microchiroptera, 147, 158
eyes, 56
Migration, **372**
in artiodactyls, 372
in bats, 145, 372
in seals, 372
in whales, 184, 189
Mink. See Mustelidae
Mississippian, 4
Molariform teeth. See premolars and molars
Molars, 17-24
Mole. See Talpidae
Mole, golden. See Chrysochloridae
Mole, marsupial. See Notoryctidae
tooth types, 23
Molossidae, 147, *155-157,* 158
migration, 145
Molts, *30*-31
Moluscivores. See also diets
Mongoose. See Viverridae
Monkey. See Callimiconidae, Cebidae and Cercopithecidae
Monodontidae, 182, 189, *190,* 196
Monotremata, 102, 106-107, **109-112**. See also Prototheria.
bifurcate glans penis, 62
dentition, 17, 24
gestation, 69
mammae, 70
pectoral girdle, 43
reproduction, 66-67
spurs, 42
testes, abdominal, 62
Mormoopidae, 147, *150,* 158
Mortality, 73
Mouse. See Myomorpha
Mouse, pocket. See Heteromyidae
†Multituberculata, 18-*19,* 101 107
plagiaulacoid teeth, 113
Muridae, 219, 223-224, *230-232, 234,* 247-248
mammae, *Mastomys,* 70
Muroidea, 248
Mustelidae, 199-202, *211-212,* 213
bone growth and age determination, 293-294
diet studies, 298
fossorial behavior, badgers, 50
rooting, skunks, 309
scents, 33
seasonal color changes, *Mustela,* 30-31
semiaquatic adaptations, otters, 51-52
special study skin tail preparation, skunks, 338, 343
teeth and diet, sea otter, 23
waterproofing, otter, 33
Myocastoridae, 223, *243,* 249
mammae, placement, 70-71
special scaly tail preparation techniques, 343
Myomorpha, 222, 224, 247-248
age determination, 292-293
jaw musculature and associated structures, 221-*222*
Myomorphi, 247
Myrmecobiidae, 116-117, *119,* 122
angular process of dentary weak, 113
tooth number, 25
Myrmecophagidae, 10, *171-176*
compared to echidnas and pangolins, 112, 178
Mystacinidae, 147, 151, 158
Mysticeta, 102, 107-108, **181-188**
aquatic adaptations, 53
baleen and age determination, 294
ear plugs and age determination, 294
gestation, 69
lactation, 71
measurements, *337*
migrations, 372
recognition of individuals, 372
sweat glands absent, 33
testes, position of, 62
Myzopodidae, 147, *157*-158

Nails, *41*-42
Nasal length, 13
Nasal suture length, 13
Nasal width, 13
Natalidae, 147, *155,* 158
Natality, 73
Nearctic Faunal Region, 3
Nectivore. See also diets and tooth types
Neotropical Faunal Region, *3*
Nests. See dens
Niche, 302-303
Noctilionidae, 147, *149,* 158
Nomadism, 372
Nomen nudum, 90
Nomenclature, 81, 89-93
alternate systems, 93
availability of names, 90-91
binomial, 89-90
synonymy, 92
types, 91-92
Notoryctemorphia, 116
Notoryctidae, 116-117, *120,* 122
epipubic bones vestigial, 114
fully fossorial, 114
molars, 20
resemble Insectivora, 139
Notoryctoidea, 116
†Notoungulata, 102
Numbat. See Myrmecobiidae
Nutria. See Myocastoridae
Nycteribiidae, 354, *357*
Nycteridae, 147, *152-153,* 158

Oceania, *3*
Ochotonidae, *215-217*
bone growth and age determination, 293-294
harvesting vegetation, 309
litter size, 69
resemble hyraxes, 256
Octodontidae, 223, *242-244,* 249
Octodontoidea, 249
Odobenidae, 214
Odontoceta, 10, 15, 102, 107-108, **181-184, 188-197**
aquatic adaptations, *53*
gestation, 69
measurements, 337
migrations, 372
recognition of individuals, 372
sweat glands absent, 33
teeth, 23, 25
teeth used in age determination, 293
testes, position of, 62
Oestridae, 353-354, *356-357, 359,* 362
Okapi. See Giraffidae
Omnivore. See also diets
tooth types, 21
Oogenesis, 63
Opossum. See Didelphidae; See also possum
Orangutan. See Pongidae
Orcaelidae, 197
Order, 83
Ordovician, 4
Oriental Faunal Region, **2**
Ornithodelphia. See Monotremata
Ornithorynchidae, **109-112**
reproduction, 66-67
semiaquatic adaptations, 51-52
Orycteropodidae. See Tubulidentata
Os clitoris, 61-62
Os penis. See Baculum
Otariidae, 199-*204,* 213-214
canines, walrus, 17
cheek teeth, walrus, 23
delayed implantation, 66
limb structure, 52-53
migrations, 372
territories, 371
waterproofing, 33
Otter. See Mustelidae
Ovaries, *59-60, 63,* 66
in bats, 147
Ovulation, 64

†Palaeadonta, 174, 176, 178
†Palaeomerycidae, 266
Palatal length, 13

Palearctic Faunal Region, 3
Paleozoic, 4
Palitar length, 13
Panda, giant. See Ailuropodidae
Panda, red or lesser. See Procyonidae
Pangea, 5-6
Pangolin. See Pholidota
†Pantodonta, 102
†Pantotheria, *18-19*, 101-102, 107
†Paramyidae, 224
Parapatric, *82*
Parasites, 314, 334, 337, 353
  collecting, 353, 360-362
  identification of, 353-360, 362
Parturition, 67
Patagia
  aerial, 55-*56*, 145-*146*
  gliding, *55*, 143-144
Patronym, 89
Paucituberculata, 102, 106, 113, 115-116, **123-124**
  similarity to Diprotodonta, 131
Peccary. See Tayassuidae
Pecora, 272
Pectoral girdle, *43*
  of monotremes, 110
Pectoral limb, *43-44*
  bat wing, 145-*146*
  of Sirenia, 257
  of whales, 181-*182*
Pedetidae, *219*, 222, 224, *236*, 247-248
Pelage, 31
  of Edentata, 172
  of Insectivora, 133
Pelvic girdle, *44*. See also limb structure
  rudimentary, 181, 257
Pelvic limb, 44-45. See also limb structure
  rudimentary, 181, 257
Penis, *62*
  Chiroptera, 147
  Lagomorpha, 216
  marsupials, 114
  Monotremata, 110
  Primates, 160
  whales, 182
Pennsylvanian, 4
Peramelemorphia, 116
Peramelidae, 116, *122*-123
  limb structure, pig-footed bandicoot, 48
Peramelina, 102-106, 113, 115-117, 122-123
  hopping locomotion, 114
  placenta, 114
  syndactylous digits, 114
Perameloidea, 116
Perissodactyla, 102, 107, 259-262
  bicornuate uterus, 61
  horns, 35
  mesaxonic limbs, *47-48*
  precocial young, 69
  testes, position of, 62
Permian, 4
Pes, 45. See also feet
Petauridae, 116, 125, *129, 131*
  arboreal, 114
Petromuridae, 249
Petromyidae, 223, *243*, 249
Phalangeridae, 116, 125, **130-131**
  arboreal, 114
  economic importance, 115
  Gymnobelideus rarity, 115
Phalangeroidea, 116
Phascolarctidae, 116, 125, *128*, 131
  angular process of dentary weak, 113
  arboreal, 114
  caecum complex, 125
  marsupium, 67
Phascolomidae, 116
Phenetics, 84, 86-87
Pheromone, 33-34
Phloeomyidae, 247
Phocidae, 199-202, *204-205*, 213-214
  delayed implantation, 66
  lactation, 71
  limb structure, 52-*53*
  teeth, crab eating seal, *23*
  territories, 371
Phocoenidae, 189-190, *193-195*, 196-197
  skeleton, *53*
Pholidota, 102, 106, 174, 176, **177-179**
  compared to echidnas and anteaters, **112**, 176
  diet, 24
  placenta, 67
  scales, 27-28
  special study skin preparation, 343
Phyllostomatidae, 147, *149-151*, 158
  delayed development, 66
  shelter construction, 309
  vampire incisors, 16, *149*
Physeteridae, *181, 188*-190, *192*, 196
Pika. See Ochotonidae
Pinniped
  animals. See Phocidae and Otariidae
  limbs, 52-53, 200-201
Pinnipedia, 108, 213. See also Otariidae and Phocidae
Piscivore. See also diets
  odontocetes, 188
  tooth types, 23
Placenta, 67-*68*
  in marsupials, 114, 122
Placental scar, 67, 69
Placentata, 107
Platacanthomyidae, 223, *233-234*, 248
Platanistidae, 189-190, **194**, 196-197
Plate tectonics, 5-6
Platypus. See Ornithorynchidae
Platyrrhini, 160-*161*
Poison glands in monotremes, 42, 110
†Polydolopidae, 123
Polyprotodontia, 115-116
Pongidae, 159-161, *168*-169
  group home ranges, 370
  sleeping platforms, 308
Population
  census data recording, 317
  determination, 301
  estimation of abundance and density, 375-382
  general, 73-79
  growth, 73-78
  regulation, 78
Populations in statistics, 276, 282
Porcupine. See Erethizontidae and Hystricidae
Porpoise. See Phocoenidae
Possum. See Burramyidae, Petauridae, Phalangeridae, and Tarsipedidae; See also opossum
Postcanine teeth. See premolars and molars
Postorbital constriction, 12
Postpalatal length, 13
Potamogalidae, 140
Precambrian, 4
Precocial young, 69, 215
Premolars, **17-24**
Preservation
  data on, 314-315
  diet analysis reference samples, 298-301
  ectoparasites, 362
  embryos, 342
  histological material, 352
  scats, 300
  skeletons, 347-351
  stomachs and contents, 298, 342
  study skins, 233-347
  whole animals and internal organs, 351-352
Primates, 102, 106, 140, 144, 159-170
  arboreal adaptations, 54
  baculum in most, 62
  mammae, 70
  placenta, 67
  scent marking, 33
  testes, position of, 62
  uteri, 61
Probability, 281-282
Proboscidea, 102, 107, 251, 253-254
  graviportal limb, *50*
  incisors, 16
  testes, position of, 62
  tooth replacement, 15-*16*
Procaviidae. See Hyracoidea
Procyonidae, 10, 199-202, *205, 210, 213*-214
  paraxonic limb, 47
Pronghorn. See Antilocapridae.
Prosimii, 169
Prostigmata, 355, *361*
Protelidae, 214
Protoeutheria, 140
Prototheria, 5-6, 107, 109-112
  evolution of, 101-102
  reproduction, 59
  uterus type, 60
Protoungulate, 251
Pteropodidae. See Megachiroptera
Pyralidae, 354, *357*
†Pyrotheria, 102

Quaternary, 4

Rabbit. See Leporidae
Raccoon. See Procyonidae
Range (in statistics), 279, 281-282
Rat. See Myomorpha and many families of Hystricomorpha
Rat, kangaroo. See Heteromyidae
Reingestation, 215
Reproductive
  condition, 314
  effort, 78
  strategies, 70, 75-78
  system, female, 59-61, 63-69
  system, male, 61-62, 65-66
Rhinocerotidae, 259-262
  horns, 35, **39**
Rhinolophidae, 147, *153-154*, 158
  ear, *146*
  hibernate, migrate, 145
  nose, *148*
Rhinopomatidae, 147, *154*, 158
  nose, *148*
Rhizomyidae, 223-224, *228*, 248
Ringtail. See Procyonidae
Rodentia, 102, 106-107
  in Australian Faunal Region, 115
  accessory corpora lutea, 64
  baculum in most, 62
  baits to attract, 326
  copulation plugs, 62

duplex uterus, 59
feeding residues, 309
food in captivity, 330
gliding membranes compared to colugos, 144
horned fossils, 35
incisors, 15
insemination, 65
mammae, 70
placenta, 67
postpartum estrus, 66
resemblance to lagomorphs, 217
resemble aye aye, 168
resemble hyraxes, 256
similarities to marsupials, 120
tooth eruption and age determination, 291
urine scent marking, 33
vaginal openings, 61
Rorqual. See Balaenopteridae
Rostral breadth, 12
Rove beetles. See Staphylinidae
Ruminantia, 263, 272
incisors, 15
stomach, *265*
Runways, 307

Saharo-Sindian Faunal Region, 3
Scales, **27-28**
Dasypodidae, *171*-172
glyptodonts, *174*
Pholidota, 177-178
Scandentia, 140
Scansorial, claws, 41
Scarabaeidae, 354-*355*
Scats, 300, 306-307
Scatter diagram
three-dimensional, 288
two-dimensional, 87-*88*
Scent gland, **33-34**
Scents. See baits and scent glands
Sciuridae, 10, 222, 224, *226*, 248
arboreal and gliding adaptations, **54-55**
estrus, 65
food caches, 309
fossorial behavior, 50-51
runways, ground squirrel, 307
territories, prairie dog, 371
timing trapping activity, 326
Sciurognathi, 247
Sciuroidea, 248
Sciuromorpha, 222, 247-248
jaw musculature and associated structures, 221-*222*
Sciuromorphi, 247
Scrotum, *62*. See also testes
Lagomorpha, 216
marsupials, 114
Sea cow, Steller's, 256-258
Seal. See Phocidae and Otariidae
Seal, earless/hair. See Phocidae
Seal, eared/fur. See Otariidae
Sebaceous gland, 28, **33**
Secondary sex characters. See also sexual dimorphism
antlers, 38
general, **71-72**
Selviniidae, 223-224, *230*, 248
Semen, 62, 65
Sex determination, 334
Sex ratio, 73
Sexual dimorphism. See also secondary sex characters
horns and antlers, 35-36, 294
reproductive anatomy, 59-72
Sheep. See Bovidae
Sign, **305-310**
Silken fungus beetles. See Cryptophagidae
Silurian, 4
Siphonaptera, 354, *357*, 362
Sirenia, 102, 107, 256-258
sweat glands absent, 33
testes, position of, 62
Shrew. See Soricidae
Shrew, elephant. See Macroscelididae
Shrew, otter. See Tenrecidae
Shrew, tree. See Tupaiidae
Skin, **27-28**
Skull, **7-14**
greatest length of, 12
modified in whales, *182-183*
nomenclature in rodents, *220*-221
Skunk. See Mustelidae
Sloth,
ground, 174
tree. See Bradypodidae
Small carrion beetles. Leptodiridae
Soft ticks. See Argasidae
Solenodontidae, 134, *135-136*, 140
rarity, 133
Sonar. See echolocation
Soricidae, *133*-134, *135*, 139-140
commensal species, 133
dental formulas, 25
feeding residues, 309
incisors, 16
pitfall traps, 323, 325
semiaquatic adaptations, *51-52*
Soriciomorpha, 140
Soricoidea, 140
Spalacidae, 221-222, 224, *228*, 248
burrows, 307
Speciation, 82
Species, **82**
name, 89
sibling, 82
type species, 91
Specimens. See also collecting, data recording, and preservation
care of, 6, 330
Spines, 30
echidnas, 109-*110*
Spurs in monoteremes, 42, *110*
Squirrel. See Sciuridae
Squirrel, scaly-tailed. See Anomaluridae
Standard deviation, 279-282
Standard error of the mean, 280-281
Stapes, *1*
Staphylinidae, 354-*355*, *357*
Stenidae, 196
Stenodelphidae, 197
Stomach
Artiodactyla, 264-*265*
Carnivora, 201
Diprotodonta, 125
fistula operation, 297
Marsupicarnivora, 117
Paucituberculata, 123
Peramelina, 122
Perissodactyla, 260
preservation techniques, 298
Streblidae, 354, *356-357*
Strepsirhini, 159-160
Subgenus name, 89
Subspecies, **82**, 88
name, 89
Subungulate, 251, 253-258. See also Hyracoidea, Proboscidea, and Sirenia
nails-hoofs, 42
Sucking lice, See Anoplura
Suidae, 263-265, 271-272
bipartite uterus, 61
canines, 17
insemination, 65
mammae, 70
rooting, 309
tooth wear and age determination, wart hog, 291
Suiformes, 263, 265, *267*
Suina, 272
Suoidea, 272
Survivorship, 75
Sweat gland, **33**
†Symmetrodonta, *18-19*, 101-102, 107
Sympatric, *82*
†Synapsida, 4
Syndactyla, 115
Syndactylous digits
hyraxes, 255
marsupials, *114*, 115-116, 122, 125
otter shrew, 116
sloths, *172*
Synonym, 91
Synonymy, **92**
Systematics, 81-89, **275-288**
interpretations, 87-89
methods, 86-87
philosophies, 83-86

Tachyglossidae, *109-112*
gestation, 69
reproduction, 66-67
resemble insectivores, 139
skulls compared to anteaters and pangolins, 176, 178
spines, 30
sweat glands absent, 33
†Taeniodonta, 102
Tail
in Chiroptera, 145-*146*
in fossorial mammals, 50
in fully aquatic mammals, 53, 181, 257
in Lagomorpha, 215-216
in Pholidota, 177-178
prehensile, 54, 159, 168, 172, 177
in Rodentia, 221
in saltatores, 49
in scansorial mammals, 54
in Sirenia, 257
in sloths, 172
Talpidae, 134, *139*-140
burrows, *307*
economic importance, 133
fossorial modifications, 50-*51*
mole traps, 322-*323*, *325*
similarity to Notoryctidae, 120
sweat glands absent, 33
Tapiridae, 259-262
Tarsii, 160
Tarsiidae, 159-161, *162-163*, 168-169
Tarsiiformes, 169
Tarsioidea, 169
Tarsipedidae, 116, 125, *127*, 131
angular process of dentary weak, 113
arboreal, 114
caecum absent, 125
jugal does not form part of mandibular fossa, 113
urogenital sinus, 59-*60*
Tarsipedoidea, 116
Taxonomy, **81-89**
Tayassuidae, 264-*266*, 271-272
canines, 17
musk, 33
scat, *306*

Teeth, 15-26
Artiodactyla, 263-264, 272
basic anatomy, *15*
Carnivora, *200-201*
cheek tooth evolution, 17-21, 24
cusp nomenclature, 17, 20-21
dental formulas, 24-26
Dermoptera, 133-134
Edentata, 171-173
Hyracoidea, 255
Insectivora, 133
kinds of, 16-17, 21-24
Lagomorpha, 216
Monotremata, 109-110
Odontoceta, 188
Perissodactyla, 260
Proboscidea, 253-*254*
replacement, 15-16, 253, 257
Rodentia, 219, 221
Sirenia, 257
tooth growth and age determination, 293
tooth wear and age determination, 291
Tubulidentata, 251-*252*
Tenrecidae, 134, *136*, 138-140
molars, *21*
spines, 30
Tenrecoidea, 140
Tenrecomorpha, 140
Territory, 309, *370-371*
Tertiary, 4
Testes, *61, 63, 65*. See also scrotum
Artiodactyla, 265
Carnivora, 201
Chiroptera, 147
Dermoptera, 143
Edentata, 173
Hyracoidea, 255
Insectivora, 134
Lagomorpha, 216
measuring, 342
Monotremata, 110
Perissodactyla, 260
Pholidota, 178
Primates, 160
Proboscidea, 254
Rodentia, 221
Sirenia, 257
Tubulidentata, 252
whales, 182
†Therapsida, 4, 101
Theria, 102
Theridomyomorpha, 248
Thermoregulation
antlers, 71
color, 32
hair, 30
hair waterproofing in fur seals and otters, 33
Thryonomyidae, 223, *242*, 248
Thryonomyoidea, 249
Thylacinidae, 116-117, *120*, 122
epipubic bones vestigial, 114
limb structure, 48
†Thylacoleonidae, 125-*126*
Thylacomyidae, 116, 122-*123*
Thryropteridae, 147, *155*, 158
†Tillodontia, 102
Tongue
of aardvark, 251-252
of anteater, *171*-173
of honey opossum, *127*
of numbat, *119*
of pangolin, *177*
Tracks, 305-306
Tragulidae, *264*-266
Tragulina, 272
Traguloidea, 271
Trails. See runways
Transect. See trap placement
Traps
for estimating population density, 375-377, 379-381
for home range studies, 237, 373
kill, 320-323
live, 297, 323-325
placement of, 325, 375-377
trap nights, 379
trapping success, 326
when to trap, 326
Triassic, 4
†Tribospena, 107
†Tribospenata, 107
Trichechidae, *256-258*
cervical vertebrae, 172
tooth numbers, 25
tooth replacement, 15-16
Triconodonta, *18*, 101, 107
Trinomen, 89
†Trituberculata, 101-102, 107
t-test, 283-284
Tubulidentata, 102, 107, **251-253**
formerly in Edentata, 176
Tupaiidae, 134, *137*, 169
molars, *21*
Tupaioidea, 140, 169
Tusks. See incisors and canines
Tylopoda, 263, 265, 272
Tympanic bullae length, 13
Tympanic bullae width, 13
Type
-genus, 91
holotype, 91
lectotype, 91
neotype, 91
paratype, 91
-species, 91
syntype, 91
topotype, 91

Underhairs, 30
Ungulate hoofs, 42, 259-260, 263-264. See also Artiodactyla and Perissodactyla
Uninominal, 89
Urşidae, *199*-202, *205-206*, 213-214
delayed implantation, 66
estrus, 65
marking, 309
scent to attract, 326
semiaquatic adaptations, polar bear, 51-52
teeth and age determination, 293
Uterus
Artiodactyla, 265
Carnivora, 201
Chiroptera, 147
Dermoptera, 143
Edentata, 173
egg migration in, 66
Hyracoidea, 255
Insectivora, 134
Lagomorpha, 216
marsupials, 114
Monotremata, 110
Perissodactyla, 260
Pholidota, 178
Primates, 160
Rodentia, 221
Sirenia, 257
structure and kinds, *59-60*, 61
Tubulidentata, 252
whales, 182

Vagina, 59-*60*, 61
Variance, 279-280
analysis of (ANOVA), 285-286
Vasectomy, 65
Vertebrae
cervical
in manatees, 172, 257
in sloths, 172
in whales, 182-*183*
xenarthrous, 173
Vespertilionidae, 147, *156*, 158
delayed fertilization, 66
ears, *146*
feeding residues, pallid bat, 309
hibernation and migration, 145
nose leaf, *148*
Vibrissae, *29*
Viverridae, 199-202, *206-208*, 213-214
musk, 34
Vole. See Cricetidae
Vombatidae, 116, 125-*126*, 131
fossil form, giant, 115
incisors, 15, 113
semifossorial, 114
similarity to rodents, 247
skeleton, *115*
Vombatoidea, 116

Wallaby. See Macropodidae
Wallace's line, *3*
Walrus. See Otariidae
Weasel. See Mustelidae
Whales. See Mysticeta, Odontoceta and appropriate family names
Wolf. See Canidae
Wombat. See Vombatidae
†Wynyardiidae, 125

Xenarthra, 174, 176, 178
†Xenungulata, 102

†Zalambdodonta, 107, 140
Zapodidae, 223-224, *236*, 248
ricochetal locomotion, 49
Zebra. See Equiidae
Ziphiidae, *188*-190, *191*, 193, 196
teeth used in age determination, *293*
Zygomatic breadth, 12